P9-EFG-213

Practical Handbook of
Microbiology

Edited by
William M. O'Leary, Ph.D.
Professor
Department of Microbiology
Cornell University Medical College
New York, New York

CRC Press, Inc.
Boca Raton, Florida

Library of Congress Cataloging-in-Publication Data

Practical handbook of microbiology / editor, William M. O'Leary.
p. cm.
Condensed version of: CRC handbook of microbiology / editors,
Allan I. Laskin, Hubert A. Lechevalier. 2nd ed. 1977.
Includes bibliographies and index.
ISBN 0-8493-3704-6
1. Microbiology--Handbooks, manuals, etc. I. O'Leary, William M.
(William Michael), 1928- . II. CRC handbook of microbiology.
[DNLM: 1. Microbiology--handbooks. QW 39 P895]
QR72.5.P73 1989
576--dc19
DNLM/DLC
for Library of Congress 89-955
CIP

Direct all inquiries to CRC Press, Inc., 2000 Corporate Blvd., N.W., Boca Raton, Florida, 33431.

International Standard Book Number 0-8493-3704-6

Library of Congress Card Number 89-955
Printed in the United States

PREFACE

The second edition of CRC Press's *Handbook of Microbiology* is a major reference source in this field, now consisting of nine volumes and nearly 7000 pages. While this "main frame" publication is essential to major libraries and institutions, CRC Press realizes that its size and cost limit its availability to individual investigators and students.

Accordingly, this one-volume *Practical Handbook of Microbiology* has been prepared as a condensed version of the most generally useful portions of the larger work. It also includes a table of contents for all the information contained in the volumes of the *Handbook of Microbiology* in their entirety, so that the needful reader can be guided to the wealth of data in the thousands of pages from which this abbreviated volume has been gleaned.

Wm. M. O'Leary
New York, 1989

ACKNOWLEDGMENTS

Except for "Major Culture Collections and Sources", the contents of this volume were excerpted from Volumes I (pages 1 through 248), II (pages 251 through 348), IV (pages 349 through 453), and VI (pages 455 through 641) of the nine-volume *CRC Handbook of Microbiology*, 2nd ed., Laskin, A. I. and Lechevalier, H. A., Eds., published by CRC Press, Inc. between 1977 and 1988.

THE EDITOR

William M. O'Leary, Ph.D., Professor of Microbiology at the Cornell University Medical College, received a B.S. in Bacteriology in 1952, an M.S. in Immunology in 1953, and a Ph.D. in Microbial Biochemistry in 1957, all from the University of Pittsburgh.

He then spent several years at the Division of Biology and Medicine of the Argonne National Laboratory, working on devising techniques for studying the use of radioactive isotopes in investigations of microbial metabolism. After this, he moved to New York City and joined the Faculty of the Department of Microbiology of Cornell's Medical College.

Dr. O'Leary is a member of many scientific societies, including the American Academy of Microbiology, the American Society of Biological Chemists, the New York Academy of Medicine, and the Harvey Society. His membership in honorary groups includes Sigma Xi and Phi Beta Kappa.

His long-time major research interest has been the nature, biosynthesis, and function of microbial lipids, although he has also been active in such diverse fields as high-speed instrumental identification of pathogens, microbial causes of infertility, and modes of acton of antibiotics.

Dr. O'Leary is the author or co-author of many research papers and books. Currently he is the Editor of CRC's *Critical Reviews in Microbiology,* the Editor of this *Practical Handbook of Microbiology,* and was a contributor to and an advisor for CRC's multi-volume ''main frame'' *Handbook of Microbiology.*

TABLE OF CONTENTS

INTRODUCTION TO THE BACTERIA

H. A. Lechevalier

Bacteria are prokaryotic organisms (Table) that, if photosynthetic, do not produce oxygen. Most bacteria are quite small, being rods, cocci, or filaments that range from 0.5 to 1 μm in diameter. Since the resolution of the light microscope is of the order of 0.2 to 0.3 μm, it is easily understandable that no great progress was made in the cytology of bacteria before the introduction of the electron microscope and the development of allied methods of shadowing, thin-sectioning, and staining.

Like any other group of organisms, bacteria can be classified in various ways. In the following pages, an effort has been made to relate the various chapters to the corresponding parts of the 8th edition of *Bergey's Manual,*[1] to which readers are referred for further detail.

In the following presentation, some of the groups of bacteria are discussed by specialists in the field; other groups are simply represented by tables giving a summary of the information presented in *Bergey's Manual.*

Table 1
PROPERTIES OF PROKARYOTIC AND EUKARYOTIC CELLS

	Prokaryotes	Eukaryotes
Nuclear membrane	–	+
Histones	–	+
Sterols	–[a]	+
Mitochondria	–	+
Ribosomes	70S	80S[b]
Murein	+	–
Flagella	1 fibril	20 fibrils

[a] Sterols are present in a few bacteria, such as mycoplasmas.
[b] 70S ribosomes are present in mitochondria and in chloroplasts.

REFERENCE

1. **Buchanan, R. E. and Gibbons, N. E. (Eds.),** *Bergey's Manual of Determinative Bacteriology,* 8th ed., Williams & Wilkins, Baltimore, 1974.

Editor's note: The 8th edition of *Bergey's Manual* is a respected and widely used prime reference source. A new version edited by J. G. Holt and P. A. Sneath, entitled *Bergey's Manual of Systematic Bacteriology,* appeared in 1986, and was published by Williams & Wilkins. The interested reader may wish also to consult this publication.

THE SPIROCHAETALES*

R. M. Smibert

Members of the order Spirochaetales are slender, flexuous, unicellular, helically coiled organisms 5 to 500 μm long and 0.1 to 3 μm wide; they have one or more complete turns in the helix. The organisms are Gram-negative, but are best observed by dark-field microscopy or phase-contrast microscopy. They are motile, with a rapid whirling about the long axis of the cell, flexuous, and have movement in a corkscrew or serpentine fashion. Spirochetes consist of an outer cell envelope and an inner protoplasmic cylinder. Between the cell envelope and the protoplasmic cylinder are axial fibrils that are inserted into the cylinder wall at each end of the cell and extend along the cell toward the opposite end of the cell. Bizarre forms of spirochetes may be seen. Bullae, for example, are swellings of the cell envelope. In old cultures, spirochetes may coil up, forming spheres or coccoid forms that may break up into granules. The coccoid forms and granules are also found in cultures treated with chemicals such as penicillin.[1] Spirochetes are both aerobic and anaerobic. They are found free-living or saprophitic in nature or are parasitic. Some species are pathogenic for man and animals.

The genera of Spirochaetales may be divided into two groups on the basis of their relationship to oxygen. Anaerobic genera include *Spirochaeta, Borrelia, Treponema,* and *Cristispira;* the obligate aerobic genus is *Leptospira.*

Spirochaeta are found free-living in nature. Some are anaerobes, others are facultative anaerobes. They are not very strict anaerobes.

Borrelia cause relapsing fever in man and are transmitted by lice and ticks. They are not very strict anaerobes.

Treponema are found in the intestines, oral cavity, and genital tract of man and animals. Some are pathogens. They are anaerobes and range from strict anaerobes to those that will tolerate some oxygen.

Cristispira are found in the intestinal tract of mollusks. They are the largest of the spirochetes and have a characteristic lateral ridge ("crista") made by a large bundle – fifty to several hundred – of axial fibrils. They have not been cultured.

Leptospira are thin, tightly coiled spirochetes that usually have one or both ends hooked. Some are pathogenic to man and animals, others are saprophytes. Leptospires are the only spirochetes that are obligate aerobes.

Morphology and Ultrastructure

The different genera of spirochetes have a structural similarity that makes it desirable to consider their morphology from a comparative viewpoint.[2-32] All spirochetes are helically coiled; there is a note in the literature, however, stating that *Treponema pallidum* may be a flat wave twisted into one to five different planes.[33]

Spirochetes have an outer cell envelope – usually made up of three layers – that is from 80 to 140 Å thick, a space between the envelope and the protoplasmic cylinder, a protoplasmic cylinder made up of three layers about 80 to 100 Å thick, and an axial fibril (filament) system inserted by terminal bulbs into each end of the protoplasmic cylinder. The terminal bulbs of the axial fibrils are composed of a hook region and of discs that are structurally similar to those found on bacterial flagella. The axial fibril system runs along the cell from one end to the other in the space between the inner surface of the cell envelope and the outer surface of the protoplasmic cylinder.

Spirochetes differ in the structure and number of fibrils in the system. The axial fibril

* Part 5 in *Bergey's Manual.*

in the leptospires has also been called an axostyle, and the diameter of the structure appears to be wider than that of fibrils of other spirochetes. A cross section of leptospira axial filaments viewed under electron microscopy appears to have a fiber bundle containing 12 to 15 individual fibers;[2] the axial fibril is encased in a sheath; other papers on the structure of *Leptospira* do not show many fibers encased in a sheath. *Spirochaeta, Treponema, Borrelia,* and *Cristispira* have one or more loose axial fibrils without an outer covering that binds them all together. *Leptospira* have only one set of axial fibrils, *Treponema* have 1 to 10 sets, and *Borrelia* have 15 to 30 sets.

Intracytoplasmic microtubules have been found in most species of *Treponema* studied.

The outer cell envelope or outer cell membrane of *Treponema, Spirochaeta, Borrelia,* and nonvirulent *Leptospira* is three-layered, whereas virulent *Leptospira* has a five-layered outer cell envelope. These structures are now being isolated and studied.

Pathogenic treponemes, i.e., *T. pallidum, T. pertenue,* and *T. paraluis-cuniculi,* have been reported to have pointed ends, and the axial fibrils are inserted into the protoplasmic cylinder subterminally. The species of nonpathogenic treponemes have blunt ends, and the axial fibrils are inserted into the protoplasmic cylinder terminally. The pathogenic treponemes also seem to have a bulb on the end of the cell, which is not seen on nonpathogenic treponemes.

SPIROCHAETA

Spirochaeta are 5 to 500 μm long and 0.2 to 0.75 μm wide. They are motile, free-living in nature, and are usually found in H_2S-containing fresh- or sea-water mud as well as in sewage and polluted waters. Those species cultivated are either strict or facultative anaerobes and do not require animal sera in the media. The guanidine-plus-cytosine (G + C) content of their DNA ranges from 50 to 67%. In the known species, two axial fibrils are usually found – one inserted at each end of the protoplasmic cylinder.[34] So far, five species have been recognized in the genus.

Spirochaeta plicatilis is the type species of the genus. It is 0.5 to 0.75 μm wide and 50 to 250 μm long; it has not been cultivated. *Spirochaeta plicatilis* may be either an anaerobe that can tolerate small amounts of oxygen or a microaerophile that can also grow anaerobically.[33]

Spirochaeta stenostrepta is 0.2 to 0.3 μm wide and 15 to 45 μm long. Colonies measure 2 mm in diameter; they are subsurface, white, spherical, and fluffy.[34,36] The organism is strictly anaerobic, catalase-negative, and grows at temperatures from 15 to 40°C. Optimal temperature is between 35 and 37°C; optimal pH is between 7.0 and 7.5. Carbohydrates are fermented by the Embden-Meyerhof pathway and act as energy sources; amino acids are not fermented. An unidentified growth factor is present in yeast extract. Biotin, riboflavine, and vitamin B_{12} are either required or stimulatory; carbon dioxide is not required for growth. Generation time is 6 hr. The HL antigen, which is genus-specific for leptospires, is not present. The G + C content of the DNA is 60.2%. A rubredoxin has been isolated from *S. stenostrepta* that contains one atom of iron per molecule and no inorganic sulfide; its molecular weight is around 6,000.[35]

Spirochaeta zuelzerae is 0.2 to 0.35 μm wide and 8 to 16 μm long.[34-38] Colonies are subsurface, white, fluffy, and spherical; they have a tendency to spread in the agar medium. The organism is strictly anaerobic, catalase-negative, and grows at 20°C, but not at 45°C. Optimal temperature is between 37 and 40°C; optimal pH is between 7 and 8. Carbon dioxide is required for growth. A fermentable carbohydrate is needed as an energy source. Cells contain a protein antigen that reacts with syphilitic serum. The G + C content of the DNA is 56.1%.

Spirochaeta aurantia is 0.3 μm wide and 5 to 35 μm long.[34,39] When grown anaerobically, colonies measure 1 mm in diameter and are subsurface, white, and fluffy.

When grown aerobically, colonies range from 2 to 4 mm in diameter and are partially subsurface, colored yellow to orange, and round with slightly irregular edges. The pigment in the latter case is a carotenoid of two fractions; one has been tentatively identified as *trans*-lycopene, and the other as an isomer of lycopene. According to another paper, the major pigment of *S. aurantia* strain J1 was found to be 1′,2′-dihydro-1′-hydroxytorulene. Another facultative anaerobic spirochete, strain RS1, contained as a major pigment 4-keto-1′,2-dihydro-1′hydroxytorulene (deoxyflexixanthin), and a minor pigment similar to that found in *S. aurantia* J1.[217] The organism *S. aurantia* is a facultative anaerobe, catalase-positive, reduced nitrate to nitrite, and grows at 15°C, but not at 37°C. Optimal temperature is 30°C; optimal pH is between 7.0 and 7.3. The organism uses carbohydrates, but not amino acids, as energy sources. There is no growth without a carbohydrate in the medium. Thiamine and biotin are required; adenine, guanine, and uracil are not required. Generation time is 3.8 hr. Leptospiral HL antigen is not present. The G + C content of the DNA is 66.8%. Anaerobic fermentation of glucose is shown in Table 2. Aerobic oxidation of carbohydrates by *S. aurantia* yields CO_2, acetate, pyruvate, and lactate. A cytochrome b and cytochrome O were found in *S. aurantia.*[40] A rubredoxin and an unstable ferredoxin have also been found in this organism.[37]

Spirochaeta litoralis is 0.4 to 0.5 μm wide and 5.5 to 7 μm long.[49] Colonies are 1 to 5 mm in diameter, subsurface, cream-colored, spherical, and fluffy. The organism is obligately anaerobic and obligately marine. Optimal concentration of NaCl is between 0.2 and 0.3*M* Optimal temperature is 30°C; optimal pH is between 7 and 7.5. Fermentable carbohydrates are required as an energy source. Glucose is catabolized by the Embden-Meyerhof pathway, and a clostridial-like clastic reaction was found in degrading pyruvate to acetyl-coenzyme A, CO_2, and H_2.[42] A rubredoxin was found, but no ferredoxin. The G + C content of the DNA is 50.6%.[41]

Carbohydrates fermented by species of *Spirochaeta* are shown in Table 1. End products of glucose fermentation are presented in Table 2. Enzymes found in cell extracts of *S. stenostrepta* are listed in Table 3.

Chemical Composition

Chloroform-methanol-extractable lipids or *Spirochaeta zuelzerae* consist of 9% neutral lipids, 37% glycolipids, and 54% phospholipids.[43] Glycolipids consist of 27% glucosyldiglyceride and 7% lysoglucosyldiglyceride. The phospholipids consist of 31% phosphatidylglycerol, 16% cardiolipin, and 5% lysocardiolipin. An unidentified glycolipid (3%) and an unidentified phospholipid (2%) were found. The neutral lipids contained 7% fatty aldehydes and 1% fatty acids; the remaining 1% was an unidentified fraction. Of the fatty acids of *S. zuelzerae,* 30% were straight-chain saturated fatty acids, 47% were branched-chain fatty acids, and 23% were monounsaturated acids. The saturated and unsaturated straight-chain acids were C_{12}:0, C_{14}:0, $C_{14}:1\Delta^5$, $C_{14}:1\Delta^7$, C_{15}:0, C_{16}:0, $C_{16}:1\Delta^7$, $C_{16}:1\Delta^9$, $C_{18}:1\Delta^9$, and $C_{18}:1\Delta^{11}$. Branched-chain acids were iso-C_{13}:0, iso-C_{14}:0, iso-C_{15}:0, and iso-C_{16}:0. Fatty aldehydes present were similar to the fatty acid spectrum. Slightly different results on fatty acids present in *S. zuelzerae* have been reported in Reference 44.

According to another report,[45] *S. litoralis* had phosphatidyl glycerol, cardiolipin, and phosphatidic acid; *S. stenostrepta* also contained phosphatidyl serine; *S. zuelzerae* and *S. aurantia* contained phosphatidyl glycerol and cardiolipin. All contained a monoglycosyldiglyceride; the carbohydrate portion of the glycolipid of *S. aurantia* and *S. zuelzerae* was glucose, *S. stenostrepta* had galactose, and *S. litoralis* had mannose.

These authors[45] also reported that the fatty acids of *S. aurantia* were iso-13:0, 14:0, iso-15:0, 16:0, 16:1, iso-17:0, 18:0, and 18:1; *S. litoralis* had 14:0, iso-15:0, 16:0, and iso-17:0; *S. stenostrepta* had iso-14:0, 14:0, either iso-15:0 or anteiso-15:0, iso-16:0,

Table 1
CARBOHYDRATES FERMENTED BY SPECIES OF *SPIROCHAETA*[36–41]

Carbohydrate	*S. stenostrepa*	*S. aurantia*	*S. zuelzerae*	*S. litoralis*
Glucose	+	+	+	+
Galactose	+	+	+	+
Mannose	+	+	+	+
Fructose	+	+	–	+
Sucrose	+	+	–	+
Lactose	+	+	–	+
Maltose	+	+	+	+
Cellobiose	+	+	+	+
Ribose	+	–		–
Arabinose	+	+	+	+
Xylose	+	+	+	+
Trehalose		+	+	+
Starch			+	
Dextrin		+		
Rhamnose		+	–	+
Raffinose		–	–	+
Inulin		+	–	+
Mannitol		+	–	–
Sorbitol		–	–	–
Sorbose		–	–	–
Glycerol		+		–
Dulcitol		–		–
Fucose				+
Lactate		–		+
Pyruvate		–		+
Acetate		–		

Code: + = sugar fermented; – = sugar not fermented; blank space = no information available.

Note: Ethanol, allantoin, uric acid, succinate, orotic acid, fumarate, and α-ketoglutarate are not attacked by *S. aurantia* and *S. litoralis*.

Table 2
END PRODUCTS OF GLUCOSE FERMENTATION BY SPECIES OF *SPIROCHAETA*[34–42]

Product	*S. stenostrepta*	*S. aurantia*	*S. litoralis*	*S. zuelzerae*	Strain Z4
Acetate	20.4	69.2	37.5	+	94.8
Lactate	8.2	1.0	6.5	+	56.8
Formate	–	5.2	2.8	–	10.7
Pyruvate	–	–	0.3	–	–
Succinate	–	–	–	+	26.3
Acetoin	–	tr	–	–	–
Diacetyl	–	tr	–	–	–
Ethanol	146.2	151.0	109.5	–	10.5
CO_2	187.5	165.3	127.5	+	72.7
H_2	27.2	107.7	74.10	+	186.9

Code: The values given represent μmoles product/100 μg glucose; tr = product present in trace amounts; + = product present; – = product not found.

Note: Strain Z4 is similar to *S. stenostrepta*, but not identical.

Table 3
ENZYMES FOUND IN CELL EXTRACTS OF *SPIROCHAETA STENOSTREPTA*[36]

Enzyme		Enzyme	
Hexokinase	+	Pyruvate kinase	+
Glucosephosphate kinase	+	Phosphogluconate dehydrogenase	+
Phosphofructokinase	+	Glucose-6-phosphate dehydrogenase	–
Fructosediphosphate aldoase	+	Gluconokinase	–
Glyceraldehydephosphate dehydrogenase	+	Glycerol kinase	–
Triosephosphate isomerase	+	Glycerolphosphate dehydrogenase	–
Phosphoglyceromutase	+	Glyceroldehydrogenase	–
Phosphopyruvate hydratase	+		

Code: + = enzyme present – = enzyme not present.

16:0, 16:1, either iso-17:0 or anteiso-17:0, and 18:0; and *S. zuelzerae* had 12:0, iso-13:0, iso-14:0, 14:0, either iso-15:0 or anteiso-15:0, iso-16:0, 16:0, 16:1, either iso-17:0 or anteiso-17:0, 18:0, and 18:1.

Spirochaeta zuelzerae grows in a fatty-acid-free medium. It can synthesize all its fatty acids, fatty aldehydes, and lipids *de novo* from glucose or acetate.[43] Monoenoid fatty acids are synthesized by the anaerobic pathway, in which β-hydroxy fatty acylthioesters of medium chain length undergo β,γ-dehydration to yield Δ^3 monoenoic acylthioesters that are then chain-elongated.[43,46] All species of *Spirochaeta* can synthesize fatty acids *de novo,* and some can synthesize unsaturated fatty acids. None of the *Spirochaeta* species contain phosphatidyl choline or phosphatidyl ethanolamine.[45]

The peptidoglycan of *S. stenostrepta* and *S. litoralis* was isolated, and the composition was determined.[47] Hydrolysates of the isolated peptidoglycan from these two species contained glucosamine, muramic acid, glutamic acid, alanine, and L-ornithine. Diaminopimilic acid (DAP) was not found. Cell walls of *Spirochaeta, Treponema,* and *Borrelia* contain ornithine but no DAP; *Leptospira* cell walls have DAP, but not ornithine.[48]

BORRELIA

Borrelia are spirochetes that are 5 to 20 μm long and 0.2 to 0.5 μm wide. They are motile and have 12 to 15 axial fibrils at each end of the cell. There are three to ten spirals to the cells, the average being five to seven spirals. The organisms are loosely coiled; the amplitude of their spiral is about 1 μm. *Borrelia* cause relapsing fever in man, and a similar condition in animals; they are transmitted by lice or ticks.

The taxonomy and classification of *Borrelia* is poor and confusing. Until they are cultivated in vitro and systematically studied, the classification will remain in its present state. Classification of *Borrelia* is now accomplished mainly by naming the species of *Borrelia* after the species of host vector. Species of *Borrelia* and their vectors are listed in Table 4. Type species of the genus is *B. anserina.*

Cultivation

Until recently, *Borrelia* species have not been cultivated in vitro. Most culture media used by early investigators maintained the organisms, but did not allow unlimited subculturing. In 1971, Kelly[49] reported the cultivation of *B. hermsii* in artificial medium; *B. parkerii* and *B. turicatae* were also cultivated by Kelly. These results were confirmed in the present author's laboratory, where the original medium was modified by the addition of glutamine and asparagine (2.5 g/l of each). Kelly[50] has also reported media for cultivation of *B. hipanica* and *B. recurrentis.* The medium for *B. hermsii* is unsuitable for *B. hispanica* and *B. recurrentis.* The medium for *B. hispanica* supports

Table 4
CLASSIFICATION OF *BORRELIA*

Organism[a]	Vector[a]	Reservoir	Geographical distribution
	LOUSE-BORNE *BORRELIA*		
B. recurrentis	*Pediculus humanus*		
[obermeieri]			Cosmopolitan
[berbera]			North Africa
[carteri]			India
[novyi]			U.S.A.
[kocki]			North America
[aegyptica]			
	TICK-BORNE *BORRELIA*		
B. hispanica [subsp. *maraccana*] [subsp. *mansouria*]	*Ornithodoros erraticus erraticus*		North Africa, Spain, Portugal
B. crocidurae group	*Ornithodoros erraticus sonrae*	Rodents	Central Africa, Turkey, Middle East
B. microti	*Ornithodoros erraticus sonrae*	Rodents	Middle East, Iran
B. merionesi	*Ornithodoros erraticus sonrae*	Rodents	West Africa
B. dipodilli	*Ornithodoros erraticus sonrae*	Gerbils	East Africa, Kenya
B. duttonii	*Ornithodoros moubata saviguyi*		Tropical East and South Africa Madagascar, Sengal, Arabia
B. graingeri	*Ornithodoros graingeri*		Kenya
B. persica	*Ornithodoros*		
	tholozani	Rodents	Eastern Mediterranean
[uzbekistana]	*papillipis*	Rodents	Arabian peninsula, Iran
[sogdiana]	*crossi*	Rodents	Central Asia
[babylonensis]	*asperns*	Rodents	
B. caucasica	*Ornithodoros verrucosus*	Rodents	Caucasus
B. latyschewii	*Ornithodoros tartakauskyi*	Rodents, dogs	Caucasus
B. venezuelensis [neotropicalis]	*Ornithodoros rudis [venezuelensis]*	Rodents	Central and South America
B. turkmenica	*Ornithodoros cholodkouskyi*		
B. mazzotti	*Ornithodoros talaje*	Rodents	Central and South America, Texas
B. parkerii	*Ornithodoros parkeri*	Rodents	Western U.S.A., Canada
B. turicatae	*Ornithodoros turicata*	Rodents	Central and South America, Western U.S.A., Texas, Canada
B. hermsii	*Ornithodoros hermsi*	Rodents	Western U.S.A.
B. brasiliensis	*Ornithodoros brasiliensis*	Rodents	South America
B. dugesii	*Ornithodoros dugesi*	Rodents	Central America, Mexico
B. tillae	*Ornithodoros*	Rodents	South Africa
Unnamed species	*Ornithodoros talaje*		Central and South America, Western U.S.A., Canada

Table 4 (continued)
CLASSIFICATION OF *BORRELIA*

Organism[a]	Vector[a]	Reservoir	Geographical distribution
	TICK-BORNE ANIMAL-PATHOGENIC *BORRELIA*		
B. harveyi	Unknown		
B. theileri	*Rhipicephalus evertsi*		
B. anserina	*Argas miniatus*		
	percisus		
	reflexus		

[a] Names in brackets are possible subspecies or synonyms.

Note: Blank spaces = unknown, or information not available.

growth through many passages, but with low yields (7×10^6/ml); the new medium devised for *B. recurrentis* also supports growth with low yields (5×10^6/ml). Table 5 shows the composition of Kelly's media for *Borrelia.* In preparing the media, certain precautions are essential, in order to obtain media that are not toxic to the organisms. Glassware must be cleaned by methods satisfactory for tissue culture. The media must be filter-sterilized by pressure filtration, using nitrogen, and filtered only once, using a membrane filter; other methods of filtration make the media incapable of supporting growth. All ingredients should be made with double-distilled water or with distilled water passed through an ion-exchange resin.

Instructions for preparing the culture media described in Table 5 are given below. Double-distilled water is used for all solutions. Glassware must be cleaned to meet the requirements for tissue cultures. Serum does not have to be inactivated. The final pH of the media should be between 7.6 and 7.8.

Medium A

1. Prepare the sodium bicarbonate solution by dissolving 4.5 g of $NaHCO_2$ in 95.5 of distilled water. Make the solution fresh each time medium is made.
2. Prepare the bovine albumin solution by adding 10 g of bovine serum albumin (fraction V) to 90 ml of distilled water. Adjust the pH to 7.8.
3. Prepare the gelatin solution by dissolving 7 g of Bacto-gelatin in 93 ml of distilled water. Autoclave the solution.
4. Prepare a 0.5% solution of phenol red.
5. To 80 ml of basal solution add 34 ml of albumin solution, 4 ml of bicarbonate solution, 0.7 ml of phenol red solution, and 1.3 ml of distilled water. With the exception of the gelatin solution and serum, mix all the ingredients required for the culture medium. Sterilize the mixture by pressure filtration through a 0.22-μ membrane filter. Do not use vacuum filtration.
6. Fill 6-ml aliquots of the above medium into 13×100-mm screw-cap tubes with Teflon® liner.
7. Add 2 ml of warm gelatin solution to each tube.
8. Add 0.5 ml of sterile rabbit serum to each tube. Mix by inversion.
9. When inoculating the medium, leave 0.5 ml of air space, closing the cap tightly. Store the medium at room temperature and use within 30 days.

Table 5
CULTURE MEDIA[a] FOR *BORRELIA*[49,50]

Constituents[b]	Kelly A	Kelly B	Kelly C
Basal Medium			
Bacto-Yeastolate		–	
Hanks BSS 10 X without $NaHCO_3$, with phenol red	–	13.5 ml	–
$Na_2HPO_4 \cdot 7H_2O$	26.52	–	20 mg
$NaH_2PO_4 \cdot H_2O$	1.03	–	1.03
$KH_2 \cdot PO_4$	–	–	12 mg
NaCl	1.20	–	1.38
KCl	0.85	–	0.85
$MgCl_2 \cdot 6H_2O$	0.68	–	0.68
Glucose	12.75	0.3	550 mg
Bacto-proteose peptone No. 2	5.95	0.33	1.3
Bacto-proteose peptone No. 4	–	0.3	–
Bacto-tryptone	2.55	–	–
Sodium pyruvate	1.06	0.08	–
Sodium citrate dihydrate	0.47	0.07	34 mg
N-Acetyl-D-glucosamine	0.53	0.04	–
Bovine serum albumin	–	3	4.4
Glutamine	–	0.05	–
Choline chloride		0.01	3 mg
Asparagine	–	–	11 mg
$CaCl_2 \cdot 2H_2O$	–	–	33 mg
Distilled water	1 liter	81.5 ml	100 ml

[a] Medium A is for *B. hermsii, B. parkerii,* and *B. turicatae*; medium B is for *B. hispanica*; medium C is for *B. recurrentis.*
[b] Unless otherwise specified, amounts given represent g/l.

Medium B:

1. To 95 ml of basal solution add 5 ml of freshly prepared 2% $NaHCO_3$ solution.
2. Filter-sterilize through a 0.22-μ membrane filter.
3. Disperse 5 ml of the medium into sterile 9-ml screw-cap tubes.
4. Add 0.5 ml of sterile rabbit serum, 0.1 ml of L-cystine solution (5 mg/ml, pH 1.0), and 2.4 ml of sterile distilled water.
5. Readjust pH visually to about 7.8 with sterile 1*N* NaOH.

Medium C

1. To 100 ml of basal medium add 10 ml of sterile solution 2 (freshly prepared, 1.4 g of $NaHCO_3$ and 0.11 g of ascorbic acid dissolved in 100 ml distilled water).
2. Add 4 ml of medium to sterile 9-ml screw-cap tubes.
3. Add 1.5 ml of warm 10% gelatin solution, 0.5 ml of rabbit serum, and 2.5 ml of sterile distilled water.
4. Mix by inversion.

A spirochete called Agent 277F, isolated from *Haemophysulis temporispalustris,* was cultivated in medium 10, which consisted of pooled chorioallantoic and amniotic fluids from 12-day-old chicken embryos, fortified with 0.2% casamino acids, and crystallized bovine plasma albumin.[51] The medium was adjusted to pH 6.5 and filter-sterilized. Red blood cells (5% final concentration) from 13-day-old chicken embryos were added to the medium; the red cells, however, were found not to be essential for growth.

Laboratory animals that can be infected by different species of *Borrelia* are listed in Table 6. Mice are susceptible to *Borrelia* that infect man; young mice are usually more susceptible than old mice. Monkeys are susceptible to most *Borrelia.* Young rabbits may be infected by some species, but are more useful for antibody production. Embryonating chicken eggs have been shown to cultivate several species of *Borrelia,* such as *B. recurrentis, B. duttonii, B. anserina,* and Agent 277F.

The wild-animal reservoir of *B. hermsii* has been studied, and it was shown that *B. hermsii* can infect laboratory mice, rats, guinea pigs, and monkeys, as well as rodents found in nature.[56] Table 7 offers additional information regarding this and other species of *Borrelia.*

Borrelia anserina infects ducks, turkeys, chickens, and geese. Turkey poults (1 to 3 weeks old), inoculated intramuscularly or subcutaneously, had spirochetes in their blood in 24 to 48 hr.[57] Young turkeys (10 to 16 weeks old) rarely had spirochetes in their blood before 48 hr. Adult turkeys (7 months old) had spirochetes in their blood in 48 hr; the organisms could be found there for 4 days. Spirochaetemia usually last only about 3 to 4 days in birds of any age; 6-month-old chickens and 2-year-old hens infected with *B. anserina* have a spirochaetemia that usually clears up in 5 to 6 days with the disappearance of the organism from blood and tissues.[58,59] In general, spirochetes appear in the blood of chickens 24 to 48 hr after inoculation, depending on their age and on the amount of inoculum. Their numbers increase rapidly, so that many cells are present in the blood by 72 to 120 hr; after 120 hr the spirochetes decline rapidly in number and disappear.

Vaccines against human *Borrelia* have been studied. They protected mice only against the homologous strain. *Borrelia anserina* vaccines protected chickens against challenge and were 98.5% effective.[59]

Table 6
LABORATORY ANIMALS CAPABLE OF BEING INFECTED BY *BORRELIA*[51-55]

Organism	Mice	Rats	Guinea pigs
B. recurrentis	Young only; disease mild; blood positive for 3 to 5 days.	Young only; blood positive for 1 week.	Some young; adults not infected.
B. hispanica	Young only; disease mild; blood positive for 2 to 5 days.	Young only, disease mild; blood positive for 1 to 2 weeks.	Disease severe.
B. crocidurae group	Young only; disease of long duration.	Young only; disease of long duration.	Usually not infected.
B. duttonii	Old only; disease of long duration.	Old only, disease of long duration.	Young only; disease of long duration.
B. persica	Long incubation required; disease mild.	Long incubation required; disease mild.	Disease severe.
B. latyshevyi	Sick; only few cells in blood.	Not infected.	Not infected.
B. venezuelensis	Disease mild; blood positive for 1 to 2 weeks.	Disease mild; blood positive for 1 to 2 weeks.	Disease mild.
B. mazzottii	Disease mild; blood positive for 1 to 2 weeks.	Disease mild; blood positive for 1 to 2 weeks.	Not infected.
B. turicatae	Disease mild; few cells in blood.	Disease mild; few cells in blood.	Young only
B. parkerii	Disease mild; few cells in blood.	Disease mild; few cells in blood.	Young only.
B. hermsii	Disease mild.	Disease mild.	Disease mild.
B. brasiliensis	Disease mild; blood positive for 1 to 2 weeks.	Disease mild; blood positive for 1 to 2 weeks.	Disease mild; blood positive for 1 to 2 weeks.
B. anserina	Not infected.	Not infected.	Not infected.
Agent 277F	Not infected.	Not infected.	Not infected.

Metabolism

Biochemical activities of *Borrelia* are not well understood. Only a few reports are available; studies have been hampered by the inability to cultivate *Borrelia* in vitro. Cell-free extracts of *B. recurrentis* utilized glucose via the Embden-Meyerhof pathway.[60,61] *Borrelia novyi* was found to ferment glucose with formation of lactic acid (65%) and CO_2 (10%). *Borrelia recurrentis* fermented glucose with production of lactic acid. Hexokinase, phosphoglucoisomerase, phosphofructokinase, aldolase, phosphoglyceraldehyde dehydrogenase, triose-phosphoglycerate kinase, phosphoglyceromutase, enolase, pyruvate kinase, and a DPN-dependent lactic dehydrogenase were found in cell-free extracts of *B. recurrentis.*[61-64] Current studies in the present author's laboratory show that *B. hermsii, B. parkeri,* and *B. turicatae* ferment glucose with production of lactic acid, the latter being the only acid end product; in addition, *B. hermsii* and *B. parkeri* ferment maltose, trehalose, dextrin, starch, and glycogen, and *B. turicatae* ferments raffinose and dextrin.

Borrelia hermsii, B. parkeri, and *B. turicatae* utilize lysophosphatidylcholine from serum-containing culture medium.[50,65] Phosphatidylcholine and other phospholipids in serum were not used. *Borrelia* have phospholipase B (lysolecithinase), which yields one fatty acid and L-α-glycerophosphoryl choline; the latter is acted upon by glycerophosphorylcholine diesterase to yield choline and L-α-glycerophosphate, and this in turn, is

Table 7
WILD ANIMALS INFECTED BY *BORRELIA* EITHER NATURALLY OR IN THE LABORATORY[53-56]

Animal	*B. recurrentis*	*B. duttonii*	*B. hispanica*	*B. crocidurae* group	*B. persica*	*B. turicatae*	*B. hermsii*
Gerbil	+			+			
Dog		+	+	+		+	
Horse		+		+			
Goat		+		+			
Sheep		+		+	+		
Hedgehog		–	+	+	+		
Rat			+				
Jackal			+				
Fox			+			+	
Bat			+	+	±		
Weasel			+				
Porcupine			+				
Wild mouse			+	+	+		
Pig			–			+	
Donkey			–				
Cat			–			+	
Shrew				+			
Hamster				+			
Bandicoot				+			
Vole							+
Flying squirrel							–
Cotton rat						+	
Squirrel	+						
Wild rat					+		
Pine squirrel							+
Chipmunk							+
Ground squirrel							–
Wood rat							–
Deer mouse							–

Code: + = animal infected either naturally or in the laboratory; – = animal not infected; blank space = no information

acted upon by acid phosphatase to yield inorganic phosphate and glycerol. Lipases acting on glycerides were not found nor were phospholipase A, D, and C. *Borrelia* do not grow in lipid-free media, but will grow with both saturated and unsaturated long-chain fatty acids in medium with lipid-free serum albumin. *Borrelia* get their fatty acids from free fatty acids in serum-containing medium or from hydrolysis of lysophosphatidylcholine.

Immunity

Man and animals develop immunity against *Borrelia* during the first attack of relapsing fever. Relapse strains or variants develop in vivo, causing relapse attacks. The relapse variants are antigenically different from the attack strain, virulence of a relapse strain is usually lower than that of the original attack strain. How and why *Borrelia* can change their antigenic characteristics remains a mystery, but it seems that the relapse occurring in the infection is an immunological phenomenon resulting from the genetic capability of the organism to undergo antigenic variations. The number of antigenic variants differs from strain to strain and from species to species. *Borrelia recurrentis* has been reported to have nine phases, designated A to I;[55] phase A was the attack variant, B the first relapse phase, and phases C to I were found in the second and additional relapses. In man, phases A and B are the ones most frequently found.

Studies in rats showed *B. hermsii* to have four major serotypes, namely, O, A, B, and C.[57,66] In rats infected with serotype O, the first relapse contained serotype A, the second contained serotype B, and the third contained serotype C; thus, the relapse serotypes appeared in sequential order of A, B, and C. When a relapse serotype (A, B, or C) was used to initiate infection of rats, there was a tendency in the relapse to revert to the serotype that preceded them in infections started with serotype O.

Antigens

Antigens of some species of *Borrelia* have been studied. Three antigenic factors were isolated from *B. turicatae* and *B. parkeri* and were labeled A, B, and C.[67] Antigen B was shared by both species of *Borrelia* and may be considered genus-specific; antigens A and C were strain- and relapse-specific. For more detailed information in reviews on *Borrelia* see References 53 to 55, and 68.

TREPONEMA

Treponemes are 5 to 20 μm long and 0.09 to 0.5 μm wide. They have one or more axial fibrils inserted at each end of the cell. Members of this genus are chemoorganotrophs with a fermentative metabolism; some ferment glucose, others ferment amino acids. They are anaerobes and are catalase- and oxidase-negative. Those that can be cultured require serum or one or more volatile fatty acids added to the culture medium. Treponemes are found in the oral cavity, in the intestinal tract, and in the genital regions of man and animals. Some species are pathogenic. The G + C content of the DNA of cultivatable species ranges from 32 to 50 moles %. The type species of the genus is *Treponema pallidum.*

Classification of treponemes is difficult. At our current state of knowledge they may be divided into two groups. The first group consists of pathogens propagated in laboratory animals, but not cultured in vitro; the second group comprises organisms that have been cultivated in vitro.

The pathogens in the first group are *T. pallidum,* the cause of syphilis; *T. pertenue,* the cause of yaws; *T. carateum,* the cause of pinta; *T. paraluis-cuniculi* (*T. cuniculi* is an illegitimate name), the cause of rabbit syphilis, which is reviewed in Reference 69; and two organisms without names. One of the unnamed species is the cause of non-venereal endemic syphilis in man, the other (treponeme FB) causes a natural disease in primates. A classification system of these non-cultivatable pathogens is based on the lesions produced in laboratory animals (Table 8). It has been suggested that the treponeme of endemic syphilis is a subspecies or variant of *T. pallidum* and that treponeme FB is a subspecies or variant of *T. pertenue.*[75] The only pathogenic treponeme that has been cultivated on artificial medium is *T. hyodysenteriae,*[79] which is associated with swine dysentery and is found in the mucosal lining, luminal surface, and mucosal crypts of the large intestine of infected swine.[128,129]

The cultivatable treponemes can be grown in prereduced media, and their phenotypic characteristics can be studied like those of any bacteria. Table 9 shows the classification and some characteristics of these treponemes.[76-81] Gas chromatography is very useful in determining fatty acids and alcohols that are end products of the metabolism of treponemes (Table 10). *Treponema phagedenis* comprises the Reiter treponeme, English Reiter, and all the Kazan strains. *Treponema refringens* comprises the Noguchi and avirulent cultivated Nichols strains as well as strains previously labeled *T. calligyrum. Treponema denticola* (the correct designation for *T. microdentium,* which is an illegitimate name) includes strains formerly named *T. microdentium, T. commondonii,* and *T. ambiguum.*

Colonies

The capability of growing microorganisms as isolated colonies is necessary for obtaining pure cultures that can be used for taxonomic, biochemical, and antigenic studies. Socransky et al.[82] were able to colonize some oral treponemes both on streaked and on poured plates. Others have also reported successful colonization of treponemes.[76,77,83-85]

Treponemes require either short-chain volatile fatty acids or long-chain fatty acids incorporated into culture media. Short-chain acids can be provided by rumen fluid from cattle and other ruminants or by artificial mixtures of fatty acids. Long-chain fatty acids are supplied to treponemes from animal sera. Rabbit, horse, sheep, and cattle sera have been used in culture media at a concentration of 10 to 12%. Rabbit serum is more satisfactory than other animal sera. Bovine serum albumin can replace whole serum.[86] However, oral treponemes called *T. dentium (T. denticola)* were found to be able to grow in a medium supplemented with α-globulin or α^2-globulin.[87] Albumin gave poor growth, as did β-globulin, γ-globulin, transferrin, and ceruloplasmin.

Most treponemes require long-chain fatty acids in albumin. The albumin acts as a detoxifying carrier for the long-chain fatty acids, which by themselves are toxic at concentrations necessary for growth. The Reiter and Kazan strains of *T. phagedenis* were found to use a variety of long-chain fatty acids (Table 11). Some results were probably influenced by contamination of the albumin with lipid. In another study, using lipid-poor

Table 8
SOME BIOLOGIC CHARACTERISTICS OF PATHOGENIC TREPONEMES[70-74,80,81]

Organism	Rabbit	Hamster	Mouse	Guinea pig	Chimpanzee
T. pallidum	+	–	–	+/–	+
T. pertenue	+	+	–	–	+
T. carateum	–	–	–	–	+
T. paraluis-cuniculi	+	–	–	+	+
Endemic syphilis	+	+	–	+	+
Treponeme FB	+	+			+

Code: + = produces skin lesions; – = no skin lesions; +/– = only rarely produces skin lesions.

Table 9
SOME CHARACTERISTICS OF TREPONEMES[76-81]

Species	Glucose	Lactose	Fructose	Sucrose	Mannitol	Galactose	Cellobiose	Maltose	Mannose	Trehalose	Indol	H_2S	1% Glycine, gr.	Lactate used	Esculin hydrolysis	Cell diameter, μm
T. phagedenis, biotype Reiter	+	+	+	–	+	v	–	–	+	v	+	w	+	–	–	0.25–0.35
T. phagedenis, biotype Kazan	+	+	+	–	+	+	–	–	+	v	+	w	+	–	+	0.25–0.35
T. refringens, biotype *refringens*	–	–	–	–	–	–	–	–	–	–	+	+	–	–	+	0.25–0.35
T. refringens, biotype *calligyrum*	–	–	–	–	–	–	–	–	–	–	+	+	+	–	+	0.25–0.35
T. denticola, biotype *denticola*	–	–	–	–	–	–	–	–	–	–	+	+	–	–	+	0.15–0.25
T. denticola, biotype *comondonii*	–	–	–	–	–	–	–	–	–	–	–	+	v	–	+	0.15–0.25
T. oralis	–	–	–	–	–	–	–	–	–	–	+	+		+		0.15–0.25
T. scoliodontum	–	–	–	–	–	–	–	–	–	–	–	–	–	–	–	0.10–0.15
T. macrodentium	+	–	+	+	–	v	v	+	–	–	–	+		–	–	0.15–0.25
T. vincentii	–	–	–	–	–	–	–	–	–	–	+	+	–	–	–	0.25–0.35
T. hyodysenteriae[a]	+	+	+	–	–	–	–	+	–	–	+	–			–	0.35–0.45

[a] Associated with swine dysentery.

Code: + = positive reaction or weak acid formation without gas; – = negative reaction or no acid formation; v = variable results (some strains +, some –); w = weak reaction; blank space = no data available.

albumin thioglycollate medium, the Reiter and Kazan strains of *T. phagedenis* were shown to require a pair of fatty acids for growth (Table 12); one fatty acid was saturated, and the other was unsaturated. The saturated fatty acid needed a chain length of at least 14 carbon atoms, whereas the unsaturated fatty acid needed a chain length of 15 or more carbon atoms and 1, 2, or 3 double bonds. The pair of fatty acids could be replaced only with *trans*-18-carbon monounsaturated fatty acid. Long-chain fatty acids were incorporated unchanged into the lipid of the Kazan-5 strain, so that the fatty acids in the medium ended up unchanged in the lipid of serum-requiring treponemes. Kazan-5 and Reiter strains could not synthesize fatty acids and could not modify the chain length, nor could they reduce or desaturate long-chain fatty acids.[43,90] The combination of a saturated and a *cis*-unsaturated fatty acid probably provides the right "fluidity" for the cell membrane of the flexuous treponemes. Treponemes requiring short-chain volatile fatty acids include human oral isolates as well as strains from the intestinal tract and rumen of animals (Table 13).

Composition and Metabolism

The chemical composition of some treponemes has been investigated. Most information was obtained on the Reiter and Kazan strains of *T. phagedenis*, and the least information was obtained on *T. pallidum.* Cellular fatty acids of treponemes that have been examined indicate that the main cellular fatty acids of treponemes are C-16:0, C-18:1, C-18:2, and C-18:0.[44,97] Other acids found ranged from C-10:0 to C-20:0. Another study showed Reiter treponeme to contain as major cell lipids C-16:0, C-18:0, C-18:1 Δ^9, and C-18:2 Δ^9, Δ^{12} fatty acids.[96] The fatty aldehydes in Reiter treponeme corresponded to the fatty acids present in cells. Fatty aldehydes were in rather high concentration, at 25 μM/100 mg lipid.

Lipids of treponemes have been studied[43,45,94,97-100] and are listed in Table 14. In Kazan-15, lipids comprise 18 to 20% of the dry weight of the cells, which is a larger amount of lipid than is found in most bacterial cells. Glycolipid and phospholipid make up 90 to 95% of the total lipid, and free fatty acids and aldehydes make up the remaining 5 to 10%. The monogalactosyldiglyceride found in Kazan-5 treponeme was identified as 1-(O-α-D-galactopyranosyl)-2,3-diglyceride by Livermore and Johnson;[99] another group of investigators[100] stated that the compound isolated from the Reiter treponeme had the β-linkage.

Recent work using lipid defined medium shows that *T. phagedenis* has phosphatidylcholine (PC), phosphatidylethanolamine (PE), cardiolipin (CL), and phosphatidylglycerol (PG); *T. refringens* contained PC, PS, PG, and CL; *T. denticola* and *T.*

Table 10
END PRODUCTS OF FERMENTATION OF TREPONEMES[76-78,80,81]

Fatty acid	*T. phagedenis*	*T. refringens*	*T. denticola*	*T. oralis*	*T. scoliodontum*	*T. macrodentium*	*T. vincentii*	*T. hyodysenteriae*
Acetic acid	+	+	+	+	+	+	+	+
Propionic acid	±	t	±	+	+	–	–	–
Isobutyric acid	–	–	–	–	+	–	–	–
N-Butyric acid	+	t	t	–	–	–	+	+
Ethanol	±	–	–	–	–	–	±	–
n-Butanol	±	–	–	–	–	–	±	–
n-Propanol	±	–	–	–	–	–	+	–
Lactic acid	±	±	+	–	t	+	t	–
Succinic acid	t	t	±		t		t	–

Code: + = major end product; ± = minor end product that is usually, but not always, found; t = trace amounts sometimes found; – = not found; blank spaces = no data available

Table 11
EFFECT OF FATTY ACIDS ON GROWTH OF SOME SERUM-REQUIRING TREPONEMES[86,88,89]

Fatty acid[a]	Growth	Fatty acid[a]	Growth
Caprylic acid	±	Heptylic acid	–
Lauric acid	±	Pelargonic acid	–
Tridecylic acid	±	Capric acid	–
Palmitic acid	±	Undecylic acid	–
10-Methylhexadecanoic acid	±	Myristic acid	–
Stearic acid	±	Pentadecylic acid	–
cis-6-Octadecanoic acid	+	Margaric acid	–
trans-6-Octadecanoic acid	+	2-Methylhexadecanoic acid	–
cis-8-Octadecanoic acid	+	Nonadecylic acid	–
trans-8-Octadecanoic acid	+	Arachidic acid	–
Elaidic acid	+	Acrylic acid	–
cis-10-Octadecanoic acid	+	Erotonic acid	–
Vaccenic acid	+	Sorbic acid	–
Linoleic acid	+	Undecylic acid	–
Linolenic acid	+	Undecylenyl alcohol	–
8-Octadecynoic acid	+	Myristyl alcohol	–
9-Octadecynoic acid	+	Cetyl alcohol	–
10-Octadecynoic acid	±	Oleyl alcohol	–
Erucic acid	±	Tween 80	+
Ricinoleic acid	+	Sodium oleate	+
Oleic acid methyl ester	+	TEM-4T[b]	+
Oleic acid ethyl ester	+	TEM-4C[c]	–
Arachidonic acid methyl ester	±	TEM-4S[d]	–
Formic acid	–	C_{16}:0[e]	–
Acetic acid	–	*cis*-9-C_{18}:1[e]	–
Propionic acid	–	*trans*-9-C_{18}:1[e]	–
Butyric acid	–	*trans*-9-C_{18}:1[e]	+
Valeric acid	–	*trans*-11-C_{18}:1[e]	+
Caproic acid	–	*trans*-9, 12-C_{18}:2[e]	–

[a] The strain tested on all compounds was the Reiter strain of *T. phagedenis*. The medium contained bovine serum albumin.
[b] TEM-4T = diacetyl tartaric acid ester of tallow monoglycerides.
[c] TEM-4C = diacetyl tartaric acid ester of cotton seed oil monoglycerides.
[d] TEM-4S = diacetyl tartaric acid ester of soybean oil monoglycerides.
[e] Also tested against the Kazan-5 strain.

Code: + = growth; ± = slight growth; – = no growth.

scoliodontum had PC, PE, PG, and CL; *T. vincentii* contained PC, PG, and CL.[45] According to another paper, treponemes from human oral cavity and pig feces that required short-chain fatty acids had PG, bisphosphatidylglycerol, and traces of phosphatidic acid.[94]

Glycolipids of treponemes were monoglycosyldiglycerides.[45,96] The glycolipids of *T. phagedenis, T. scoliodontum, T. vincentii,* and of the oral and fecal strains that require short-chain fatty acids contained galactose; *T. refringens* and *T. denticola* strains contained a mixture of glucose- and galactose-containing glycolipids.

The main fatty acids of these species grown in lipid defined medium were 16:0 and 18:1; *T. vincentii* and one strain of *T. denticola* also had 14:0 fatty acid, and several strains of *T. refringens* had some 14:0, 15:0, 16:1, 18:0, 18:1, and 18:3, and larger amounts of 16:0 and 18:2 fatty acids.[45] The major fatty acids found in the strains

Table 12
EFFECT OF COMBINATIONS OF FATTY ACIDS ON TREPONEMES[89]

Fatty acids	Growth: Reiter[a]	Growth: Kazan-5[a]
C_{16}:0	–	–
cis-9-C_{18}:1	–	–
C_{4}:0 + *cis*-9-C_{18}:1	–	–
C_{5}:0 + *cis*-9-C_{18}:1	–	–
C_{6}:0 + *cis*-9-C_{18}:1	–	–
C_{7}:0 + *cis*-9-C_{18}:1	–	–
C_{8}:0 + *cis*-9-C_{18}:1	–	–
C_{9}:0 + *cis*-9-C_{18}:1	–	–
C_{10}:0 + *cis*-9-C_{18}:1	–	–
C_{11}:0 + *cis*-9-C_{18}:1	–	–
C_{12}:0 + *cis*-9-C_{18}:1	–	–
C_{13}:0 + *cis*-9-C_{18}:1	–	–
C_{14}:0 + *cis*-9-C_{18}:1	+	+
C_{15}:0 + *cis*-9-C_{18}:1		+
C_{16}:0 + *cis*-9-C_{18}:1	+	+
C_{17}:0 + *cis*-9-C_{18}:1		+
C_{18}:0 + *cis*-9-C_{18}:1		+
cis-9-C_{14}:1 + C_{16}:0	–	–
cis-9-C_{16}:1 + C_{16}:0	+	+
cis-9-C_{18}:1 + C_{16}:0	+	+
cis-11-C_{18}:1 + C_{16}:0	+	+
cis-9,12-C_{18}:2 + C_{16}:0	+	+
cis-9,12,15-C_{18}:3 + C_{16}:0	+	+
cis-5,8,11,14-C_{20}:4 + C_{16}:0	–	–
trans-9-C_{18}:1		+
trans-9-C_{18}:1 + C_{8}:0		+
trans-9-C_{18}:1 + C_{10}:0		+
trans-9-C_{18}:1 + C_{12}:0		+
trans-9-C_{18}:1 + C_{14}:0		+
trans-9-C_{18}:1 + C_{15}:0		–
trans-9-C_{18}:1 + C_{16}:0		–
trans-9-C_{18}:1 + C_{17}:0		–
trans-9-C_{18}:1 + C_{18}:0		–
trans-9-C_{18}:1 + *cis*-9-C_{18}:1		+

[a] Medium contained 2% lipid-poor bovine albumin.

Code: + = growth; – = no growth; blank spaces = no data available.

Table 13
EFFECT OF FATTY ACIDS ON GROWTH OF RUMEN FLUID-REQUIRING TREPONEMES[90–96]

Fatty acid(s)	Growth
Acetic acid	–
Propionic acid	–
Isobutyric acid	+[a]
n-Butyric acid	–
Isovaleric acid	–
n-Valeric acid	–
DL-2-Methylbutyric acid	–
Caproic acid	–
Caprylic acid	–
2-Methyl valerate	–
Oleic acid	–
Linoleic acid	–
Lauric acid	–
Palmitic acid	–
Formic acid	–
Isobutyric + *n*-valeric acid	+[b]
Isobutyric + isovaleric acid	–
Isobutyric + acetic acid	–
Isobutyric + propionic acid	–
Isobutyric + caproic acid	–
Acetic + propionic acid	–
n-Butyric + isobutyric acid	–
n-Butyric + *n*-valeric acid	–
n-Butyric + acetic acid	–
n-Butyric + propionic acid	–
n-Butyric + caproic acid	–

[a] Strains of *T. macrodentium*.[90,91]
[b] Isolates from animal intestinal tract or rumen or from human oral cavity.

Code: + = growth; – = no growth.

Table 14
LIPIDS FOUND IN *TREPONEMA* AND *SPIROCHAETA*[43,45,94,97–100]

Lipid	Reiter[43]	Reiter[100]	Reiter[98]	Kazan-5[98]	Noguchi[98]	Nicholas[a98]	*S. zuelzerae*[43]
Phosphatidylcholine	+	+		+			–
Lysophosphatidylcholine	±	–		–			–
Phosphatidylglycerol	±	+		–			+
Cardiolipin	±	+		±			+
Lysocardiolipin	±	–		–			±
Unidentified phospholipid	±	±		–			±
Monogalactosyldiglyceride	+	+	+	+	+	+	–
Lysogalactosyldiglyceride	±	–		–			–
Monoglucosyldiglyceride	–	–		–			+
Lysoglucosyldiglyceride	–	–		–			+
Unidentified glycolipid	±	±		±			±
Phosphatidylethanolamine	–	–		+			–

[a] Avirulent cultivated Nichols strain.

Code: + = present; ± = small amount found; – = none found; blank space = not reported.

Note: Neutral lipids are also present, as well as free fatty acids and aldehydes.

requiring short-chain fatty acids were iso-14:0, 14:0, 15:0, iso-16:0, and 16:0 fatty acids.[94] For other information on lipids, see Reference 96.

The Reiter treponeme contains 1.8% DNA and 10.2% RNA.[105] Guanine + cytosine content of the DNA was 38 to 40% in the Reiter treponeme, 37 to 40% in *T. refringens,* 36 to 37% in *T. denticola,* 39% in *T. macrodentium,* and 37% in *T. oralis.*[34,77,78,105]

The metabolism of amino acids has been investigated in a few strains (Table 15).

Enzymes

Some enzyme systems of treponemes have been investigated.[101-104,106-117] The Reiter treponeme has been reported to have cytochromes.[106,107] *Treponema denticola* ferments glucose by the Embden-Meyerhof pathway;[104] amino acids were also fermented, and they serve as the major energy source for the organism. Only a small amount of the end products were produced from glucose. The Reiter, Kazan-2, Kazan-4, and Kazan-5 strains had acetokinase, phosphotransacetylase, and β-galactosidase activity, but no aceto-CoA-kinase activity. The pathogenic Nichols strain of *T. pallidum* did not have acetokinase, aceto-CoA-kinase, phosphotransacetylase, or β-galactosidase activity.[108,113,114] Proteolytic activity of treponemes has been reported in Table 16.[115-117]

Antigens

The antigenic relationships of treponemes have received some study;[118-127] they are shown in Table 17. The Reiter and Kazan treponemes are very similar, but not identical. The Kazan-2 strain has been shown to have two phenol-water-extractable

Table 15
AMINO ACIDS USED BY TREPONEMES[101-104]

(Used as Energy Source)

Amino acid	Reiter	*T. denticola*	Dissimilated[a] Reiter
Lysine	–	–	–
Histidine	+	–	+
Arginine	+	–	+
Aspartic acid	–	–	–
Threonine	+	–	+
Serine	+	+	
Glutamate	+	–	+
Proline	–	–	–
Glycine	–	+	–
Alanine	–	+	–
Valine	–	–	–
Methionine	–	–	–
Isoleucine	–	–	–
Leucine	–	–	–
Tyrosine	–	–	–
Phenylalanine	–	–	–
Cysteine	–	+	+

[a] Deaminated or decarboxylated by cell-free extracts.

Code: + = fermented or dissimilated by cell-free extracts; – = not used; blank space = no data available.

Table 16
PEPTIDASE ACTIVITY OF THE REITER TREPONEME[110,112,115-117]

Enzyme	Activity	Enzyme	Activity
Hyaluronidase	+	N-Acetyl-DL-glutamate	–
Proteolysis	+	N-Acetyl-DL-leucine	–
Glycylglycine	+	N-Acetyl-DL-methionine	–
Glycyldiglycine	+	L-Leucyl-β-naphthylamide	+
Glycyltriglycine	+	L-Tyrosyl-β-naphthylamide	+
Glycyl-L-leucine	+	L-Arginyl-β-naphthylamide	+
Glycyl-L-phenylalanine	+	L-Alanyl-β-naphthylamide	+
Glycyl-L-tyrosine	+	L-Pyrrolidonyl-β-naphthylamide	+
L-Leucylglycine	+	α-L-Glutamyl-β-naphthylamide	+
L-Prolylglycine	+	γ-L-Glutamyl-α-naphthylamide	+
L-Glutamine	+	Glycyl-β-naphthylamide	+
L-Asparagine	+	L-Histidyl-β-naphthylamide	+
Glutathione	+	L-Cystinyldi-β-naphthylamide	+
γ-DL-Glutamylglycylglycine	+	L-Propyl-β-naphthylamide	+

Code: + = activity found; – = activity not found.

Note: *T. denticola* strains and *T. vincentii* have been reported as not having hyaluronidase or chondroitinase activity; another report states that oral treponemes had hyaluronidase activity.

Table 17
SEROLOGICAL RELATIONSHIPS OF TREPONEMES[118-125]

	Serogroup or organism					
Serogroup or organism	***T. phagedenis*** **Reiter Kazan Kazan-2–8**	***T. refringens*** **Nichols Noguchi** ***T. calligyrum***	**Kroó**	***T. denticola*** **MRB FM N-9**	***T. vincentii*** **N-9**	***T. minutum***
T. phagedenis	+	–		–	–	∓
Reiter	+	–	–	–	–	∓
English Reiter	+	–		–	–	∓
Kazan	+	–	–	–	–	
Kazan-2	+	–		–	–	
Kazan-4	+	–		–	–	
Kazan-5	+	–		–	–	
Kazan-8	+	–		–	–	
T. refringens	–	+		–	–	∓
Nichols	–	+	–	–	–	
Noguchi	–	+	–	–	–	
T. calligyrum	–	+		–	–	∓
T. minutum	–	–		–	–	+
Kroó	–	–	+	–	–	
T. denticola MRB	–	–		+	–	
T. denticola FM	–	–		+	–	
T. denticola N-39	–	–		+	–	
T. vincentii N-9	–	–		–	+	

Code: + = cross reaction; ∓ = weak cross reaction; – = no cross reaction; blank spaces = no data available.

Note: Most strains will agglutinate each others' sera, due to possession of a common antigen on cells. Most of the data given above were derived from sera absorbed with Reiter treponen to remove the common antigen.

polysaccharide antigens not possessed by the Reiter strain, as well as three other antigens found in sonically disrupted cells.[120] *Trzponema phagedenis* and the Reiter treponeme are also very closely related, having at least six antigens in common;[122,123] the Reiter strain has at least one antigen not found in *T. phagedenis. Treponema refringens* has four or five antigens in common with *T. calligyrum,* but both organisms also have some individual antigens not found in the other. The Nichols and Noguchi strains are identical. *Treponema minutum* shares a weak antigen with *T. refringens, T. calligyrum,* and the Reiter strain. *Treponema pallidum* has only one antigen in common with all of the cultivated treponemes;[122-124] this common antigen is probably the same one that is shared by all treponemes and by *Spirochaeta zuelzerae.* Larger amounts of this antigen are found on the Reiter strain and on *T. phagedenis;* smaller amounts are found on *T. refringens, T. calligyrum,* and *T. minutum.*[122,123]

Cell Envelope

Treponemes are covered with a three-layered outer cell envelope of outer membrane. The outer envelope has been removed from the cell by 1.4 mM of sodium dodecyl sulfate (SDS).[130] The "solubilized" envelope could be reaggregated by a divalent cation, Mg^{2+}, and the reaggregated membrane can be resolubilized by ethylenediaminetetraacetic acid or SDS.

Treponema pallidum

The increase in the number of cases of syphilis has started a renewed effort to cultivate the organism in vitro and to understand the immunity involved in syphilis.

Using radiolabeled amino acids and radioautography, it was found that a temperature of 34°C and a pH of 7.6 was optimal for protein synthesis under anaerobic conditions in an atmosphere of CO_2, H_2, and N_2.[131] Serine and valine were the amino acids most actively incorporated into protein. Radiolabeled carbon sources were also investigated;[132] of 22 carbon sources examined, glucose and pyruvate were degraded. End products were CO_2 and acetate.

In 1974, an interesting article reported the oxygen uptake by *T. pallidum.*[133] This raised the question whether *T. pallidum* is an aerobe or an anaerobe. Maximal metabolic activity was found at 10 to 20% 0_2 levels; metabolic activity was much less under anaerobic conditions.[134] Cyanide was not effective in inhibiting metabolic activity. Motility was vigorous in an oxygen atmosphere and sluggish in anaerobic conditions. The question whether this organism is aerobic or anaerobic will be answered when *T. pallidum* is cultivated.

Cultivation of *T. pallidum* is being investigated. Conventional culture methods for bacteria have not proved successful. Tissue culture methods have given some interesting results. Increased retention of motility and virulence of *T. pallidum* have been reported using anaerobic coincubation with rat glial cells and anaerobic incubation in glial-cell spent culture medium.[135,136] Another report claims growth and subculture of *T. pallidum* in BHK-21 baby hamster kidney cell cultures under aerobic conditions.[137]

The subject of immunity and the kind of immunity developed in syphilis has also come under investigation in recent years. Although man has little innate immunity, an acquired immunity is developed to the disease. There is some evidence that cell-mediated immunity is suppressed in the early stages of syphilis, but becomes activated when latency begins.[138] The infection progresses despite the development of circulating antibodies. However, passive transfer of some immunity has been reported in rabbits treated with large doses of hyperimmune serum;[139-141] when large doses of immune serum were given rabbits 48 hr prior to challenge with *T. pallidum* and daily after challenge, the appearance of skin lesions was suppressed as long as the treatment was continued.[140] Although new information is available, the question concerning the kind and nature of acquired immunity in syphilis still remains to be answered.

Reaction to Antibiotics

Antibiotic susceptibility of treponemes has received some attention.[118,142-147] The sensitivity of pathogenic treponemes has been established by observing the immobilization of cells in the presence of the antibiotic and comparing the percentage of immobilized cells with a non-antibiotic control. The relationship between immobilization of cells and inhibition of growth may be fallacious; a better index of susceptibility might be the use of a cultivated treponeme. The minimal inhibitory concentrations of some cultivated treponemes are listed in Tables 18 and 19. All strains were resistant to cycloserine (500 to 1,000 μg/ml), polymyxin B (500 to 1,000 units/ml), nitrofurazone (100 to 1,000 μg/ml), sulfathiazole (1,000 μg/ml), sulfaquinoxaline (1,000 μg/ml), sulfadiazine (1,000 μg/ml), succinylsulfathiazole (1,000 μg/ml), nalidixic acid (500 to 1,000 μg/ml), methenamine mandelate (500 to 1,000 μg/ml), 5-aminouracil (1,000 μg/ml), 5-iodouracil (1,000 μg/ml), and lysostaphin (1,000 μg/ml). The bactericidal

Table 18
INHIBITORY CONCENTRATIONS[a] OF ANTIBIOTICS FOR SOME STRAINS OF TREPONEMES[142,143]

	T. phagedenis				*T. denticola*		*T. vincentii*	Rumen fluid	
Antibiotic	Reiter	Kazan	*T. refringens*	Nichols	FM	T-32	N-9	Oral	Intestinal
P	1	0.1	0.1	1	0.1	0.1	0.1	100	10
A	1	1	1	0.1	0.1	0.1	1	100	10
N	1	1	10	1	1	0.1	1	100	10
Ox	1	1	10	0.1	0.1	0.1	1	100	10
Cl	1	1	1	0.1	0.1	0.1	0.1	100	10
KP	1	0.1	0.1	1	0.1	0.1	0.1	500	10
Ce	10	10	0.01	0.1	0.1	0.1	0.1	1	1
No	100	10	100	500	100	10	1	100	1
Van	1	1	1	0.1	1	10	1	10	100
Bac	1	0.1	0.1	0.1	1	0.1	0.1	1	10
E	0.1	0.1	0.1	0.01	0.01	0.01	0.1	0.01	0.1
Ty	100	10	10	1	100	10	1	1	10
Lin	100	10	1	1	10	10	10	10	10
Tet	1	1	1	1	1	1	1	1	100
Chl	1	10	10	10	10	1	1	10	100
Oxt	10	10	1	1	1	1	1	1	100
DM	1	1	1	1	1	0.1	0.1	1	100
Doxy	1	0.1	0.1	1	1	0.1	0.1	1	100
Me	1	1	0.1	1	1	0.1	0.1	1	100
Chlo	500	500	100	100	100	100	100	100	100
S	100	500	10	10	10	100	10	500	1,000
DHS	500	100	1	1	1	100	100	100	1,000
K	1,000	1,000	100	1	100	100	100	100	100
Gen	500			1	10				
Neo	500	500	10	1	10	10	100	100	100
Vio	1,000+	1,000+	100	10	10	100	1,000	100	500
Tyr	500	100	100	10	100	100	500	10	500

[a] Concentrations are given in μg/antibiotic/ml medium.

Code: P = penicillin G; A = ampicillin; N = nafcillin; Cl = cloxacillin; KP = K-phenoxymethyl penicillin; Ce = cephalothin; No = novobiocin; Van = vancomycin; Bac = bacitracin; E = erythromycin; Ty = tylosin; Lin = lincomycin; Tet = tetracycline; Chl = chlorotetracycline; Oxt = oxytetracycline; DM = demethylchlorotetracycline; Doxy = doxycycline; ME = methacycline; Chlo = chloramphenicol; S = streptomycin; DHS = dihydrostreptomycin; K = kanamycin; Gen = gentamycin; Neo = neomycin; Vio = viomycin; Tyr = tyrothricin; blank spaces = not tested.

Note: Clindamycin is not inhibitory at 0.5 μg/ml; higher concentrations were not tested.

concentrations of the penicillins are 10 to 1,000 times higher than the inhibitory concentrations.[143] Synergism of antibiotics against treponemes shows that there is lowering of the bactericidal concentrations for combinations of erythromycin and penicillin, erythromycin and tetracycline, erythromycin and vancomycin, cephalothin and tetracyclin, and bacitracin in combination with erythromycin, cephalothin, penicillin, tetracyclin, or vancomycin.[144]

Natural or induced resistance of treponemes to penicillin has not been found in syphilis or in the laboratory.[143] L-forms of treponemes have been reported by some workers;[20,148] others have not been able to induce stable cultivatable L-forms.[1]

For current reviews and texts on treponemes, see References 69, 75, 119, 126, 127, 147, and 149 to 152.

LEPTOSPIRA

Leptospires are aerobic spirochetes, whose cells are flexuous, motile, tightly coiled, and have a single axial fibril (axostyle). Some are pathogenic for man and animals, and others are saprophytes found in water.

Classification is based on serologic analysis of the antigens of leptospires and divides the pathogenic leptospires into serogroups, serotypes, and subserotypes (Table 20). Recent ideas on the taxonomy of leptospires have resulted in reducing all species names previously used to the taxonomic level of serotypes and retaining only two species names: *L. interrogans* for the pathogens, and *L. biflexa* for the water forms or saprophytes. Others recognize one species – *L. interrogans.* Twenty-eight water leptospires were grouped by their antigens into 16 serogroups.[153]

Some very interesting work[154] that has a great influence on the taxonomy of leptospires has been done with G + C ratios and DNA homology (Table 21). Serologic relationships of the strains studied did not correlate with homology groups, but sensitivity to 8-azaguanine and 2,6-diaminopurine, growth at 13°C, and lipase production did correlate with the four homology groups (Table 22). Additional DNA–DNA homology work has shown three groups in the pathogenic complex, three in the saprophytic complex, and one in the *illini* group.[155] The pathogenic complex consisted of serotypes *bataviae, javanica,* and *ranarum,* and the saprophytic complex consisted of serotypes *patoc, codice,* and A-183. The groups in the pathogenic complex were sensitive

Table 19
INHIBITORY CONCENTRATIONS[a] OF CHEMICALS FOR SOME STRAINS OF TREPONEMES[142,143]

	T. phagedenis				*T. denticola*			Rumen fluid[b]	
Antibiotic	Reiter	Kazan	*T. refringens*	Nichols	FM	T-32	*T. vincentii*	Oral	Intestinal
Fur	100	100	100	100	100	100	10	100	500
USNIC	1,000	1,000	100	100	100	100	500	100	10
5-FL	500	1,000	1,000	1,000	100	100	100	100	1,000
Tell	100	100	100	100	100	10	10	100	100
Thall	100	500	100	500	500	100	100	100	100
BG	500	500	100	100	100	100	100	100	100
CV	100	500	100	100	500	500	100	100	100

[a] Concentrations are given in μg chemical/ml medium.
[b] Oral strains require fluid for growth; intestinal strains require rumen fluid for growth.

Code: Fur = furazolidone; USNIC = usnic acid; 5-FL = 5-fluorouracil; Tell = K-tellurite; Thall = thallium acetate; BG = brilliant green; CV = crystal violet.

Table 20
SEROGROUPS AND SEROTYPES OF PATHOGENIC LEPTOSPIRES OF THE SPECIES *LEPTOSPIRA INTERROGANS*

Serogroup	Serotype	Serogroup	Serotype	Serogroup	Serotype
icterohaemorrhigiae	icterohaemorrhagiae	celledoni	celledoni	hebdomadis	hebdomadis
	copenhageni		whitcombi		nona
	mankarso				kambale
	naam	cynopteri	cynopteri		kremastos
	mwogolo		canalzonae		worsfoldi
	dakota		butembo		jules
	birkini	autumnalis	autumnalis		maru
	smithi		rachmati		borincana
	ndambari		fort-bragg		kabura
	budapest		sumatrana		mini
	sarmin		bulgarica		szwajizak
	weaveri		bangkinang		georgia
			erinacei-auriti		perameles
javanica	javanica		mooris		hardjo
	poi		sentot		recreo
	sorex-jalna		louisiana		medanesis
	coxi		orleans		wolffi
	sofia		djasiman		trinidad
			gurungi		sejroe
canicola	canicola				balcanica
	bafani	australis	australis		polonica
	kamituga		lora		saxkoebing
	jonsis		muenchen		nero
	sumneri		jalna		haemolytica
	broomi		bratislava		nicardi
	bindjei		fugis		
	schueffneri		bangkok	bataviae	bataviae
	benjamin		peruviana		paidjan
	malaya		pina		djatzi
			nicaragua		kobbe
ballum	ballum				balboa
	castellonis	pomona	pomona		claytoni
	arboreae		kennewicki		brasiliensis
			monjakov		
pyrogenes	pyrogenes		mozdak	tarassovi	tarassovi
	zanoni		tropica		bakeri
	myocastoris		proechimys		atlantae
	abramis				guidae
	biggis	grippotyphosa	grippotyphosa		kisuba
	hamptoni		valbuzzi		bravo
	alexi				atchafalaya
	robinsoni				chagres
	manilae				rama
					gatuni
				panama	panama
				shermani	shermani
				semaranga	semaranga
					patoc
					sao-paulo
				andamana	andamana

Note: See *W.H.O. Tech. Rep. Ser.*, No. 380, 1967, and Joint FAO/WHO Expert Committee on Zoonoses, third report, FAO Agricultural Studies, No. 74, *W.H.O. Tech. Rep. Ser.*, No. 378, 1967.

to 8-azaguanine, copper ions, and $NaHCO_3$; the others were resistant to these chemicals. *Bataviae* and *javanica* did not grow at 13°C, but *ranarum* and the other serotypes did. *Bataviae, patoc, codice,* and A-183 were lipase-positive, and *javanica* lipase-negative. The G + C content of their DNA is as follows: *bataviae,* 35.3%; *javanica,* 39.9%; *ranarum,* 41.2%; *patoc,*38.3%; *codice,* 38.0%; A-183, 36.0%; and *illini,*53.0%.

Cultivation

Leptospires are usually cultured in dilute media supplemented with animal sera or albumin and Tween 80. Various formulations of culture media and supplements are commercially available. Pyruvate has been found to be a good addition to Tween 80-albumin medium in promoting growth from a small number of cells.[156,157] In addition to pyruvate, acetate was found to permit full use of fatty acids, and glycerol shortened generation time.[157] Tween 80 could be rendered nontoxic for use in medium without albumin by passage through an anion exchange column; Fe^{2+} was also found to eliminate the toxicity of Tween 80.

Nutrition and Metabolism

Horse, cattle, fetal-calf, sheep, pig and guinea pig sera have been used, but the most successful is pooled rabbit serum. Some rabbit sera have been found to be inhibitory for leptospires.[158,159] The inhibitory fraction, a so-called natural antibody, was found to be a β-macroglobulin (IgM), which is heat-labile (2 hr at 65°C) and is reduced by

Table 21
PERCENT G + C IN THE DNA OF PATHOGENIC AND *BIFLEXA* LEPTOSPIRES

Complex	Serotype	Strain	% G + C as determined by: Tm	Buoyant density
Pathogens (*L. interrogans*)	*autumnalis*	Akiyami A	35.5 ± 1.3[a]	35.4 ± 0.95
	australis	Ballico	36.7 ± 0.72	35.5 ± 0.95
	ballum	Mus 127		39.0 ± 1.57
	bataviae	Van Tienen	36.6 ± 0.58	35.3 ± 1.35
	canicola	Hond Utrecht	36.7 ± 0.9	
	hyos	Mitis Johnson	40.2 ± 0.73	
	copenhageni	M-20	36.6 ± 0.70	35.4 ± 0.69
	icterohaemorrhagiae	RGA	35.5 ± 1.3	35.4 ± 0.95
	javanica	Veldrat Bataviae 46	40.4 ± 0.75	39.9 ± 1.32
	javanica	TR-73[b]		37.7 ± 1.0
	celledoni	Celledoni		38.3 ± 0.30
	pomona	Pomona	36.0 ± 1.38	36.0 ± 0.79
	pomona	Cornelli CB	36.0 ± 1.09	35.4 ± 1.85
	pyrogenes	Salinem		34.2 ± 2.04
Unclassified		Turtle strain A-183[c]		36.0 ± 1.8
		Turtle strain A-284[c]		35.0 ± 0.46
L. biflexa	*andamana*	Correo	39.1 ± 1.03	39.4 ± 0.32
	patoc	Patoc I	39.0 ± 0.7	38.3 ± 1.0
	sao-paulo	Sao Paulo		37.8 ± 0.87
	undetermined	CDC		38.0 ± 0.59

[a] Twice the standard deviation of the mean.
[b] Isolated from rodents in Thailand.
[c] Isolated from cloacae of turtles in Illinois by L. E. Hanson.

From Haapala, D. K., Rogul, M., Evans, L. B., and Alexander, A. D., *J. Bacteriol.*, 98, 421, 1969. With permission.

Table 22
SELECTED PHENOTYPIC CHARACTERISTICS OF GENETIC GROUPS OF *LEPTOSPIRA*

Genetic group	Strains used in DNA duplex studies	% G + C	Pathogenicity	Growth at 13°C[a]	DAP sensitivity[b]	8-Aza-guanine-sensitivity[b]	Lipase production[a]
I	*australis* (Ballico), *bataviae* (Van Tienen), and *pomona* (Pomona)	36 ± 1	+	–	–	–	+
II	*javanica* (Veldbat Bataviae 46), *celledoni* (Celledoni), and *hyos* (Mitis Johnson)	39 ± 1	+	–	+	–	–
III	*patoc* (Patoc I), *sao-paulo* (Sao Paulo), and *andamana* (Correo)	39 ± 1	–	+	+	+	+
IV	*biflexa* type (CDC)	39 ± 1	–	+	+	+	+

[a] Data of Johnson and Harris,[20,21] Kmety and Bakoss,[25] Fuzi and Czoka,[17] and Parnas et al.,[39] published in the reference cited below.
[b] DAP (2,6-diaminopurine) concentration is 50 to 10 μg/ml; 8-azaguanine concentration is 200 μg/ml.

From Haapala, D. K., Rogul, M., Evans, L. B., and Alexander, A. D., *J. Bacteriol.*, 93, 421, 1969. With permission.

2-mercaptoethanol. This serum fraction seems to act in conjunction with complement and lysozyme. Formalin-treated cells have a Z-antigen, which absorbs the toxic serum fraction. The Z-antigen is formalin-stable and may be associated with virulence.[158]

The active growth-promoting fraction in serum is albumin. Globulin fractions are inactive; however, the globulin fraction of serum combined with serum ultrafiltrate and either soluble starch or the weakly basic resin Amberlite IR-45 is active in promoting growth. Amberlite IR-45 alone did not support growth. The weakly acidic resin Amberlite ICR-50, the strongly acidic resin Dowex 50, the strongly basic resin Dowex-1, and charcoal did not support growth when combined with globulin and serum ultrafiltrate.[160] The PPLO serum fraction supports growth, as does albumin with oleic acid or with Tween 80 and Tween 60.

Although the active fraction of serum is albumin, it has been shown that leptospires actually require long-chain fatty acid contained in the serum and that serum protein (albumin) acts to inhibit the toxicity of the fatty acids to the organism.[161-164] Fatty acid can be used as the sole source of carbon energy. There is no growth when lipid-free albumin is used to supplement media.

Table 23 shows the lipids and fatty acids that can be used by leptospires or that support growth of these organisms in media containing albumin.[163,165-169] Saprophitic leptospires (serotype *semaranga*) grew in albumin medium containing saturated and unsaturated fatty acids with chain lengths ranging from C_{12} to C_{18}.[169] Pathogenic serotypes *canicola* and *ballum* grew in media containing only unsaturated fatty acids that had 14 carbon atoms or in media containing saturated fatty acids that had 12 to 18 carbon atoms.

Cellular lipids of some parasitic and saprophytic leptospires have been investigated; they make up 18 to 26% of the dry weight of the organisms.[170] The total lipid content was composed of 60 to 70% phospholipids, the remainder consisting of free long-chain fatty acids. The phospholipids found were phosphatidylethanolamine, phosphatidylglycerol, and diphosphatidylglycerol. Traces of lysophosphatidylethanolamine were often found. Glycolipids were not found in *Leptospira*.[45,170] The major fatty acids were hexadecanoic, hexadecenoic, and octadecanoic acids.

Leptospires cannot chain-elongate fatty acids, but they are capable of β-oxidation of fatty acids.[167,169,171] They can desaturate fatty acids. Usually they do not require an unsaturated fatty acid in the medium and can use saturated fatty acids. Water *Leptospira* strain B-16 grown in a synthetic medium without amino acids produced CO_2 and acetate from oxidation of oleic acid. This strain was able to grow in a medium with acetate as the sole carbon and energy source.[173]

Various enzymes of leptospires have been investigated; they are listed in Table 24. Of newest interest is the utilization of urea by *Leptospira*.[172] Virulent serotypes as well as an avirulent strain grew in a medium when urea replaced an ammonium salt as nitrogen source. Urea-grown cells of virulent serotypes had urease activity; however, urease could not be detected in the saprophytic strain. This could be associated with the colonization of *Leptospira* in the kidney of infected animals and resulting kidney damage.

Baseman and Cox[175] have investigated the terminal electron transport system in leptospires. Spectral evidence showed cytochromes of the a, c, and c_1 types in serotypes *pomona* and *shueffneri* and in the water isolate B-16. Cytochrome b was not found. Absorption peaks showed that cytochrome oxidase of the O type was present in all strains; cytochrome oxidase of the a_1 or a_3 type and a pigment were found only in the two pathogenic strains.

Serotype *pomona* was reported to have its growth increased by addition of glucose to the medium.[184] However, Baseman and Cox[185] investigated the metabolic pathway for serotypes *pomona* and *shueffneri* and for the water isolate B-16. Figure 1 shows their proposed pathway of *Leptospira* for the β-oxidation of oleic acid. Glucose was not used.

The leptospires do not require preformed purines or pyrimidines in culture media. Some encouraging work has been done with purine analogues. Johnson and Harris[186] found that leptospires could be divided into three groups on the basis of purine analogue sensitivity and lipase activity. Group 1, the parasitic leptospires, cannot grow in media containing 10 μg of 2,6-diaminopurine or 200 μg of 8-azaguanine/ml; the organisms have lipase (triolein) activity. Group 2, also parasitic leptospires, grows in media containing 2,5-diaminopurine, but not 8-azaguanine; these organisms do not have lipase or triolein activity. Group 3 consists of saprophytic leptospires that have lipase activity and grow in media containing both purine analogues. *Leptospira biflexa* was insensitive to 8-azaguanine at concentrations varying from 25 to 600 μg/ml. Twenty pathogenic serotypes were sensitive to 8-azaguanine.[187] Another purine analogue, 6-mercaptopurine, also inhibited pathogenic strains. The pyrimidine analogues 5-fluorouracil and 5-bromouracil

Table 23
LIPIDS AND FATTY ACIDS UTILIZED OR HYDROLYZED BY SAPROPHYTIC AND PATHOGENIC LEPTOSPIRES[163,165–169]

Lipid or fatty acid(s)	Concentration	Used[a]	Lipid or fatty acid(s)	Concentration	Used[a]
2-Octadecanoic acid	200 μg/ml	–	Monoolein	2 mg/100 ml	+/–
3-Octadecanoic acid	200 μg/ml	+	Olive oil	2 mg/100 ml	+
4-Octadecanoic acid	200 μg/ml	+	Peanut oil	2 mg/100 ml	+
5-Octadecanoic acid	200 μg/ml	–	TEM-4T[c]	5 mg/100 ml	+
6-Octadecanoic acid	200 μg/ml	+	Tween 60	14 mg/100 ml	+/–
7-Octadecanoic acid	200 μg/ml	–	Tween 80	28 mg/100 ml	+
9-Octadecanoic (oleic) acid	200 μg/ml	+			
10-Octadecanoic acid	200 μg/ml	–	Aleuric acid		–
11-Octadecanoic acid	200 μg/ml	+	Capric acid		–
12-Octadecanoic acid	200 μg/ml	–	Caproic acid		–
13-Octadecanoic acid	200 μg/ml	–	Caprylic acid		–
14-Octadecanoic acid	200 μg/ml	–	Heptadecanoic acid		+
15-Octadecanoic acid	200 μg/ml	+	Methylacetic acid		–
16-Octadecanoic acid	200 μg/ml	+	Methylcholesterol		–
17-Octadecanoic acid	200 μg/ml	–	Methyloleic acid		–
2- + 9-Octadecanoic acid	200 μg/ml	+	Methylpalmitic acid		–
3- + 9-Octadecanoic acid	200 μg/ml	+	Myristic acid		+/–
			Myristoleic acid		+/–
Arachidic acid	1–4 mg/100 ml	–	Palmitic acid		+/–
Arachidonic acid	1–4 mg/100 ml	–	Palmitoleic acid		+
Butyric acid	1–4 mg/100 ml	–	Pentadecanoic acid		+/–
Capric acid	1–4 mg/100 ml	–	Stearic acid		+/–
Caproic acid	1–4 mg/100 ml	–	Tridecanoic acid		+/–
Caprylic acid	1–4 mg/100 ml	–	Triglycerides of butyric acid		–
Lauric acid	1–4 mg/100 ml	+/–[b]	Triolein		+/–
Linoleic acid	1–4 mg/100 ml	–	Tween 20		+/–
Linolenic acid	1–4 mg/100 ml	–	Tween 40		+/–
Propionic acid	1–4 mg/100 ml	–			
Ricinoleic acid	1–4 mg/100 ml	–	Acetic + capric acid		–
Undecanoic acid	1–4 mg/100 ml	–	Acetic + caproic acid		–
			Acetic + caprylic acid		–
Egg lecithin	2 mg/100 ml	+	Acetic + lauric acid		–
Egg yolk fat	2 mg/100 ml	+			

[a] Where two results are shown, data were obtained from different reports.
[b] +/– = saprophyte/pathogen; for differences in fatty acid utilization by saprophytic and pathogenic leptospires, see Reference 169.
[c] TEM-4T = diacetyl tartaric ester of tallow monoglycerides.

did not inhibit either pathogenic or saprophytic leptospires; 5-fluorouracil has been used successfully in isolation media to inhibit contaminating bacteria.

The outer envelope or sheath of *Leptospira interrogans* serotype *canicola* was solubilized and removed by 0.02% sodium dodecyl sulfate;[188] the isolated envelope was then reaggregated, after removal of the SDS, by concentration. The isolated outer envelope did not contain muramic acid, but the protoplasmic cylinder left after envelope removal did. The outer envelope was removed from serotype *pomona* by suspending the cells in phosphate buffer;[189] purification of the isolated envelope was done by gradient centrifugation. The envelope contained 47% protein, 27% carbohydrate, and 23% lipid. The carbohydrates found were hexose, pentose, 6-deoxyhexose, and hexosamine. Muramic acid and 2-keto-3-deoxyoctonate were not found. Lipids found were phosphatidylethanolamine and traces of lysophosphatidylethanolamine. Fatty acids were octadecanoic, octadecenoic, hexadecanoic, and hexadecenoic acids. The isolated outer envelope protected hamsters from challenge when used as a vaccine.[190]

Antigens

Antigens of leptospires have been studied. Extraction of leptospires with 70% ethanol yields two water-soluble antigens; the P-antigen is type-specific, whereas the S-antigen is genus-specific.[191] Extraction of serotype *pomona* with 50% ethanol and 0.15*M* NaCl also yielded two fractions,[192,193] an alcohol fraction and saline S fraction.

Cox[194] discovered the HL-antigen (hemolytic-lysis antigen), which is genus-specific for leptospires. HL-antigen-sensitized sheep erythrocytes are hemolyzed in serum containing leptospiral antibodies in the presence of complement. A hemagglutination and a hemagglutination inhibition test using the HL-antigen have been developed.[195] The hemagglutination test has been evaluated as a sensitive and simple test for human leptospiral infection.[196] The HL-test is genus-specific, and therefore a good screening test for leptospirosis in man and animals.

Table 24
ENZYMES OF LEPTOSPIRES[171,174-184]

Enzyme	Present	Enzyme	Present
Catalase	+	Succinate dehydrogenase	+
Hemolysin	+	Acyl thiolase	+
Oxidase	+	Acyl CoA synthetase	+
Lipase	+	Glucokinase	–
Transaminase	+	Phosphoglucoisomerase	–
Fumarase	+	Glucose-6-phosphate dehydrogenase	–
Enolase	+	Pyruvate kinase	–
Aconitase	+	Lactate dehydrogenase	–
NADH oxidase	+	Acetokinase	–
Phosphoglyceromutase	+	Phosphotransacetylase	–
Phosphofructokinase	+	Alcohol dehydrogenase	–
α-Ketoglutarate dehydrogenase	+	Phosphoriboisomerase	+
Malate dehydrogenase	+	D-Ribulose-5-phosphate 3-epimerase	+
Transaldolase	+	Phosphatase	+
Fructose-1,6-diphosphate aldolase	+	Malic dehydrogenase	+
Condensing enzyme	+	α-Glycerophosphate dehydrogenase	+
Acyl CoA dehydrogenase	+	6-Phosphogluconic dehydrogenase	+
Triosephosphate isomerase	+	Naphthylamidase	+
Isocitrate dehydrogenase	+	Esterase	+
Glyceraldehyde-3-phosphate dehydrogenase	+	Phospholipase C	+
Acyl dehydrogenase	+	Aminopeptidase	+
Enoyl hydrase	+	Transketolase	+
β-Hydroxyacyl dehydrogenase	+		

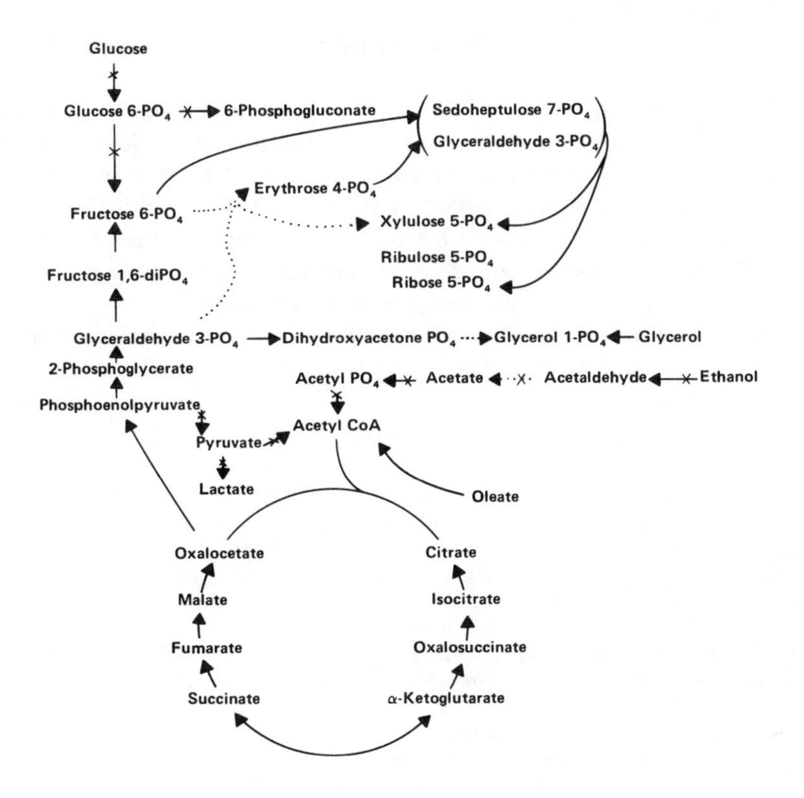

FIGURE 1. Proposed metabolic pathway of *Leptospira*. → = enzyme present in leptospiral extract; – x → = enzyme absent from leptospiral extract; ••→ = enzyme not assayed, but presence of enzyme suspected; •• x ••→ = enzyme not assayed, but presence of enzyme doubted. (From Baseman, J. B. and Cox, C. D., *J. Bacteriol.*, 97, 992, 1969. With permission.)

An erythrocyte-sensitizing substance (ESS) similar to the HL-antigen has been isolated from leptospires. The sensitized-erythrocyte lysis (SEL) test detects ESS antibodies in people with leptospirosis. It has broad reactivity against most serogroups of leptospires.[197]

A leptospire type-specific antigen (TM) has been isolated by extraction with 90% phenol, enzyme treatment, and fractionation by ethanol precipitation.[198] It had only one antigenic component.

Work on soluble antigens of *Leptospira* is now in progress.[199-201] It has now been shown that extraction of serotype *biflexa* with 0.2% trypsin yielded three precipitating antigens by immunodiffusion (axial filament antigen, antigen d, and antigen e).[199] Using axial filament antigen, eight groups were found that showed some correlation with gentic homology data, but little with the classical agglutination-absorption classification system.[200]

Colonies

Leptospires can be grown as colonies on solid agar medium.[202] They grow into the agar medium, giving several types of colonies.[202-208]

For additional information on Leptospires, see References 209 to 211.

CRISTISPIRA

Cristispira have very coarse spirals. They are 28 to 120 μm long and 0.5 to 3 μm wide. Cells have fifty to several hundred axial fibrils, which form a lateral ridge of "crista" that may be seen by dark-field or phase microscopy. *Cristispira* are found in mollusks. They have not been cultured. The type species of the genus is *C. pectinis* Gross, 1910.[212] Another organism found in mollusks, *C. hartmanni,* is smaller than *C. pectinis,* has fewer axial fibrils, and may be considered a treponeme. Very little is known about *Cristispira.* For more information on *Cristispira,* see Reference 15.

Recent texts and reviews on spirochetes (*Spirochaeta, Treponema, Borrelia,* and *Leptospira*) are found in References 151, 152, and 213 to 216.

REFERENCES

1. **Abramson, I. J.,** Ph.D. thesis, Effect of Antimicrobial Agents on Treponemes, Virginia Polytechnic Institute and State University, Blacksburg, Va., 1971.
2. **Ritchie, A. E. and Ellinghausen, H. C.,** *J. Bacteriol.,* 89, 223, 1965.
3. **Nauman, R. K., Holt, S. C., and Cox, C. D.,** *J. Bacteriol.,* 98, 264, 1969.
4. **Anderson, D. L. and Johnson, R. C.,** *J. Bacteriol.,* 95, 2293, 1968.
5. **Pillot, J. and Ryter, A.,** *Ann. Inst. Pasteur Paris,* 108, 791, 1965.
6. **Listgarten, M. A. and Socransky, S. S.,** *Arch. Oral Biol.,* 10, 127, 1965.
7. **Jackson, S. and Black, S. H.,** *Arch. Mikrobiol.,* 76, 308, 1971.
8. **White, F. H. and Simpson, C. F.,** *J. Infect. Dis.,* 115, 123, 1965.
9. **Simpson, C. F. and White, F. H.,** *J. Infect. Dis.,* 109, 243, 1961.
10. **Bharier, M. A. and Rittenberg, S. C.,** *J. Bacteriol.,* 105, 413, 1971.
11. **Holt, S. C. and Canale-Parola, E.,** *J. Bacteriol.,* 96, 822, 1968.
12. **Bladen, H. A. and Hampp, E. G.,** *J. Bacteriol.,* 87, 1180, 1964.
13. **Listgarten, M. A. and Socransky, S. S.,** *J. Bacteriol.,* 88, 1087, 1964.
14. **Jepsen, O. B., Hovind Hougen, K., and Birch-Anderson, A.,** *Acta Pathol. Microbiol. Scand.,* 74, 241, 1968.
15. **Ryter, A. and Pillot, J.,** *Ann. Inst. Pasteur Paris,* 109, 552, 1965.
16. **Bharier, M. A. and Rittenberg, S. C.,** *J. Bacteriol.,* 105, 422, 1971.
17. **Bharier, M. A. and Rittenberg, S. C.,** *J. Bacteriol.,* 105, 430, 1971.
18. **Jackson, S. and Black, S. H.,** *Arch. Mikrobiol.,* 76, 308, 1971.
19. **Hovind Hougen, K. and Birch-Anderson, A.,** *Acta Pathol. Microbiol. Scand.,* 79, 37, 1971.
20. **Oveinnikov, N. M. and Delktorskij, V. V.,** *Br. J. Vener. Dis.,* 47, 315, 1971.
21. **Yanagihara, Y. and Mifuchi, I.,** *J. Bacteriol.,* 95, 2403, 1968.
22. **Birch-Anderson, A., Hovind Hougen, K., and Borg-Peterson, C.,** *Acta Pathol. Microbiol. Scand. Sect. B.,* 81, 665, 1973.
23. **Wiegand, S. E., Strobel, P. L., and Glassman, L. H.,** *J. Invest. Dermatol.,* 58, 186, 1972.
24. **Hovind Hougen, K.,** *Acta Pathol. Microbiol. Scand. Sect. B,* 82, 495, 1974.
25. **Hovind Hougen, K.,** *Acta Pathol. Microbiol. Scand. Sect. B,* 82, 329, 1974.
26. **Hovind Hougen, K.,** *Acta Pathol. Microbiol. Scand. Sect. B,* 83, 91, 1975.
27. **Hovind Hougen, K.,** *Acta Pathol. Microbiol. Scand. Sect. B,* 80, 297, 1972.
28. **Hovind Hougen, K.,** *Acta Pathol. Microbiol. Scand. Sect. B,* 81, 15, 1973.
29. **Hovind Hougen, K., Birch-Anderson, A., and Jorgen Skovgard Jensen, H.,** *Acta Pathol. Microbiol. Scand. Suppl.,* 215, 23, 1970.
30. **Hovind Hougen, K.,** *Acta Pathol. Microbiol. Scand. Sect. B,* 82, 799, 1974.
31. **Joseph, R. and Canale-Parola, E.,** *Arch. Mikrobiol.,* 81, 146, 1972.
32. **Breznak, S. A. and Canale-Parola, E.,** *Arch. Mikrobiol.,* 83, 261, 1972.
33. **Cox, C. D.,** *J. Bacteriol.,* 109, 943, 1972.
34. **Canale-Parola, E., Udris, Z., and Mandel, M.,** *Arch. Mikrobiol.,* 63, 385, 1968.
35. **Blakemore, R. P. and Canale-Parola, E.,** *Arch. Mikrobiol.,* 89, 273, 1973.
36. **Hespell, R. B. and Canale-Parola, E.,** *J. Bacteriol.,* 103, 216, 1970.
37. **Johnson, P. W., and Canale-Parola, E.,** *Arch. Mikrobiol.,* 89, 341, 1973.
38. **Veldhamp, H.,** *Antonie van Leeuwenhoek J. Microbiol. Serol.,* 26, 103, 1960.
39. **Breznak, J. A. and Canale-Parola, E.,** *J. Bacteriol.,* 97, 386, 1968.
40. **Breznak, S. A. and Canale-Parola, E.,** *Arch. Mikrobiol.,* 83, 278, 1972.
41. **Hespell, R. B. and Canale-Parola, E.,** *Arch. Mikrobiol.,* 74, 1, 1970.

42. **Hespell, R. B. and Canale-Parola, E.,** *J. Bacteriol.,* 116, 931, 1973.
43. **Meyer, H. and Meyer, F.,** *Biochim. Biophys. Acta,* 231, 93, 1971.
44. **Cohen, P. G., Moss, C. W., and Farshtchi, D.,** *Br. J. Vener. Dis.,* 46, 10, 1970.
45. **Livermore, B. and Johnson, R. C.,** *J. Bacteriol.,* 120, 1268, 1974.
46. **Meyer, H. and Meyer, F.,** *Biochim. Biophys. Acta,* 176, 202, 1969.
47. **Joseph, R., Holt, S. C., and Canale-Parola, E.,** *J. Bacteriol.,* 115, 426C, 1973.
48. **Johnson, R. C.,** in *The Biology of the Parasitic Spirochetes,* Johnson, R. C., Ed., Academic Press, New York, 1976, p. 39.
49. **Kelly, R.,** *Science,* 173, 443, 1971.
50. **Kelly, R. T.,** in *The Biology of Parasitic Spirochetes,* Johnson, R. C., Ed., Academic Press, New York, 1976, p. 87.
51. **Pickens, E. G., Gerhoff, R. K., and Burgdorfer, W.,** *J. Bacteriol.,* 95, 291, 1968.
52. **Coffey, E. M. and Eveland, W. C.,** *J. Infect. Dis.,* 117, 29, 1967.
53. **Southern, P. M. and Sanford, J. P.,** *Medicine,* 48, 129, 1969.
54. **Felsenfeld, O.,** *Bacteriol. Rev.,* 29, 46, 1965.
55. **Felsenfeld, O.,** *Borrelia Strains, Vectors, Human and Animal Borreliosis,* Warren H. Green, St. Louis, Missouri, 1971.
56. **Burgdorfer, W. and Mavros, A. J.,** *Infect. Immun.,* 2, 256, 1970.
57. **McNeil, E., Hinshaw, W. R., and Kissling, R. E.,** *J. Bacteriol.,* 57, 191, 1949.
58. **Dhanhov, I., Soumrov, I., Lozera, T., and Penev, P.,** *Zentralbl. Veterinaermed. Reihe B,* 17, 544, 1970.
59. **Packchanian, H. and Smith, J. B.,** *Tex. Rep. Biol. Med.,* 28, 287, 1970.
60. **Jepson, W. F.,** *Nature,* 160, 874, 1947.
61. **Fulton, J. D. and Smith, P. J. C.,** *J. Biochem.,* 76, 491, 1960.
62. **Smith, P. J. C.,** *J. Biochem.,* 76, 500, 1960.
63. **Smith, P. J. C.,** *J. Biochem.,* 76, 508, 1960.
64. **Smith, P. J. C.,** *J. Biochem.,* 76, 514, 1960.
65. **Pickett, J. and Kelly, R.,** *Infect. Immun.,* 9, 279, 1974.
66. **Coffey, E. M. and Eveland, W. C.,** *J. Infect. Dis.,* 117, 23, 1967.
67. **Felsenfeld, O., Decker, W. J., Wohlhieter, J. A., and Rafyi, Z.,** *J. Immunol.,* 94, 805, 1965.
68. **Davis, G. E.,** *Annu. Rev. Microbiol.,* 2, 305, 1943.
69. **Smith, J. L. and Pesitsky, B. R.,** *Br. J. Vener. Dis.,* 43, 117, 1967.
70. **Paris-Hamelin, A., Vaisman, A., and Dunoyer, F.,** *Bull. W.H.O.,* 38, 308, 1968.
71. **Vaisman, A., Paris-Hamelin, A., Dunoyer, F., and Dunoyer, M.,** *Bull. W.H.O.,* 36, 339, 1967.
72. **Fribourg-Blanc, A. and Mallaret, H. H.,** *WHO/VDT/Res.,* 68, 135, 1968.
73. **Sepetjian, M., Tissot Guerraz, F., Salassola, D., Thirolet, T., and Monnier, J. C.,** *Bull. W.H.O.,* 40, 141, 1969.
74. **Kuhn, U. S. G., III, Media, R., Cohen, P. G., and Vegas, M.,** *Br. J. Vener. Dis.,* 46, 311, 1970.
75. Treponematoses Research, *W.H.O. Tech. Rep. Ser.,* No. 455, 1970.
76. **Holdeman, L. V. and Moore, W. E. C. (Eds.),** *Anaerobe Laboratory Manual,* Virginia Polytechnic Institute and State University, Blacksburg, Va., 1972.
77. **Smibert, R. M.,** *WHO/VDT/Res.,* 71, 242, 1971.
78. **Socransky, S. S., Listgarten, M. A., Hubersak, C., Cotmore, J., and Clark, H.,** *J. Bacteriol.,* 98, 878, 1969.
79. **Harris, D. L., Glock, R. D., Christensen, C. R., and Kinvon, J. M.,** *Vet. Med. Small Anim. Clin.,* 67, 61, 1972.
80. **Smibert, R. M.,** in *The Biology of the Parasitic Spirochetes,* Johnson, R. C., Ed., Academic Press, New York, 1976, p. 121.
81. **Smibert, R. M.,** in *Bergey's Manual of Determinative Bacteriology,* 8th ed., Buchanan, R. E. and Gibbons, N. E., Eds., Williams & Wilkins, Baltimore, 1974.
82. **Socransky, S. S., MacDonald, J. B., and Sawyer, S.,** *Arch. Oral. Biol.,* 1, 171, 1959.
83. **Hardy, P. H., Lee, Y. C., and Nell, E. E.,** *J. Bacteriol.,* 86, 616, 1963.
84. **Christiansen, A. H.,** *Acta Pathol. Microbiol. Scand.,* 60, 234, 1964.
85. **Hanson, A. W. and Cannefax, G. R.,** *Br. J. Vener. Dis.,* 41, 163, 1965.
86. **Oyama, V. I., Steinman, H. G., and Eagle, H.,** *J. Bacteriol.,* 65, 609, 1953.
87. **Socransky, S. S. and Hubersak, C.,** *J. Bacteriol.,* 94, 1795, 1967.
88. **Power, D. A. and Pelczar, M. J.,** *J. Bacteriol.,* 77, 789, 1959.
89. **Johnson, R. C. and Eggerbraten, L. M.,** *Infect. Immun.,* 3, 723, 1971.
90. **Socransky, S. S., Loesche, W. J., Hubersak, C., and MacDonald, J. B.,** *J. Bacteriol.,* 88, 200, 1964.
91. **Hardy, P. H. and Munro, C. O.,** *J. Bacteriol.,* 91, 27, 1966.
92. **Wegner, G. H. and Foster, E. M.,** *J. Bacteriol.,* 85, 53, 1963.

93. **Sachan, D. S. and Davis, C. L.,** *J. Bacteriol.*, 98, 300, 1969.
94. **Livermore, B. P. and Johnson, R. C.,** *Can. J. Microbiol.*, 21, 1877, 1975.
95. **Smibert, R. M. and Claterbaugh, R. L., Jr.,** *Can. J. Microbiol.*, 18, 1073, 1972.
96. **Smibert, R. M.,** in *The Biology of the Parasitic Spirochetes*, Johnson, R. C., Ed., Academic Press, New York, 1976, p. 49.
97. **Vaczi, L., Kiraly, K., and Rethy, H.,** *Acta Microbiol. Acad. Sci. Hung.*, 13, 79, 1966.
98. **Johnson, R. C., Livermore, B. P., Jenkins, H. M., and Eggerbraten, L. M.,** *Infect. Immun.*, 2, 606, 1970.
99. **Livermore, B. P. and Johnson, R. C.,** *Biochim. Biophys. Acta*, 210, 315, 1970.
100. **Coulon-Morelec, M. J., Dupouey, P., and Marechal, J.,** *C. R. Acad. Sci. Paris*, 269, 854, 1969.
101. **Allen, G. L., Johnson, R. C., and Peterson, D.,** *Infect. Immun.*, 3, 727, 1971.
102. **Barban, S.,** *J. Bacteriol.*, 68, 493, 1954.
103. **Barban, S.,** *J. Bacteriol.*, 69, 274, 1955.
104. **Hespell, R. B. and Canale-Parola, E.,** *Arch. Mikrobiol.*, 78, 234, 1971.
105. **Rathlev, T. and Pfau, C. J.,** *Arch. Biochem. Biophys.*, 106, 343, 1964.
106. **Kawata, T.,** *J. Gen. Appl. Microbiol.*, 13, 405, 1967.
107. **Kawata, T.,** *Jpn. J. Bacteriol.*, 22, 590, 1967.
108. **Ajello, F.,** *G. Microbiol.*, 15, 17, 1967.
109. **Bucca, M. A.,** *J. Vener. Dis. Inf.*, 32, 16, 1951.
110. **Hussey, M. S. and Nowminski, W. W.,** *Tex. Rep. Biol. Med.*, 7, 73, 1949.
111. **Tauber, H., Cannefax, G. R., Hanson, A. W., and Russell, H.,** *Exp. Med. Surg.*, 20, 324, 1962.
112. **Berger, U.,** *Zentralbl. Bakteriol. Parasitenkd. Infektionskr. Hyg. Abt. 1 Orig.*, 165, 563, 1956.
113. **Ajello, F.,** *G. Microbiol.*, 17, 107, 1969.
114. **Ajello, F.,** *WHO/VDT/Res.*, 71, 240, 1971.
115. **Hampp, E. G., Mergenhagen, S. E., and Omata, R. R.,** *J. Dent. Res.*, 38, 979, 1959.
116. **Omata, R. R. and Hampp, E. G.,** *J. Dent. Res.*, 40, 171, 1961.
117. **Szewezuk, A. and Metzger, M.,** *Arch. Immunol. Ther. Exp.*, 18, 643, 1970.
118. **Turner, T. B. and Hollander, D. H.,** Biology of the treponematoses, *W.H.O. Monogr. Ser.*, No. 35, 1957.
119. **Meyer, P. E. and Hunter, E. F.,** *J. Bacteriol.*, 93, 784, 1967.
120. **Eagle, H. and Germuth, E. G., Jr.,** *J. Immunol.*, 60, 223, 1948.
121. **Christiansen, A. H.,** *Acta Pathol. Microbiol. Scand.*, 60, 123, 1964.
122. **Dupouey, P.,** *Ann. Inst. Pasteur Paris*, 105, 725, 1963.
123. **Dupouey, P.,** *Ann. Inst. Pasteur Paris*, 105, 949, 1963.
124. **Kiraly, K., Jobbagy, A., and Kovats, L.,** *J. Invest. Dermatol.*, 48, 98, 1967.
125. **Pillot, J., Dupouey, P., and Faure, M.,** *Ann. Inst. Pasteur Paris*, 98, 734, 1960.
126. **Wallace, A. L. and Harris, A.,** Reiter treponeme: a review of the literature, *Bull. W.H.O.*, 36, Suppl. 2, 1967.
127. **Willcox, R. R. and Guthe, R.,** *Treponema pallidum:* a bibliographical review of the morphology, culture and survival of *T. pallidum* and associated organisms, *Bull. W.H.O.*, 35, 1960.
128. **Harris, D. L. and Kinyon, J. M.,** *Am. J. Clin. Nutr.*, 27, 1297, 1974.
129. **Glock, R. D., Harris, D. L., and Kluge, J. B.,** *Infect. Immun.*, 9, 167, 1974.
130. **Johnson, R. C., Wachter, M. S., and Ritzi, D.,** *Infect. Immun.*, 7, 249, 1973.
131. **Baseman, J. B. and Hayes, N. S.,** *Infect. Immun.*, 10, 1350, 1974.
132. **Nichols, J. C. and Baseman, J. B.,** *Infect. Immun.*, 2, 1044, 1975.
133. **Cox, C. D. and Barber, M. K.,** *Infect. Immun.*, 10, 123, 1974.
134. **Baseman, J. B., Nichols, J. C., and Hayes, N. S.,** *Infect. Immun.*, 13, 704, 1976.
135. **Graves, S. R., Sandok, P. L., Jenkins, H. M., and Johnson, R. C.,** *Infect. Immun.*, 12, 1116, 1975.
136. **Sandok, P. L., Jenkins, H. M., Graves, S. R., and Knight, S. T.,** *J. Clin. Microbiol.*, 3, 72, 1976.
137. **Jones, R. H., Finn, M. A., Thomas, J. J., and Folger, C.,** *Br. J. Vener. Dis.*, 52, 18, 1976.
138. **Schell, R. F. and Musher, D. M.,** *Infect. Immun.*, 9, 658, 1974.
139. **Perine, P. L., Weiser, R. S., and Klebanoff, S. J.,** *Infect. Immun.*, 8, 787, 1973.
140. **Weiser, R. S., Erickson, D., Perine, P. L., and Pearsall, N. N.,** *Infect. Immun.*, 13, 1402, 1976.
141. **Turner, T. B., Hardy, P. H., Jr., Newman, B., and Nell, E. E.,** *Johns Hopkins Med. J.*, 133, 241, 1973.
142. **Abramson, I. J. and Smibert, R. M.,** *Br. J. Vener. Dis.*, 47, 407, 1971.
143. **Abramson, I. J. and Smibert, R. M.,** *Br. J. Vener. Dis.*, 47, 413, 1971.
144. **Abramson, I. J. and Smibert, R. M.,** *Br. J. Vener. Dis.*, 48, 113, 1972.
145. **Fitzgerald, R. J. and Hampp, E. G.,** *J. Dent. Res.*, 31, 20, 1952.
146. **Berger, U.,** *Arch. Hyg. Bakteriol.*, 140, 605, 1956.

147. **Rosebury, T.,** *Microorganisms Indigenous to Man,* Blakiston Division, McGraw-Hill, New York, 1962.
148. **Ovcinnikov, N. M., Delektorskij, V. V., and Ustimenko, L. M.,** *Vestn. Dermatol. Venerol.,* 44, 53, 1970.
149. **Miller, J. N., Falcone, V. H., Golden, B., Israel, C. W., Kuhn, U. S. G., and Smibert, R. M.,** *Spirochetes in Body Fluids and Tissues: Manual of Investigative Methods,* Charles C Thomas, Springfield, Ill., 1971.
150. **Smith, J. L.,** *Spirochetes in Late Seronegative Syphilis, Penicillin Notwithstanding,* Charles C Thomas, Springfield, Ill., 1969.
151. **Smibert, R. M.,** *Crit. Rev. Microbiol.,* 2, 491, 1973.
152. **Johnson, R. C. (Ed.),** *The Biology of Parasitic Spirochetes,* Academic Press, New York, 1976.
153. **Henneberry, R. C. and Cox, C. D.,** *J. Bacteriol.,* 96, 1419, 1968.
154. **Haapala, D. K., Rogul, M., Evans, L. B., and Alexander, A. D.,** *J. Bacteriol.,* 98, 421, 1969.
155. **Brendle, J. J., Rogul, M., and Alexander, A. D.,** *Int. J. Syst. Bacteriol.,* 24, 205, 1974.
156. **Johnson, R. C., Walby, J., Henry, R. A., and Auran, N. E.,** *Appl. Microbiol.,* 26, 118, 1973.
157. **Staneck, J. L., Henneberry, R. C., and Cox, C. D.,** *Infect. Immun.,* 7, 886, 1973.
158. **Rhu, E.,** *Bull. Inst. Zool. Acad. Sin.* (Taipei), 3, 1, 1964.
159. **Faine, S. and Carter, J. N.,** *J. Bacteriol.,* 95, 280, 1968.
160. **Johnson, R. C. and Wilson, J. B.,** *J. Bacteriol.,* 80, 406, 1960.
161. **Ellinghausen, H. C. and McCullough, W. G.,** *Am. J. Vet. Res.,* 26, 45, 1965.
162. **Ellinghausen, H. C. and McCullough, E. G.,** *Am. J. Vet. Res.,* 26, 39, 1965.
163. **Helprin, J. J. and Hiatt, C. W.,** *J. Infect. Dis.,* 100, 136, 1957.
164. **Stalheim, O. H. V.,** *J. Bacteriol.,* 92, 946, 1966.
165. **Jenkin, H. M., Anderson, L. E., Halman, R. T., Ismaie, I. A., and Tunstone, F. P.,** *J. Bacteriol.,* 98, 1026, 1969.
166. **Johnson, R. C. and Gary, N. D.,** *J. Bacteriol.,* 85, 976, 1963.
167. **Stalheim, O. H. V. and Wilson, J. B.,** *J. Bacteriol.,* 88, 55, 1964.
168. **Bertok, L. and Kemens, F.,** *Acta Microbiol. Acad. Sci. Hung.,* 7, 251, 1960.
169. **Johnson, R. C., Harris, V. G., and Walby, J. K.,** *J. Gen. Microbiol.,* 55, 399, 1969.
170. **Johnson, R. C., Livermore, B. P., Walby, J. K., and Jenkin, H. M.,** *Infect. Immun.,* 2, 286, 1970.
171. **Henneberry, R. C. and Cox, C. D.,** *Can. J. Microbiol.,* 16, 41, 1970.
172. **Kadis, S. and Pugh, W. L.,** *Infect. Immun.,* 10, 793, 1974.
173. **Henneberry, R. C., Baseman, J. B., and Cox, C. D.,** *Antonie van Leeuwenhoek J. Microbiol. Serol.,* 36, 489, 1970.
174. **Rao, P. J., Larson, A. D., and Cox, C. D.,** *J. Bacteriol.,* 88, 1045, 1964.
175. **Baseman, J. B. and Cox, C. D.,** *J. Bacteriol.,* 97, 1001, 1969.
176. **Patel, V., Goldberg, H. S., and Blendon, D. C.,** *J. Bacteriol.,* 88, 877, 1964.
177. **Green, S. S. and Goldberg, H. S.,** *J. Bacteriol.,* 93, 1739, 1967.
178. **Chorvath, B. and Fried, M.,** *J. Bacteriol.,* 102, 879, 1970.
179. **Berg, R. N., Green, S. S., Goldberg, H. S., and Blenden, D. C.,** *Appl. Microbiol.,* 17, 467, 1969.
180. **Green, S. S., Goldberg, H. S., and Blenden, D. C.,** *Appl. Microbiol.,* 15, 1104, 1967.
181. **Burton, G., Blenden, D. C., and Goldberg, H. S.,** *Appl. Microbiol.,* 19, 586, 1970.
182. **Stalheim, O. H. V.,** *Am. J. Vet. Res.,* 32, 843, 1971.
183. **Markovetz, A. J. and Larson, A. D.,** *Proc. Soc. Exp. Biol. Med.,* 101, 638, 1959.
184. **Ellinghausen, H. C.,** *Am. J. Vet. Res.,* 29, 191, 1969.
185. **Baseman, J. B. and Cox, C. D.,** *J. Bacteriol.,* 97, 992, 1969.
186. **Johnson, R. C. and Harris, V. G.,** *Appl. Microbiol.,* 16, 1584, 1968.
187. **Johnson, R. C. and Rogers, P.,** *J. Bacteriol.,* 88, 1618, 1964.
188. **Auran, N. E., Johnson, R. C., and Ritzi, D. M.,** *Infect. Immun.,* 5, 968, 1972.
189. **Zeigler, J. A. and VanEseltine, W. P.,** *Can. J. Microbiol.,* 21, 1102, 1975.
190. **Bey, R. F., Auran, N. E., and Johnson, R. C.,** *Infect. Immun.,* 10, 1051, 1974.
191. **Rothstein, N. and Hiatt, C. W.,** *J. Immunol.,* 77, 257, 1956.
192. **Schricker, R. L. and Hanson, L. E.,** *Am. J. Vet. Res.,* 24, 854, 1963.
193. **Schricker, R. L. and Hanson, L. E.,** *Am. J. Vet. Res.,* 24, 861, 1963.
194. **Cox, C. D.,** *Proc. Soc. Exp. Biol. Med.,* 90, 610, 1955.
195. **Baker, L. A. and Cox, C. D.,** *Appl. Microbiol.,* 25, 697, 1973.
196. **Sulzer, C. R. and Jones, W. L.,** *Appl. Microbiol.,* 26, 655, 1973.
197. **Sharp, C. F.,** *J. Pathol. Bacteriol.,* 77, 349, 1958.
198. **Shinagawa, M. and Yanagawa, R.,** *Infect. Immun.,* 5, 12, 1972.
199. **Graves, S. and Faine, S.,** *Aust. J. Exp. Biol. Med. Sci.,* 52, 615, 1974.
200. **Chang, A., Faine, S., and Williams, W. T.,** *Aust. J. Exp. Biol. Med. Sci.,* 52, 549, 1974.

201. **Chang, A. and Faine, S.,** *Aust. J. Exp. Biol. Med. Sci.,* 52, 569, 1974.
202. **Cox, C. D. and Larson, A. D.,** *J. Bacteriol.,* 73, 587, 1957.
203. **Smibert, R. M.,** *Can. J. Microbiol.,* 15, 127, 1968.
204. **Stalheim, O. H. V. and Wilson, J. B.,** *J. Bacteriol.,* 86, 482, 1963.
205. **Fujikura, T.,** *Jpn. J. Vet. Res.,* 28, 63, 1966.
206. **Armstrong, J. C. and Goldberg, H. S.,** *Am. J. Vet. Res.,* 21, 311, 1960.
207. **Fujikura, T.,** *Jpn. J. Vet. Res.,* 28, 297, 1966.
208. **Fujikura, T.,** *Jpn. J. Microbiol.,* 10, 79, 1966.
209. **Turner, L. H.,** *Trans. R. Soc. Trop. Med. Hyg.,* 61, 842, 1967.
210. **Turner, L. H.,** *Trans. R. Soc. Trop. Med. Hyg.,* 62, 880, 1968.
211. **Turner, L. H.,** *Trans. R. Soc. Trop. Med. Hyg.,* 64, 623, 1970.
212. **Kuhn, D. A.,** *Int. J. Syst. Bacteriol.,* 20, 301, 1970.
213. **Hunter, E. F.,** *Crit. Rev. Microbiol.,* 5, 315, 1975.
214. **Canale-Parola, E.,** in *Methods in Microbiology,* Norris, J. R. and Ribbons, D. W., Eds., Academic Press, New York, 1973, p. 61.
215. **Stalheim, O. H. V.,** *Crit. Rev. Microbiol.,* 2, 423, 1973.
216. **Breznak, J. A.,** *Crit. Rev. Microbiol.,* 2, 457, 1973.
217. **Greenberg, E. P. and Canale-Parola, E.,** *J. Bacteriol.,* 123, 1006, 1975.

VIBRIOS AND SPIRILLA*

R. R. Colwell

Short curved or straight cells, single or united into spirals, that grow well and rapidly on the surfaces of standard culture media can be readily isolated from salt- and fresh-water samples. These heterotrophic organisms vary in their nutritional requirements; some occur as parasites and pathogens for animals and for man. These short curved, asporogenous, Gram-negative rods, members of the genera *Vibrio* and *Spirillum,* are most commonly encountered in the marine or fresh-water habitat. Distinguishing species of the genus *Vibrio* from those of *Spirillum* can often be accomplished by examining Gram and flagella stains of carefully prepared specimens; *Spirillum* species are frequently seen as rigid, helical cells with a single or several turns, motile by means of biopolar polytrichous flagella, whereas *Vibrio* species are short rods with a curved axis, motile by means of a single polar flagellum. However, *Vibrio* species may be short straight rods (1.5–3.0 μm × 0.5 μm), or they may be S-shaped or spiral-shaped when individual cells are joined. Possession of two or more flagella in a polar tuft has also been demonstrated in *Vibrio* species, as have lateral flagella. Thus, to identify and classify vibrios and spirilla, physiological and biochemical taxonomic tests should be made.[1] *Vibrio* species are facultatively anaerobic, with both a respiratory (oxygen-utilizing) and a fermentative metabolism. *Spirillum* species are aerobic or microaerophilic, with a strictly respiratory metabolism (oxygen is the terminal electron acceptor). Hylemon et al.[2] proposed a division of the genus *Spirillum* into three genera – *Spirillum, Aquaspirillum,* and *Oceanospirillum;* this was supported by the work of Carney et al.[3] Marine vibrios and spirilla require 1 to 3% NaCl for growth.

In the 8th edition of *Bergey's Manual,*[1] published in 1974, the genus *Vibrio* is placed among the Gram-negative facultatively anaerobic rods in the family Vibrionaceae, whereas the spirilla are grouped as the spiral and curved bacteria in the family Spirillaceae, containing the genera *Spirillum* and *Campylobacter.* The Spirillaceae are defined as simple cells, helically curved rods that, after transverse division, frequently remain attached to each other to form chains of spirally twisted cells. The cells are described as rigid, usually motile by means of a single flagellum (rarely two) or a fascicle of several polar flagella, Gram-negative, and most frequently isolated from water, although some species are pathogenic for higher animals and man.

The borderline between the genus *Vibrio* and the genus *Pseudomonas* has been considered in the past to be somewhat vague. However, evidence that sharpens the demarcation between the two genera has been gathered. *Pseudomonas* species are oxidative and possess overall DNA base compositions in the range from 57 to 67% G + C, whereas *Vibrio* species are fermentative, producing acid in carbohydrate-containing media without formation of gas, and possess a DNA with G + C in the range from 39 to 49 moles %. Separation of *Spirillum* and *Pseudomonas* species, unfortunately, remains less clear-cut, and those *Spirillum* species with G + C ranging from 57 to 67% may prove to be appropriately grouped with *Pseudomonas.* DNA/DNA homology studies should prove useful in establishing the relationships among the genera *Spirillum, Aquaspirillum, Oceanospirillum, Pseudomonas* and *Vibrio.*

GENUS *VIBRIO* PACINI, 1854

The type species for the genus *Vibrio* is *Vibrio cholerae* Pacini, 1854. A total of five species of *Vibrio* are listed in *Bergey's Manual.*[1] The strictly anaerobic species have been

* Parts 6 and 8 in *Bergey's Manual.*

removed from the genus, since no cultures of obligately anaerobic *Vibrio* species are extant, and the microaerophilic species, including *Vibrio fetus* and *Vibrio bubulus,* have been transferred to the genus *Campylobacter* in the family Spirillaceae by Sebald and Véron,[4] who observed the DNA base composition of *Campylobacter* species to be in the range from 30 to 34% G + C.

Of the species of *Vibrio* listed in the 7th edition of *Bergey's Manual,*[5] those not attacking carbohydrates were transferred to the genus *Commamonas.* Davis and Park[6] described *Commamonas* as Gram-negative rod-like bacteria that give no reaction on carbohydrate and an alkaline reaction in the Hugh and Leifson test.[7] Other test results described were as follows: negative in indole, cholera red, methyl red, phenylalanine deaminase, Moeller's lysine, arginine, ornithine, and Voges-Proskauer tests; positive in urease, catalase, and oxidase tests; no liquefaction of gelatin, no utilization of citrate, no evidence of growth in KCN medium, no pigment or fluorescence under ultraviolet light, and possession of lophotrichous flagella. Confirming the observations of Davis and Park,[6] Sebald and Véron[4] assigned, on the basis of DNA G + C composition (64%), *Vibrio percolans, Vibrio cyclosites, Vibrio neocistes,* and *Vibrio alcaligenes* to the genus *Commamonas,* which in the 8th edition of *Bergey's Manual* is now listed among the species of *Pseudomonas.* Colwell and Liston[8] observed that *Vibrio cuneatus* produced a green-fluorescent pigment, and subsequent DNA studies[9] confirmed the conclusion that this species should be assigned to the genus *Pseudomonas.*

The C27 organisms of Ferguson and Henderson,[10] previously assigned to *Aeromonas* by Ewing et al.,[11] to *Plesiomonas* by Habs and Schubert[12] and by Eddy and Carpenter,[13] and to *Fergusonia* by Sebald and Véron,[4] were suggested to belong to the genus *Vibrio* by Hendrie et al.[14] However, in the 8th edition of *Bergey's Manual* these organisms are grouped in the genus *Plesiomonas.* Thus, the number of characterized and defined species resident in the genus *Vibrio* has been markedly reduced.

The description of the genus *Vibrio*, as amended by the Subcommittee on Taxonomy of Vibrios, International Committee on Nomenclature of Bacteria, is concise and provides a good working definition:

> "Gram-negative, asporogenous rods which have a single, rigid curve or which are straight. Motile by means of a single, polar flagellum. Produce indophenol oxidase and catalase. Ferment glucose without gas production. Acidity is produced from glucose by the Embden-Meyerhof glycolytic pathway. The quanine plus cytosine in the DNA of *Vibrio* species is within the range of 40 to 50 moles per cent."*

The type species of the genus, *Vibrio cholerae* Pacini 1854, can be succinctly described as follows: producing L-lysine and L-ornithine decarboxylases; not producing L-arginine dihydrolase and hydrogen sulfide (Kligler iron agar). The G + C content in the DNA of *Vibrio cholerae* is approximately 48 ± 1%. *Vibrio cholerae* includes strains that may or may not elicit the cholera-red (nitroso-indole) reaction, may or may not be hemolytic, may or may not be agglutinated by Gardner and Venkatraman O group I antiserum, and may or may not be lysed by Mukerjee *Vibrio cholerae* bacteriophages I, II, III, IV, and V.[15]

Vibrio cholerae strains possess a common H antigen and can be serologically grouped into 39 serotypes according to their O antigens, as described in Reference 16. Strains agglutinated by Gardner and Venkatraman O group I antiserum are in serotype I and are the principal cause of cholera in man. An M antigen can obscure the agglutinability of mucoid strains of *Vibrio cholerae.* Furthermore, *Vibrio cholerae* strains in the R (rough) form cannot be serotyped. General reviews on cholera are provided in References 16 to 19. Isolation and diagnosis procedures are outlined in these reviews and in References 20 and 21.

* From Hugh, R. and Feeley, J. C., *Int. J. Syst. Bacteriol.*, 22, 123, 1972. With permission.

In the past, many cholerae-like vibrios have been given separate species status because they were isolated from patients suffering diarrhea, not cholera, or from water and foods. *Vibrio proteus, Vibrio metschnikovii, Vibrio berolinensis, Vibrio albensis,* and *Vibrio paracholerae* can be considered to be biotypes of *Vibrio cholerae.* The biotypes of *Vibrio cholerae* recognized in the 8th edition of *Bergey's Manual* are *cholerae, eltor, proteus,* and *albensis.* The so-called non-agglutinable (NAG) or non-cholera (NCV) vibrios, including *Vibrio eltor,* have been shown by Citarella and Colwell,[22] using the techniques of DNA/DNA reassociation measurements, to be related to *Vibrio cholerae* at the species level.

Since its initial isolation by Fujino and Fukumi[23] in 1953, *Vibrio parahaemolyticus* has received considerable attention. *Vibrio parahaemolyticus* (syn. *Pasteurella parahaemolytica, Pseudomonas enteritis, Oceanomonas parahaemolytica*) is the causative agent of food poisoning arising from ingestion of contaminated seafood and can be isolated from the marine environment. A review of its identification and classification is provided in References 24 to 26.

The species *Vibrio alginolyticus,* although previously considered by some investigators to be synonymous with *Vibrio parahaemolyticus,* is considered to be a separate species. Nevertheless, in the 8th edition of *Bergey's Manual* it is listed as a biotype of *Vibrio parahaemolyticus.*

Other species of *Vibrio* that should be considered are the following: *Vibrio anguillarum* (syn. *Vibrio piscium, Achromobacter ichthyodermis, Pseudomonas ichthyodermis, Vibrio piscium* var. *japonicus, Vibrio ichthyodermis*), isolated from diseased conditions in marine- and fresh-water fish; *Vibrio fischeri* and *Vibrio marinus,* found in sea water and associated with marine animals;[27,28] and *Vibrio costicola,* a species tolerating salt concentrations from 2 to 23%, with an optimal concentration of 6 to 12%.

Distinctions can be made among the species of the genus *Vibrio,* and useful differentiating characteristics are given in Table 1. Some species of *Vibrio* may demonstrate sheathed flagella – i.e., a flagellum with a central core and an outer sheath – and, under certain conditions of growth, peritrichous flagella.[29] "Round bodies" or sphaeroplasts are commonly present during various stages of growth,[30,31] and fimbriae (pili) have been observed in strains of *Vibrio cholerae.*[32]

Separation of *Vibrio* from *Aeromonas, Plesiomonas, Photobacterium,* and *Lucibacterium* may, on occasion, prove to be difficult. Features useful for such differentiation are given in Table 2.

GENUS *SPIRILLUM* EHRENBERG, 1832

The genus *Spirillum* has been less sharply defined than the genus *Vibrio. Spirillum* Ehrenberg, 1832, as described in the 7th edition of *Bergey's Manual,*[5] included crescent-shaped to spiral cells that are frequently united into spiral chains of cells, which are not embedded in zoogloeal masses. In the 8th edition of *Bergey's Manual*[1] the genus is described as including spiral cells, usually motile by means of polar flagellation, i.e., possessing a tuft of polar flagella at one or both ends of the cells. The cells form either long screws or portions of a turn. Intracellular granules of polyhydroxybutyrate are present in most species. Spirilla are either aerobic, growing well on ordinary culture media, or microaerophilic.

Nine species were listed in the 7th editon of *Bergey's Manual,*[5] but at least 30 species have been described in the literature. Species described in the 8th edition of *Bergey's Manual*[1] include 19 *Spirillum* species. In a numerical taxonomy study carried out by Colwell,[24] *Spirillum itersonii,* ATCC strain 11331, was found to be a green-fluorescent, pigment-producing organism clustering with *Pseudomonas* species, including *Pseudomonas fluorescens.* The overall G + C content of this organism and that of *Spirillum serpens,*

Table 1
FEATURES USEFUL IN DIFFERENTIATING AND CHARACTERIZING SPECIES OF THE GENUS *VIBRIO*

Characteristic	*Vibrio cholerae*	*Vibrio parahaemolyticus*	*Vibrio anguillarum*	*Vibrio marinus*	*Vibrio costicolus*
Rod shape	+	+	+	+	+
Motility	+	+	+	+	+
Single polar flagellum	+	+	+	v	+
Lophotrichous flagella	–	–	–	v	–
Gram reaction	–	–	–	–	–
Diffusible pigment	–	–	–	–	–
Luminescence	–	–	–	–	–
Pathogenicity for man or animals	+	+	+	–	–
DNA base composition (% G + C)	46–49	44–46	44–45	40–44	50
Indole reaction	+	+	+	–	–
Methyl-red reaction	+	+	+	+	–
Voges-Proskauer reaction	+	v	+	–	+
Citrate utilization	+	+	–	–	–
Citrulline utilization	–	–	+	–	nt
Sensitivity					
0/129[a]	+	+	+	+	v
novobiocin, 10 μg	+	+	+	+	+
penicillin, 10 units	+	–	–	–	nt
polymyxin, 300 units	v	–	v	+	nt
streptomycin, 10 μg	+	–	–	+	nt
Growth					
in 0% NaCl	+	–	+	+	–
in 1% NaCl	+	+	+	+	–
in 7% NaCl	v	+	v	+	+
in 10% NaCl	–	+	–	–	+
at 5°C	–	–	+	+	+
at 20°C	+	+	+	+	+
at 37°C	+	+	v	–	v
at 42°C	+	+	–	–	–
Acid production					
from arabinose	–	+	–	–	–
from inositol	–	–	–	v	–
from mannitol	+	+	+	+	v
from mannose	+	+	+	+	+
from salicin	–	–	–	v	–
from sucrose	+	v	+	v	+
Gelatin liquefaction	+	+	+	+	v
Hydrolysis					
casein	+	+	+	+	–
starch	+	+	+	v	–
Tween 80	+	+	+	+	+
H_2S production (on lead acetate agar)	–	–	–	–	–
Lecithinase (egg yolk)	+	+	nt	nt	v
Arginine dihydrolase	–	–	+	–	+
Lysine decarboxylase	+	+	–	+	–
Ornithine decarboxylase	+	+	–	–	–
Hemolysis	+	+	v	–	–

[a] 2,4-Diamino-6,7-diisopropylpteridine.

Symbol code: + = positive or present; – = negative or absent; v = reaction varied among the strains tested; nt = not tested.

Data compiled from several sources, including References 14, 24, 27, 28 and 33, and from unpublished data obtained by the author.

Table 2
DIFFERENTIATION OF RELATED GENERA FREQUENTLY ISOLATED FROM THE SAME SOURCE IN NATURE

Characteristic	*Vibrio*	*Aeromonas*	*Plesiomonas*	*Photobacterium*	*Lucibacterium*	*Pseudomonas*	*Spirillum*	*Campylobacter*
Morphology	Straight or curved rod	Straight rod	Straight rod	Straight rod	Straight or curved rod	Straight rod	Helical	Spirally curved rod
Diffusible pigment	None	None[a]	None	None	None	None or green-fluorescent	None or green-fluorescent	None
Motility	+	+	+	+	+	+	+	+
Flagella	Polar	Polar	Lophotrichous	Polar	Peritrichous (usually)	Polar	Lophotrichous	Polar
Carbohydrate metabolism	Fermentative	Fermentative	Fermentative	Fermentative	Fermentative	Respiratory or not metabolized	Respiratory	Not metabolized
Gas production from carbohydrates	–	v	–	+	–	–	–	–
Luminescence	v	–	–	+	+	–	–	–
Oxidase	+	+	+	v	+	+	+	+
0/129 sensitivity	+	–	+	+	–	–	–	–
"Round bodies" or "cysts" produced	+	–	–	–	–	–	+	+

[a] Species of *Aeromonas* may produce a brown pigment.

Symbol code: + = positive or present; – = negative or absent; v = variable.

subsp. *serpens,* ATCC strain 11330, both 64%, suggest that green-fluorescent, pigment-producing strains of *Spirillum* with a G + C content of more than 50% may, in fact, be strains of *Pseudomonas* with spiral-shaped (as opposed to straight-rod) morphology. Furthermore, the work of Hylemon et al.,[2] as well as that of Carney et al.,[3] indicates that there are four genera that should be included in the family Spirillaceae: *Spirillum, Aquaspirillum, Oceanospirillum,* and *Campylobacter.* However, the genus *Campylobacter* may eventually be placed in a family other than the Spirillaceae.

In general terms, the genus can be described as comprising rigid, helical, Gram-negative cells, motile with bipolar polytrichous flagella. Spirilla possess a strictly respiratory metabolism, with oxygen as the terminal electron acceptor; they are oxidase- and catalase-positive, and usually phosphatase-positive. H_2S is usually produced from cysteine, but indole, sulfatase, and amylase are not produced. Other reactions usually negative are hydrolysis of casein, hippurate, and gelatin, and production of urease and acid from sugars. Though the species cannot utilize sugars, various organic acids, alcohols, or amino acids can suffice as sole carbon sources. Most *Spirillum* species do not demonstrate amino acid, vitamin, purine, or pyrimidine requirements, and ammonium ion can usually be utilized as a sole nitrogen source. Growth below 10°C or above 45°C is uncommon. Salt requirements for marine species range from 1 to 3% NaCl.

Table 3 lists differentiating characteristics of the spirilla. Species of *Spirillum* and related genera other than these have been described elsewhere (see References 2, 3, and 33 to 37). It must be emphasized that the taxonomy of the genus *Spirillum* is presently under study in several laboratories and will doubtless be further altered in the near future.

DIFFERENTIATION FROM OTHER BACTERIA

Characteristics useful for distinguishing vibrios and spirilla from several genera with which they are frequently confused are given in Table 2.

Genera of uncertain affiliation include *Desulfovibrio, Butyrivibrio, Succinivibrio, Succinimonas, Bdellovibrio, Microcyclus, Pelosigma,* and *Brachyarcus.* Of these, the *Bdellovibrio, Microcyclus, Pelosigma,* and *Brachyarcus* spp. are included with the spiral and curved bacteria in the 8th edition of *Bergey's Manual.*[1]

Bdellovibrio spp. are, in general, small, curved, Gram-negative motile rods. They are parasitic for other bacteria; that is they attach to and penetrate into the host cell – the outstanding characteristic of the genus. Various Gram-negative and Gram-positive bacteria may serve as hosts, with a specific host range often characteristic of given species of *Bdellovibrio.* All strains of *Bdellovibrio* isolated from nature are parasitic, but host-independent strains frequently are developed in the laboratory. *Bdellovibrio* spp. possess a polar sheathed flagellum, usually monotrichous. The G + C content of the DNA of the type species, *Bdellovibrio bacteriovorus,* is 50 ± 1%, whereas that of the other species, *Bdellovibrio stolpii* and *Bdellovibrio starrii,* is 42 to 43 moles %. The catalase and protease reactions and sensitivity to the vibriostat 0/129 are used to speciate the *Bdellovibrio.*[1]

The genera *Microcyclus, Pelosigma,* and *Brachyarcus* are, as yet, incompletely described. They are found in aquatic environments, with *Microcyclus* being noted as curved aerobic rods, on occasion forming rings. *Pelosigma* and *Brachyarcus* spp. have not been isolated in pure culture.

Table 3
DIFFERENTIATING CHARACTERISTICS OF SELECTED *SPIRILLUM* AND RELATED SPECIES

Characteristic	*Spirillum volutans*	*Aquaspirillum anulus*	*Aquaspirillum itersonii*	*Aquaspirillum bengal*[a]	*Aquaspirillum serpens*	*Oceanospirillum linum*	*Campylobacter fetus*
Size	1.4–1.7 μm[b] × 14–60 μm	1.4–1.5 μm × 7–15 μm	0.4–0.6 μm × 2.0–7.0 μm	0.9–1.2 μm × 5.2–22 μm	0.5–1.0 μm × 5- 35 μm	0.4–0.6 μm × 4–30 μm	0.2–0.5 μm × 1.5–5.0 μm
Motility (lophotrichous)	+	+	+	+	+	+	+
Fluorescent pigment	–	–	+	–[c]	v	+	–
Oxidase	+	+	+	+	+	+	+
Catalase	+	+	+	+	+	+	+
Phosphatase	+	+	+	+	+	+	v
Urease	–	–	–	–	–	–	–
Phenylalanine deaminase	–	–	–	–	–	–	–
DNAse	–	+	+	v	+	–	–
RNAse	–	+	+	–	+	v	–
Gelatin hydrolysis	–	–	–	–	–	–	nt
Esculin hydrolysis	–	–	+	–	–	–	–
Nitrate reduction	–	–	+	–	–	–	+
Acid from carbohydrates	–	–	v	–	–	–	–
Growth							
in 1% bile	–	–	+	+	+	+	+
in 3% NaCl	–	–	–	–	–	+	–
at 10°C	–	–	–	–	–	v	v
at 30–32°C	+	+	+	+	+	+	+
at 42°C	–	–	–	+	–	–	–(v)
Oxygen requirements	Obligately microaerophilic	Strict aerobe	Strict aerobe	Aerobic to microaerophilic	Strict aerobe	Strict aerobe	Microaerophilic
DNA base composition	38%	58–59%	62–64%	51%	64%[d]	48%	32–35%

[a] From Kumar et al., *Int. J. System. Bacteriol.*, 24, 453, 1974.
[b] Diameter; the other measurements represent width by length.
[c] Water-soluble pigments from tyrosine and tryptophan.
[d] Reported as 50% by Hylemon et al.[2]

Symbol code: + = positive or present; – = negative or absent; v = variable among the strains tested; nt = not tested.

REFERENCES

1. **Buchanan, R. E. and Gibbons, N. E.,** *Bergey's Manual of Determinative Bacteriology,* 8th ed., Williams & Wilkins, Baltimore, 1974.
2. **Hylemon, P. B., Wells, J. S., Krieg, N. R., and Jannasch, H. W.,** *Int. J. Syst. Bacteriol.,* 23, 340, 1973.
3. **Carney, J. F., Wan, L., Lovelace, T. E., and Colwell, R. R.,** *Int. J. Syst. Bacteriol.,* 25, 38, 1975.
4. **Sebald, M. and Véron, M.,** *Ann. Inst. Pasteur Lille,* 105, 897, 1963.
5. **Breed, R. S., Murray, E. G. D., and Smith, N. R.,** *Bergey's Manual of Determinative Bacteriology,* 7th ed., Williams & Wilkins, Baltimore, 1957.
6. **Davis, G. H. G. and Park, R. W. A.,** *J. Gen. Microbiol.,* 27, 101, 1962.
7. **Hugh, R. and Leifson, E.,** *J. Bacteriol.,* 66, 24, 1953.
8. **Colwell, R. R. and Liston, J.,** *J. Bacteriol.,* 82, 1, 1961.
9. **Colwell, R. R. and Mandel, M.,** *J. Bacteriol.,* 87, 1412, 1964.
10. **Ferguson, W. W. and Henderson, N. D.,** *J. Bacteriol.,* 54, 178, 1947.
11. **Ewing, W. H., Hugh, R., and Johnson, J. G.,** *Studies on the Aeromonas Group,* Communicable Disease Center, Atlanta, 1961.
12. **Habs, H. and Schubert, R. H. W.,** *Zentralbl. Bakteriol. Parasitenkd. Infektionskr. Hyg. Abt. 1 Orig.,* 186, 316, 1962.
13. **Eddy, B. R. and Carpenter, K. P.,** *J. Appl. Bacteriol.,* 27, 96, 1964.
14. **Hendrie, M. S., Shewan, J. M., and Véron, M.,** *Int. J. Syst. Bacteriol.,* 21, 25, 1971.
15. **Hugh, R. and Feeley, J. C.,** Report (1966–1970) of the Subcommittee on Taxonomy of Vibrios to the International Committee on Nomenclature of Bacteria, *Int. J. Syst. Bacteriol.,* 22, 123, 1972.
16. **Sakazaki, R., Kazunichi, T., Gomez, C. Z., and Sen, R.,** *Jpn. J. Med. Sci. Biol.,* 23, 13, 1970.
17. **Pollitzer, R.,** *Cholera* (W.H.O. Monograph Series, No. 43), World Health Organization, Geneva, 1959, p. 1019.
18. **Felsenfeld, O.,** *Bacteriol. Rev.,* 28, 72, 1964.
19. **Felsenfeld, O.,** *Bull. W.H.O.,* 34, 161, 1966.
20. **Burrows, W. and Pollitzer, R.,** *Bull. W.H.O.,* 18, 275, 1958.
21. **Carpenter, K. P., Hart, J. M., Hatfield, J., and Wicks, G.,** Identification methods for microbiologists (Gibbs, B. M. and Shapton, D. A., Eds.), *Soc. Appl. Bacteriol. Tech. Ser.,* 2, 8, 1968.
22. **Citarella, R. V. and Colwell, R. R.,** *J. Bacteriol.,* 104, 434, 1970.
23. **Fujino, T. and Fukumi, H. (Eds.),** *Vibrio parahaemolyticus,* 2nd ed., Naya Shoten, Tokyo, 1967 (in Japanese).
24. **Colwell, R. R.,** *J. Bacteriol.,* 104, 410, 1970.
25. **Vanderzant, C., Nickelson, R., and Parker, J. C.,** *J. Milk Food Technol.,* 33, 161, 1970.
26. **Colwell, R. R.,** *Microbiology 1974,* American Society for Microbiology, Washington, D.C., 1974.
27. **Colwell, R. R. and Morita, R. Y.,** *J. Bacteriol.,* 88, 831, 1964.
28. **Bianchi, M. A. G.,** *Arch. Mikrobiol.,* 77, 127, 1971.
29. **Baumann, P., Baumann, L., and Mandel, M.,** *J. Bacteriol.,* 107, 268, 1971.
30. **Felter, R. A., Kennedy, S. F., Colwell, R. R., and Chapman, G. B.,** *J. Bacteriol.,* 102, 552, 1969.
31. **Kennedy, S. F., Colwell, R. R., and Chapman, G. B.,** *Can. J. Microbiol.,* 16, 1027, 1970.
32. **Tweedy, J. M., Park, R. W. A., and Hodgkiss, W.,** *J. Gen. Microbiol.,* 51, 235, 1968.
33. **Williams, M. A.,** *Int. Bull. Bacteriol. Nomencl. Taxon.,* 9, 137, 1959.
34. **Watanabe, N.,** *Bot. Mag.* (Tokyo), 72, 77, 1959.
35. **Terasaki, Y.,** *Bull. Suzugamine Womens Coll. Nat. Sci.,* Suppl. 8–9, 1, 1962.
36. **Wells, J. S.,** Ph.D. thesis, Virginia Polytechnic Institute, Blacksburg, Va., 1970.
37. **Hylemon, P. B.,** Ph.D., thesis, Virginia Polytechnic Institute, Blacksburg, Va., 1971.

GRAM-NEGATIVE AEROBIC RODS AND COCCI*

H. A. Lechevalier

Generic name (G + C, molar %) [Number of species in *Bergey's Manual*]	Morphology	Physiology	Ecology	Examples of species
Pseudomonas (58 to 70) [29]	Straight or curved rods with polar flagellation; 0.5–1 by 1.5–4 μm.	Never fermentative; respiratory metabolism; some are facultative chemolithotrophs, using H_2 or CO as source of energy; some can denitrify and respire anaerobically.	Found in soil and in water bodies; active in mineralization of organic matter. Some species are pathogenic.	*P. aeruginosa.* Type species; produces fluorescent pigments; may be an opportunistic pathogen of man, animals, and plants; cause of "blue pus"; produces the blue phenazine pigment pyocyanin. *P. fluorescens.* Produces fluorescent pigments; often associated with spoilage of foods and isolated from clinical specimens. *P. pseudomallei.* Soil organisms; opportunistic pathogen causing melioidosis. *P. mallei.* Cause of glanders and farcy of horses and donkeys.
Xanthomonas (64 to 69) [5]	Straight rods, with single polar flagellum; 0.2–0.8 by 0.6–2 μm; usually yellow.	Never fermentative; respiratory metabolism; oxidase reaction negative or weak; catalase-positive; nitrates not reduced.	Plant pathogens found in association with plant or plant materials.	*X. campestris.* Type species; causes a vascular and parenchymatous disease of some crucifers. *S. fragariae.* Causes a leaf spot disease of strawberry.

* Part 7 in *Bergey's Manual.*

Generic name (G + C, molar %) [Number of species in *Bergey's Manual*]	Morphology	Physiology	Ecology	Examples of species
Zoogloea (–) [2]	Rods 0.5–1.0 by 1.0–3.0 μm; have a single polar flagellum when young; when older, aggregate into firm macroscopic flocs with finger-like or dendritic outgrowths; extracellular fibrils interlace the cells in the flocs.	Chemoorganotrophs; respiratory metabolism; do not hydrolyze cellulose and corn starch; not proteolytic; litmus milk not peptonized; indole and H_2S not produced; nitrates not reduced; cytochrome oxidase and catalase positive; vitamin B_{12} required; biotin stimulatory.	Found in natural water and sewage.	*Z. ramigera*. Type species.
Gluconobacter (60 to 64) [1]	Ellipsoidal to rod-shaped cells, 0.6–0.8 by 1.5–2.0 μm; occur singly, in pairs, or in chains; if motile, with 1 to 8 polar flagella.	Chemoorganotrophs; respiratory metabolism; produce acetic acid from ethanol; do not oxidize acetic acid and lactate to CO_2; usually produce ketones and acids from sugars; catalase-positive.	Found in flowers, fruits, and vegetables, in fermented beverages, in vinegar, and in soil.	*G. oxydans*. Type species.
Acetobacter (55 to 64) [3]	Ellipsoidal to rod-shaped cells, 0.6–0.8 by 1.0–3.0 μm; may be slightly curved; occur singly, in pairs, or in chains; if motile, by peritrichous flagellation.	Chemoorganotrophs; respiratory metabolism; produce acetic acid from ethanol; oxidize acetic acid and lactate to CO_2; do not hydrolyze dextrin and starch.	Same as above.	*A. aceti*. Type species; catalase-positive, produces 5-ketoglucomate. *A. pasteurianus*. Catalase-positive; does not produce 5-ketogluconate. *A. peroxydans*. Catalase-negative; does not produce 5-ketogluconate.

Generic name (G + C, molar %) [Number of species in *Bergey's Manual*]	Morphology	Physiology	Ecology	Examples of species
Azotobacter (63 to 66) [4]	Large ovoid cells, more than 2 μm in diameter; occur singly, in pairs, or in irregular clumps; form thick-walled cysts; if motile, have peritrichous flagella; most strains are capsulated and produce copious slime.	Fix atmospheric nitrogen; in cultures, at least 10 mg/g carbohydrate consumed; molybdenum required for nitrogen fixation, but may be replaced by vanadium; not proteolytic; catalase-positive; may grow under reduced oxygen tension.	Found in soil and in water bodies.	*A. chroococcum*. Type species; no fluorescent pigment; motile; starch and mannitol utilized. *A. beijerinckii*. No fluorescent pigment; non-motile; starch and mannitol not utilized. *A. vinelandii*. Green fluorescent pigment; motile; mannitol utilized, but not starch. *A. paspali*. Green fluorescent pigment; motile; starch and mannitol not utilized; in soil, found particularly on the root surface of *Paspalum* grass.
Azomonas (53 to 59) [3]	Large ovoid cells, more than 2 μm in diameter; occur singly, in pairs, or in irregular clumps; do not form cyst; motile, with polar or peritrichous flagella.	Same as above.	Same as above.	*A. agilis*. Type species; white fluorescent pigment; peritrichous flagellation. *A. insignis*. No fluorescent pigment; lophotrichous flagellation. *A. macrocytogenes*. White fluorescent pigment; monotrichous flagellation.

Generic name (G + C, molar%) [Number of species in *Bergey's Manual*]	Morphology	Physiology	Ecology	Examples of species
Beijerinckia (55 to 59) [4]	Straight, slightly curved, or pear-shaped rods with rounded ends, 0.5–1.5 by 1.7–4.5 μm; large, highly refractile bodies contain poly-β-hydroxybutyrate at each end of the cells; if motile, by peritrichous flagellation; cysts, enclosing one cell, or capsules, enclosing several cells, may be formed.	Molybdenum required for nitrogen fixation, but cannot be replaced by vanadium.	Same as above.	*A. indica.* Type species; widely distributed in acidic tropical soils.
Derxia (70) [1]	Rod-shaped cells with rounded ends, 1.0–1.2 by 3.0–6.0 μm; occur singly or in short chains; motile by a short polar flagellum; older cells have large refractile bodies distributed throughout.	Same as above, but catalase-negative.	Same as above.	*D. gummosa.* Type species.
Rhizobium (59 to 66) [6]	Rods, 0.5–0.9 by 1.2–3.0 μm; pleomorphic under adverse conditions; motile by 1 to 6 peritrichous flagella; produce	Chemoorganotrophs; respiratory metabolism; may grow under reduced oxygen tension; casein not	Able to invade root hairs of legumes and incite the production of root nodules;	*R. leguminosarum.* Type species; forms nodules in peas, vetches, and lentils; motile by 2 to 6 flagella.

Generic name (G + C, molar %) [Number of species in *Bergey's Manual*]	Morphology	Physiology	Ecology	Examples of species
Rhizobium (continued)	nodules on roots of legumes; pleomorphic in nodules (bacteroids).	hydrolyzed; cellulose and starch not utilized; some strains require water-soluble vitamins; slime produced when grown on carbohydrate-containing media; nitrogen fixed within root nodules.	all strains show host specificity.	*R. phaseoli.* Forms nodules in beans; motile by 2 to 6 flagella. *R. japonicum.* Forms nodules in *Glycine* and *Lupinus* spp.; motile by a polar or subpolar flagellum.
Agrobacterium (60 to 63) [4]	Rods, 0.8 by 1.5–3.0 μm; motile by 1 to 4 peritrichous flagella; produce hypertrophies on many plants.	Same as above, but nitrogen not fixed.	Initiate stem or root hypertrophies on diverse plants wherein the bacteria are intracellular parasites; enter host through existing lesions; strains do not show host specificity.	*A. tumefaciens.* Type species; cause galls in plants belonging to more than 40 families. *A. radiobacter.* An avirulent form of *A. tumefaciens.*
Methylomonas (52) [3]	Cells straight, curved, or branched; 0.5–1.0 by 1–4 μm; motile by polar flagellum.	Respiratory metabolism; utilize only one-carbon organic compounds, as source of carbon; methane and methanol are the only known sources of carbon and energy; organic factors not required.	Found in mud and water from ponds, rivers, streams, and ditches; also present in soil.	*M. methanica.* Type species.

Generic name (G + C, molar %) [Number of species in *Bergey's Manual*]	Morphology	Physiology	Ecology	Examples of species
Methylococcus (63) [1]	Cells spherical; usually occur in pairs.	Same as above.	Same as above.	*M. capsulatus.* Type species.
Halobacterium (Major 66 to 68, minor 57 to 60) [2]	Rods, 0.6–1.0 by 1–6 μm; if motile, by polar flagella; colonies red, pink, orange, vermilion, or mauve.	Require about 12% NaCl for growth; sodium, chloride, and magnesium ions are required to maintain cell integrity; cell wall does not contain diaminopimelic acid or muramic acid, but is mainly lipoproteinaceous; metabolism respiratory; amino acids are mainly used as sources of energy; contain carotenoids, mainly bacterioruberin; DNA with 2 components.	Found in salt lakes, salterns, and salted products.	*H. salinarium.* Type species.
Halococcus (Major 67, minor 59) [1]	Cocci, 0.6–1.5 μm diameter; occur in pairs, tetrads, or irregular clusters of tetrads; nonmotile.	Same as above.	Same as above.	*H. morrhuae.* Type species.

Generic name (G + C, molar %) [Number of species in *Bergey's Manual*]	Morphology	Physiology	Ecology	Examples of species
Alcaligenes (58 to 70) [4]	Cocci to rods, 0.5–1.2 by 0.5–2.6 μm; motile, with 1 to 8 peritrichous flagella.	Chemoorganotrophs; respiratory metabolism; some may respire anaerobically in the presence of nitrate or nitrite; oxidase-positive.	Saprophytic inhabitants of the intestinal tract of vertebrates; occur in dairy products, rotting foods, soil, and water bodies.	*A. faecalis.* Type species; some strains have a characteristic strawberry-like odor.
Brucella (56 to 58) [6]	Coccobacilli to short rods, 0.5–0.7 by 0.6–1.5 μm, non-motile.	Chemoorganotrophs; respiratory metabolism; require thiamine, niacin, and biotin; catalase-positive; growth from small inocula may require CO_2 and serum; cultures prone to spontaneous dissociation.	Mammalian parasites and pathogens; facultatively intracellular; can infect man, causing malta fever (undulant fever), which is debilitating, but usually self-limiting; in animals, have an affinity for reproductive and lactating organs, often causing contagious abortion (Bang's disease).	*B. melitensis.* Type species; usually pathogenic for goats and sheep, but can also affect other species, including cattle and man. *B. abortus.* Usually pathogenic for cattle, causing abortion, but can also affect other species, including man. *B. suis.* Usually pathogenic for pigs, but can also affect hares, reindeer, and other species, including man. *B. neotomae.* Affects desert wood rat. *B. ovis.* Pathogenic for sheep, causing epididymitis in rams and abortions in ewes. *B. canis.* Causes epididymitis in dogs and abortion in bitches.

Generic name (G + C, molar %) [Number of species in *Bergey's Manual*]	Morphology	Physiology	Ecology	Examples of species
Bordetella (58 to 70) [3]	Coccobacilli, 0.2–0.3 by 0.5–1.0 μm; arranged singly, in pairs, or – rarely – in short chains; if motile, by lateral polytrichous flagella.	Chemoorganotrophs; respiratory metabolism; catalase-positive; require nicotinic acid, cysteine, and methionine.	Pathogens of the respiratory tract of mammals.	*B. pertussis.* Type species; causes whooping cough; non-motile; does not utilize citrate. *B. parapertussis.* Causes a whooping-cough-like disease; non-motile; utilizes citrate. *B. bronchiseptica.* Causes bronchopneumonia; motile; utilizes citrate.
Francisella (33 to 36) [2]	Coccoid to ellipsoid pleomorphic rods. 0.2–0.5 by 0.2–1.7 μm; non-motile.	Growth on media enriched with egg yolk, blood, serum, or animal tissues; acid, but no gas may be formed from carbohydrates; strictly aerobic.	Found in many species of wild animals, especially the jack rabbit, and in natural waters.	*F. tularensis.* Type species; causes tularemia, a disease of wild rodents, especially rabbits; transmitted to man by contact with infected tissues, and thus most common in hunters; bites of ticks and deer flies can also transmit the infection to man, where it takes many forms, such as ulceroglandular, oculoglandular, pneumonic, or typhoidal; the latter two forms are the most serious, with mortality rates of about 30% in untreated patients; streptomycin is the drug of choice.

Generic name (G + C, molar %) [Number of species in *Bergey's Manual*]	Morphology	Physiology	Ecology	Examples of species
Francisella (continued)				*F. novicida.* Isolated from a water sample; produces lesions similar to those of tularemia in mice, guinea pigs, and hamsters; does not affect rabbits and white rats.
Thermus (64 to 67) [1]	Rods, 0.5–0.8 by 5–10 μm, forming filaments up to 200 μm; spheres 10–20 μm formed by association of individual cells produced in old cultures; non-motile.	Chemoorganotrophs; obligately aerobic; optimal temperature 70–72° C; usually yellow to orange.	Common inhabitants of thermal aquatic habitats, both natural and man-made; bright orange tufts of inter-woven hyphae can be seen macro-scopically.	*T. aquaticus.* Type species.

*PSEUDOMONAS**

N. J. Palleroni

INTRODUCTION

Members of the genus *Pseudomonas* are very common in nature and can be isolated from a large variety of natural materials. A number of strains are notorious for their nutritional versatility towards organic compounds of low molecular weight in media totally devoid of organic growth factors; this capacity, combined with a fast growth rate, allows them to predominate in the microflora growing in natural media that have a reaction close to neutrality and some organic matter in solution.

The basic morphological features common to almost all species are the straight rod shape and the presence of one or several polar flagella. No spores are produced, and the Gram reaction is negative. These morphological attributes define the "pseudomonads", but admission to the genus *Pseudomonas* requires some additional physiological properties as well. These include an energy metabolism purely respiratory, and nutrition of the chemoorganotrophic type, either absolute or facultative (some species can live autotrophically in contact with an atmosphere containing hydrogen, oxygen, and carbon dioxide).

In the 8th edition of *Bergey's Manual of Determinative Bacteriology,*[1] detailed descriptions of 29 species of *Pseudomonas* are presented, and the names of 236 other organisms assigned to the genus, but not as extensively characterized, are given in appendixes. An index based on some distinctive characters supplements the taxonomic treatment. However, the authors did not claim to have compiled a comprehensive list; it is, indeed, a difficult task to present a complete list of species for a genus that has suffered considerable nomenclatural hypertrophy, and very likely the grand total of species names is substantially larger than 265.

The species included in *Bergey's Manual* have a DNA base composition ranging from 57 to 69 mol% guanine plus cytosine (G + C). No internal discontinuities in this range are useful for the separation of subgeneric categories; in fact, some of the natural groups within the genus cover a G + C span almost as wide as that of the whole genus. The genus is clearly heterogeneous, and by appropriate methods several internal subdivisions can be outlined, which perhaps correspond to separate genera. These various genera, however, share all the important phenotypic properties, and their circumscription in a clear-cut manner can only be achieved by application of rather sophisticated techniques, which are still beyond the reach of the practical taxonomist; at present it appears wise, therefore, to keep all the species under a single generic designation.[2]

BASIC CRITERIA FOR THE INTERNAL SUBDIVISION OF THE GENUS PSEUDOMONAS

A list of the properties that have been found useful for internal subdivision of the genus as presently defined is presented in Table 1. The properties listed in the left-hand column of the table are of basic importance and have been studied in all the well-characterized species; those on the right belong to special categories on which modern approaches to the phylogeny of the genus can be attempted. Due to the amount of labor involved, or due to the limited applicability, relatively few cases have thus far been thoroughly analyzed for these special properties.

* Part 7 in *Bergey's Manual.*

Table 1
CHARACTERS OF *PSEUDOMONAS* STRAINS USED FOR INTERNAL SUBDIVISION OF THE GENUS

Morphology (shape, size of cells)
Flagella (number, insertion, wavelength)
Reserve materials:
  poly-β-hydroxybutyrate
  glucose polysaccharide
Pigments
Temperature relationships
Growth factor requirements
Oxidase reaction
Denitrification
Nitrate reduction
Hydrolytic reactions:
  gelatin
  starch
  poly-β-hydroxybutyrate
  Tween 80 (lipase)
  lecithinase (egg yolk reaction)
Arginine dihydrolase
Ring fission mechanisms
Utilization of carbon sources
Utilization of nitrogen sources
DNA base composition
Nucleic acid hybridization
Metabolic pathways
Regulatory mechanisms
Immunological studies
Amino acid sequence of proteins
Genetic studies

The appropriate methodology for the study of many of these characters has been thoroughly discussed by Stanier et al.[3] and by Palleroni and Doudoroff.[2] It is important to keep in mind that the methods are very important and that no unique set of methods can be recommended for the classification of all groups of bacteria. Thus, the methods popular in the field of the Enterobacteriaceae are inadequate for revealing the taxonomically important characters of *Pseudomonas* species; therefore, a brief discussion of the most important properties and of some of the procedures for their study seems pertinent here. Some of these properties for the various species are summarized in Table 2.

Morphology

Pseudomonas cells are typically straight rods, but curved rods are frequently observed in many cultures. The rods can be very short, almost coccoid, but in some strains (for instance, in some plant-pathogenic species) they can be very long. Average dimensions are 0.5 to 1 μm by 1.5 to 4 μm.

Only one species of the genus, *P. mallei,* is non-flagellated. In all other species, flagella are present, although non-flagellated strains can be isolated from time to time. The flagella are typically polar, and their number per cell (normally expressed as "one" or "several") is taxonomically important. It must be emphasized, however, that in a typically monotrichous species, such as *P. aeruginosa,* cells with more than one flagellum can be observed, and the reverse is also true for multitrichous species. Ideally, the number of flagella should be expessed on a statistical basis from observations of cells that have grown under well-defined conditions.[4] Two species, *P. diminuta* and *P. vesicularis,* have flagella of very short wavelength. Lateral flagella have been observed in some strains; these flagella have short wavelengths, are more easily shed, and are probably controlled by different genetic determinants than polar flagella.

Reserve Materials

Poly-β-hydroxybutyrate (PHB) is by far the most common reserve material found among pseudomonads and has been identified in 17 of the 29 well-characterized species. The capacity for PHB synthesis is an excellent taxonomic character; the synthesis can be stimulated by growth of the cells in nitrogen-deficient media.

In one species, *P. vesicularis,* a glucose polysaccharide has been detected as carbon reserve material.[5]

Table 2
SELECTED PHENOTYPIC CHARACTERS OF *PSEUDOMONAS* SPECIES

			Pigments							Growth at		Hydrolysis of					
Species	**Number of flagella**	**Poly-β-hydroxybutyrate as reserve material**	**Fluorescent**	**Phenazine**	**Carotenoids**	**Growth factor requirements**	**Denitrification**	**Oxidase reaction**	**Arginine dihydrolase**	**4°C**	**41°C**	**Gelatin**	**Poly-β-hydroxy-butyrate**	**Starch**	**Cleavage of diphenols**	**Chemolithotrophic growth with hydrogen**	**G + C**
P. aeruginosa	1	–	+	+	–	–	+	+	+	–	+	+	–	–	*o*	–	67
P. putida	>1	–	+	–	–	–	–	+	+	v	–	–	–	–	*o*	–	60–63
P. fluorescens	>1	–	+	–	–	–	v	+	+	+	–	+	–	–	*o*	–	59–61
P. chlororaphis	>1	–	+	+	–	–	+	+	+	+	–	+	–	–	*o*	–	63
P. aureofaciens	>1	–	+	+	–	–	–	+	+	+	–	+	–	–	*o*	–	63
P. syringae	>1	–	+	–	–	–	–	–	–	v	–	v	–	–	*o*	–	59–61
P. cichorii	>1	–	+	–	–	–	–	+	–		–	–	–	–	*o*	–	59
P. stutzeri	1	–	–	–	–	–	+	+	–	–	v	–	–	+	*o*	–	61–66
P. mendocina	1	–	–	–	+	–	+	+	+	–	+	–	–	–	*o*	–	63–64
P. alcaligenes	1	–	–	–	v	–	–	+	+	–	+	+	–	–		–	66–68
P. pseudoalcaligenes	1	v	–	–	–	–	–	+	v	–	+	v	–	–		–	62–64
P. pseudomallei	>1	+	–	–	–	–	+	+	+	–	+	+	+	+	*o*	–	69
P. mallei	0	+	–	–	–	–	v	+	+	–	+	+	+	v	*o*	–	69
P. caryophylli	>1	+	–	–	–	–	v	+	+	–	+	–	–	–	*o*	–	65
P. cepacia	>1	+	–	v	–	–	–	+	–	–	v	+	–	–	*o*	–	67–68
P. marginata	>1	+	–	–	–	–	–	+	–	–	+	+	–	–	*o*	–	68
P. lemoignei	1	+	–	–	–	–	–	+	–	–	+	–	+	–		–	58
P. testosteroni	>1	+	–	–	–	–	–	+	–	–	–	–	v	–	*m*	–	62
P. acidovorans	>1	+	–	–	–	–	–	+	–	–	–	–	–	–	*m*	–	67
P. delafieldii	1	+	–	–	–	–	–	+	–	–	–	–	+	–	*m*	–	65–66
P. pickettii	1	+	–	–	–	–	+	+	–	–	+	–	–	–	*o*	–	64
P. solanacearum	>1	+	–	–	–	–	v	+	–		–	–	–	–		–	66–68
P. facilis	1	+	–	–	–	–	–	+	–		–	+	+	–	*m*	+	62–64
P. saccharophila	1	+	–	–	–	–	–	+	–		–	–	–	+	*m*	+	69
P. ruhlandii	1	+	–	–	–	–	–	+	–		–	–	–	–		+	
P. flava	1	+	–	–	+	–	–	+	–		–	–	–	–	*m*	+	67
P. palleronii	1	+	–	–	+	–	–	+	–		–	–	–	–	*m*	+	67
P. maltophilia	>1	–	–	–	–	+	–	–	–	–	–	+	–	–		–	67
P. vesicularis	1	+	–	–	+	+	–	±	–	–	–	–	–	–		–	66
P. diminuta	1	+	–	–	–	+	–	+	–	–	v	–	–	–		–	66–67

Code: + = positive for all or most strains; – = negative for all or most strains; ± = weak reaction; v = positive for a variable number of strains; *m* = *meta* type of cleavage; *o* = *ortho* type of cleavage.

Pigments

In the early taxonomic treatments of *Pseudomonas,* pigment production was a character of primary importance at the generic level. Later, many non-pigmented species were included in the genus, but the character of pigmentation retained an important place among the diagnostic traits of some species. The type species of the genus, *P. aeruginosa,* is capable of producing a wide variety of pigments.[2]

Usually, pigment production can be induced or enhanced in special culture media, but repeated transfer of strains in the laboratory sometimes results in total loss of pigment production. In fact, pigmentation is one of the most erratic of all phenotypic traits.

Diffusible pigments that fluoresce under ultraviolet radiation of short wavelength (ca. 254 nm) are collectively called "fluorescent pigments" and characterize important species now placed in the so-called fluorescent group.

Other common pigments (blue, red, yellow, or green) are those belonging to the chemical family of the phenazines; some of these are insoluble, and others diffuse into the medium. Some soluble phenazine pigments fluoresce under long-wavelength ultraviolet radiation (ca. 350 nm), an important distinction with respect to the "true" fluorescent pigments described earlier.

Some species of *Pseudomonas* produce carotenoid pigments. The production of these compounds is not too useful for the circumscription of species groups within the genus; thus, among the many species of the fluorescent group, only *P. mendocina* and one strain of *P. alcaligenes* are known to produce carotenoids.

Temperature Relationships

Growth at 4°C and at 41°C has been determined for a large number of strains. *Pseudomonas* species are typically mesophilic; some, such as *P. fluorescens,* can grow at 4°C and can occasionally be identified in the psychrophilic flora responsible for food spoilage. *Pseudomonas aeruginosa* does not grow at 4°C, but can grow at 41°C, and many strains still do well at 44°C. Determination of these extreme growth temperatures requires the use of well-regulated water baths; air incubators are very unreliable for this purpose.

Growth Factor Requirements

Pseudomonas maltophilia, P. diminuta, and *P. vesicularis* require some amino acids as growth factors. All other well-characterized species of the genus are independent of these nutritional requirements.

Nitrate Reduction and Denitrification

Most *Pseudomonas* species can utilize nitrate as a nitrogen source, and this implies the possession of a reductive system for the conversion of nitrate into amino compounds. The first step of the reduction probably gives nitrate in all cases; this step is the basis for the reaction called reduction of nitrate, which is assayed by testing for the appearance of nitrite in the nitrate medium. The reduction-of-nitrate reaction was performed in very few of the *Pseudomonas* strains of the Berkeley collection, because it was thought that the reaction provided little taxonomically useful information; besides, the accumulation of nitrite probably depends on the activity of the components of the reduction system, so that nitrite may be undetectable when its further reduction is not a limiting step of the process.

Some species of *Pseudomonas* can carry the reduction of nitrate to nitrogen gas or nitrous oxide, a process known as denitrification. Denitrifiers can grow anaerobically, using nitrate in place of oxygen as the electron acceptor. Nitrogen gas escapes to the atmosphere and represents a wasteful process in nature.

Molecular nitrogen cannot be utilized as a nitrogen source by *Pseudomonas* species, and some reports in the literature could not be confirmed when rigorous tests were applied. Recently, however, a new species *P. glathei*, has been thoroughly characterized,[6] which is capable of growth in nitrogen-deficient media. Even though this suggests a capacity for nitrogen fixation, acetylene reduction (a classical test for fixation) could not be demonstrated with cells of *P. glathei* grown under the above conditions.

Extracellular Polysaccharides

Levan formation from sucrose is taxonomically important. It can be determined in various media with a high concentration (4%) of the disaccharide. A number of *Pseudomonas* species can use sucrose without production of levan or dextran; with the exception of *P. saccharophila* (which is capable of phosphorolytic cleavage of sucrose), the metabolism of the disaccharide has not been extensively explored.

Oxidase Reaction and Cytochrome Composition

The oxidase reaction is positive for most *Pseudomonas* species; only *P. maltophilia* and the fluorescent plant pathogen *P. syringae* are negative, and *P. vesicularis* gives a weak reaction (see Table 2). There is a positive correlation between this reaction and the presence of cytochrome C in the cells.

Production of Hydrolytic Enzymes

A number of substrates (gelatin, starch, poly-β-hydroxybutyrate, Tween 80, egg yolk) have been used for the detection of various hydrolytic activities. Gelatinases are of widespread occurrence in the genus; a large number of strains also hydrolyze sorbitan monooleate polyoxyethylene (Tween 80) and manifest strong lecithinase action in the egg yolk reaction. Starch and poly-β-hydroxybutyrate are far less commonly hydrolyzed (see Table 2).

Arginine Dihydrolase Reaction

This reaction is carried out under anaerobic conditions by some *Pseudomonas* species, and it can be tested by direct chemical determination of arginine disappearance or ornithine appearance, and also, less specifically, by pH changes. All strains that have a positive arginine dihydrolase reaction can grow with arginine as the sole carbon source; however, many strains capable of growth with arginine do not decompose the amino acid anaerobically. As a taxonomic character, the reaction is very satisfactorily correlated with many other phenotypic properties, and therefore appears to be very valuable.

Ring Fission Mechanisms

The mechanisms for the degradation of aromatic compounds have been studied extensively in some members of the genus.[7] In these pathways, hydroxylation of the compound precedes its cleavage, and this last step is carried out in a manner that is characteristic of every group of organisms. Taxonomic conclusions, however, should be drawn only when different organisms are compared under the same conditions of induction.[8] In Table 2, the two main methods of cleavage are indicated as *o* and *m* (*ortho* and *meta* respectively).

Nutritional Properties

The basic nutritional requirements for most *Pseudomonas* species are quite simple, since they can grow in mineral media supplemented with a single organic compound as the source of carbon and energy. The number of organic compounds that can support growth varies within wide limits; *P. lemoignei* can use about 10 out of 160, whereas strains of *P. cepacia* can grow at the expense of more than 100. What is more important from the taxonomic point of view is the fact that the list of compounds used by strains of a given species is qualitatively characteristic. Thus, the few compounds used by *P. lemoignei* are organic acids, whereas the *P. cepacia* spectrum covers the carbohydrates, acids, alcohols, aromatic compounds, amino acids, and amines. Some versatile species, such as *P. acidovorans* or *P. testosteroni,* typically do not grow at the expense of aldo sugars, and so on. Some of the diagnostically important nutritional properties of *Pseudomonas* species are presented in Table 3.

The ease with which the nutritional analysis can be performed and the wealth of taxonomic information that can be obtained have been the main reasons for the preferential attention given to this aspect of *Pseudomonas* physiology in modern taxonomic treatments of the genus. The mass of clear-cut nutritional data obtainable is ideally suited for numerical analysis.

A methodology useful for the analysis of a large number of strains has been described

Table 3
SOME DIAGNOSTICALLY IMPORTANT NUTRITIONAL PROPERTIES OF *PSEUDOMONAS* SPECIES

Species	D-Fucose	Glucose	Trehalose	Cellobiose	Starch	*m*-Inositol	Geraniol	2-Ketogluconate	Maleate	Glycollate	Lactate
P. aeruginosa	–	+	–	–	–	–	+	+	–	–	+
P. putida	–	+	–	–	–	–	–	+	–	(–)	+
P. fluorescens	–	+	+	–	–	+	–	+	(–)	(–)	+
P. chlororaphis	–	+	+	–	–	+	–	+	–	–	+
P. aureofaciens	–	(+)	(+)	–	–	+	–	(+)	–	–	+
P. syringae	–	+	–	–	–	(+)	–	–	–	–	(–)
P. cichorii	–	+	–	–	–	+	–	–	–	–	+
P. stutzeri	–	(+)	–	–	(+)	–	–	–	–	(+)	+
P. mendocina	–	+	–	–	–	–	+	–	–	+	+
P. alcaligenes	–	–	–	–	–	–	–	–	–	–	+
P. pseudoalcaligenes	–	–	–	–	–	–	–	–	–	–	+
P. pseudomallei	+	+	+	+	+	+	–	+	–	–	+
P. mallei	+	+	+	+	(+)	(+)	–	(+)	–	–	+
P. caryophylli	+	+	(–)	+	–	+	–	+	–	+	+
P. cepacia	+	+	(+)	+	–	+	–	+	–	(+)	+
P. marginata	(+)	+	+	+	–	+	–	+	–	–	+
P. lemoignei	–	–	–	–	–	–	–	–	–	–	–
P. testosteroni	–	–	–	–	–	–	(–)	–	–	–	–
P. acidovorans	–	–	–	–	–	(–)	–	–	+	+	+
P. delafieldii	–	+	–	–	–	–	–	+	(+)	–	+
P. pickettii	–	+	–	–	–	–	(–)	+	(+)	+	+
P. solanacearum	–	+	(+)	–	–	(+)	–	–	–	(–)	(+)
P. facilis	–	+	–	–	–	–	–	–	–	–	+
P. saccharophila	–	+	+	+	+	–	–	–	–	–	+
P. ruhlandii	–	+	–	–	–	–	–	–	–	–	+
P. flava	–	+	+	+	–	+		–	–	–	+
P. palleronii	–	+	–	–	–	+		–	–	+	+
P. maltophilia	–	+	+	+	–	–	–	–	–	–	+
P. vesicularis	–	+	–	+	–	–	–	–	–	–	–
P. diminuta	–	–	–	–	–	–	–	–	–	–	–

Code: + = positive for all strains; – = negative for all strains; (+) = positive for 50% or more of the strains (–) = negative for 50% or more of the strains; $+^{m}$ = growth after mutation.

by Stanier et al.,[3] and some additional comments on the procedure have been presented by Palleroni and Doudoroff.[2]

The nutritional analysis of *Pseudomonas* has been very valuable for the clear circumscription of several species and species groups. However, it is only fair to admit that in some cases this approach has been insufficient for unraveling the formidable taxonomic complexities of some groups. Thus, several biotypes of the species *P. fluorescens* are still in highly unsatisfactory taxonomic condition, giving the impression that the number of internal subdivisions will increase when a larger collection is analyzed. A similar situation holds for *P. cepacia* and *P. stutzeri,* species of marked internal heterogeneity, for which the nutritional analysis has done little to define clear internal subdivisions. However, other approaches to the phylogeny of these entities have also met

Table 3 (continued)
SOME DIAGNOSTICALLY IMPORTANT NUTRITIONAL PROPERTIES OF *PSEUDOMONAS* SPECIES

Species	PHB	Adipate	*m*-Hydroxybenzoate	Testosterone	Acetamide	Arginine	Valine	Norleucine	D-Tryptophan	Betaine	Pentothenate	Ethylene glycol
P. aeruginosa	–	+	–	–	+	+	+	–	–	+	–	–
P. putida	–	–	(–)	(–)	(–)	+	+	–	–	+	(–)	(–)
P. fluorescens	–	(–)	(–)	(–)	–	+	+	–	–	+	(–)	–
P. chlororaphis	–	–	(–)	(–)	–	+	(+)	–	–	+	–	–
P. aureofaciens	–	–	–	–	–	+	+	–	–	+	–	–
P. syringae	–	–	–	–	–	(+)	–	–	–	(+)	–	–
P. cichorii	–	–	–	–	–	+	–	–	–	+	–	–
P. stutzeri	–	(–)	–	–	–	–	(+)	–	–	(–)	–	+
P. mendocina	–	–	–	–	–	+	+	–	–	+	–	(+)
P. alcaligenes	–	–	–	–	–	(+)	–	–	–	–	–	–
P. pseudoalcaligenes	–	–	–	–	–	+	–	–	–	(+)	–	(–)
P. pseudomallei	+	+	–	–	–	+	+	–	–	+	–	(–)
P. mallei	+	(+)	–	–	–	+	(+)	–	–	+	–	–
P. caryophylli	–	–	–	–	–	+	(–)	–	–	+	–	–
P. cepacia	–	+	+	(+)	(+)	+	(+)	–	+m	+	–	–
P. marginata	–	+	–	–	–	+	+	(–)	–	+	–	–
P. lemoignei	+	–	–	–	–	–	–	–	–	–	–	–
P. testosteroni	(–)	+	+	+	–	–	–	+	–	–	–	–
P. acidovorans	–	+	+	–	+	–	–	+	+	–	–	–
P. delafieldii	+	+	–	–	–	–	–	–	–	–	–	–
P. pickettii	–	+	–	–	–	–	+	–	(+)	–	–	–
P. solanacearum	–	–	–	–	–	–	–	–	–	(–)	–	–
P. facilis	+	–	–	–	–	–	–	–	–	–	–	–
P. saccharophila	–	+m	–	–	–	–	–	–	–	–	–	+m
P. ruhlandii	–	(+)	–	–	–	–	(–)	(–)	–	–	–	(–)
P. flava	–	–	+	–	–	–	–	–	–	–	–	–
P. palleronii	–	+	+	–	–	–	–	–	–	–	–	–
P. maltophilia	–	–	–	–	–	–	–	–	–	–	–	–
P. vesicularis	–	–	–	–	–	–	–	–	–	–	–	–
P. diminuta	–	–	–	–	–	–	–	–	–	–	+	–

with little success so far. A clear understanding of the phylogeny in these cases may require reevaluation of the concept of a bacterial species, since it does not appear likely that experimental approaches other than those now in use will completely resolve their complexity.

Production of Acids from Sugars and Sugar Alcohols

The tests for production of acids from various carbohydrates have been quite popular in bacteriology for determinative purposes. In *Pseudomonas,* these characters have remained largely unexplored, due to ambiguities in the interpretation of the results and to the possibility of redundancy, since several sugars may be oxidized by the same enzyme.

In Vitro Nucleic Acid Hybridization

Nucleic acid hybridization results have been particularly rewarding in understanding the internal complexity of the genus *Pseudomonas.* Deoxyribonucleic acid (DNA–DNA) homology studies have generally supported the conclusions based on the phenotypic properties of the various species. In addition, the DNA–DNA hybridization studies also revealed internal heterogeneity in groups that appeared rather homogeneous on phenotypic grounds. DNA homology groups were therefore outlined,[9] including species linked directly or indirectly to one another by some detectable level of DNA homology.

Ribosomal RNA–DNA experiments have shed additional light on the relationships of some of the natural groups within the genus.[10] As expected from the conservative nature of the ribosomal RNA genes, the homology values are higher than those obtained by means of DNA–DNA hybridization, and the RNA homology groups, therefore, occasionally include species not showing any detectable DNA homology. In the discussion below, the various RNA homology groups will be examined succinctly. Various other approaches to the phylogeny of the genus, presented in Table 1, will not be discussed here, and the reader is referred to a recent review.[11]

INTERNAL SUBDIVISION OF THE GENUS *PSEUDOMONAS*: THE RNA HOMOLOGY GROUPS

The *P. fluorescens* RNA Homology Group

Pseudomonas fluorescens occupies a central position in this group. Both fluorescent and non-fluorescent pseudomonads are included. At present, the members of this group are *P. aeruginosa* (the type species of the genus), *P. putida* (two biotypes), *P. fluorescens* (five biotypes), *P. chlororaphis, P. aureofaciens, P. syringae, P. cichorii, P. stutzeri, P. mendocina, P. alcaligenes,* and *P. pseudoalcaligenes.* The natural relationships of the species in the group are suggested by the results of nucleic acid homology and by the similarity in some phenotypic characters, biochemical pathways, immunological properties of enzymes, and regulatory mechanisms. The group is, however, the most heterogeneous cluster within the genus, with few phenotypic traits shared by all members. The different species within the group show various degrees of uniformity. *Pseudomonas aeruginosa* and *P. mendocina* appear to be internally quite homogeneous, whereas *P. stutzeri* and *P. fluorescens* can be subdivided into subgroups or biotypes, since they are clearly heterogeneous.

Pseudomonas aeruginosa has received considerable attention as an opportunistic pathogen. The interested reader will find excellent discussions on important aspects of the biology of this species in a recent book edited by Clarke and Richmond.[12] *Pseudomonas aeruginosa* is common in soils, from which it can be isolated by direct streaking on plates of nutrient agar, or, more conveniently, after enrichment under denitrification conditions with appropriate carbon sources. Several pigments can be produced by strains of the species;[2,9] pyocyanine is one of the most characteristic; in addition, fluorescent pigment is produced by most strains.

The diagnosis of *P. aeruginosa* is fairly simple: monotrichous flagellation, pigment production, capacity for growth at 37°C, ability to denitrify, and growth with geraniol and acetamide as carbon sources. This constellation of characters is very useful, even though strains isolated from nature may lack one or several of the properties listed. The species is quite isolated from others in the group. The DNA homology among strains is high (usually 90% or more by the competition technique); the homology is considerably lower with other fluorescent pseudomonads.

Pseudomonas aeruginosa can be further subdivided into "types" by serological, phage, or pyocin typing. This internal classification of the species is useful for epidemiological purposes because it facilitates tracing the origin of infections. The typing methods are capable of detecting very small differences among strains, and even a homogeneous species such as *P. aeruginosa* presents a serious task to any single typing method. It is now accepted that typing gives the best results when at least two typing approaches are used simultaneously.

Pseudomonas fluorescens have been subdivided into a number of biotypes;[3] two of these (D and E) have recently been restored to their original species level.[1] The internal subdivision of *P. fluorescens* can be based on characters such as levan formation from sucrose, denitrification, pigment production, and various nutritional characters. The neotype of the species, originally proposed by Rhodes,[13] has been allocated by Stanier et al.[3] to biotype A, which is characterized by the capacity for levan formation and the inability to denitrify.

Pseudomonas putida can be differentiated from *P. aeruginosa* and *P. fluorescens* by means of a constellation of negative characters, namely, the inability to liquefy gelatin, to denitrify, and to produce levan from sucrose, the absence of lipase and lecithinase activity, and the incapacity for growth at 41°C and (with few exceptions) at 4°C. The strains of *P. putida* can be grouped into two biotypes, but it is now considered that only biotype A may represent typical *P. putida,* since the strains of biotype B appear to be more closely related to *P. fluorescens* than to biotype A.[10] The strains of biotype A are rather homogeneous in phenotypic properties, although, surprisingly, they have considerable heterogeneity in DNA homology.

Many plant-pathogenic fluorescent pseudomonads can be assigned to one of two species, *P. syringae* or *P. cichorii,* and these two species constitute a group that is considered to represent a phylogenetically separate branch within the *P. fluorescens* RNA homology group. There are other phytopathogenic fluorescent species outside of the *syringae-cichorii* group; *P. aeruginosa* and some strains of *P. fluorescens* biotype B *(P. marginalis)* can also be agents of plant diseases. The group constituted by *P. syringae* and *P. cichorii* can be separated from other organisms of the *P. fluorescens* group on the basis of their negative arginine dihydrolase reaction (see Table 2). *Pseudomonas syringae* is a collective species including an enormous number of nomen species,[1] which have been named according to the host of origin. *Pseudomonas syringae* includes oxidase-negative strains, whereas *P. cichorii* is the name reserved for the oxidase-positive. The present state of the nomenclature of the fluorescent phytopathogens is very confusing, and our knowledge of the phylogenetic relationships among the various types is still very limited. It is thought that some of the species now considered to be synonyms of *P. syringae* deserve independent species rank.

The non-fluorescent members of the *P. fluorescens* RNA homology group are *P. stutzeri, P. mendocina, P. alcaligenes,* and *P. pseudoalcaligenes. Pseudomonas stutzeri* is the species that has the broadest range of DNA base composition of the whole genus (from 61 to 66 mol% G + C), and, as expected, it is also heterogeneous in DNA homology.[14] The number of carbon compounds that can be used by various strains can vary considerably.[2] The general phenotypic properties of the species, however, are readily identifiable: wrinkled appearance of the colonies, capacity for denitrification, and use of starch and ethylene glycol for growth. The colonies may lose their wrinkled appearance after repeated transfers in laboratory media, but in some cases the original type can be recovered after subcultivation under conditions of denitrification.

Pseudomonas mendocina is a species somewhat similar to *P. stutzeri;*[14] the strains are denitrifiers and can use ethylene glycol for growth. In contrast to most members of the *P. fluorescens* RNA group, they produce carotenoid pigments.

Pseudomonas alcaligenes and *P. pseudoalcaligenes* are non-fluorescent members of the *P. fluorescens* RNA homology group. *Pseudomonas alcaligenes* is still a poorly defined species, due to the low similarity among the three strains that have been described.[15] One of these strains is particularly aberrant and produces carotenoid pigments. *Pseudomonas pseudoalcaligenes* strains are phenotypically heterogeneous, but the DNA homology is relatively high among the 15 strains known at present. Outside of the species, the homology is higher with the strains of *P. mendocina* than with *P. alcaligenes.* Some members of *P. pseudoalcaligenes* accumulate poly-β-hydroxybutyrate as carbon reserve material, a property uniformly negative in all other members of the RNA homology group.

The *pseudomallei-cepacia* RNA Homology Group

Most of the species included in this natural group are pathogenic; all species accumulate poly-β-hydroxybutyrate and have several polar flagella per cell (see Table 2); the only exception is *P. mallei,* which is permanently non-motile. *Pseudomonas pseudomallei* and *P. mallei* are the agents of melioidosis and of glanders respectively. A discussion of these two species as agents of diseases transmissible from animals to man has been recently presented.[16] *Pseudomonas pseudomallei* is a free-living organism that can be isolated from many tropical soils.[17]

Pseudomonas cepacia, P. marginata (syn. *P. alliicola* and probably *P. gladioli*[18]), and *P. caryophylli* are plant pathogens. *Pseudomonas cepacia* has also been found to be an occasional human pathogen.[19] This species is rather remarkable, because it is the nutritionally most versatile of all *Pseudomonas*[3,20] and has been isolated from a large variety of materials, including clinical specimens.

The only species of the *pseudomallei-cepacia* RNA homology group that has not yet been accused of pathogenic inclinations is *P. pickettii,* described by Ralston et al.[21] DNA–DNA hybridization experiments reveal a relationship of *P. pickettii* to other members of the group and also to *P. solanacearum,* which, until the discovery of *P. pickettii,* appeared to be a member of the RNA group not having any direct DNA homology with any of the other species.

Pseudomonas solanacearum is an important plant pathogen and a complex species, which has been subdivided into a number of biotypes.[22] DNA homology experiments performed among the members of the various biotypes can only define two clusters.[23]

The *acidovorans* RNA Homology Group

This RNA homology complex comprises three separate small DNA homology groups. *Pseudomonas acidovorans* and *P. testosteroni* were placed close to one another by Stanier et al.[3] on the basis of clear phenotypic resemblances. Although they differ in their DNA base composition, some level of DNA homology can be demonstrated between the two species. Members of the two species can use for growth higher dicarboxylic acids, norleucine, and *m*-hydroxybenzoate, but not aldose sugars or amines. Another distinctive property is the mode of cleavage of the aromatic ring: the cleavage of protocatechuate (an intermediate in the degradation of *p*-hydroxybenzoate) is of the *"meta"* type.

Pseudomonas delafieldii and *P. facilis* constitute a second small DNA homology group within the *acidovorans* RNA group. *Pseudomonas facilis* is one of the hydrogen pseudomonads (see below), whereas *P. delafieldii* is unable to live autotrophically.

Pseudomonas saccharophila, the remaining species of the *acidovorans* RNA homology group, is unrelated by DNA homology to other species of the genus *Pseudomonas.* The species (reduced until recently to a single strain) was studied for many years by Doudoroff and his collaborators, particularly in relation to its sugar metabolism. *Pseudomonas saccharophila* is a hydrogen pseudomonad.

The *diminuta* RNA Homology Group

Two species, *P. diminuta* and *P. vesicularis,* are included in this small RNA homology group. The group can be distinguished from all others by an important morphological character, namely, the short wavelength of the flagella. The two species have been extensively characterized by Ballard et al.,[5] who defined their organic growth factor requirements.

The *P. maltophilia-Xanthomonas* RNA Homology Group

Pseudomonas maltophilia is a very distinct species of the genus. Its oxidase reaction is negative, methionine is required as growth factor, nitrate cannot be used as a nitrogen source, and lactose can serve as growth substrate. Its distribution in nature seems to be very wide, since the reported sources of the known strains have been water, soil, milk, and clinical specimens. Its relation to *Xanthomonas* can only be demonstrated by ribosomal RNA–DNA hybridization,[11] and the meaning of this unexpected relationship is at present obscure.

Miscellaneous Species Not Yet Assigned to RNA Homology Groups

Species of *Pseudomonas* for which the phylogenetic relationships have not yet been established include *P. lemoignei, P. ruhlandii, P. flava,* and *P. palleronii.* The last three are hydrogen pseudomonads.[24]

Pseudomonas lemoignei is an interesting species, represented at present by only one strain. It was singled out for study by Delafield et al.[25] because of the production of a very active extracellular enzyme that catalyzes the hydrolysis of poly-β-hydroxybutyrate. As mentioned before, the nutritional spectrum of the species is very limited; out of about 160 organic compounds, it can only use acetate, butyrate, valerate, succinate, β-hydroxybutyrate, poly-β-hydroxybutyrate, and pyruvate. Propionate, L-malate, α-ketoglutarate, and citrate are used very poorly.

The Hydrogen Pseudomonads

Hydrogen bacteria are facultative chemolithotrophs that can live in a purely mineral medium in equilibrium with an atmosphere containing hydrogen, oxygen, and carbon dioxide. Several known hydrogen bacteria are now placed in the genus *Pseudomonas.* The rejection of the genus *Hydrogenomonas* has been proposed by Davis et al.[26] because of its clearly artifical nature as a taxon. The genus included bacteria of various morphological types: flagellated rods (peritrichous, polar, subpolar, or degenerately peritrichous), non-motile rods, and coccoid types. The hydrogen pseudomonads are themselves quite heterogeneous as a group, since, as we have seen, the species now assigned to *Pseudomonas* belong to different natural clusters within the genus, and some bear a closer resemblance to purely heterotrophic species than to other hydrogenomonads.

Few properties are shared by all hydrogen pseudomonads, aside from the capacity for chemoorganotrophic life; all can live heterotrophically at the expense of various organic compounds, and all are capable of accumulating poly-β-hydroxybutyrate as reserve material. *Pseudomonas saccharophila* and *P. facilis* are the two best-known species of hydrogen pseudomonads. They are now placed in the same RNA homology group, although they do not share any DNA homology. The relationships of the remaining species (*P. ruhlandii, P. flava,* and *P. palleronii*) are unclear at present; limited DNA hybridization experiments with *P. palleronii* demonstrate no relationship to other *Pseudomonas* species.[27]

REFERENCES

1. **Doudoroff, M. and Palleroni, N. J.,** in *Bergey's Manual of Determinative Bacteriology,* 8th Ed., Buchanan, E. R. and Gibbons, N. E., Eds., Williams & Wilkins, Baltimore, 1974, p. 217.
2. **Palleroni, N. J. and Doudoroff, M.,** *Annu. Rev. Phytopathol.,* 10, 73, 1972.
3. **Stanier, R. Y., Palleroni, N. J., and Doudoroff, M.,** *J. Gen. Microbiol.,* 43, 159, 1966.
4. **Lautrop, H. and Jessen, O.,** *Acta Pathol. Microbiol. Scand.,* 60, 588, 1964.
5. **Ballard, R. W., Doudoroff, M., Stanier, R. Y., and Mandel, M.,** *J. Gen. Microbiol.,* 53, 349, 1968.
6. **Zolg, W. and Ottow, J. C. G.,** *Z. Allg. Mikrobiol.,* 15, 287, 1975.
7. **Clarke, P. H. and Ornston, L. N.,** in *Genetics and Biochemistry of Pseudomonas,* Clarke, P. H. and Richmond, M. H., Eds., John Wiley & Sons, London, 1975, pp. 191 and 263.
8. **Feist, C. F. and Hegeman, G. D.,** *J. Bacteriol.,* 100, 869, 1969.
9. **Palleroni, N. J., Ballard, R. W., Ralston, E., and Doudoroff, M.,** *J. Bacteriol.,* 110, 1, 1972.
10. **Palleroni, N. J., Kunisawa, R., Contopoulou, R., and Doudoroff, M.,** *Int. J. Syst. Bacteriol.,* 23, 333, 1973.
11. **Palleroni, N. J.,** in *Genetics and Biochemistry of Pseudomonas,* Clarke, P. H. and Richmond, M. H., Eds., John Wiley & Sons, London, 1975, p. 1.
12. **Clarke, P. H. and Richmond, M. H. (Eds.),** *Genetics and Biochemistry of Pseudomonas,* John Wiley & Sons, London, 1975.
13. **Rhodes, M. E.,** *J. Gen. Microbiol.,* 21, 221, 1959.
14. **Palleroni, N. J., Doudoroff, M., Stanier, R. Y., Solanes, R. E., and Mandel, M.,** *J. Gen. Microbiol.,* 20, 215, 1970.
15. **Ralston, E.,** Some Contributions to the Taxonomy of the Genus *Pseudomonas,* (thesis), University of California, Berkeley, 1972.
16. **Redfearn, M. S. and Palleroni, N. J.,** in *Diseases Transmissible from Animals to Man,* Hubbert, W. T., McCulloch, W. F., and Schnurrenberger, P. I., Eds., Charles C Thomas, Springfield, Ill., 1975, p. 110.
17. **Redfearn, M. S., Palleroni, N. J., and Stanier, R. Y.,** *J. Gen. Microbiol.,* 43, 293, 1966.
18. **Hildebrand, D. C., Palleroni, N. J., and Doudoroff, M.,** *Int. J. Syst. Bacteriol.,* 23, 433, 1973.
19. **Sinsabaugh, H. A. and Howard, G. W.,** *Int. J. Syst. Bacteriol.,* 25, 187, 1975.
20. **Ballard, R. W., Palleroni, N. J., Doudoroff, M., Stanier, R. Y., and Mandel, M.,** *J. Gen. Microbiol.,* 60, 199, 1970.
21. **Ralston, E., Palleroni, N. J., and Doudoroff, M.,** *Int. J. Syst. Bacteriol.,* 23, 15, 1973.
22. **Hayward, A. C.,** *J. Appl. Bacteriol.,* 27, 265, 1964.
23. **Palleroni, N. J. and Doudoroff, M.,** *J. Bacteriol.,* 107, 690, 1971.
24. **Davis, D., Stanier, R. Y., Doudoroff, M., and Mandel, M.,** *Arch. Mikrobiol.,* 70, 1, 1970.
25. **Delafield, F., Doudoroff, M., Palleroni, N. J., Lusty, C., and Contopoulou, R.,** *J. Bacteriol.,* 90, 1455, 1965.
26. **Davis, D., Doudoroff, M., Stanier, R. Y., and Mandel, M.,** *Int. J. Syst. Bacteriol.,* 19, 376, 1969.
27. **Ralston, E., Palleroni, N. J., and Doudoroff, M.,** *J. Bacteriol.,* 109, 465, 1972.

GRAM-NEGATIVE FACULTATIVELY ANAEROBIC RODS*

H. A. Lechevalier

Generic name (G + C, molar %) [Number of species in *Bergey's Manual*]	Morphology	Physiology	Ecology	Examples of species
Enterobacteriaceae (39 to 59) [12 genera]	Small rods; motile by peritrichous flagella.	Chemoorganotrophs; respiratory and fermentative metabolism; acid produced from the fermentation of glucose, other carbohydrates, and alcohols; catalase-positive; oxidase-negative; nitrate reduced to nitrities.	Found in the intestinal tracts of animals, and in feces, urine, food, water, and soil; some species are pathogenic to animals, others to plants; many are opportunistic pathogens.	*Escherichia coli.* The most studied bacterium; found in the lower part of the intestine of warm-blooded animals. *Salmonella typhi.* Causes typhoid fever; transmitted by water or food contaminated by human excreta. *Shigella dysenteriae.* Causes bacterial dysentery; normally found in the intestinal tract of man and higher monkeys. *Klebsiella pneumoniae.* Found in soil, in water, and in the intestinal tract of man and animals; may be pathogenic. *Serratia marcescens.* Found in water, soil, and food; may be pathogenic; famous for the production of the red pigment prodigiosin. *Proteus mirabilis.* Found in soil, sewage, fecal matter of animals, and human clinical material. *Yersinia pestis.* Causative agent of plague in man and rodents; transmitted from rat to man by fleas. *Erwinia amylovora.* Causes fire-blight of *Rosaceae.*

* Part 8 in *Bergey's Manual.*

Generic name (G + C, molar %) [Number of species in *Bergey's Manual*]	Morphology	Physiology	Ecology	Examples of species
Vibrionaceae (39 to 63) [5 genera]	Rods, straight or curved; polarly flagellated.	Chemoorganotrophs; respiratory and fermentative metabolism; oxidase-positive.	Found in aquatic habitats, and occasionally in fish or man.	*Vibrio cholerae.* Found in water and foods, and in the intestinal tract of humans and animals; presumably causes cholera and diarrhea. *Aeromonas hydrophila.* Causes diseases of cold-blooded animals and occasional human infections. *Photobacterium phosphoreum.* Luminescent; found in sea water and in marine animals, especially in the luminous organs of the latter.
Zymomonas (47 to 48) [2 species]	Plump rods, 1–2 by 2–5 μm, occurring singly or in pairs; lophotrichous flagellation.	Chemoorganotrophs; ferment glucose or fructose to ethanol and CO_2; produce acetaldehyde catalase and H_2S; gelatin not liquefied; nitrate not reduced to nitrite; methyl red test negative; anaerobes somewhat oxygen-tolerant.	Found in tainted cider and beer, and in fermented plant juices.	*Z. mobilis.* Type species; found in tropical countries in fermenting plant juices. *Z. anaerobia.* Found in tainted cider and beer.
Chromobacterium (63 to 72) [2 species]	Rods with rounded ends, 0.6–1.2 by 1.5–6 μm; motile by 1 polar and 1 to 4 lateral flagella; form violet colonies.	Chemoorganotrophs; respiratory or fermentative metabolism; acid, but no gas formed from some carbohydrates; catalase-positive; produce the indole pigment violacein (UV max. in ethanol at 579 nm, min. at 430 nm)	Found in soil and in water bodies; occasionally cause infections of animals or spoilage of food.	*C. violaceum.* Type species; mesophil; grows at 37°C, but not at 4°C; produces hydrogen cyanide and acid from trehalose; no hydrolysis of esculin; predominates in the tropics. *C. lividum.* Psychrophil; grows at 4°C, but not at 37°C; no production of hydrogen cyanide, nor of acid from trehalose; hydrolysis of esculin; predominates in temperate regions.

(G + C, molar %) [Number of species in *Bergey's Manual*]	Morphology	Physiology	Ecology	Examples of species
Flavobacterium (30 to 42; also 63 to 70) [12 species]	Coccobacilli to slender rods, 0.4–1.0 by 0.7–6 μm; if motile, by peritrichous flagella; yellow, orange, and red to brown pigments not soluble in media.	Chemoorganotrophs; respiratory metabolism; fastidious; require B vitamins; some strains are halotolerant.	Found in soil and in water bodies, including sewage and sea, in vegetables, and in dairy products; some strains are pathogenic to man and animals.	*F. aquatile.* Type species; found in fresh-water bodies. *F. meningosepticum.* Pathogenic to infants, causing septicemia and meningitis.
Haemophilus (38 to 42) [14 species]	Coccobacilli to rods, 0.2–0.3 by 0.5–2.5 μm; may be pleomorphic and form filaments; non-motile.	Strict parasites of vertebrates; require growth factors of blood, including hemin (X-factor), phosphopyridine nucleotide (V-factor), diphosphothiamine, or adenosine; some species are hemolytic, either of the α-type (green) or of the β-type (colorless); some species are pathogenic.	Found in lesions and secretions of vertebrates, as well as on their normal mucous membranes.	*H. influenzae.* Type species; normal inhabitant of the nasopharynx of man; may cause various infections.
Pasteurella (37 to 43) [4 species]	Ovoid or rod-shaped cells, 0.3–0.5 by 1.0–1.8 μm; non-motile; bipolar staining common.	Chemoorganotrophs; fermentative metabolism; fastidious; grow best in the presence of blood; glucose and other carbohydrates attacked with the production of acid, but no gas; catalase- and oxidase-positive; nitrates reduced.	Parasites of animals, including man.	*P. multocida.* Type species; causes fowl cholera and hemorrhagic septicemia of ruminants, horses, rabbits, rats, sheep, dogs, cats, and numerous other animals; it is also found associated with pneumonia of various animals.
Actinobacillus (41 to 42) [2 species]	Cocci to rods, 0.3–0.5 by 0.6–1.4 μm; cocci and bacilli often give a morse code appearance; non-motile; cultures very sticky.	Chemoorganotrophs; fermentative metabolism; nitrates reduced to nitrite; urease-positive; acid, but no gas produced from fermented carbohydrates.	Parasites of animals.	*A. lignieresii.* Type species; pathogenic for cattle and sheep. *A. equirli.* Pathogenic for horses and pigs.

Generic name (G + C, molar %) [Number of species in *Bergey's Manual*]	Morphology	Physiology	Ecology	Examples of species
Cardiobacterium (62) [1 species]	Rods, 0.5–0.75 by 1–3 μm; pleomorphic; non-motile.	Chemoorganotroph; fermentative metabolism; acid, but no gas produced from fermented carbohydrates; catalase-negative; nitrates not reduced.	Found in human nose and throat; may produce endocarditis in man.	*C. hominis.*
Streptobacillus (24 to 26) [1 species]	Rods, 0.3–0.7 by 1–5 μm; form chains that may be 150 μm long; pleomorphic; non-motile; conversion to L-phase common.	Chemoorganotroph; fermentative metabolism; fastidious; requires serum, ascitic fluid, or blood for growth.	Associated with animals.	*S. moniliformis.* Causes streptobacillary rat-bite fever of man and joint diseases of various animals.
Calymmatobacterium [1 species]	Rods, 1–2 μm in length; pleomorphic; non-motile.	Requires fresh egg yolk for isolation.	Causes granuloma inguinale of man.	*C. granulomatis.*

ENTEROBACTERIACEAE*

C. F. Clancy

Some Enterobacteriaceae exist as saprophytes in nature, others are parasitic for plants, causing blights and soft rots; many exist as commensals in the intestinal tracts of man and animals, and still others may cause serious infectious disease.

The group includes motile organisms with peritrichous flagella as well as non-motile varieties. Most species ferment glucose with production of acid, reduce nitrate to nitrite, and lack a cytochrome C oxidase. Their biochemical and serological characteristics have been studied intensively in efforts to arrive at a logical classification. Five tribes and about 12 genera are currently recognized as distinguishable entities; however, so many intermediate strains exist that it is difficult to express them in a single formal classification.

Many of the genera to be discussed are susceptible to lysis by specific bacteriophages, and many fundamental principles of phage–host-cell relationships have been elucidated by studies with *Escherichia coli* and its various phages. Another characteristic of these enteric bacteria is the ability to produce bacteriocins, metabolites that are lethal for other strains of the same or closely related species. Bacteriocins are named after the producing strain; for example, colicins are produced by *Escherichia coli.* They may play a role in the maintenance of normal flora in the intestinal tract, although in vivo activity has not been thoroughly investigated. Endotoxins, composed of lipopolysaccharides, occur in the cell wall of nearly all the enteric bacteria. The endotoxins of *E. coli, Salmonella, Shigella,* and *Serratia* have been studied intensively and will be discussed in a subsequent volume of the *Handbook.*

Whenever these Gram-negative rods are implicated as the etiologic agent of infection, the question arises as to the best antimicrobial agent, if any, to be used for therapy. With the exception of typhoid fever and shigellosis, no reference will be made in the following pages to specific chemotherapy for each individual infectious agent. Most are sensitive, in varying degrees, to ampicillin, cephalothin, chloramphenicol, colistin, gentamycin, tetracycline, sulfonamide, etc., but in any specific instance the isolated strain must be tested to determine its particular antimicrobial-sensitivity pattern.

The more important biochemical reactions of these organisms, which enable them to be separated into 12 genera, are shown in Table 1. Antigenic analysis and serotyping are also used as taxonomic tools and will be discussed under the several genera for which they are most valuable. The DNA base composition within this family ranges from 36 to 57 mol % guanosine and cytosine; however, if one considers a single genus, established on phenotypic characteristics, the ratio may lie within a very close range. For example, in the genus *Shigella* the base compositions determined for various species are between 49 and 53 mol % G + C.

ESCHERICHIA

The only important species of this genus, *Escherichia coli* is a common inhabitant of the lower bowel of man and animals. It is often present in man in concentrations of 10^7 or more viable organisms per gram of fecal material. It is commonly looked for in water supplies and food as an indicator of fecal pollution. It may be motile or non-motile, always ferments glucose, and about 90% of strains ferment lactose in 1 to 2 days. When studied by the IMViC (indol, methyl red, Voges-Proskauer, and citrate) reactions, it

* Part 8 in *Bergey's Manual.*

Table 1
DIFFERENTIATION OF ENTEROBACTERIACEAE BY BIOCHEMICAL TESTS

TEST or SUBSTRATE	ESCHERICHIEAE		EDWARDSIELLEAE	SALMONELLEAE			KLEBSIELLEAE									PROTEEAE					
	Escherichia	Shigella	Edwardsiella	Salmonella	Arizona	Citrobacter	Klebsiella	Enterobacter						Serratia	Pectobacterium	Proteus				Providencia	
								cloacae	aerogenes	hafniae		liquefaciens				vulgaris	mirabilis	morganii	rettgeri	alcelifaciena	stuartii
										37 C	22 C	37 C	22 C		25 C						
INDOL	+	- or +	+	-	-	-	- or +	-	-	-	-	-	-	-	- or +	+	-	+	+	+	+
METHYL RED	+	+	+	+	+	+	-	-	-	+ or -	-	+ or -	- or +	- or +	+ or -	+	+	+	+	+	+
VOGES - PROSKAUER	-	-	-	-	-	-	+	+	+	+ or -	+	- or +	+ or -	+	- or +	-	- or +	-	-	-	-
SIMMONS'S CITRATE	-	-	-	d	+	+	+	+	+	(+) or -	d	+	+	+	d	d	+ or (+)	-	+	+	+
HYDROGEN SULFIDE (TSI)	-	-	+	+	+	+ or -	-	-	-	-	-	-	-	-	-	+	+	-	-	-	-
UREASE	-	-	-	-	-	d^{w}	+	+ or -	-	-	-	d	-	d^{w}	d^{w}	+	+	+	+	-	-
KCN	-	-	-	-	-	+	+	+	+	+	+	+	+	+	+ or -	+	+	+	+	+	+
MOTILITY	+ or -	-	+	+	+	+	-	+	+	+	+	d	+	+	+ or -	+	+	+	+	+	+
GELATIN (22 C)	-	-	-	-	(+)	-	-	(+) or -	- or (+)		-		+	+	+ or (+)	+ or (+)	+	-	-	-	-
LYSINE DECARBOXYLASE	d	-	+	+	+	-	+	-	+	+	+	+ or -	+	+	-	-	-	-	-	-	-
ARGININE DIHYDROLASE	d	- or (+)	-	(+) or +	+ or (+)	d	-	+	-	-	-	-	-	-	- or +	-	-	-	-	-	-
ORNITHINE DECARBOXYLASE	d	d$^{(1)}$	+	+	+	d	-	+	+	+	+	+	+	+	-	-	+	+	-	-	-
PHENYLALANINE DEAMINASE	-	-	-	-	-	-	-	-	-	-	-	-	-	-	-	+	+	+	+	+	+
MALONATE	-	-	-	-	+	d	+	+ or -	+ or -	+ or -	+ or -	-	-	-	- or +	-	-	-	-	-	-
GAS FROM GLUCOSE	+	-$^{(1)}$	+	+	+	+	+	+	+	+	+	+	+	+ or -$^{(3)}$	- or +	+ or -	+	d	- or +	+ or -	-
LACTOSE	+	-$^{(1)}$	-	-	d	d	+	+	+	- or (+)	- or (+)	d	(+)	- or (+)	d	-	-	-	-	-	-
SUCROSE	d	-$^{(1)}$	-	-	-	d	+	+	+	d	d	+	+	+	+	+	d	-	d	d	d
MANNITOL	+	+ or -	-	+	+	+	+	+	+	+	+	+	+	+	+	-	-	-	+ or -	-	d
DULCITOL	d	d	-	d$^{(2)}$	-	d	- or +	- or +	-	-	-	-	-	-	-	-	-	-	-	-	-
SALICIN	d	-	-	-	-	d	+	+ or (+)	+	d	d	+	+	+	+	d	d	-	d	-	-
ADONITOL	-	-	-	-	-	-	+ or -	- or +	+	-	-	d	d	d	-	-	-	-	d	+	-
INOSITOL	-	-	-	d	-	-	+	d	+	-	-	+	+	d	-	-	-	-	+	-	+
SORBITOL	+	d	-	+	+	+	+	+	+	-	-	+	+	+	-	-	-	-	d	-	d
ARABINOSE	+	d	-	+$^{(2)}$	+	+	+	+	+	+	+	+	+	-	+	-	-	-	-	-	-
RAFFINOSE	d	d	-	-	-	d	+	+	+	-	-	+	+	-	+ or (+)	-	-	-	-	-	-
RHAMNOSE	d	d	-	+	+	+	+	+	+	+	+	-	-	-	d	-	-	-	+ or -	-	-

(1) Certain biotypes of *Shigella flexneri* produce gas: *S. sonnei* cultures ferment lactose and sucrose slowly and decarboxylate ornithine. (2) *Salmonella typhi, S. cholerae-suis, S. enteritidis* bioser. Paratyphi A and Pullorum, and a few others ordinarily do not ferment dulcitol promptly; *S. cholerae-suis* does not ferment arabinose. (3) Gas volumes produced by cultures of *Serratia, Proteus,* and *Providencia* are small. (From Edwards, P. R. and Ewing, W. H., *Identification of Enterobacteriaceae,* 3rd ed., Burgess Publ. Co., Minneapolis, 1972. With permission of Elsevier Science Publishers, Amsterdam.)

produces indol from tryptophan and is methyl-red-positive. Acetyl methyl carbinol is not formed, and citrate is not utilized as a sole source of carbon.

The results of serological studies on *E. coli* have shown it to have three different classes of antigens.[1] The "O antigens" are somatic antigens, stable to heat at 100°C or at 121°C. "K antigens" are somatic antigens that occur as sheaths, envelopes, or capsules and may act to inhibit "O" agglutination; they are designated as L, A, or B, depending on their heat-lability at 100°C or at 120°C. The "H antigens" are flagellar and are heat-labile at 100°C. By application of serotyping, as done with *Salmonella,* it has been possible to characterize many strains of *E. coli* in relation to these three antigens.

Evidence has now accumulated to show that many outbreaks of diarrheal disease in which no *Salmonella* or *Shigella* can be isolated are associated with certain serotypes of *E. coli.* A particular serotype of *E. coli* (O 11:B 4) was first noted in 1945 by Bray[2] to be associated with infantile diarrhea; many subsequent reports confirm that one of several serotypes of *E. coli* may be isolated from the stool of infants with epidemic diarrhea of the newborn. Commercial typing sera that contain appropriate antibody against the O (somatic) antigen and B (envelope) antigen are available for the identification of ten different serotypes known to be associated with infantile diarrhea. Some of these same serotypes are also found in diarrhea of adults.

Occasionally biotypes of *E. coli* are encountered that are anaerogenic and non-motile. They were once thought to be *Shigella,* but more recently have been referred to as the *Alkalescens-Dispar* group. They do have O antigens in common with *Escherichia* and are now included in this genus.

Escherichia coli is the etiologic agent of a variety of other types of human and animal infections. It is one of the most common causes of cystitis and other infections of the urinary tract. It is also one of the organisms most commonly recovered from peritonitis following rupture of the appendix or other bowel perforation, and from pneumonia following aspiration of intestinal contents. It may cause pyogenic wound infections, especially in lesions that are fecally contaminated.

Escherichia coli has also served as a "model" organism for many types of studies in bacterial physiology, cytology by electron microscopy, genetics, chromosome mapping, etc. The transfer of genetic material by conjugation, transduction, and transformation has been demonstrated by certain strains of this species. It grows well on simple media, has a generation time of about 20 min, can be grown in large quantities, and generally lends itself to laboratory manipulation.

SHIGELLA

Members of this genus, the etiologic agents of bacillary dysentery, are widespread throughout the world, especially in tropical areas. They are non-motile Gram-negative rods, anaerogenic, and ferment dextrose, but not lactose. Their reactions in the IMViC tests are the same as those of *Escherichia coli,* except that not all strains produce indol. Some species also ferment mannitol, a characteristic that is useful in the identification of the species (Table 2). Group A (*S. dysenteriae*) organisms do not ferment mannitol, whereas Group B (*S. flexneri*), Group C (*S. boydii*), and Group D (*S. sonnei*) are nearly all fermenters of mannitol. Group A has 10 serotypes based on O antigens that are serologically related to each other, and Groups B and C are composed of 12 to 15 serotypes. Group D (*S. sonnei*) has only one serotype.

It should be noted[3] that the O antigens of 10 *Shigella* serotypes are identical to those of *Escherichia coli* or intermediate coliform bacilli; furthermore, 17 other *Shigella* serotypes also bear reciprocal O antigen relationships with *E. coli* and coliform intermediates. This is of some practical importance in the identification of *Shigella*

Table 2
GENUS *SHIGELLA*

Serologic group	Mannitol	Species
A (1)	Negative	*S. dysenteriae*
A (2–10)	Negative	*S. schmitzii, A. arabinotarda,* etc.
B	Positive	*S. flexneri* and miscellaneous serotypes
C	Positive	*S. boydii* and miscellaneous serotypes
D	Positive	*S. sonnei*

species, especially if one is dealing with an isolate that is actually an anaerogenic, slow lactose-fermenting *E. coli.* It is important to analyze carefully the results of both biochemical and serological tests.

The natural habitat of *Shigella* is the lower bowel of man. Isolation of the organisms from monkeys has been reported, but they are rarely isolated from other animal species. Transmission of bacillary dysentery usually occurs directly by contact with a human case or carrier or by ingestion of food or water that is fecally contaminated. Dysentery is characterized by diarrhea, fever, cramps, and sometimes vomiting; the watery stools characteristically contain blood, mucus, and many pus cells. The incubation period is usually less than 7 days. In contrast to typhoid fever, the organisms in *Shigella* infections are localized in the lower bowel. They cannot be cultured from blood, but are readily isolated from feces druing illness and for several weeks afterward. The carrier state may then persist for months, and occasionally for a year or more. Clinical severity of the disease may be affected by the age, state of nutrition, and general well-being of the patient. Infants and debilitated elderly persons may be severely affected. In temperate zones the disease may be self-limiting, whereas in the tropics, under conditions of crowding, severe epidemics occur.

An attack of dysentery does not confer protection against subsequent attacks, although persons living in areas where the disease is endemic do not continue to have recurrent attacks. Antibody formation after an attack of dysentery cannot be demonstrated by conventional serologic techniques, but if measured by the hemagglutination technique,[4] titers of 1/40 or higher may be demonstrated.

Shigella dysenteriae type 1 (the Shiga bacillus) produces a disease that is considerably more severe than that caused by other organisms in the group. It has been shown to produce a powerful exotoxin, which damages the blood vessels of the central nervous system of rabbits, producing symptoms of flaccid paralysis. In rats and hamsters, no neurological symptoms appear, but the blood vessels of various other organs are affected. The true role of the toxin in human disease is not known. Disease due to the Shiga bacillus has been quite rare for many years, except for endemic foci in Asia, but in 1969 an epidemic, first noted in Guatemala,[5] spread throughout Central America.

Specific treatment of bacillary dysentery with chloramphenicol or tetracycline results in a rapid decline in symptoms. Earlier attempts at chemotherapy with sulfonamides, however, resulted in bacteriostasis of the organisms, with subsequent emergence of drug-resistant strains. Studies on the antibiotic sensitivity of *Shigella* strains in Japan have resulted in the discovery of the RTF (resistance transfer factor). It was found that, by means of transfer of an episome through conjugation, a *Shigella* strain could acquire resistance to an antibiotic from a resistant strain of *Escherichia coli.*

SALMONELLA

Salmonella organisms are usually motile and produce both acid and gas from glucose.

Table 3
ANTIGENIC STRUCTURES OF REPRESENTATIVE *SALMONELLA* AS DEPICTED IN THE KAUFFMANN-WHITE SCHEMA

Group	Species	O antigens	H antigens	
			Phase 1	Phase 2
A	*S. paratyphi* A	1, 2, 12	a	
B	*S. paratyphi* B (*schottmuelleri*)	1, 4, 5, 12	b	1, 2
	S. typhimurium	1, 4, 5, 12	i	1, 2
	S. derby	1, 4, 5, 12	f, g	
C_1	*S. cholerae-suis*	6, 7	c	1, 5
C_2	*S. newport*	6, 8	e, h	1, 2
D	*S. typhi*	9, 12	d	
	S. enteritidis	1, 9, 12	g, m	
	S. pullorum[a]	9, 12		

[a] *S. pullorum,* one of the few non-motile species, has no flagella, and hence no H antigens.

Lactose, sucrose, and salicin are not fermented, but hydrogen sulfide is usually produced. The organisms do not produce acetyl methyl carbinol, rarely liquefy gelatin, and do not hydrolyze urea. They are found in man and in lower animals, especially in the intestinal tract. Members of this group cause typhoid fever, gastroenteritis, and sepsis in man, as well as similar infections in lower animals.

The typhoid bacillus, now designated as *Salmonella typhi,* was the first member of this group to be described and isolated.[6,7] As bacteriologists began to study the etiology of enteric fevers, it became apparent that many of the organisms isolated had the common characteristics mentioned above, but attempts to identify and classify these organisms by the usual biochemical tests failed. By use of antigenic analysis, White[8] and Kauffmann[9] established some taxonomic order; they showed that each organism was a serologic entity with characteristic somatic and flagellar antigens, of which some were common to other members of the group.

The O (*ohne,* German for "without") or somatic antigens are present both in motile and in non-motile bacilli and are resistant to boiling at 100°C and to treatment with alcohol. The H (*Hauch,* German for "haze") or flagellar antigens are only found in motile cultures and are destroyed at 100°C or by treatment with alcohol or dilute acids. At least two species, *S. typhi* and *S. paratyphi* C, have, in addition to O and H antigens, an antigen designated as Vi. This substance is an envelope antigen and was orginally called Vi to denote association with a virulent strain. The virulence concept has not been borne out; however, presence of this antigen in a freshly isolated culture of *S. typhi* may interfere with its agglutination by antiserum for the specific O-antigenic factors.

The Kauffmann-White scheme is a systematic tabulation of the exact antigenic structure of the thousand-odd species of *Salmonella* now described. Somatic (O) antigens are designated with arabic numerals, as may be seen in Table 3; for example, *S. typhimurium* cells contain O-antigenic factors 1, 4, 5, and 12.

The O antigen, a complex lipopolysaccharide, is a constitutive part of the bacterial cell wall and is closely associated with endotoxin. It consists of three portions the first is a core polysaccharide, to which lipid A is attached; the second is a glycophospholipid; the third portion, the O-specific chain of repeating units of monosaccharides, is also attached to the core polysaccharide. The O-specific chains consist of distinctive combinations of ordinary sugars, such as galactose, mannose, rhamnose, etc. The specific O antigens of *S. typhi* (factors 9 and 12) are thus a reflection of the distinctive arrangement of mannose, rhamnose, and tyvelose in the O-specific chain. The endotoxin appears to reside in the lipid fraction, whereas the specific antigenic determinants are in the polysaccharide portion of the molecule.

Flagellar H antigens are divided into specific (Phase 1) factors, designated by lower-case letters, and nonspecific (Phase 2) factors, designated by arabic numerals. These two entities differ in that Phase 1 antigens are shared by only a few serotypes, whereas the nonspecific antigens are common to many different serotypes. Thus, it may be noted (Table 3) that the flagellar antigens of *S. typhimurium* are as follows: Phase 1 – i; Phase 2 – 1, 2. The specificity of the flagellar antigens is dependent on the proteins present in the flagella.

The Kauffmann-White schema, as developed, has become an aid in the serologic identification of organisms isolated from disease in man and animals. Before the multitude of serotypes was comprehended, each was given a species name, usually the name of the area of isolation (e.g., *S. rutgers*). It now seems apparent that spontaneous mutations occur in nature and that genetic recombinations may occur by transduction and conjugation. This may explain the multiplicity of serotypes and the overlapping patterns of antigenicity.

*Salmonella typhi,** the etiologic agent of typhoid fever, is found only in man or in food or water fecally contaminated by a human carrier. The disease is an acute febrile illness of about 3 weeks' duration. The cellular response to *S. typhi* infection, a neutropenia, is unlike that of most bacterial diseases. *Salmonella typhi* may be isolated from the blood, urine, and feces of the patients at various stages in the disease. Such isolates are often subjected to bacteriophage typing for epidemiological purposes to correlate with isolations from carriers or other cases of typhoid.

Chloramphenicol is the drug of choice for patients with typhoid fever. It has been found that treatment with other drugs (e.g., tetracycline), though showing activity in vitro against *S. typhi,* appeared to exert little or no effect on the course of the disease. More recently, ampicillin has been shown to be useful in the therapy of typhoid fever. After an attack of typhoid fever, the patient may become a carrier of the organism for weeks and months; occasionally the carrier state continues for years, and these people serve as a reservoir for the maintenance of the disease. Typhoid fever now occurs usually in small, localized epidemics.

A vaccine for immunization against *Salmonella* infection has been available since the early days of prophylactic immunization. The usual preparation, a mixture, contains killed cells of *S. typhi, S. paratyphi* A, and *S. paratyphi* B. Although widely used for many years, no experimental proof of its efficacy can be demonstrated; however, it is recommended for persons living in areas where typhoid fever occurs.

Several other organisms (*S. paratyphi* A, *S. paratyphi* B) produce a disease clinically similar to typhoid fever, but less severe. These organisms, like the typhoid bacillus, are harbored almost exclusively by man and are spread by carriers to susceptible individuals.

Gastroenteritis due to *Salmonella* is probably the most commonly occurring form of infection caused by these organisms. The course of the disease is 3 to 7 days and is usually self-limited. Patients are carriers for weeks after the infection, but eventually most clear spontaneously. The strains of *Salmonella* causing gastroenteritis are usually animal pathogens or organisms maintained in animal species. For example, the *Salmonella* most commonly isolated from cases of human gastroenteritis is *S. typhimurium,* the causative agent of mouse typhoid. Some of the more common sources of *Salmonella* dissemination are listed following:

* *S. typhi, S. pullorum,* and *S. gallinarum* ferment glucose, but are anaerogenic.

1. Contaminated water and food, especially poultry products, such as eggs, egg powder, improperly cooked meat (e.g., turkey roll), etc.
2. Animal reservoirs; direct contamination of food, such as fecal contamination by mice carrying *S. typhimurium,* or direct contact with animals (e.g., handling of infected pet turtles)
3. Animal feed prepared from contaminated meat scraps, tankage, etc.
4. Human cases, subclinical human infections, and carriers.

Certain Group C *Salmonella,* especially *S. cholerae-suis,* are known to cause disease unrelated to the gastrointestinal tract. Abscesses may develop anywhere in the body, sometimes resulting in bacteremia. The source of infection in these patients is not clear. *Salmonella cholerae-suis* is found in swine, associated with hog cholera, but not the causal agent; it is rarely isolated from the feces of humans or animals.

A simplified system of nomenclature for *Salmonellae* is recommended by Edwards and Ewing.[1] They suggest having only three species; *S. cholerae-suis* and *S. typhi* would be the names applied to two specific organisms; the remaining 1400 or more existing species would then be designated as serotypes or bioserotypes of *S. enteritidis.* Thus, *S. typhimurium* would be called *S. enteritidis* ser. typhimurium, *S. pullorum* would become *S. enteritidis* bioser. Pullorum, etc.

MISCELLANEOUS ENTEROBACTERIACEAE

A transitional group of Gram-negative rods should be mentioned, which has certain biochemical and serologic characteristics in common with *Salmonella,* with *Shigella,* and also with other members of the Enterobacteriaceae. They include *Arizona, Citrobacter, Enterobacter hafniae,* and *Providencia* (Table 1), as well as a more recently described genus, *Edwardsiella.*[12] This somewhat heterogeneous group of organisms was formerly referred to as the "paracolon group," a term of convenience that had no taxonomic standing.

These organisms are of importance mainly because they complicate the process of isolating *Salmonella* and *Shigella* from stool, since they either do not ferment lactose or ferment it only slowly. Thus, on stool culture media, which depend on non-fermentation of lactose to indicate suspect colonies, they mimic the appearance of the true pathogens.

The *Arizona* strains are quite closely related to *Salmonella* and have numerous O and H antigens in common with the latter. They are found in lower animals, especially reptiles, and occasionally produce *Salmonella*-like gastroenteritis in man.

Representative *Citrobacter* strains (formerly called *Escherichia freundii*) ferment lactose and produce large amounts of hydrogen sulfide. However, many strains attack this key carbohydrate slowly or not at all. This group of bacteria, which also has some antigens in common with *Salmonella,* has also been termed "Bethesda-Ballerup."

Enterobacter hafniae is included in the tribe Klebsielleae. It does not produce hydrogen sulfide and is mentioned here because it usually does not ferment lactose. The *Providencia* strains are closely related to *Proteus* in their biochemical reactions.

In summary, this heterogeneous group of organisms is of nuisance value in the search for *Salmonella* and *Shigella* from cases of enteric fever. Although the true role of these bacteria in the etiology of gastroenteritis is uncertain, they may cause urinary-tract or other infections if they somehow gain access to a part of the body normally sterile; e.g., *Edwardsiella tarda*[13] has been shown to have caused a case of bacterial meningitis.

Klebsiellae

Members of the tribe Klebsiellae are widespread in their biologic activities; they are found as human pathogens, as commensals in man and animals; in soil and water, and as

Table 4
DIAGNOSTIC ANTIGENIC SCHEMA OF THE GENUS *KLEBSIELLA*

O group	Capsule type
1	1, 2, 3, 7, 8, 10, 12, 16, 19, 20, 21, 22, 23, 24, 26, 29, 30, 32, 34, 37, 39, 41, 44, 45, 46, 47, 62
2	2, 3, 4, 5, 6, 8, 27, 28, 35, 43, 59
3	11, 25, 31, 33, 48, 49, 50, 51, 53, 54, 55, 58
4	15, 42
5	57, 61
Ungrouped	9, 13, 14, 17, 18, 36, 38, 40, 52, 56, 60, 63–72

Reprinted with permission from Kauffman, F., *The Bacteriology of Enterobacteriaceae,* Williams & Wilkins, Baltimore, 1966. Copyright by Munksgaard International Publishers, Ltd., Copenhagen, Denmark.

plant pathogens. Most genera ferment glucose, lactose, and sucrose, and utilize citrate, but do not produce indol, hydrogen sulfide, or phenylalanine deaminase. Those in the genus *Klebsiella* are non-motile, whereas most of the *Enterobacter* strains are motile.

Classification of these organisms into an orderly scheme has been difficult. Historically, organisms arising from the respiratory tract were designated as *Klebsiella,* whereas those from urine, stool, soil, water, etc., were called *Aerobacter aerogenes.* As time went on, it became apparent that application of critical biochemical tests showed some strains of *A. aerogenes* to be indistinguishable from *Klebsiella.* Julianelle,[14] using a capsular antigen–antibody reaction, showed that at least three serotypes of *Klebsiella* (A, B, and C) could be recognized on the basis of the specificity of the polysaccharide in the capsule. Subsequently, Kauffmann[15] prepared a schematic representation (Table 4) of the antigenic structure of *Klebsiella* based on their somatic O and capsular K antigens.

It may be seen in Table 4 that 72 serotypes have been described. Identification of organisms on this basis is not practical for the rcutine diagnostic laboratory, but Eichoff et al.[16] studied 257 strains of *Klebsiella* from human sources and found that serotypes 1, 3, 4, and 5 were most frequently encountered in the respiratory tract, whereas type 2 and untypable strains were more often found in urine. In the same study, the motile species *Enterobacter cloacae* and *E. aerogenes* were more commonly, but not exclusively, isolated from urine. These latter two species are distinct from *Klebsiella* in being motile, as are *Serratia* and some strains of *Enterobacter liquefaciens.*

Klebsiella is the causative agent of Friedlander's pneumonia, a fulminating type of disease that has a high case fatality rate if untreated. Since the advent of chemotherapy, however, the disease has been less serious. It occurs primarily in chronic alcoholics or in physically debilitated individuals. Those patients recovering from the disease may have chronic abscesses and cavities that persist over a period of slow convalescence.

Serratia

Serratia marcescens has long been used as a laboratory stock culture to hand out to students in introductory microbiology. It is a small Gram-negative rod or coccobacillus that does not ferment lactose, or does so only slowly. It is found in water, in soil, and occasionally in humans, but not nearly as constantly as *Enterobacter aerogenes.* The colonies have orange-red pigmentation, which varies in color intensity and shade, depending on conditions of cultivation. The red pigment, prodigiosin, has been characterized chemically and will be discussed in a subsequent volume of the *Handbook.*

Serratia marcescens has been used as a tracer organism to follow the natural spread of bacteria from person to person, instruments to person, etc., because the colonies are easily recognized on subculture. The organism was once considered a harmless saprophyte, but it is now recognized that *Serratia* can be the cause of infections in various parts of the human body. Some strains are non-pigmented, and it is probable that in the past many of these were reported as *Aerobacter.* It is now believed that this organism may spread in epidemic form, causing nosocomial infections in hospitals, similar to that of penicillin-resistant staphylococci in the newborn and to that of *Salmonella derby* gastroenteritis.[17]

Enterobacter agglomerans

A group of organisms that were described at one time in the genus *Erwinia,* and later as *Pectobacterium,* are now classified as *Enterobacter agglomerans.* They are found in soil and invade plant tissues, causing dry necroses, galls, wilts, and soft roots. Certain members of the species produce pectinolytic enzymes, which is manifested in their ability to produce plant diseases. The biochemical reactions of these organisms are listed in Table 1, under the heading *Pectobacterium.* The genus *Erwinia* is still favored for classification by some microbiologists, but on the basis of phenotypic characteristics and DNA base homology these organisms can be considered members of the genus *Enterobacter.*

Occasionally, these bacilli are isolated from human sources.[18,19] Their role as human pathogens is doubtful. In one instance, however, in which they were accidentally introduced into the blood stream of patients via contaminated bottles of intravenous glucose solution, they apparently caused serious illness and death.

The *Proteus* Group

Organisms of this genus are motile; they ferment dextrose, but not lactose. They produce a strong urease and phenylalanine deaminase. Four species (Table 1) are recognized; two of these, *P. vulgaris* and *P. mirabilis,* produce large amounts of hydrogen sulfide, liquefy gelatin, and are so motile that they swarm over the surface of a moist agar plate. *Proteus morganii* and *P. rettgerii* do not liquefy gelatin, nor do they produce hydrogen sulfide; they are motile, but do not tend to swarm. It is interesting that the DNA base composition of *P. morganii* is 50 to 53 mol % G + C; this is somewhat remote from the 38 to 41 mol % G + C value that has been measured for other species in the genus. The habitat of *Proteus* is the intestinal tract of man and of lower animals; in nature, the organisms occur in locales that are fecally contaminated.

The O, H, and K antigens of the *Proteus* group have been studied, but they are of little importance, since identification on the basis of biochemical reactions is usually satisfactory. Of more practical interest is the fact that certain strains of *Proteus* (OX19, OX2, and OXK) have an O-antigenic component in common with certain rickettsiae. This cross reaction is used in the Weil-Felix reaction, in which the serum of patients convalescing from epidemic typhus agglutinates *Proteus* OX19 antigen. Similarly, sera of patients who have had scrub typhus agglutinate the OXK antigen. The phenomenon is interesting, but not as specific for diagnostic purposes as the complement fixation test with purified rickettsial antigens.

Proteus species are commonly isolated from urinary-tract infections; they are sometimes difficult to eliminate because of their resistance to the more commonly used antimicrobial agents. Like other Enterobacteriaceae, they may cause pyogenic infections in other parts of the body when accidentally introduced.

REFERENCES

1. **Edwards, P. R. and Ewing, W. H.,** *The Identification of Enterobacteriaceae,* 3rd ed., Burgess Publishing Co., Minneapolis, 1972.
2. **Bray, J.,** Isolation of antigenically homogeneous strains of *Bact. coli neapolitanum* from summer diarrhea of infants, *J. Pathol. Bacteriol.,* 57, 239, 1945.
3. **Ewing, W. H.,** Serological relationships between *Shigella* and coliform cultures, *J. Bacteriol.,* 66, 333, 1953.
4. **Neter, E., Westphal, O., Lüderitz, O., and Gorzynski, E. A.,** The bacterial hemagglutination test for the demonstration of antibodies to Enterobacteriaceae, *Ann. N.Y. Acad. Sci.,* 66, 141, 1956.
5. **Mata, L. J., Gangarosa, E. J., Carceres, A., Perera, D. R., and Mejicanos, M. L.,** Epidemic Shiga bacillus dysentery in Central America. I. Etiologic investigations in Guatemala, 1969, *J. Infect. Dis.,* 122, 170, 1970.
6. **Eberth, C. J.,** Die Organismen in den Organen bei Typhus abdominalis, *Virchow's Arch.,* 81, 58, 1880.
7. **Gaffky, G.,** *Mitt. Kaiserl. Gesundh. Amtes,* 2, 372, 1884.
8. **White, P. B.,** *Med. Res. Counc. G. B. Spec. Rep. Ser.,* No. 103, 1926.
9. **Kauffmann, F.,** Der Antigenaufbau der Typhus-Paratyphus Gruppe, *Z. Hyg. Infektionskr.,* 111, 740, 1930.
10. **Kauffmann, F., Lüderitz, O., Stierlin, H., and Westphal, O.,** Zur Chemie der O-Antigene von Enterobacteriaceae. I. Analyse der Zuckerbausteine von *Salmonella* O-Antigenen, *Zentralbl. Bakteriol. Parasitenkd. Infektionskr. Hyg. Abt. I Orig.,* 178, 442, 1960.
11. **Lüderitz, O., Staub, A. M., and Westphal, O.,** Immunochemistry of O and R antigens of *Salmonella* and related Enterobacteriaceae, *Bacteriol. Rev.,* 30, 192, 1966.
12. **Ewing, W. H., McWhorter, A. C., Escobar, M. R., and Lubin, A. H.,** *Edwardsiella,* a new genus of Enterobacteriaceae, based on a new species, *E. tarda, Int. Bull. Bacteriol. Nomencl. Taxon.,* 149, 33, 1965.
13. **Sonnenwirth, A. C. and Bozena, A. K.,** Meningitis due to *Edwardsiella tarda.* First report of meningitis caused by *E. tarda, Am. J. Clin. Pathol.,* 49, 92, 1968.
14. **Julianelle, L. A.,** A biological classification of *Encapsulatus pneumoniae* (Friedlander's bacillus), *J. Exp. Med.,* 44, 113, 1926.
15. **Kauffman, F.,** *The Bacteriology of Enterobacteriaceae,* Williams & Wilkins, Baltimore, 1966.
16. **Eickhoff, T. C., Steinhauer, B. W., and Finland, M.,** The *Klebsiella-Enterobacter-Serratia* division. Biochemical and serologic characteristics and susceptibility to antibiotics, *Ann. Intern. Med.,* 65, 1163, 1966.
17. **Sweeney, F. J. and Randall, E. L.,** Clinical and epidemiological studies of *Salmonella derby* infections in a general hospital, in *Proceedings of the National Conference on Salmonellosis, 1964,* U.S. Department of Health, Education and Welfare, Washington, D.C., 1964, pp. 130–139.
18. **Ewing, W. H. and Fife, M. A.,** *Enterobacter agglomerans,* U.S. Department of Health, Education and Welfare, Washington, D.C., 1971.
19. **von Gravenitz, A. and Strouse, A.,** Isolation of *Erwinia* spp. from human sources, *Antonie van Leeuwenhock J. Microbiol. Serol.,* 32, 429, 1966.

GRAM-NEGATIVE ANAEROBIC BACTERIA*

H. A. Lechevalier

Generic name (G + C, molar %) [Number of species in *Bergey's Manual*]	Morphology	Physiology	Ecology	Examples of species
Bacteroides (40 to 55) [22]	Rods up to 10 μm long, somewhat pleomorphic; if motile, with peritrichous flagella.	Chemoorganotrophs; obligate anaerobes; metabolize carbohydrates or peptone; fermentation products include succinic, lactic, acetic, formic, and propionic acids; butyric acid is not a major product; catalase usually not produced.	Found in cavities of man and other animals, in infections of soft tissues, and in sewage.	*B. fragilis.* Type species; part of the intestinal flora; may cause appendicitis, peritonitis, heart valve infections, etc. *B. ruminicola.* Part of the rumen population of most ruminants. *B. oralis.* Part of the oral flora; may be found in oral, upper respiratory, and genital tract infections. *H. hypermegas.* Found in the intestinal tract of poultry. *B. termitidis.* Found in the intestinal tract of termites. *B. nodosus.* Causes foot rot in sheep. *B. melaninogenicus.* Pathogenic, usually associated with other organisms; found in various types of infections.
Fusobacterium (26 to 34) [16]	Rods up to 10 μm long, which may be fusiform and pleomorphic; if motile, with peritrichous flagella.	Same as above, but butyric acid is a major product.	Same as above.	*F. nucleatum.* Type species; found in the mouth and in infections of the upper respiratory tract. *F. varium.* Found in human feces, in the intestinal contents of cockroach, and in various types of infections. *F. necrophorum.* Found in natural cavities of man and animals and in various types of necrotic lesions, particularly liver abscesses of cattle and pigs. *F. bullosum.* Found in the feces of man and in actinomycotic infections.

* Part 9 in *Bergey's Manual.*

Generic name (G + C, molar %) [Number of species in *Bergey's Manual*]	Morphology	Physiology	Ecology	Examples of species
Leptotrichia (32 to 34) [1]	Rods that may be slightly curved, 1–1.5 by 5–15 μm; many cells have pointed ends; filaments up to 200 μm may be formed in old cultures; non-motile.	Heterotrophic; fastidious; require 5% CO_2 for optimal growth; lactic acid and acetic acid are the main fermentation products; butyric acid not formed.	Found in oral cavities of man.	*L. buccalis.*
Desulfovibrio (42 to 65) [5]	Curved rods up to 10 μm long; motile by polar flagella.	Chemoorganotrophs; reduce sulfate to H_2S; lactate, pyruvate, and malate oxidized to acetate and CO_2; strict anaerobes.	Found in fresh- or salt-waters, particularly in polluted waters rich in H_2S, and in soils rich in organic matter and waterlogged.	*D. desulfuricans.* Type species.
Butyrivibrio (–) [1]	Curved rods, 0.3–0.8 by 1–5 μm; motile by a polar or subpolar flagellum.	Chemoorganotrophs; strict anaerobes; glucose fermented mainly to butyrate.	Found in the rumen of most ruminants, and also in the intestines of other animals.	*B. fibrisolvens.*
Succinivibrio (–) [1]	Curved rods, 0.3–0.7 by 1–7 μm; with pointed ends; motile by a single polar flagellum.	Chemoorganotrophs; strict anaerobes; glucose fermented mainly to succinate, acetate, and formate.	Found in the rumen of cattle and sheep.	*S. dextrinosolvens.*
Succinimonas (–) [1]	Short to coccoid rods, 1–1.5 by 1–3 μm; motile by a single polar flagellum.	Chemoorganotrophs; strict anaerobes; glucose fermented with production of succinate and some acetate; catalase not produced.	Found in the rumen of cattle.	*S. amylolytica.*

Generic name (G + C, molar %) [Number of species in *Bergey's Manual*]	Morphology	Physiology	Ecology	Examples of species
Lachnospira (–) [1]	Curved rods, 0.4–0.6 by 2–4 μm; motile by a lateral to subterminal flagellum; young cultures weakly Gram-positive.	Chemoorganotrophs; not proteolytic; glucose fermented with production of ethanol, lactic, formic, and acetic acids, and CO_2.	Found in bovine rumen.	*L. multiparis*
Selenomonas (53–61) [2]	Rods curved or helical, 0.8–1.1 by 2–7 μm; usually kidney- to crescent-shaped cells; motile by a tuft of lateral flagella located on the concave side of the cells.	Chemoorganotrophs; fermentation of glucose yields acetate, propionate, CO_2, and/or lactate; organic growth factors required; strict anaerobes; catalase not produced.	Found in the gastrointestinal tract of animals and in polluted fresh-water.	*S. putigena*. Type species; found in the mouth of man. *S. ruminantium*. Found in the rumen of ruminants.

GRAM-NEGATIVE COCCI AND COCCOBACILLI*

H. A. Lechevalier

Generic name G + C molar % [Number of species in *Bergey's Manual*]	Morphology	Physiology	Ecology	Examples of species
Neisseria[1] (47 to 52) [6]	Cocci, 0.6–1.0 μm, single or in pairs, with occasional formation of tetrads; non-motile.	Chemoorganotrophic; complex nutritional requirements; catalase produced; aerobic or facultatively anaerobic; temperature optimum about 37°C.	Found associated with man.	*N. gonorrhoeae.* Type species; causes gonorrhea and other infections of man. *N. meningitidis.* Causes epidemic cerebrospinal fever and other infections of man; may be found in the nasopharynx of healthy individuals.
Branhamella[1] (40 to 45) [1]	Cocci, in pairs; non-motile.	Chemoorganotrophic; blood not required for growth; acid not produced from carbohydrates; aerobic; temperature optimum about 37°C.	Associated with mucous membranes of mammals; may cause inflammations of the membranes.	*B. catarrhalis.*
Moraxella[1] (40 to 46) [5]	Short, plump rods, 1.0–1.5 by 1.5–2.5 μm, mainly in pairs and short chains; no flagella, but may "twitch" on solid surfaces.	Chemoorganotrophic; oxidative metabolism; fastidious growth requirements; oxidase-positive; usually catalase-positive; strictly aerobic.	Associated with the mucous membranes of man and of warm-blooded animals.	*M. lacunata.* Type species; may cause human conjunctivitis. *M. bovis.* Causes "pink-eye" of cattle.

* Part 10 in *Bergey's Manual.*

Generic name (G + C, molar %) [Number of species in *Bergey's Manual*]	Morphology	Physiology	Ecology	Examples of species
Acinetobacter[1] (39 to 47) [1]	As for *Moraxella* in the log phase; cocci in the stationary phase.	Chemoorganotrophic; oxidative metabolism; no specific growth requirement; oxidase-negative; catalase-positive; strictly aerobic.	Found in soil, water, and associated with man and animals; may be pathogenic to debilitated subjects.	*A. calcoaceticus.*
Paracoccus (64 to 67) [2]	Cocci, 0.5–1.1 μm; non-motile.	Chemoorganotrophic; respiratory metabolism; aerobic, except in the presence of nitrate; oxidase- and catalase-positive.	Found in soil or in brines.	*P. denitrificans.* Type species; facultatively chemolithotrophic; able to use the oxidation of H_2 as source of energy; found in soil. *P. halodenitrificans.* Halophilic; found in brines; not chemolithotrophic.
Lampropedia (61) [1]	Coccoid to cubical, 1.0–2.5 μm, in pairs, tetrads, and tablets of cells; the cells in the tablets are enclosed within a complex structured envelope; non-motile, but groups of cells flicker.	Chemoorganotrophic; respiratory metabolism; obligate aerobe; energy source limited to Krebs cycle intermediates; growth from 10–35°C at pH 6–8.5.	Isolated from stagnant water, decomposing organic material and gastrointestinal content of herbivorous animals.	*L. hyalina.*

REFERENCES

1. Hendriksen, S. D., *Moraxella, Neisseria, Branhamella,* and *Acinetobacter, Annu. Rev. Microbiol.*, 30, 63–83, 1976.

NEISSERIACEAE

C. F. Clancy

There are four genera in the family Neisseriaceae, namely, *Neisseria, Branhamella, Moraxella,* and *Acinetobacter.*

The *Neisseria* are Gram-negative "biscuit"-shaped diplococci; one side of each cell of the pair is flattened against the side of the other cell, and the pair always appears as one unit. In cultures 3 to 4 days old or older they may "balloon out" in size and have a tendency to become Gram-positive; in smears of body fluids and young cultures they usually have the typical diplococcus arrangement and are uniformly Gram-negative. All species are found in the human body, with the exception of *N. caviae,* which has been found in the pharynx of guinea pigs. All *Neisseria* produce cytochrome-oxidase that reacts with dimethyl- or tetramethyl-*p*-phenylene diamine; when flooded with the reagent, colonies turn pink to purple to black because of the oxidation reaction.

Branhamella catarrhalis, formerly named *Neisseria catarrhalis,* is commonly found in sputum and in throat cultures of normal individuals. The colony is dirty white in color, grows to a diameter of 2 to 3 mm in 2 days, and often is rough. *Niesseria sicca* has the same habitat, but colonies grow in a rather lacy fashion, likened by some to cartwheels. Several other species (e.g., *N. flava*) have dirty-yellow pigmentation and may also be readily isolated from sputa or from throat swabs. These different species are similar to one another in having rough colonies that are difficult to emulsify. They are a well-recognized entity in the oral flora and are not associated with any disease processes. They grow at 22°C. *Moraxella* and *Acinetobacter* species are occasionally the cause of human infections.

NEISSERIA MENINGITIDIS

Neisseria meningitidis (formerly *intracellularis*), the meningococcus, is the cause of a bacterial meningitis and meningococcemia. This Gram-negative diplococcus is maintained in the population in the throats of healthy carriers and produces its disease only when conditions are favorable for the development of the disease in a susceptible host. Colonies of this species grow to be 2 to 3 mm in diameter and are usually gray and mucoid on primary isolation. Good growth occurs on blood agar or chocolate agar at 37°C and is enhanced by incubation in an atmosphere of 5% carbon dioxide. When grown in a special cystine-casein semisolid agar with appropriate CHO substrate, glucose and maltose are fermented, but not sucrose. The amount of acid produced is not large.

Meningococci are separated into four serologic groups (A, B, C, and D) on the basis of antigenically specific polysaccharides present in the capsules. A rapid slide agglutination test may be carried out with a freshly isolated culture by use of appropriate antisera that have been absorbed, to ensure group specificity. Encapsulated cells demonstrate the quellung reaction when mixed with homologous antiserum. Serologic studies show that group A organisms are usually isolated from epidemic cases, whereas group B and group C strains are usually recovered from sporadic cases; group D organisms are rarely encountered.

A potent endotoxin can be extracted from meningococcal cells; it is believed to play a part in the pathogenesis of the infection. When injected into experimental animals, the endotoxin produces vascular damage; the lesions resemble those seen in meningococcemia. Analysis[1] of endotoxin from an organism of the serologic group C showed it to contain 20% lipid, composed of cephalin, fatty acids, and plasmalogen; the polysaccharide constituents were identified as galactose, glucose, glucosamine, and sialic acid. The lipids, when separated from the endotoxin, fail to show lethal activity and are nonpyrogenic.

Meningococcal Disease

Man is the only natural host for meningococci, and the disease is maintained and spread by carriers harboring the organism in their nasopharynx. The relationship between the carrier state and clinical disease is not understood, since studies have shown a carrier rate of 25% in a population where only sporadic cases of the disease may occur. Meningococcal disease attacks children and young adults, and epidemics may break out in institutions or military induction camps. Development of the disease seems to be related in some way to physical stress; in military units the carrier rate may rise well above 50% during epidemic periods.

Invasion of the blood by the organism may produce chronic or acute meningococcemia; this may result in the development of petechiae in the skin, with subsequent progress to purpurae several centimeters in diameter, due to hemorrhage. Positive cultures may be recovered from the hemorrhaged areas, and it is even possible to see the Gram-negative diplococci in smears of blood from the purpurae. Cultures of blood drawn at this time are usually positive.

Invasion of the central nervous system from the blood results in development of the meningeal form of the disease. Smears of spinal fluid sediment reveal the presence of Gram-negative diplococci, and cultures made at this time should be positive. It should be emphasized that the meningococcus undergoes autolysis rather quickly, and bacteriological studies should, therefore, be carried out immediately after collection of the specimen.

Chemotherapy and Prophylaxis

Initially, all strains of meningococci were believed to be sensitive to sulfonamides. Clinical cures were reported, and mass prophylaxis with sulfonamides was shown to curb epidemics in closed communities, such as military installations. However, the situation is different today; more than 70% of the cases are caused by sulfonamide-resistant meningococci.[2]

Penicillin is effective in the treatment of clinical cases, but on a large scale it does not lend itself to prophylaxis of the carrier state. It has been shown[3] that application of oral rifampin, a semisynthetic antibiotic, was effective in clearing the carrier state. Another interesting approach to the carrier problem has been the immunization[4] of potential carriers with a high-molecular-weight polysaccharide prepared from *N. meningitidis*, group C. Follow-up studies[5] showed the results of the vaccination to have a degree of efficiency in the prevention of subsequent meningococcal disease.

NEISSERIA GONORRHOEAE

Neisseria gonorrhoeae, the gonococcus, is the causative agent of gonorrhea, a venereal disease, and of several other conditions, which are usually sequela to the venereal infection. Like the meningococcus, the organism only infects man, and transmission of the disease is practically always by person-to-person contact.

Microscopically, the organism is indistinguishable from many other *Neisseria;* it is, however, more fastidious in its growth requirements. The gonococcus will not grow on blood agar, but it will grow on chocolate agar to which a yeast supplement has been added; the latter provides glutamine, cocarboxylase, and glutathione, which have been shown to be essential for growth. Certain amino acids in peptone and agar are somewhat toxic for the gonococcus, but the serum in the chocolate agar adsorbs or neutralizes these toxic factors. For primary isolation, appropriate concentrations of crystal violet in the chocolate agar will reduce the growth of Gram-positive organisms that might be present in urethral or cervical swabs. It is also necessary to incubate plates for primary isolations from clinical material in an atmosphere of about 5% added carbon dioxide. Colonies, about 1 to 2 mm in diameter, grow in 2 days, are colorless to gray, shiny, and smooth. The oxidase test has been used to identify *Neisseria* colonies on plates from urethral or cervical swabs, where contaminating organisms make recognition of gonococcal colonies difficult. The gonococcus ferments glucose, but not maltose and sucrose; the sugar fermentations are of value in confirming the identity of an isolate.

Laboratory Diagnosis

The gonococcus is even more delicate than the meningococcus and will not withstand adverse conditions, such as drying; hence, it is important to be sure that smears and cultures are made promptly after collection of specimens. Cultures should be made on chocolate agar, of course, as discussed above. Smears from urethral discharge are prepared by rolling the swab onto a glass slide, to avoid destruction of leukocytes. The finding of Gram-negative intracellular diplococci in such smears is almost specific identification for gonococcus, provided that the history and clinical findings are coincident.

In the past, because of the presence of a multiplicity of other organisms, the demonstration of gonococci in cervical swabs was so difficult that laboratory studies were often not carried out and the diagnosis was made on a clinical basis. Since the advent of fluorescent-antibody techniques for identification of microorganisms, appropriately labeled antigonococcal serum has also been used for identification of *N. gonorrhoeae.* Slides for fluorescent-antibody studies may be prepared directly from the swab or from a culture.

Gonococcal Disease

Gonorrhea usually appears in males 2 to 8 days after exposure and is manifested by urethritis with purulent discharge. In females the onset of disease may also be urethritis, but the infection often progresses to involvement of the ovaries and other organs and is then called "pelvic inflammatory disease" (P.I.D.).

Among the complications of gonorrhea are gonococcal arthritis and ophthalmia neonatorum. Prophylaxis for the latter, a disease of newborns, consists of washing the eyes of the infant with a solution of penicillin immediately after birth.

When sulfonamides first became available, they were used successfully in the treatment of gonorrhea. During World War II, certain strains of the gonococcus appeared, which were resistant to the drug. Later, penicillin became the drug of choice, and many strains were sensitive to 0.06 units/ml or less.[6] Subsequent studies showed that the sensitivity to penicillin is decreasing each year; as of 1969,[7] more than 65% of routine isolations were resistant to penicillin doses larger than 0.05 units/ml, and some even to doses larger than 3.0 units/ml. Penicillin is still the drug used most often; when given in high doses, with addition of probenecid to slow down excretion, it provides an effective therapeutic measure.

In spite of the fact that effective antimicrobial agents have been available since about 1940, the incidence of gonorrhea is increasing at an alarming rate. From 1954 to 1962 the total of reported cases in the United States was around 250,000 per year; this has been rising constantly, so that by 1971 a total of almost 700,000 cases was reported.

REFERENCES

1. **Mergenhagen, S. E., Martin, G. R., and Schiffman, E.,** Studies on an endotoxin of a group C *Neisseria meningitidis, J. Immunol.*, 90, 312, 1963.
2. *Meningococcal Infections, Morbidity and Mortality Weekly Reports,* Vol. 18, Communicable Disease Center, Atlanta, Ga., 1969, p. 135.

3. **Deal, W. B. and Sanders, E.,** Efficacy of rifampin in treatment of meningococcal carriers, *N. Engl. J. Med.*, 281, 641, 1969.
4. **Gotschlich, E. C., Goldschneider, I., and Artenstein, M. S.,** Human immunity to the meningococcus. V. The effect of immunization with meningococcal group C polysaccharide on the carrier state, *J. Exp. Med.*, 129, 1385, 1969.
5. **Artenstein, M. S., Gold, R., Zimmerly, J. G., Luyle, F. A., Schneider, H., and Harkins, C.,** Prevention of meningococcal disease by group C polysaccharide vaccine, *N. Engl. J. Med.*, 282, 417, 1970.
6. **Love, B. D. and Finland, M.,** Susceptibility of *Neisseria gonorrhoeae* to eleven antibiotics and sulfadiazine, *Arch. Intern. Med.*, 95, 66, 1955.
7. **Martin, J. E., Jr., Lester, A., Price, E. V., and Schmale, J. D.,** Comparative study of gonococcal susceptibility to penicillin in the United States, 1965–1969, *J. Infect. Dis.*, 122, 459, 1970.

NONMOTILE, NONSPORULATING, GRAM-NEGATIVE ANAEROBIC COCCI*

Veillonellaceae

H. A. Lechevalier

Generic name (G + C, molar %) [Number of species in *Bergey's Manual*]	Morphology	Physiology	Ecology	Examples of species
Veillonella (40 to 44) [2]	Cocci, 0.35–0.5 μm, in pairs or short chains.	Chemoorganotrophic; complex nutritional requirements; CO_2 required; carbohydrates and polyols not fermented; acetate, propionate, CO_2, and H_2 produced from lactate; CO_2 and propionate produced from succinate; growth good at 30–37°C and pH 6.5–8.0.	Parasitic in the mouth and in the intestinal and respiratory tracts of animals and man.	*V. parvula.* Type species; catalase-negative. *V. abcalescens.* Catalase-positive.
Acidaminococcus (55 to 57) [1]	Cocci, 0.6–1.0 μm, in pairs.	Chemoorganotrophic; complex nutritional requirements; certain amino acids are required and can serve as sole energy source; lactate, succinate, fumarate, malate, citrate, and pyruvate not used; weak saccharolastic activity; growth good at 30–37°C and pH 6.5–7.5.	Found in the intestinal tract of the pig and of man.	*A. fermentans.*
Megasphaera (53 to 54) [1]	Cocci, 2 μm or larger, in pairs or occasional short chains.	Chemoheterotrophic; complex nutritional requirements; carbohydrates fermented.	Found in the rumen of cattle and sheep and in the cecum of pigs.	*M. elsdenii.*

* Part 11 in *Bergey's Manual.*

ANAEROBIC NONMOTILE GRAM-POSITIVE COCCI*

Peptococcaceae

H. A. Lechevalier

Generic name (G + C, molar %) [Number of species in *Bergey's Manual*]	Morphology	Physiology	Ecology	Examples of species
Peptococcus (36 to 37) [6]	Cocci, 0.5–1.0 μm, single, in pairs, in tetrads, or in irregular masses; short chains may be formed.	Chemoorganotrophic; can use protein hydrolysates as sole energy source; gas may be produced with or without carbohydrates; volatile fatty acids, CO_2, H_2, and NH_3 are produced from amino acids; lactate is not a major product of glucose fermentation; lactate and malate are not fermented; optimal temperature, 35–37°C; optimal pH, 7–8.	Isolated from various locations in the human body and in animal bodies, with or without pathological manifestations; some strains have been isolated from mud.	*P. niger.* Type species. *P. asaccharolyticus.* Ferments glutamate, but not glycine; gelatin not hydrolyzed. *P. activus.* Ferments glutamate, but not glycine; gelatin hydrolyzed. *P. anaerobius.* Ferments glycine, but not glutamate; gelatin not hydrolyzed.
Peptostreptococcus (33 to 34) [4]	Coccoid to ovoid, 0.3–1.0 μm, in pairs or in short or long chains.	Chemoorganotrophic; peptones and amino acids are main sources of energy; optimal temperature, 35–37°C; optimal pH, 7–7.5.	Isolated from normal or pathological locations in man and in animals; may be pathogenic.	*P. anaerobius.* Type species; produces gas from pyruvate. *P. micros.* Very small; does not produce gas from pyruvate.

* Part 14 in *Bergey's Manual.*

Generic name (G + C, molar %) [Number of species in *Bergey's Manual*]	Morphology	Physiology	Ecology	Examples of species
Ruminococcus (40 to 45) [2]	Coccoid to ovoid, 0.7–1.2 μm, single, in pairs, or in short chains; colonies may be white, light tan, yellow, or orange.	Chemoorganotrophs with complex nutritional requirements; ammonia is the main source of nitrogen; cellulose is usually digested, and carbohydrates are fermented; good growth at 30–37°C; optimal pH, 6.5.	Widely distributed in the rumen and other parts of the digestive tract of herbivorous animals.	*R. flavefaciens.* Type species.
Sarcina (28 to 31) [2]	Cocci, 1.8–3 μm, in packets of eight or more.	Chemoorganotrophs; ferment carbohydrates; from glucose, CO_2, H_2, and acetic acid are the main products; some amino acids and vitamins are required for growth; temperature optimum, 30–37°C; optimal pH range, 1–9.8.	Isolated from soil, from plants, especially the coat of cereal seeds, and from the stomach contents of man and of animals.	*S. ventriculi.* Type species.

AEROBIC OR FACULTATIVELY ANAEROBIC GRAM-POSITIVE COCCI*

H. A. Lechevalier

Generic name (G + C, molar %) [Number of species in *Bergey's Manual*]	Morphology	Physiology	Ecology	Examples of species
Micrococcus (66 to 75) [3]	Cocci, 0.5–3.5 μm, single, in pairs, in irregular clusters, in tetrads, or in cubical packets; usually non-motile.	Chemoorganotrophs; respiratory metabolism; glucose oxidized to acetate or to CO_2 and H_2O; catalase produced; nutritional requirements variable; aerobic; optimal temperature, 25–30°C.	Found in fresh water, in soils, and on the skin of man and of animals.	*M. luteus.* Type species; yellow. *M. roseus.* Pink to red.
Staphylococcus (30 to 40) [3]	Cocci, 0.5–1.5 μm, single, in pairs, or in irregular clusters; non-motile.	Chemoorganotrophs; respiratory or fermentative metabolism; catalase produced; carbohydrates utilized with the production of acid; anaerobically, glucose fermented to lactic acid; in the presence of air, acetic acid is formed; facultative anaerobes; optimal temperature, 35–40°C; pH optimum, 7.0–7.5.	Found associated with the skin, skin glands, and mucous membranes of warm-blooded animals, may be pathogenic.	*S. aureus.* Type species; may cause boils, abscesses, meningitis, pyemia, osteomyelitis, and food poisoning; does not require biotin for growth. *S. epidermidis.* Requires biotin for growth; very common on the skin and in the membranes of warm-blooded animals; may cause diseases, including contagious impetigo of swine and bacterial endocarditis of man.

* Part 14 in *Bergey's Manual.*

Generic name (G + C, molar %) [Number of species in *Bergey's Manual*]	Morphology	Physiology	Ecology	Examples of species
Planococcus (48 to 52) [1]	Cocci, 1–1.2 μm, single, in pairs, in threes, or in tetrads; motile with one or two flagella.	Chemoorganotrophs; respiratory metabolism; strict aerobes; catalase-positive; grow at temperatures from 20–37°C.	Found in sea water.	*P. citreus.*
Streptococcus (33 to 42) [21]	Coccoid to ovoid, less than 2 μm, in pairs or in chains; may be motile.	Chemoorganotrophs; fermentative metabolism; produce dextrorotary lactic acid from glucose; homofermentative; facultative anaerobes; nutritional requirements generally complex; temperature optimum, about 37°C.	Found in the buccal cavity of man and of animals, in their respiratory tract, and in blood in cases of septicemia; associated with numerous diseases.	*S. pyogenes.* Type species; associated with man. *S. equi.* The cause of equine strangles. *S. dysgalactiae.* A cause of bovine mastitis and lamb polyarthritis. *S. sanguis.* Found in dental plaques; may cause endocarditis. *S. faecalis.* Found in human and animal feces. *S. lactis.* A common contaminant of milk and dairy products. *S. pneumoniae.* Found in the upper respiratory tract of man and of animals.
Leuconostoc (38 to 44) [6]	Spherical to lenticular, in pairs and in chains; non-motile.	Chemoorganotrophs; complex nutritional requirements; glucose fermented with the production of D(−) lactic acid, ethanol, and CO_2; catalase-negative; facultative anaerobes; optimal temperature, 20–30°C.	Found in sugar solutions, fruits, vegetables, milk, and milk products; nonpathogenic to man and animals.	*L. mesenteroides.* Type species.

Generic name (G + C, molar %) [Number of species in *Bergey's Manual*]	Morphology	Physiology	Ecology	Examples of species
Pediococcus (34 to 44) [5]	Cocci, in pairs or in tetrads; non-motile.	Fermentative metabolism; homolactic fermentation; DL-lactic acid produced; complex nutritional requirements; microaerophilic; catalase-negative.	Found in fermenting plant material, especially spoiled beer; rare in dairy products.	*P. cerevisiae.* Type species
Aerococcus (36 to 40) [1]	Cocci, 1–2 μm, often in tetrads; non-motile.	Chemoorganotrophs; acid, but no gas from glucose; microaerophilic; catalase-negative.	Found in air, in meat brine, and in vegetables.	*A. viridans.*
Gemella (31 to 35) [1]	Cocci, about 0.5 μm, single or in pairs; non-motile.	Chemoorganotrophs; fermentative metabolism; complex nutritional requirements; aerobic to facultative anaerobic; optimal temperature, 37°C.	Found in bronchial secretions and in slime from the respiratory tract of mammals.	*G. haemolysans.*

STAPHYLOCOCCUS

C. F. Clancy

The staphylococci are commonly found in and on the human and animal body. *Staphylococcus epidermidis* (formerly *albus*) is usually found on the skin and on the external nares; it is usually not considered pathogenic, although it may cause such diseases as subacute bacterial endocarditis, meningitis, etc., after accidental introduction into the body. The colonies on agar are white, shiny, butyrous, and about 3 mm in diameter after an incubation period of 48 hr. The organism grows at room temperature, but more rapidly at 37°C.

Staphylococcus aureus is the cause of a wide variety of human and animal infections. It is the common cause of furunculosis and pyogenic wound infections, and may produce an infection in almost any part of the body into which it is introduced. It is one of the common causes of mastitis in cattle. Colonies on agar are initially a dirty-white color that may develop into a golden color after 2 to 3 days of incubation; it should be noted that the pigmentation is not the lemon-yellow color seen with some of the other micrococci. There is considerable variation from strain to strain in the amount of golden pigmentation that develops, even after several days of incubation.

Differentiation between *S. epidermidis* and *S. aureus* is not too difficult, since the former has a white colony, does not ferment mannitol, nor produce the enzyme coagulase. *Staphylococcus aureus,* on the other hand, ferments mannitol and produces coagulase. Occasional strains may deviate from the usual pattern.

TOXINS AND ENZYMES

Staphylococcus aureus, when grown on blood agar plates, usually produces a zone of clear hemolysis after 2 days of incubation. The size of the zone is enhanced by cooling the plates to 5°C for several hours. Four antigenically distinct hemolysins, designated as *alpha-, beta-, gamma-,* and *delta-,* have been shown to be involved in this red-cell lysis.

The *alpha*-hemolysin, a protein with a molecular weight of about 44,000, is active against rabbit and sheep erythrocytes. It produces necrosis on intracutaneous injection into rabbits, and small amounts are lethal for rabbits and mice. This hemolysin is produced mainly by strains isolated from human sources. It should be noted that the *alpha*-hemolysin produces a clear type of hemolysis that should not be confused with the reaction produced on blood agar by *alpha*-streptococci. The *beta*-hemolysin is active against sheep erythrocytes and is produced mainly by strains from animal sources. In vitro activity is best demonstrated by preliminary incubation of blood broth tubes at 37°C, then holding overnight at 4°C, the so-called "hot–cold lysis." This hemolysin is much less toxic on injection into mice or rabbits. The *gamma*- and *delta*-hemolysins are active against erythrocytes of a wider species of animals, including man. They are less toxic for experimental animals than the *alpha*-hemolysin. It is believed that these four toxins play a part in the pathogenesis of staphylococcal disease, but their true role is unknown.

Staphylococcal enterotoxin[1] is produced by relatively few strains of *S. aureus,* usually those in phage-lytic group II. There are three types of enterotoxin, each antigenically distinct from the others and from the hemolysins.

Leucocidin, a substance produced by many strains of *S. aureus,* can be shown to destroy white blood cells of a variety of animal species. It is distinct from the other metabolites of staphylococci, proteinaceous in nature, and antigenic.

A number of enzymes are produced by staphylococci, such as hyaluronidase (spreading factor), staphylokinase, proteinases, lipases, coagulase, and penicillinase.

Coagulase can be demonstrated in cultures of nearly all strains of *S. aureus* that are freshly isolated from human infections. This enzyme has a thrombokinase-like activity that initiates the conversion of fibrinogen into fibrin. Coagulase does not require calcium ions to initiate the conversion of fibrinogen into fibrin; however, it does require a factor called CRF (coagulase reacting factor), which is present in high concentrations in horse, human, and rabbit plasma. The coagulase test, as performed in the laboratory, is usually carried out using human or rabbit plasma in a tube to which a broth culture of the organism is added; after a suitable incubation period, a typical fibrin clot forms if the enzyme is present. A slide test may be carried out by adding plasma to a uniform suspension of the organism and rocking the mixture for a few minutes; the bacterial cells become coated with fibrin as it is formed, then clump together, appearing much the same as in a rapid slide agglutination test.

The production of the enzyme penicillinase, by staphylococci has considerable clinical importance. The enzyme acts by hydrolyzing the unstable β-lactam ring of penicillin G, thereby yielding inactive penicilloic acid. Penicillinase-producing strains are resistant to penicillin G when tested in vitro, and infections caused by them do not respond to treatment with penicillin G. Penicillinase is much less active against the semisynthetic penicillins, due to the steric effect of the bulky side chains substituted in the acyl group.

SEROLOGIC RELATIONSHIPS

Identification of staphylococci by serologic tests has been described.[2] There are two main groups: the pathogenic group A, and the nonpathogenic group B. The specificity depends on the polysaccharides in the cell wall, which are precipitated by antiserum prepared against the whole cells. Subsequent work has shown that immune serum prepared from pathogenic staphylococci agglutinates many strains of *S. aureus,* but not *S. epidermidis.* Many strains are inagglutinable, however, and serologic studies seem, therefore, of little value.

BACTERIOPHAGE TYPING

One of the first descriptions of bacteriophage lysis was that of Twort,[3] who noted the moth-eaten appearance of colonies of bacteria (presumably *S. epidermidis*) cultured from calf vaccine. Six years later, d'Herelle[4] reported on the isolation of the lytic principle from *S. aureus* isolated from a human boil. A system of phage typing[5] has been developed that enables the characterization of cultures of *S. aureus* according to their lysis by 20 to 30 different phage types. Phage typing is useful for epidemiological studies, especially in tracking down healthy carriers of *S. aureus* in institutional outbreaks, such as occur in maternity and newborn units. It is now recognized that certain phage types, such as 80/81 and 52A/79, are epidemic types and are repeatedly isolated from staphylococcal infections in many parts of the world.

THE NATURE OF STAPHYLOCOCCAL DISEASE

The reservoir that maintains staphylococcal disease in the population is man, either as hosts with actual infections or as healthy carriers. *Staphylococcus epidermidis* can be isolated routinely from the skin and nares of most people, but occasionally, for reasons not known, some individuals harbor *S. aureus* in the nares.

The most common type of infection is that of suppurative wound infection following injury. This type of infection may occur in any part of the body; the source of the organism is not always clear. Staphylococcal infections tend to become walled-off abscesses, which eventually drain or are incised by the surgeon's scalpel. A particularly annoying type of infection is recurrent furunculosis, which usually occurs in youths and young adults. Furuncles develop without apparent break in the skin, proceed to drain or are incised, and finally heal; subsequently, another furuncle develops at a different site. This condition may continue for weeks and months. At one time vaccination with a killed autogenous vaccine (prepared from the patient's own organism) was tried, but proved to have little therapeutic effect. Application of chemotherapy also seems to have no effect in controlling the disease.

Chronic osteomyelitis may be a sequela to pyogenic infection with *S. aureus*. It often persists for years and is especially refractory to chemotherapy.

In recent years, staphylococcal gastroenteritis has been found to occur occasionally in patients who receive intensive antibiotic therapy. It is believed that, when the normal flora of the lower bowel are altered, an antibiotic-resistant staphylococcus may be established in the gut, resulting in gastroenteritis. The disease is often severe, especially in patients who already have another disease.

Staphylococcal food poisoning is an intoxication due to ingestion of food that contains preformed enterotoxin. Cream-filled pastries, puddings, and other foods that are only lightly heated in cooking often are the vehicle, although almost any food contaminated by a carrier and improperly refrigerated may cause this disease. The incubation period is short, 2 to 6 hr, and the course of the disease is limited to about 24 hr. The patient develops nausea, then violent diarrhea and vomiting, which is followed by prostration. Recovery occurs in 1 to 2 days, except in debilitated persons.

REFERENCES

1. **Dack, G. M.,** *Food Poisoning,* 3rd ed., University of Chicago Press, Chicago, 1956.
2. **Julianelle, L. A. and Wieghard, C. W.,** The immunological specificity of staphylococci. I. The occurrence of serological types, *J. Exp. Med.*, 62, 11, 1935.
3. **Twort, F. W.,** An investigation on the nature of ultramicroscopic viruses, *Lancet,* 2, 1241, 1915.
4. **d'Herelle, F.,** *Le Bacteriophage: Son Role dans l'Immunité. Monographie de l'Institut Pasteur,* Masson et Cie, Paris, 1921.
5. **Wentworth, B. B.,** Bacteriophage typing of staphylococci, *Bacteriol. Rev.*, 27, 253, 1963.

STREPTOCOCCUS

C. F. Clancy

The streptococci are Gram-positive cocci that appear under the microscope in pairs and in chains. Chains can best be visualized in smears from fluid media: strains that grow "rough" usually have longer chains than smooth-growing strains. A basic character for classification is the reaction produced on the surface of blood agar plates:[1] *alpha*-hemolytic streptococci turn the medium (actually the hemoglobin) green in the vicinity of the colony; *beta*-hemolytic streptococci lyse the red blood cells, creating a clear zone around the colony; *gamma*-hemolytic streptococci have no effect, and the medium remains unchanged. Four basic groups of streptococci are now recognized: (1) hemolytic streptococci, (2) viridans streptococci, (3) enterococci, and (4) lactic streptococci.

HEMOLYTIC STREPTOCOCCI

These organisms are, for the most part, human and animal pathogens. Most appear as *beta* hemolytic streptococci when grown on blood agar. *Streptococcus pyogenes* was recognized very early in microbiology as a causative agent of human infection. It is maintained in the throats of carriers, usually children. Although streptococcal disease is not as rampant as in the pre-antibiotic era, it continues to smolder and breaks out as clinical disease from time to time.

Streptococcus pyogenes grows only poorly in ordinary media, but addition of blood gives much better growth. Colonies on blood agar are of three different types – mucoid, matt, and glossy – and about 1 to 2 mm in diameter. The hemolysin produced diffuses into the medium and lyses the red cells, creating a clear area surrounding each colony. In broth there is usually a floccular or granular sediment, due to long chains of cocci settling to the bottom of the tube. The classification of hemolytic streptococci was considerably simplified when Lancefield[2] showed that an antigen, now called the C carbohydrate, could be extracted from cells of *S. pyogenes.* The antigen was specific and antigenically identical in all strains studied. Thus, the human strains from scarlet fever, streptococcal sore throat, erysipelas, puerperal sepsis, wound infections, etc., became designated as serologic group A. It is now recognized that the C carbohydrate is in the cell wall. Elaboration of studies of the C carbohydrate of streptococci from many sources has shown at least 13 different serologic groups.

Another antigenically active component of streptococci is the M protein, present in the cell wall. Within group A, more than 40 antigenically distinct serotypes are recognized (on the basis of the M antigen) and are used for epidemiologic purposes in studying the spread of streptococcal disease.

Streptococcus pyogenes Metabolites

Erythrogenic toxin is produced by many group A strains and is responsible for the rash of cases of scarlet fever. Injection of toxin into the skin (Dick test) produces an erythematous reaction in susceptible children, whereas in convalescents it is negative, presumably due to neutralization of the toxin by antibody. It is of interest that erythrogenic strains are lysogenic,[3] similar to toxigenic *Corynebacterium diphtheriae.*

Streptolysin O lyses red blood cells, but does so only under anaerobic conditions. It is antigenic and stimulates antibody production during infection. This reaction has clinical application in that a significant increase in Streptolysin O antibody level indicates a recent streptococcal infection.

Streptolysin S is stable in air and is cell-bound although it can be extracted from the bacterial cells. It is apparently non-antigenic.

Streptokinase (originally called fibrinolysin) is an enzyme that initiates the lysis of fibrin. It acts to convert plasminogen to plasmin, a protease found in blood plasma. It has had some use in human medicine as an agent to lyse fibrin deposits, but is of limited value because of its immunogenicity.

Hyaluronidase, originally called "spreading factor," is an enzyme that acts on hyaluronic acid; the latter is found in the streptococcal capsule and is very similar to the hyaluronate in the ground substance of connective tissue. Culture filtrates injected intradermally into animals enhance the spreading of such inert substances as India ink.

STREPTOCOCCAL DISEASE

The usual streptococcal infections have been listed above; however, these organisms can produce infection and disease in any part of the body to which they gain entrance. Two diseases that are sequelae of streptococcal infection are acute glomerulonephritis and rheumatic fever. The pathogenesis of these conditions is essentially unknown, except that they are known to follow an acute streptococcal infection at a time when the patient has developed antibodies against the disease.

Acute glomerulonephritis follows a streptococcal infection by about 1 week. The glomeruli of the kidney are damaged, but the patients usually undergo spontaneous recovery. It is interesting that the great majority of cases of this disease follow infection by serotype 12, group A streptococci. In acute rheumatic fever, the latent period after streptococcal infection is about 3 weeks. Patients, usually children, show carditis, migratory polyarthritis, and fever for many weeks; there is great danger of permanent damage to the heart. In this disease there may be recurrent attacks following subsequent streptococcal infections; prophylactic penicillin is often recommended to ward off further streptococcal infections. Both of these conditions may be an expression of autoimmune phenomenon.

LABORATORY DIAGNOSIS

Streptococcus pyogenes grows out readily from throat or wound swabs when planted on blood agar plates. The colonies and hemolytic activity are most typical after 2 days of incubation. Positive identification may be accomplished in 5 to 6 hr by inoculation of a swab into broth, followed by preparation of a smear with appropriate fluorescent-antibody and microscopic examination; fluorescent-antibody studies may even be done on the initial swab, if large numbers of organisms are present. Another simple test, though less specific, is to subculture the isolate on a blood agar plate, apply a filter paper disk containing the antibiotic bacitracin, and look for a zone of inhibition of growth. Most strains of *S. pyogenes* are sensitive to bacitracin, whereas most other *beta*-hemolytic streptococci are bacitracin-resistant. Serologic tests for grouping streptococci are too cumbersome for routine diagnostic purposes.

Streptococcus pyogenes is very sensitive to penicillin, and apparently no antibiotic-resistant mutants have appeared, even though the drug has been vigorously applied for nearly 30 years. Most strains are also sensitive to the tetracyclines, chloramphenicol, etc., as well as to sulfonamides.

Other Hemolytic Streptococci

Streptococcus agalactiae of the Lancefield group B streptococci is indigenous to cattle and a common cause of mastitis, as are *S. dysgalactiae,* a group C organism, and *S. uberis,* a viridans streptococcus. Some group B and C strains cause human infections, others infect horses (*S. equi*) and a wide variety of domestic animals.

VIRIDANS STREPTOCOCCI

The viridans streptococci are usually *alpha*-hemolytic, and occasionally non-hemolytic. Some species (*S. salivarius, S. mitis*) are a part of the normal human oral flora; others have been found in milk and in the bovine alimentary tract. The human species are usually commensals, being found in the oral cavity. When they are accidentally introduced by trauma (e.g., vigorous tooth brushing, tooth extraction, etc.) into the blood stream of persons with valvular heart disease, these benign organisms can attach themselves to the heart valves, grow, and produce vegetation, causing development of subacute bacterial endocarditis (SBE) in the patient. At one time a uniformly progressive, fatal disease, endocarditis can now usually be treated effectively with penicillin or other antimicrobials. Viridans streptococci do not have a C carbohydrate antigen in the cell wall, and therefore cannot be classified on a serologic basis.

Diplococci

Streptococcus pneumoniae, the common cause of lobar pneumonia in man, appears as lanceolate-shaped diplococci in which the pair comprises the unit. The organisms are, of course, Gram-positive, but they may be readily decolorized by the use of too much ethanol or acetone in the Gram-staining procedure. Their usual habitat is the upper respiratory tract of healthy carriers. Pneumococci are delicate, die off readily away from the body, and often cannot be subcultured if they are held longer than 2 or 3 days in the incubator. They will not grow on ordinary culture media unless blood is added; in blood broth they grow with a light uniform turbidity. On primary isolation on a blood agar plate, they appear after 1 day as tiny colonies, which enlarge after 2 days to about 1 mm in diameter. The typical appearance is flat and glassy, with a central depression. Growth is much better in a partial carbon dioxide atmosphere, Some strains show a marked tendency to autolyze; if left in the incubator for several days, the entire colony may disappear.

Pneumococci possess a polysaccharide capsule, which accounts for the glassy appearance of the colonies. Certain serotypes (types 3 and 8) have unusally large capsules, a property that is reflected in large colonies. Serological studies on capsular polysaccharide have made it possible to establish a system of classification based on the serologic specificity of the capsular material. As initially developed, there were only a few serotypes; but as time went on, new ones were discovered, until more than 100 serotypes have been described. Prior to the antimicrobial era, serotherapy of pneumococcal infection was successfully used; however, the therapeutic antiserum had to be of the same serotype as the organism that infected the patient.

PNEUMOCOCCAL DISEASE

The pneumococcus is the most common cause of lobar pneumonia, and, as a complication, may produce empyema or pericarditis. It also may cause mastoiditis and otitis media, and is a common agent of bacterial meningitis.

The laboratory diagnosis of pneumococcal infection depends on demonstration of the organism by smear and on confirmation by culture. A smear of body fluid (e.g., spinal fluid) and a Gram stain may be sufficient to show the presence of Gram-positive lanceolate-shaped diplococci. The most specific test that can be done to identify the pneumococcus is the Neufeld quellung reaction. If an encapsulated pneumococcus is mixed with its homologous serum (i.e., antibodies against the specific capsular polysaccharide), a precipitation reaction occurs, and the capsule can then be seen surrounding the organism. Visualization under the microscope is enhanced by adding a drop of methylene blue and a cover slip. Commercially prepared sera are available for the serotyping procedure.

Identification of the culture depends on the appearance of typical flat, glassy colonies on blood agar plates in 1 to 2 days. The blood in the agar is turned green, an effect similar to that produced by *alpha*-hemolytic streptococci. Confirmation tests that may be done are fermentation of inulin and bile solubility tests. The pneumococcus ferments inulin, and the cells are readily lysed by 10% sodium taurocholate. However, a simple test, and one that is easier to interpret, is the cuprein hydrochloride (Optochin®) susceptibility test. This compound is active against pneumococci, but not against *alpha*-hemolytic streptococci. Disks impregnated with cuprein hydrochloride may be placed on a plate that has been freshly streaked with the unknown organism; after incubation, there will be a zone of inhibition around the disk if the organism is a pneumococcus, whereas *alpha*-hemolytic streptococci will not be inhibited. The Neufeld quellung reaction also may be used for identification of young cultures of pneumococci on primary isolation.

Pneumococci are sensitive to many antimicrobial agents. Penicillin is usually the drug of choice, since the pneumococcus is extremely sensitive to this antibiotic.

ENTEROCOCCI

The enterococci, such as *S. faecalis,* are quite different from the other streptococci in many ways. They are so consistently a part of the bowel flora of humans and animals that they have sometimes been used as indicators of fecal pollution in the bacteriological examination of water. Some strains grow so smoothly that it is difficult to demonstrate the presence of chains. Colonies on agar plates are 2 to 3 mm in diameter in 2 days, colorless to gray, and usually butyrous. Hemolytic activity is not critical for identification, but two of the four species are *beta*-hemolytic and the other two are non-hemolytic. Enterococci differ from other human streptococci in their ability to grow at 10°C in broth that contains 6.5% sodium chloride. They have a specific C carbohydrate antigen and are designated as Lancefield group D.

Enterococci have the same capabilities to cause subacute bacterial endocarditis as some of the viridans group organisms, often as a sequela to prostatectomy or bowel surgery. They are penicillin-resistant, but sensitive to some degree to a wide spectrum of other antibiotics. Enterococci are also a cause of cystitis and other urinary-tract infections.

LACTIC STREPTOCOCCI

Streptococcus lactis and *S. cremoris* are found in milk and in milk products. These organisms play an important role in the souring of milk and are often used as starter cultures for the preparation of some cheeses. Some strains of *S. lactis* produce an antibiotic, nisin, that acts against other Gram-positive bacteria.

REFERENCES

1. **Brown, J. H.,** *The Use of Blood Agar for the Study of Streptococci,* Monograph No. 9, Rockefeller Institute for Medical Research, New York, 1919.
2. **Lancefield, R. C.,** A serologic differentiation of human and other groups of hemolytic streptococci, *J. Exp. Med.,* 57, 571, 1933.
3. **Zabriskie, J. B.,** The role of temperate bacteriophage in the production of erythrogenic toxin by group A streptococci, *J. Exp. Med.,* 119, 761, 1964.

ENDOSPORE-FORMING RODS AND COCCI*

H. A. Lechevalier

Generic name (G + C molar %) [Number of species in *Bergey's Manual*]	Morphology	Physiology	Ecology	Examples of species
Bacillus (32 to 62) [48] See pp. 319–336 for further details.	Rods, 0.3–2.2 by 1.2–7 μm; usually motile by peritrichous flagella; endospores, formed one to a cell, may be variously located within the sporangial cell; usually Gram-positive.	Chemoorganotrophs; respiratory or fermentative metabolism; aerobic to facultative anaerobes; usually catalase-positive.	Widely distributed in nature, mainly in soil.	*B. subtilis*. Type species. *B. anthracis*. Causative agent of anthrax. *B. thurigiensis*. Produces a crystalline product that is toxic larvae of *Lepidoptera*.
Sporolactobacillus (39 to 47) [1]	Rods, 0.7–0.8 by 3–5 μm; motile by peritrichous flagella; Gram-positive.	Chemoorganotrophs; fermentative metabolism, producing lactic acid; microaerophilic.	Original isolate was from chicken feed.	*S. inulinus*.
Clostridium (23 to 43) [61] See pp. 337–345 for further details.	Rods, 0.3–1.5 by 1.3–14 μm; usually motile by peritrichous flagella; endospores usually distend the sporangial cell; usually Gram-positive.	Chemoorganotrophs; fermentative metabolism; some species fix nitrogen; do not reduce sulfate; usually strict anaerobes.	Found in soil, in marine and fresh water sediments, and in the intestinal tract of man and animals.	*C. butyricum*. Type species. *C. pasteurianum*. Fixes nitrogen. *C. botulinum*. Found in soil and in food; causes botulism. *C. perfringens*. Found in soil; causes gas gangrene. *C. tetani*. Found in soil; causes tetanus.

* Part 15 in *Bergey's Manual*.

Generic name (G + C molar %) [Number of species in *Bergey's Manual*]	Morphology	Physiology	Ecology	Examples of species
Desulfotomaculum (42 to 46) [3]	Rods, 0.3–1.5 by 3–6 μm, may be curved; motile by peritrichous flagella; endospores cause slight swelling of the sporangial cells; black colonies produced on lactate-sulfate agar containing ferrous salts; Gram-negative.	Chemoorganotrophs; respiratory metabolism; reducible sulfur compounds reduced to H_2S; limited range of substrates utilized; strict anaerobes; tend to form acetate and CO_2.	Found in soil, in fresh water, in the intestines of insects, and in the rumen of ruminants.	*D. nigricans.* Type species.
Sporosarcina (40–43) [1]	Cocci, 1.2–2.5 μm in diameter; occur in tetrads or packets; if motile, by randomly spaced flagella; Gram-positive; endospores 0.8–1.0 μm, born centrally.	Chemoorganotrophs; respiratory metabolism; strict aerobes.	Found in soil, in urine, and in sea water.	*S. ureae.* Converts urea to ammonium carbonate.
Oscillospira (–) [1]	Long rods or filaments, 3–6 by 10–40 μm; divided by closely spaced cross walls; motile by peritrichous flagella; endospores 2.5 by 4 μm, not always formed; Gram-negative.	Have not been grown in pure culture; presumably anaerobic; endospores have not been observed to germinate.	Found in the alimentary tract of herbivorous animals.	*O. guillermondi.*

THE GENUS *BACILLUS**

R. E. Gordon

The rod-shaped bacteria that aerobically form refractile endospores are assigned to the genus *Bacillus.* One spore appears in the spore-bearing cell, although some large rods observed in the intestines of tadpoles, but never cultivated in vitro, were described as forming two spores in the sporangium and assigned to the genus *Bacillus.*[1] The endospores of the bacilli, like those of the clostridia and a few other taxa,[2] are more resistant than the vegetative cells to heat, drying, disinfectants, and other destructive agents, and thus may remain viable for centuries. The basis of the spore's resistance and longevity, its formation, morphology, composition, and stages of germination continue to be subjects of many investigations.[3,4]

With a few exceptions, strains of the genus *Bacillus* form catalase, which, in addition to the aerobic production of spores, distinguishes bacilli from clostridia. The production of catalase also differentiates bacilli from strains of *Sporolactobacillus.*[5] The exceptions, i.e., those strains that produce no catalase or only trace amounts, are strains of *B. larvae, B. lentimorbus, B. popilliae,* and some strains of *B. stearothermophilus.*

The genus *Bacillus* encompasses a great diversity of strains. Some species are strictly aerobic, others are facultatively anaerobic. Strains of *B. polymyxa* fix atmospheric nitrogen.[6-8] Strains of some species grow well in a solution of glucose, ammonium phosphate, and a few mineral salts; others need additional growth factors or amino acids; still others have increasingly complex nutritional requirements.[9,10] Strains of *B. fastidiosus* grow only when uric acid or allantoin is available.[11] Although a pH of 7.0 is suitable for growth of most bacilli, a pH of 9.0 to 10.0 was described[12] as a growth prerequisite for *B. alcalophilus,* and *B. acidocaldarius* was described[13] as growing at pH values of 2.0 to 6.0, with optimal growth at pH 3.0 to 4.0. The bacilli also exhibit great variation in temperatures of growth; some thermophiles grow from a minimum temperature of 45°C to a maximum temperature of 75°C or higher, and some psychrophiles grow at temperatures from −5°C to 25°C. Undoubtedly because of their special requirements for growth, many strains that exist in nature are unknown in the laboratory. On the other hand, because a microorganism has the capacity to adapt to unusual conditions of growth, counterparts of a strain isolated from an unusual source could possibly be found in our culture collections.

The strains of bacilli with ellipsoidal or cylindrical spores that do not appreciably distend the sporangia are generally Gram-positive. Strains with ellipsoidal or round spores that swell the sporangia, however, may be Gram-positive, -variable, or -negative. Strains may be motile or non-motile. Although a few motile strains have been reported as having polar flagella,[14,15] the flagella of the bacilli are generally peritrichous. For a very small number of representative strains the surface configuration of the spores was described as distinctive for some species.[16]

According to the examination of a limited number of strains of various species, the base composition of deoxyribonucleic acid (moles % guanine + cytosine) of the bacilli ranges from 32 to 66.[17,18] A more recent report of an examination of 114 strains in one laboratory disclosed a range of 35 to 57 moles % G + C.[19] Although sporulation and other growth characteristics complicated the techniques of paper chromatography, a study of 75 mesophilic strains indicated that strains might be differentiated specifically by the amino acids of their whole cells.[20] For taxonomic purposes, characteristic constituents of the cell lipids,[21-23] the peptidoglycan types of the cell walls,[24] and the

* Part 15 in *Bergey's Manual.*

presence of hippurate hydrolyase, detected by thin-layer chromatography[25] may also be useful. But the promise of these and other studies of the chemical composition of the cells can only be realized by comparative examination of many strains. In some species, spore antigens and bacteriophages have provided information that correlates with other properties, but much more work is required before full applications can be determined.[26,27]

Because the bacteria of the genus *Bacillus* are widely distributed in soil, water, and air, and because their spores are so resistant, their control is of great economic concern in the food processing industry and in the preparation of all sterile products. Although *B. anthracis,* which played a memorable role in the history of microbiology, is generally regarded as the only species of the genus that is virulent for man and animals, evidence of infections due to strains of species hitherto accepted as "nonpathogenic" is accumulating.[28-39] At least five species of the genus (*B. thuringiensis, B. larvae, B. popilliae, B. lentimorbus,* and *B. sphaericus*) are pathogenic for insects. *Bacillus thuringiensis* insecticides, effective against many pests of agricultural crops, forests, and stored food, are produced commercially in several countries.[40]

DESCRIPTIONS OF SPECIES

The concept of a microbial species is admittedly man-made, imperfect, and difficult to define.[41] Nevertheless, the specific naming of groups of strains with certain similarities does provide a means of communication among microbiologists. Each name points to a description or definition. Because the rules of nomenclature affix the name of a species to the strain first named and described, old strains as well as freshly isolated ones must be examined, and their more stable properties (properties that persist after years of cultivation in vitro) must be incorporated into the species' descriptions. Microorganisms possess many characteristics that are of varying degrees of stability. Therefore, to establish the reliability of a characteristic for describing a species, many strains must be examined.

The most intensively studied species of the genus *Bacillus* are those that are important in medicine or industry and those that are abundantly represented in nature and easily cultivated in vitro. As a result, knowledge of the species has accumulated unevenly, and much more work on comparative description of the species remains to be done. With some small modifications in nomenclature, the species presented here are those recognized in the 8th edition of *Bergey's Manual of Determinative Bacteriology.* Histories of the strains and descriptions of media and methods for the determination of properties listed in Tables 1 through 6 and in Keys 1 and 2 are available elsewhere[27,42] (also see Appendix). The species are arranged in three groups according to shape of the spore and swelling of the sporangium by the spore;[27] the first group is subdivided by diameter of the rod and appearance of its protoplasm (Key 2).

The spores of strains of the species assigned to Group 1 are generally ellipsoidal or cylindrical and do not appreciably distend the sporangia. The cells of *B. megaterium* and *B. cereus* (Table 1) are usually wider than the cells of *B. licheniformis, B. subtilis, B. pumilus, B. firmus,* and *B. coagulans* (Table 2). In addition, the cells of *B. megaterium* and *B. cereus* (Figures 1 and 2), when grown on glucose agar (18 to 24 hr) and lightly stained, are filled with unstained globules, whereas the cells of other species are not (Figure 3).

Of the properties listed in Table 1, anaerobic growth, Voges-Proskauer and egg yolk reactions, resistance to lysozyme, and acid production from mannitol are the most useful in separating *B. megaterium* and *B. cereus.* Some characteristics used by other investigators[10,26,43] to delineate these two species are growth in an inorganic ammonium basal solution with glucose, the methyl-red test, and production of urease and

FIGURE 1. *Bacillus megaterium.* A free spore, cells grown on glucose agar, and a sporangium.

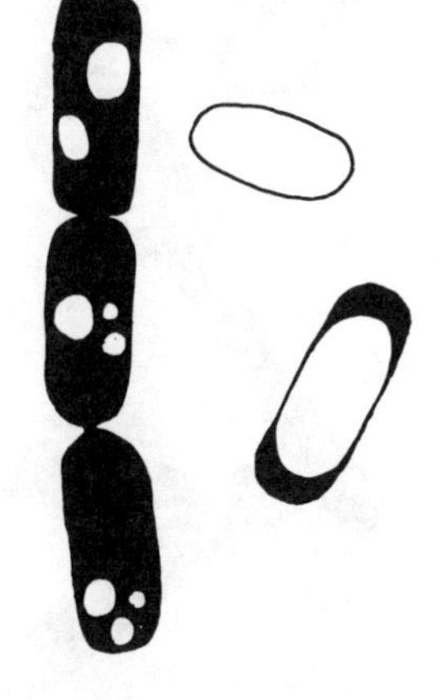

FIGURE 2. *Bacillus cereus.* Cells grown on glucose agar, a free spore, and a sporangium.

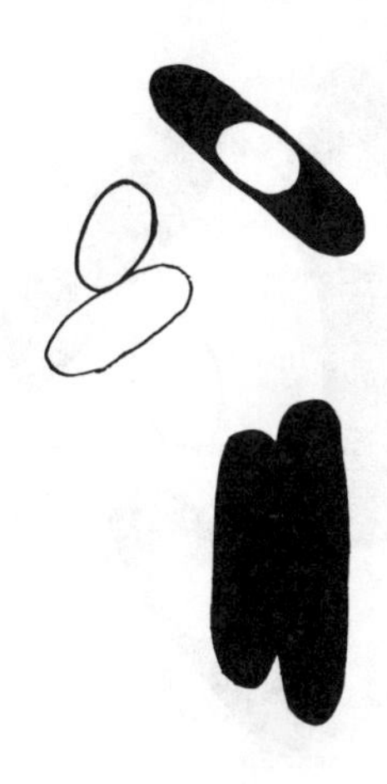

FIGURE 3. *Bacillus subtilis.* A sporangium, free spores, and cells grown on glucose agar.

of acid from raffinose and inulin. An egg yolk and polymyxin medium and a fluorescent-antibody procedure were proposed for presumptive identification in foods of *B. cereus,* a recognized cause of food poisoning.[44,45]

The assignment of *B. anthracis* and *B. thuringiensis* to varietal status (Table 1) is controversial. We undertook this assignment for the following reasons: (1) our data[27,42] do not provide any stable correlating properties separating *B. anthracis, B. mycoides,* and *B. thuringiensis* from *B. cereus;* (2) data of other investigators support our contention that avirulent strains of *B. anthracis* and *B. thuringiensis* are indistinguishable from strains of *B. cereus;*[17,46-50] and (3) we believe that a strain should always bear the same species name, despite its loss of unstable properties. In the laboratories of human and veterinary medicine and of insect pathology, however, recognition of virulent strains is admittedly far more important than problems of nomenclature. Criteria for the identification of virulent strains are plentiful.[26,46,51-53] Although the rhizoid strains of *B. mycoides* are currently regarded as unworthy of varietal status,[43,54] in Table 1 the rhizoid strains are presented as a varietal group to demonstrate the similarity between their characteristics and those of the anthrax bacilli.

The strains of the *B. subtilis* group, or the *B. subtilis* spectrum (*B. licheniformis, B. subtilis,* and *B. pumilus*), are morphologically similar and share many physiological properties. In addition to the properties of these three species listed in Table 2, decomposition of chitin and amygdaline, spore antigens, the anaerobic production of gas from nitrate, arginine dihydrolase, reduction of nitrite, and the utilization of acetate and tartrate were reported as useful for the separation of *B. licheniformis* and *B. subtilis.*[55-59] The third member of the *B. subtilis* group, *B. pumilus,* is admittedly very similar to *B. subtilis* and has been given varietal status.[26] However, in addition to the characteristics that separate *B. pumilus* from *B. subtilis* given in Table 2, examination of the nutritional requirements of 21 strains of *B. pumilus* disclosed that all strains required biotin for growth in a solution of glucose and ammonium phosphate, whereas 26 of 27 strains of *B. subtilis* and 12 strains of *B. licheniformis* grew without biotin.[10] The presence of urease has also been used to differentiate between *B. subtilis* and *B. pumilus.*[43]

Despite all attempts to establish the identity of *B. subtilis,* the type species of the genus,[58-61] a Michigan strain of *B. cereus* bearing the label *B. subtilis* is still extant – a

FIGURE 4. *Bacillus coagulans.* Sporangia, a cell, and a free spore.

FIGURE 5. *Bacillus circulans.* A free spore, sporangium, and cells.

FIGURE 6. *Bacillus alvei.* Spores and cells.

warning to all microbiologists against acceptance of a tube's label as identification of the culture.

Although morphologically similar to strains of the *B. subtilis* group, strains of *B. firmus* are distinguishable from this group by their negative V-P reaction, their inability to utilize citrate, and their sensitivity to acid (Table 2). Cultures of *B. firmus* form acid from glucose, but are so sensitive to acid that growth ceases at pH 6.0 or above. Strains of *B. lentus* have these same characteristics, but they have been differentiated from strains of *B. firmus*[62] by their failure to decompose casein and gelatin and by their ability to produce urease. In a study of their nutritional requirements,[9] 20 strains of *B. firmus* and 6 strains of *B. lentus* showed a gradation in their requirements; the most exacting strains of *B. firmus* and the least exacting strains of *B. lentus* required the same growth factors. Since both *B. firmus* and *B. lentus* are poorly represented in culture collections, examination of many more strains of each is needed.

In contrast to *B. firmus,* strains of *B. coagulans* (Table 2) are aciduric and produce a low pH (4.0 to 5.0) in media containing utilizable carbohydrates. Spoilage of acid foods, such as canned tomatoes, is usually caused by *B. coagulans.* Strains of *B. coagulans* are morphologically variable. The rods of some strains resemble those of the *B. subtilis* group, whereas those of other strains are longer and slender, and the sporangia may or may not distend the spores (Figure 4). In the presentation of species in Key 2, therefore, *B. coagulans* is intermediate between Groups 1 and 2. Strains of *B. coagulans* also vary in their nutritional requirements, depending on strain, temperature of incubation, and basal medium.[63-65] Separation of *B. coagulans* into two types (Type A with a negative V-P reaction and growth at 65°C, and Type B with a positive V-P reaction and no growth at 65°C)[26] is confirmed. The strains of *B. coagulans* described in Key 1, Tables 2 and 4, belong to Type B. *Bacillus fastidiosus* is one of the two recognized species[54] whose special requirements for growth exclude them from many comparative observations with other recognized species. Because of the limitation of *B. fastidiosus* to media containing uric acid or allantoin, data on this species are not included in Tables 1 to 6. The rods of *B. fastidiosus* are described[54] as large (1.5 to 2.5 by 3 to 6μm), motile, and Gram-positive in the early stages of growth; the spores are oval to cylindrical, terminal or subterminal, and do not distend the sporangia appreciably. Cultures grown on uric acid agar are aerobic, catalase-positive, and mesophilic.

Among the species whose sporangia are swollen by ellipsoidal spores (Figure 5) and whose strains hydrolyze starch (Table 3), *B. polymyxa* is most easily distinguished. Its spores generally have heavily ribbed surfaces; the ribs are longitudinal and parallel; and cross sections of the spores are star-shaped.[16,66,67] The presence of lipase, absence of chitinase, spore antigens, and bacteriophage have been successfully used for identification of its strains.[55,68,69] If all the species of the genus *Bacillus* were as distinctive as *B. polymyxa,* the taxonomy of the bacilli would be further advanced.

Although *B. macerans* resembles *B. polymyxa* in some of its properties, including the morphology of its spores[16] and its active formation of gas from carbohydrates, it is easily separated from *B. polymyxa* (Table 3). At present, however, only gas production and, less satisfactorily, the formation of crystalline dextrins divide *B. macerans* from *B. circulans.* The latter species, aptly labeled a complex rather than a species,[70] encompasses a group of morphologically, nutritionally, physiologically, and chemically heterogeneous strains. The heterogeneity of the complex is well illustrated by an investigation of 136 strains of *B. circulans* that divided them into two main groups and five subgroups on the basis of their spore precipitinogens and of some of their biochemical properties.[71]

Bacillus stearothermophilus, a thermophilic species, is also heterogeneous, as demonstrated by the separation of 230 strains into three principal groups and ten subgroups by their spore agglutinogens and by some of their physiological characteristics.[72] The spores formed by a culture of *B. stearothermophilus* may vary in size. However, in cultures grown on soil extract agar[42] at 45 to 55°C for 3 days, spores with diameters wider than the rods predominated. *Bacillus stearothermophilus* is, therefore, assigned to Group 2, Key 2.

In an arrangement (Table 4) of some species according to temperatures of growth of representative strains,[27] *B. stearothermophilus,* the cause of "flat-sours" in canned foods, and *B. coagulans* grow at the higher temperatures. The division between *B. stearothermophilus* and *B. coagulans* is not as clear-cut as the data in Tables 2 to 4 indicate, because of the strains assigned to Type A of *B. coagulans.*[26]

The nomenclatural type strain of *B. alvei,* named and described in 1885, was isolated from foul brood of the honeybee, and for a time *B. alvei* was mistakenly thought to be the cause of the disease.[73] Smears of cultures grown on agar usually show the spores arranged side by side in rows (Figure 6). Typical strains of *B. alvei,* as well as some strains of *B. circulans,* are actively motile and may form motile colonies on agar. Although strains of *B. alvei* have been isolated from soil as well as in association with diseases of honeybees, their numbers in culture collections are usually small. Not only are the comparative data on *B. alvei* (Table 3) based on too few strains, but the occurrence of strains seemingly intermediate between *B. alvei* and *B. circulans* makes the reliability of the species description of *B. alvei* uncertain.

Bacillus laterosporus (Table 5) is also sparsely represented in culture collections. A distinctive morphological characteristic of this species is a canoe-shaped[74] or C-shaped[16] parasporal body attached to the spore, with resulting lateral position of the spore in the sporangium. The parasporal body is easily stained and can be seen on the spore after lysis of the sporangium (Figure 7).

In the early 1940s, gramicidin, one of the first antibiotics, was isolated from strains of *B. brevis.* The strains of this species usually form ellipsoidal central spores that distend the sporangia into spindle shapes; they resemble strains of *B. laterosporus* in their inability to hydrolyze starch (Table 5). When only inorganic nitrogen is available, cultures of *B. brevis* (86% of 92 strains) form acid from glucose (under the conditions of the test); when organic nitrogen is present, production of ammonia masks any action on glucose.

The strains of *B. larvae, B. popilliae,* and *B. lentimorbus* are catalase-negative, grow slowly, and require special media[42] for growth and observation (Table 6). Cultures of *B. larvae,* the cause of American foul brood of the honeybee, grow and sporulate satisfactorily in vitro. Cultures of *B. popilliae* and *B. lentimorbus,* causal agents of milky

FIGURE 7. *Bacillus laterosporus.* Spores and a sporangium.

FIGURE 8. *Bacillus sphaericus.* A cell and spores.

diseases in the larvae of the Japanese beetle and European chafer, can be maintained indefinitely by serial transfer in vitro.[75] Infective spores (after ingestion), abundantly produced in the larvae, have not, however, been produced in vitro. Insecticides of *B. popilliae* and *B. lentimorbus,* therefore, are not economically feasible.

The spores of *B. sphaericus* are generally round (Figure 8) and terminal, resulting in drumstick-like sporangia (Group 3, Key 2). Examination of 21 strains[10] disclosed that 19 strains grew in a casein basal medium plus thiamine (6 strains) or plus thiamine and biotin (13 strains). Growth is not enhanced by urea or ammonia. Among the round-spored species, *B. sphaericus* is generally the best represented species in culture collections, and the strains of *B. sphaericus* are quite readily recognized (Table 5). However, since other round-spored strains, including psychrophilic strains,[76] have been reported[54] as differing from *B. sphaericus* in one or more of the distinguishing properties of the species (Table 5), further investigation of more strains may reveal another taxonomically difficult complex.

Bacillus pasteurii is the second recognized species[54] that requires special conditions for growth. The cultures are described as requiring alkaline media containing ammonia (approximately 1% NH_4Cl) or urea (1%). The author of the species[77] characterized freshly isolated strains as decomposing 3 g of urea per liter per hour during their period of maximal growth. The rods of *B. pasteurii* are reported[54] as measuring 0.5 to 1.2 by 1.3 to 4μm, motile, and Gram-positive. Spores are generally spherical, terminal or subterminal, and swell the sporangia. With the exception of the inability to hydrolyze starch, the physiological properties of the strains are described as variable.

In addition to the species recognized here, many other species names appear in the literature and in our culture collections. Representative strains of a large number of these names have either been lost, found to belong to previously described and named species (see Indexes 1 and 2 in Reference 42 for species names in the synonymy of earlier established species), are believed to be unlike the original strains of their respective species (misnamed), or are temporarily listed as species *incertae sedis* (Table 7). This designation does not imply that the strains bearing these names do not represent distinct species. Some of them undoubtedly do, but the number of available representative strains is insufficient to establish their stable, reliable characteristics. In other instances, time has not permitted a comparison of the representative strains with significant numbers of other *Bacillus* species. Information on the comparative examination of strains of some of the species *incertae sedis* (Table 7) is available.[42] The reliability of each characteristic for species description, however, is not established. Despite continued attempts since 1936 to obtain newly described strains and strains bearing species names not represented in our collection, there are unquestionably some we do not have. We should gratefully welcome any representative strains of other *Bacillus* species.

IDENTIFICATION OF STRAINS

For the identification of strains of the genus *Bacillus,* the necessity to compare the unknown strains with a goodly number of named strains of the genus cannot be overstressed. No matter how carefully media and methods are described, the results of a test can be affected by some inadvertently omitted step in technique, by changes in the composition of media, by regional differences in the water supply, and by other factors. By subjecting known strains to a specific test, the investigator may quickly learn whether the test as conducted in his laboratory gives results comparable to those found in the literature. If the results are comparable, the test may be applied to unknown strains; if they are not comparable, the test may be discarded, modified, or replaced by another, provided the usefulness of the modified or new test is established by application to a large number of reference strains. If the investigator cannot spend the time necessary for making a comparison of his unknown strains with available reference strains, he should seek the assistance of someone having a background of experience with the taxon.

For the microbiologist who wishes to acquire a background of knowledge about the taxonomy of the genus *Bacillus,* a mechanical key (Key 1) is presented, to aid in tentatively identifying strains. Although on a much smaller scale, this key resembles *A Guide to the Identification of the General of Bacteria*[78] and may be similarly used. In common with other taxonomic keys, Key 1 applies only to the more typical strains of each species.

A more conventional key (Key 2) for the tentative identification of unknown strains is also presented. Key 2 contains more morphological and physiological criteria than Key 1, and the species are arranged in approximately the same order as in the comparative Tables 1 through 6. After tentative identification of an unknown strain as, for example, *B. licheniformis,* observations of the significant properties listed for *B. licheniformis* in Table 2 should then be made. Known reference strains should be used as both positive and negative controls of each observation. The results of these observations would then confirm or disprove the tentative identification of the unknown strain.

Although there undoubtedly remain representative strains of new species to be isolated, described, and named, there are a great many strains belonging to previously established species that were first characterized and named as "new species," a fact that workers should constantly keep in mind.

APPENDIX

The media and methods used in preparing the following keys and tables are given in Reference 42 with the following modification and additions.

For determining resistance to lysozyme, 0.1 g of lysozyme (6000 to 10,000 units/mg, from Nutritional Biochemicals Corp., Cleveland, Ohio) was dissolved in 100 ml of sterile distilled water in a 100-ml volumetric flask, and the solution was sterilized by filtration; 1 ml of the lysozyme solution was mixed with 99 ml of sterile nutrient broth and dispensed in 2.5-ml amounts into sterile plugged tubes. A tube of the 0.001% lysozyme and a tube of nutrient broth were inoculated with a loopful (small loop) of a broth culture. Growth in the two tubes was recorded at 7 days and at 14 days. For the fastidious insect pathogens, 1 ml of the 0.1% lysozyme solution was added to 99 ml of sterile semisolid J-agar. A tube of the 0.001% solution of lysozyme and a tube of semisolid J-agar without lysozyme were inoculated with two or three drops of a culture in semisolid J-agar. Growth was recorded at 7 days and at 14 days.

Growth in ammonia basal medium plus glucose (3%, w/v) was demonstrated according to the method described in Reference 10. Cultures that grew in ten serial transfers in the ammonia basal medium with glucose were accepted as capable of utilizing ammonia nitrogen.

Liquefaction of nutrient gelatin, used in the separation of strains of *B. licheniformis* from strains of *B. subtilis,* was determined in the following medium: beef extract, 3 g; peptone, 5 g; gelatin, 120 g; distilled water, 1000 ml; pH 7.0. The gelatin was slowly suspended in cold water, then heated and dissolved. After addition of the other ingredients and adjustment of the pH, the medium was tubed and sterilized by autoclaving. Stab cultures were made in the gelatin, hardened by cooling, and incubated at 20°C. The amount of liquefaction was measured after 2 weeks of incubation.

Key 1
TENTATIVE IDENTIFICATION OF TYPICAL STRAINS OF *BACILLUS* SPECIES

Numbers on the right indicate the number (on the left) of the next test to be applied until the right-hand number is replaced by a species name.

1. Catalase: positive → 2
   negative → 16

2. Voges-Proskauer: positive → 3
   negative → 9

3. Growth in anaerobic agar: positive → 4
   negative → 8

4. Growth at 50°C: positive → 5
   negative → 6

5. Growth in 7% NaCl: positive . . . *B. licheniformis*
   negative . . . *B. coagulans*

6. Acid and gas from glucose (inorganic N_2): positive . . . *B. polymyxa*
   negative → 7

7. Growth at pH 5.7: positive . . . *B. cereus*
   negative . . . *B. alvei*

8. Hydrolysis of starch: positive . . . *B. subtilis*
   negative . . . *B. pumilus*

9. Growth at 65°C: positive . . . *B. stearothermophilus*
   negative → 10

10. Hydrolysis of starch: positive → 11
    negative → 14

11. Acid and gas from glucose (inorganic N_2): positive . . . *B. macerans*
    negative → 12

12. Utilization of citrate: positive . . . *B. megaterium*
    negative → 13

13. pH in V-P broth <6.0: positive . . . *B. circulans*
    negative . . . *B. firmus*

14. Growth in anaerobic agar: positive . . . *B. laterosporus*
    negative → 15

15. Acid from glucose (inorganic N_2): positive . . . *B. brevis*
    negative . . . *B. sphaericus*

16. Growth at 65°C: positive . . . *B. stearothermophilus*
    negative → 17

17. Decomposition of casein: positive . . . *B. larvae*
    negative → 18

18. Parasporal body in sporangium: positive . . . *B. popilliae*
    negative . . . *B. lentimorbus*

Key 2
TENTATIVE IDENTIFICATION OF TYPICAL STRAINS OF *BACILLUS* SPECIES

Group 1

Sporangia not definitely swollen; spores ellipsoidal or cylindrical, central to terminal; Gram-positive

A. Unstained globules demonstrable in protoplasm of lightly stained cells grown on glucose agar
- 1. Strictly aerobic; acetoin not produced: *B. megaterium*
- 2. Facultatively anaerobic; acetoin produced: *B. cereus*
  - a. Pathogenic to insects: *B. cereus* var. *thuringiensis*
  - b. Rhizoid growth: *B. cereus* var. *mycoides*
  - c. Causative agent of anthrax: *B. cereus* var. *anthracis*

B. Unstained globules not demonstrable in protoplasm of lightly stained cells grown on glucose agar
- 1. Growth in 7% NaCl; acid not produced in litmus milk
  - a. Growth at pH 5.7; acetoin produced
    - (1) Starch hydrolyzed; nitrates reduced to nitrites
      - (a) Facultatively anaerobic; propionate utilized: *B. licheniformis*
      - (b) Aerobic;[a] propionate not utilized: *B. subtilis*
    - (2) Starch not hydrolyzed; nitrates not reduced to nitrites: *B. pumilus*
  - b. No growth at pH 5.7; acetoin not produced: *B. firmus*
- 2. No growth in 7% NaCl; acid produced in litmus milk: *B. coagulans*[b]

Group 2

Sporangia swollen by ellipsoidal spores; spores central to terminal; Gram-positive, -negative, or -variable

A. Gas formed from carbohydrates
- 1. Acetoin produced; dihydroxyacetone formed from glycerol: *B. polymyxa*
- 2. Acetoin not produced; dihydroxyacetone not formed: *B. macerans*

B. Gas not formed from carbohydrates
- 1. Starch hydrolyzed
  - a. Indole not formed
    - (1) No growth at 65°C: *B. circulans*
    - (2) Growth at 65°C: *B. stearothermophilus*
  - b. Indole formed: *B. alvei*
- 2. Starch not hydrolyzed
  - a. Catalase-positive; survives serial transfer in nutrient broth
    - (1) Facultatively anaerobic; pH of cultures in glucose broth less than 8.0: *B. laterosporus*
    - (2) Aerobic; pH of cultures in glucose broth 8.0 or higher: *B. brevis*
  - b. Catalase-negative; fails to survive serial transfer in nutrient broth
    - (1) Nitrates reduced to nitrites; casein decomposed: *B. larvae*
    - (2) Nitrates not reduced to nitrites; casein not decomposed
      - (a) Sporangium contains a parasporal body; growth in 2% NaCl: *B. popilliae*
      - (b) Sporangium does not contain a parasporal body; no growth in 2% NaCl: *B. lentimorbus*

Group 3

Sporangia swollen; spores generally spherical, terminal to subterminal; Gram-positive, -negative, or -variable

A. Starch not hydrolyzed; urea or alkaline pH not required for growth: *B. sphaericus*

[a] Do not growth anaerobically in BBL anaerobic agar without glucose or Eh indicator.

[b] Swelling of the sporangia by the spores is a variable property of strains of this species and makes the species intermediate between Group 1 and Group 2.

Table 1
COMPARISON OF *BACILLUS* SPECIES

[Group 1A, Key 2]

Property	*B. megaterium*	*B. cereus*	*B. cereus* var. *thuringiensis*	*B. cereus* var. *mycoides*	*B. cereus* var. *anthracis*
Rods					
width, μm	1.2–1.5	1.0–1.2	1.0–1.2	1.0–1.2	1.0–1.2
length, μm	2–5	3–5	3–5	3–5	3–5
Gram reaction	+	+	+	+	+
Unstained globules in the protoplasm	+	+	+	+	+
Spores					
ellipsoidal	+	+	+	+	+
round	v	–	–	–	–
central or paracentral	+	+	+	+	+
swelling the sporangium	–	–	–	–	–
Crystalline parasporal bodies	–	–	a	–	–
Motility	a	a	a	–	–
Catalase	+	+	+	+	+
Anaerobic growth	–	+	+	+	+
V-P reaction	–	+	+	+	+
pH in V-P broth	4.5–6.8	4.3–5.6	4.3–5.6	4.5–5.6	5.0–5.6
Temperature of growth, °C					
maximum	35–45	35–45	40–45	35–40	40
minimum[a]	3–20	10–20	10–15	10–15	15–20
Egg yolk reaction	–	+	+	+	+
Growth in					
0.001% lysozyme	–	+	+	+	+
7% NaCl	+	+	+	a	+
media at pH 5.7	+	+	+	+	+
ammonia glucose medium	+	–	–	–	–
Acid from					
glucose	+	+	+	+	+
arabinose	a	–	–	–	–
xylose	a	–	–	–	–
mannitol	+	–	–	–	–
Hydrolysis of starch	+	+	+	+	+
Use of citrate	+	+	+	a	b
Reduction of NO_3 to NO_2	b	+	+	+	+
Deamination of phenylalanine, 1 week	a	–	–	–	–
Decomposition of					
casein	+	+	+	+	+
tyrosine	a	+	+	a	–

[a] The lowest temperature tested was 3°C.

Symbol code: + = 85 to 100% of the strains positive; a = 50 to 84% of the strains positive; b = 15 to 49% of the strains positive; – = 0 to 14% of the strains positive; v = character inconstant.

Table 2
COMPARISON OF *BACILLUS* SPECIES

[Group 1B, Key 2]

Property	*B. licheniformis*	*B. subtilis*	*B. pumilus*	*B. firmus*	*B. coagulans*
Rods					
width, μm	0.6–0.8	0.7–0.8	0.6–0.7	0.6–0.9	0.6–1
length, μm	1.5–3	2–3	2–3	1.2–4	2.5–5
Gram reaction	+	+	+	+	+
Unstained globules in protoplasm	–	–	–	–	–
Spores					
ellipsoidal or cylindrical	+	+	+	+	+
central or paracentral	+	+	+	v	v
subterminal or terminal	–	–	–	v	v
swelling the sporangium	–	–	–	–	v
Motility	+	+	+	a	+
Catalase	+	+	+	+	+
Anaerobic growth	+	–	–	–	+
V-P reaction	+	+	+	–	a
pH in broth					
V-P	5.0–6.5	5.4–8.0	4.8–5.5	6.0–6.8	4.2–4.8
anaerobic glucose	5.0–5.6	6.2–8.2			
Temperature of growth, °C					
maximum	50–55	45–55	45–50	40–45	55–60
minimum	15	5–20	5–15	5–20	15–25
Egg yolk reaction	–	–	–	–	–
Growth in					
0.001% lysozyme	–	b	a	–	–
media at pH 5.7	+	+	+	–	+
7% NaCl	+	+	+	+	–
0.02% azide[a]	–	–			+
Acid from					
glucose	+	+	+	+	+
arabinose	+	+	+	b	a
xylose	+	+	+	b	a
mannitol	+	+	+	+	b
Hydrolysis of					
starch	+	+	–	+	+
hippurate, 4 weeks	–	–	+		
Use of					
citrate	+	+	+	–	b
propionate	+	–	–	–	–
Reduction of NO_3 to NO_2	+	+	–	+	b
Decomposition of					
casein	+	+	+	+	b
tyrosine	–	–	–	b	–
Liquefaction of nutrient gelatin, 20°C, 2 weeks	<1.0 cm	1.0 cm or more			

[a] Only strains growing at 55°C or higher were tested.

Symbol code: + = 85 to 100% of the strains positive; a = 50 to 84% of the strains positive; b = 15 to 49% of the strains positive; – = 0 to 14% of the strains positive; v = character inconstant.

Table 3
COMPARISON OF *BACILLUS* SPECIES

[Groups 2A and 2B(1), Key 2]

Property	*B. polymyxa*	*B. macerans*	*B. circulans*	*B. stearothermophilus*	*B. alvei*
Rods					
width, μm	0.6–0.8	0.5–0.7	0.5–0.7	0.6–1.0	0.5–0.8
length, μm	2–5	2.5–5	2–5	2–3.5	2–5
Gram reaction	v	v	v	v	v
Spores					
ellipsoidal	+	+	+	+	+
central or paracentral	v	–	v	–	v
subterminal or terminal	v	+	v	+	v
swelling the sporangium	+	+	+	+	+
Motility	+	+	a	+	+
Catalase	+	+	+	a	+
Anaerobic growth	+	+	a	–	+
V-P reaction	+	–	–	–	+
pH in V-P broth	4.5–6.8	4.5–5.0	4.5–6.6	4.8–5.8	4.6–5.2
Temperature of growth, °C					
maximum	35–45	40–50	35–50	65–75	35–45
minimum	5–10	5–20	5–20	30–45	15–20
Growth in					
0.001% lysozyme	a	–	b	–	+
media at pH 5 7	+	+	b	–	–
0.02% azide				–	
5% NaCl	–	–	a	b	b
10% NaCl	–	–	–	–	–
Acid from					
glucose	+	+	+	+	+
arabinose	+	+	+	b	–
xylose	+	+	+	a	–
mannitol	+	+	+	b	–
Gas from fermented carbohydrates	+	+	–	–	–
Hydrolysis of starch	+	+	+	+	+
Use of citrate	–	b	b	–	–
Reduction of NO_3 to NO_2	+	+	b	a	–
Formation of					
crystalline dextrins	–	+	–		
dihydroxyacetone	+	–	–	–	+
indole	–	–	–	–	+
Decomposition of					
casein	+	–	b	a	+
tyrosine	–	–	–	–	b

Symbol code: + = 85 to 100% of the strains positive; a = 50 to 84% of the strains positive; b = 15 to 49% of the strains positive; – = 0 to 14% of the strains positive; v = character inconstant.

Table 4
GROWTH TEMPERATURES OF SEVERAL SPECIES OF *BACILLUS*

Species	Number of strains	Number of strains growing at								
		28°C	33°C	37°C	45°C	50°C	55°C	60°C	65°C	70°C
B. stearo-thermophilus	87	0	10	73	81	87	87	87	87	45
B. coagulans	73	53	73	73	73	72	66	23	0	
B. subtilis	154	154	154	154	150	105	17	0		
B. brevis	57	57	57	57	38	16	7	0		
B. circulans	55	55	55	51	18	6	1	0		
B. pumilus	65	65	65	65	64	43	0			
B. macerans	13	13	13	13	13	9	0			
B. cereus	50	50	50	50	23	0				
B. sphaericus	42	42	42	42	15	0				

Table 5
COMPARISON OF *BACILLUS* SPECIES

[Groups 2B(2a) and 3, Key 2]

Property	*B. laterosporus*	*B. brevis*	*B. sphaericus*
Rods			
width, μm	0.5–0.8	0.6–0.9	0.6–1
length, μm	2–5	1.5–4	1.5–5
Gram reaction	v	v	v
Spores			
ellipsoidal	+	+	–
round	–	–	+
central or paracentral	+	v	–
subterminal or terminal	–	v	+
with "C"-shaped rims	+	–	–
swelling the sporangium	+	+	+
Motility	+	+	+
Catalase	+	+	+
Anaerobic growth	+	–	–
V-P reaction	–	–	–
pH in V-P broth	5.0–6.0	8.0–8.6	7.4–8.6
Temperature of growth, °C			
maximum	35–50	40–60	30–45
minimum	15–20	10–35	5–15
Growth in			
0.001% lysozyme	+	b	a
media at pH 5.7	–	b	b
5% NaCl	a	–	+
10% NaCl	–	–	–
0.02% azide		–	
Acid from			
glucose	+	+	–
arabinose	–	–	–
xylose	–	–	–
mannitol	+	a	–
Hydrolysis of starch	–	–	–
Use of citrate	–	b	b

Table 5 (continued)
COMPARISON OF *BACILLUS* SPECIES

[Groups 2B(2a) and 3, Key 2]

Property	*B. laterosporus*	*B. brevis*	*B. sphaericus*
Reduction of NO_3 to NO_2	+	a	–
Formation of			
dihydroxyacetone	–	–	–
indole	a	–	–
Deamination of phenylalanine, 3 weeks	–	–	+
Decomposition of			
casein	+	+	a
tyrosine	+	+	–

Symbol code: + = 85 to 100% of the strains positive; a = 50 to 84% of the strains positive; b = 15 to 49% of the strains positive; – = 0 to 14% of the strains positive; v = character inconstant.

Table 6
COMPARISON OF *BACILLUS* SPECIES

[Group 2B(2b), Key 2]

Property	*B. larvae*	*B. popilliae*	*B. lentimorbus*
Rods			
width, μm	0.5–0.6	0.5–0.8	0.5–0.7
length, μm	1.5–6	1.3–5.2	1.8–7
Gram reaction	+	–	–
Spores			
ellipsoidal	+	+	+
round	–	–	–
central or paracentral	v	v	v
subterminal or terminal	v	v	v
swelling the sporangium	+	+	+
Parasporal body in sporangium		+	–
Motility	a	a	–
Catalase	–	–	–
Anaerobic growth	+	+	+
V-P reaction	–	–	–
pH in V-P broth	5.5–6.2	5.7–6.2	5.9–6.9
Temperature of growth, °C			
maximum	40	35	35
minimum	25	20	20
Growth in			
0.001% lysozyme	+	+	+
media at pH 5.7	–	–	–
nutrient broth	–	–	–
2% NaCl	+	+	–
Acid from			
glucose	+	+	+
trehalose	+	+	+

Table 6 (continued)
COMPARISON OF *BACILLUS* SPECIES

[Group 2B(2b), Key 2]

Property	*B. larvae*	*B. popilliae*	*B. lentimorbus*
arabinose	–	–	–
xylose	–	–	–
mannitol	b	–	–
Hydrolysis of starch	–	–	–
Use of citrate	–	–	–
Reduction of NO_3 to NO_2	a	–	–
Formation of			
dihydroxyacetone	–	–	–
indole	–	–	–
Deamination of phenylalanine, 3 weeks	–	–	–
Decomposition of			
casein	+	–	–
gelatin	+	–	–
tyrosine	–	–	–

Symbol code: + = 85 to 100% of the strains positive; a = 50 to 84% of the strains positive; b = 15 to 49% of the strains positive; – = 0 to 14% of the strains positive; v = character inconstant.

Table 7
SPECIES *INCERTAE SEDIS* OF THE GENUS *BACILLUS*

B. acidocaldarius Darland and Brock
B. alcalophilus Vedder
B. aminovorans den Dooren de Jong
B. aneurinolyticus Aoyama
B. anthracoides Flügge
B. apiarius Katznelson
B. badius Batchelor
B. cirroflagellosus ZoBell and Upham
B. cubensis Stührk
B. epiphytus ZoBell and Upham
B. filicolonicus ZoBell and Upham
B. freudenreichii (Miquel) Chester
B. globisporus Larkin and Stokes
B. insolitus Larkin and Stokes
B. laevolacticus Nakayama and Yanoshi
B. lentus Gibson
B. lubinskii Kruse
B. macquariensis Marshall and Ohye
B. macroides Bennett and Canale-Parola
B. maroccanus Delaporte and Sasson
B. medusa Delaporte
B. pacificus Delaporte
B. pantothenticus Proom and Knight
B. psychrophilus Larkin and Stokes
B. psychrosaccharolyticus Larkin and Stokes
B. pulvifaciens Katznelson
B. racemilacticus Nakayama and Yanoshi
B. rarerepertus Schieblich
B. thiaminolyticus Kuno
B. virgula Trevisan
Krusella cascainesis Castellani

REFERENCES

1. **Delaporte, B.,** *Ann. Inst. Pasteur Paris,* 107, 845, 1964.
2. **Cross, T.,** *J. Appl. Bacteriol.,* 33, 95, 1970.
3. **Halvorson, H. O., Hanson, R., and Campbell, L. L. (Eds.),** *Fifth International Spore Conference, Fontana, Wisconsin, 1971,* American Society for Microbiology, Washington, D.C., 1972.
4. **Gould, G. W. and Hurst, A. (Eds.),** *The Bacterial Spore,* Academic Press, London and New York, 1969.
5. **Kitahara, K. and Suzuki, J.,** *J. Gen. Appl. Microbiol.,* 9, 59, 1963.
6. **Kalininskaya, T. A.,** *Mikrobiologiya* (USSR), 37, 923, 1968.
7. **Grau, F. H. and Wilson, P. W.,** *J. Bacteriol.,* 83, 490, 1962.
8. **Bredemann, G.,** *Zentralbl. Bakteriol. Parasitenkd. Infektionskr. Hyg. Abt. 2,* 22, 44, 1909.

9. **Proom, H. and Knight, B. C. J. G.,** *J. Gen. Microbiol.,* 13, 474, 1955.
10. **Knight, B. C. J. G. and Proom, H.,** *J. Gen. Microbiol.,* 4, 508, 1950.
11. **den, Dooren de Jong, L. E.,** *Zentralbl. Bakteriol. Parasitenkd. Infektionskr. Hyg. Abt. 2,* 79, 344, 1929.
12. **Vedder, A.,** *Antonie van Leeuwenhoek Ned. Tijdschr. Hyg. Microbiol. Serol.,* 1, 141, 1934.
13. **Darland, G. and Brock, T. D.,** *J. Gen. Microbiol.,* 67, 9, 1971.
14. **Mann, E. W.,** *Southwest. Nat.,* 13, 349, 1968.
15. **Delaporte, B., and Sasson, A.,** *C. R. Acad. Sci. Paris Sér. D,* 264, 2344, 1967.
16. **Bradley, D. E. and Franklin, J. G.,** *J. Bacteriol.,* 76, 618, 1958.
17. **Normore, W. M.,** in *Handbook of Microbiology: Microbial Composition,* Laskin, A. I. and Lechevalier, H. A., Eds., CRC Press, Cleveland, 1973, p. 585.
18. **Jones, D. and Sneath, P. H. A.,** *Bacteriol. Rev.,* 34, 40, 1970.
19. **Bonde, G. J.,** *Dan. Med. Bull.,* 22, 41, 1975.
20. **Jayne-Williams, D. J. and Cheeseman, G. C.,** *J. Appl. Bacteriol.,* 23, 250, 1960.
21. **Bulla, L. A., Jr., Bennett, G. A., and Shotwell, O. L.,** *J. Bacteriol.,* 104, 1246, 1970.
22. **Shen, P. Y., Coles, E., Foote, J. L., and Stenesh, J.,** *J. Bacteriol.,* 103, 479, 1970.
23. **Kaneda, T.,** *J. Bacteriol.,* 93, 894, 1967.
24. **Schleifer, K. H. and Kandler, O.,** *Bacteriol. Rev.,* 36, 407, 1972.
25. **Ottow, J. C. G.,** *J. Appl. Bacteriol.,* 37, 15, 1974.
26. **Wolf, J. and Barker, A. N.,** in *Identification Methods for Microbiologists,* Gibbs, B. M. and Shapton, D. A., Eds., Academic Press, London, 1968, p. 93.
27. **Smith, N. R., Gordon, R. E., and Clark, F. E.,** *Agriculture Monograph No. 16,* U.S. Department of Agriculture, Washington, D.C., 1952.
28. **Reller, L. B.,** *Am. J. Clin. Pathol.,* 60, 714, 1973.
29. **Goepfert, J. M., Spira, W. M., and Kim, H. U.,** *J. Milk Food Technol.,* 35, 213, 1972.
30. **Coonrood, J. D., Leadley, P. J., and Eickhoff, T. C.,** *J. Infect. Dis.,* 123, 102, 1971.
31. **Leffert, H. L., Baptist, J. N., and Gidez, L. I.,** *J. Infect. Dis.,* 122, 547, 1970.
32. **Pearson, H. E.,** *Am. J. Clin. Pathol.,* 53, 506, 1970.
33. **Allen, B. T. and Wilkinson, H. A., III,** *Johns Hopkins Med. J.,* 125, 8, 1969.
34. **Melles, Z., Nikodémusz, I., and Ábel, A.,** *Zentralbl. Bakteriol. Parasitenkd. Infektionskr. Hyg. Orig.,* 212, 174, 1969.
35. **Le Lourd, R., Domec, L., and Le Lourd, F.,** *Sem. Hop.* (Paris), 43, 2729, 1967.
36. **Curtis, J. R., Wing, A. J., and Coleman, J. C.,** *Lancet,* 1, 136, 1967.
37. **Lázár, J. and Juresák, L.,** *Zentralbl. Bakteriol. Parasitenkd. Infektionskr. Hyg. Abt. I Orig.,* 199, 59, 1966.
38. **Stopler, T., Camuescu, V., and Voiculescu, M.,** *Microbiol. Parazitol. Epidemiol.,* 9, 457, 1964; English translation, *Rum. (Bucaresti) Med. Rev.,* 19, 7, 1965.
39. **Farrar, W. E., Jr.,** *Am. J. Med.,* 34, 134, 1963.
40. **Norris, J. R.,** in *The Bacterial Spore,* Gould, G. W. and Hurst, A., Eds., Academic Press, London and New York, 1969, p. 485.
41. **Cowan, S. T.,** *J. Gen. Microbiol.,* 67, 1, 1971.
42. **Gordon, R. E., Haynes, W. C., and Pang, C. H.-N.,** *Agriculture Handbook No. 427,* U.S. Dept. of Agriculture, Washington, D.C., 1973.
43. **Lemille, F., de Barjac, H., and Bonnefoi, A.,** *Ann. Inst. Pasteur Paris,* 116, 808, 1969.
44. **Kim, H. U. and Goepfert, J. M.,** *Appl. Microbiol.,* 22, 581, 1971.
45. **Kim, H. U. and Goepfert, J. M.,** *Appl. Microbiol.,* 24, 708, 1972.
46. **Krieg, A.,** *J. Invertebr. Pathol.,* 15, 313, 1969.
47. **Dowdle, W. R. and Hansen, P. A.,** *J. Infect. Dis.,* 108, 125, 1961.
48. **Lamanna, C. and Eisler, D.,** *J. Bacteriol.,* 79, 435, 1960.
49. **Bennett, E. O., Peterson, G. E., and Williams, R. P.,** *Antibiot. Chemother.,* 9, 115, 1959.
50. **Burdon, K. L.,** *J. Bacteriol.,* 71, 25, 1956.
51. **Weaver, R. E., Brachman, P. S., and Feeley, J. C.,** in *Diagnostic Procedures for Bacterial, Mycotic, and Parasitic Infections,* 5th ed., Bodily, H. L., Updyke, E. L., and Mason, J. O., Eds., American Public Health Association, New York, 1970, p. 354.
52. **de Barjac, H. and Bonnefoi, A.,** *C. R. Acad. Sci. Paris Sér. D,* 264, 1811, 1967.
53. **Heimpel, A. M.,** *Annu. Rev. Entomol.,* 12, 287, 1967.
54. **Gibson, T. and Gordon, R. E.,** in *Bergey's Manual of Determinative Bacteriology,* 8th ed., Williams & Wilkins, Baltimore, 1974.
55. **de Barjac, H. and Cosmao-Dumanoir, V.,** *Ann. Microbiol. Paris,* 126, 83, 1975.
56. **Hánáková-Bauerová, E., Kocur, M., and Martinec, T.,** *J. Appl. Bacteriol.,* 28, 384, 1965.
57. **Kundrat, W.,** *Zentralbl. Veterinaermed. Reihe B,* 10, 418, 1963.
58. **Gibson, T.,** *J. Dairy Res.,* 13, 248, 1944.

59. **Lamanna, C.,** *J. Bacteriol.*, 44, 611, 1942.
60. *Report of Proceedings, Second International Congress for Microbiology, London, 1936,* St. John-Brooks, R., Ed., 1937, p. 28.
61. **Conn, H. J.,** *J. Infect. Dis.*, 46, 341, 1930.
62. **Gibson, T.,** *Zentralbl. Bakteriol. Parasitenkd. Infektionskr. Hyg. Abt. 2,* 92, 364, 1935.
63. **Marshall, R. and Beers, R. J.,** *J. Bacteriol.*, 94, 517, 1967.
64. **Campbell, L. L., Jr., and Sniff, E. E.,** *J. Bacteriol.*, 78, 267, 1959.
65. **Humphreys, T. W. and Costilow, R. N.,** *Can. J. Microbiol.*, 3, 533, 1957.
66. **Murphy, J. A. and Campbell, L. L.,** *J. Bacteriol.*, 98, 737, 1969.
67. **Holbert, P. E.,** *J. Biophys. Biochem. Cytol.*, 7, 373, 1960.
68. **Davies, S. N.,** *J. Gen. Microbiol.*, 5, 807, 1951.
69. **Francis, A. E. and Rippon, J. E.,** *J. Gen. Microbiol.*, 3, 425, 1949.
70. **Gibson, T. and Topping, L. E.,** *Proc. Soc. Agric. Bacteriol.*, p. 43, 1938.
71. **Wolf, J. and Chowdhury, M. S. U.,** in *Spore Research 1971,* Barker, A. N., Gould, G. W. and Wolf, J., Eds., Academic Press, London, 1971, p. 227.
72. **Walker, P. D. and Wolf, J.,** in *Spore Research 1971,* Barker, A. N., Gould, G. W., and Wolf, J., Eds., Academic Press, London, 1971, p. 247.
73. **Bailey, L.,** *Nature,* 180, 1214, 1957.
74. **Hannay, C. L.,** *J. Biophys. Biochem. Cytol.*, 3, 1001, 1957.
75. **Haynes, W. C. and Rhodes, L.,** *J. Bacteriol. Proc.*, p. 10, 1963.
76. **Larkin, J. M. and Stokes, J. L.,** *J. Bacteriol.*, 94, 889, 1967.
77. **Miquel, P.,** *Ann. Microgr.*, 1, 506, 552, 1889; 2, 13, 1889.
78. **Skerman, V. B. D.,** *A Guide to the Identification of the Genera of Bacteria,* 2nd ed. Williams & Wilkins, Baltimore, 1967.

THE CLOSTRIDIA*

L. DS. Smith

The genus *Clostridium* is composed of spore-forming anaerobic rods. Spore-forming anaerobic cocci – the genus *Sporosarcina* – have been described, but we know very little about them. The genus *Desulfotomaculum* also contains anaerobic sporing rods, but these may be differentiated from the clostridia by a combination of three characteristics.[3] (1) Gram-negativity; (2) DNA base composition, G + C = 41 to 46%; and (3) cytochrome of the protoheme class. Some members of *Clostridium*, especially the cellulose fermenters, are Gram-negative; others have DNA with G + C = 43 to 46%; in still others, cytochromes have been found. However, the most important characteristic shared by organisms in *Desulfotomaculum* is their ability to reduce sulfate to sulfide, a characteristic uncommon among strains of *Clostridium*. Of the three species of *Desulfotomaculum*, *D. nigrificans* is thermophilic. The two mesophilic species can be differentiated by the ability of *D. ruminis* to grow in formate and sulfate and by the inability of *D. orientis* to do so.

The clostridia vary in tolerance to oxygen, in nutritional requirements, and in optimal and limiting temperature for growth, and in many ways it is difficult to generalize concerning them. *Clostridium carnis, C. histolyticum,* and *C. tertium,* for example, are aerotolerant and will form colonies on freshly poured blood agar medium incubated aerobically. *Clostridium haemolyticum,* on the other hand, is a strict anaerobe[12] and will not grow even on reduced media if the oxygen tension is greater than 0.5%. Some species such as *C. butyricum,* can grow with ammonia as the nitrogen source and biotin as the only vitamin,[5] whereas others, such as *C. perfringens,* require more than 20 amino acids and vitamins. Some of the psychrophiles, such as *C. putrefaciens,* will not grow at temperatures above 30°C; some of the thermophiles, such as *C. thermosaccharolyticum,* will hardly grow below 50°C. For more information concerning the clostridia than can be given here, see References 19, 23, and 25, as well as *J. Appl. Bacteriol.*, 28, No. 1, 1965.

Although the clostridia are usually considered as retaining Gram's stain, not all of them do so equally. Many of the species with terminal spores lose the stain readily, and Gram-positive cells are seldom seen in overnight cultures. Several species with subterminal spores, such as *C. haemolyticum,* also lose stain completely and appear as large Gram-negative rods. Cells of *C. chauvoei* do so only partially and appear mottled or spotted.

It is sometimes necessary to determine whether a strain is an aerotolerant *Clostridium* or a facultative *Bacillus.* Clostridia produce spores anaerobically; bacilli do so aerobically. In addition, clostridia rarely produce catalase, and then only in trace amounts. If spores can be demonstrated in anaerobic culture, the strain should be considered as belonging to *Clostridium,* especially if it does not produce catalase.

The demonstration of spores in clostridia is usually not difficult, because most strains form spores when incubated in a good medium under strictly anaerobic conditions and 3 to 8°C below their optimal temperature for growth. The best single medium for demonstrating spore formation is agar made from cooked-meat medium, slanted in tubes, and prepared anaerobically. If a fluid medium is desired, cooked-meat medium may be used.[20] For spore formation by most species, the sporulation medium should not contain fermentable sugar. A few saccharolytic species, however, will not sporulate unless a fermentable carbohydrate is present. A few other species, such as *C. perfringens* and *C. paraperfringens* (*barati*) will not sporulate unless special media are used.

If a strain does not readily form spores, it may often be stimulated to do so by

* Part 15 in *Bergey's Manual.*

heat-shocking. Heat a freshly inoculated tube of fluid cooked-meat medium at 80°C for 10 min, then incubate at a temperature below the optimum for growth. Some clostridial spores, such as those of some strains of *C. botulinum* type E, will not withstand 10 min at 80°C; for such strains, 10 min at 70°C or treatment with 50% ethanol for 1 hr at room temperature will serve.[10] Although such treatment will usually, but not always, stimulate the production of spores by a strain, it should not be assumed that if a little heat is good, more is better. Many strains will withstand heating at 100°C for an hour or more, but spore production from ensuing cultures will be less than if the heat treatment had been milder, just enough to kill off the vegetative cells. (For a review of some of the factors involved in the production of clostridial spores, see References 18 and 20. For the effect of heat on the properties of cultures, see Reference 17.)

In the identification of the clostridia, the shape – spherical or oval – and the position – terminal or subterminal – of the spore is used. Determination of the shape and position of the spores should be made only with cultures showing fully mature spores, taking no stain and highly refractile. Spores that will, in their mature state, be truly and unmistakably terminal are, in some species, first formed in a subterminal position. As the spore matures, the vegetative material surrounding it disappears, and the spore attains its position at the end of the rod. Subterminal spores, however, do not do this; they remain in their subterminal to central position until the vegetative portion of the cell is lysed. A few species, such as *C. carnis,* consist of cells that have subterminal spores and cells that have terminal spores, giving the appearance of a mixed culture. For microscopic demonstration of spores, Gram-staining is sufficient. Spore stains are of no real advantage when searching for spores. A phase microscope is better than Gram-staining when mature spores are present, but is less good when the spores are immature and not refractile. Little difficulty will be experienced in determining whether spores are oval or spherical if the spores are mature. Almost all species forming spherical spores do so in the terminal position, giving rise to an unmistakable drumstick configuration.

If strains of clostridia are to be resurrected from old spore suspensions, they should be inoculated into fresh infusion broth containing 0.2% starch, heat-shocked for 10 min at 80°C (except for *C. botulinum* type E strains, which should be heated at 70°C), and incubated in an atmosphere containing at least 10% carbon dioxide at a temperature somewhat below the optimum for growth. Thioglycollate should not be used in such media, for it is inhibitory to the outgrowth of some strains;[9,16] its inhibiting effect is much less if the medium contains glucose. The factors governing recovery of severely heated or otherwise damaged spores are complex, for spores of different strains of the same species may behave differently, and spores of the same strain produced in different media may differ in response to the same recovery produce and the same recovery medium.[21]

When clostridia are to be isolated, it should be remembered that they cover the entire range in their need for anaerobiosis and that some of them are quite fastidious. Consequently, relatively rigorous means of anaerobiosis should be used, or such fastidious strains will be missed. This is important even when handling some of the clostridia usually not considered highly sensitive, for the most toxigenic variants of some species are fastidious as far as oxygen is concerned, whereas variants of lower toxigenicity are not. Consequently, if inadequate anaerobic methods are used with such strains, only cultures of low or no toxigenicity will be obtained. This seems especially to be the case with *C. botulinum* types C and D.

Two methods of obtaining anaerobiosis can be recommended: the roll tube method of Hungate as modified by Moore,[10] and the plastic anaerobic glove box of Freter.[1] Freshly poured blood agar plates, incubated in anaerobe jars, may also be successfully used, but if clostridia sensitive to oxygen are to be isolated, the agar should be streaked as soon as it has hardened. Oxygen-sensitive clostridia that will not form colonies on freshly poured and solidified agar may usually be grown by "sloppy streaking", i.e., by streaking the fluid blood agar just before it solidifies. Blood agar serves as an excellent solid medium for the clostridia, particularly if 0.05% cysteine and 0.03% dithiothreitol, dithioerythritol, or sodium formaldehyde sulfoxylate is added. For an investigation into the cultivation of exacting species, see Reference 4.

Several media have been devised for the isolation of clostridia from a variety of sources (for an excellent description of methods and media, see Reference 22). For the isolation of clostridia from food, semiselective media containing sulfite and iron are most commonly used. The clostridia reduce sulfite to sulfide, thus causing the blackening of the colonies.[15] Various other substances – inhibitory for other organisms, but with little effect on the clostridia – have also been used in efforts to make media that will grow clostridia, but no other organisms. Unfortunately, as Gibbs and Fraeme[6] have pointed out, no really satisfactory selective medium for the clostridia is known. Even for an organism as hardy as *C. perfringens* it has not been possible to devise a medium that suppresses the outgrowth of contaminant bacteria, but permits the quantitative recovery of *C. perfringens.*[7,8] If a substance is inhibitory for all the facultative bacteria, or even for most of them, it is also inhibitory for at least some of the clostridia. Consequently, media containing crystal violet, sorbic acid, sulfadiazine, neomycin, cycloserine, etc., cannot be trusted for the quantitative isolation of all species of *Clostridium.* When such media are used, nonselective media should be used in parallel. If the limitations of selective media are understood, however, they may be very useful for special purposes. It is well to have a strain of *C. haemolyticum* on hand and occasionally to try obtaining isolated colonies of this organism. If isolated colonies of *C. haemolyticum* can be obtained, the medium and the isolation procedures are adequate for the other clostridia. If isolated colonies of this organism cannot be obtained, some factor is unsatisfactory.

Whatever procedure of incubation and whatever medium is used, after incubation the colonies should be scrutinized with a dissection microscope to make certain that they are well isolated and that the plate is not covered by a thin film of one of the swarming clostridia. Often this film may be just barely perceptible if one is looking carefully for it. When this happens, isolation should be repeated, using a shorter incubation time or increasing the concentration of the agar to 4%. The edge of a swarming area should not be picked with the hope of getting a pure culture of the swarming strain, for bacteria that are swarming often carry other bacteria along with them. An isolated colony should be picked to a tube of cooked-meat/glucose medium, incubated overnight, and used to inoculate the diffrential media listed in Table 1 for identification.

The cultural characteristics of the most commonly occurring species are listed in Tables 2 and 3. Among the species that will be encountered most often will be *C. perfringens, C. bifermentans, C. sordelli, C. sporogenes, C. innocuum,* and *C. subterminale.* If pathologic specimens from animals are being examined, *C. novyi, C. chauvoei,* and *C. septicum* will also be encountered fairly frequently. It will be convenient to be familiar with the morphology, colonial aspects, and outstanding cultural characteristics of these organisms.

In addition to the determination of the usual cultural characteristics, it is most helpful to determine, with the aid of a gas chromatograph, the principal fermentation products that each strain produces. This may be done from the culture in peptone/yeast-extract/glucose medium[10] after incubation for 3 days or longer. The major products of fermentation in this medium are listed in Tables 4 and 5. Determination of the fermentation products is a valuable aid when quick identification of a strain is desired, and is always helpful in confirming conclusions arrived at by consideration of other data.

The fluorescent-antibody technique has also been found to be of considerable value in the rapid identification of some of the pathogenic species. *Clostridium chauvoei, C. septicum, C. novyi, C. tetani,* and *C. botulinum* have been studied by this method.[2]

Table 1
DIFFERENTIAL MEDIA FOR THE CLOSTRIDIA

Medium	Purpose
Glucose broth	Fermentation, chromatography, toxin
Lactose broth	Fermentation
Maltose broth	Fermentation
Mannitol broth	Fermentation
Mannose broth	Fermentation
Salicin broth	Fermentation
Sucrose broth	Fermentation
Trehalose broth	Fermentation
Gelatin	Liquefaction
Milk	Digestion
Cooked-meat medium	Motility, indol, digestion
Cooked-meat slant	Sporulation (incubate at 30°C)
Molten deep agar	Aerobic or anaerobic growth
Blood agar plate	Colony description, hemolysis
Blood agar plate, aerobic	Aerotolerance
Egg yolk agar plate	Lecithinase, lipase

Clostridium chauvoei strains apparently form a single antigenic group when tested by this method, whereas *C. septicum* strains form two groups. Strains of *C. novyi* of whatever type share antigens. However, antiserum prepared against *C. haemolyticum* (*C. novyi* type D) and absorbed with *C. novyi* strains of the other types has been reported as satisfactory for identifying *C. haemolyticum* in tissue smears.[13] The indirect-fluorescence method has also been used to demonstrate spores of *C. haemolyticum* in biopsy specimens.[24]

TOXIN PRODUCTION AND TESTING

For the identification of some species, and of types within some species, it is necessary to determine the production of toxin and to identify the toxin that is produced. For example, it is not possible to distinguish, on cultural characteristics, between strains of *C. sporogenes* and proteolytic strains of *C. botulinum* of types A, B, and F. *Clostridium botulinum* forms a potent paralytic neurotoxin, whereas *C. sporogenes* does not, and the demonstration of such a toxin is sufficient to remove a strain from *C. sporogenes*. Determination of a type within the species *C. botulinum, C. novyi,* or *C. perfringens* can be carried out only by identifying the toxin by neutralization with type-specific antitoxin.

All of the pathogenic clostridia produce toxin. All clostridia toxins are produced in the interior of the cell, but some, the "extracellular" toxins, readily diffuse out and are found in maximum amount in the culture fluid shortly after the end of the log phase of growth. Others, the "protoplasmic" toxins, do not readily diffuse out and reach their maximum in the culture fluid only after lysis has occurred. The toxins of all types of *C. botulinum,* the neurotoxin of *C. tetani,* and the alpha toxin of *C. novyi* are all protoplasmic toxins. Almost all the other clostridial toxins are extracellular. The enteropathogenic toxin of *C. perfringens* differs from the rest; it is formed intracellularly and is released only at the time of sporulation.

Some toxins appear to be synthesized by the clostridia in fully active form. Others are synthesized as less active prototoxins and attain full toxicity only after exposure to certain proteolytic enzymes. In some cases proteolytic enzyme is also synthesized by the organism producing the prototoxin, and activation is spontaneous; in other cases the bacteria do not synthesize proteolytic enzyme, or not enough of it to fully activate the

Table 2
CULTURAL CHARACTERISTICS OF CLOSTRIDIA WITH SUBTERMINAL SPORES

Species	Gelatin digestion	Milk digestion	Lecithinase production	Lipase production	Indol production	Glucose fermentation	Lactose fermentation	Maltose fermentation	Mannitol fermentation	Mannose fermentation	Salicin fermentation	Sucrose fermentation	Toxin production
C. sordelli[a]	+	+	+	–	+	+	–	+	–	v	–	–	+
C. bifermentans[a]	+	+	+	–	+	+	–	+	–	v	–	–	–
C. sporogenes[b]	+	+	–	+	–	+	–	v	–	–	–	–	–
C. botulinum ABF (proteolytic)	+	+	–	+	–	+	–	v	–	–	–	–	+
C. hystolyticum[c]	+	+	–	–	–	–	–	–	–	–	–	–	+
C. subterminale[c]	+	+	–	–	–	–	–	–	–	–	–	–	–
C. limosum	+	+	+	–	–	–	–	–	–	–	–	–	∓
C. perfringens	+	–	+	–	–	+	+	+	–	+	–	+	+
C. novyi A[b]	+	–	+	+	–	+	–	+	–	–	–	–	+
C. novyi B	–	+	+	–	–	+	–	+	–	+	–	–	+
C. haemolyticum	+	–	+	–	+	+	–	–	–	v	–	–	+
C. botulinum CD[b]	+	–	–	+	–	+	–	+	–	+	–	–	+
C. botulinum BEF[b] (nonproteolytic)	+	–	–	+	–	+	–	+	–	+	–	+	+
C. difficile	+	–	–	–	–	+	–	–	+	+	v	–	+
C. septicum	+	–	–	–	–	+	+	+	–	+	+	–	+
C. chauvoei	+	–	–	–	–	+	+	+	–	+	–	+	+
C. paraperfringens	–	–	+	–	–	+	+	+	–	+	+	+	–
C. fallax	–	–	–	–	–	+	+	+	+	+	+	+	–
C. butyricum	–	–	–	–	–	+	+	+	–	+	+	+	–
C. carnis[d]	–	–	–	–	–	+	v	+	–	+	+	+	–
C. beijerinckii	–	–	–	–	–	+	+	+	–	+	–	+	–

[a] *C. sordelli* produces urease; *C. bifermentans* does not.
[b] Toxin tests are required for identification.
[c] *C. histolyticum* is aerotolerant; *C. subterminale* is not.
[d] *C. carnis* is aerotolerant; spores are subterminal to terminal.

prototoxin, and treatment with trypsin or some similar enzyme is required for full toxic activity.

The neurotoxins of all serological types of *C. botulinum* seem to be synthesized as prototoxins, but only that of type E and those of some strains of other nonproteolytic types of this species require trypsin activation. The situation with regard to *C. tetani* neurotoxin is not clear. The epsilon and iota toxins of *C. perfringens* definitely require activation for full toxin activity. Trypsin activation cannot be carried out routinely with all cultures, for all clostridial toxins can be inactivated by proteolytic enzymes if the treatment is sufficiently rigorous.

If a strain under investigation produces easily demonstrable toxin, its identification is relatively easy. More work is required if toxin is not easily demonstrable, for it may be necessary to incubate for different lengths of time and to examine culture fluid that has been treated with trypsin as well as culture fluid that has not been so treated. Both young (6 to 12 hr incubation) and old (3 to 5 days incubation) cultures should be examined. A portion of each should be used for toxicity tests (0.5 ml injected intraperitoneally into

Table 3
CULTURAL CHARACTERISTICS OF CLOSTRIDIA WITH TERMINAL SPORES

Species	Gelatin digestion	Milk digestion	Lecithinase production	Lipase production	Indol production	Glucose fermentation	Lactose fermentation	Maltose fermentation	Mannitol fermentation	Mannose fermentation	Salicin fermentation	Sucrose fermentation	Trehalose fermentation	Toxin production
C. cadaveris	+	+	–	–	+	+	–	–	–	–	–	–	–	–
C. lentoputrescens	+	+	–	–	+	–	–	–	–	–	–	–	–	–
C. putrificum	+	+	–	–	–	+	–	–	–	–	–	–	–	–
C. tetani	+	–	–	–	v	–	–	–	–	–	–	–	–	+
C. malenominatum	–	–	–	–	+	–	–	–	–	–	–	–	–	–
C. tertium[a]	–	–	–	–	–	+	+	+	+	+	+	+	+	–
C. carnis[a]	–	–	–	–	–	+	v	+	–	+	+	+	–	+
C. paraputrificum[a]	–	–	–	–	–	+	+	+	–	+	+	+	–	–
C. ramosum[a]	–	–	–	–	–	+	+	+	v	+	+	+	+	–
C. innocuum	–	–	–	–	–	+	–	–	+	+	+	+	+	–
C. cochlearium	–	–	–	–	–	–	–	–	–	–	–	–	–	–

[a] *C. tertium* and *C. carnis* are aerotolerant; *C. paraputrificum* and *C. ramosum* are not.

Table 4
FERMENTATION PRODUCTS OF CLOSTRIDIA WITH SUBTERMINAL SPORES

Major products	Species
Acetic acid	*C. histolyticum, C. limosum*
Acetic and butyric acids; smaller amounts of propionic acid	*C. perfringens, C. septicum, C. chauvoei, C. botulinum* types B, E, F (nonproteolytic), *C. butyricum, C. fallax, C. paraperfringens, C. carnis, C. beijerinckii*
Propionic and butyric acids, with or without acetic acid	*C. botulinum* types C, D, *C. novyi, C. haemolyticum*
Acetic, isobutyric, and isovaleric acids	*C. subterminale*
Acetic, isobutyric, sometimes butyric, isovaleric, and isocaproic acids	*C. sordelli, C. bifermentans*
Acetic and butyric acids, with smaller amounts of propionic, isobutyric, isovaleric, and valeric acids; with or without valeric acid and ethanol, butyanol, isoamyl alcohol	*C. sporogenes, C. botulinum* types A, B, F (proteolytic), *C. difficile*

each of two mice) without treatment with trypsin, and another portion after trypsin treatment. Trypsin treatment is done by adjusting the pH of the culture fluid to a level between 6.0 and 6.2, adding a solution of commercial trypsin (Difco 1:250, for example) to a final concentration of 0.1%, then incubating at 37°C for 1 hr. Too long or too rigorous treatment with trypsin will first activate and then inactivate prototoxin; consequently, it is best to have the hydrogen ion concentration of the fluid below neutrality, where the action of trypsin is slower than under alkaline conditions.

When species- or type-specific antitoxin is available, 0.2 ml of antitoxin should be

Table 5
FERMENTATION PRODUCTS OF CLOSTRIDIA WITH TERMINAL SPORES

Major products	Species
Acetic and formic acids	*C. ramosum*
Butyric acid	*C. cochlearium, C. lentoputrescens*
Acetic and butyric acids	*C. innocuum*
Acetic and butyric acids; smaller amount of propionic acid	*C. malenominatum, C. terium, C. carnis, C. paraputrificum*
Acetic and butyric acids, butanol; with or without ethanol	*C. tetani, C. cadaveris*
Acetic, isobutyric, butyric, and isovaleric acids	*C. putrificum*
Acetic, propionic, isobutyric, and isovaleric acids, and ethyl, propyl, butyl, and isoamyl alcohols	*C. glycolicum*

added to 1.3 ml of culture fluid, mixed, and allowed to stand for 30 min at room temperature before injecting 0.5 ml intraperitoneally or 0.2 ml intravenously into each of two mice. Culture fluid that has been mixed with normal serum in similar fashion should be injected into a second pair of mice. The mice should be observed for 2 days. If the mice receiving the mixture of culture fluid and antitoxin die after the control mice do, the experiment should be repeated, diluting the culture fluid 1:10 with broth. With most toxigenic strains the control mice will die overnight; only occasionally will death occur on the second day. It is not necessary to sterilize clostridial culture fluid by filtration when intraperitoneal or intravenous injection is used, for thorough centrifuging suffices for the removal of the great majority of the organisms, and infection is not encountered.

The pathogenic clostridia and their *lethal* toxins are listed in Table 6.

Table 6
LETHAL CLOSTRIDIAL TOXINS

Species	Designation	Mode of action
C. botulinum	Neurotoxin	Prevents release of acetylcholine at motor end plate
C. carnis	None	Hemolytic; lethal activity unknown
C. chauvoei	Alpha	Necrotizing;[a] hemolytic; leucocidic
C. difficile	None	Lethal activity unknown
C. haemolyticum	Identical to the beta toxin of *C. novyi*	Hemolytic: necrotizing: phospholipase
C. histolyticum	Alpha	Lethal activity unknown
	Beta	Collagenase
C. novyi	Alpha	Necrotizing
	Beta	Necrotizing; hemolytic; phospholipase
C. perfringens	Alpha	Necrotizing; hemolytic; phospholipase
	Beta	Necrotizing
	Epsilon	Necrotizing
	Iota	Necrotizing; increases capillary permeability
	Enterotoxin	Diarrheal; erythemal
C. septicum	Alpha	Necrotizing; hemolytic; leucocidic
C. sordelli	None	Lethal activity unknown
C. tetani	Tetanospasmin	Neurotoxic; action in central nervous system not known
	Tetanolysin	Hemolytic; edema-inducing

[a] Necrotizing properties were determined by intracutaneous inoculation of guinea pigs. The true lethal action of toxins designated as necrotizing is not known.

REFERENCES

1. **Aranki, A., Syed, S. A., Kenney, E. B., and Freter, R.,** Isolation of anaerobic bacteria from human gingiva and mouse cecum by means of a simplified glove box procedure, *Appl. Microbiol.,* 17, 568, 1969.
2. **Batty, I. and Walker, P. D.,** Colonial morphology and fluorescent-labelled antibody staining in the identification of species of the genus *Clostridium, J. Appl. Bacteriol.,* 28, 112, 1965.
3. **Campbell, L. L. and Postgate, J. R.,** Classification of the spore-forming sulfate-reducing bacteria, *Bacteriol. Rev.,* 29, 359, 1965.
4. **Collee, J. G., Rutter, J. M., and Watt, B.,** The significantly viable particle: a study of the subculture of an exacting sporing anaerobe, *J. Med. Microbiol.,* 4, 271, 1971.
5. **Cummins, C. S. and Johnson, J. L.,** Taxonomy of the clostridia: wall composition and DNA homologies in *Clostridium butyricum* and other butyric acid-producing clostridia, *J. Gen. Microbiol.,* 67, 33, 1971.
6. **Gibbs, B. M. and Fraeme, B.,** Methods for the recovery of clostridia from foods, *J. Appl. Bacteriol.,* 28, 95, 1965.

7. **Harmon, S. M., Kautter, D. A., and Peeler, J. T.,** Comparison of media for the enumeration of *Clostridium perfringens, Appl. Microbiol.,* 21, 922, 1971.
8. **Harmon, S. N., Kautter, D. A., and Peeler, J. T.,** Improved medium for enumeration of *Clostridium perfringens, Appl. Microbiol.,* 22, 688, 1971.
9. **Hibbert, H. R. and Spencer, R.,** An investigation of the inhibitory properties of sodium thioglycollate in media for the recovery of clostridial spores, *J. Hyg.,* 68, 131, 1970.
10. **Holdeman, L. V. and Moore, W. E. C.,** *Anaerobic Laboratory Manual,* Virginia Polytechnic Institute Educational Foundation, Blacksburg, Va., 1972.
11. **Johnson, R., Harmon, S., and Kautter, D.,** Method to facilitate the isolation of *Clostridium botulinum* type E, *J. Bacteriol.,* 88, 1521, 1964.
12. **Loesche, W. J.,** Oxygen sensitivity of various anaerobic bacteria, *Appl. Microbiol.,* 19, 723, 1969.
13. **McCain, C. S.,** Isolation and identification of *Clostridium haemolyticum* in cattle in Florida, *Am. J. Vet. Res.,* 28, 878, 1967.
14. **Moore, W. E. C., Cato, E. P., and Holdeman, L. V.,** Fermentation characteristics of *Clostridium* species, *Int. J. Syst. Bacteriol.,* 16, 383, 1966.
15. **Mossel, D. A. A.,** Enumeration of sulfite-reducing clostridia occurring in foods, *J. Sci. Food Agric.,* 10, 662, 1959.
16. **Mossel, D. A. A. and Beerens, H.,** Studies on the inhibitory properties of sodium thioglycollate on the germination of wet spores of clostridia, *J. Hyg.,* 66, 269, 1968.
17. **Nishida, S., Yamagishi, T., Tamai, K., Sanada, I., and Takahashi, K.,** The effects of heat selection on toxigenicity, cultural properties, and antigenic structures of clostridia, *J. Infect. Dis.,* 120, 507, 1969.
18. **Perkins, W. E.,** Production of clostridial spores, *J. Appl. Bacteriol.,* 28, 1, 1965.
19. **Prévot, A. R., Turpin, A., and Kaiser, P.,** *Les Bactéries Anaérobies,* Dunod, Paris, 1967.
20. **Roberts, T. A.,** Sporulation of mesophilic clostridia, *J. Appl. Bacteriol.,* 30, 430, 1967.
21. **Roberts, T. A.,** Recovering spores damaged by heat, ionizing radiation, or ethylene oxide, *J. Appl. Bacteriol.,* 33, 74, 1970.
22. **Shapton, D. A. and Board, R. G.,** *Isolation of Anaerobes,* Academic Press, New York and London, 1971.
23. **Smith, L. DS. and Holdeman, L. V.,** *The Pathogenic Anaerobic Bacteria,* Charles C Thomas, Springfield, Ill., 1968.
24. **Van Kampen, K. R. and Kennedy, O. C.,** Experimental bacillary hemoglobinuria: intrahepatic detection of spores of *Clostridium haemolyticum* by immunofluorescence in the rabbit, *Am. J. Vet. Res.,* 29, 2173, 1968.
25. **Willis, A. T.,** *Clostridia of Wound Infection,* Butterworths, London, 1969.

GRAM-POSITIVE ASPOROGENOUS ROD-SHAPED BACTERIA*

H. A. Lechevalier

Generic name (G + C, molar %) [Number of species in *Bergey's Manual*]	Morphology	Physiology	Ecology	Examples of species
Lactobacillus (33 to 54) [25]	Rods, 0.5–2.0 by 1.0–9 μm, often forming chains; usually non-motile; if motile, by peritrichous flagella; become Gram-negative with age; rarely pigmented; if so, yellow or orange to reddish.	Fermentative metabolism; anaerobic to facultatively aerobic; saccharoclastic, accumulating lactate; gelatin not liquefied; casein not digested; catalase-negative; complex nutritional requirements characteristic for each species; grow best below pH 6.0.	Found in dairy products, water, sewage, fruits and juices, pickled products, and other foods; also found in the natural cavities and gastrointestinal tract of animals, including man; usually non-pathogenic.	*L. delbrueckii.* Type species. *L. lactis.* Used in the manufacture of cheeses. *L. helveticus.* Used in the manufacture of Emmental and Gruyère cheeses. *L. casei.* From Tilsit cheese. *L. trichodes.* Growth forms a tangled mass in fortified wines.
Listeria (38 to 56) [4]	Rods, 0.4--0.8 by 0.5–2.5 μm, with a tendency to form chains and produce V and Y cells; become Gram-negative with age; motile by peritrichous flagella, if grown at 20–25°C.	Fermentative and oxidative metabolism; aerobic to microaerophilic; acid, but no gas from glucose; esculin hydrolyzed; gelatin and casein not hydrolyzed; usually catalase-positive.	Found in the feces of animals and man; parasitic on animals, including man; widely distributed in nature: decaying vegetation, soil, silage, sewage, etc.	*L. monocytogenes.* Type species; widely distributed; causes numerous different types of animal and human infections, depending on the species and on the portal of entry; meningitis is one of the infections caused by this species.
Erysipelothrix (36) [1]	Rods, 0.2–0.4 by 0.5–2.5 μm, with a tendency to form filaments; non-motile; old cultures have a tendency to become Gram-negative.	Aerobic, growing best in presence of 5–10% CO_2; acid, but no gas from glucose; esculin not hydrolyzed; catalase-negative; alpha-hemolysin on blood agar.	Widely distributed in nature; parasitic on mammals, birds, and fish.	*E. rhusiopathiae.* Causes swine erysipelas; may be pathogenic to man, causing an erysipeloid infection, which can be caught by handling infected fish.

* Parts 16 and 17 in *Bergey's Manual.*

Generic name (G + C, molar %) [Number of species in *Bergey's Manual*]	Morphology	Physiology	Ecology	Examples of species
Caryophanon (–) [1]	Rods or filaments, 3 by 6–30+ μm; divided by numerous cross walls, which are often incomplete; motile by peritrichous flagella.	Chemoorganotrophs; respiratory metabolism; strict aerobes.	Found in cow feces and in decaying fungi.	*C. latum*.
Corynebacterium diphtheriae-type (52 to 68) [9]	Rods, 0.3–0.8 by 1–8 μm, which may be slightly curved or have club-shaped swellings; snapping division produces angular or palisade arrangements of cells; usually non-motile; not acid-fast.	Chemoorganotrophs; respiratory and fermentative metabolism; aerobic to facultatively anaerobic; catalase-positive; cell walls of Type IV (*meso*-DAP + arabinose and galactose).	Found on the skin and in mucous membranes of animals; often pathogenic to animals.	*C. diphtheriae.* Type species; found in the nasopharynx of man; causes diphtheria. *C. pseudotuberculosis.* Causes abscesses of animals. *C. equi.* Causes various infections in horses and cattle. *C. bovis.* Found in cow udders; may cause mastitis.
Plant pathogens (65 to 75) [12] For more information see pp. 351–359.	Same as above, but snapping division is less evident, and motility is more common.	Same as above, but cell wall composition is variable.	Associated with plant diseases.	*C. fascians.* Causes a fasciation disease of plants. *C. tritici.* Causes a gumming disease of wheat. *C. michiganense.* Causes wilt, canker, and leaf spots of tomatoes and relatives. *C. flaccumfaciens.* Causes a wilt of bean.
Arthrobacter (60 to 72) [7]	Old (2 to 7 days) cells are cocci, 0.4 to 0.9 μm in diameter; young cells are coryneform rods, 0.4–0.8 by 1–6 μm; if motile, by one subpolar flagellum or by a few peritrichous flagella.	Chemoorganotrophs; respiratory metabolism; strict aerobes; no acid from glucose; catalase-positive; some species require terregens factor, vitamin B_{12}, or other growth factors.	Abundantly distributed in soil.	*C. globiformis.* Type species.

Generic name (G + C, molar %) [Number of species in *Bergey's Manual*]	Morphology	Physiology	Ecology	Examples of species
Brevibacterium (47 to 71)	Organisms have the morphology of either *Arthrobacter* or coryneforms.	Most cultures presently in this genus will probably be transferred to other genera upon further studies.		*B. linens.* Type species.
Microbacterium (58 to 69)	Coryneform rods; non-motile; aerobic.	Same as above; L(+)-lactic acid produced from carbohydrates; catalase-positive.	Found in dairy products and on dairy utensils.	*M. lacticum.* Type species.
Cellulomonas (72 to 73) [1]	Irregular rods, 0.5 by 0.7–2 μm; may occasionally show rudimentary branching; motile by one polar or subpolar flagellum or by a few peritrichous flagella; a part of the cells may be Gram-negative; not acid-fast.	Chemoorganotrophs; respiratory metabolism; aerobic to anaerobic; acid produced from glucose; catalase-positive; cellulose is decomposed.	Found in soil.	*C. flavigena.*
Kurthia (37 to 38)[a] [1]	Old (3 to 7 days) cells are cocci or short rods; young cells are rods, 0.8 by 2–8 μm; motile by peritrichous flagella.	Chemoorganotrophs; respiratory metabolism; strict aerobes; no acid from glucose; catalase-positive.	Found in manure, in stagnant water, in meat products, and in milk.	*K. zopfii.*
Propionibacterium (59 to 66) [8]	Rods, up to 6 μm long; pleomorphic; may become coccoid, bifid, or branched; non-motile; Gram-positive; either not pigmented, or pink, yellow, or orange.	Chemoorganotrophs; anaerobic to aerotolerant; ferment carbohydrates, peptone, pyruvate, or lactate, accumulating mainly propionic and acetic acids; acid always formed from glucose.	Found in dairy products or on the skin and in the intestinal tract of man and animals.	*P. freudenreichii.* Type species; found in dairy products, including Swiss cheese. *P. acidi-propionici.* Found in dairy products. *P. avidum.* Found in feces and in various infections. *P. acnes.* Found on normal skin and in various infections.

[a] Shaw, S. and Keddie, R. M., personal communication.

Generic name (G + C, molar %) [Number of species in *Bergey's Manual*]	Morphology	Physiology	Ecology	Examples of species
Eubacterium (–) [28]	Rods, up to 9 μm long; may be pleomorphic; may be motile.	Chemoorganotrophs; obligate anaerobes; ferment sugars or peptone with production of butyric, acetic, or formic acid, or may not be fermentative; usually catalase-negative.	Found in cavities of animals, including man, in plant products, and in soil; some species may be pathogenic.	*E. foedans.* Type species. *E. cellulosolvens.* Found in the rumen of sheep and cows. *E. parvum.* Found in feces and in various infections.

THE CORYNEFORM BACTERIA*

C. S. Cummins

The generic name *Corynebacterium* means "club-shaped bacterium," and it is indeed characteristic of *C. diphtheriae*, the type species of the genus, that the organisms appear in stained smears as Gram-positive, irregularly stained, curved rods that frequently have club-shaped swellings. The less specific terms "coryneform" and "diphtheroid" mean that the organisms to which they are applied have a club-like shape (coryneform) or resemble the diphtheria bacillus (diphtheroid). As such, these two terms are valuable shorthand descriptions of a particular type of morphology in Gram-positive bacteria, and they are generally used interchangeably.

A little reflection will show, however, that – apart, perhaps, from organisms in the genera *Clostridium, Bacillus,* or *Lactobacillus* – the terms "coryneform" or "diphtheroid" can be applied to almost all Gram-positive organisms in the orders Eubacteriales or Actinomycetales at some stage of their growth cycle. This type of morphology, therefore, is very commonly encountered, and the implication of some kind of "special relationship" to the type species, which these two terms carry, has been the cause of much taxonomic confusion.

The coryneform bacteria have been the subject of several intensive studies and reviews in recent years, mostly by numerical taxonomy (see References 2, 13, 14, 19, 24, 26, 52, 58, 61, 62, 63, and 64). However, all of these studies have been wide-ranging, covering in some cases representatives of up to 20 named genera of Gram-positive organisms. The present contribution will be confined to those organisms traditionally assigned to the genus *Corynebacterium,* and even here it is evident that there are several different groups, each probably worthy of generic rank. It is not proposed to give details of isolation procedures, nor detailed keys for strain identification, but to indicate some of the characters that may be useful in assigning strains to the correct group. The treatment, therefore, is deliberately broad, with (it is hoped) sufficient references to enable those interested to follow up specific points.

CORYNEBACTERIUM

The organisms to which the generic name *Corynebacterium* is usually applied can be broadly arranged in four groups on the basis of cell wall composition and other properties, as indicated in Table 1. Of these groups, three are essentially parasites and pathogens in man and animals, whereas the plant-pathogenic corynebacteria have a rather different ecological background and were at one time classified in a different genus, *Phytomonas.*

1. *Corynebacterium diphtheriae* and Related Organisms

These organisms are what might legitimately be called the "classical" corynebacteria, since they share many properties with the type species, *C. diphtheriae.* Basically they are aerobic, Gram-positive, non-motile, non-sporing rods, characterized by having an arabinogalactan polysaccharide in their cell walls, with *meso*-DAP as the diamino acid of the peptidoglycan component. This "arabinose-galactose-*meso*-DAP" pattern of components is also characteristic of mycobacteria and nocardias (see page 365) and is probably associated with a common cell wall polysaccharide antigen found in all these organisms.[8] Several of the corynebacteria of this group have also been found to resemble

* Part 17 in *Bergey's Manual.*

Table 1
CELL WALL COMPONENTS, DNA COMPOSITION (% G + C), AND OTHER PROPERTIES IN VARIOUS GROUPS OF CORYNEBACTERIA

Group of strains	Cell wall components: Sugars	Cell wall components: Amino acids in peptidoglycan	Presence of corynomycolic acids	Motility	DNA % G + C
1. *C. diphtheriae* and most other human- and animal-pathogenic corynebacteria	Arabinose, galactose (mannose and glucose may also be present)	ala, glu, *meso*-DAP[a]	+	–	58–68
2. Plant pathogenic corynebacteria:					
i. *C. fascians*	Arabinose, galactose	ala, glu, *meso*-DAP	+	–	68
ii. *C. poinsettiae*	Rhamnose, mannose, galactose				
C. betae	Rhamnose, mannose, fucose	gly, glu, homoser, ala, orn	–	+	71
C. flaccumfaciens	Glucose, mannose				
iii. *C. tritici*	Xylose, glucose, mannose				
C. insidiosum	Fucose, rhamnose, mannose, galactose	gly, glu, DAB,[b] ala	–	–[c]	75
C. sepedonicum	Rhamnose, mannose, galactose				
3. *C. pyogenes*	Rhamnose, glucose	ala, glu, lys	–	–	58–59
C. haemolyticum	Rhamnose				
4. *C. acnes* and "anaerobic diphtheroids"	Galactose and/or glucose and/or mannose (not all strains have all 3 sugars)	ala, glu, gly, L-DAP (occasionally a strain with *meso*-DAP)	–	–	58–63

[a] DAP = α,ϵ-diaminopimelic acid.
[b] DAB = diaminobutyric acid.
[c] Motility has been reported in *C. tritici.*

Adapted from data in Reference 4.

the mycobacteria and nocardias in having mycolic acids as a characteristic feature of their lipids, although the corynomycolic acids are of lower molecular weight (c. C_{32}) than those found in either nocardias (c. C_{56}) or mycobacteria (c. C_{88}).[6,27,28,41,60]

Although in Table 1 *C. fascians* has been placed with the other plant-pathogenic corynebacteria, it is evident that it too has the "arabinose-galactose-*meso*-DAP" pattern of cell wall components, and therefore falls into the same group as the type species. Strains of *C. fascians* possess the common cell wall antigen mentioned above,[8] and it has recently been reported that they contain a mycolic acid.[6] They also show some relationship to *Nocardia* and *Mycobacterium,* since they may be acid-fast under some conditions.[26,29]

The literature on *C. diphtheriae* is very extensive, since it is a major pathogen of man: for reviews on the organism and its exotoxin see References 1 and 5. Literature on most other organisms in the group is scantier, and in some cases almost confined to the paper containing the original description.

Table 2 contains a list of corynebacterial species (other than *C. fascians*) that have the arabinose-galactose-*meso*-DAP pattern of cell wall components. The different cultural varieties of *C. diphtheriae* (*gravis, mitis* and *intermedius,* see Reference 31) have been omitted, as have strains often referred to as *C. ulcerans* and *C. belfanti,* which seem to be intermediates between *C. diphtheriae* and *C. pseudotuberculosis* (see, e.g., References 1, 18, and 23).

2. Plant-pathogenic Corynebacteria

As can be seen from Table 1, this is a heterogeneous group. As noted above, it seems certain that strains of *C. fascians* do not belong here; taxonomically they appear to have a good deal in common with *C. diphtheriae,* but they also show some relationships to *Nocardia* and *Mycobacterium.*

The other plant-pathogenic corynebacteria appear to fall into two groups, as indicated in sections 2.ii and 2.iii of Table 1. Strains of *C. tritici* are reported to be motile, but otherwise motility appears to be confined to the organisms in section 2.iii. Besides being motile, these plant-pathogenic coryneforms differ from the type species in several other respects. They have a variety of sugars in the cell wall polysaccharides, but none contain the arabinogalactan common to the animal pathogens. They also have unusual amino acid patterns in peptidoglycan; in group 2.iii, diaminobutyric acid replaces DAP, and in group 2.ii, the place of DAP is taken by homoserine, which in fact is not a diamino acid. For details of these unusual peptidoglycan structures see References 17 and 49.

The plant pathogens also seem to have a considerably higher guanine + cytosine content in their DNA than the type species; up to 76% G + C has been reported for strains of *C. sepedonicum,* and 78% for *C. insidiosum,*[3,7,50,59,61,62] although most results quoted are rather lower (see Table 3).

Some of the plant-pathogenic corynebacteria are of considerable economic importance. For example, *C. michiganense* (tomato canker), *C. insidiosum* (wilt of lucerne), and *C. sepedonicum* (ring rot in potatoes) may all cause major reduction of yield in the crops they attack. A list of the most common named species of plant pathogens and the diseases they produce is given in Table 3.

3. *Corynebacterium pyogenes* and *C. haemolyticum*

Both of these organisms are Gram-positive and coryneform in morphology, but otherwise very different from the type species. For example, they contain lysine instead of DAP as the diaminoacid of peptidoglycan (Table 1), and they do not have arabinogalactan or mycolic acids. They are also negative, or at most weakly positive, in the catalase test.[9,42]

Table 2
C. DIPHTHERIAE AND RELATED ORGANISMS

Species	Presence of corynomycolic acids	Sensitivity to *C. diphtheriae* phage	DNA G + C content[a] (%)	Ecology and disease production
C. diphtheriae	+	+	58	Causes diphtheria in man. Found in the nasopharynx of cases and of healthy carriers, and occasionally elsewhere.
C. pseudotuberculosis (*C. ovis*)	+	+	67	Causes ulcerative lymphangitis and abscesses in sheep, goats, horses, etc., and occasionally in man.
C. xerosis	+	–	57–59	Found on the skin and in the mucous membranes of man. Nonpathogenic.
C. pseudodiphtheriticum (*C. hofmannii*)	+	–	56–58	Found in the mucous membranes of man.
C. equi	+	–		Causes pneumonia in foals and pseudotuberculous lesions in swine.
C. renale	+	–	54–58	Causes pyelitis and cystitis in cattle.
C. bovis	+	–	63–64	Isolated from cows milk.
C. kutscheri (*C. murium*)	–	–		Parasite of rats and mice.
C. paurometabolum				Isolated from bed bugs.
C. minutissinum				Isolated from lesions of erythrasma in man.
Lipophilic diphtheroids from skin			58–61	Found on the skin of man.

[a] Figures for G + C content have been rounded up to the nearest whole number.

Code: – = no effect; blank space = not determined.

Compiled from information in References 1, 3, 6, 7, 10, 21, 22, 25, 27, 28, 33, 41, 48, 51, 52, 56, 57, 61, and 62.

Table 3
COMMON NAMED SPECIES OF PLANT-PATHOGENIC CORYNEBACTERIA

Organism	Cell wall group (see Table 1)	DNA % G + C[a]	Disease caused
C. fascians	2.i	70	Hypertrophic disease in many plants
C. poinsettiae	2.ii	70–73	Canker and leaf spot in poinsettias
C. flaccumfaciens[b]	2.ii	68–74	Vascular wilt of beans
C. betae	2.ii	70–73	Vascular wilt of beet
C. oortii	?	?	Vascular wilt of tulips
C. michiganense	2.ii	67–75	Tomato canker
C. insidiosum	2.iii	73–78	Vascular wilt and stunting of luceme
C. sepedonicum	2.iii	70–72	Vascular wilt and ring rot of potatoes
C. tritici	2.iii	70	Gumming disease of wheat
C. rathayi	?	?	Gumming disease of cocksfoot grass

[a] Figures for G + C content have been rounded up to the nearest whole number.
[b] Two variants, producing different colored pigments, have been named *C. flaccumfaciens* var. *aurantiacum* and *C. flaccumfaciens* var. *violaceum*.

Compiled from information in References 3, 4, 7, 30, 32, 33, 50, 59, and 62.

Corynebacterium pyogenes is a very frequent cause of purulent infections in cattle, pigs, and sheep, and less frequently in other animals, either alone or in combination with other organisms, e.g., *Fusobacterium necrophorus.*[43–47,53]

Corynebacterium haemolyticum was originally isolated from throat, ear, skin infections in military personnel in the Pacific area during World War II, but has subsequently been found in human infections in the U.S.[20] and elsewhere (see, for example, References 38, 54, and 55).

The two organisms appear to be related,[1,53] but their exact taxonomic relationship to other coryneform organisms is still uncertain.

4. Anaerobic Coryneforms

Traditionally, the acne bacillus is placed in the genus *Corynebacterium* along with a number of other named species of anaerobic diphtheroids, such as *C. diphtheroides, C. avidum, C. parvum, C. liquefaciens, C. granulosum,* and *C. lymphophilum.*[39,40] On the basis of gross morphology, these organisms are undoubtedly coryneform. However, they lack the characteristic arabinogalactan in their walls, and the diaminoacid of the peptidoglycan is usually L-DAP instead of the *meso*-isomer. Also, they are anaerobic, have a propionic acid type of metabolism,[15,34] and the principal fatty acids in the cell lipids are C_{15} branched-chain compounds.[35–37] The corynomycolic acids (C_{32-36}), which seem to be characteristic of the classical corynebacteria (see above), have not been found in the anaerobic diphtheroids.[16] In all of these respects the anaerobic coryneforms resemble the propionibacteria instead of the corynebacteria, *sensu stricto,* as defined above in subsection 1. It seems more logical, therefore, to classify them in the genus *Propionibacterium,* as has been done in the latest edition of *Bergey's Manual of Determinative Bacteriology.*[4]

As a result of combined cell wall analysis and DNA homology studies, it has been possible to divide the anaerobic diphtheroids into three species, *P. acnes, P. avidum,* and *P. granulosum,* as shown in Table 4. These species can also be distinguished by physiological and serological tests.[11,12,25] In homology tests, *P. acnes* and *P. avidum* strains show approximately 50% homology to each other, but only 12 to 15% homology to strains of *P. granulosum.*[25] All three groups show relatively low homology, about 15%, to strains of the classical propionibacteria.[25]

MEDIA AND CONDITIONS OF GROWTH FOR CORYNEBACTERIA

With a group of such diverse origins it is obviously unlikely that a single medium will be satisfactory for all members. Moreover, little is known about the nutritional requirements of many strains. Without going with details of special media for the isolation and identification of certain species, e.g., *C. diphtheriae,* Table 5 below gives an idea of the broad requirements of each of the four groups described in Table 1.

Table 4
ANAEROBIC CORYNEFORMS

***P. acnes* (*C. acnes*) and Related Organisms**

Organism	Serological type	Cell wall components		DNA G + C content (%)	DNA homology relationships, % homology to reference strains		
		Sugars[a]	DAP-isomer		0389 acnes	0575 avidum	0507 granulosum
P. acnes	acnes type I	Galactose, glucose, mannose	L-	58–60	97	51	16
	acnes type II	Glucose, mannose	L- or *meso-*				
P. avidum	avidum type I	Galactose, glucose, mannose	L-	62–63	50	90	17
	avidum type II	Glucose, Mannose	L- or *meso-*				
P. granulosum	only one type identified	Galactose, mannose (glucose, none or trace)	L-	61–63	12	15	95

[a] In all cases except *P. granulosum,* galactosamine is also a constituent of the cell wall polysaccharide in these organisms.

Compiled from information in References 11, 12, and 25.

Table 5
NUTRITIONAL AND GROWTH REQUIREMENTS OF CORYNEBACTERIA

Group	Suitable media	Conditions for good growth
1. *C. diphtheriae* and animal pathogens	Media containing blood or serum	Aerobic; often grow best as a surface pellicle, at or near 37°C.
2. Plant-pathogenic corynebacteria	Media containing soil extract[a]	Aerobic; grow best at 25–30°C.
3. *P. acnes* and related organisms	Media containing Tween® 80 or some other source of oleic acid	Anaerobic; grow best at or near 37°C.
4. *C. pyogenes* and *C. haemolyticum*	Media containing blood or serum	Aerobic or anaerobic; grow best at or near 37°C.

[a] Soil extract is generally prepared by autoclaving garden soil with tap water, filtering the extract, and adding to some basal medium.[26]

REFERENCES

1. **Barksdale, W. L.,** *Bacteriol. Rev.,* 34, 378, 1970.
2. **Bousfield, I. J.,** *J. Gen. Microbiol.,* 71, 441, 1972.
3. **Bowie, I. S., Grigor, M. R., Dunckley, G. C., Loutit, M. W., and Loutit, J. S.,** *Soil Biol. Biochem.,* 4, 397, 1972.
4. **Buchanan, R. E. and Gibbons, N. E. (Eds.),** *Bergey's Manual of Determinative Bacteriology,* 8th ed., Williams & Wilkins, Baltimore, 1974.
5. **Collier, R. J.,** *Bacteriol. Rev.,* 39, 54, 1975.
6. **Collins, M. D., Goodfellow, M., and Minnikin, D. E.,** *Proc. Soc. Gen. Microbiol.,* 3, 98, 1976.
7. **Crombach, W. H. T.,** *Antonie van Leuweenhoek J. Microbiol. Serol.,* 38, 105, 1975.
8. **Cummins, C. S.,** *J. Gen. Microbiol.,* 28, 35, 1962.
9. **Cummins, C. S.,** *Can. J. Microbiol.,* 17, 1001, 1971.
10. **Cummins, C. S.,** *J. Bacteriol.,* 105, 1227, 1971.
11. **Cummins, C. S.,** *J. Clin. Microbiol.,* 2, 104, 1975.
12. **Cummins, C. S. and Johnson, J. L.,** *J. Gen. Microbiol.,* 80, 433, 1974.
13. **DaSilva, G. A. N. and Holt, J. G.,** *J. Bacteriol.,* 90, 921, 1965.
14. **Davis, G. H. G. and Newton, K. G.,** *J. Gen. Microbiol.,* 56, 195, 1969.
15. **Douglas, H. C. and Gunter, S. E.,** *J. Bacteriol.,* 52, 15, 1946.
16. **Etemadi, A. H.,** *Bull. Soc. Chim. Biol.,* 45, 1420, 1963.
17. **Fiedler, F. and Kandler, O.,** *Arch. Mikrobiol.,* 89, 51, 1973.
18. **Gundersen, W. B.,** *Acta Pathol. Scand.,* 47, 65, 1959.
19. **Harrington, B. J.,** *J. Gen. Microbiol.,* 45, 31, 1966.
20. **Hermann, G. J.,** *Am. J. Med. Technol.,* 27, 61, 1961.
21. **Honda, E. and Yanagawa, R.,** *Int. J. System. Bacteriol.,* 23, 226, 1973.
22. **Jayne-Williams, D. J. and Skerman, T. M.,** *J. Appl. Bacteriol.,* 29, 72, 1966.
23. **Jebb, W. H.,** *J. Pathol. Bacteriol.,* 60, 403, 1948.
24. **Jensen, H. L.,** *Annu. Rev. Microbiol.,* 6, 77, 1952.
25. **Johnson, J. L. and Cummins, C. S.,** *J. Bacteriol.,* 109, 1047, 1972.
26. **Jones, D.,** *J. Gen. Microbiol.,* 87, 52, 1975.
27. **Kato, M.,** *J. Bacteriol.,* 101, 709, 1970.
28. **LaCave, C., Asselineau, J., and Toubiana, R.,** *Eur. J. Biochem.,* 2, 37, 1967.
29. **Lacey, M. S.,** *Trans. Br. Mycol. Soc.,* 38, 49, 1955.
30. **Lelliott, R. A.,** *J. Appl. Bacteriol.,* 29, 114, 1966.
31. **McLeod, J. W.,** *Bacteriol. Rev.,* 7, 1, 1943.
32. **Marmur, J. and Doty, P.,** *J. Mol. Biol.,* 5, 109, 1962.
33. **Marmur, J., Falkow, S., and Mandel, M.,** *Annu. Rev. Microbiol.,* 17, 329, 1963.
34. **Moore, W. E. C. and Cato, E. P.,** *J. Bacteriol.,* 85, 870, 1963.
35. **Moss, C. W. and Cherry, W. B.,** *J. Bacteriol.,* 95, 241, 1968.

36. Moss, C. W., Dowell, V. R., Lewis, V. J., and Schekter, M. A., *J. Bacteriol.*, 94, 1300, 1967.
37. Moss, C. W., Dowell, V. R., Farshtchi, D., Raines, L. J., and Cherry, W. B., *J. Bacteriol.*, 97, 561, 1969.
38. Patocka, F., Mara, M., Soucek, A., and Souckova, A., *J. Hyg. Epidemiol. Microbiol. Immunol.*, 6, 1, 1962.
39. Prévot, A. R. and Fredette, V., *Manual for the Classification and Determination of the Anaerobic Bacteria,* Lea and Febiger, Philadelphia, 1966.
40. Prévot, A. R., Turpin, A., and Kaiser, P., *Les Bactéries Anaérobies,* Dunod, Paris, 1967.
41. Pudles, J. and Lederer, E., *Biochim. Biophys. Acta,* 11, 163, 1953.
42. Reid, J. D. and Joya, M. A., *Int. J. System. Bacteriol.*, 19, 273, 1969.
43. Roberts, D. S., *Br. J. Exp. Pathol.*, 48, 665, 1967.
44. Roberts, D. S., *Br. J. Exp. Pathol.*, 48, 674, 1967.
45. Roberts, R. J., *Vet. Rec.*, 79, 346, 1966.
46. Roberts, R. J., *J. Pathol. Bacteriol.*, 95, 127, 1968.
47. Roberts, R. J., *Res. Vet. Sci.*, 9, 350, 1968.
48. Schildkraut, C. L., Marmur, J., and Doty, P., *J. Mol. Biol.*, 4, 430, 1962.
49. Schleifer, K. H. and Kandler, O., *Bacteriol. Rev.*, 36, 407, 1972.
50. Schuster, M. L., Vidaver, A. K., and Mandel, M., *Can. J. Microbiol.*, 14, 423, 1968.
51. Skerman, T. M. and Jayne-Williams, D. J., *J. Appl. Bacteriol.*, 29, 167, 1966.
52. Skyring, G. W. and Quadling, C., *Can. J. Microbiol.*, 16, 95, 1970.
53. Sorensen, G. H., *Acta Vet. Scand.*, 15, 544, 1974.
54. Soucek, A., Mara, M., and Souckova, A., *J. Hyg. Epidemiol. Microbiol. Immunol.*, 9, 67, 1965.
55. Soucek, A., Souckova, A., Mara, M., and Patocka, F., *J. Hyg. Epidemiol. Microbiol. Immunol.*, 6, 13, 1962.
56. Sukapure, R. S., Lechevalier, M. P., Reber, H., Higgins, M. C., Lechevalier, H. A., and Prauser, H., *Appl. Microbiol.*, 19, 527, 1970.
57. Tomlinson, A. J. H., *J. Appl. Bacteriol.*, 29, 131, 1966.
58. Veldkamp, H., *Annu. Rev. Microbiol.*, 24, 209, 1970.
59. Vidaver, A. K. and Mandel, M., *Int. J. System. Bacteriol.*, 24, 482, 1974.
60. Welby-Guisse, M., Lanéelle, M. A., and Asselineau, J., *Eur. J. Biochem.*, 13, 164, 1970.
61. Yamada, K. and Komagata, K., *J. Gen. Appl. Microbiol.*, 16, 103, 1970.
62. Yamada, K. and Komagata, K., *J. Gen. Appl. Microbiol.*, 16, 215, 1970.
63. Yamada, K. and Komagata, K., *J. Gen. Appl. Microbiol.*, 18, 399, 1972.
64. Yamada, K. and Komagata, K., *J. Gen. Appl. Microbiol.*, 18, 417, 1972.

THE ACTINOMYCETALES*

INTRODUCTION

H. A. Lechevalier and L. Pine

Actinomycetes are filamentous, branching bacteria that are widely distributed in nature. They are mainly found in soil, where they play a major role in the decomposition of organic matter. Some species cause diseases of animals or plants; a few species are parasitic and have been isolated only from their animal hosts. Some actinomycetes are of medical and industrial importance as producers of antibiotics, and others are used industrially as agents of chemical transformations. For general reviews on actinomycetes see References 1 to 9.

Although branched filaments may also be observed in some budding bacteria, mycoplasmata, and highly pleomorphic organisms found in the genera *Actinobacillus* and *Streptobacillus,*[10] these are not classified in the Actinomycetales. Basically, members of the Actinomycetales may be considered to be those bacteria that form a well-defined coherent mycelium. But this structure may be so rudimentary and transient in the genera *Actinomyces* and *Mycobacterium* that it may go unnoticed or be essentially nonexistent. Indeed, in the genus *Bifidobacterium* no mycelium has been described. Thus, no unambiguous morphological feature serves to unite all the members of this widely diversified group.

Actinomycetales may form a substrate mycelium only, or both aerial and substrate mycelia, or an aerial mycelium only (*Sporichthya*). Although some species are holocarpic, eucarpic members may show highly complex mycelial structures with conidia and sporangia.[11] Motility, when observed, is due to flagella. Conidia and sporangiospores may be variously arranged and/or variously structured with surface hairs, spines, or ridges.

Although generic separation may be made on morphological differences in the case of highly evolved types, classification of the most primitive forms rests primarily on oxygen requirements, fermentative abilities, formation of catalase and respiratory cytochrome systems, and cell wall and lipid composition. Most actinomycetes are Gram-positive, but part of their thallus (depending on the stages of growth) may be Gram-negative; members of the genus *Mycoplana* are Gram-negative.

Phylogenetically, there is very little reason to classify organisms on the basis of ecology. However, a realistic appraisal of research pertinent to the actinomycetes shows that two major groups may be recognized: (1) parasitic forms found in the mucosal cavities of man and animals, and (2) soil forms. It is recognized that among the so-called soil forms parasitic actinomycetes may be found; similarly, some soil forms may be closely related to the animal parasites.

As a group, the parasitic actinomycetes are presently contained within the family Actinomycetaceae and are unified by the following characteristics: they are Gram-positive; they are anaerobic to aerobic, but generally microaerophilic; they ferment glucose under anaerobic or aerobic conditions to form acids, which can account for 50 to 100% of the substrate carbon; they form a loosely bound or transient mycelium, but do not sporulate or form motile cells; at best, they have a limited oxidative system, by which oxygen is utilized for growth.

* Part 17 in *Bergey's Manual.*

Conversely, the soil actinomycetes may be viewed as Gram-positive, sometimes acid-fast organisms having an aerobic metabolism that mediates growth and does not result in large accumulation of acid from carbohydrate substrates. Well-recognizable coherent mycelium is usually formed, often with specialized structures and spores.

As a practical device for presentation of the various genera, the following review on actinomycetes is divided into two sections: soil or oxidative actinomycetes, and parasitic or fermentative actinomycetes. At one extreme of the soil actinomycetes are oxidative mycelial organisms, whereas at the opposite pole of the parasitic actinomycetes are fermentative diphtheroids. Between these two extremes there are no voids, but a population of intermediate forms.

REFERENCES

1. **Lechevalier, H. A. and Lechevalier, M. P.,** *Annu. Rev. Microbiol.*, 21, 71, 1967.
2. **Prauser, H.,** *The Actinomycetales,* Gustav Fischer, Jena, Germany, 1970.
3. **Waksman, S. A.,** *The Actinomycetes,* Ronald Press, New York, 1967.
4. **Williams, S. T., Davies, F. L., and Cross, T.,** in *Identification Methods for Microbiologists,* Part B, Gibbs, B. M. and Shapton, D. A., Eds., Academic Press, London, 1968, p. 111.
5. **Pine, L. and George, L. K.,** *Int. J. Syst. Bacteriol.*, 15, 143, 1965.
6. **Pine, L.,** *Int. J. Syst. Bacteriol.*, 20, 445, 1970.
7. **Lechevalier, H. A., Lechevalier, M. P., and Gerber, N. N.,** Chemical composition as a criterion in the classification of actinomycetes, *Adv. Appl. Microbiol.*, 14, 47–72, 1971.
8. **Sykes, G. and Skinner, F. A. (Eds.),** *Actinomycetales: Characteristics and Practical Importance,* Academic Press, London, 1973.
9. **Goodfellow, M., Brownell, G. H., and Serrano, J. A. (Eds.),** *The Biology of the Nocardiae,* Academic Press, London, 1976.
10. **Lechevalier, H. A. and Pramer, D.,** *The Microbes,* J. B. Lippincott, Philadelphia, 1971.
11. **Lechevalier, H. A., Lechevalier, M. P., and Holbert, P. E.,** *J. Bacteriol.*, 92, 1228, 1966.

THE ACTINOMYCETALES*

SOIL OR OXIDATIVE ACTINOMYCETES

H. A. Lechevalier

Analyses of cell wall preparations of soil or oxidative actinomycetes, some of which are animal pathogens (see Tables 7, 8, and 9), reveal that they fall into four main groups, as indicated in Tables 1 to 5.

Strains of *Oerskovia, Agromyces,* and *Mycoplana* do not produce aerial mycelia. Oerskoviae and mycoplanae form branched filaments, which break up into motile segments. Strains of *Agromyces* are microaerophilic to aerobic catalase-negative actinomycetes with no special morphological features, but they are abundantly distributed in soil.

It is possible to determine by one-way chromatography of whole-cell hydrolysates whether an actinomycete contains major amounts of L- or of *meso*-DAP.[1] On the basis of paper chromatography of sugars present in whole-cell hydrolysates of actinomycetes containing major amounts of *meso*-DAP, it is possible to separate them into four groups.[2,3] As can be seen in Table 6, whole-cell sugar pattern A indicates the presence of Type IV wall, and whole-cell sugar pattern D indicates the presence of Type II wall. Organisms with Type III wall can be separated into two groups, depending on the presence or absence of madurose (see Tables 4 and 6).

A distinction between the genera *Mycobacterium* and *Nocardia* can be made on the basis of the type of mycolic acid produced.[4] Mycobacteria contain true mycolic acids, giving C_{22} to C_{26} fatty acids upon pyrolysis. Nocardiae contain related nocardomycolic acids, which are smaller molecules and give smaller fatty acids (C_{12} to C_{18}) upon pyrolysis.

A tentative phylogenetic sketch of some of the better-known aerobic actinomycetes is given in Figure 1. Chemical criteria are incorporated in the scheme. Table 7 lists species of actinomycetes of special importance, following the same generic order as in Tables 2 to 5. Tables 8 and 9 list properties of zoopathogenic members of the genera *Nocardia, Actinomadura, Streptomyces,* and *Mycobacterium.*

* Part 17 in *Bergey's Manual.*

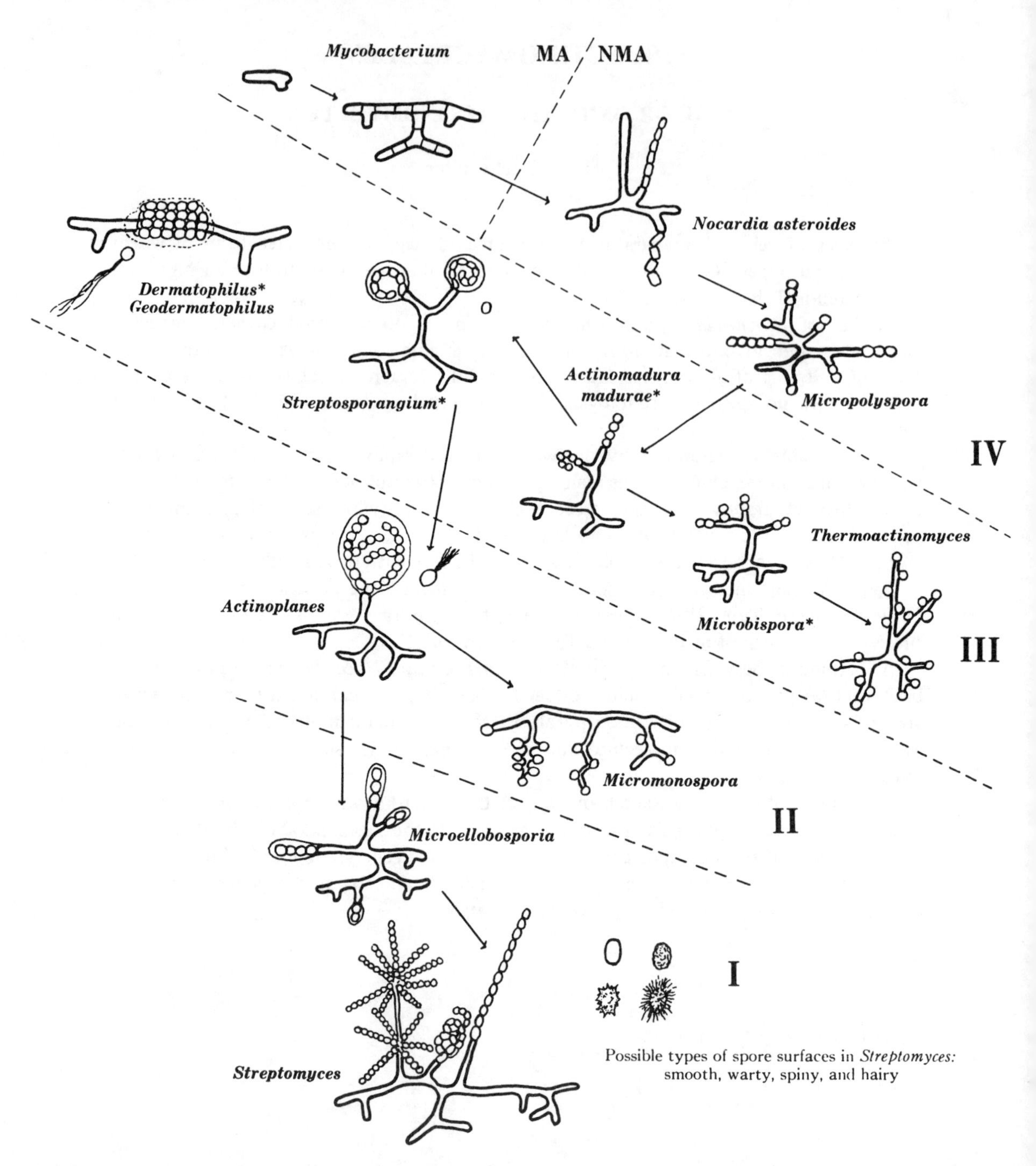

FIGURE 1. Phylogenetic sketch of some aerobic actinomycetes. Code: Roman numerals refer to cell wall types (see Table 1); * = madurose-containing; MA = contains mycolic acid; NMA = contains nocardomycolic acid. (From *Laval Méd.*, 38, 740, 1967. Reproduced by permission of the copyright owners, *Laval Médical*, Ecole de Médicine, Université Laval, Quebec, Canada.)

Table 1
MAJOR CONSTITUENTS OF CELL WALLS OF SOIL ACTINOMYCETES[a]

Cell wall type	Major constituents	Reference
Streptomyces or Type I	L-DAP,[b] glycine	5, 6
Micromonospora or Type II	*meso*-DAP,[c] glycine; hydroxy-DAP may also be present	5, 6
Actinomadura or Type III	*meso*-DAP	5, 6
Nocardia or Type IV	*meso*-DAP, arabinose, galactose	5, 6
Oerskovia	Lysine, aspartic acid, galactose	7
Agromyces	DAB,[d] glycine	8
Mycoplana[e]	*meso*-DAP; also many amino acids	9

[a] All cell wall preparations contain major amounts of alanine, glutamic acid, glucosamine, and muramic acid.
[b] DAP = 2,6-diaminopimelic acid.
[c] No differentiation is made between *meso*-DAP and D-DAP.
[d] DAB = 2,4-diaminobutyric acid.
[e] These microorganisms are Gram-negative.

Table 2
MORPHOLOGIC CHARACTERISTICS OF ACTINOMYCETALES WITH A TYPE I CELL WALL[a]

Generic name	Morphologic characteristics
Streptomyces[b]	Aerial mycelium with chains (usually long) of non-motile conidia.
Streptoverticillium[c]	Same as *Streptomyces,* but the aerial mycelium bears verticils consisting of at least three side branches, which may be chains of conidia or hold sporulating terminal umbels.
Nocardioides	Both substrate and aerial mycelia fragment into rod- and coccus-shaped elements.
Chainia	Same as *Streptomyces,* but sclerotia are also formed.
Actinopyenidium	Same as *Streptomyces,* but pycnidia-like structures are also formed.
Actinosporangium	Same as *Streptomyces,* but spores accumulate in drops.
Elytrosporangium	Same as *Streptomyces,* but merosporangia are also formed.
Microellobosporia	No chains of conidia; merosporangia with non-motile spores are formed.
Sporichthya	No substrate mycelium is formed; aerial chains of motile, flagellated conidia are held to the surface of the substratum by holdfasts.[12]
Intrasporangium	No aerial mycelium; substrate mycelium forms terminal and subterminal vesicles.[13]

[a] All aerobic.
[b] This genus includes the most common soil forms and most of the important producers of antibiotics. See Reference 10 for methods of characterization.
[c] See Reference 11.

Table 3
MORPHOLOGIC CHARACTERISTICS OF ACTINOMYCETALES WITH A TYPE II CELL WALL[a]

Generic name	Morphologic characteristics
Micromonospora	Aerial mycelium absent, conidia single.
Actinoplanes	Globose to lageniform sporangia; globose spores with one polar tuft of flagella.
Amorphosporangium	Same as *Actinoplanes*, but the sporangia are often very irregular; sporangiospores are usually non-motile.
Ampullariella	Lageniform to globose sporangia; rod-shaped spores with one polar tuft of flagella.
Dactylosporangium	Claviform sporangia, each with one chain of spores with one polar tuft of flagella.

[a] All aerobic, except for some micromonosporae; the cell wall composition of anaerobic micromonosporae is not known.

Table 4
MORPHOLOGIC CHARACTERISTICS OF ACTINOMYCETALES WITH A TYPE III CELL WALL[a]

Generic name	Morphologic characteristics
Actinomadura[a]	Short chains of conidia on the aerial mycelium.
Nocardiopsis	Very long chains of conidia on the aerial mycelium.[2,26]
Microbispora[b]	Longitudinal pairs of conidia on the aerial mycelium.
Thermoactinomyces	Single spores are formed on the aerial and substrate mycelia; the spores are heat-resistant endospores.
Thermomonospora	Single spores are formed on the aerial mycelium or on both the aerial and substrate mycelia; the spores are not heat-resistant endospores.
Actinobifida	Same as *Thermomonospora*, but the sporophores are dichotomously branched.
Streptosporangium[b]	Globose sporangia containing non-motile spores.
Spirillospora[b]	Globose sporangia with rod-shaped spores, each with a subpolar tuft of flagella.
Planomonospora[b]	Cylindrical sporangia, each containing one motile spore with one polar tuft of flagella.
Planobispora[b]	Cylindrical sporangia, each containing two motile spores with peritrichous flagella.
Dermatophilus[b]	Hyphae dividing in all planes, forming packets of cocci motile by means of a tuft of flagella; pathogenic to animals.
Geodermatophilus	Same as *Dermatophilus*, but a soil form.[14]

[a] All aerobic.
[b] Madurose-containing organisms (see Table 6).

Table 5
MORPHOLOGIC CHARACTERISTICS OF ACTINOMYCETALES WITH A TYPE IV CELL WALL[a]

Generic name	Morphologic characteristics
Mycobacterium	Filamentation is usually limited, and aerial mycelium is usually not formed; filaments fall easily apart into rods and cocci.
Nocardia	Filamentation is abundant, and aerial mycelium is often formed; chains of conidia may be formed.
Micropolyspora	Short chains of globose conidia are formed on both the aerial and the substrate mycelia.
Pseudonocardia	Long, cylindrical conidia in chains on the aerial mycelium.
Saccharomonospora	Single spores, mainly on the aerial mycelium.
Saccharopolyspora[25]	Morphology similar to that of *Actinomadura dassonvillei.*

[a] All aerobic.

Table 6
CELL WALL TYPES AND WHOLE-CELL SUGAR PATTERNS OF AEROBIC ACTINOMYCETES CONTAINING *meso*-DIAMINOPIMELIC ACID[a]

Cell wall[5,6]		Whole-cell sugar pattern[2,3]	
Type	Distinguishing major constituents[b]	Type	Diagnostic sugars
II	Glycine	D	Xylose, arabinose
III	None	B	Madurose[c]
		C	None
IV	Arabinose, galactose	A	Arabinose, galactose

[a] No differentiation is made between *meso*-DAP and D-DAP.

[b] All cell-wall preparations contain major amounts of alanine, glutamic acid, glucosamine, and muramic acid.

[c] Madurose = 3-O-methyl-D-galactose.[15]

Table 7
IMPORTANT SPECIES OF AEROBIC ACTINOMYCETES

Organism	Importance	Reference
Streptomyces		
antibioticus	Production of actinomycin.	16
aureofaciens	Production of chlortetracycline and tetracycline.	16
erythreus	Production of erythromycin.	16
fradiae	Production of neomycin.	16
griseus	Production of streptomycin, cycloheximide, and candicidin.	16
griseus (scabies)	Cause of potato scab.	17
nodosus	Production of amphotericin B.	16
noursei	Production of nystatin.	16
rimosus	Production of oxytetracycline.	16
somaliensis	Found in mycetomas.	18
venezuelae	Production of chloramphenicol.	16
Micromonospora		
echinospora	Production of gentamicin.	19
purpurea	Production of gentamicin.	19
Actinomadura		
madurae	Found in mycetomas.	18
pelletieri	Found in mycetomas.	18
Thermoactinomyces		
vulgaris	Cause of pneumonitis (farmer's lung).	20
Dermatophilus		
congolensis	Cause of streptothricosis of animals.	18
Mycobacterium		
avium	Cause of tuberculosis.	21
bovis	Cause of tuberculosis.	21
farcinogenes	Cause of bovine farcy.	22
fortuitum	Causes abscesses in man; grows more rapidly and is more filamentous than most pathogenic mycobacteria.	21
kansasii	Cause of tuberculosis.	21
leprae	Associated with leprosy; has never been cultured on laboratory media.	21
marinum	Cause of skin lesions in man.	21
paratuberculosis	Cause of Johne's disease of cattle.	23
tuberculosis	Cause of tuberculosis.	21
Nocardia		
asteroides	Cause of deep-seated nocardiosis; also found in mycetomas.	18
brasiliensis	Cause of deep-seated nocardiosis; also found in mycetomas.	18
caviae	Cause of deep-seated nocardiosis; also found in mycetomas.	18
farcinica	See *Mycobacterium farcinogenes*.	
Micropolyspora		
faeni	Cause of pneumonitis (farmer's lung).	24

Table 8
DISTINCTIVE PROPERTIES OF MYCOBACTERIAL SPECIES

CODE

Column 1: *M. abscessus*
Column 2: *M. avium*
Column 3: *M. bovis*
Column 4: *M. flavescens*
Column 5: *M. fortuitum*
Column 6: *M. gastri*
Column 7: *M. intracellulare*
Column 8: *M. kansasii*
Column 9: *M. marianum (scrofulaceum)*
Column 10: *M. marinum*
Column 11: *M. terrae*
Column 12: *M. tuberculosis*
Column 13: *M. ulcerans*
Column 14: *M. xenopei*
Column 15: "Aquae" strains
Column 16: "V" *(M. triviale)*
Column 17: other Group IV[a]

Property	1	2	3	4	5	6	7	8	9	10	11	12	13	14	15	16	17
Specimen Source																	
See footnote b	+								+	+			+				
Growth Rate[c]																	
At 45°C	–	+			–		–							S			V
At 37°C	R	S	S	M	R	S	S	S	S	∓	S	S	–	S	S	S	R
At 31°C										M			S				
At 24°C	R	±	–	M	R	S	±	S	S	M	S	–	–	–	S	S	R
Colony																	
Always rough on 7H10	–	–	+	–	–	–	–	–	–	–	–	+	+	–	–	+	–
Usually smooth, thin		+					+		+								
Branched filamentous extension	–			–	+									+			V
Pigment																	
Photochromogen	–	–	–		–	–	–	+		+	–	–	–			–	–
Scotochromogen	–	–	–	+	–	–	–	–	+	–	–	–	–	+	+	–	V
Tests																	
Niacin	V	–	–			–	–				–	+	+			–	
Susceptible to isoniazid (1 μg/ml)	–	–	+	+	–	+	–	+	–	+	–	+	–	+	–	–	–
Susceptible to T_2H (10 μg/ml)		–	+			–	–				–	–	–			–	
Agglutination tests available	+	+			+		+	+	+	+							
Tween® hydrolysis (5 days)	–	–		+	∓	+	–	+	–	+	+		–	–	+	+	+
Tellurite reduction (3 days)	–	+		–	+	–	+		–		–			–	–	–	V
Nitrate reduction	–	–	–	+	+	–	–	+	–	–	+	+	–	–	–	+	+
Arylsulfatase (3 days)	+			+	+	–					–					+	–
MacConkey agar	±			–	+	–					–					–	–
Iron uptake				–	+	–			–		–			–	–	–	+
68°C catalase	+	+	–	+	+	–	+				+	–	+	+		+	+
>45 mm catalase	+	–		+	+	–	–	+	+		+			–		+	+
5% NaCl tolerance	+			±	+	–					–					+	+

[a] Group IV of Runyon = growing fully in less than one week.
[b] A plus sign indicates that the nature of the source of the specimen (as from a superficial body area) is information contributing to the identification.
[c] S = slow; M = moderate; R = rapid; V = variable.
[d] Thiophene-2-carboxylic acid hydrazide.

From Blair, J. E., Lennette, E. H., and Truant, J. P. (Eds.), *Manual of Clinical Microbiology,* American Society for Microbiology, Washington, D.C., 1970, 122. Reproduced with permission of the copyright owners.

Table 9
PHYSIOLOGICAL CHARACTERISTICS OF PATHOGENIC *NOCARDIA*, *ACTINOMADURA*, AND *STREPTOMYCES* SPECIES

Species	Decomposition[a] of					BCP[b] milk	Urea	Acid from arabinose and xylose
	Casein	Tyrosine	Xanthine	Starch	Gelatin			
Nocardia								
asteroides	–	–	–	–[c]	–[d]	–[e]	+	–
brasiliensis	+	+	–	–[c]	+	+	+	–
caviae	–	–	+	–[c]	–	–[e]	+	–
farcinica	–	–	–	–	–	–	–	
Actinomadura								
madurae	+	+	–	+	+	+	–	+
pelletieri	+	+	–	–	+	+	–	–
Streptomyces								
paraguayensis	+	+	+	–	+	+	+	
somaliensis	+	+	–	±	+	+	–	–

[a] Within 2 weeks at 27°C.
[b] BCP = bromcresol purple.
[c] About 50% of strains give a positive test result when a different method is used.
[d] Some strains reportedly liquefy certain gelatin media.
[e] Usually turns alkaline.

From Blair, J. E., Lennette, E. H., and Truant, J. P. (Eds.), *Manual of Clinical Microbiology*, American Society for Microbiology, Washington, D.C., 1970, 138. Reproduced with permission of the copyright owners.

REFERENCES

1. **Becker, B., Lechevalier, M. P., Gordon, R. E., and Lechevalier, H. A.,** *Appl. Microbiol.*, 12, 421, 1964.
2. **Lechevalier, H. A. and Lechevalier, M. P.,** A critical evaluation of the genera of aerobic actinomycetes, in *The Actinomycetales*, Prauser, H., Ed., Gustav Fischer, Jena, Germany, 1970, pp. 393–405.
3. **Lechevalier, M. P.,** *J. Lab. Clin. Med.*, 71, 934, 1968.
4. **Lechevalier, M. P., Horan, A. C., and Lechevalier, H. A.,** *J. Bacteriol.*, 105, 313, 1971; **Lechevalier, M. P., Lechevalier, H., and Horan, A. C.,** *Can. J. Microbiol.*, 19, 965, 1973.
5. **Becker, B., Lechevalier, M. P., and Lechevalier, H. A.,** *Appl. Microbiol.*, 13, 236, 1965.
6. **Yamaguchi, T.,** *J. Bacteriol.*, 89, 444, 1965.
7. **Sukapure, R. S., Lechevalier, M. P., Reber, H., Higgins, M. L., Lechevalier, H. A., and Prauser, H.,** *Appl. Microbiol.*, 19, 527, 1970.
8. **Gledhill, W. E. and Casida, L. E.,** *Appl. Microbiol.*, 18, 340, 1969.
9. **Higgins, M. L., Lechevalier, M. P., and Lechevalier, H. A.,** *J. Bacteriol.*, 93, 1446, 1967.
10. **Shirling, E. B. and Gottlieb, D.,** *Int. J. Syst. Bacteriol.*, 16, 313, 1966.
11. **Locci, R., Baldacci, E., and Petrolini-Baldan, B.,** *G. Microbiol.*, 17, 1, 1969.
12. **Lechevalier, M. P., Lechevalier, H. A., and Holbert, P.E.,** *Ann. Inst. Pasteur*, 114, 277, 1968.
13. **Lechevalier, H. and Lechevalier, M. P.,** *J. Bacteriol.*, 100, 522, 1969.
14. **Luedemann, G. M.,** *J. Bacteriol.*, 96, 1848, 1968.
15. **Lechevalier, M. P. and Gerber, N. N.,** *Carbohyd. Res.*, 13, 451, 1970.
16. **Waksman, S. A. and Lechevalier, H. A.,** *Antibiotics of Actinomycetes*, Williams & Wilkins, Baltimore, 1962.
17. **Waksman, S. A.,** *The Actinomycetes*, Ronald Press, New York, 1967.
18. **Gordon, M. A.,** in *Manual of Clinical Microbiology*, Blair, J. E., Lennette, E. H., and Truant, J. P., Eds., American Society for Microbiology, Bethesda, Maryland, 1970, p. 137.
19. **Lechevalier, H. A. and Lechevalier, M. P.,** *Annu. Rev. Microbiol.*, 21, 71, 1967.
20. **Wenzel, F. J., Emanuel, D. A., and Lawton, B. R.,** *Am. Rev. Resp. Dis.*, 95, 652, 1967.

21. **Runyon, E. H., Kubica, G. P., Morse, W. C., Smith, C. R., and Wayne, L. G.,** in *Manual of Clinical Microbiology,* Blair, J. E., Lennette, E. H., and Truant, J. P., Eds., American Society for Microbiology, Bethesda, Maryland, 1970, p. 112.
22. **Asselineau, J., Lanéele, M. A., and Chamoiseau, G.,** *Rev. Elev. Med. Vet. Pays Trop.,* 22, 205, 1969.
23. **Siegmund, O. H. and Eaton, L. G. (Eds.),** *The Merck Veterinary Manual,* 3rd ed., Merck and Company, Inc., Rahway, N.J., 1967.
24. **Lacey, J.,** *J. Gen. Microbiol.,* 41, 406, 1965.
25. **Lacey, J. and Goodfellow, M.,** *J. Gen. Microbiol.,* 88, 75, 1975.
26. **Meyer, J.,** *Int. J. System. Bacteriol.,* 26, 487, 1976.

THE ACTINOMYCETALES*

PARASITIC OR FERMENTATIVE ACTINOMYCETES

L. Pine

Twelve species of organisms distributed among five genera have been proposed for inclusion in the family Actinomycetaceae[1,72] (see Tables 1 and 2). Of the species listed, the taxonomical status of *Actinomyces eriksonii* remains to be firmly established. On the basis of glucose fermentation, it may well belong in the genus *Bifidobacterium,* although lysine, not ornithine, is a major component of its cell wall. The genus *Agromyces*[2] is microaerophilic and fermentative and lacks both catalase and a cytochrome system. It belongs to the Actinomycetaceae, but it is discussed as one of the soil actinomycetes (see page 153). All species listed in Table 2, except *Actinomyces humiferus,* have been isolated from human or animal sources, and several cause invasive disease. The actinomycete group and species summarized here are described individually and in complete detail by Slack and Gerencser.[72]

In general, there are two microcolony types. Colonies showing soft, smooth convex surfaces with entire edges are associated with *Actinomyces bovis, A. eriksonii, A. odontolyticus,* or *Bifidobacterium bifidum.* All other species form rough-edged mycelial microcolonies that are composed of loosely attached branching cells or of true mycelium giving a spider-like appearance to the colony. The macrocolony, which may take 7 to 14 days to develop, may exhibit a wider range of structures. Regardless of the microcolony type, the macrocolony developing from it may vary from a smooth-surfaced convex colony to one resembling a cauliflower or a classical molar tooth. For excellent photographs of the various morphological forms, the reader is directed to References 3 to 8 and 72. Diagnostically, the nature of the disease, material, and cellular or colonial morphology of the actinomycete within it do not identify genera; nevertheless, they are strongly suggestive, and often highly characteristic.

Differentiation between species can be made on the basis of the physiological characteristics listed in Table 3. When acid is formed, it is not accompanied by the formation of gas. The amount of acid produced depends on the rate of growth and on the period of incubation. In view of the slow growth rates of some strains, fermentation tubes should be incubated no less than 14 days. *Arachnia propionica,* for example, does not ferment glycerol readily, but some strains may do so in incubation periods longer than 7 days. Similarly, the formation of red colonies by some strains of *Actinomyces odontolyticus* requires up to 21 days on blood agar.

Serologically, each of the *Actinomyces* species may be identified by the specific-fluorescent-antibody technique; *A. bovis, A. israelii, A. viscosus,* and *A. odontolyticus,* each show two relatively specific species serotypes.[1,9-12,72]

Relationships between the effects of carbon dioxide and oxygen on growth are complex (see Table 4). In general, carbon dioxide is required for growth, and substantial amounts may be fixed into the fermentation products. Some species show a twofold increase in cell yields when grown aerobically in shake culture. Regardless of their ability to grow aerobically, cultures of *Actinomyces israelii, A. naeslundii,* and *Arachnia propionica* are maintained anaerobically. Of the species that have no catalase, only one, *Actinomyces humiferus,* reportedly requires oxygen for growth. *Actinomyces viscosus, Bacterionema matruchotii,* and *Rothia dentocariosa,* which are catalase-positive, are also maintained aerobically. Anaerobic *Bacterionema* species have been described.[13,14]

The anaerobic and aerobic fermentations of glucose by the various species of the

* Part 17 in *Bergey's Manual.*

family Actinomycetaceae are given in Table 5, with fermentations reported for closely related genera. The fermentations are of three types: a propionic acid fermentation (*Arachnia* and *Bacterionema*),[5,15-17] a homolactic acid fermentation (*Actinomyces* and *Rothia*),[18-20] and a heterolactic acid fermentation (*Bifidobacterium*).[21-23]

Just as the species are divided into groups forming lactic acid or propionic acid (see Table 1), they may also be divided into two groups based on cell wall composition (see Table 6). Thus, the two genera fermenting glucose to propionic acid are the only members having diaminopimelic acid in the cell walls; all other species have lysine or lysine + ornitihine as the dibasic component of the cell wall murein. Within this latter group, a further division can be made; species can be divided into those having glucose and rhamnose or those primarily having galactose in the polysaccharide of the cell wall. If one considers oxygen relationship, cell wall composition, and morphology, the actinomycetes exhibit a phylogenetic progression from a strictly homofermentative, catalase-negative, anaerobic bifurcated type of organism (*Actinomyces bovis*) that has lysine in its cell wall to a strict aerobe that has catalase and a cytochrome system,[24,25] diaminopimelic acid in its cell wall, and a specialized mycelial structure that produces spores (Streptomycetes). An examination of the G + C ratios of the DNA of various actinomycetes (see Volume II of the first edition of the *Handbook*) also shows a general progression from 57% for some Actinomycetaceae to about 75% for the Streptomycetes. Although the percent G + C of *Arachnia propionica* (63 to 65%) is intermediate to the percent G + C exhibited by groups of coryneforms and by species of *Propionibacterium* (57 to 68%), the DNA of several strains of *A. propionica* showed no homology with these groups.[71]

Table 1
CHARACTERISTICS DELINEATING GENERA OF ACTINOMYCETACEAE[a]

Oxygen relationship	Catalase	Cell wall	Fermentation	Major products formed from glucose
		ACTINOMYCES		
Anaerobe, facultative anaerobe	+,–	Lysine + ornithine Lysine + aspartic acid	Homolactic	Formate + acetate + lactate + succinate
		ARACHNIA		
Facultative anaerobe	–	LL-Diaminopimelic acid + glycine	Propionic	Acetate + propionate
		BACTERIONEMA		
Aerobe, facultative anaerobe	+,–	*meso*-Diaminopimelic acid DL-Diaminopimelic acid + glycine	Propionic	Lactate + propionate
		BIFIDOBACTERIUM		
Anaerobe	–	Ornithine	Heterolactic	Acetate + lactate
		ROTHIA		
Aerobe	+	Lysine	Homolactic	Lactate

[a] Members of the family Actinomycetaceae are Gram-positive, non-acid-fast, non-motile, non-spore-forming, diphtheroid or branched rod-like bacteria that ferment carbohydrates, producing volatile and non-volatile acids.

Table 2
GENERA AND SPECIES OF FERMENTATIVE ACTINOMYCETES (FAMILY ACTINOMYCETACEAE)

Organism	Synonym	Habitat	Disease	References
Actinomyces				
bovis		Cattle	Lumpy jaw, actinomycosis	26–31
eriksonii		Man	Pulmonary and subcutaneous abscess	32
humiferus		Soil	None	33
israelii		Oral cavity, dental calculus, tonsillar crypts of man	Lacrimal canaliculitis, actinomycosis	6, 8, 26, 31, 33
naeslundii		Oral cavity, dental calculus, tonsillar crypts	Actinomycosis	6, 35–37
odontolyticus		Human dental caries	None	38, 39
suis		Swine	Swine mammary actinomycosis	40, 41
viscosus	Hamster organism *Odontomyces viscosus*	Oral cavity of hamster, human dental calculus	Subgingival plaque, peridontal disease	20, 42–45
Arachnia				
propionica	*Actinomyces propionicus*	Human tissues	Actinomycosis, lacrimal canaliculitis	5, 17, 36, 46, 47
Bacterionema				
matruchotii	*Leptotrichia*	Oral cavity of man and primates	None	13, 14, 16, 48, 49
Bifidobacterium				
bifidum	*Lactobacillus bifidus, Actinomyces parabifidus, Lactobacillus parabifidus, Actinobacterium bifidum*	Stool of milk-fed infants, adult stool, rat feces, bovine rumen, turkey liver	None	50–56
Rothia				
dentocariosa	*Nocardia salivae*	Oral cavity of man	None	4, 57–59

Table 3
FERMENTATIVE AND METABOLIC CHARACTERISTICS OF ACTINOMYCETACEAE

Organism	Catalase	Nitrate reduction	Casein and gelatin hydrolysis	Starch hydrolysis	Glucose	Starch	Mannitol	Ribose	Xylose	Glycerol	Color on blood agar	Oxygen requirement	% G + C composition of DNA[71,72]
Actinomyces													
bovis	–	–	–	+	A	A	–	–	A(–)[a]	–(A)[a]	White	$+CO_2$ An	57–63
eriksonii	–	–	–	+	A	A	A	A	A	–	White	An	62
humiferus	–	–	+[b]	+	A[c]	A	A	–	A	±	White	Ae	73
israelii	–	+(–)[a]	–	–	A	–	A	A	A	–	White	$+CO_2$ An	57–60
naeslundii	–	+	–	–	A	–	–	–	–	–	White	$+CO_2$ An	63–64
odontolyticus	–	+	–	–	A	–	–	–	A(–)[a]	A	Dark red	An	62
suis	–	NR	–	+	A	A	A(–)[a]	NR	A	A	Brown to red on blood	An	NR
viscosus	+	+	–	+	A	±	–	–	–	–	White	$+CO_2$ An	63–70
Arachnia													
propionica	–	+	–	–	A	–	A	–	–	–(A)[a]	White	An	63–65
Bacterionema													
matruchotii	+	+	–	+(–)[a]	A	NR	–	NR	–	–	White	Ae	50–56
Bifidobacterium													
bifidum	–	–	–	–	A	–	–	NR	–	–	White	An	57–64
Rothia													
dentocariosa	+	+	–	–	A	–	–	–	–	A	White	Ae	47–53

Note: A = acid produced; An = anaerobically; Ae = aerobically; NR = not reported.

[a] A few strains of the species give the reaction shown in parentheses.
[b] Strong + for casein, ± for gelatin.[33]
[c] Fermentation of glucose is limited; fructose is readily fermented.[33]

Compiled from Slack, J. M., *Report to the International Committee on Bacterial Nomenclature, Subgroup on Taxonomy of Microacrophilic Actinomycetes,* Department of Microbiology, West Virginia University, Morgantown, West Virginia, 1969.

Table 4
RELATIONSHIPS BETWEEN ANAEROBIOSIS, CARBON DIOXIDE, OXYGEN, AND GROWTH IN SPECIES OF THE FAMILY ACTINOMYCETACEAE

CODE

Column 1: *Actinomyces bovis*
Column 2: *Actinomyces eriksonii*
Column 3: *Actinomyces humiferus*
Column 4: *Actinomyces israelii*
Column 5: *Actinomyces naeslundii*
Column 6: *Actinomyces viscosus*
Column 7: *Arachnia propionica*
Column 8: *Bacterionema matruchotii*
Column 9: *Bifidobacterium bifidum*
Column 10: *Rothia dentocariosa*

Conditions of growth	1	2	3	4	5	6	7	8	9	10
Net fixation of CO_2	+	+	?	+	+	+	–	–	±	–
CO_2 required for anaerobic growth	+	?	?	+	+	+	–	–	+	na
Anaerobic growth increased by CO_2	+	–	–	+	+	+	–	?	–	na
Aerobic growth	+	–	+	+	+	+	+	+	–	+
Aerobic growth requires CO_2	+	na	–	+	–	+	–	–	na	–
Aerobic growth (in CO_2) is two or more times greater than anaerobic growth	–	na	?	+	+	+	–	+	na	+
Aerobic growth changes fermentation, signifying complete or partial oxidation of substrate	na	na	+	+	+	+	+	+	na	+
Aerobic formation of CO_2	na	na	?	+	+	+	+	+	na	+
Catalase	–	–	–	–	–	+	–	+	–	+
Maintenance of stock cultures	an	an	ae	an	an	ae	an	ae	an	ae

Note: na = not applicable; an = anaerobic; ae = aerobic.

Data compiled from References 5, 7, 16, 20, 33, 60, 61, and 62.

Table 5
ANAEROBIC AND AEROBIC FERMENTATIONS OF ACTINOMYCETACEAE AND RELATED ORGANISMS

Organisms	Fermentations
	ANAEROBIC CULTURES
Actinomyces bovis, israelii, naeslundii, viscosus	1 glucose → 2 lactate 3.5 glucose + 3 CO_2 → 3 formate + 3 acetate + 3 succinate + 1 lactate 1 glucose + 2 malate → 2 formate + 2 acetate + 2 succinate
Arachnia propionica	18 glucose → 10 CO_2 + 2 formate + 10 acetate + 23 propionate + 1 succinate + 1 lactate
Bifidobacterium bifidum	10 glucose + CO_2 → 1 formate + 13 acetate + 10 lactate + 1 succinate 10 glucose → 15 acetate + 10 lactate
Propionibacterium acnes	10 glucose → 5 CO_2 + 4 acetate + 12 propionate + 2 succinate + 1 lactate
Propionibacterium pentosaceum	10 glucose → 6 CO_2 + 2 acetate + 14 propionate + 2 succinate
	ANAEROBIC MICROAEROPHILIC CULTURES
Actinomyces humiferus	1 glucose → 2 lactate
Bacterionema matruchotii	10 glucose → 6 propionate + 13 lactate
	AEROBIC SHAKE CULTURES
Actinomyces israelii, naeslundii *Arachnia propionica* *Propionibacterium acnes, arabinosum*	1 glucose + O_2 → 2 acetate + 2 CO_2
Actinomyces viscosus	10 glucose → 10 lactate + 2 succinate + 1 formate + 1 acetate
Bacterionema matruchotii	10 glucose → 3 acetate + 4 propionate + 5 lactate
Rothia dentocariosa	10 glucose → 15 lactate + traces of formate, acetate, and succinate

Data compiled from References 5, 18, 33, 36, 60, 61, 62, 63, and 64.

Table 6
CELL WALL COMPOSITION OF ACTINOMYCETACEAE AND RELATED SPECIES[a]

Organism	Catalase	Aspartic acid	Lysine	Ornithine	Glycine	DAP	Galactose	Mannose	Glucose	Rhamnose	Fucose	Deoxytalose	Arabinase
Actinomyces													
bovis													
ATCC 13683	–	+	+					+	+	+	+	+	
P2R	–		+					+	+	+	+	+	
13R	–		+					+	+	+	+	+	
eriksonii	–	+	+				+				+		
humiferus	–	+	+	+					±	+	±		
israelii													
ATCC 12102, Type 1	–		+	+			+						
ATCC 12102, Type 2	–	–	+	+			+			+			
israelii (296)													
ATCC 10049	–	±	+	?			+	+		+			
naeslundii (297)													
ATCC 12104	–	±	+	+				+	+	+	+	+	
odontolyticus	–		+	+			+		+	±	±	±	
viscosus	+		+	+			+	+	+				
Arachnia													
propionica	–				+	L	+		+				
Bacterionema													
matruchotii	+					*meso*	+	+	+				+
Bifidobacterium													
bifidum	–			+			+		+				
corynebacterium													
Propionibacterium													
species	+				+	L	+		+				
acnes	+				+	L	+		+				
diphtheriae	+					*meso*	+						+
Rothia													
dentocariosa	+		+				+						

[a] All strains have alanine and glutamic acid.

Data compiled from References 2, 5, 26, 35, 57, 65–72.

REFERENCES

1. **Slack, J. M.,** *Report to the International Committee on Bacterial Nomenclature, Subgroup on Taxonomy of Microaerophilic Actinomycetes,* Department of Microbiology, West Virginia University, Morgantown, West Virginia, 1969.
2. **Gledhill, W. E. and Casida, L. E.,** *Appl. Microbiol.,* 18, 340, 1969.
3. **Brock, D. W. and Georg, L. K.,** *J. Bacteriol.,* 97, 589, 1969.
4. **Brown, J. M., Georg, L. K., and Waters, L. C.,** *Appl. Microbiol.,* 17, 150, 1969.
5. **Buchanan, B. B. and Pine, L.,** *J. Gen. Microbiol.,* 28, 305, 1962.
6. **Howell, A., Jr., Murphy, W. C., Paul, F., and Stephan, R. M.,** *J. Bacteriol.,* 78, 82, 1959.
7. **Pine, L., Howell, A., Jr., and Watson, S. J.,** *J. Gen. Microbiol.,* 23, 403, 1960.
8. **Slack, J. M., Langried, S., and Gerencser, M. A.,** *J. Bacteriol.,* 97, 873, 1969.
9. **Blank, C. H. and Georg, L. K.,** *J. Lab. Clin. Med.,* 71, 283, 1968.
10. **Brock, D. W. and Georg, L. K.,** *J. Bacteriol.,* 97, 581, 1969.
11. **Lambert, F. W., Brown, J. M., and Georg, L. K.,** *J. Bacteriol.,* 94, 1287, 1967.
12. **Slack, J. M., and Gerencser, M. A.,** *J. Bacteriol.,* 103, 266, 1970.
13. **Gilmour, M. N.,** *Bacteriol. Rev.,* 25, 142, 1961.
14. **Gilmour, M. N. and Beck, P. H.,** *Bacteriol. Rev.,* 25, 152, 1961.
15. **Allen, S. H. G., Kellermeyer, R. W., Stjernholm, R. L., and Wood, H. G.,** *J. Bacteriol.,* 87, 171, 1964.

16. **Howell, A., Jr. and Pine, L.,** *Bacteriol. Rev.*, 25, 162, 1961.
17. **Pine, L. and Hardin, H.,** *J. Bacteriol.*, 78, 165, 1959.
18. **Buchanan, B. B. and Pine, L.,** *J. Gen. Microbiol.*, 46, 225, 1967.
19. **Buyze, G., van den Hamer, J. A., and DeHaan, P. G.,** *Antonie van Leeuwenhoek J. Microbiol. Serol.*, 23, 345, 1951.
20. **Howell, A., Jr. and Jordan, H. V.,** *Sabouraudia,* 3, 93, 1963.
21. **DeVries, W. and Stouthamer, A. H.,** *J. Bacteriol.*, 93, 574, 1967.
22. **DeVries, W., Gerbrandy, S. J., and Stouthamer, A. H.,** *Biochem. Biophys.*, 136, 415, 1967.
23. **Scardovi, V., and Trovatelli, L. D.,** *Ann. Microbiol. Enzimol.*, 15, 19, 1965.
24. **Domnas, A. and Grant, N. G.,** *J. Bacteriol.*, 101, 652, 1970.
25. **Hein, A. H., Silver, W. S., and Birk, Y.,** *Nature,* 180, 608, 1957.
26. **Cummins, C. S.,** *J. Gen. Microbiol.*, 28, 35, 1962.
27. **Erikson, D.,** *Med. Res. Counc. G.B. Spec. Rep. Ser.*, No. 240, 1, 1940.
28. **Harz, C. O.,** *Dsch. Z. Tiermed.*, 5, 125, 1879.
29. **Silberschmidt, W.,** *Zentralbl. Bakteriol. Parasitenkd. Infektionskr. Hyg. Abt. I. Orig.*, 27, 486, 1900.
30. **Silberschmidt, W.,** *Z. Hyg. Infektionskr.*, 37, 345, 1901.
31. **Thompson, L.,** *Proc. Staff Meet. Mayo Clin.*, 25, 81, 1950.
32. **Georg, L. K., Roberstad, G. W., Brinkman, J. A., and Hicklin, M. D.,** *J. Infect. Dis.*, 115, 88, 1965.
33. **Gledhill, W. E. and Casida, L. E.,** *Appl. Microbiol.*, 18, 114, 1969.
34. **Wolff, M. and Israel, J.,** *Virchows Arch. Abt. A Pathol. Anat.*, 126, 11, 1891.
35. **Pine, L. and Boone, C. J.,** *J. Bacteriol.*, 94, 875, 1967.
36. **Pine, L., Hardin, H., Turner, L., and Roberts, S. S.,** *Am. J. Ophthalmol.*, 49, 1278, 1960.
37. **Thompson, L. and Lovestadt, S. A.,** *Proc. Staff Meet. Mayo Clin.*, 26, 169, 1951.
38. **Batty, I.,** *J. Pathol. Bacteriol.*, 75, 455, 1958.
39. **Georg, L. K. and Coleman, R. M.,** Comparative pathogenicity of various *Actinomyces* species, in *The Actinomycetales,* Prauser, H., Ed., Gustav Fischer, Jena, Germany, 1970.
40. **Grasser, R.,** *Zentralbl. Bakteriol. Parasitenkd. Infektionskr. Hyg. Abt. I. Orig.*, 184, 478, 1962.
41. **Grasser, R.,** *Zentralbl. Bakteriol. Parasitenkd. Infektionskr. Hyg. Abt. I. Orig.*, 188, 251, 1963.
42. **Georg, L. K., Pine, L., and Gerencser, M. A.,** *Int. J. Syst. Bacteriol.*, 19, 291, 1969.
43. **Gerencser, M. A. and Slack, J. M.,** *Appl. Microbiol.*, 18, 80, 1969.
44. **Howell, A., Jr.,** *Sabouraudia,* 3, 81, 1963.
45. **Jordan, H. V. and Keyes, P. H.,** *Arch. Oral Biol.*, 9, 401, 1967.
46. **Gerencser, M. A. and Slack, J. M.,** *J. Bacteriol.*, 94, 109, 1967.
47. **Pine, L. and Georg, L. K.,** *Int. J. Syst. Bacteriol.*, 19, 267, 1969.
48. **Cock, D. J. and Bowen, W. H.,** *J. Periodontal Res.*, 2, 36, 1967.
49. **Gilmour, M. N., Howell, A., Jr., and Bibby, B. J.,** *Bacteriol. Rev.*, 25, 131, 1961.
50. **Harrison, A. P. and Hansen, J. A.,** *J. Bacteriol.*, 60, 543, 1954.
51. **Orla-Jensen, J., Orla-Jensen, A. D., and Winther, O.,** *Zentralbl. Bakteriol. Parasitenkd. Infektionskr. Hyg. Abt. II. Naturwiss.*, 93, 321, 1936.
52. **Pine, L. and Georg, L. K.,** *Int. J. Syst. Bacteriol.*, 15, 143, 1965.
53. **Prévot, A. R.,** *Traité de systématique bactérienne,* Vol. 2, Dunnod, Paris, (1961).
54. **Puntoni, V.,** *Ann. Ig. Sper.*, 47, 157, 1937.
55. **Weiss, J. E. and Rettger, L. F.,** *J. Bacteriol.*, 35, 17, 1938.
56. **Weiss, J. E. and Rettger, L. F.,** *J. Infect. Dis.*, 62, 115, 1938.
57. **Davis, G. H. C. and Freer, J. H.,** *J. Gen. Microbiol.*, 23, 163, 1960.
58. **Onisi, M.,** *Shikagaku Zasshi,* 6, 273, 1949.
59. **Roth, G. D. and Thurn, A. N.,** *J. Dent. Res.*, 41, 1279, 1962.
60. **Buchanan, B. B. and Pine, L.,** *Sabouraudia,* 3, 26, 1963.
61. **Ichikawa, Y.,** *Chem. Abstr.*, 51, 11458, 1957.
62. **Pine, L. and Howell, A., Jr.,** *J. Gen. Microbiol.*, 15, 428, 1956.
63. **Moore, W. E. C. and Cato, E. P.,** *J. Bacteriol.*, 85, 870, 1963.
64. **Wood, H. G. and Werkman, C. H.,** *Biochem. J.*, 30, 618, 1936.
65. **Becker, B., Lechevalier, M. P., and Lechevalier, H. A.,** *Appl. Microbiol.*, 13, 236, 1965.
66. **Cummins, C. S.,** in *The Actinomycetales,* Prauser, H., Ed., Gustav Fischer, Jena, Germany, 1970, p. 29.
67. **Cummins, C. S. and Harris, H.,** *J. Gen. Microbiol.*, 14, 583, 1956.
68. **Cummins, C. S. and Harris, H.,** *J. Gen. Microbiol.*, 18, 173, 1958.
69. **Cummins, C. S., Glendenning, O. M., and Harris, H.,** *Nature,* 180, 337, 1957.
70. **Deweese, M. S., Gerencser, M. A., and Slack, J. M.,** *Appl. Microbiol.*, 16, 1713, 1968.
71. **Johnson, J. L. and Cummins, C. S.,** *J. Bacteriol.*, 109, 1047, 1972.
72. **Slack, J. M. and Gerencser, M. A.,** *Actinomyces, Filamentous Bacteria: Biology and Pathogenicity,* Burgess Publ. Co., Minneapolis, 1975.

THE RICKETTSIALES*

P. Fiset, W. F. Myers, and C. L. Wisseman, Jr.

The order Rickettsiales comprises small rod-shaped or coccoid organisms parasitic of man or animals. They are considered Gram-negative, although the Gram stain is not the stain of choice. Some are obligate intracellular parasites, others may grow in a cell-free environment. Rickettsiales have both RNA and DNA. They exist in natural vertebrate hosts and are usually transmitted from host to host by more or less specific arthropod vectors. Most are capable of causing inapparent infections in their vertebrate and arthropod hosts.

The seventh edition of *Bergey's Manual* lists four families in the order Rickettsiales: Rickettsiaceae, Chlamydiaceae, Bartonellaceae, and Anaplasmataceae. The next edition will probably classify the family Chlamydiaceae in a new, monofamilial order, the Chlamydiales, which are discussed by Dr. L. A. Page on pages 187 to 194 of this volume. The major characteristics of the Rickettsiales are described in Table 1.

For reasons of convenience the present section will be divided into two parts: the family Rickettsiaceae, and the families Bartonellaceae and Anaplasmataceae.

RICKETTSIACEAE

The family Rickettsiaceae comprises the following important genera: *Rickettsia, Coxiella, Rochalima,* and *Ehrlichia.* All except *Ehrlichia* are potential human pathogens. All except *Rochalima* are obligate intracellular parasites. All are transmitted to their natural hosts by arthropods, although *Coxiella* is usually transmitted by aerosol. All replicate by binary fission, and all have a cell wall. Although most of the Rickettsiaceae are considered obligate intracellular parasites, they are capable of a large variety of independent metabolic activities. Except for *Ehrlichia,* all genera present certain hazards to laboratory workers and should be handled carefully. All Rickettsiaceae are susceptible to varying but clinically attainable concentrations of chloramphenicol and tetracycline, and to a lesser degree to erythromycin.

Chemical Composition

Most of the chemical-composition data available were derived from studies of a limited number of members of this group, primarily *Rickettsia prowazeki, R. typhi (R. mooseri),* and *Coxiella burnetii.* However, there is no reason to believe that other members of this group are different. From the data available, all Rickettsiaceae would seem to contain DNA, RNA, proteins, polysaccharides, lipids, phospholipids, and coenzymes. In addition, their cell walls are similar to those Gram-negative bacteria and contain amino acids, muramic acid, diaminopimelic acid, hexoses, and hexosamines.

Morphology

Rickettsiaceae vary in morphology from small rods to coccoid forms, ranging in size from approximately 0.5 to 2.0 μm in length and 0.2 to 0.6 μm in width. Under certain, as yet poorly defined, conditions some may assume filamentous form.

Although very few electron-microscopic studies have been carried out, the data available indicate that the ultrastructures of rickettsiae resemble that of bacteria. The rickettsial cell wall and cytoplasmic membrane appear to be trilamellar structures. In addition, an amorphous capsular substance has been demonstrated in *Rickettsia prowazeki;* there is also evidence for the presence of cytoplasmic membranous organelles.

* Part 18 in *Bergey's Manual.*

Ultrathin sections of other rickettsiae suggest the presence of ribosome-like structures and nuclear material similar to that of bacteria. In *Coxiella burneti,* typical bacterial ribosomes have been demonstrated.

Biological Properties

Tables 2 to 6 present detailed descriptions of the biological properties of all Rickettsiaceae other than *Ehrlichia. Ehrlichia* was excluded from the tables because of the paucity of information available on the biological characteristics of this group of organisms. Suffice it to say that the genus *Ehrlichia* comprises three species: *E. bovis, E. ovina,* and *E. canis. Ehrlichia canis* is the only species that has been studied to any significant extent in recent years. It causes tropical canine pancytopenia; it is prevalent in Southeast Asia, Central America, the Caribbean, Florida, and Texas. The usual laboratory animals are refractory, but experimental infection is readily achieved in dogs (beagles and German shepherds), though clinical manifestations vary with the breed. The organism has not been grown in chick embryos or in tissue cultures other than canine monocytes. The dog tick, *Rhipicephalus sanguineus,* seems to be the natural vector, in which the organism undergoes transovarial and transtadial transmission. Although the natural vertebrate hosts are not really known, foxes, coyotes, and jackals are suspected.

The Rickettsiaceae are grouped on the basis of antigenic relationships.

BARTONELLACEAE AND ANAPLASMATACEAE

The family Bartonellaceae includes the genera *Bartonella, Haemobartonella, Eperythrozoon,* and *Grahamella;* the family *Anaplasmataceae* contains only the genus *Anaplasma.* The genus *Bartonella* is the only one of the group that infects man. The other Bartonellaceae produce infections in lower vertebrates, frequently with few or no clinical symptoms. *Anaplasma marginale* produces a severe infectious anemia of cattle, anaplasmosis, a disease of economic importance in the USA and elsewhere. A common characteristic of all these agents is their location on or within erythrocytes.

Human bartonellosis is manifested in two distinctly different clinical forms: Oroya fever, an acute hemolytic anemia, and verruga peruana, a relatively benign, chronic type of disease. The latter disease syndrome seems to reflect a partial immunity to the agent and is the usual form of bartonellosis seen in cases of reinfection. The asymptomatic carrier state is also probably an expression of a partial immunity. Latent infections seem to be a common characteristic of both human and animal bartonellosis. Persisting nonapparent infections and partial immunity seem to be a common theme for all the Bartonellaceae and Anaplasmataceae.

Only *Bartonella* and *Grahamella* have been cultured in cell-free media and possess true cell walls. Arthropod vectors are involved in the transmission of most of these agents. Virtually nothing is known about their metabolism or chemical and antigenic composition. The classification of this group rests on somewhat tenuous grounds, and the status of these organisms will undoubtedly change as more is learned about their biological properties. Tables 7 and 8 describe the known biological properties of the Bartonellaceae and Anaplasmataceae.

Table 1
GENERAL CHARACTERISTICS OF RICKETTSIALES

Families and genera	Characteristics
Rickettsiaceae	Most are human pathogens. All have a true cell wall. They replicate by binary fission, mostly within endothelial and reticuloendothelial cells. They are not associated with erythrocytes.
1. *Rickettsia*	Most show a close association with arthropod vectors, which are required for transmission to natural vertebrate hosts, including man. All are potential human pathogens. All are labile and are readily inactivated outside the natural host. All are considered obligate intracellular parasites. Several species are known.
2. *Coxiella*	Similar to the genus *Rickettsia*, except: (a) they are highly resistant to physical and chemical agents in an extracellular environment; (b) arthropod vectors are not usually involved in natural transmission to vertebrate hosts; (c) they undergo an antigenic phase variation. Only one species is known.
3. *Rochalima*	Similar to the genus *Rickettsia*, except that these organisms are not obligate intracellular parasites. Man is the only known vertebrate host. Only one species is known.
4. *Ehrlichia*	They have many of the characteristics of the genus *Rickettsia*, except that they are nonpathogenic for man. Three species are known, the most important being *E. canis*, the agent of tropical canine pancytopenia.
Bartonellaceae	Only one genus is pathogenic for man. Some lack a true cell wall. They replicate by binary fission on or inside erythrocytes.
1. *Bartonella*	Human pathogens growing in the endothelial cells of man, the only known vertebrate host. They grow in cell-free medium. The organisms have cell walls. Only one species is known.
2. *Haemobartonella*	These organisms are nonpathogenic for man. Multiple species are defined on the basis of vertebrate-host association. They are closely associated with erythrocytes, but are rarely found free in plasma. No demonstrable growth occurs in host tissue other than erythrocytes. The organisms rarely produce disease in natural hosts without splenectomy. They have not been grown in cell-free medium. *Haemobartonella* have no cell wall. There is no differentiated nuclear structure, although both RNA and DNA are reported to be present.
3. *Eperythrozoon*	Similar to *Haemobartonella*. Differentiation between these two genera is difficult and somewhat arbitrary. There are minor differences in morphology, ring forms being predominant with *Eperythrozoon*. Although associated with red cells, numerous organisms are found free in plasma.
4. *Grahamella*	Morphologically similar to *Bartonella*. They grow inside erythrocytes. Multiple species are defined on the basis of host association. Although parasitemia can be demonstrated, no clinical disease is manifested in natural vertebrate hosts, even after splenectomy. The organisms grow in cell-free medium. They possess true cell walls. *Grahamella* are not infectious to man.
Anaplasmataceae	Nonpathogenic for man. The organisms lack a true cell wall. They replicate inside erythrocytes within a vacuole (marginal body).
1. *Anaplasma*	The only genus in this family. The growth cycle is different from that of all other members of the Rickettsiales; it resembles that of the Chlamydiales. Three species are known to infect cattle and sheep. Anaplasmosis in cattle varies from asymptomatic infection to severe anemia. The organisms have not been grown in cell-free medium.

Table 2
HUMAN DISEASES AND GEOGRAPHIC DISTRIBUTION OF RICKETTSIACEAE

Species	Human diseases	Geographic distribution
	TYPHUS GROUP: GENUS *RICKETTSIA*	
R. prowazcki	Epidemic typhus: the only epidemic rickettsial disease; outbreaks occur under conditions of overcrowding and louse infestation.	Worldwide; endemic on all continents except Australia, with occasional epidemic outbreaks in Africa, Central and South America, Eastern Europe, and Asia.
	Brill-Zinsser disease: recrudescence of infection in individuals who had typhus years earlier.	
R. typhi (mooseri)	Murine typhus: less severe than epidemic typhus.	Worldwide.
R. canada	Serologic evidence for a Rocky Mountain spotted feverlike disease; no isolation from man at this time.	Unknown (only one isolation reported in Ontario, Canada, from rabbit ticks).
	SPOTTED FEVER GROUP: GENUS *RICKETTSIA*	
R. rickettsi	Rocky Mountain spotted fever (RMSF); severe.	Western hemisphere; the organisms are transmitted by different species of ixodid ticks in different areas.
R. conori	Boutonneuse fever, Kenya tick typhus, South African tick typhus, Indian tick typhus (may be caused by other species): less severe than RMSF.	Mediterranean area, East and South Africa, Pakistan, and India; the organisms are transmitted by different species of ixodid ticks in these different areas.
R. sibirica	Siberian or North Asian tick typhus: less severe than RMSF.	Siberia, Armenia, and Central Asian Republics of USSR, possibly also in Czechoslovakia.
R. australis	Queensland tick typhus: less severe than RMSF.	Australia.
R. akari	Rickettsial pox: mild.	Urban infections in USA and USSR (house mouse); sylvatic infections in Korea (field mouse).
	SCRUB TYPHUS: GENUS *RICKETTSIA*	
R. tsutsugamushi (orientalis)	Scrub typhus: mild to severe, depending on the strain.	Soviet Far East, Japan, Chinese Mainland, Pacific islands, Southeast Asia to Northern Australia, and westward to West Pakistan.
	Q-FEVER: GENUS *COXIELLA*	
C. burnetii	Unrecognized infections: mostly mild. Q-fever: moderately severe. Hepatitis: moderately severe. Subacute endocarditis: severe.	Worldwide.
	TRENCH FEVER: GENUS *ROCHALIMA*	
R. quintana	Trench fever, probably many unrecognized infections.	Probably worldwide; same distribution as epidemic typhus.

Table 3
NATURAL VERTEBRATE HOSTS AND VECTORS OF RICKETTSIACEAE

Species	Natural vertebrate hosts	Vectors transmitting within the natural cycle	Vectors transmitting to man	Infection in vectors
		TYPHUS GROUP: GENUS *RICKETTSIA*		
R. prowazeki	Man	*Pediculus humanus humanus* (human body louse)	*Pediculus humanus humanus*	Growth in gut epithelium; lice die 8 to 10 days after infection; organisms excreted in feces.
R. typhi (mooseri)	*Rattus rattus* and *R. norvegicus*	*Xenopsylla cheopis* (rat flea) *Polyplax sinulosa* (rat louse)?	*Xenopsylla cheopis* *Pulex irritans* (human flea)? *Pediculus humanus humanus?*	Permanent infection; fleas do not die of infection, growth in gastric epithelium; organisms excreted in feces.
R. canada	Unknown	*Haemaphysalis leporispalustris* (rabbit tick)	Unknown	Experimental infection of *Haemaphysalis leporispalustris* indicates transovarial and transtadial transmission.
		SPOTTED FEVER GROUP: GENUS *RICKETTSIA*		
R. rickettsi	Various rodents; rabbits; dog?	Northwest USA: *Dermacentor andersoni* Eastern USA and Canada: *Dermacentor variabilis* *Haemaphysalis leporispalustris* South and Southwest USA: *Amblyoma americanum* Mexico and South America: *Rhipicephalus sanguineus* *Amblyoma cajenense* *Ixodes* sp. ?	Same as in the zoonotic cycle, except for *Haemaphysalis leporispalustris*, which does not bite man.	Persistent infection in ticks without overt pathology (?); transovarial and transtadial transmission; growth in all tissues of ticks; organisms excreted in saliva and feces.
R. conori	Primarily dogs, but also several species of rodents; (rabbits ?)	Mediterranean area: *Rhipicephalus sanguineus* Kenya: *Haemaphysalis leachi* *Rhipicephalus simus* South Africa: *Amblyoma herbreum* *Haemaphysalis aegypticum* Other species in other areas	Same as in the zoonotic cycle.	Same as for *R. rickettsi.*

Table 3 (continued)
NATURAL VERTEBRATE HOSTS AND VECTORS OF RICKETTSIACEAE

Species	Natural vertebrate hosts	Vectors transmitting within the natural cycle	Vectors transmitting to man	Infection in vectors
R. sibirica	Domestic animals; rodents	Various species of ticks: *Dermacentor nuttali*, *D. sylvarum*, *D. marginatus*, *D. pictus*, *Haemaphysalis concinna*, *H. punctata*	Same as in the zoonotic cycle.	Same as for *R. rickettsi.*
R. australis	Several species of marsupials	*Ixodes holocyclus* (tick of marsupials).	Same as in the zoonotic cycle.	Same as for *R. rickettsi.*
R. akari	*Mus musculus* (house mouse); *Microtus fortis pellicus* (sylvatic in Korea)	The mite *Allodermanyssus sanguineus*	Same as in the zoonotic cycle.	Transmission by biting; transovarial transmission.
		SCRUB TYPHUS: GENUS *RICKETTSIA*		
R. tsutsugamushi (orientalis)	Various rodents and other small mammals	The trombiculid mites: *Leptotrombidium akamushi*, *L. deliense*, *L. pallidum*, *L. scutellare*, others	Same as in the zoonotic cycle.	Distribution of the parasites in the vector is not well known; transmission by biting; transovarial and transtadial transmission.
		Q-FEVER: GENUS *COXIELLA*		
C. burnetii	Various domestic animals: sheep, goats, cattle; various other animals: birds, rodents, marsupials.	Direct transmission by aerosol; several species of ticks may be involved in some animal cycles.	Aerosol	In ticks, transovarial and transtadial transmission; organisms are excreted in saliva and feces.
		TRENCH FEVER: GENUS *ROCHALIMA*		
R. quintana	Man	*Pediculus humanus humanus*	*Pediculus humanus humanus*	Lice do not die of infection, but seem to remain infected for life; growth is extracellular in the gut; organisms are excreted in feces.

Table 4
EXPERIMENTAL INFECTIONS IN LABORATORY ANIMALS AND SEROLOGIC REACTIONS OF RICKETTSIACEAE

Species	Experimental infections		Serologic reactions	
	Guinea pig	Mouse	With rickettsial antigens	With *Proteus*[a]
		TYPHUS GROUP: GENUS *RICKETTSIA*		
R. prowazeki	High susceptibility to infection; mild disease with fever; scrotal swelling (rare).	Low susceptibility to infection; acute toxic death with large inoculum.	A "soluble" antigen that is group-specific in complement fixation reaction is released by ether treatment; after ether treatment, rickettsial bodies exhibit species-specific activity in complement fixation, agglutination, and fluorescent-antibody tests.	OX-19, OX-2
R. typhi (mooseri)	Organisms more virulent than *R. prowazeki*, usually causing scrotal swelling.	High-dose infection causes death in 3 to 8 days; acute toxic death with large inoculum.		OX-19, OX-2
R. canada	Subclinical infection shown by serologic conversion.	Subclinical infection shown by serologic conversion; acute toxic death with large inoculum.		Not determined
		SCRUB TYPHUS: GENUS *RICKETTSIA*		
R. tsutsugamushi (orientalis)	Except for a few strains, the organisms usually cause no apparent infection.	High susceptibility to infection; strains vary from almost avirulent to lethal; death occurs 7 to 14 days after inoculation with lethal strains; acute toxic death has been demonstrated with only one strain.	At least three serotypes are recognized, but there are probably many more; the rickettsial surface behaves like a mosaic with predominant "type-specific' antigens; a group-specific. ether-released, water-soluble antigen has been described.	OX-K
		Q-FEVER: GENUS *COXIELLA*		
C. burnetii	Animal of choice for primary isolation; considerable variation in virulence with different strains; fever and enlargement of spleen.	High susceptibility to infection; acute toxic death has not been produced.	Only one serotype; two antigenic phases; phase I is found in nature; surface phase I polysaccharide is lost on adaptation to chick embryos → phase II.	Negative
		SPOTTED FEVER GROUP: GENUS *RICKETTSIA*		
R. rickettsi	High susceptibility to infection; most strains are virulent, causing scrotal swelling and necrosis as well as death.	Low susceptibility to infection; acute toxic death with large inoculum.		

Table 5
CULTIVATION AND STABILITY OF RICKETTSIACEAE

Species	Embryonated hen's egg (yolk sac)	Cell culture[a]	Optimal temperature	Stability
		TYPHUS GROUP: GENUS *RICKETTSIA*		
R. prowazeki	Generation time about 6 hours; peak titer just prior to death; rickettsial growth stops after embryo death; optimal yield about $10^9 - 10^{10}$ infectious units per egg.	Plaques (1 mm) in 10 days with primary chick embryo cell culture; other cultures employed: BS-C-1, guinea pig kidney, chick entodermal cells, *Hyalomma* tick cells, L-cells, and colubrid snake cells.	32°C for plaques, 35°C for optimal growth	Labile; readily inactivated at 56°C; unstable at room temperature, but may survive for months in dried vector feces in a cool, dry climate; stable at −76°C and under lyophilization; labile between −5°C and −20°C; stabilized in SPG (sucrose phosphate-glutamate) medium.
R. typhi (mooseri)	Same as *R. prowazeki.*	Plaques (1 mm) in 10 days with primary chick embryo cell culture; other cultures employed: *Hyalomma* tick cells.		Same as *R. prowazeki.*
R. canada	Grows well in yolk sac after adaptation.	Plaques (0.75 mm) in 7 days with primary chick embryo cell culture.	35°C	Same as *R. prowazeki.*
		SPOTTED FEVER GROUP: GENUS *RICKETTSIA*		
R. rickettsi	Embryo may die early due to toxicity; rickettsial multiplication continues about 2 days after death.	Plaques (2 mm) in 6 days with primary chick embryo cell culture; other cultures employed: 14 pf (rat fibroblast) and guinea pig scrotal tissue explant.	32°C	Labile; similar to the organisms of the typhus group, except not stabilized by SPG; some stability is conferred by glutathione and by microaerophilic conditions.
R. conori	Same as *R. rickettsi.*	Plaques (2 mm) in 5 days with primary chick embryo cell culture; other cultures employed: guinea pig kidney, pig embryo kidney, and *Hyalomma* tick cells.	32°C	Same as *R. rickettsi.*
R. sibirica	Same as *R. rickettsi.*	Plaques (2 mm) in 5 days with primary chick embryo cell culture.	32°C	Same as *R. rickettsi.*
R. australis	Same as *R. rickettsi.*	Plaques (2 mm) in 5 days with primary chick embryo cell culture.	32°C	Same as *R. rickettsi.*
R. akari	Same as *R. rickettsi.*	Plaques (3 mm) in 5 days with primary chick embryo cell culture; other cultures employed: *Hyalomma* tick cells.	32°C	Same as *R. rickettsi.*

Table 4 (continued)
EXPERIMENTAL INFECTIONS IN LABORATORY ANIMALS AND SEROLOGIC REACTIONS OF RICKETTSIACEAE

	Experimental infections		Serologic reactions	
Species	Guinea pig	Mouse	With rickettsial antigens	With *Proteus*[a]
R. conori	Organisms less virulent than *R. rickettsi*, usually causing fever and scrotal swelling, but not scrotal necrosis and death.	Low susceptibility to infection; acute toxic death with large inoculum.	Group-specific "soluble" antigens used in complement fixation tests; species-specific rickettsia-associated antigens used in agglutination and fluorescent-antibody tests.	OX-19, OX-2
R. sibirica	Infection similar to that of *R. rickettsi.*	Low susceptibiliby to infection; acute toxic death with large inoculum.		OX-19, OX-2
R. australis	Infection similar to that of *R. conori.*	Infection produced by intraperitoneal route; acute toxicity has not been demonstrated.		OX-19, OX-2
R. akari	Infection similar to that of *R. conori.*	High susceptibility to infection; acute toxicity has not been demonstrated.		Negative
		TRENCH FEVER: GENUS *ROCHALIMA*		
R. quintana	Infection not produced; experimental infections have been produced only in man and in monkeys.	Infection not produced; acute toxic death has not been produced.	Only one recognized serotype; no ether-released "soluble" antigen.	Negative

[a] Weil & Felix reaction (*Proteus* agglutination).

Table 5 (continued)
CULTIVATION AND STABILITY OF RICKETTSIACEAE

Species	Embryonated hen's egg (yolk sac)	Cell culture[a]	Optimal temperature	Stability
		SCRUB TYPHUS: GENUS *RICKETTSIA*		
R. tsutsugamushi (orientalis)	Peak titer just prior to death; titer declines rapidly after death; poor growth and low yield.	Plaques (1 mm) in 10 to 17 days with primary chick embryo cell culture; other cultures employed: 14 pf, guinea pig kidney, pig embryo kidney, MB-III, and L-929.	35°C	Very labile; inactivated at 37°C and at room temperature in several hours; stabilized by SPG, albumin, or protein, stable at −70°C; inactivated by lyophilization.
		Q-FEVER: GENUS *COXIELLA*		
C. burnetii	Slow growth rate, may require three passages; generation time about 12 hours; yield about $10^{10}-10^{11}$ infectious particles per egg.	Plaques (0.75 mm) in 16 days with primary chick embryo cell culture; other cultures employed: L-cells, guinea pig kidney, pig embryo kidney, chick fibroblast, *Antheraea* (moth) and *Aedes* (mosquito) cell lines, Detroit-6, H-Ep-2, and human amnion.	35°C	Very stable; survives for months in dried materials at room temperature; may survive heating at 60°C for 30 min.
		TRENCH FEVER: GENUS *ROCHALIMA*		
R. quintana	Poor growth; will grow in cell-free medium (blood agar with 5% carbon dioxide atmosphere).	Has been grown with some difficulty; will grow in cell-free medium.	32–34°C	Stability similar to that of *R. prowazeki;* organisms remain viable for several months in dried louse feces in a cool, dry climate.

[a] Plaque size may vary, depending on the type of cell used in the culture, and on the medium used for growth, and on the temperature of incubation.

Table 6
ENERGY PRODUCTION AND BIOSYNTHESIS OF RICKETTSIACEAE

Species	Energy production	Biosynthesis
	TYPHUS GROUP: GENUS *RICKETTSIA*	
R. prowazeki	Oxygen uptake with glutamate, glutamine, pyruvate, and succinate; demonstration of dicarboxylic portion of citric acid cycle; adenosine diphosphate → adenosine triphosphate; glutamic dehydrogenase.	Glutamate-oxaloacetate transaminase; glycine-^{14}C, methionine-^{14}C, or valine-^{14}C + amino acid mixture → protein-^{14}C; acetate-^{14}C → lipid-^{14}C.
R. typhi (mooseri)	Oxygen uptake with glutamate, glutamine, pyruvate, and succinate; demonstration of dicarboxylic portion of citric acid cycle; glutamine → glutamate; asparagine → aspartic acid; glutamic hydrogenase.	Glutamate-oxaloacetate transaminase; methionine-^{14}C + amino acid mixture → protein-^{14}C; *Chlorella* protein hydrolysate-^{14}C → protein-^{14}C.
R. canada	Unknown	Unknown
	SPOTTED FEVER GROUP: GENUS *RICKETTSIA*	
R. rickettsi	Carbon dioxide-^{14}C from all carbon atoms of glutamate; carbon dioxide-^{14}C production reduced by addition of unlabeled citric acid cycle intermediates.	Unknown
R. conori	Unknown	Unknown
R. sibirica	Unknown	Unknown
R. australis	Unknown	Unknown
R. akari	Unknown	Unknown
	SCRUB TYPHUS: GENUS *RICKETTSIA*	
R. tsutsugamushi (orientalis)	Unknown	Unknown
	Q-FEVER: GENUS *COXIELLA*	
C. burnetii	Oxygen uptake with pyruvate, glutamate, succinate, α-ketoglutarate, oxaloacetate, fumarate, malate, and serine, glutamate, malate, and isocitrate dehydrogenase; citrate synthase; hexokinase; glucose-6-phosphate dehydrogenase, adenosine diphosphatase, adenosine triphosphatase.	Glycine + formaldehyde → serine; ornithine + carbamyl phosphate → citrulline; aspartate + carbamyl phosphate → ureidosuccinate; glutamate-oxaloacetate transaminase; DNA-dependent RNA polymerase; leucine-^{14}C or phenylalanine-^{14}C + amino acid mixture → protein-^{14}C; algal hydrolysate-^{14}C → protein-^{14}C.
	TRENCH FEVER: GENUS *ROCHALIMA*	
R. quintana	Oxygen uptake with succinate, glutamine, glutamate, α-ketoglutarate, pyruvate; demonstration of dicarboxylic portion of citric acid cycle.	Glutamate-oxaloacetate transaminase.

Table 7
NATURAL INFECTION AND HOST RANGE, MODE OF TRANSMISSION, AND GEOGRAPHIC DISTRIBUTION OF BARTONELLACEAE AND ANAPLASMATACEAE

Species	Natural infection and host range	Mode of transmission	Geographic distribution
Bartonellaceae			
Bartonella bacilliformis[a]	Man is the only known host. Clinical manifestations: acute hemolytic anemia (Oroya fever), and subacute benign wart-like skin eruptions (verruga peruana). In both diseases the parasite is found in reticuloendothelial cells. In Oroya fever, the parasite is found on the surface of red blood cells; at the peak of the disease, 90% of the red blood cells are parasitized. If untreated, Oroya fever has a mortality of about 40%. In endemic areas, asymptomatic carriers represent 5 to 10% of the population.	Known vectors are *Phlebotomus verrucarum* and *P. columbianus* (sand flies). The geographic distribution of the vectors agrees with that of the diseases. Sand flies feed at night. The status of the parasite in the vectors is unknown.	The mountainous regions of Peru, Ecuador, and Columbia at altitudes from 2,500 to 8,000 but not elsewhere.
Haemobartonella muris[b,c]	Asymptomatic infection in apparently normal rats. Acute anemia develops after splenectomy, with the appearance of large numbers of the parasites on erythrocytes; this may result in death. Surviving animals are resistant to reinfection.	Transmission by *Polyplax spinulosa* (rat louse). Infection is induced by biting; feces are apparently noninfective.	Not determined, but probably the same as that of rats.
felis	Infectious anemia of cats occurs without splenectomy; loss of weight, weakness, and dyspnea. In untreated cases the mortality is high.	Transmission probably by cat bite. No vectors have been incriminated.	Probably worldwide.
Eperythrozoon coccoides[b]	Asymptomatic infection in apparently normal mice. Splenectomy in infected mice leads to moderate anemia. Several species of wild rats and mice are susceptible, but the common laboratory mouse is most susceptible.	*Polyplax serrata* (mouse louse) is the natural vector.	Worldwide.
Grahamella talpae[b,c]	Infection without clinical symptoms in a variety of animals, including shrews, voles, mice, desert rats, gerbils, hamsters, and squirrels. Splenectomy has virtually no effect.	Transmission by fleas; contamination of the bite by flea feces causes the infection.	Worldwide.
Anaplasmataceae			
Anaplasma marginal[d]	Anaplasmosis in cattle varies from nonapparent infection to severe anemia; it is more severe in older animals. In deer the infection is widespread and asymptomatic. Deer seem to be a natural source of infection for cattle and are probably important in maintaining the parasite in nature.	Many species of ticks have been experimentally infected, but *Dermacentor andersoni* and *D. occidentalis* seem to be the most important vectors in the USA. Both transtadial and transovarial transmission occur. Horse flies seem to be important insect vectors, but transmission is strictly mechanical. Mosquitoes may be involved.	Probably worldwide.

[a] The only species in the genus *Bartonella*.
[b] Prototype species; many species have been described, based on host relationships.
[c] Definite speciation is not well established.
[d] Prototype species; other species, of lesser importance, are *A. centrale* (cattle) and *A. ovis* (sheep).

Table 8
EXPERIMENTAL INFECTIONS, CULTIVATION, AND MORPHOLOGY OF BARTONELLACEAE AND ANAPLASMATACEAE

Species	Experimental infections	Cultivation	Morphology
Bartonellaceae			
Bartonella bacilliformis	The experimental host range is limited to primates. Clinical manifestations are localized verrugas following intradermal or subcutaneous inoculation. Experimental Oroya fever has been produced in man only.	Growth becomes visible in semisolid agar enriched with serum and hemoglobin in 10 days at 28°C. The organisms are obligate aerobes. No hemolysin has been demonstrated. In tissue cultures, growth is intra- and extracellular. In chick embryos, growth occurs in chorioallantoic fluid and in the yolk sac, with no erythrocyte association.	The parasites are found on the surface of red cells, with as many as ten organisms per cell at the peak of Oroya fever. They are extremely pleomorphic and may exist as spheres, rods, and ring-like bodies, 0.3–3.0 × 0.2–0.5 μm. In culture, there is a predominance of rods in the early growth phase. Cell walls and flagella have been demonstrated, but the latter are found in culture only.
Haemobartonella muris	Experimental infection can only be demonstrated in splenectomized rats.	Several reports of growth in cell-free media and in chick embryos have been confirmed.	Coccoid bodies, 0.3–0.5 μm, are found in blood; under light microscopy, chains of cocci may resemble rods. The parasites are found attached to or within the red blood cells. No cell wall has been demonstrated; there is a single limiting membrane. The organisms appear to divide by binary fission.
felis	The organisms are transmitted only to cats. Recovery from the disease leads to a carrier state.	Reports of growth in cell-free media are conflicting	Ring forms, 0.5–1.0 μm. Four to twelve parasites, sometimes in chains, are found on the surface of red cells. No cell wall has been observed.
Eperythrozoon coccoides	Transmission to intact mice leads to subclinical infection and a carrier state. Infection of splenectomized mice leads to moderate anemia and parasitemia. Concomitant infection with mouse hepatitis virus leads to fatal hepatitis.	No growth has been reported in cell-free media. Adaptation is required for growth on chick embryos.	The organisms have a single limiting membrane; no cell wall has been demonstrated. They are ring forms with clear centers, 0.5 μm in diameter; some are rods, 0.6–1.2 μm in length.
Grahamella talpae	Persistent parasitemia may be produced in a large variety of animals without overt clinical manifestation.	The organisms will grow in semisolid medium containing blood or hemoglobin. They are obligate aerobes. Growth occurs in about 10 days at temperatures between 20°C and 37°C. In rat embryo cell cultures, growth occurs both intra- and extracellularly.	In blood smears, the parasites appear as bacilliform bodies or, more rarely, as coccoids within the erythrocytes; the rods measure 0.5–1.0 × 0.2 μm. The organisms are nonmotile and appear to divide by binary fission. They possess true cell walls.

Table 8 (continued)
EXPERIMENTAL INFECTIONS, CULTIVATION, AND MORPHOLOGY OF BARTONELLACEAE AND ANAPLASMATACEAE

Species	Experimental infections	Cultivation	Morphology
Anaplasmataceae			
Anaplasma marginale	The usual laboratory animals are refractory to infection. A large variety of ungulates are susceptible to experimental infection with infected blood, washed erythrocytes, or organ suspensions administered by inoculation.	No growth has been demonstrated in a wide variety of cell-free media, in cell cultures, or in chick embryos.	The organisms grow within the erythrocyte. Dense spherical bodies, known as marginal bodies (0.3–1.0 μm in diameter) and possessing a limiting membrane, contain subunits (initial bodies). The latter are 0.3–0.4 μm and are enclosed in a double membrane and, possibly, an outer envelope-like membrane. Dense aggregates of fine granular material are seen within the initial bodies. The initial bodies are the infectious units. They divide by binary fission.

SELECTED REFERENCES

1. **Hubbert, W. T., McCulloch, W. F., and Schnorrenberger, P. R.,** *Diseases Transmitted from Animals to Man,* 6th ed., Charles C Thomas, Springfield, Ill., 1975.
2. **Huxsoll, D. L., Hildebrandt, P. K., Nims, R. M., and Walker, J. S.,** Tropical canine pancytopenia, in *Current Veterinary Therapy,* Vol. 4, Kirk, R. W., Ed., W. B. Saunders, Co., Philadelphia, 1971, pp. 677–679.
3. **Kobayashi, Y., Nagai, K., and Tachibana, N.,** Purification of complement-fixing antigens of *Rickettsia orientalis* by ether extraction, *Am. J. Trop. Med. Hyg.,* 18, 942, 1969.
4. **McDade, J. E., Stakebake, J. R., and Gerone, P. J.,** Plaque assay system for several species of rickettsiae, *J. Bacteriol.,* 99, 910, 1969.
5. **Ormsbee, R. A.,** Q-fever rickettsia, in *Viral and Rickettsial Infections of Man,* 4th ed., Horsfall, F. L. and Tamm, I., Eds., J. B. Lippincott, Philadelphia, 1965, pp. 1144–1160.
6. **Ormsbee, R. A.,** Rickettsiae (as organisms), *Annu. Rev. Microbiol.,* 23, 275, 1969.
7. **Paretsky, D.,** Biochemistry of rickettsiae and their infected hosts, with special reference to *Coxiella burneti, Zentralbl. Bakteriol. Parasitenkd. Infektionskr. Hyg. Abt. Orig.,* 206, 283, 1968.
8. **Peters, D. and Wigand, R.,** Bartonellaceae, *Bacteriol. Rev.,* 19, 150, 1955.
9. **Philip, C. B.,** The Rickettsiales, in *Bergey's Manual of Determinative Bacteriology,* 7th ed., William & Wilkins, Baltimore, 1957.
10. **Smadel, J. E. and Elisberg, B. L.,** Scrub typhus rickettsiae, in *Viral and Rickettsial Infections of Man,* 4th ed., Horsfall, F. L. and Tamm, I., Eds., J. B. Lippincott, Philadelphia, 1965, pp. 1130–1143.
11. **Synder, J. C.,** Typhus fever rickettsiae, in *Viral and Rickettsial Infections of Man,* 4th ed., Horsfall, F. L. and Tamm, I., Eds., J. B. Lippincott, Philadelphia, 1965, pp. 1059–1094.
12. **Stoker, M. G. P. and Fiset, P.,** Phase variation of the nine-mile and other strains of *Rickettsia burneti, Can. J. Microbiol.,* 2, 310–321, 1956.
13. **Tanaka, H., Hall, W. T., Sheffield, J. B., and Moore, D. H.,** Fine structure of *Haemobartonella muris* as compared with *Eperythrozoon coccoides* and *Mycoplasma pulmonis, J. Bacteriol.,* 90, 1735, 1965.
14. **Vimson, J. W. and Fuller, H. S.,** Studies on trench fever. I. Propagation of *Rickettsia*-like organisms from a patient's blood, *Pathol. Microbiol.,* 24 (Suppl.), 152–166, 1961.
15. **Weinman, D. and Ristic, M.,** *Infectious Blood Diseases of Man and Animals,* Academic Press, New York, 1968.
16. **Weiss, E.,** Comparative metabolism of rickettsiae and other host-dependent bacteria, *Zentralbl. Bakeriol. Parasitenkd. Infektionskr. Hyg. Abt. Orig.,* 206, 292, 1968.
17. **Weiss, E. and Moulder, J. W.,** Taxonomy of the rickettsiae, in *Bergey's Manual of Determinative Bacteriology,* 8th ed., Buchanan, R. E. and Gibbons, N. E., Eds., Williams & Wilkins, Baltimore, 1973.
18. **Wike, D. A., Tallent, G., Peacock, M. G., and Ormsbee, R. A.,** Studies of the rickettsial plaque assay technique, *Infect. Immun.,* 5, 715, 1972.
19. **Wisseman, C. L., Jr.,** Some biological properties of rickettsiae pathogenic for man, *Zentralbl. Bakteriol. Parasitenkd. Infektionskr. Hyg. Abt. Orig.,* 206, 299, 1968.
20. **Woodward, T. E. and Jackson, E. B.** Spotted fever rickettsiae, in *Viral and Rickettsial Infections of Man,* 4th ed., Horsfall, F. L. and Tamm, I., Eds., J. B. Lippincott, Philadelphia, 1965, pp. 1095–1129.
21. **Zdrodovskii, P. F. and Golinevich, H. M.,** *The Rickettsial Diseases,* Pergamon Press, New York, 1960, p. 629.

SELECTED REFERENCES

THE CHLAMYDIAE*

L. A. Page

Chlamydiae are pathogenic bacteria that multiply only within the cytoplasm of vertebrate host cells by a developmental cycle that is unique among microorganisms. They are Gram-negative, non-motile, coccoidal organisms, 0.2 to 1.5 μm in diameter, classified in the genus *Chlamydia,* family Chlamydiaceae, order Chlamydiales. The intracellular developmental cycle begins after phagocytosis of the small (0.2 to 0.5 μm), infectious form of the organism, called elementary body (EB). Within a cytoplasmic vesicle, the EB reorganizes into a larger (0.8 to 1.5 μm), noninfectious form, called initial body, that multiplies by fission. Numerous daughter cells are formed, and these again reorganize, becoming small, electron-dense EB's, which, when released from damaged host cells, survive extracellularly to repeat the cycle in other host cells.

Elementary bodies contain compact nuclear material (RNA/DNA ratio is 1:1) and ribosomes and are bounded by a rigid trilaminar cell wall that is chemically similar to that of Gram-negative bacteria.[1] Initial bodies contain a loose network of nuclear fibrils with an RNA/DNA ratio of 2:1 and have a more fragile cell wall.[2] Because of this nucleic fibrilar network, they are sometimes called reticulate bodies. The initial bodies are the intracellular vegetative form of the organism and have limited ability to survive extracellularly.

METABOLIC CHARACTERISTICS AND CLASSIFICATION

Chlamydiae depend on host cells for certain growth factors. Though chlamydiae are capable of independent enzymatic activities – such as catabolism of glucose, pyruvate, and glutamate – when provided with essential organic and inorganic cofactors, they have no apparent ability to produce high-energy compounds (e.g., ATP) for energy storage and utilization.[3] Thus, they have been characterized as "energy parasites".[4]

Some strains are capable of synthesizing folates and glycogen, thereby providing biochemical criteria for distinguishing two species: *C. trachomatis* and *C. psittaci.*[5,6] Strains of *C. trachomatis* synthesize folates, and hence their growth is inhibited by sulfonamide compounds; they also produce a glycogen matrix surrounding the organisms in their intracytoplasmic microcolony. On the other hand, strains of *C. psittaci* fail to synthesize either folates or glycogen, and their growth is not inhibited by sulfonamides. This division of the chlamydial strains into two species is supported by evidence that hybridization of DNA strands between the two species is 10% or less (see Table 1).[7]

ANTIBIOTIC SENSITIVITIES AND DISINFECTION

Multiplication of chlamydiae in host cells is inhibited by tetracyclines, chloramphenicol, erythromycin, and 5-fluorouracil. Some strains are sensitive to penicillin and D-cycloserine. Most strains appear to be insensitive to streptomycin, vancomycin, kanamycin, neomycin, bacitracin, and gentamycin. Infectivity is rapidly destroyed by dilute solutions of quaternary ammonium detergents and lipid solvents, but is destroyed less rapidly by dilute acids, alkalies, alcohols, phenol, creosyl, or oxidants such as hypochlorite or permanganate.[8] The organisms are moderately resistant to low-energy ultraviolet treatment, to drying, and to lyophilization. They are inactivated in 5 to 30 min at 56°C when suspended in beef heart infusion broth as a 10% infected tissue homogenate.[9]

* Part 18 in *Bergey's Manual.*

Table 1
BIOCHEMICAL PROPERTIES DIFFERENTIATING TWO SPECIES OF *CHLAMYDIA*

Chlamydia trachomatis	*Chlamydia psittaci*
1. Forms compact intracytoplasmic microcolonies that produce glycogen and phospholipids, which stain differentially with iodine-potassium iodide solution.	1. Intracytoplasmic microcolonies are less compact, with developing microorganisms distributed throughout the cytoplasm of the host cell. Glycogen or phospholipids detectable by iodine staining of infected cultures are apparently not formed.
2. Synthesizes essential folates, and growth in chicken embryos is therefore inhibited by sodium sulfadiazine at the level of 1 mg per embryo.	2. Growth in chicken embryos is not sensitive to sodium sulfadiazine at the level of 1 mg per embryo.
3. Degree of DNA hybridization with strains of *C. psittaci* is 10% or less.	3. Degree of DNA hybridization with strains of *C. trachomatis* is 10% or less.

Note: Strains of chlamydiae now considered to be members of the single species *C. psittaci* were formerly known as any of eight species of *Miyagawanella* (*Bergey's Manual of Determinative Bacteriology,* 7th ed., 1957). They have also been variously labeled as *Bedsonia, Rickettsiformis, Rakeia,* and *Ehrlichia.*[4]

DISEASES CAUSED BY CHLAMYDIAE

Chlamydiae produce cytopathology and are the etiologic agents of a variety of diseases of man and other animals. Strains of *C. trachomatis* cause well-known diseases of the ocular and urogenital tracts in humans[10] (see Table 2), and strains of *C. psittaci* cause numerous diseases of man and animals, manifested primarily as pneumonitis, arthritis, placentitis (leading to abortion), or enteritis[11] (see Table 3). Chlamydiae are widely distributed in nature.[12] They have occasionally been isolated from mites and ticks, but arthropod transmission of chlamydial disease has not been proven.[13]

ANTIGENIC STRUCTURE

Every chlamydial strain possesses common group antigens, which are lipopolysaccharides. These antigens are resistant to heat, phenol, proteases, and nucleases, but are inactivated by periodate and lecithinase. They are solubilized in aqueous solutions by treatment of chlamydiae suspensions with deoxycholate or sodium lauryl sulfate.[14,15] Some of these antigens are common to all strains, but certain groups of specific antigens present in a number of strains indicate whether the strain's natural host was man, other mammals, or birds.[16] Clusters of strain-specific antigens are found in chlamydial cell walls.

TINCTORIAL CHARACTERISTICS

Chlamydiae, like rickettsiae, stain red with the Gimenez or Macchiavello procedure, or purple with Giemsa stain. The Gimenez procedure is preferred for staining chlamydiae propagated in yolk sacs of infected chicken embryos.[17] Giemsa stain or Macchiavello's method may be preferred for staining chlamydiae in infected animal tissues or exudates.

Table 2
DISEASES AND INFECTIONS CAUSED BY VARIOUS STRAINS OF *CHLAMYDIA TRACHOMATIS*

Disease	Natural host	Principal effects and transmission routes
Trachoma	Man	Progressive conjunctivitis, primarily of upper eyelid, with hyperemia, exudation, follicular hypertrophy; neovascularization of cornea may lead to opacity and pannus formation. Transmitted by contamination of conjunctiva with infectious exudate.
Inclusion conjunctivitis	Man	Conjunctivitis, primarily of lower eyelid, which tends to heal spontaneously; organisms may spread to mucous membranes of genitalia. Transmitted by contamination of eyelid with infectious material.
Non-gonococcal urethritis, proctitis	Man	Inflammation and exudation in urethra of males and in cervix and anus of females; infection may spread by contact contamination to conjunctiva. Transmitted venereally.
Lymphogranuloma venereum	Man	Effects on lymphatic tissue of iliac and inguinal region result in lymphoadenopathy with suppuration, occasional elephantiasis of penis or scrotum, or rectal strictures in females. Transmitted venereally.
Murine pneumonitis	Mouse	Mild pneumonitis, rarely fatal; endemic in some mouse colonies. Probably transmitted via aerosol of respiratory ejecta.

Intracellular chlamydiae may be clearly observed in fresh wet-mounts of infected cell suspension or exudates by using a microscope equipped with phase contrast optics. Purified suspensions of chlamydiae, free of host cells, are Gram-negative, but various forms of intracellular chlamydiae stain irregularly by Gram's method.

DIAGNOSIS OF CHLAMYDIOSIS

Whatever the host species, conclusive proof of chlamydial infection or disease usually rests upon isolation and identification of the etiologic agent. Clinical history, evidence of typical lesions, and positive serology (or skin test, where appropriate) greatly assist confirmation of frank chlamydiosis, but they are not conclusive by themselves. Overt clinical trachoma may be a diagnostic exception. Evidence of a fourfold or greater rise in serologic titer between acute and convalescent phases of infection is usually diagnostic of recent infection, but in cases of primary brucellosis or Q-fever this evidence may occasionally be misleading. The serology of viral diseases in animals may be obscured by chlamydial antibody titer rises due to incidental subclinical intestinal infection with chlamydiae of low virulence. On the other hand, concurrent salmonellosis or trichomoniasis in birds greatly enhances otherwise benign chlamydial infection, producing high mortality. In domestic herbivores, chlamydiae of modest virulence combined with pasteurellae, mycoplasmata, or parainfluenza viruses may cause a "shipping fever" syndrome.

ISOLATION AND IDENTIFICATION OF CHLAMYDIAE

Choice of tissue for examination depends on the site of principal damage caused by chlamydial infection (see Table 4). Contaminated specimens may be treated with antibiotics known not to affect chlamydial growth.

Table 3
DISEASES AND INFECTIONS CAUSED BY VARIOUS STRAINS OF *CHLAMYDIA PSITTACI*

Disease	Natural hosts	Principal effects, epidemiology, and transmission route
Psittacosis, ornithosis	Wild and domestic birds	Lethargy, hyperthermia, anorexia, abnormal excretions, lowered egg production. Fibrinous airsacculitis, pericarditis, peritonitis, perihepatitis, splenomeglay, hepatopathy. Mortality 0 to 40%, depending on the virulence of the organism. Endemic in psittacine and columbine species. Transmitted by air-borne route to domestic birds and man.
Psittacosis	Man	Malaise, headache, hyperthermia, anorexia, cough. Pneumonitis, splenitis, occasional meningitis. Mortality is less that 1% in antibiotic-treated cases, and 20% in untreated cases. World-wide distribution. Transmitted by air-borne route from birds to man and from man to man.
Pneumonitis	Cats, sheep, cattle, goats, pigs, horses, rabbits	Conjunctival and/or nasal mucopurulent discharge, lethargy, anorexia, labored breathing, hyperthermia, Conjunctivitis, pneumonitis. Rarely fatal, unless complicated by concurrent infection with viruses or other bacteria, especially *Mycoplasma* and *Pasteurella.*
Polyarthritis	Lambs, calves, pigs	Lameness, swollen carpal, tarsal, and stifle joints, hyperthermia, anorexia, lethargy. Fibrinous synovitis, tendonitis, occasional hepatopathy. Mortality is variable. Widespread among lambs and calves in the western U.S. and among pigs in Europe.
Placentitis (leading to abortion)	Cattle, sheep, pigs, goats, rabbits, mice	Transient hyperthermia, chlamydemia, inflammation and necrosis of placentome, abortion of fetus late in gestation. Fetal hepatopathy, edema, ascites, vascular congestion, tracheal petechia. Periodically epidemic in California and Oregon cattle. Endemic in sheep throughout the world.
Encephalomyelitis	Calves	Lethargy, incoordination, weakness, hyperthermia, anorexia, diarrhea, paralysis. Fibrinous perihepatitis, pericarditis, ascites. Endemic in cattle in the midwestern and western U.S.
Conjunctivitis	Sheep, cattle, pigs, cats, guinea pigs	Vascular congestion and edema of conjunctiva, mucopurulent discharge, hyperthermia in cats. Follicular conjunctivitis, keratitis, pannus formation. May be related to pneumonitis strains. Probably transmitted by air-borne route and by contact contamination.
Fatal enteritis	Snowshoe hare, muskrat	Bizarre behavior, diarrhea. Enteritis, splenomegaly, focal necrosis in liver. High mortality in hares caused extensive deaths in snowshoe hares in Canada between 1959 and 1961. Transmission route is unknown; muskrats may be reservoirs.

Table 3 (continued)
DISEASES AND INFECTIONS CAUSED BY VARIOUS STRAINS OF *CHLAMYDIA PSITTACI*

Disease	Natural hosts	Principal effects, epidemiology, and transmission route
Enteritis	Cattle, sheep	Diarrhea, weakness, and death in newborns. Enteritis. Epidemiology in relation to widespread subclinical infection of adult animals is not known.
Subclinical intestinal infection	Cattle, sheep	No clinical signs in adults; may cause transient hyperthermia and diarrhea in newborns. Widespread in cattle and sheep throughout the U.S. Probably transmitted by ingestion of feces-contaminated feed.

Note: Clinical signs and gross lesions vary widely in cases of mild or chronic chlamydial infections. Variations in effects may also be caused by differences in natural resistance of individuals, virulence of the organisms, dosages, presence of other pathogens, and other stress factors.

Table 4
PREFERRED SPECIMENS, MICROSCOPY, AND HOSTS FOR ISOLATION OF CHLAMYDIAE

(CE = chicken embryo; CC = cell culture[a])

Disease	Preferred specimen[b]	Microscopy, stain, or method[c]	Principal (alternative) isolation hosts; incubation temperature, °C
Trachoma	Conjunctival scraping	Giemsa stain, iodine stain, fluorescent antibody	CE (CC); 35.Primates, conjunctival scarification
Conjunctivitis; urethritis or cervicitis	Conjunctival scraping; urethral or cervical swab	Fluorescent antibody, Giemsa stain, iodine stain	CE (CC); 35. Primates, conjunctival scarification
Lymphogranuloma venereum	Bubos aspirate	Inconclusive	CE (CC); 35. Mice, intracerebral route
Human pneumonitis	Sputum, blood; lungs spleen at autopsy	Inconclusive	CE (CC); 37–39. Mice, intraperitoneal route
Avian pneumonitis	Airsac or pericardial exudate, spleen, liver	Giemsa stain, Macchiavello stain, phase contrast microscopy of fresh exudate	CE (CC); 37–39. Mice, intraperitoneal route
Mammalian pneumonitis	Tracheal exudate, lung	Giemsa stain, Macchiavello stain	CE; 37–39. Guinea pig, intraperitoneal route
Mammalian polyarthritis	Synovial fluid of carpal, tarsal, or stifle joints	Phase contrast microscopy of fresh exudate, fluorescent antibody	CE; 37–39. Guinea pig, intraperitoneal route
Mammalian placentitis	Hyperemic areas of placenta, placentomes, fetal liver	Giemsa stain, Macchiavello stain	CE; 37–39. Guinea pig, intraperitoneal route
Mammalian enteritis	Surface mucus of formed feces, hyperemic areas of intestinal wall	Inconclusive	CE; 37–39 after centrifugation and antibiotic treatment to remove contaminating bacteria

[a] McCoy, "L", or HeLa cell lines.

[b] For isolation of chlamydiae, all specimens should be treated with a combination of antibiotics, e.g., streptomycin sulfate, Vancomycin®, and kanamycin sulfate, 1 mg each per ml of phosphate-buffered saline.

[c] Positive specimens contain mononuclear cells with intracytoplasmic microcolonies of chlamydiae.

All strains of chlamydiae may be isolated in developing chicken embryos inoculated by the yolk-sac route on the 6th to 7th day of development. Many strains of *C. trachomatis* are sensitive to temperatures above 37°C, and embryos inoculated with homogenates of specimens from humans should, therefore, be incubated at 35°C. Some chlamydial strains from humans are difficult to isolate in chicken embryos, because they multiply slowly and require numerous passages before they are fully adapted to growth in avian cells. Embryos inoculated with specimens from infected birds or mammals may be incubated at 37 to 39°C. The latter temperature enhances the growth rate of most *C. psittaci* strains.[18] Depending on the virulence and growth rate, chlamydiae multiply to numbers sufficient to kill the embryo in 4 to 12 days. If the inoculum contains few organisms, several blind passages of infectious material in eggs may be necessary to establish an embryo death pattern.

Alternative isolation hosts are listed in Table 4. These species include primates, mice, guinea pigs, and lung cell cultures ("L" cell line, McCoy cell line). Lesions observed in these hosts are described in detail elsewhere.[19,20]

Identification of chlamydiae in yolk sacs of chicken embryos or in tissues of other experimental hosts depends on demonstration of the presence of chlamydial group antigen. Infected yolk sacs are triturated in phosphate-buffered saline (pH 7.2) to make a 20 to 25% suspension and boiled for 30 min; when the suspension has cooled, phenol is added to a final concentration of 0.5%. The suspension should then be reacted in serial dilution in Veronal® buffer against a constant dilution of chlamydial antiserum in a CF test. The dilution of antiserum should contain four to eight CR units of antibody per chlamydial group antigen. If the yolk sac suspension has a CF titer of 1:32 or more against antiserum and is negative against normal serum, the yolk sac can then be assumed to contain organisms of the genus *Chlamydia.* It has been calculated[6,21] that approximately 10,000 embryo LD_{50} of chlamydiae will fix one unit of complement against four units of antiserum in a CF test.* Confirmatory examination of infected yolk sacs may be made after staining yolk sac smears by the Gimenez method. Chlamydiae stain red, whereas other bacteria stain blue against a greenish background.

Chlamydiae in cell cultures and animal tissues may also be identified microscopically by using specific fluorescent-antibody (FA) methods, but reliable FA preparations are not available for general use. Most FA preparations are products of individual laboratory effort and are used experimentally. FA preparations that are specific for certain strains of organisms, e.g., those that cause trachoma, have been used successfully to distinguish serotypes.[22]

Chlamydial species are identified on the basis of characteristics described in Table 1. The test for glycogen production is performed on infected cell monolayer cultures (and uninoculated control cultures) by staining them with 5% iodine-potassium iodide solution after chlamydial growth has occurred.[5] Iodine-positive microcolonies of chlamydiae are dark tan against a light tan background. The test for sulfonamide sensitivity is performed by inoculating decimal dilutions of a suspension of chlamydiae by the yolk sac route into 6-day incubated embryos, using at least 12 embryos per dilution. Half of the embryos in each series of dilutions should be given a second yolk sac inoculation, using a solution containing sufficient sodium sulfadiazine so that each embryo receives 1 mg. The eggs are then incubated for 14 days, the deaths are recorded, and the LD_{50} is calculated for each series. If the chlamydiae are sensitive to sulfadiazine, there should be at least a 2-log difference between the LD_{50} calculation for the series with sulfadiazine and that for the series without sulfadiazine.[6]

* Yolk sacs containing large numbers of bacteria of the genera *Herellea* or *Bacterioides* may fix complement with chlamydial antisera also.

SEROLOGY

The method most commonly used at present for detecting antibodies in sera of infected individuals is the CF test.[16] For this test, group- and strain-specific antigens have been prepared from yolk-sac- or cell-culture-propagated organisms. However, more rapid and simpler serologic methods are now being used, namely, the agar gel precipitin method[23,24] and immunofluorescence[19] tests for antibodies.

Rapid increases in circulating chlamydial antibodies generally signify current infection. Incidences of 50% or more of serologic titers of 1:64 among a herd or flock of domestic animals also indicate current infection. The incidence of seropositive tests among apparently healthy populations of humans, cattle, or sheep may range from 25 to 70%, reflecting the ubiquity of chlamydiae in nature and the possibility of widespread natural infections that stimulate production of antibodies, but escape clinical notice because of their mildness.

REFERENCES

1. **Manire, G. P. and Tamura, A.,** Preparation and chemical composition of the cell walls of mature infectious dense forms of meningopneumonitis organisms, *J. Bacteriol.,* 94, 1178, 1967.
2. **Tamura, A. and Manire, G. P.,** Preparation and chemical composition of the cell wall membranes of developmental reticulate forms of meningopneumonitis organisms, *J. Bacteriol.,* 94, 1184, 1967.
3. **Weiss, E. and Wilson, N. N.,** Role of exogenous adenosine triphosphate in catabolic and synthetic activities of *Chlamydia psittaci, J. Bacteriol.,* 97, 719, 1969.
4. **Moulder, J. W.,** The relation of the psittacosis group (chlamydiae) to bacteria and viruses, *Annu. Rev. Microbiol.,* 20, 107, 1969.
5. **Gordon, F. B. and Quan, A. L.,** Occurrence of glycogen in inclusions of the psittacosis-lymphogranuloma venereum-trachoma agents, *J. Infect. Dis.,* 115, 86, 1965.
6. **Page, L. A.,** Proposal for the recognition of two species in the genus *Chlamydia* Jones, Rake and Stearns, 1945, *Int. J. Syst. Bacteriol.,* 18, 51, 1968.
7. **Kingsbury, D. T. and Weiss, E.,** Lack of deoxyribonucleic acid homology between species of genus *Chlamydia, J. Bacteriol.,* 96, 1421, 1968.
8. **Nabli, B. and Tarizzo, M. L.,** The effect of antiseptics and other substances on TRIC agents, *Am. J. Ophthalmol.,* 63, 1441, 1967.
9. **Page, L. A.** Thermal inactivation studies on a turkey ornithosis virus, *Avian Dis.,* 3, 67, 1959.
10. **Jawetz, E.,** Agents of trachoma and inclusion conjunctivitis, *Annu. Rev. Microbiol.,* 18, 301, 1964.
11. **Storz, J.,** *Chlamydia and Chlamydia-Induced Diseases,* Charles C Thomas, Springfield, Ill., 1971.
12. **Meyer, K. F.,** The host spectrum of psittacosis-lymphogranuloma venereum (PL) agents, *Am. J. Ophthalmol.,* 63, 1224, 1967.
13. **Eddie, B., Radovsky, F. J., Stiller, D., and Kumada, D.,** Psittacosis-lymphogranuloma venereum (PL) agents *(Bedsonia, Chlamydia)* in ticks, fleas, and native mammals in California, *Am. J. Epidemiol.,* 90, 449, 1969.
14. **Jenkin, H. M.,** Preparation and properties of cell walls of the agent of meningopneumonitis, *J. Bacteriol.,* 80, 639, 1960.
15. **Benedict, A. A. and McFarland, C.,** Direct complement fixation tests of ornithosis in turkeys, *Proc. Soc. Exp. Biol. Med.,* 92, 768, 1956.
16. **Fraser, C. E. O.,** Analytical serology of the Chlamydiaceae, in *Analytical Serology of Microorganisms,* Vol. I, John Wiley & Sons, New York, 1969, p. 257.
17. **Gimenez, D. F.,** Staining rickettsiae in yolk sac cultures, *Stain Technol.,* 39, 135, 1964.
18. **Page, L. A.,** Influence of temperature on multiplication of chlamydiae in chicken embryos (Proceedings, International Trachoma Conference, Boston, 1970), *Excerpta Med. Int. Congr. Ser.,* p. 273, 1971.

19. **Wang, S. P.,** A microimmunofluorescence method: study of antibody response to TRIC organisms in mice (Proceedings, International Trachoma Conference, Boston, 1970), *Excerpta Med. Int. Congr. Ser.,* No. 223, p. 273, 1971.
20. **Page, L. A.,** Interspecies transfer of psittacosis-LGV-trachoma agents: pathogenicity of two avian and two mammalian strains for eight species of birds and mammals, *Am. J. Vet. Res.,* 27, 397, 1966.
21. **Schachter, J.,** Recommended criteria for the identification of trachoma and inclusion conjunctivitis agents, *J. Infect. Dis.,* 122, 105, 1970.
22. **Wang, S. P. and Grayson, J. T.,** Classification of TRIC and related strains with microimmunofluorescence (Proceedings, International Trachoma Conference, Boston, 1970), *Excerpta Med. Int. Congr. Ser.,* No. 223, 305, 1971.
23. **Collins, A. R. and Barron, A. L.,** Demonstrations of group- and species-specific antigens of chlamydial agents by gel diffusion, *J. Infect. Dis.,* 121, 1, 1970.
24. **Page, L. A.,** Application of an Agar Gel Precipitin Test to the Serodiagnosis of Avian Chlamydiosis, Proceedings, 17th Annual Meeting of the American Association of Veterinary Laboratory Diagnosticians, 1974, pp. 51–61.

THE MOLLICUTES*

MYCOPLASMAS

J. G. Tully and S. Razin

Mycoplasmas form a distinct group of minute, filterable prokaryotic microorganisms that lack a cell wall. Although approaching the larger viruses in size, their ability to reproduce independently of any other living cell and the fact that they contain both DNA and RNA distinguish them from viruses. Their small size, the ability to pass through bacteriological filters, and the absence of a cell wall separates them from the true bacteria. Thus, mycoplasmas are the smallest and perhaps simplest free-living organisms known. Mycoplasmas have been recovered from almost all animal species surveyed, and within recent years evidence has accumulated to verify their occurrence in a variety of plants and insects (spiroplasmas). No less intriguing is the recent finding of prokaryotes with similar characteristics in very specialized habitats, such as acidic coal refuse piles (thermoplasmas) or the strictly anaerobic bovine and ovine rumen (anaeroplasmas).

The discussion here and in subsequent sections will use the general trivial term "mycoplasmas" to refer to all members of the class Mollicutes.

MORPHOLOGY AND REPRODUCTION OF MYCOPLASMAS

Although published work on the morphology and ultrastructure of mycoplasmas has been extensive, much of this work has been carried out on a relatively small number of species.[1] Thus, many members of the class Mollicutes have not been adequately characterized, both morphologically and ultrastructurally, especially in relationship to recent information that the osmolarity of fixatives[2] and of specific buffers[3] may drastically alter the actual size and shape of these wall-free and plastic organisms. Ideally, the morphology and ultrastructure of mycoplasmas should be based upon close correlation of the appearance of the organisms under phase contrast or dark-field microscopy with their appearance in the electron microscope. Additional comments and references on technical aspects of morphologic studies can be found in several recent publications or reviews.[1,4,160]

Mycoplasmas generally have been found to vary in shape from spherical structures (300 to 800 nm in diameter) to slender branched filaments of uniform diameter (100 to 800 nm), ranging in length from a few μm to 150 μm.[5-7] Much of the earlier information that led to the concept of very small viable cells, down to 100 nm in diameter, was based upon sizing data obtained by filtration or by electron microscopy. The limitations of these techniques with regard to plastic organisms have been noted.[5,8] There is growing evidence that the diameter of the smallest mycoplasma cell capable of reproduction is in the size range of 300 nm.[9,161]

Electron microscopy of thin sections of mycoplasma cells reveals a very simple ultrastructure. Essentially, the mycoplasma cell is built of only three organelles: the cell membrane, the ribosomes, and the characteristic prokaryotic genome. There is no evidence of any intracellular membranous structures, such as mesosomes. In several species, however, specialized organelles or structures have been observed. The pear-shaped *Mycoplasma gallisepticum* cells have a terminal bleb structure.[4] A special structure, consisting of a dense central rod-like core surrounded by a lucent space, has been described at the tip of filamentous *Mycoplasma pneumoniae* cells.[10] A similar tip

* Part 19 in *Bergey's Manual.*

structure has also been found in *Mycoplasma pulmonis*.[11] These terminal structures appear to play a role in the attachment of the mycoplasmas to the surface of respiratory epithelial cells as well as to glass and plastic surfaces.[12] The tip structures also appear to be directly involved in the motility or so-called gliding movement of these three *Mycoplasma* species, as observed on liquid-covered surfaces. The leading direction of the movement is always oriented tip first.[13]

An additional intracytoplasmic structure, termed rho fibers, has been observed in variants of *M. mycoides* subsp. *mycoides* and subsp. *capri*, and in several other caprine and bovine mycoplasmas.[14] The cells contain a seemingly rigid, rod-like striated fiber that extends axially throughout the cell and terminates at the plasma membrane in a knob-like structure.[15] The culture medium is an important factor in the selection and expression of rho forms,[15] and their presence correlates with the synthesis of two proteins within the cell.[16] The function and significance of these organelles in mycoplasmas is unknown, but they appear to occur in vivo as well as in artificial media.[17]

The cell membrane of mycoplasmas has become a most useful model in biological membrane research, largely due to its easy isolation, freedom from other types of membranes, and the ease by which controlled alterations in its composition can be made. Extensive reviews have been published on the preparation and characterization of mycoplasma membranes,[5,18-22] their reconstitution,[23] and their immunochemistry.[24] The existence of a slime layer made of galactan on the surface of *M. mycoides* subsp. *mycoides* cells has long been known.[25] More recently, a hexosamine polymer has been found to cover *Acholeplasma laidlawii* cells. This polymer appears to be closely associated with the membrane, since a considerable amount of it remains attached to the membrane even after extensive washing.[26] Capsule-like material has also been observed by electron microscopy on *M. meleagridis*[27] and *M. dispar*[28] cells.

The mode of reproduction of mycoplasmas has been a matter of dispute. However, study of the molecular biology of these organisms showed clearly that replication of their prokaryotic genome, which must precede cell division, follows the same pattern as with other prokaryotes dividing by binary fission.[29] For binary fission to occur, however, cytoplasmic division must be fully synchronized with genome replication, which is not always the case with mycoplasmas. The lack of a cell wall may well be responsible for the poor coordination of the two processes. In many mycoplasmas cytoplasmic division lags behind genome replication, resulting in the formation of multinucleate filaments. The subsequent division of the cytoplasm leads to the formation of characteristic chains of beads, which later fragment to give single cells.[6] Budding, frequently seen in mycoplasma cultures,[30,31] may also be regarded as a form of binary fission in which the cytoplasm is not equally divided between the daughter cells. Perhaps the best experimental evidence for this mode of reproduction was provided by phase contrast cinematography of mycoplasma broth cultures.[7] Classical binary fission was clearly seen to occur side by side with budding and fragmentation of filaments.

CLASSIFICATION OF MYCOPLASMAS

The current basis for nomenclature and classification of mycoplasmas was proposed in 1956,[32] and this taxonomic system has received general acceptance.[33] Subsequent progress in the taxonomy of these organisms has developed through modification or extension of this scheme.[34] In 1967, the ICSB Subcommittee on the Taxonomy of Mycoplasmatales recommended[35,36] that the Mycoplasmatales be placed in a separate new class (Mollicutes) to recognize the major distinctions between mycoplasmas and bacteria. A fundamental characteristic of the Mollicutes is the absence of a cell wall, which is associated with the inability to synthesize the peptidoglycan polymer or its precursors. It should be noted here that stable L-phase variants of bacteria also lack a cell wall, and though some of these organisms still maintain the presence of peptidoglycan or its precursors, other L-phase variants are apparently devoid of these components.[37] The absence of a cell wall in mycoplasmas is a property of outstanding importance, and one to which the organisms owe many of their peculiarities – for example, their plasticity and morphological instability, their osmotic sensitivity, their ability to grow within the interstices of the fibrilar agar gel, their susceptibility to lysis by detergents, alcohol, or specific antibody and complement, and their resistance to penicillin or other antibiotics known to inhibit peptidoglycan synthesis and polymerization.

The Mollicutes are generally Gram-negative, except for *Spiroplasma citri,* which has been reported to give a Gram-positive staining reaction.[38] Dependence of growth on sterol is also an outstanding characteristic of the Mollicutes, and – although not a property of all mycoplasmas – this growth requirement is sufficiently distinct to exclude these organisms from members of the class Schizomycetes.

Additional support for the separation of mycoplasmas from bacteria has also come through analyses of the DNA of this group. The guanine + cytosine ratio (G + C) of the DNA of most members of the class is at the lower limit of the ratios known for bacteria.[39-51] In addition, the genome size of mycoplasmas is only one fifth to one half the size recorded for most bacteria.[46,47,49-53]

Recent extensions to the classification scheme (Table 1) have involved subdivision of the order Mycoplasmatales into two families (Mycoplasmataceae and Acholeplasmataceae) and establishment of corresponding genera to provide for differences in sterol requirement for growth,[54,55] establishment of a separate genus (*Ureaplasma*) within the Mycoplasmataceae to provide a separate status for a group of the urea-hydrolyzing mycoplasmas, formerly known as T-mycoplasmas,[56] and proposals to establish a separate genus (*Spiroplasma*)[38] and family (Spiroplasmataceae)[57] within the order Mycoplasmatales for the new plant and insect mycoplasmas. Two additional groups of organisms

Table 1
TAXONOMY OF THE CLASS MOLLICUTES

Class: Mollicutes
- Order: Mycoplasmatales
  - Family I: Mycoplasmataceae
    1. Sterol required for growth
    2. Genome size about 5.0×10^8 daltons
    3. NADH oxidase localized in cytoplasm
    - Genus I: *Mycoplasma* (about 50 species current)
      1. Do not hydrolyze urea
    - Genus II: *Ureaplasma* (single species with serotypes)
      1. Hydrolyzes urea
  - Family II: Acholeplasmataceae
    1. Sterol not required for growth
    2. Genome size about 1.0×10^9 daltons
    3. NADH oxidase localized in the membrane
    - Genus I: *Acholeplasma* (6 species current)
  - Family III: Spiroplasmataceae (proposed)
    1. Helical organisms during some phase of growth
    2. Sterol required for growth
    3. Genome size about 1.0×10^9 daltons
    4. NADH oxidase localized in cytoplasm
    - Genus I: *Spiroplasma* (1 species current)
  - Genera of uncertain taxonomic position:
    - *Thermoplasma* (single species)
    - *Anaeroplasma* (2 species)

given proposed generic designations, *Thermoplasma*[58] and *Anaeroplasma,*[59] are presently considered to be members of the Mollicutes, but of uncertain taxonomic position within the class. Additional information on the occurrence and characteristics of these two latter groups will be necessary before their proper relationship to other classified mycoplasmas can be established.

Distinctions at the subgeneric level involve a series of biochemical and serological properties. To improve the standards of published descriptions of new species, the ICSB Subcommittee on the Taxonomy of Mycoplasmatales published a proposal for the minimal standards for description of new species within the order.[60] These recommendations state the properties to be determined in establishing differences from existing species, including structure, cultural and biochemical characteristics, and serological distinctions. More specific details of these procedures will be given in subsequent sections.

ISOLATION AND IDENTIFICATION OF MYCOPLASMAS

Table 2 gives a rough comparison of some characteristics of mycoplasmas with those of other groups of microorganisms. Recommended steps in the identification of mycoplasmas, as proposed by the ICSB Subcommittee on the Taxonomy of Mycoplasmatales,[60] are outlined in Figure 1. General principles for the isolation and detection of mycoplasmas, and specifically comments on tissue preparation,[61-64] use of various culture media,[61-63] cell culture procedures and tissue culture contamination,[62,65-68] and recovery of mycoplasmas from various animal tissues,[61,62,69,70] have been published in recent years.

Most mycoplasmas are facultative anaerobes, but will grow aerobically. Since a few species require anaerobic environments, and since isolations from primary tissue specimens frequently grow out only under anaerobic conditions, an atmosphere of 95% N + 5% CO_2 is preferred. The "fried egg" colony, consisting of an opaque, granular central zone that grows down into the medium and a flat, translucent peripheral zone on the

Table 2
CHARACTERISTICS OF MYCOPLASMAS AND SOME OTHER PROKARYOTIC ORGANISMS

Property	Mycoplasmas	Schizomycetes	Chlamydiae	Rickettsiae	Viruses
Growth on cell-free medium	+	+[a]	–	–[b]	–
Absence of cell wall or cell wall peptidoglycan	+	–	–	–	+
Generation of metabolic energy	+	+	–	+	–
Dependent on host cell nucleic acid for multiplication	–	–	–	–	+
Can synthesize proteins by own enzymes	+	+	+	+	–
Sterol requirement	+[c]	–	–	–	–
Visible in optical microscope (1500 ×)	+	+	+	+	–
Filterability through 450-nm pore size filters	+	–[d]	+	+	+
Contain both DNA and RNA	+	+	+	+	–
Growth inhibited by antibody alone	+	–	+	+	+
Growth inhibited by antibiotics acting on protein synthesis	+	+	+	+	–

[a] With few exceptions, such as *Treponema pallidum* and *Mycobacterium leprae.*
[b] Except *Rochalimaea quintana* and *Bartonella* species.
[c] Except *Acholeplasma* species.
[d] With few exceptions.

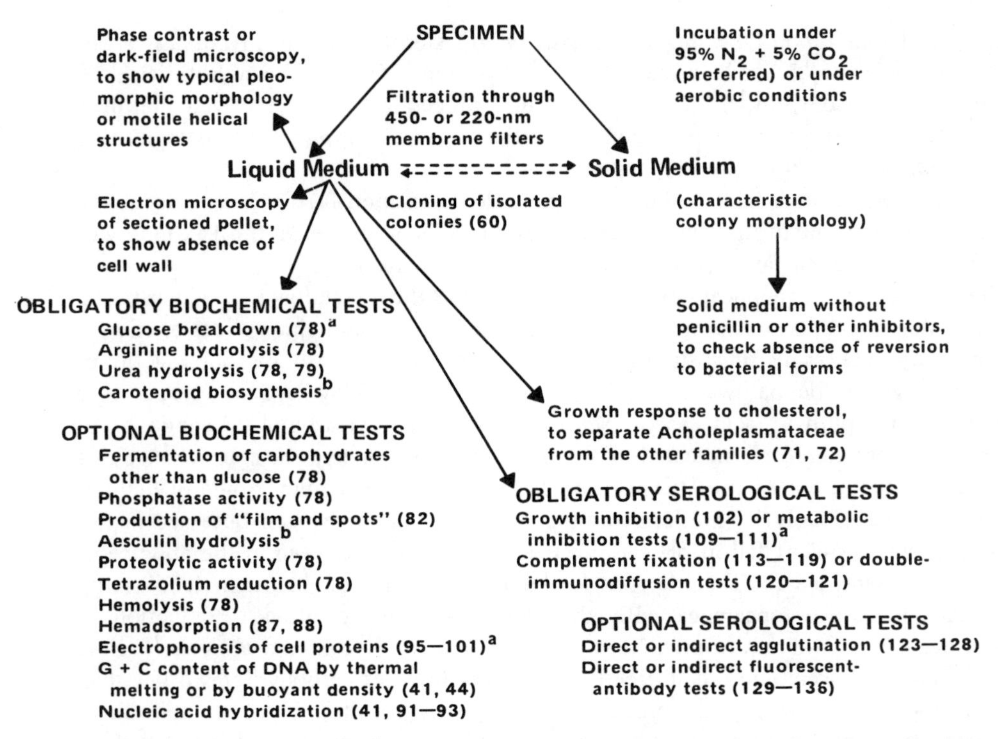

[a] **Recommended techniques for these procedures are also contained in documents prepared by the Board for the WHO/FAD Program on Comparative Mycoplasmology. See text for further information.**

[b] **See chapter on *Acholeplasma* species.**

FIGURE 1. Recommended steps in the identification of mycoplasmas. (Slightly modified from recommendations of the Subcommittee on the Taxonomy of Mycoplasmatales, 1972.[60] Additional references to test procedures are given within parentheses and in Reference 162.)

medium surface, is an important characteristic of mycoplasmas. The colonies are usually very small, their diameters ranging from 50 to 600 μm, but occasionally, when well isolated, they may approach 4 mm in diameter. A "fried egg" colony is also typical of the L-phase variants of bacteria. These variants, however, which are usually induced in the laboratory by a single exposure to antibiotics affecting cell wall synthesis, revert to the bacterial phase and lose the "fried egg" colonial form when the antibiotic is omitted from the growth medium. It is, therefore, recommended to subculture each new isolate suspected of being a mycoplasma at least five consecutive times on media not containing penicillin or other antibacterial agents, to check for non-reversion to a bacterial form.

Another important step in mycoplasma identification is cloning of the culture to establish its purity. For this purpose, an agar medium is inoculated with a filtrate of a broth culture of the organism. Filtration through a membrane filter with the smallest pore diameter possible breaks up any clumps of organisms and results in colonies originating from single cells. An isolated colony is then transferred to the broth, and the cloning procedure is repeated at least twice more. The cloned cultures are examined by light and electron microscopy to verify the characteristic mycoplasma or spiroplasma morphology and the absence of a cell wall.

Growth response to cholesterol,[71,72] combined with cell morphology, should be the minimal determinations made to place any new isolate within one of the three families in the order Mycoplasmatales. In addition to the two recommended direct tests for establishing a requirement for cholesterol, several indirect tests have been developed, which are based upon the higher sensitivity of sterol-requiring mycoplasmas to polyanetholsulfonate,[150–153] lysolecithin,[154] digitonin,[153,155,156] amphotericin,[157] Filipin®, or lucensomycin.[158] The digitonin disc procedure[153] has received the most extensive evaluation and appears to combine reliability with simplicity in distinguishing *Acholeplasma* species from other mycoplasmas.

Classification of an isolate within the family depends upon further biochemical and serological tests. Some of these are designated by the Subcommittee on the Taxonomy of Mycoplasmatales as obligatory; others are optional (see Figure 1).

The obligatory biochemical tests include tests for the breakdown of glucose and arginine. Most mycoplasmas utilize either glucose or arginine as major sources of energy. The carbohydrate-fermenting strains generally catabolize glucose by homolactic or heterolactic glycolytic pathways, the major end products being lactic acid and, to a smaller extent, pyruvic acid, acetic acid, and acetylmethylcarbinol. Breakdown of carbohydrates is, therefore, indicated by the production of acid and the color change of a pH indicator incorporated into the carbohydrate-containing medium. Some difficulties with certain non-glycolytic (nonfermentative) mycoplasmas have been observed in this procedure, since a slight fall in pH can occur in control cultures even in the absence of glucose. More sensitive and specific tests for the determination of glucose metabolism by measuring glucose disappearance with the glucose–oxidase reaction,[73] detection of acid-fermentation products from radioactive glucose or of hexokinase activity,[74] have been proposed to overcome this difficulty. The breakdown of carbohydrates other than glucose, such as mannose, mannitol, lactose, xylose, sorbitol, glycerol, cellobiose, saccharose, salicin, fructose, and galactose, is of some diagnostic value, and these determinations were accordingly included among some of the optional tests. In the vast majority of species, the respiratory pathways seem to be flavine-terminated, so that heme compounds (cytochromes, catalase, etc.) are absent.[18] Exceptions have been noted, however.[75] Most, but not all of the non-glycolytic mycoplasmas contain the arginine dihydrolase pathway:

(a) arginine $\xrightarrow{\text{arginine deiminase}}$ citrulline + NH_3

(b) citrulline + Pi $\xrightleftharpoons{\text{ornithine transcarbamylase}}$ ornithine + carbamyl phosphate

(c) carbamyl phosphate + ADP $\xrightleftharpoons[\text{Mg 2+}]{\text{carbamyl phosphokinase}}$ ATP + NH_3 + CO_2

This pathway can supply the organism with ATP,[76] but not all arginine-utilizing mycoplasmas use this pathway as a major energy source.[77] The liberated ammonia raises the pH of the medium, furnishing the basis for the detection of arginine hydrolysis in the laboratory test.[78]

Hydrolysis of urea is a property of all *Ureaplasma* species and, as a consequence, becomes an obligatory biochemical test to separate these organisms from other mycoplasmas.[56,78,79]

The optional biochemical tests listed in Figure 1 are very useful in the classification of the mycoplasmas, as may be seen in subsequent sections, where summaries of the biochemical and physiological characteristics of *Mycoplasma, Ureaplasma, Acholeplasma, Spiroplasma, Thermoplasma,* and *Anaeroplasma* species are tabulated and discussed. Some of the data presented in these tables might not be totally representative of species characteristics, since possible strain variations do occur, and in a few instances tests were performed on only the type strain, or a single isolate has been used to establish the species (monotypy).

The tetrazolium reduction test is based on the ability of many mycoplasmas to reduce

2,3,5-triphenyltetrazolium chloride. More strains are capable of reducing tetrazolium under aerobic than under anaerobic conditions. Phosphatase activity is determined according to the ability or inability of the mycoplasmas to hydrolyze phenolphthalein diphosphate incorporated in the growth medium,[78,80] or by the ability of washed-cell suspensions to hydrolyze *p*-nitrophenylphosphate.[81] Film and spot reaction is indicative of the intensity of the lipolytic activity of the organisms, a property that has some diagnostic value. During the growth of certain mycoplasmas on media containing horse serum or egg yolk emulsion, a characteristic wrinkled, pearly film appears on the medium surface, together with tiny black spots beneath and around the colonies. The film contains cholesterol and phospholipids, whereas the spots consist of calcium and magnesium salts of fatty acids liberated by the mycoplasma lipases.[82] Proteolytic activity, usually determined by the ability to liquefy gelatin or coagulated serum, may serve as another criterion for distinguishing mycoplasmas.[78,83]

The optional tests also include hemolysis of sheep and guinea pig red blood cells by mycoplasmas, which can be tested by covering colonies with a thin layer of blood agar[78] or by inoculating concentrated suspensions of organisms onto blood agar.[84] Weak to strong *alpha* and *beta* hemolysis was shown almost throughout to result from the production of peroxide by the organisms.[84-86] Since hemolytic activity is shared by most mycoplasmas, and since its degree and type seem to depend on minor differences in the technique used, demonstration of the property *per se* is of secondary importance. Of greater diagnostic value is the ability of certain mycoplasmas to adsorb erythrocytes or other types of animal cells. This property can be tested microscopically by determining the adsorption of erythrocytes or other animal cells to mycoplasma colonies.[87-89]

More sophisticated methods for mycoplasma classification are those based on molecular genetics. Determination of the G + C content of the mycoplasma DNA by thermal melting, buoyant density, or chemical analysis may be of great diagnostic value.[45,48] The data presented in subsequent sections show that the G + C content ranges from 23 to 46%, reflecting the marked genetic heterogeneity of mycoplasmas. The most typical aspect is the low G + C content of mycoplasma DNA, which in many strains is as low as 23 to 24%, much less than in bacterial DNA's.[90] Since a similar base composition does not always imply genetic identity, the nucleotide sequence in the DNA strands must also be determined. Various nucleic acid hybridization techniques have helped to establish the identity of unknown mycoplasma strains.[40,45,91-94] A simpler approach to genetic classification, though less direct, is the comparison of the electrophoretic patterns of cell proteins in polyacrylamide gels.[95-101] Since the synthesis of cell proteins is genetically directed, the electrophoretic patterns are likely to reflect the genetic identity or non-identity of microorganisms.

Serological tests form an essential part of any identification and classification of a new mycoplasma isolate. The new isolate should be compared serologically with other named species – ideally with all of them. The minimum requirement is that it should differ antigenically from all species having the same habitat and sharing the same general biological properties. The obligatory serological tests, according to the recommendations of the Subcommittee (see Figure 1), include either the growth or the metabolism inhibition test together with any one of the less specific complement fixation and/or double-immunodiffusion tests. The growth inhibition test, based on inhibition of growth on agar around discs saturated with the specific antiserum,[102-108] is the most specific serological test, but it requires highly potent sera, which are not always available. The metabolism inhibition test is much more sensitive than the growth inhibition procedure. The metabolism inhibition technique is based upon determining the amount of growth inhibition by antibody directed to the inhibition of certain metabolic activities of the mycoplasmas, specifically to glucose fermentation or arginine and urea hydrolysis.[109-112] Complement fixation tests with whole-cell antigens or cell extracts[113-119] and immunodiffusion tests with extracts of mycoplasma cells[113,120-122] have the advantage of showing antigenic relationships between strains, and one of them should, therefore, be included in the battery of tests to supplement the growth or metabolism inhibition test.

Additional optional serological tests include the direct and indirect agglutination[123-128] and fluorescent-antibody tests.[129,130] The direct identification of mycoplasma colonies on agar by specific fluorescent antibodies[62,108,131,163] is perhaps the most useful and rapid test for diagnosis. It has proved to be very specific and is the only test capable of distinguishing between a mixture of colonies of different serotypes on the same plate – a most important feature, since clinical material very frequently contains more than one species or serotype. This test has also been modified by using agar blocks containing colonies or mycoplasmas grown on glass cover slips in order to reduce the amount of conjugated antiserum required,[132-136] or by using conjugated antiglobulin to eliminate the need for a battery of specific conjugated antisera.[132]

It is important to note here that antigenic diversity exists with certain *Mycoplasma* species, especially *M. hominis,*[122,137,138] *M. pulmonis,*[119] and *M. arginini.*[139] Such antigenic heterogeneity has been noted with these species in a variety of serological tests and appears to be correlated with differences between membrane antigens or membrane proteins of individual strains. The serological tests recommended by the Subcommittee (see Figure 1) provide suitable selections for distinguishing different species within the Mycoplasmatales or for grouping species within a genus. Additional information and discussion on the preparation of mycoplasma antisera and on the respective values of various serological procedures in the classification and identification of mycoplasmas can be found in several recent reviews.[108,140,141,159,162]

Reference antisera and seed to 27 *Mycoplasma* and *Acholeplasma* species[141] is currently available in limited amounts from the FAO/WHO Collaborating Center for Animal Mycoplasmas, Institute of Medical Microbiology, University of Aarhus, DK8000 Aarhus, Denmark, or from Research Resources Branch, National Institute of Allergy and Infectious Diseases, National Institutes of Health, Bethesda, MD 20014. Type strains of most established species within the Mollicutes are also available from the American Type Culture Collection, 12301 Parklawn Drive, Rockville, MD 20852, and from the National Collection of Type Cultures, Public Health Laboratory, Colindale Ave., London, NW9 5EQ, England. Recommended techniques and reviews of several aspects of mycoplasma characterization (determination of glucose fermentation, gel electrophoretic techniques, metabolism inhibition and growth inhibition tests, and lyophilization procedures) prepared under auspices of the WHO/FAO Program on Comparative Mycoplasmology[142] are available from the Veterinary Public Health, Division of Communicable Diseases, World Health Organization, 1211 Geneva 27, Switzerland.

More comprehensive information on the biology and pathogenicity of mycoplasmas may be found in several books or recent monographs.[20,143-149]

REFERENCES

1. **Boatman, E. S.,** *Ann. N.Y. Acad. Sci.,* 225, 172–180, 1973.
2. **Lemcke, R. M.,** *J. Bacteriol.,* 110, 1154–1162, 1972.
3. **Cole, R. M., Tully, J. G., Popkin, T. J., and Bove, J. M.,** *J. Bacteriol.,* 115, 367–386, 1973.
4. **Maniloff, J. and Morowitz, H. J.,** *Bacteriol. Rev.,* 36, 263–290, 1972.
5. **Razin, S.,** *Ann. Rev. Microbiol.,* 23, 317–356, 1969.
6. **Freundt, E. A.,** in *The Mycoplasmatales and the L-Phase of Bacteria,* Hayflick, L., Ed., Appleton-Century-Crofts, New York, 1969, pp. 281–315.
7. **Bredt, W., Heunert, H. H., Hofling, K. H., and Milthaler, B.,** *J. Bacteriol.,* 113, 1223–1227, 1973.

8. **Lemcke, R. M.,** *Nature,* 229, 492–493, 1971.
9. **Maniloff, J.,** *J. Bacteriol.,* 100, 1402–1408, 1969.
10. **Biberfeld, G. and Biberfeld, P.,** *J. Bacteriol.,* 102, 855–861, 1970.
11. **Richter, C. B.,** in *Morphology of Experimental Respiratory Carcinogenesis,* Vol. 21, Nettesheim, P., Hanna, M. G., Jr., and Deatherage, J. W., Jr., Eds., USAEC Symposium Series, Oak Ridge, Tenn., 1970, pp. 365–380.
12. **Collier, A. M. and Clyde, W. A., Jr.,** *Am. Rev. Respir. Dis.,* 110, 765–773, 1974.
13. **Bredt, W.,** in *Les Mycoplasmes,* Vol. 33, Bove, J. M. and Duplan, J. F., Eds., Colloques INSERM, Paris, 1974, pp. 47–52.
14. **Rodwell, A. W., Peterson, J. E., and Rodwell, E. S.,** *Ann. N.Y. Acad. Sci.,* 225, 190–200, 1973.
15. **Peterson, J. E., Rodwell, A. W., and Rodwell, E. S.,** *J. Bacteriol.,* 115, 411–425, 1973.
16. **Rodwell, A. W., Peterson, J. E., and Rodwell, E. S.,** *J. Bacteriol.,* 122, 1216–1229, 1975.
17. **Rodwell, A. W., Peterson, J. E., and Rodwell, E. S.,** in *Les Mycoplasmes,* Vol. 33, Bove, J. M. and Duplan, J. F., Eds., Colloques INSERM, Paris, 1974, pp. 43–46.
18. **Razin, S.,** *Ann. N.Y. Acad. Sci.,* 143, 115–129, 1967.
19. **Razin, S., Rottem, S., Hasin, M., and Gershfeld, N. L.,** *Ann. N.Y. Acad. Sci.,* 225, 28–37, 1973.
20. **Razin, S.,** in *Advances in Microbial Physiology,* Vol. 10, Rose, A. H. and Tempest, D. W., Eds., Academic Press, New York, 1973, pp. 1–80.
21. **Razin, S.,** in *Progress in Surface and Membrane Science,* Vol. 9, Cadenhead, D. A., Danielli, J. F., and Rosenberg, M. D., Eds., Academic Press, New York, 1975, pp. 257–312.
22. **Razin, S. and Rottem, S.,** in *Biochemical Methods in Membrane Studies,* Maddy, A. H., Ed., Chapman Hall, London, 1976, pp. 3–26.
23. **Razin, S.,** *J. Supramol. Struct.,* 2, 670–681, 1974.
24. **Razin, S., Kahane, I., and Kovartovsky, J.,** in *Pathogenic Mycoplasmas* (Ciba Symposium), Elliott, K. and Birch, J., Eds., Elsevier, Amsterdam, 1973, pp. 93–122.
25. **Gourlay, R. N. and Thrower, K. J.,** *J. Gen. Microbiol.,* 54, 155–159, 1968.
26. **Gilliam, J. M. and Morowitz, H. J.,** *Biochim. Biophys. Acta,* 274, 353–363, 1972.
27. **Green, F. and Hanson, R. P.,** *J. Bacteriol.,* 116, 1011–1018, 1973.
28. **Howard, C. J. and Gourlay, R. N.,** *J. Gen. Microbiol.,* 83, 393–398, 1974.
29. **Morowitz, H. J. and Wallace, D. C.,** *Ann. N.Y. Acad. Sci.,* 225, 62–73, 1973.
30. **Anderson, D. R. and Barile, M. F.,** *J. Bacteriol.,* 90, 180–192, 1965.
31. **Whitescarver, J. and Furness, G.,** *J. Med. Microbiol.,* 8, 349–355, 1975.
32. **Edward, D. G. ff. and Freundt, E. A.,** *J. Gen. Microbiol.,* 14, 197–207, 1956.
33. **Freundt, E. A.,** *Ann. N.Y. Acad. Sci.,* 225, 7–13, 1973.
34. **Edward, D. F. ff.,** in *Les Mycoplasmes,* Vol. 33, Bove, J. M. and Duplan, J. F., Eds., Colloques INSERM, Paris, 1974, pp. 13–18.
35. ICSB Subcommittee on the Taxonomy of Mycoplasmata, *Int. J. Syst. Bacteriol.,* 17, 105–109, 1967.
36. **Edward, D. G. ff. and Freundt, E. A.,** *Int. J. Syst. Bacteriol.,* 17, 267–268, 1967.
37. **Gilpin, R. W., Young, F. E., and Chatterjee, A. N.,** *J. Bacteriol.,* 113, 486–499, 1973.
38. **Saglio, P., L'Hospital, M., Lefleche, D., Dupont, G., Bove, J. M., Tully, J. G., and Freundt, E. A.,** *Int. J. Syst. Bacteriol.,* 23, 191–204, 1973.
39. **Neimark, H. C. and Pene, J. J.,** *Proc. Soc. Exp. Biol. Med.,* 118, 517–519, 1965.
40. **McGee, Z. A., Rogul, M., and Wittler, R. G.,** *Ann. N.Y. Acad. Sci.,* 143, 21–30, 1967.
41. **Neimark, H. C.,** *Ann. N.Y. Acad. Sci.,* 143, 31–37, 1967.
42. **Bak, A. L. and Black, F. T.,** *Nature,* 219, 1044–1045, 1968.
43. **Kelton, W. H. and Mandel, M.,** *J. Gen. Microbiol.,* 56, 131–135, 1969.
44. **Williams, C. O., Wittler, R. G., and Burris, C.,** *J. Bacteriol.,* 99, 341–343, 1969.
45. **Neimark, H. C.,** *J. Gen. Microbiol.,* 63, 249–263, 1970.
46. **Allen, T. C.,** *J. Gen. Microbiol.,* 69, 285–286, 1971.
47. **Black, F. T., Christiansen, C., and Askaa, G.,** *Int. J. Syst. Bacteriol.,* 22, 241–242, 1972.
48. **Askaa, G., Christiansen, C., and Ernø, H.,** *J. Gen. Microbiol.,* 75, 283–286, 1973.
49. **Howard, C. J., Gourlay, R. N., Garwes, D. J., Pocock, D. H., and Collins, J.,** *Int. J. Syst. Bacteriol.,* 24, 373–374, 1974.
50. **Christiansen, C., Freundt, E. A., and Black, F. T.,** *Int. J. Syst. Bacteriol.,* 25, 99–101, 1975.
51. **Searcy, D. G. and Doyle, E. K.,** *Int. J. Syst. Bacteriol.,* 25, 286–289, 1975.
52. **Morowitz, H. J., Bode, H. R., and Kirk, R. G.,** *Ann. N.Y. Acad. Sci.,* 143, 110–113, 1967.
53. **Bak, A. L., Black, F. T., Christiansen, C., and Freundt, E. A.,** *Nature,* 224, 1209–1210, 1969.
54. **Edward, D. G. ff. and Freundt, E. A.,** *J. Gen. Microbiol.,* 57, 391–395, 1969.
55. **Edward, D. G. ff. and Freundt, E. A.,** *J. Gen. Microbiol.,* 62, 1–2, 1970.

56. **Shepard, M. C., Lunceford, C. D., Ford, D. K., Purcell, R. H., Taylor-Robinson, D., Razin, S., and Black, F. T.,** *Int. J. Syst. Bacteriol.*, 24, 160–171, 1974.
57. **Skripal, I. G.,** *Mikrobiol. Zh. Akad. Nauk Ukr. SSR*, 36, 462–466, 1974.
58. **Darland, G., Brock, T. D., Samsonoff, W., and Conti, S. F.,** *Science*, 170, 1416–1418, 1970.
59. **Robinson, I. M. and Allison, M. J.,** *Int. J. Syst. Bacteriol.*, 25, 182–186, 1975.
60. ICSB Subcommittee on the Taxonomy of Mycoplasmatales, *Int. J. Syst. Bacteriol.*, 22, 184–188, 1972.
61. **Fallon, R. J. and Whittlestone, P.,** in *Methods in Microbiology*, Vol. 3B, Norris, J. R. and Ribbons, D. W., Eds., Academic Press, New York, 1970, pp. 211–267.
62. **Barile, M. F. and Del Giudice, R. A.,** in *Pathogenic Mycoplasmas* (Ciba Foundation Symposium), Elliott, K. and Birch, J., Eds., Elsevier, Amsterdam, 1973, pp. 165–181.
63. **Barile, M. F.,** in *Les Mycoplasmes*, Vol. 33, Bove, J. M. and Duplan, J. F., Eds., Colloques INSERM, Paris, 1974, pp. 135–142.
64. **Mardh, P.-A. and Taylor-Robinson, D.,** *Med. Microbiol. Immunol.*, 158, 259–266, 1972.
65. **Barile, M. F., Hopps, H. E., Grabowski, M. W., Riggs, D. B., and Del Giudice, R. A.,** *Ann. N.Y. Acad. Sci.*, 225, 251–264, 1973.
66. **Hopps, H. E., Meyer, B. C., Barile, M. F., and Del Giudice, R. A.,** *Ann. N.Y. Acad. Sci.*, 225, 265–276, 1973.
67. **Barile, M. F.,** in *Contamination in Tissue Culture*, Fogh, J., Ed., Academic Press, New York, 1973, pp. 131–172.
68. **Kenny, G. E.,** in *Contamination in Tissue Culture*, Fogh, J., Ed., Academic Press, New York, 1973, pp. 107–129.
69. **Frey, M. L., Thomas, G. B., and Hale, P. A.,** *Ann. N.Y. Acad. Sci.*, 225, 334–346, 1973.
70. **Whittlestone, P.,** in *Les Mycoplasmes*, Vol. 33, Bove, J. M. and Duplan, J. F., Eds., Colloques INSERM, Paris, 1974, pp. 143–152.
71. **Razin, S. and Tully, J. G.,** *J. Bacteriol.*, 102, 306–310, 1970.
72. **Edward, D. G. ff,** *J. Gen. Microbiol.*, 69, 205–210, 1971.
73. **Edward, D. G. ff, and Moore, W. B.,** *In Vitro* (CSSR), 2, 148–155, 1973.
74. **Cirillo, V. P. and Razin, S.,** *J. Bacteriol.*, 113, 212–217, 1973.
75. **Smith, P. F.,** *The Biology of Mycoplasmas*, Academic Press, New York, 1971, pp. 174–176.
76. **Schimke, R. T., Berlin, C. M., Sweeney, E. W., and Carroll, W. R.,** *J. Biol. Chem.*, 241, 2228–2236, 1966.
77. **Hahn, R. G. and Kenny, G. E.,** *J. Bacteriol.*, 117, 611–618, 1974.
78. **Aluotto, B. B., Wittler, R. G., Williams, C. O., and Faber, J. E.,** *Int. J. Syst. Bacteriol.*, 20, 35–58, 1970.
79. **Black, F. T.,** *Ann. N.Y. Acad. Sci.*, 225, 131–143, 1973.
80. **Black, F. T.,** *Int. J. Syst. Bacteriol.*, 23, 65–66, 1973.
81. **Makki, M. A.,** *J. Hyg. Epidemiol. Microbiol. Immunol.*, 15, 417–423, 1971.
82. **Fabricant, J. and Freundt, E. A.,** *Ann. N.Y. Acad. Sci.*, 143, 50–58, 1967.
83. **Watanabe, T., Mishima, K., and Horikawa, T.,** *Jpn. J. Microbiol.*, 17, 151–153, 1973.
84. **Cole, B. C., Ward, J. R., and Martin, C. H.,** *J. Bacteriol.*, 95, 2022–2030, 1968.
85. **Cohen, G. and Somerson, N. L.,** *J. Bacteriol.*, 98, 543–551, 1969.
86. **Johnson, D. W. and Muscoplat, C. C.,** *Am. J. Vet. Res.*, 33, 2593–2595, 1972.
87. **Sobeslavsky, O., Prescott, B., and Chanock, R. M.,** *J. Bacteriol.*, 96, 695–705, 1968.
88. **Manchee, R. J. and Taylor-Robinson, D.,** *J. Gen. Microbiol.*, 50, 465–478, 1968.
89. **Manchee, R. J. and Taylor-Robinson, D.,** *Br. J. Exp. Pathol.*, 50, 66–75, 1969.
90. **Bak, A. L.,** *Curr. Top. Microbiol. Immunol.*, 61, 89–149, 1973.
91. **Reich, P. R., Somerson, N. L., Rose, J. A., and Weissman, S. M.,** *J. Bacteriol.*, 91, 153–160, 1966.
92. **Reich, P. R., Somerson, N. L., Hybner, C. J., Chanock, R. M., and Weissman, S. M.,** *J. Bacteriol.*, 92, 302–310, 1966.
93. **Somerson, N. L., Reich, P. R., Walls, B. E., Chanock, R. M., and Weissman, S. M.,** *J. Bacteriol.*, 92, 311–317, 1966.
94. **Somerson, N. L. and Weissman, S. M.,** in *The Mycoplasmatales and the L-Phase of Bacteria*, Hayflick, L., Ed., Appleton-Century-Crofts, New York, 1969, pp. 201–218.
95. **Razin, S. and Rottem, S.,** *J. Bacteriol.*, 94, 1807–1810, 1967.
96. **Razin, S.,** *J. Bacteriol.*, 96, 687–694, 1968.
97. **Razin, S., Valdesuso, J., Purcell, R. H., and Chanock, R. M.,** *J. Bacteriol.*, 103, 702–706, 1970.
98. **Zola, H., Baxendale, W., and Sayer, L. J.,** *Res. Vet. Sci.*, 2, 397–399, 1970.
99. **Theodore, T. S., Tully, J. G., and Cole, R. M.,** *Appl. Microbiol.*, 21, 272–277, 1971.
100. **Daniels, M. J. and Meddins, B. M.,** *J. Gen. Microbiol.*, 76, 239–242, 1973.
101. **Wreghitt, T. G., Windsor, G. D., and Butler, M.,** *Appl. Microbiol.*, 28, 530–533, 1974.

102. **Clyde, W. A., Jr.,** *J. Immunol.*, 92, 958–965, 1964.
103. **Stanbridge, E. and Hayflick, L.,** *J. Bacteriol.*, 93, 1392–1396, 1967.
104. **Dighero, M. W., Bradstreet, C. M. P., and Andrews, B. E.,** *J. Appl. Microbiol.*, 33, 750–757, 1970.
105. **Black, F. T.,** *Appl. Microbiol.*, 25, 528–533, 1973.
106. **Ernø, H. and Jurmanova, K.,** *Acta Vet. Scand.*, 14, 524–537, 1973.
107. **Jordan, F. T. W.,** *Vet. Sci.*, 14, 387–389, 1973.
108. **Freundt, E. A.,** in *Les Mycoplasmes,* Vol. 33, Bove, J. M. and Duplan, J. F., Eds., Colloques INSERM, Paris, 1974, pp. 161–168.
109. **Taylor-Robinson, D., Purcell, R. H., Wong, D. C., and Chanock, R. M.,** *J. Hyg.*, 64, 91–104, 1966.
110. **Purcell, R. H., Taylor-Robinson, D., Wong, D. C., and Chanock, R. M.,** *Am. J. Epidemiol.*, 84, 51–66, 1966.
111. **Purcell, R. H., Taylor-Robinson, D., Wong, D. C., and Chanock, R. M.,** *J. Bacteriol.*, 92, 6–12, 1966.
112. **Woode, G. N. and McMartin, D. A.,** *J. Gen. Microbiol.*, 75, 43–50, 1973.
113. **Taylor-Robinson, D., Somerson, N. L., Turner, H. C., and Chanock, R. M.,** *J. Bacteriol.*, 85, 1261–1273, 1963.
114. **Lemcke, R. M.,** *J. Hyg.*, 62, 199–219, 1964.
115. **Kenny, G. E.,** *Ann. N.Y. Acad. Sci.*, 143, 676–681, 1967.
116. **Somerson, N. L., James, W. D., Walls, B. E., and Chanock, R. M.,** *Ann. N.Y. Acad. Sci.*, 143, 384–389, 1967.
117. **Frey, M. L. and Hanson, R. P.,** *Avian Dis.*, 13, 185–197, 1969.
118. **Kenny, G. E.,** *Infect. Immun.*, 3, 510–515, 1971.
119. **Forshaw, K. A. and Fallon, R. J.,** *J. Gen. Microbiol.*, 72, 501–510, 1972.
120. **Lemcke, R. M.,** *J. Gen. Microbiol.*, 38, 91–100, 1965.
121. **Kenny, G. E.,** *J. Bacteriol.*, 98, 1044–1055, 1969.
122. **Hollingdale, M. R. and Lemcke, R. M.,** *J. Hyg.*, 68, 469–477, 1970.
123. **Edward, D. G. ff and Kanarek, A. D.,** *Ann. N.Y. Acad. Sci.*, 79, 696–702, 1960.
124. **Adler, H. E. and Damassa, A. J.,** *Proc. Soc. Exp. Biol. Med.*, 116, 608–610, 1964.
125. **Morton, H. E.,** *J. Bacteriol.*, 92, 1196–1205, 1966.
126. **Krogsgaard-Jensen, A.,** *Appl. Microbiol.*, 22, 756–759, 1971.
127. **Lind, K.,** *Acta Pathol. Microbiol. Scand.*, 73, 459–472, 1968.
128. **Lam, G. T. and Morton, H. E.,** *Appl. Microbiol.*, 27, 356–359, 1974.
129. **Clark, H. W., Bailey, J. S., Fowler, R. C., and Brown, T. M.,** *J. Bacteriol.*, 85, 111–118, 1963.
130. **Tully, J. G.,** *J. Infect. Dis.*, 115, 171–185, 1965.
131. **Del Giudice, R. A., Robillard, N. F., and Carski, T. R.,** *J. Bacteriol.*, 93, 1205–1209, 1967.
132. **Rosendal, S. and Black, F. T.,** *Acta Pathol. Microbiol. Scand. Sect. B*, 80, 615–622, 1972.
133. **Ertel, P. Y., Ertel, I. J., Somerson, N. L., and Pollack, J. D.,** *Proc. Soc. Exp. Biol. Med.*, 134, 441–446, 1970.
134. **Al-Aubaidi, J. M. and Fabricant, J.,** *Cornell Vet.*, 61, 519–542, 1971.
135. **Baas, E. J. and Jasper, D. E.,** *Appl. Microbiol.*, 23, 1097–1100, 1972.
136. **Lehmkuhl, H. D. and Frey, M. L.,** *Appl. Microbiol.*, 27, 1170–1171, 1974.
137. **Taylor-Robinson, D., Ludwig, W. M., Purcell, R. H., Mufson, M. A., and Chanock, R. M.,** *Proc. Soc. Exp. Biol. Med.*, 118, 1073–1083, 1965.
138. **Lin, J.-S. and Kass, E. H.,** *Infect. Immun.*, 10, 535–540, 1974.
139. **Thirkill, C. E. and Kenny, G. E.,** *J. Immunol.*, 114, 1107–1111, 1975.
140. **Lemcke, R. M.,** *Ann. N.Y. Acad. Sci.*, 225, 46–53, 1973.
141. **Freundt, E. A., Ernø, H., Black, F. T., Krogsgaard-Jensen, A., and Rosendal, S.,** *Ann. N.Y. Acad. Sci.*, 225, 161–171, 1973.
142. FAO/WHO Program on Comparative Mycoplasmology, *Vet. Rec.*, 95, 457–461, 1974.
143. **Hayflick, L. (Ed.),** *The Mycoplasmatales and the L-Phase of Bacteria,* Appleton-Century-Crofts, New York, 1969.
144. **Sharp, J. T. (Ed.),** *The Role of Mycoplasmatales and L-Forms of Bacteria in Disease,* Charles C Thomas, Springfield, Ill., 1970.
145. **Smith, P. F.,** *The Biology of Mycoplasmas,* Academic Press, New York, 1971.
146. **Kenny, G. E., Lemcke, R. M., and Clyde, W. A., Jr. (Eds.),** Workshop on Mycoplasmatales as agents of disease, *J. Infect. Dis.*, 127, (Suppl.), 1971.
147. **Elliott, K. and Birch, J. (Eds.),** *Pathogenic Mycoplasmas* (Ciba Foundation Symposium), Elsevier, Amsterdam, 1973.
148. **Maramorosch, K. (Ed.),** Mycoplasma and Mycoplasma-like agents of human, animal, and plant diseases, *Ann. N.Y. Acad. Sci.*, 225, 1–532, 1973.

149. **Bove, J. M. and Duplan, J. F. (Eds.),** *Les Mycoplasmes,* Vol. 33, Colloques INSERM, Paris, 1974.
150. **Kunze, M.,** *Zentralbl. Bakteriol. Parasitenkd. Infektionskr. Hyg. Abt. 1 Orig.,* 216, 501–505, 1971.
151. **Kunze, M. and Flamm, H.,** *Zentralbl. Bakteriol. Parasitenkd. Infektionskr. Hyg. Abt. 1 Orig.,* 220, 203–206, 1972.
152. **Andrews, B. E. and Kunze, M.,** *Med. Microbiol. Immunol.,* 157, 175, 1972.
153. **Freundt, E. A., Andrews, B. E., Ernø, H., Kunze, M., and Black, F. T.,** *Zentralbl. Bakteriol. Parasitendk. Infektionskr. Hyg. Abt. 1 Orig.,* 225, 104–112, 1973.
154. **Mardh, P.-A. and Taylor-Robinson, D.,** *Med. Microbiol. Immunol.,* 158, 219–226, 1973.
155. **Smith, P. F. and Rothblat, G. H.,** *J. Bacteriol.,* 80, 842–850, 1960.
156. **Razin, S. and Shafer, Z.,** *J. Gen. Microbiol.,* 58, 327–339, 1969.
157. **Rottem, S.,** *Appl. Microbiol.,* 23, 659–660, 1972.
158. **Grabowski, M. W., Rottem, S., and Barile, M. F.,** *J. Clin. Microbiol.,* 3, 110–112, 1976.
159. **Kenny, G. E.,** in *The Antigens,* Vol. 3, Sela, M., Ed., Academic Press, New York, 1975, pp. 449–478.
160. **Robertson, J., Gomersall, M., and Gill, P.,** *J. Bacteriol.,* 124, 1019–1022, 1975.
161. **Robertson, J., Gomersall, M., and Gill, P.,** *J. Bacteriol.,* 124, 1017–1018, 1975.
162. **Freundt, E. A., Ernø, H., and Lemcke, R. M.,** in *Methods in Microbiology,* Vol. 10, Norris, J. R. and Ribbons, D. W., Eds., Academic Press, New York, in press.
163. **Bradbury, J. M., Oriel, C. A., and Jordan, F. T. W.,** *J. Clin. Microbiol.,* 3, 449–452, 1976.

THE MOLLICUTES*

MYCOPLASMAS AND UREAPLASMAS

J. G. Tully and S. Razin

INTRODUCTION

Mycoplasma and *Ureaplasma* species within the family Mycoplasmataceae possess the morphology and general ultrastructural features described for organisms in the class Mollicutes. Nutritional requirements are usually met with a complex medium containing either whole animal serum (equine, bovine, porcine, or avian) or specialized bovine serum fractions.[1-3] All species within each genus require cholesterol or related sterols for growth. The genome size of species examined to date is about 5×10^8 daltons (see preceding chapter).

GENUS *MYCOPLASMA*

The biochemical and physiological characteristics of species in the genus *Mycoplasma* are summarized in Table 1. Most species utilize glucose or arginine as major energy sources, with a few species and strains within some species capable of utilizing both substrates.[4] In a few instances (*M. agalactiae, M. bovis, M. bovigenitalium,* and *M. verecundum*),[5,12,79] neither arginine nor glucose appear to be utilized. The major separation of *Mycoplasma* and *Ureaplasma* species is based primarily upon the inability of *Mycoplasma* species to hydrolyze urea.

Habitat and pathogenicity of *Mycoplasma* species are reviewed in Table 2. Mycoplasmal infections have been most frequently associated with diseases of the respiratory and urogenital tracts, especially tissue sites containing mucous membranes, and with joints. Little definitive information is available on the mechanisms of mycoplasmal pathogenicity. More detailed discussion and extensive bibliographies on the pathogenicity of mycoplasmas and possible mechanisms involved can be found in several recent reviews[6-11] or in texts and monographs referred to in the previous chapter.

GENUS *UREAPLASMA* (T-MYCOPLASMAS)

Morphology and Classification of Ureaplasmas

As noted above, the morphology of ureaplasmas is basically similar to that of *Mycoplasma* species, although long filamentous forms have not been observed in a number of ureaplasmas.[270-274] Electron and phase contrast microscopy of organisms grown in liquid medium show coccoid cells (0.25 to 1.0 μm in diameter), mostly single or in pairs, with occasional short filaments up to 2 μm in length.[272-275]

The terms "T-strains" and "T-mycoplasmas" (T denoting tiny colony size) were originally intended as temporary designations for urea-hydrolyzing mycoplasmas initially isolated from man.[276] With increasing information on their biological and serological properties, and on the basis of their wide distribution in animal hosts (see Table 4), it became apparent that a more formal taxonomic structure was needed for this group. Although the demonstration of urea hydrolysis[277,278] and of cholesterol requirement[275] for T-mycoplasmas of human origin was considered sufficient to justify assigning these organisms to a new genus *(Ureaplasma)* within the family Mycoplasmataceae, the ICSB Subcommittee on the Taxonomy of Mycoplasmatales

* Part 19 in *Bergey's Manual.*

recommended[279] that serologically distinct subdivisions of the type species be given "serotype" designations rather than individual species epithets. This recommendation was offered primarily because serologically distinct members within the human ureaplasmas did not otherwise exhibit distinguishing characteristics, and there would be definite disadvantages to establishing a classification system based solely upon serological properties. The proposed taxonomic framework for ureaplasmas of human origin[280] provides for a single named species *(Ureaplasma urealyticum)* and a numbered serotype system, currently containing eight serotypes (see Table 4). Proposals for extension of this system to provide for classification of additional ureaplasmas of human origin as well as of those that might be characterized from isolations made from other animal hosts have been given.[280]

Biochemistry and Physiology of Ureaplasmas

The basic biological properties of the ureaplasmas are given in Table 3. The most extensive comparative biochemical studies have been performed on the eight serotypes of *U. urealyticum,*[281] although information on the biochemical properties of a number of strains from bovine,[282] canine, and primate hosts[283] is also available. Current information indicates that the basic biochemical and physiological characteristics of various animal ureaplasmas are similar to those ureaplasmas isolated from man. The G + C content of bovine ureaplasma DNA, however, is somewhat higher (29 to 29.8 moles %) than that of the human ureaplasmas (27 to 28 moles %), suggesting that the two groups of organisms might represent different species.[291] Electrophoretic analysis of cell proteins of a number of human ureaplasmas showed marked similarity,[292,293] and though this type of study has not been extended to a large number of different human and animal ureaplasmas, limited examination of a number of isolates from man, monkey, and canines[283] confirmed the similarity of the protein patterns.

The property used to distinguish ureaplasmas from other mycoplasmas is their ability to hydrolyze urea. The physiological function of this activity is uncertain for ureaplasmas, as it is for many other urease-positive plants and microorganisms.[287] Ureaplasmas, like other urease-positive organisms, appear to be able to grow and reproduce in the absence of urea.[288] Yet, the addition of urea enables growth of ureaplasmas in a dialyzed-serum medium.[278,284] This effect of urea is apparently not specific, since putrescine and some other diamines could replace the requirement for urea in the dialyzed medium.[288] Urease activity has been localized in the cytoplasm of the organism rather than in membrane preparations.[285,302]

Cultivation of Ureaplasmas

The most critical factor in the cultivation of ureaplasmas on artificial media is their extremely low cell yield, as compared to other mycoplasmas. Although the first part of the growth curve of ureaplasmas resembles that of other mycoplasmas (i.e., a logarithmic phase with a generation time of about 3 hr), this phase comes to an end once the titer reaches 10^6 to 10^7 colony-forming units (CFU) per ml.[280] When urea is present and the medium is poorly buffered, there is a very rapid loss of viability at the point where liberated ammonia raises the pH of the medium to about 8.0. However, when the pH of the medium is kept in the acid range, either by the removal of ammonia[305,386] or by efficient buffering, such as by HEPES,[297] L-histidine,[306] or by incubation under 100% CO_2,[272,304] a rapid decline in viability can be prevented. Yet, the peak titer under these conditions rarely exceeds 10^7 CFU/ml. The extremely low cell yields in liquid media are also reflected in the small colony size on solid media. Ureaplasma colonies, when originally observed on an agar medium,[294] were found to be extremely small (10 ± 5 μm in diameter), and they did not show the usual "fried egg" colony shape characteristic of mycoplasmas. However, adjustment of the reaction of the medium to pH 6.0 ± 0.5,[280] and inclusion of HEPES,[297] phosphate,[300,301] or 100% CO_2 atmosphere[272] to maintain the pH in the acid range, results not only in an increase in colony size (up to 200 or 300 μm in diameter), but also in the formation of the characteristic "fried egg" colony. Thus, *Ureaplasma* colony shape and size under these conditions is not different from that of other *Mycoplasma* species.

The reason for the abrupt cessation in the reproduction of the organisms once the culture reaches a titer of about 10^7 CFU/ml is not clear. A plausible hypothesis may be based on the accumulation of a toxic product[286] or on the exhaustion of an essential nutrient. Dialysis cultures of ureaplasmas[289] improved growth somewhat, enabling a peak titer of about 10^8 CFU/ml. The low cell yield of ureaplasmas also impedes enzymological studies, which explains the scarcity of data on their metabolism and antigenic structure.

Differential media for the separation of the ureaplasmas from all other mycoplasmas in clinical specimens utilize the detection of urease activity by ureaplasmas, either by means of an indicator color change or by the precipitation of MnO_2 on the colonies, due to the ammonia liberated by urea hydrolysis.[298,299,380] Since ureaplasmas are more sensitive to thallium acetate than other mycoplasmas, this selective antibacterial agent should be omitted from their growth medium.[282,295,296]

Serological Characteristics of Ureaplasmas

The ureaplasmas of human origin, forming the species *Ureaplasma urealyticum,* fall into eight serotypes on the basis of metabolism inhibition,[277,280,311] growth inhibition, and indirect fluorescent-antibody tests[293,311,312] as well as by a complement-dependent mycoplasmacidal test.[313-315] Limited comparison of the serologic properties of various ureaplasmas of animal origin have been carried out. Bovine ureaplasmas were first thought to comprise a rather large heterogenous group.[316-318] More recent findings, utilizing the indirect immunofluorescence test[319] or the immune-inactivation procedure,[320] indicate that bovine ureaplasmas might be separated into at least eight serotypes.

Few studies have been performed to determine the chemical nature of the ureaplasma antigens. The presence of glycolipids in a ureaplasma has been noted,[321] and this component apparently participates in the complement fixation test.[322] However, the extremely small cell yields and the inevitable heavy contamination of the washed-cell or membrane pellets with precipitated medium components, mostly serum lipoproteins, impedes immunological studies on ureaplasmas.[304,323,324]

Habitat and Pathogenicity of Ureaplasmas

The known distribution of ureaplasmas in man and other animal hosts, and their possible role in disease, are summarized in Table 4. The sensitivity of ureaplasmas to antibiotics, compared to that of other genital mycoplasmas, is presented in Table 5.

Table 1
BIOCHEMICAL AND PHYSIOLOGICAL CHARACTERISTICS OF MYCOPLASMAS

(Ae = aerobic; An = anaerobic; NT = not tested or not known)

Mycoplasma species	% G + C of DNA	Glucose catabolism	Mannose catabolism	Arginine hydrolysis	Tetrazolium reduction (Ae/An)	Phosphatase	Film and spot reaction	Gelatin hydrolysis	Coagulated-serum digestion	Hemadsorption	Preferred atmosphere	Rate of growth	Colony morphology	Special growth factors	References
M. agalactiae	33.5–34.2	–	–	–	+/+	+	+ or –	–	–	+	Ae	Moderate	Regular[a]	None	12–14
M. alkalescens	25.9	–	NT	+	–/–	+	–	NT	–	–	Ae	Moderate	Regular	None	12, 17–18, 360
M. alvi	26.4	+	NT	+	–/+	NT	–	NT	NT	NT	An	Moderate	Regular	None	363
M. anatis	NT	+	+ or –	–	–/+	+	+	NT	NT	–	Ae	Slow	Regular	None	19–22
M. arginini	27.6–28.6	–	–	+	–/+	–	–	NT	–	–	Ae	Moderate	Regular	None	12, 23, 360
M. arthritidis	30.0–33.7	–	–	+	–/–	+	–	+	–	–	Ae	Moderate	Regular	None	24–27
M. bovigenitalium	28.0–32.0	–	–	–	–/+	+	+	–	–	+	Ae	Moderate	Regular	DNA on primary isolation	12, 27
M. bovirhinis	24.5–25.7	+	–	–	+/+	+ or –	–	–	+ or –	+ or –	Ae	Moderate	Regular	None	12, 28, 29, 360
M. bovis (formerly *M. agalactiae* subsp. *bovis* or *M. bovimastitidis*)	32.7–32.9	–	–	–	+/+	+	+ or –	NT	–	–	Ae	Moderate	Regular	None	12, 15, 16, 360, 366
M. bovoculi	29.0	+	NT	+ or –	+/+	+ or –	+	–	–	+	Ae	Moderate	Regular	None	12, 17, 30
M. buccale (formerly *M. orale* 2)	24–28	–	–	+	–/+	+	–	–	–	–	An	Slow to moderate	Small; no central nipple	None	31, 32
M. canis	28.5–29.1	+	–	–	–/+	–	–	+ or –	–	+	Ae	Moderate	Regular	None	27, 33, 34
M. canadense	29	–	–	+	–/+	NT	–	NT	NT	–	Ae	Slow	Regular	None	35, 36
M. capricolum	25.5	+	+	+ or –	+/+	+	–	NT	+	–	Ae	Rapid	Regular	None	37
M. caviae	NT	+	NT	–	NT	NT	NT	NT	NT	NT	Ae	Moderate	Regular	None	38
M. conjunctivae	NT	+	+	–	+/+	–	–	NT	–	–	Ae	Moderate	Regular	None	39
M. cynos	NT	+	+	–	±/+	+	+	NT	NT	+	Ae	Moderate	Regular	None	34, 40
M. dispar	28.5–29.3	+	NT	–	NT	NT	NT	NT	NT	NT	An	Slow	Lacy; no central nipple	Exacting; reported inhibition by penicillin G, but not by ampicillin	41, 42
M. edwardii	29.2	+	–	–	–/+	–	+ (slow)	–	–	+ or –	Ae	Moderate	Regular	None	33, 34, 43

[a] Regular colony morphology means the typical "fried egg" colony.

Table 1 (continued)
BIOCHEMICAL AND PHYSIOLOGICAL CHARACTERISTICS OF MYCOPLASMAS

(Ae = aerobic; An = anaerobic; NT = not tested or not known)

Mycoplasma species	% G + C of DNA	Glucose catabolism	Mannose catabolism	Arginine hydrolysis	Tetrazolium reduction (Ae/An)	Phosphatase	Film and spot reaction	Gelatin hydrolysis	Coagulated-serum digestion	Hemadsorption	Preferred atmosphere	Rate of growth	Colony morphology	Special growth factors	References
M. equirhinis	NT	–	NT	+	–/–	NT	+	NT	NT	NT	Ae	Moderate	Medium to large-granular	None	44
M. faucium (formerly *M. orale* 3)	NT	–	–	+	–/–	–	NT	NT	NT	+ (chick RBC only)	An	Moderate	Small	Growth stimulation by L cysteine	32, 45
M. feliminutum	NT	+	+	–	–/+	–	+ or –	NT	NT	–	An	Slow	Small	None	34, 46 (Tully, unpublished)
M. felis	25.0–25.4	+	–	–	–/+	+	+	–	–	–	Ae	Moderate	Regular	None	44, 46–48
M. fermentans	27.5–29.1	+	–	+	–/+	– or +	+	–	–	–	An	Moderate	Regular	None	27
M. flocculare	NT	+	NT	–	–/?	NT	–	NT	NT	NT	An	Very slow	Minute; no central nipple	Reported sensitivity to penicillin G	49, 50
M. gallinarum	26.3–28.0	–	–	+	+/+	–	+	–	–	–	Ae	Moderate	Regular	None	20, 21, 27, 51
M. gallisepticum	31.6–35.7	+	+	–	+/+	–	–	–	–	+	Ae	Moderate to rapid	Small; central nipple	None	21, 51
M. gateae	28.4–28.6	–	–	+	–/±	–	–	–	–	NT	Ae	Moderate	Regular	None	46, 47
M. hominis	27.3–29.3	–	–	+	–/–	–	–	–	–	– (+ for some human RBC's)	Ae	Moderate	Regular	None	27, 52
M. hyopneumoniae/*M. suipneumoniae*	NT	+	–	–	NT	NT	NT	NT	NT	NT	An	Very slow	Minute; no central nipple	Exacting; reported inhibition by penicillin G	53–56
M. hyorhinis	27–28	+	–	–	+/±	+	–	–	–	–	Ae	Moderate	Regular	None	57
M. hyosynoviae	NT	–	–	+	–/–	–	+	NT	–	–	Ae	Slow to rapid	Regular	Growth enhanced by mucin	58, 188
M. iners	28.9–29.6	–	–	+	–/–	–	+	–	–	–	Ae	Moderate	Regular	None	51
M. lipophilum	NT	–	–	+	NT	NT	+	NT	NT	–	An	Moderate	Regular	None	59
M. maculosum	26.5–29.6	–	–	+	–/+	+	+	–	–	–	Ae	Moderate	Regular	None	33, 34, 60

Table 1 (continued)
BIOCHEMICAL AND PHYSIOLOGICAL CHARACTERISTICS OF MYCOPLASMAS

(Ae = aerobic; An = anaerobic; NT = not tested or not known)

Mycoplasma species	% G + C of DNA	Glucose catabolism	Mannose catabolism	Arginine hydrolysis	Tetrazolium reduction (Ae/An)	Phosphatase	Film and spot reaction	Gelatin hydrolysis	Coagulated-serum digestion	Hemadsorption	Preferred atmosphere	Rate of growth	Colony morphology	Special growth factors	References
M. meleagridis	28.0–28.5	–	–	+	–/+	+	–	NT	NT	– or +	Ae	Moderate	Regular	None	61
M. moatsii	NT	+	NT	+	–/–	NT	NT	NT	NT	NT	Ae	Moderate	Regular	None	62
M. molare	NT	+	+	–	+/+	–	+	NT	NT	NT	Ae	Moderate	Regular	None	63
M. mycoides subsp. *capri*	23.6–25.8	+	+	–	+/+	–	–	+	+	–	Ae	Rapid	Large	None	60, 64, 65
M. mycoides subsp. *mycoides*	26.1–26.8	+	+	–	+/+	–	–	+	+ or –	–	Ae	Moderate to rapid	Regular	None	12, 65, 66
M. neurolyticum	22.8–26.5	+	+	–	–/+	–	–	–	–	–	Ae	Moderate to rapid	Regular	None	24, 67
M. opalescens	NT	–	–	+	–/–	+	+	NT	NT	NT	Ae	Moderate	Regular	None	68, 124, 125
M. orale (formerly *M. orale* 1)	24.0–28.2	–	–	+	–/–	–	–	–	–	+ (chick RBC's only)	An	Moderate	Regular	None	69, 70
M. ovipneumoniae	27–29	+	NT	–	+/+	NT	–	NT	NT	NT	An	Moderate to rapid	No central nipple; lacy	None	269, 368
M. pneumoniae	38.6–40.8	+	+	–	+/+	–	–	–	–	+	Ae	Slow to moderate	No central nipple on primary isolation	None	71–73
M. primatum	28.6	–	–	+	–/–	+	–	–	–	–	An	Moderate	Regular	None	74
M. pulmonis	27.5–28.3	+	+	–	–/+	–	+	–	–	+	Ae	Moderate	Lacy; central nipple less defined	None	24, 27, 75
M. putrefaciens	28.9	+	+	–	+/+	+	–	NT	–	+	Ae	Rapid	Regular	None	37
M. salivarium	27.0–31.5	–	–	+	–/±	–	+	–	–	–	An	Moderate	Regular	None	27, 60
M. spumans	28.4–29.1	–	–	+	–/–	+	–	–	–	+	Ae	Moderate	Coarse on primary isolation	None	33, 34
M. synoviae	34.2	+	NT	NT	NT	NT	+	NT	NT	+ or –	Ae	Slow	Regular	NADH required for growth, and some stimulation by cysteine	21, 76–78
M. verecundum	27.0–29.2	–	–	–	–/?	NT	+ (slow)	NT	NT	NT	Ae	Moderate	Regular	None	79

Table 2
HABITAT AND PATHOGENICITY OF MYCOPLASMAS

Type strain (Reference collection numbers)[a]	Natural host	Site of recovery (Material for isolation)	Disease manifestations	Experimental pathology	References
M. agalactiae PG2 (NCTC 10123)	Goat, sheep	Mammary glands, lymph nodes, joints (Milk, synovial fluid)	Contagious agalactia, characterized by arthritis, mastitis, keratitis, and vulvovaginitis	Goats and sheep susceptible to experimental infections by subcutaneous inoculation. Inflammatory lesions localized in udders of females and in joints (10–20% of cases). Vulvovaginitis also experimentally produced.	64, 80–83
M. alkalescens D12 (NCTC 10135)	Cattle	Nasal cavity and serum	Not known	Not studied.	12, 28
M. alvi Ilsley (NCTC 10157)	Cattle	Intestinal and urogenital tracts (Feces)	Not known	Not studied.	363
M. anatis 1340 (ATCC 25524)	Duck	Sinuses and air sac (the latter with single isolation	Not known	Not studied.	19, 21
M. arginini (NCTC 10129)	Cattle, sheep, goats, chamois, wild felines	Brain, joints, eye, lung, and kidney of a variety of animals; also contaminated cell cultures	Not known	Not pathogenic to mice.	23, 48, 93–98, 266

[a] ATCC = American Type Culture Collection; NCTC = National Collection of Type Cultures.

Table 2 (continued)
HABITAT AND PATHOGENICITY OF MYCOPLASMAS

Type strain (Reference collection numbers)[a]	Natural host	Site of recovery (Material for isolation)	Disease manifestations	Experimental pathology	References
M. arthritidis PG6 (ATCC 19611)	Rat, non-human primates	Joint fluid, infected tissues	Polyarthritis of rats, submandibular abscesses, ocular lesions, middle-ear infections, purulent rhinitis, and lung lesions	Generalized infections characterized by suppurative polyarthritis, conjunctivitis, and urethritis can be produced by intravenous inoculation of virulent strains. Subcutaneous inoculation with agar produces localized abscesses and septicemia in rats. Localized arthritis is produced by inoculation into the footpads. Organisms are nonpathogenic to monkeys, rabbits, and guinea pigs; they may or may not be pathogenic to mice.	99–109, 372
M. bovigenitalium PG11 (B2) (ATCC 19852) (NCTC 10122)	Cattle Dog? Horse? Swine?	Common inhabitant of the lower genital tract in both male and female cattle	Bovine mastitis; possible cause of vaginal disorders and seminal vesiculitis	Mastitis and seminal vesiculitis characterized by eosinophilic cell response can be experimentally produced. Intravenous inoculation into calves produced general infection with low fever, mild diarrhea, and arthritic lesions. Marked cytopathic effects in a variety of cell cultures.	34, 66, 110–115
M. bovirhinis PG43 (5M331)(ATCC 19884) (NCTC 10118)	Cattle	Common inhabitant of the upper respiratory tract or joints (Lung material, milk, and synovial fluid)	Possible etiologic agent in respiratory disease and mastitis in cattle	Mastitis experimentally produced by some strains Cytopathic for bovine and canine embryo kidney cell cultures.	64, 66, 98, 116

Table 2 (continued)
HABITAT AND PATHOGENICITY OF MYCOPLASMAS

Type strain (Reference collection numbers)[a]	Natural host	Site of recovery (Material for isolation)	Disease manifestations	Experimental pathology	References
M. bovis (formerly *M. agalactiae* subsp. *bovis* or *M. bovimastitidis*) Donetta (ATCC 25523) (NCTC 10131)	Cattle	Udder, joints, and respiratory tract (Milk or exudate from udder, semen, and synovial fluid)	Bovine mastitis and arthritis; primary role in calf pneumonia suspected, but not confirmed	Mastitis experimentally produced by inoculation of organisms into udder. Endometritis and salpingo-oophoritis, sometimes associated with impaired fertility, have been produced by inoculation of cultures into the heifer uterus. Intravenous inoculation into cows or calves produced persistent joint infections. Histopathologic examination shows characteristic eosinophilic cell response. Some strains cytopathic for bovine embryo tissue culture. Endobronchial inoculation of infected tissue suspensions produced pneumonic lesions and arthritis in calves.	66, 84–92, 264, 268, 374
M. bovoculi M165/69 (NCTC 10141)	Cattle	Eye and associated fluids	Possible etiologic agent in bovine infectious keratoconjunctivitis	Not studied.	30, 117, 118, 365
M. buccale (formerly *M. orale* 3) CH20247 (ATCC 23636)	Monkey, man	Oropharynx (common in monkeys, rare in man) and vagina (monkeys)	Apparently not pathogenic	Not studied.	31, 119–122
M. canis PG14 (C55) (ATCC 19525)	Dog Non-human primates?	Common inhabitant of the upper respiratory tract and genital tract of dogs	Apparently not pathogenic	Not studied.	33, 122–128

Table 2 (continued)
HABITAT AND PATHOGENICITY OF MYCOPLASMAS

Type strain (Reference collection numbers)[a]	Natural host	Site of recovery (Material for isolation)	Disease manifestations	Experimental pathology	References
M. canadense 275C (NCTC 10152)	Cattle	Joint, urogenital tract (Milk (mastistis), semen, and vaginal mucus)	Possible etiologic agent in bovine mastitis	Clinical mastitis produced by intramammary inoculation in cows. Symptoms of swollen quarter in bovine udder and leukocytosis noted.	35, 36
M. capricolum California kid (ATCC 27343) (NCTC 10154)	Goat	Arthritic joint and eye (Synovial fluid, blood and various exudates)	Associated with septicemia, acute polyarthritis, and some conjunctivitis (older goats).	Experimental infections reproduced in goats and sheep.	37, 129, 130
M. caviae G122 (ATCC 27108) (NCTC 10126)	Guinea pig	Nasopharynx and genital tract	Pathogenicity unknown	Not studied.	38, 131
M. conjunctivae HRC581 (ATCC 25834)	Goat, sheep, chamois, cattle	Conjunctival tissue	Associated with "pink eye" in goats, sheep, and chamois	Apparently not capable of producing experimental keratoconjunctivitis in calves in the absence of other microbial flora.	39, 266, 364
M. cynos H831 (ATCC 27544) (NCTC 10142)	Dog	Conjunctivae, upper respiratory tract, lung, and genital mucosa	Pathogenicity unknown	Not studied.	40, 126
M. dispar 462/2 (ATCC 27140) (NCTC 10125)	Cattle	Normal habitat not fully defined; isolated from pneumonic lungs of calves	Associated with pneumonia in calves, but pathogenicity not fully defined	Experimental endobronchial inoculations into conventionally reared calves produced pneumonia. Also pathogenic for bovine mammary gland under experimental conditions.	41, 132–134, 377–379
M. edwardii PG24 (C21) (ATCC 23462) (NCTC 10132)	Dog	Throat, upper respiratory tract, and genital tract (Pneumonic lung tissue)	Significance in respiratory or genital disease not clear	Not studied.	33, 123–127

Table 2 (continued)
HABITAT AND PATHOGENICITY OF MYCOPLASMAS

Type strain (Reference collection numbers)[a]	Natural host	Site of recovery (Material for isolation)	Disease manifestations	Experimental pathology	References
M. equirhinis M432/72 (NCTC 10148)	Horse	Nasopharynx and trachea	Not studied.	Not studied.	44, 367
M. faucium formerly *M. orale* 3) DC333 (ATCC 25293)	Man, non-human primates	Oropharynx (rare)	Apparently not pathogenic	Not studied.	45, 119
M. feliminutum Ben (ATCC 25749)	Dog, cat	Pneumonic lung of dogs and oral cavity of cats	Associated with pneumonia in dogs, but pathogenicity not defined	Not studied.	34, 46, 126
M. felis CO (ATCC 23391)	Cat, wild felines, horse	Very common inhabitant of oral and nasal cavity, conjunctivae, and lower genital tract of felines; also found in nasopharynx and trachea of horses	Possibly some association with feline conjunctivitis	Young kittens are susceptible to experimental inoculation, with development of conjunctivitis and serological response. Some histologic evidence of interstitial pneumonia, with isolations of agent from trachea and lung. Synergism with other microbial agents as yet not excluded.	46, 48, 97, 135, 136
M. fermentans PG18 (G) (ATCC 19989) (NCTC 10117)	Man, non-human primates	Genitourinary tract, oropharynx (relatively rare) (Urethral or cervical scrapings)	Not known	Certain strains can produce chromosomal aberrations in cell cultures and a leukemoid disease in mice.	119, 137–143
M. flocculare Ms42 (ATCC 27399) (NCTC 10143)	Swine	Lungs (pneumonic)	Associated with enzootic pneumonia in pigs (relationship to *M. hyopneumoniae*/*M. suipneumoniae* uncertain at present	Experimental aerosol infections produced mild histologic changes in the lungs of pigs. Symptoms are consistent with enzootic pneumonia, but no overt disease is evident	49
M. gallinarum PG16 (Fowl) (ATCC 19708) (NCTC 10120)	Chicken	Upper respiratory tract	Not pathogenic	Not studied.	21

Table 2 (continued)
HABITAT AND PATHOGENICITY OF MYCOPLASMAS

Type strain (Reference collection numbers)[a]	Natural host	Site of recovery (Material for isolation)	Disease manifestations	Experimental pathology	References
M. gallisepticum PG31 (X95) (ATCC 19610) (NCTC 10115)	Chicken, turkey, peafowls	Respiratory system, air sac, ovaries, and eggs	Chronic respiratory disease in chickens, infectious sinusitis in turkeys, encephalitis associated with polyarthritis of cerebral arteries in turkeys	Chronic respiratory disease in chickens experimentally reproduced. Neurological disease and characteristic pathology produced in turkeys by intravenous inoculation of high doses of washed cells, but not of cell-free filtrates. Presence of cell-bound toxin is suspected. Experimental infections in mice and monkeys produced lesions in heart, lung, and liver.	21, 144–150
M. gateae CS (ATCC 23392)	Cat, dog	Common inhabitant of the upper respiratory tract, conjunctivae, and genital mucosa of cats; less frequent in the throat of dogs	Not known	Not studied.	34, 46, 47, 97
M. hominis PG21 (H50) (ATCC 23114) (NCTC 10111)	Man, non-human primates	Human genitourinary tract (common) and oropharynx (less common), oropharynx of monkeys, oropharynx, vagina, and rectum of chimpanzees (Urethral or cervical exudates, prostatic secretions, rectum, urinary sediments, products of abortion, pleural fluid, abscesses, throat swabs, blood, and sputum)	Usually considered commensal, but may be potential pathogen under some circumstances; may act in concert with other parts of microbial flora	Produces chromosomal aberrations in cell cultures.	7, 9, 120–122, 137, 143, 151–158, 205, 267

Table 2 (continued)
HABITAT AND PATHOGENICITY OF MYCOPLASMAS

Type strain (Reference collection numbers)[a]	Natural host	Site of recovery (Material for isolation)	Disease manifestations	Experimental pathology	References
M. hyopneumoniae/M. suipneumoniae[b] 11 (ATCC 25617) (NCTC 10127) J (ATCC 25934) (NCTC 10110)	Swine	Respiratory tract (Pneumonic lung and semen)	Etiological agent of at least one form of enzootic pneumonia of pigs	Pneumonia experimentally produced by intranasal inoculation of organisms into both conventional and gnotobiotic pigs.	10, 159–169
M. hyorhinis BTS-7 (ATCC 17981) (NCTC 10130)	Swine	Common inhabitant of nasal cavity (Nasal secretions and semen; contaminated cell cultures)	May play primary role in some forms of swine pneumonia; septicemia associated with arthritis-polyserositis reported in young pigs	Young pigs susceptible to experimental infection, which is characterized by fever, arthritis, and polyserositis. Challenged chick embryos show irregular mortality rates, with some pericarditis and peritoneal lesions. Most strains are cytopathic to a variety of primary animal cell cultures.	169–181, 263, 375
M. hyosynoviae S16 (ATCC 25591)	Swine	Respiratory tract and joints (Synovial fluid, nasal secretions, tonsillar material, lungs of pigs with catarrhal pneumonia)	Produces acute synovitis and arthritis in swine	Experimental production of acute arthritis produced by intravenous challenge. Gross lesions consist of serofibrinous synovial fluid, hyperemia of synovium. Large numbers of organisms are required to produce cytopathic effect in swine synovial cell cultures.	58, 182–190
M. iners PG30 (M) (ATCC 19705)	Chicken, turkey	Respiratory tract	Not known	Induces joint lesions when inoculated into chick embryos	21
M. lipophilum MaBy (ATCC 27104)	Man	Oral cavity	Not known	Not studied.	59

[b] Type strains of *M. hyopneumoniae* and *M. suipneumoniae* are serologically identical. Proper taxonomic designation and type strain have not been established as yet.

Table 2 (continued)
HABITAT AND PATHOGENICITY OF MYCOPLASMAS

Type strain (Reference collection numbers)[a]	Natural host	Site of recovery (Material for isolation)	Disease manifestations	Experimental pathology	References
M. maculosum PG15 (C27) (ATCC 19327)	Dog	Genitourinary tract and upper respiratory tract	Not known	Not studied.	33, 123, 124, 126, 128, 191
M. meleagridis 17529 (N) (ATCC 25294)	Turkey	Respiratory and urogenital tract; also present in semen, ovaries, and eggs	Air sacculitis in turkeys	Air sacculitis experimentally produced in turkeys. Mixed infections with bacteria, may produce purulent pneumonia, pericarditis, or peritonitis. Turkey embryos surviving experimental inoculation may later show ascites and skeletal deformities.	192–201
M. moatsii MK405 (ATCC 27625)	Monkey	Oropharynx and genitourinary tract (vagina and prepuce)	Not known	Not studied.	62
M. molare H542 (ATCC 27746) (NCTC 10144)	Dog	Pharynx	Not known	Not studied.	63
M. mycoides subsp. *capri* PG3 (NCTC 10137)	Goat, sheep	Lung, spleen, and liver (Exudate, synovial fluid, and blood)	Caprine pleuropneumonia and/or polyarthritis, cellulitis, and septicemia	Pneumonia produced experimentally in goats by intratracheal inoculation of aerosols. Organisms inhibit ciliary activity of tracheal organ cultures. Mice are susceptible when the organisms are suspended in mucin.	64, 202–206

Table 2 (continued)
HABITAT AND PATHOGENICITY OF MYCOPLASMAS

Type strain (Reference collection numbers)[a]	Natural host	Site of recovery (Material for isolation)	Disease manifestations	Experimental pathology	References
M. mycoides subsp. *mycoides* PG1 (NCTC 10114)	Cattle, sheep, goats	Respiratory tract (Pleural exudates and infected lung material)	Bovine pleuropneumonia; etiological role in caprine and ovine diseases not established, but may play some part in pleuropneumonia	Cattle experimentally infected by nasal instillation of lung material. Goats and sheep are susceptible to subcutaneous inoculation. Subcutaneous lesions are induced in rabbits by strains both virulent and avirulent for cattle. Mice are susceptible when cultures are mixed with mucin. Toxin is suspected to play some role in disease.	64, 82, 207–214
M. neurolyticum Type A (ATCC 19988)	Mouse	Brain, eye, nasopharynx, middle ear, and lung (Exudates or tissue washings)	"Rolling" disease and epidemic conjunctivitis in mice	Organism produces extracellular toxin with neurological properties. Intravenous challenge of mice or rats with neurotoxin produces characteristic "rolling" disease syndrome, neuropathological changes, and pulmonary hemorrhage. Toxin is produced in some tissue culture fluids after a variable number of passages. Toxin appears to be a large-molecular-weight protein (>200,000).	99, 215
M. opalescens MH5408 (ATCC 27921) (NCTC 10149)	Dog	Throat and genitourinary tract	Not known	Not studied.	68, 124, 125
M. orale (formerly *M. orale* 1) CH19299 (ATCC 23714) (NCTC 10112)	Man, non-human primates	Common inhabitant of oropharynx of man; also present in oropharynx of some subhuman primates; contaminated cell cultures	Apparently not pathogenic	Not studied.	70, 121

Table 2 (continued)
HABITAT AND PATHOGENICITY OF MYCOPLASMAS

Type strain (Reference collection numbers)[a]	Natural host	Site of recovery (Material for isolation)	Disease manifestations	Experimental pathology	References
M. ovipneumoniae Y98 (NCTC 10151)	Sheep	Nose, trachea, bronchi, lung, and eye (Exudates or washings)	Etiological agent in proliferative interstitial pneumonia in sheep	Experimental infections induced in gnotobiotic or normal lambs by intratracheal inoculation, aerosols, or contact.	216, 217, 269, 368, 369, 376
M. pneumoniae FH (ATCC 15531) (NCTC 10119)	Man	Respiratory tract, oral cavity; rare isolations from middle ear and spinal fluid (Sputum and infected lung tissue)	Atypical pneumonia, febrile upper respiratory tract infections, bullous myringitis, myocarditis, various cutaneous and neurological conditions	Pneumonitis experimentally produced in man, hamsters, guinea pigs, and cotton rats. Organism produces distinct cytopathology, including loss of ciliary activity, in hamster and chicken tracheal organ cultures.	218–234, 265
M. primatum HRC292 (ATCC 25948)	Non-human primates, man	Oral and urogenital tract of monkeys; urogenital tract of man (rare)	Not known	Not studied.	74, 119, 120, 235
M. pulmonis PG34 (Ash) (ATCC 19612)	Mouse, rat Rabbit?	Common inhabitant of the respiratory tract and joints (arthritic) of mice and rats	Infectious catarrh, sometimes complicated by otitis media and bronchopneumonia; Also associated with arthritis in mice and rats, and with fetal wastage in mice	Rhinitis, otitis media, and characteristic pneumonic lesions produced in gnotobiotic mice. Intraperitoneal inoculation into female mice shows predilection for ovaries and oviducts. Intravenous inoculation into some mouse strains produces arthritis. Rats infected by intranasal route show altered pulmonary function.	99, 236–253, 370, 371
M. putrefaciens KS-1 (ATCC 15718) (NCTC 10155)	Goat?	Single isolation, supposedly from caprine joint	Apparently not pathogenic	Single isolate not pathogenic for goats, sheep, calves, or pigs.	37, 254

Table 2 (continued)
HABITAT AND PATHOGENICITY OF MYCOPLASMAS

Type strain (Reference collection numbers)[a]	Natural host	Site of recovery (Material for isolation)	Disease manifestations	Experimental pathology	References
M. salivarium PG20 (H110) (ATCC 23064) (NCTC 10113)	Man, non-human primates	Very common inhabitant of human oral cavity, particularly in the gingival sulci; common in the oral cavity of a variety of subhuman primates, as well as in the nasal cavity and urogenital tract under less frequent circumstances	Apparently not pathogenic	Not studied.	101, 119–121
M. spumans PG13 (C48) (ATCC 19526)	Dog	Genitourinary and respiratory tract (Lung)	Not known	Not studied.	33, 123, 124, 126, 128
M. synoviae WVU 1853 (ATCC 25204) (NCTC 10124)	Chicken, turkey	Joints and respiratory tract	Infectious synovitis and air sacculitis in chickens and turkeys; pericarditis and myocarditis appears in chronic phase of disease	Natural disease manifestations can be experimentally produced by inoculation of organisms through the footpad of chickens, or through aerosols, intravenous, or air sac inoculation of turkey poults. Cytopathic effects are noted in chick fibroblast cell cultures.	21, 146, 255–262, 373
M. verecundum 107 (NCTC 10145) (ATCC 27862)	Cattle	Eye	Recovered from spontaneous outbreak of conjunctivitis in calves	Not studied.	79

Table 3
BIOLOGICAL PROPERTIES OF UREAPLASMAS

Property	Characteristic	References
Genome size, daltons	$4.1-4.8 \times 10^8$	290
Guanine + cytosine of DNA (buoyant density), moles %	26.9–29.8	290, 291
Colony morphology	Minute colonies (15–30 μm in diameter) with no peripheral growth in unbuffered media. "Fried egg" colonies (up to 300 μm in diameter) in buffered acid media.	280
Optimal pH for growth	6.0 ± 0.5	280
Temperature range for growth	22–40°C (some serotypes to 20°C)	281
Preferred atmosphere	5–15% CO_2 in air or N_2 [b]	280
Cholesterol requirement	+[a]	275, 281
Sensitivity to digitonin	+[a]	281
Urea hydrolysis	+[a]	277, 278, 281
Carbohydrate fermentation	–[a]	281
Arginine hydrolysis	–[a]	281
Tetrazolium reduction (aerobic/anaerobic)	–/–[a]	281
Phosphatase	+[a]	281
Proteolytic activity	+	307, 308
Hemolysis (guinea pig RBC's)	+ (beta)[a]	281
Hemadsorption (guinea pig RBC's)	+ (human serotype III) – (all other human serotypes)	281
Sensitivity to erythromycin	+ (0.8 to 3.0 μg/ml)	281
Sensitivity to thallium acetate	+ (1:500 to 1:2000 dilution)	281

[a] Reactions observed with eight serotypes of *Ureaplasma urealyticum*.
[b] For some strains, 100% CO_2 is superior.[272,304]

Table 4
HABITAT AND PATHOGENICITY OF UREAPLASMAS

Host	Proposed taxonomic designation	Proposed serotypes and representative strain numbers	Site of recovery	Pathogenicity	Experimental infections	Additional references
Human	*Ureaplasma urealyticum*	I 7 (ATCC 27813) II 23 (ATCC 27814) III 27 (ATCC 27815) IV 58 (ATCC 27816) V 354 (ATCC 27817) VI Pi (ATCC 27818) VII Co (ATCC 27819) VIII 960(Cx8) (ATCC 27618)	Urine, vagina, cervix, mouth, pharynx, blood, semen, urethra, fallopian tube, rectum	Possible role in non-gonoccocal urethritis in men and vaginitis in women not fully established. Ureaplasmas recovered from cases of pelvic inflammatory disease, septic abortion, and puerperal fever. Organisms also associated with infertility, abortion, and low birth weight, but direct role uncertain at present. (See References 7, 9, 280, 334, 335, 337, 339, 343–347, 381, and 382.)	Formation of bladder stones in experimental infection of rats. Murine and caprine mammary gland susceptible to challenge. Persistent infection, with no obvious damage to human fallopian-tube organ cultures. Some cytopathic effect on normal human fibroblasts in culture, and increased incidence of chromosomal abnormality in human lymphocyte cultures. Usually minimal cytopathic changes in cell cultures infected with ureaplasmas. (See References 328–330, 332, 333, 342, and 361.)	267, 271 276, 281 295, 309 310, 312, 348, 349
Bovine	None	None	Conjunctiva, lung, urethra, vagina, bladder, prepuce, semen	May be associated with pneumonia in calves and with infectious keratoconjunctivitis in cattle. (See References 330, 356, and 385.)	Mastitis produced in cattle, goats and mice. Immunity to reinfection to the homologous serotype only. Persistent infection, but no obvious tissue damage noted in bovine tracheal organ cultures or in human fallopian-tube organ cultures. More severe cytopathic effects noted in bovine oviduct organ cultures. (See References 325–328, 330, 331, 361, and 384.)	282, 320 336, 350–355
Canine	None	None	Throat, prepuce, semen	Not studied.	Not studied.	283
Feline	None	None	Throat	Not studied.	Not studied.	357, 358
Non-human primates (monkey and chimpanzee)	None	None	Throat, urogenital tract	Not studied.	Not studied.	283, 359, 362
Caprine	None	None	Genital tract	Not studied.	Not studied.	329

Table 5
SELECTED ANTIBIOTIC SENSITIVITIES OF HUMAN GENITAL MYCOPLASMAS[a]

Antibiotic	*M. hominis*	*M. fermentans*	Ureaplasmas
Ampicillin	>1000 (>1000)[b]	1000	>1000 (>1000)
Tetracycline	0.2 (0.1–0.4)	12.5	0.8 (0.4–1.6)
Erythromycin	1000 (500–1000)	125	12.5 (6.2–25)
Lincomycin	3.1 (1.6–6.2)	0.16	500 (100–>1,000)
Spectinomycin	4 (2–8)	Not tested	4 (2–16)

[a] Data principally from References 338 and 341. For additional information on antibiotic sensitivity of these organisms, see the bibliographies of the above references or References 339, 340, and 383.

[b] Median minimal inhibitory concentrations (μg/ml), followed by range values, shown in parentheses.

REFERENCES

1. **Smith, P. F. and Morton, H. E.,** *J. Bacteriol.*, 61, 395–405, 1951.
2. **Smith, P. F., Leece, J. G., and Lynn, R. J.,** *J. Bacteriol.*, 68, 627–633, 1954.
3. **Hughes, J. H., Thomas, D. C., Hamparian, V. V., and Somerson, N. L.,** *J. Med. Microbiol.*, 7, 35–40, 1974.
4. **Freundt. E. A.,** in *Bergey's Manual of Determinative Bacteriology*, 8th ed., Buchanan, R. E. and Gibbons, N. E., Eds., Williams & Wilkins, Baltimore, 1974, pp. 929–955.
5. **Edward, D. G. ff. and Moore, W. B.,** *J. Med. Microbiol.*, 8, 451–454, 1975.
6. **Whittlestone, P.,** in *Microbial Pathogenicity in Man and Animals* (Society for General Microbiology, 22nd Symposium Proceedings), Smith, H. and Pearce, J. H., Eds., Cambridge University, Press, 1972, pp. 217–250.
7. **McCormick, W. M., Braun, P., Lee, Y.-H., Klein, J. O., and Kass, E. H.,** *N. Engl. J. Med.*, 288, 78–89, 1973.
8. **Freundt, E. A.,** *Pathol. Microbiol.*, 40, 155–187, 1974.
9. **Lee, Y.-H., McCormick, W. M., Marcy, S. M., and Klein, J. O.,** *Pediatr. Clin. North Am.*, 21, 457–466, 1974.
10. **Whittlestone, P.,** in *Advances in Veterinary Science and Comparative Medicine*, Vol. 17, Brandly, C. A. and Cornelius, C. E., Eds., Academic Press, New York, 1973, pp. 1–55.
11. **Stanbridge, E.,** *Annu. Rev. Microbiol.*, 30, 169–187, 1976.
12. **Leach, R. H.,** *J. Gen. Microbiol.*, 75, 135–153, 1973.
13. **Edward, D. G. ff. and Freundt, E. A.,** *Int. J. Syst. Bacteriol.*, 23, 55–61, 1973.
14. **Al-Aubaidi, J. M. and Dardiri, A. H.,** *Int. J. Syst. Bacteriol.*, 24, 136–138, 1974.
15. **Hale, H. H., Helmboldt, C. F., Plastridge, W. N., and Stula, E. F.,** *Cornell Vet.*, 52, 582–591, 1962.
16. **Jain, N. C., Jasper, D. E., and Dellinger, J. D.,** *J. Gen. Microbiol.*, 49, 401–410, 1967.
17. **Ernø, H. and Stipkovits, L.,** *Acta Microbiol. Acad. Sci. Hung.*, 20, 305–315, 1973.
18. **Askaa, G., Christiansen, C., and Ernø, H.,** *J. Gen. Microbiol.*, 75, 283–286, 1973.
19. **Roberts, D. H.,** *Vet. Rec.*, 75, 665–667, 1963.
20. **Roberts, D. H.,** *J. Comp. Pathol. Ther.*, 74, 447–456, 1964.
21. **Fabricant, J.,** in *The Mycoplasmatales and the L-Phase of Bacteria*, Hayflick, L., Ed., Appleton-Century-Crofts, New York, 1969, pp. 621–641.
22. **Barber, T. L. and Fabricant, J.,** *Avian Dis.*, 15, 125–138, 1971.
23. **Barile, M. F., Del Giudice, R. A., Carski, T. R., Gibbs, C. J., and Morris, J. A.,** *Proc. Soc. Exp. Biol. Med.*, 129, 489–494, 1968.
24. **Sabin, A. B.,** *Bacteriol. Rev.*, 5, 1–67, 1941.
25. **Lynn, R. J. and Haller, G. J.,** *Antonie van Leeuwenhoek J. Microbiol. Serol.*, 34, 249–256, 1968.
26. **Williams, C. O., Wittler, R. G., and Burris, C.,** *J. Bacteriol.*, 99, 341–343, 1969.
27. **Aluotto, B. B., Wittler, R. G., Williams, C. O., and Faber, J. E.,** *Int. J. Syst. Bacteriol.*, 20, 35–58, 1970.
28. **Leach, R. H.,** *Ann. N.Y. Acad. Sci.*, 143, 305–316, 1967.
29. **Jurmanova, K. and Mensik, J.,** *Zentralbl. Veterinaermed.*, 18, 457–464, 1971.
30. **Langford, E. V. and Leach, R. H.,** *Can. J. Microbiol.*, 19, 1435–1444, 1973.

31. Taylor-Robinson, D., Fox, H., and Chanock, R. M., *Am. J. Epidemiol.*, 81, 180–191, 1965.
32. Freundt, E. A., Taylor-Robinson, D., Purcell, R. H., Chanock, R. M., and Black, F. T., *Int. J. Syst. Bacteriol.*, 24, 252–255, 1974.
33. Koshimizu, K. and Ogata, M., *Jpn. J. Vet. Sci.*, 36, 391–406, 1974.
34. Rosendal, S., *Acta Pathol. Microbiol. Scand. Sect. B*, 82, 25–32, 1974.
35. Langford, E. V., Ruhnke, H. L., and Onoviran, O., *Int. J. Syst. Bacteriol.*, 26, 212–219, 1976.
36. Ruhnke, H. L. and Onoviran, O., *Vet. Rec.*, 96, 203, 1975.
37. Tully, J. G., Barile, M. F., Edward, D. G. ff., Theodore, T. S., and Ernø, H., *J. Gen. Microbiol.*, 85, 102–120, 1974.
38. Hill, A., *J. Gen. Microbiol.*, 65, 109–113, 1971.
39. Barile, M. F., Del Giudice, R. A., and Tully, J. G., *Infect. Immun.*, 5, 70–76, 1972.
40. Rosendal, S., *Int. J. Syst. Bacteriol.*, 23, 49–54, 1973.
41. Gourlay, R. N. and Leach, R. H., *J. Med. Microbiol.*, 3, 111–123, 1970.
42. Andrews, B. E., Leach, R. H., Gourlay, R. N., and Howard, C. J., *Vet. Rec.*, 93, 603, 1973.
43. Tully, J. G., Barile, M. F., Del Giudice, R. A., Carski, T. R., Armstrong, D., and Razin, S., *J. Bacteriol.*, 101, 346–349, 1970.
44. Allam, N. M. and Lemcke, R. M., *J. Hyg.*, 74, 385–408, 1975.
45. Fox, H., Purcell, R. H., and Chanock, R. M., *J. Bacteriol.*, 98, 36–43, 1969.
46. Heyward, J. T., Sabry, M. Z., and Dowdle, W. R., *Am. J. Vet. Res.*, 30, 615–622, 1969.
47. Cole, B. C., Golightly, L., and Ward, J. R., *J. Bacteriol.*, 94, 1451–1458, 1967.
48. Hill, A., *Res. Vet. Sci.*, 18, 139–143, 1975.
49. Friis, N. F., *Acta Vet. Scand.*, 13, 284–286, 1972.
50. Meyling, A. and Friis, N. F., *Acta Vet. Scand.*, 13, 287–289, 1972.
51. Edward, D. G. ff. and Kanarek, A. D., *Ann. N.Y. Acad. Sci.*, 79, 696–702, 1960.
52. Freundt, E. A., *Acta Pathol. Microbiol.*, 34, 127–144, 1954.
53. Mare, C. J. and Switzer, W. P., *Vet. Med.*, 60, 841–846, 1965.
54. Mare, C. J. and Switzer, W. P., *Am. J. Vet. Res.*, 27, 1687–1693, 1966.
55. Goodwin, R. F., Pomeroy, A. P., and Whittlestone, P., *J. Hyg.*, 65, 85–96, 1967.
56. Friis, N. F., *Acta Vet. Scand.*, 12, 120–121, 1971.
57. Switzer, W. P., *Am. J. Vet. Res.*, 16, 540–544, 1955.
58. Ross, R. F. and Karmon, J. A., *J. Bacteriol.*, 103, 707–713, 1970.
59. Del Giudice, R. A., Purcell, R. H., Carski, T. R., and Chanock, R. M., *Int. J. Syst. Bacteriol.*, 24, 147–153, 1974.
60. Edward, D. G. ff., *Int. Bull. Bacteriol. Nomencl. Taxon.*, 5, 85–93, 1955.
61. Yamamoto, R., Bigland, C. H., and Ortmayer, H. B., *J. Bacteriol.*, 90, 47–49, 1965.
62. Madden, D. L., Moats, K. E., London, W. T., Matthew, E. B., and Sever, J. L., *Int. J. Syst. Bacteriol.*, 24, 459–464, 1974.
63. Rosendal, S., *Int. J. Syst. Bacteriol.*, 24, 125–130, 1974.
64. Cottew, G. S. and Leach, R. H., in *The Mycoplasmatales and the L-Phase of Bacteria*, Hayflick, L., Ed., Appleton-Century-Crofts, New York, 1969, pp. 527–570.
65. Al-Aubaidi, J. M., Dardiri, A. H., and Fabricant, J., *Int. J. Syst. Bacteriol.*, 22, 155–164, 1972.
66. Al-Aubaidi, J. M., and Fabricant, J., *Cornell Vet.*, 61, 490–518, 1971.
67. Tully, J. G., *J. Infect. Dis.*, 115, 171–185, 1965.
68. Rosendal, S., *Acta Pathol. Microbiol. Scand. Sect. B*, 83, 457–470, 1975.
69. Herdershee, D., Ruys, A. C., and van Rhijn, G. R., *Antonie van Leeuwenhoek J. Microbiol. Serol.*, 29, 157–162, 1963.
70. Taylor-Robinson, D., Canchola, J., Fox, H., and Chanock, R. M., *Am. J. Hyg.*, 80, 135–148, 1964.
71. Chanock, R. M., Hayflick, L., and Barile, M. F., *Proc. Natl. Acad. Sci. U.S.A.*, 48, 41–49, 1962.
72. Somerson, N. L., Purcell, R. H., Taylor-Robinson, D., and Chanock, R. M., *J. Bacteriol.*, 89, 813–818, 1965.
73. Eaton, M. D. and Low, I. E., *Ann. N.Y. Acad. Sci.*, 143, 375–383, 1967.
74. Del Giudice, R. A., Carski, T. R., Barile, M. F., Lemcke, R. M., and Tully, J. G., *J. Bacteriol.*, 108, 439–445, 1971.
75. Forshaw, K. A. and Fallon, R. J., *J. Gen. Microbiol.*, 72, 501–510, 1972.
76. Olson, N. O., Kerr, K. M., and Campbell, A., *Avian Dis.*, 8, 209–214, 1964.
77. Olson, N. O. and Meadows, J. K., *Avian Dis.*, 16, 387–396, 1972.
78. Windsor, G. F., Thompson, G. W., and Baker, N. W., *Res. Vet. Sci.*, 18, 59–63, 1975.
79. Gourlay, R. H., Leach, R. N., and Howard, C. J., *J. Gen. Microbiol.*, 81, 475–484, 1974.
80. Edward, D. G. ff., *Vet. Rec.*, 65, 873–875, 1953.
81. Cottew, G. S., Watson, W. A., Arisoy, F., Erdag, O., and Buckley, L. S., *J. Comp. Pathol.*, 78, 275–282, 1968.
82. Cottew, G. S., Watson, W. A., Erdag, O., and Arisoy, F., *J. Comp. Pathol.*, 79, 541–551, 1969.

83. Singh, N., Rajya, B. S., and Mohanty, G. C., *Cornell Vet.*, 64, 435–442, 1974.
84. Hirth, R. S., Plastridge, W. N., Tourtellotte, M. E., and Nielsen, S. W., *J. Am. Vet. Med. Assoc.*, 148, 277–282, 1966.
85. Hirth, R. S., Nielsen, S. W., and Plastridge, W. N., *Pathol. Vet.*, 3, 616–632, 1966.
86. Hirth, R. S., Nielsen, S. W., and Tourtellotte, M. E., *Infect. Immun.*, 2, 101–104, 1970.
87. Hirth, R. S., Tourtellotte, M. E., and Nielsen, S. W., *Infect. Immun.*, 2, 105–111, 1970.
88. Mosher, A. H., Plastridge, W. N., Tourtellotte, M. E., and Helmboldt, C. F., *Am. J. Vet. Res.*, 29, 512–522, 1968.
89. Karst, O. and Onoviran, O., *Br. Vet. J.*, 127, ix–x, 1971.
90. Hjerpe, C. A. and Knight, H. D., *J. Am. Vet. Med. Assoc.*, 160, 1414–1418, 1972.
91. Fabricant, J., *Ann. N.Y. Acad. Sci.*, 225, 369–381, 1973.
92. Jasper, D. E., Al-Aubaidi, J. M., and Fabricant, J., *Cornell Vet.*, 64, 407–415, 1974.
93. Leach, R. H., *Vet. Rec.*, 87, 319–320, 1970.
94. Barile, M. F. and Kern, J., *Proc. Soc. Exp. Biol. Med.*, 138, 432–437, 1971.
95. Jurmanova, K. and Krejci, J., *Vet. Rec.*, 89, 585–586, 1971.
96. Foggie, A. and Angus, K. W., *Vet. Rec.*, 90, 312–313, 1972.
97. Tan, R. J. S. and Miles, J. A. R., *Res. Vet. Sci.*, 16, 27–34, 1974.
98. Stipkovits, L., Bodon, L., Romvary, J., and Varga, L., *Acta Microbiol. Acad. Sci. Hung.*, 22, 45–51, 1975.
99. Tully, J. G., in *The Mycoplasmatales and the L-Phase of Bacteria*, Hayflick, L., Ed., Appleton-Century-Crofts, New York, 1969, pp. 571–605.
100. Cole, B. C., Ward, J. R., Jones, R. S., and Cahill, J. F., *Infect. Immun.*, 4, 344–355, 1971.
101. Cole, B. C., Graham, C. E., Golightly-Rowland, L., and Ward, J. R., *Can. J. Microbiol.*, 18, 1431–1437, 1972.
102. Cahill, J. F., Cole, B. C., Wiley, B. B., and Ward, J. R., *Infect. Immun.*, 3, 24–35, 1971.
103. Hannan, P. C. T. and Hughes, B. O., *Ann. Rheum. Dis.*, 30, 316–321, 1971.
104. Cole, B. C., Ward, J. R., and Golightly-Rowland, L., *Infect. Immun.*, 7, 218–225, 1973.
105. Cole, B. C. and Ward, J. R., *Infect. Immun.*, 7, 691–699, 1973.
106. Rosendal, S. and Thomsen, A. C., *Acta Pathol. Microbiol. Scand. Sect. B*, 82, 895–898, 1974.
107. Hill, A. and Dagnall, G. J. R., *J. Comp. Pathol.*, 85, 45–52, 1975.
108. Laber, G., Walzi, H., and Schutze, E., *Zentralbl. Bakteriol. Parasitenkd. Infektionskr. Hyg. Abt. 1 Orig.*, 230, 385–397, 1975.
109. Walzi, H., Schutze, E., and Laber, G., *Zentralbl. Bakteriol. Parasitenkd. Infektionskr. Hyg. Abt. 1 Orig.*, 231, 229–242, 1975.
110. Afshar, A., Stuart, P., and Huck, R. A., *Vet. Rec.*, 78, 512–519, 1966.
111. Al-Aubaidi, J. M., McEntee, K., Lein, D. H., and Roberts, S. J., *Cornell Vet.*, 62, 581–596, 1972.
112. Gois, M., Kuksa, F., and Franz, J., *Vet. Rec.*, 93, 47–48, 1973.
113. Langford, E. V., *Vet. Rec.*, 94, 528, 1974.
114. Jasper, D. E., Al-Aubaidi, J. M., and Fabricant, J., *Cornell Vet.*, 64, 296–302, 1974.
115. Parsonson, I. M., Al-Aubaidi, J. M., and McEntee, K., *Cornell Vet.*, 64, 240–264, 1974.
116. Shimizu, T., Nosaka, D., and Nakamura, N., *Jpn. J. Vet. Sci.*, 35, 535–537, 1973.
117. Langford, E. V. and Dorward, W. J., *Can. J. Comp. Med.*, 33, 275–279, 1969.
118. Nicolet, J. and Buttiker, W., *Vet. Rec.*, 95, 442–443, 1974.
119. Del Giudice, R. A., Carski, T. R., Barile, M. F., Yamashiroya, M., and Verna, J. E., *Nature*, 222, 1088–1089, 1969.
120. Madden, D. L., Hildebrandt, J., Monif, G. R. G., London, W. T., Sever, J. L., and McCullough, N. B., *Lab. Anim. Care*, 20, 467–473, 1970.
121. Cole, B. C., Ward, J. R., Golightly-Rowland, L., and Graham, C. E., *Can. J. Microbiol.*, 16, 1331–1339, 1970.
122. Martinez-Lahoz, A., Kalter, S. S., Pinkerton, M. E., and Hayflick, L., *Ann. N.Y. Acad. Sci.*, 174, 820–827, 1970.
123. Barile, M. F., Del Giudice, R. A., Carski, T. R., Yamashiroya, H. M., and Verna, J. E., *Proc. Soc. Exp. Biol. Med.*, 134, 146–148, 1970.
124. Armstrong, D., Tully, J. G., Yu, B., Morton, V., Friedman, M. H., and Steger, L., *Infect. Immun.*, 1, 1–7, 1970.
125. Armstrong, D., Morton, V., Yu, B., Friedman, M. H., Steger, L., and Tully, J. G., *Am. J. Vet. Res.*, 33, 1471–1478, 1972.
126. Rosendal, S., *Acta Pathol. Microbiol. Scand. Sect. B*, 81, 441–445, 1973.
127. Rosendal, S. and Laber, G., *Zentralbl. Bakteriol. Parasitenkd. Infektionskr. Hyg. Abt. 1 Orig.*, 225, 346–349, 1973.
128. Kirchhoff, H., Basu, A., and Loh, M., *Zentralbl. Veterinaermed. Reihe B*, 20, 474–480, 1973.
129. Cordy, D. R., Adler, H. E., and Yamamoto, R., *Cornell Vet.*, 45, 50–68, 1955.

130. Hudson, J. R., Cottew, G. S., and Adler, H. E., *Ann. N.Y. Acad. Sci.*, 143, 287–297, 1967.
131. Stalheim, O. H. V. and Matthews, P. J., *Lab. Anim. Care*, 25, 70–73, 1975.
132. St. George, T. D., Horsfall, N., and Sullivan, N. D., *Aust. Vet. J.*, 49, 580–586, 1973.
133. Pirie, H. M. and Allan, E. M., *Vet. Rec.*, 97, 345–349, 1975.
134. Ose, E. E. and Muenster, O. A., *Vet. Rec.*, 97, 97, 1975.
135. Campbell, L. H., Synder, S. B., Reed, C., and Fox, J. G., *J. Am. Vet. Med. Assoc.*, 163, 991–995, 1973.
136. Tan, R. J. S., *Jpn. J. Exp. Med.*, 44, 235–240, 1974.
137. Taylor-Robinson, D., Addey, J. P., Hare, M. J., and Dunlop, E. M. C., *Br. J. Ven. Dis.*, 45, 265–273, 1969.
138. Williams, M. H., Brostoff, J., and Roitt, I. M., *Lancet*, 2, 277–280, 1970.
139. Gabridge, M. G., Abrams, G. D., and Murphy, W. H., *J. Infect. Dis.*, 125, 153–160, 1972.
140. Plata, E. J., Abell, M. R., and Murphy, W. H., *J. Infect. Dis.*, 128, 588–597, 1973.
141. Fogh, J. and Fogh, H., *Ann. N.Y. Acad. Sci.*, 225, 311–329, 1973.
142. Windsor, G. D., Nicholls, A., Maini, R. N., Edward, D. G. ff., Lemcke, R. M., and Dumonde, D. C., *Ann. Rheum. Dis.*, 33, 70–74, 1974.
143. Thomsen, A. C., *Acta Pathol. Microbiol. Scand. Sect. B*, 83, 10–16, 1975.
144. Thomas, L., Davidson, M., and McCluskey, R. T., *J. Exp. Med.*, 123, 897–912, 1966.
145. Sun, S. C., Sohal, R. S., Chu, K. C., Colcolough, H. L., Leiderman, E., and Burch, G. E., *Am. J. Pathol.*, 53, 1073–1096, 1968.
146. Vardaman, T. H. and Yoder, H. W., *Poult. Sci.*, 49, 157–161, 1970.
147. Tripathy, S. B., Acharjyo, L. N., Singh, U., Ray, S. K., and Misra, S. K., *Br. Vet. J.*, 128, 428–431, 1972.
148. Grimes, T. M. and Rosenfeld, L. E., *Aust. Vet. J.*, 48, 113–116, 1972.
149. Kubo, N., Hashimoto, K., Sato, T., and Inaguchi, T., *Jpn. J. Vet. Sci.*, 35, 487–498, 1973.
150. Clyde, W. A., Jr. and Thomas, L., *Proc. Natl. Acad. Sci. U.S.A.*, 70, 1545–1549, 1973.
151. Mufson, M. A., Ludwig, W. M., Purcell, R. H., Cate, T. R., Taylor-Robinson, D., and Chanock, R. M., *J. Am. Med. Assoc.*, 192, 1146–1152, 1965.
152. Allison, A. C. and Paton, G. R., *Lancet*, 2, 1229–1230, 1966.
153. Tully, J. G., Brown, M. S., Sheagren, J. N., Young, V. M., and Wolf, S. M., *N. Engl. J. Med.*, 273, 648–650, 1965.
154. Tully, J. G. and Smith, L. G., *J. Am. Med. Assoc.*, 204, 827–828, 1968.
155. Harwick, H. J., Purcell, R. H., Iuppa, J. B., and Fekety, F. R., *J. Infect. Dis.*, 121, 260–268, 1970.
156. Harwick, H. J., Purcell, R. H., Iuppa, J. B., and Fekety, F. R., *Obstet. Gynecol.*, 37, 765–768, 1971.
157. Csonka, G. W., Williams, R. E. O., and Corse, J., *Ann. N.Y. Acad. Sci.*, 143, 794–798, 1967.
158. Gregory, J. E. and Cundy, K. R., *Appl. Microbiol.*, 19, 268–270, 1970.
159. Goodwin, R. F. W., Pomeroy, A. P., and Whittlestone, P., *Vet. Rec.*, 77, 1247–1249, 1965.
160. Goodwin, R. F. W. and Whittlestone, P., *Vet. Rec.*, 81, 643–647, 1967.
161. Goodwin, R. F. W., Pomeroy, A. P., and Whittlestone, P., *J. Hyg.*, 66, 595–603, 1968.
162. Hodges, R. T., Betts, A. O., and Jennings, A. R., *Vet. Rec.*, 84, 268–273, 1969.
163. Goodwin, R. F. W. and Whittlestone, P., *J. Hyg.*, 69, 391–397, 1971.
164. Goodwin, R. F. W., *Res. Vet. Sci.*, 13, 262–267, 1972.
165. Switzer, W. P., *J. Am. Vet. Med. Assoc.*, 160, 651–653, 1972.
166. Livingston, C. W., Jr., Stair, E. L., Underdahl, N. R., and Mebus, C. A., *Am. J. Vet. Res.*, 33, 2249–2258, 1972.
167. Smith, I. M., Hodges, R. T., Betts, A. O., and Hayward, A. H. S., *J. Comp. Pathol.*, 83, 307–321, 1973.
168. Roberts, D. H., *Br. Vet. J.*, 130, 68–74, 1974.
169. Schulman, A. and Estola, T., *Vet. Rec.*, 94, 330, 1974.
170. Roberts, E. D., Switzer, W. P., and Ramsey, F. K., *Am. J. Vet. Res.*, 24, 19–31, 1963.
171. Gois, M., Valicek, L., and Sovadina, M., *Zentralbl. Veterinaermed. Reihe B*, 15, 230–240, 1968.
172. Duncan, J. R. and Ross, R. F., *Am. J. Pathol.*, 57, 171–186, 1969.
173. Schulman, A., Estola, T., and Garry-Andersson, A.-S., *Zentralbl. Veterinaermed. Reihe B*, 17, 549–553, 1970.
174. Poland, J., Edington, N., Gois, M., and Betts, A. O., *J. Hyg.*, 69, 145–154, 1971.
175. Barden, J. A. and Decker, J. L., *Arthritis Rheum.*, 14, 193–201, 1971.
176. Ennis, R. S., Dalgard, D., Willerson, J. T., Barden, J. A., and Decker, J. L., *Arthritis Rheum.*, 14, 202–211, 1971.
177. Gois, M., Pospisil, Z., Cerny, M., and Mrva, V., *J. Comp. Pathol.*, 81, 401–410, 1971.
178. Ross, R. F., Dale, S. E., and Duncan, J. R., *Am. J. Vet. Res.*, 34, 367–372, 1973.

179. Barden, J. A., Decker, J. L., Dalgard, D. W., and Aptekar, R. G., *Infect. Immun.*, 8, 887–890, 1973.
180. Gois, M., Kuksa, F., and Franz, J., *Zentralbl. Veterinaermed. Reihe B*, 21, 176–187, 1974.
181. Gois, M. and Kuksa, F., *Zentralbl. Veterinaermed. Reihe B*, 21, 352–361, 1974.
182. Ross, R. F. and Duncan, J. R., *J. Am. Vet. Med. Assoc.*, 157, 1515–1518, 1970.
183. Friis, N. F., *Acta Vet. Scand.*, 11, 487–490, 1970.
184. Ross, R. F., Switzer, W. P., and Duncan, J. R., *Am. J. Vet. Res.*, 32, 1743–1749, 1971.
185. Potgieter, L. N. D., Frey, M. L., and Ross, R. F., *Can. J. Comp. Med.*, 36, 145–149, 1972.
186. Roberts, D. H., Johnson, C. T., and Tew, N. C., *Vet. Rec.*, 90, 307–309, 1972.
187. Potgieter, L. N. D. and Ross, R. F., *Am. J. Vet. Res.*, 33, 99–105, 1972.
188. Gois, M. and Taylor-Robinson, D., *J. Med. Microbiol.*, 5, 47–54, 1972.
189. Ross, R. F. and Spear, M. L., *Am. J. Vet. Res.*, 34, 373–378, 1973.
190. Furlong, S. L. and Turner, A. J., *Aust. Vet. J.*, 51, 291–293, 1975.
191. Skalka, B. and Krejcin, T., *Acta Univ. Agric. Brno Fac. Vet.*, 37, 57–64, 1968.
192. Yamamoto, R. and Bigland, C. H., *Avian Dis.*, 9, 108–118, 1965.
193. Yamamoto, R. and Bigland, C. H., *Am. J. Vet. Res.*, 27, 326–330, 1966.
194. Mohamed, Y. S. and Bohl, E. H., *Avian Dis.*, 12, 554–566, 1968.
195. Moorhead, P. D. and Saif, Y. M., *Am. J. Vet. Res.*, 31, 1645–1653, 1970.
196. Reis, R. and Yamamoto, R., *Am. J. Vet. Res.*, 32, 63–74, 1971.
197. Rhoades, K. R., *Avian Dis.*, 15, 762–774, 1971.
198. Rhoades, K. R., *Avian Dis.*, 15, 910–922, 1971.
199. Ghazikhanian, G. and Yamamoto, R., *Am. J. Vet. Res.*, 35, 417–424, 1974.
200. Ghazikhanian, G. and Yamamoto, R., *Am. J. Vet. Res.*, 35, 425–430, 1974.
201. Wise, D. R. and Evans, E. T. R., *Res. Vet. Sci.*, 18, 190–192, 1975.
202. Smith, G. R., *J. Comp. Pathol.*, 77, 21–27, 1967.
203. Smith, G. R., *J. Comp. Pathol.*, 79, 261–265, 1969.
204. Cherry, J. D. and Taylor-Robinson, D., *Infect. Immun.*, 2, 431–438, 1970.
205. Cherry, J. D. and Taylor-Robinson, D., *Ann. N.Y. Acad. Sci.*, 225, 290–303, 1973.
206. Perreau, P., Cuong, T., and Vallee, A., *Bull. Acad. Vet. Fr. Microbiol.*, 45, 109–116, 1972.
207. Laws, L., *Aust. Vet. J.*, 32, 326–329, 1956.
208. Lloyd, L. C., *J. Pathol. Bacteriol.*, 92, 225–229, 1966.
209. Karst, O., *Vet. Rec.*, 87, 506–507, 1970.
210. Lloyd, L. C., *J. Comp. Pathol.*, 80, 195–209, 1970.
211. Perreau, P., *Rev. Elev. Med. Vet. Pays Trop.*, 24, 343–348, 1971.
212. Piercy, D. W., *J. Comp. Pathol.*, 80, 549–558, 1970.
213. Ernø, H., Freundt, E. A., Krogsgaard-Jensen, A., and Rosendal, S., *Acta Vet. Scand.*, 13, 263–265, 1972.
214. Ojo, M. O., *Bull. Epizoot. Dis. Afr.*, 21, 319–323, 1973.
215. Thomas, L., *Annu. Rev. Med.*, 21, 179–186, 1970.
216. Carmichael, L. E., St. George, T. D., Sullivan, N. D., and Horsfall, N., *Cornell Vet.*, 62, 654–679, 1972.
217. Sullivan, N. D., St. George, T. D., and Horsfall, N., *Aust. Vet. J.*, 49, 57–62, 1973.
218. Liu, C., *J. Exp. Med.*, 106, 455–466, 1957.
219. Couch, R. B., Cate, T. R., and Chanock, R. M., *J. Am. Med. Assoc.*, 187, 442–447, 1964.
220. Chanock, R. M., *N. Engl. J. Med.*, 273, 1199–1206, 1257–1264, 1965.
221. Purcell, R. H., and Chanock, R. M., *Med. Clin. North Am.*, 51, 791–802, 1967.
222. Smith, C. B., Friedewald, W. T., and Chanock, R. M., *J. Am. Med. Assoc.*, 199, 353–358, 1967.
223. Collier, A. M., Clyde, W. A., Jr., and Denny, F. W., *Proc. Soc. Exp. Biol. Med.*, 132, 1153–1158, 1969.
224. Denny, F. W., Clyde, W. A., Jr., and Glezen, W. P., *J. Infect. Dis.*, 123, 74–92, 1971.
225. Clyde, W. A., Jr., *Infect. Immun.*, 4, 757–763, 1971.
226. Brunner, H., James, W. D., Horswood, R. L., and Chanock, R. M., *J. Infect. Dis.*, 127, 315–318, 1973.
227. Lerer, R. J. and Kalavsky, S. M., *Pediatrics*, 52, 658–668, 1973.
228. Lascari, A. D., Garfunkel, J. M., and Mauro, D. J., *Am. J. Dis. Child.*, 128, 254–255, 1974.
229. Lewes, D., Rainford, D. J., and Lane, W. F., *Br. Heart J.*, 36, 924–932, 1974.
230. Collier, A. M. and Clyde, W. A., Jr., *Am. Rev. Respir. Dis.*, 110, 765–773, 1974.
231. Hara, K., Izumikawa, K., Kinoshita, I., Ota, M., Ikebe, A., Koike, M., and Hamada, M., *Tohoku J. Exp. Med.*, 114, 315–337, 1974.
232. Hu, P. C., Collier, A. M., and Baseman, J. B., *Infect. Immun.*, 11, 704–711, 1975.
233. Fernald, G. W., Collier, A. M., and Clyde, W. A., Jr., *Pediatrics*, 55, 327–335, 1975.
234. Murray, H. W., Masur, H., Senterfit, L. B., and Robert, R. B., *Am. J. Med.*, 58, 229–242, 1975.
235. Thomsen, A. C., *Acta Pathol. Microbiol. Scand. Sect. B*, 82, 653–656, 1974.

236. **Organick, A. B. and Lutsky, I. I.,** *J. Bacteriol.,* 96, 250–258, 1968.
237. **Deeb, B. J. and Kenny, G. E.,** *J. Bacteriol.,* 93, 1416–1424, 1967.
238. **Barden, J. A. and Tully, J. G.,** *J. Bacteriol.,* 100, 5–10, 1969.
239. **Hannan, P. C. T.,** *J. Gen. Microbiol.,* 67, 363–365, 1971.
240. **Lindsey, J. R., Baker, H. J., Overcash, R. G., Cassell, G. H., and Hunt, C. E.,** *Am. J. Pathol.,* 64, 675–708, 1971.
241. **Kohn, D. F.,** *Lab. Anim. Sci.,* 21, 856–861, 1971.
242. **Kohn, D. F. and Kirk, B. E.,** *Lab. Anim. Sci.,* 19, 321–330, 1969.
243. **Okano, H., Homma, J. Y., Chosa, H., and Kusano, N.,** *Jpn. J. Exp. Med.,* 40, 453–465, 1970.
244. **Jones, T. C. and Hirsch, J. G.,** *J. Exp. Med.,* 133, 231–259, 1971.
245. **Whittlestone, P., Lemcke, R. M., and Olds, R. J.,** *J. Hyg.,* 70, 387–407, 1972.
246. **Somerson, N. L. Kontras, S. B., Pollack, J. D., and Weiss, H. S.,** *Science,* 171, 66–68, 1971.
247. **Harwick, H. J., Kalmanson, G. M., Fox, M. A., and Guze, L. B.,** *J. Infect. Dis.,* 128, 533–540, 1973.
248. **Ganaway, J. R., Allen, A. M., Moore, T. D., and Bohner, H. J.,** *J. Infect. Dis.,* 127, 529–537, 1973.
249. **Jersey, G. C., Whitehair, C. K., and Carter, G. R.,** *J. Am. Vet. Med. Assoc.,* 163, 599–604, 1973.
250. **Cassell, G. H., Lindsey, J. R., Overcash, R. G., and Baker, H. J.,** *Ann. N.Y. Acad. Sci.,* 225, 395–412, 1973.
251. **Lindsey, J. R. and Cassell, G. H.,** *Am. J. Pathol.,* 72, 63–84, 1973.
252. **Taylor, G., Taylor-Robinson, D., and Slavin, G.,** *Ann. Rheum. Dis.,* 33, 376–384, 1974.
253. **Taylor-Robinson, D., Rassner, C., Furr, P. M., Humber, D. P., and Barnes, R. D.,** *J. Reprod. Fertil.,* 42, 483–490, 1975.
254. **Barber, T. L. and Yedloutschnig, R. J.,** *Cornell Vet.,* 60, 297–308, 1970.
255. **Kerr, K. M. and Olson, N. O.,** *Avian Dis.,* 14, 291–320, 1970.
256. **Kerr, K. M. and Bridges, C. H.,** *Am. J. Pathol.,* 59, 399–406, 1970.
257. **Kume, K., Hayatsu, E., and Yoshioka, M.,** *Jpn. J. Vet. Sci.,* 38, 445–450, 1976.
258. **Ghazikhanian, G., Yamamoto, R., and Cordy, D. R.,** *Avian Dis.,* 17, 122–136, 1973.
259. **Kleven, S. H., Fletcher, O. J., and Davis, R. B.,** *Avian Dis.,* 19, 126–135, 1975.
260. **Gilchrist, P. T. and Cottew, G. S.,** *Aust. Vet. J.,* 50, 81, 1974.
261. **Wyeth, P. J.,** *Vet. Rec.,* 95, 208–211, 1974.
262. **Aldridge, K. E.,** *Infect. Immun.,* 12, 198–204, 1975.
263. **Gois, M., Sisak, F., Kuksa, F., and Sovadina, M.,** *Zentralbl. Veterinaermed. Reihe B,* 22, 205–219, 1975.
264. **Thomas, L. H., Howard, C. J., and Gourlay, R. N.,** *Vet. Rec.,* 97, 55–56, 1975.
265. **Cherry, J. D., Hurwitz, E. S., and Welliver, R. C.,** *J. Pediat.,* 87, 369–373, 1975.
266. **Nicolet, J. and Freundt, E. A.,** *Zentralbl. Veterinaermed. Reihe B,* 22, 302–307, 1975.
267. **McCormick, W. M., Rosner, B., Lee, Y-H., Rankin, J. S., and Lin, J-S.,** *Lancet,* 1, 596–597, 1975.
268. **Stalheim, O. H. V. and Page, L. A.,** *J. Clin. Microbiol.,* 2, 165–168, 1975.
269. **St. George, T. D. and Carmichael, L. E.,** *Vet. Rec.,* 97, 205–206, 1975.
270. **Shepard, M. C.,** *J. Bacteriol.,* 73, 161–171, 1957.
271. **Shepard, M. C.,** in *The Mycoplasmatales and the L-Phase of Bacteria,* Hayflick, L., Ed., Appleton-Century-Crofts, New York, 1969, pp. 49–65.
272. **Razin, S., Masover, G. K., Palant, M., and Hayflick, L.,** *J. Bacteriol.,* 130, 464–471, 1977.
273. **Black, F. T., Birch-Anderson, A., and Freundt, E. A.,** *J. Bacteriol.,* 111, 254–259, 1972.
274. **Whitescarver, J. and Furness, G.,** *J. Med. Microbiol.,* 8, 349–355, 1975.
275. **Rottem, S., Pfendt, E. A., and Hayflick, L.,** *J. Bacteriol.,* 105, 323–330, 1971.
276. **Shepard, M. C.,** *Am. J. Syph. Gonorrhea Vener. Dis.,* 38, 113-124, 1954.
277. **Purcell, R. H., Taylor-Robinson, D., Wong, D., and Chanock, R. M.,** *J. Bacteriol.,* 92, 6–12, 1966.
278. **Shepard, M. C. and Lunceford, C. D.,** *J. Bacteriol.,* 93, 1513–1520, 1967.
279. ICSB Subcommittee on the Taxonomy of Mycoplasmatales, *Int. J. Syst. Bacteriol.,* 24, 390–392, 1974.
280. **Shepard, M. C., Lunceford, C. D., Ford, D. K., Purcell, R. H., Taylor-Robinson, D., Razin, S., and Black, F. T.,** *Int. J. Syst. Bacteriol.,* 24, 160–171, 1974.
281. **Black, F. T.,** *Ann. N.Y. Acad. Sci.,* 225, 131–143, 1973.
282. **Taylor-Robinson, D., Williams, M. H., and Haig, D. A.,** *J. Gen. Microbiol.,* 54, 33–46, 1968.
283. **Taylor-Robinson, D., Martin-Bourgon, C., Watanabe, T., and Addey, J. P.,** *J. Gen. Microbiol.,* 68, 97–107, 1971.
284. **Ford, D. K. and MacDonald, J.,** *J. Bacteriol.,* 93, 1509–1512, 1967.
285. **Furness, G. and Coles, R. S.,** *Proc. Soc. Exp. Biol. Med.,* 150, 807–809, 1975.

286. **Furness, G.,** *J. Infect. Dis.,* 127, 9–16, 1973.
287. **Sumner, J. B.,** in *The Enzymes,* Vol. 1, Part 2, Sumner, J. B. and Myrbach, K., Eds., Academic Press, New York, 1951, pp. 873–892.
288. **Masover, G. K., Benson, J. R., and Hayflick, L.,** *J. Bacteriol.,* 117, 765–774, 1974.
289. **Masover, G. K. and Hayflick, L.,** *J. Bacteriol.,* 118, 46–52, 1974.
290. **Black, F. T., Christiansen, C., and Askaa, G.,** *Int. J. Syst. Bacteriol.,* 22, 241–242, 1972.
291. **Howard, C. J., Gourlay, R. N., Garwes, D. J., Pocock, D. H., and Collins, J.,** *Int. J. Syst. Bacteriol.,* 24, 373–374, 1974.
292. **Razin, S., Valdesuso, J., Purcell, R. H., and Chanock, R. M.,** *J. Bacteriol.,* 103, 702–706, 1970.
293. **Black, F. T. and Krogsgaard-Jensen, A.,** *Acta Pathol. Microbiol. Scand. Sect. B,* 82, 340–353, 1974.
294. **Shepard, M. C.,** *J. Bacteriol.,* 71, 362–369, 1956.
295. **Shepard, M. C.,** *Ann. N.Y. Acad. Sci.,* 143, 505–514, 1967.
296. **Lee, Y-H., Bailey, P. E., and McCormick, W. M.,** *J. Infect. Dis.,* 125, 318–321, 1972.
297. **Manchee, R. J. and Taylor-Robinson, D.,** *J. Bacteriol.,* 100, 78–85, 1969.
298. **Shepard, M. C. and Howard, D. R.,** *Ann. N.Y. Acad. Sci.,* 174, 809–819, 1970.
299. **Shepard, M. C. and Lunceford, C. D.,** *Appl. Microbiol.,* 20, 539–543, 1970.
300. **Windsor, G. D., Edward, D. G. ff., and Trigwell, J. A.,** *J. Med. Microbiol.,* 8, 183–187, 1975.
301. **Romano, N., Ajello, F., Massenti, M. F., and Scarlata, G.,** *Bol. Ist. Sieroter. Milan.,* 54, 292–295, 1975.
302. **Masover, G. K., Razin, S., and Hayflick, L.,** *J. Bacteriol.,* 130, 297–302, 1977.
303. **Razin, S., Masover, G. K., and Hayflick, L.,** *Proceedings of a Workshop on Nongonococcal Urethritis,* Hobson, D. and Holmes, K. K., Eds., American Society for Microbiology, Washington, D.C., in press.
304. **Masover, G. K., Razin, S., and Hayflick, L.,** 130, 292–296, 1977.
305. **Hendley, J. O. and Allred, E. N.,** *Infect. Immun.,* 5, 164–168, 1972.
306. **Ajello, F. and Romano, N.,** *Appl. Microbiol.,* 29, 293–294, 1975.
307. **Vinther, O. and Black, F. T.,** *Acta Pathol. Microbiol. Scand.,* 82, 917–918, 1974.
308. **Watanabe, T., Mishima, K., and Hozikawa, T.,** *Jpn. J. Microbiol.,* 17, 151–153, 1973.
309. **Purcell, R. H., Wong, D., Chanock, R. M., Taylor-Robinson, D., Canchola, J., and Valdesuso, J.,** *Ann. N.Y. Acad. Sci.,* 143, 664–675, 1967.
310. **Ford, D. K.,** *Ann. N.Y. Acad. Sci.,* 143, 501–504, 1967.
311. **Black, F. T.,** *Appl. Microbiol.,* 25, 528–533, 1973.
312. **Black, F. T.,** Proceedings, 5th International Congress on Infectious Diseases, Vienna, 1970, Vol. 1, pp. 407–411.
313. **Lin, J-S. and Kass, E. H.,** *J. Infect. Dis.,* 122, 93–95, 1970.
314. **Lin, J-S., Kendrick, M. I., and Kass, E. H.,** *J. Infect. Dis.,* 126, 658–663, 1972.
315. **Lin, J-S. and Kass, E. H.,** *Infect. Immun.,* 7, 499–500, 1973.
316. **Howard, C. J. and Gourlay, R. N.,** *Br. Vet. J.,* 128, xxxvii–xl, 1972.
317. **Howard, C. J. and Gourlay, R. N.,** *J. Gen. Microbiol.,* 78, 277–285, 1973.
318. **Howard, C. J. and Gourlay, R. N.,** *J. Gen. Microbiol.,* 79, 129–134, 1973.
319. **Howard, C. J., Gourlay, R. N., and Collins, J.,** *Int. J. Syst. Bacteriol.,* 25, 155–159, 1975.
320. **Livingston, C. W., Jr. and Gauer, B. B.,** *Am. J. Vet. Res.,* 35, 1469–1471, 1974.
321. **Romano, N., Smith, P. F., and Mayberry, W. R.,** *J. Bacteriol.,* 109, 565–569, 1972.
322. **Romano, N. and Scarlata, G.,** *Infect. Immun.,* 9, 1062–1065, 1974.
323. **Masover, G. K., Mischak, R. P., and Hayflick, L.,** *Infect. Immun.,* 11, 530–539, 1975.
324. **Massali, R. and Taylor-Robinson, D.,** *J. Med. Microbiol.,* 4, 125–138, 1971.
325. **Taylor-Robinson, D. and Carney, F. E., Jr.,** *Br. J. Vener. Dis.,* 50, 212–216, 1974.
326. **Thomas, L. H. and Howard, C. J.,** *J. Comp. Pathol.,* 84, 193–201, 1974.
327. **Gourlay, R. N., Howard, C. J., and Brownlie, J.,** *J. Hyg.,* 70, 511–521, 1972.
328. **Howard, C. J., Gourlay, R. N., and Brownlie, J.,** *J. Hyg.,* 71, 163–170, 1973.
329. **Gourlay, R. N., Brownlie, J., and Howard, C. J.,** *J. Gen. Microbiol.,* 76, 251–254, 1973.
330. **Gourlay, R. N.,** in *Les Mycoplasmes,* Vol. 33, Bove, J. M. and Duplan, J. F., Eds. Colloques INSERM, Paris, 1974, pp. 365–374.
331. **Howard, C. J., Gourlay, R. N., and Brownlie, J.,** *Infect. Immun.,* 9, 400–403, 1974.
332. **Kundsin, R. B., Ampola, M., Streeter, S., and Neurath, P.,** *J. Med. Genet.,* 8, 181–187, 1971.
333. **Friedlander, A. M. and Braude, A. I.,** *Nature,* 247, 67–69, 1974.
334. **Shurin, P. A., Alpert, S., Rosner, B., Driscoll, S. G., Lee, Y.-H., McCormick, W. M., Santamarina, B. A. G., and Kass, E. H.,** *N. Engl. J. Med.,* 293, 5–8, 1975.
335. **Foy, H., Kenny, G., Bor, E., Hammar, S., and Hickman, R.,** *J. Clin. Microbiol.,* 2, 226–230, 1975.
336. **Taylor-Robinson, D., Thomas, M., and Dawson, P. L.,** *J. Med. Microbiol.,* 2, 527–533, 1969.

337. Piot, P., *Br. J. Vener. Dis.*, 52, 266–268, 1976.
338. Braun, P., Klein, J. O., and Kass, E. H., *Appl. Microbiol.*, 19, 62–70, 1970.
339. Ford, D. K. and Smith, J. R., *Br. J. Vener. Dis.*, 50, 373–374, 1974.
340. Gnarpe, H., *Microbios*, 10, 247–252, 1974.
341. Lee, Y.-H., Alpert, S., Bailey, P. E., Duancic, A., and McCormick, W. M., *J. Am. Vener. Dis. Assoc.*, 1, 37–39, 1974.
342. Masover, G. K., Namba, M., and Hayflick, L., *Exp. Cell Res.*, 99, 363–374, 1976.
343. Shepard, M. C., in *Les Mycoplasmes*, Vol. 33, Bove, J. M. and Duplan, J. F., Eds., Colloques INSERM, Paris, 1974, pp. 375–380.
344. Lamey, J. R., Foy, H. M., and Kenny, G. E., *Obstet. Gynecol.*, 44, 703–708, 1974.
345. Matthews, C. D., Elmslie, R. G., Clapp, K. H., and Svigos, J. M., *Fertil. Steril.*, 26, 988–990, 1975.
346. Kundsin, R. B., in *Progress in Gynecology*, Vol. 6, Taymor, M. L. and Green, T. H., Jr., Eds., Grune & Stratton, New York, 1975, pp. 291–306.
347. McCormick, W. M., in *Les Mycoplasmes*, Vol. 33, Bove, J. M. and Duplan, J. F., Eds., Colloques INSERM, Paris, 1974, pp. 381–388.
348. Kundsin, R. B., Driscoll, S. G., and Ming, P.-M. L., *Science*, 157, 1573–1574, 1967.
349. Taylor-Robinson, D. and Furr, P. M., *Ann. N.Y. Acad. Sci.*, 225, 108–117, 1973.
350. Taylor-Robinson, D., Haig, D. A., and Williams, M. H., *Ann. N.Y. Acad. Sci.*, 143, 517–518, 1967.
351. Gourlay, R. N. and Thomas, L. H., *Vet. Rec.*, 84, 416-417, 1969,
352. Gourlay, R. N., *Res. Vet. Sci.*, 9, 376–378, 1968.
353. Gourlay, R. N., Mackenzie, A., and Cooper, J. E., *J. Comp. Pathol.*, 80, 575–584, 1970.
354. Livingston, C. W., Jr., *Am. J. Vet. Res.*, 33, 1925–1929, 1972.
355. Ruhnke, H. L. and van Dreumel, A. A., *Can. J. Comp. Med.*, 36, 317–318, 1972.
356. Shimizu, T., Nosaka, D., and Nakamura, N., *Jpn. J. Vet. Sci.*, 37, 121–131, 1975.
357. Tan, R. J. S. and Markham, J. G., *Jpn. J. Exp. Med.*, 41, 247–248, 1971.
358. Tan, R. J. S. and Miles, J. A. R., *Aust. Vet. J.*, 50, 142–145, 1974.
359. Kundsin, R. B., Rowell, T., Shepard, M. C., Parreno, A., and Lunceford, C. D., *Lab. Anim. Sci.*, 25, 221–224, 1975.
360. Jurmanova, K., *Zentralbl. Veterinaermed. Reihe B*, 22, 529–534, 1975.
361. Howard, C. J., Anderson, J. C., Gourlay, R. N., and Taylor-Robinson, D., *J. Med. Microbiol.*, 8, 523–529, 1975.
362. Brown, W. J., Jacobs, N. F., Arum, E. S., and Arko, R. J., *Lab. Anim. Sci.*, 26, 81–83, 1976.
363. Gourlay, R. N., Wyld, S. G., and Leach, R. H., *Int. J. Syst. Bacteriol.*, in press.
364. Pugh, G. W., Hughes, D. E., and Schulz, V. D., *Am. J. Vet. Res.*, 37, 493–495, 1976.
365. Nicolet, J., Dauwalder, M., Boss, P. H., and Anetzhofer, J., *Schweiz. Arch. Tierheilkd.*, 118, 141–150, 1976.
366. Askaa, G. and Ernø, H., *Int. J. Syst. Bacteriol.*, 26, 323–325, 1976.
367. Hooker, J. M. and Butler, M., *J. Comp. Pathol.*, 86, 87–92, 1976.
368. Leach, R. H., Cottew, G. S., Andrews, B. E., and Powell, D. G., *Vet. Rec.*, 98, 377–379, 1976.
369. Foggie, A., Jones, G. E., and Buxton, D., *Res. Vet. Sci.*, 21, 28–35, 1976.
370. Harwick, H. J., Mahoney, A. D., Kalmanson, G. M., and Guze, L. B., *J. Infect. Dis.*, 133, 103–112, 1976.
371. Organick, A. B. and Lutsky, I. I., *Lab. Anim. Sci.*, 26, 419–429, 1976.
372. Cole, B. C., Golightly-Rowland, L., and Ward, J. R., *Ann. Rheum. Dis.*, 35, 14–22, 1976.
373. Weinack, O. M. and Snoeyenbos, G. H., *Avian Dis.*, 20, 253–259, 1976.
374. Gourlay, R. N., Thomas, L. H., and Howard, C. J., *Vet. Rec.*, 98, 506–507, 1976.
375. Decker, J. L. and Barden, J. A., *Rheumatology*, 6, 338–345, 1975.
376. Jones, G. E., Foggie, A., Mould, D. L., and Livitt, S., *J. Med. Microbiol.*, 9, 39–52, 1976.
377. Gourlay, R. N. and Thomas, L. H., *Vet. Rec.*, 85, 583, 1969.
378. Howard, C. J., Gourlay, R. N., Thomas, L. H., and Stott, E. J., *Res. Vet. Sci.*, 21, 227–231, 1976.
379. Gourlay, R. N., Howard, C. J., and Brownlie, J., *Infect. Immun.*, 12, 947–950, 1975.
380. Shepard, M. C. and Lunceford, C. D., *J. Clin. Microbiol.*, 3, 613–625, 1976.
381. Fowlkes, D. M., MacLeod, J., and O'Leary, W. M., *Fertil. Steril.*, 12, 1212–1218, 1975.
382. Tafari, N., Ross, S., Naeye, R. L., Judge, D. M., and Marboe, C., *Lancet*, 1, 108–109, 1976.
383. Spaepen, M. S., Kundsin, R. B., and Horne, H. W., *Antimicrob. Agents Chemother.*, 9, 1012–1018, 1976.
384. Stalheim, O. H. V., Proctor, S. J., and Gallagher, J. E., *Infect. Immun.*, 13, 915–925, 1976.
385. Oghiso, Y., Yamamoto, K., Goto, N., Takahashi, R., Fujiwara, K., and Miura, T., *Jpn. J. Vet. Sci.*, 38, 15–24, 1976.
386. Windsor, G. D. and Trigwell, J. A., *J. Med. Microbiol.*, 9, 101–103, 1976.

THE MOLLICUTES*

ACHOLEPLASMAS, SPIROPLASMAS, THERMOPLASMAS, AND ANAEROPLASMAS

J. G. Tully and S. Razin

INTRODUCTION

The mycoplasmas discussed in this section represent an interesting and rather unique collection of agents; some of these have been only recently described and characterized. Taxonomically, most of these organisms have been accepted formally within the class Mollicutes; in several instances, however, insufficient information has delayed their further classification to order or family, and so they have simply been loosely attached to the class. The four groups of organisms discussed here also represent the extremes of diversity. Some of the acholeplasmas are among the earliest mycoplasmas isolated and described, whereas others in the group include organisms recovered from very unusual and specialized habitats and are representative of the most recent expansion of the host range of mycoplasmas. As will be seen, the groups discussed here also provide interesting and provocative variations in morphology, nutritional requirements, and pathogenicity. These variations, as exemplified among some of the newly described mycoplasmas, have important implications to a number of areas of biological research. The helical morphology and motility of the spiroplasmas raise significant questions about how these organisms maintain this structure and function in the absence of a cell wall and axial filaments. The recovery and growth of wall-less prokaryotes in the extremely acidic and thermophilic environments outlined below would seem to indicate an unusually stable organization of their cell membrane. Finally, the fact that newly recovered spiroplasmas pathogenic for vertebrates possessed nutritional needs not met by currently employed culture media strongly suggests that as yet unrecognized mycoplasmas pathogenic for man and other hosts may exist.

GENUS *ACHOLEPLASMA*

Classification of Acholeplasmas

Mycoplasmas now classified as *Acholeplasma laidlawii* were initially recovered from sewage, compost, and soil. These organisms were unique even when compared to the other mycoplasmas known at the time, since they possessed no growth requirement for serum and sterol and were able to grow at temperatures as low as 22°C. Since a sterol requirement has been for years one of the fundamental properties that distinguishes mycoplasmas from true bacteria, the strains isolated initially from soil and sewage provided a taxonomic dilemma.[1] Some resolution of the problem occurred when it was shown that another previously described *Mycoplasma* species (*M. granularum*) could grow in the absence of serum or cholesterol.[2] These studies prompted a reevaluation of the taxonomy of mycoplasmas, and a proposal was made to establish a new family (Acholeplasmataceae) and genus (*Acholeplasma*) within the order Mycoplasmatales, to give status to those species having no growth requirement for cholesterol.[3] This reasoned and logical extension of the classification system has been given additional support by the subsequent finding that *Acholeplasma* species possess a genome size in the range of 1×10^9 daltons, as opposed to a genome size of approximately 5×10^8 daltons for the *Mycoplasma* and *Ureaplasma* species (see the chapter on the Mollicutes). More recently, additional biochemical distinctions have been observed between a number of species

* Part 19 in *Bergey's Manual.*

within each family. Acholeplasmas (15 strains distributed among three species) were shown to synthesize fatty acids from acetate, whereas *Mycoplasma* species possessed very limited or no ability to utilize acetate for lipid synthesis.[4] The reduced nicotinamide adenine dinucleotide (NADH) oxidase activity of *Acholeplasma* species has been localized in their cell membrane, in contrast to its localization in the cytoplasm of *Mycoplasma* species.[5] Other studies, recently summarized,[6] developed the distinction that acholeplasmas possess lactate dehydrogenases (LDH) that are activated specifically by fructose-1,6-diphosphate. No such LDH activity has been detected in either fermentative or nonfermentative *Mycoplasma* species. Further information on differences in cholesterol requirement and on test results that are used to establish this requirement, as well as on a variety of other procedures that indirectly measure cholesterol requirement, can be found in the first chapter of this series.

The current classification scheme for acholeplasmas includes six species. Differentiation at the species level is presently limited to serological properties and to the ability of strains to ferment aesculin[7] or produce pigmented carotenoids in amounts sufficient to be recorded in a standard test.[2,8] Most of the remaining biochemical and physiological characteristics recorded for individual *Acholeplasma* species are of little value in species separation.[9] There are recent reports[10,11] that emphasize the use of arbutin fermentation as a distinguishing property of the species *A. axanthum,* although only a few strains of this species have been examined to verify the consistency of this property. Despite the need for further biochemical and physiological characteristics that will enhance the separation of *Acholeplasma* species, data obtained from DNA hybridization tests among at least three species confirm the validity of the species designations made earlier solely on serological and biochemical properties (see the review in Reference 9). Recently, a phylogenetic relationship between acholeplasmas and certain lactic acid bacteria has been proposed.[127]

Morphology and Biochemical Characteristics of Acholeplasmas

Cellular morphology and reproductive mechanisms of acholeplasmas do not differ from those described for most of the members of the class Mollicutes (see earlier chapter). The biochemical and physiological properties established for the acholeplasmas are summarized in Table 1. All acholeplasmas described to date are glycolytic, and none have been shown to hydrolyze arginine or urea. At least three *Acholeplasma* species have been found to lack the phosphoenol pyruvate-dependent phosphotransferase system present in other glycolytic *Mycoplasma* species.[12] Although acholeplasmas are able to grow at temperatures from 22 to 37°C and in the absence of sterols, rapid and heavy growth (10^9 to 10^{10} colony-forming units per ml) occurs when 5 to 20% horse serum is added to the medium and the cultures are incubated at 37°C. This attribute, combined with the absence of a cell wall and with marked susceptibility to osmotic lysis, has made the acholeplasmas a most useful model in membrane studies.[122]

No single biochemical or physiological test has been found to separate *Acholeplasma* species, and variations exist even in some strains within a given species. The lipids of some acholeplasmas, particularly in *A. axanthum,*[13] appear to be rather unique compounds, and it has been suggested that these substances and lipopolysaccharides isolated from these organisms and from mycoplasmas might offer some value in taxonomic differentiation.[14,123]

Serological Characteristics of Acholeplasmas

The acholeplasmas are serologically distinct from other mycoplasmas, although differences in the serological properties of individual *Acholeplasma* species are occasionally difficult to demonstrate. Again, no single serological test procedure possesses all the requirements of specificity, sensitivity, and ease of performance. The growth

Table 1
BIOCHEMICAL AND PHYSIOLOGICAL CHARACTERISTICS OF ACHOLEPLASMAS

(Ae = aerobic; An = anaerobic; NT = not tested or not known)

Acholeplasma species	% G + C of DNA	Glucose catabolism	Mannose catabolism	Arginine hydrolysis	Aesculin hydrolysis	Sensitivity to digitonin	Tetrazolium reduction (Ae/An)	Carotenoid synthesis test	Phosphatase	Film and spot reaction	Gelatin hydrolysis	Coagulated-serum digestion	Hemadsorption	Preferred atmosphere	Rate of growth	Colony morphology	References
A. axanthum	31.3	+	–	–	+	–	+/+	–	–	–	NT	–	–	Ae	Moderate	Regular[a]	9, 11
A. equifetale	NT	+	NT	–	±	–	±/+	NT	–	+	–	NT	–	Ae	Moderate	Regular	16, 17
A. granularum	30.5–32.4	+	–	–	–	–	±/+	+	–	–	–	–	–	Ae	Moderate	Regular, yellow	2, 9, 18
A. laidlawii	31.7–35.7	+	+ or –	–	+ or –	–	±/+	+	± or –	–	+ or –	–	–	Ae	Rapid	Regular, yellow	2, 9, 10, 19
A. modicum	29.3	+	–	–	–	–	+/+	–	–	–	NT	–	–	Ae	Moderate	Regular	9, 10, 19–21
A. oculi	NT	+	–	–	+	–	+/+	+	NT	–	NT	NT	NT	Ae	Moderate	Regular	9, 22

[a] Regular morphology means the typical "fried egg" colony.

inhibition and fluorescent-antibody tests, particularly the latter as applied to colonies on agar, probably come closest to meeting these needs (see the chapter on the Mollicutes for details). An antiserum possessing satisfactory growth-inhibiting antibody usually provides an adequate conjugate for fluorescent-antibody tests. However, preparation of potent growth-inhibiting antiserum in rabbits or other hosts is sometimes very difficult. Some of the factors that might be involved and suggested techniques have been presented.[9]

Habitat and Pathogenicity of Acholeplasmas

The habitat and possible pathogenicity of the acholeplasmas are detailed in Table 2. The acholeplasmas were initially recovered from sewage and soil more than 40 years ago, and therefore they were considered to be free-living saprophytes. However, there is no clear evidence for the prolonged survival of acholeplasmas in these environments, and their rather strict nutritional requirements and sensitivity to osmotic changes would argue against this concept. In the past 15 years, a large number of reports (summarized in a review[9]) document the direct association of acholeplasmas with a wide variety of animal hosts. *Acholeplasma laidlawii* is the most widely distributed species of *Acholeplasma* (and maybe of all mycoplasmas) and has been most frequently isolated from either the upper respiratory or the lower genital tract of most animals.[15] Isolations of *A. laidlawii* have also been reported from plants (see later comments this section). *Acholeplasma granularum* has recently been recovered from porcine feces.[126] Thus, it would appear that the close association of acholeplasmas with animal hosts provides a ready explanation of their occurrence in soil and sewage. The distribution of other *Acholeplasma* species in animals appears more restricted, but this may be only a reflection of inadequate sampling and documentation.

There is no clear evidence at present that any of the acholeplasmas are pathogenic, despite their frequent isolation from animals and, on occasion, from diseased tissue. This does not mean that these organisms play no part in the various syndromes from which they have been recovered. It can be a reflection of the present lack of controlled studies with freshly isolated strains in susceptible hosts and in hosts without evidence of complications from normal acholeplasma flora. The isolation of acholeplasmas from respiratory diseases in swine,[27] cattle,[43] sheep,[31] from cases of conjunctivitis in cattle[44] and goats,[22] and from various sites in the genital tract of cattle[29,45,46] indicates the need for further study regarding their potential for producing disease.

GENUS *SPIROPLASMA*

Morphology of Spiroplasmas

Spiroplasmas are motile, helical, wall-free prokaryotes. Cellular morphology varies from pleomorphic spherical cells measuring 200 to 300 nm in diameter to helical and branched nonhelical filaments. The helical forms, 3 to 12 μm long and 0.1 to 0.2 μm in diameter, predominate in logarithmic-phase broth cultures.[47-50] Cultures in the late logarithmic phase or in the stationary phase of growth exhibit more filaments, many of which have lost their helical appearance. There is a tendency for these filaments to become greatly elongated and to develop irregular round bodies or blebs.[47,51] Ultrastructurally, the organisms lack a cell wall and are bounded by a single triple-layered membrane. There is some evidence of a thin, fuzzy layer on the outer membrane surface of some spiroplasmas.[47] Significant amounts of a hexosamine have been identified in membrane preparations of *S. citri,*[51] but there is as yet no evidence linking this component to the fuzzy layer observed in negatively stained spiroplasmas. Bacterial peptidoglycan was found to be absent in the one *Spiroplasma* species examined.[52] Helical filaments are motile, exhibiting either rapid rotary motion or flexional movements.[53] Motility has been demonstrated in spiroplasmas from plants[53] and from insects[49,54] as well as in cultured spiroplasmas.[47,48,55-57] No evidence of flagella, axial filaments, or

Table 2
HABITAT AND PATHOGENICITY OF ACHOLEPLASMAS

Type strain Strain designation (Reference Collection Numbers)[a]	Natural host	Site of recovery; material for isolation	Disease manifestations	Experimental pathology	References
A. axanthum S-743 (ATCC 25176) (NCTC 10138)	Bovine, equine, porcine	Nasal cavity, lymph nodes, and kidney of cattle, oral cavity of horse, lung and peribronchial lymph nodes of swine; contaminated tissue cultures (from animal serum source)	Possible association with respiratory disease in swine	Experimental challenge of specific-pathogen-free piglets by intranasal route showed mild clinical symptoms with gross lung lesions and histologic changes in the lungs.	9, 11, 23–26
A. equifetale C-112	Equine	Nasopharynx and trachea	Not known	Not studied	16, 17
A. granularum BTS-39 (ATCC 19168) (NCTC 10128)	Porcine, equine	Nasal cavity and feces of swine; conjunctivae of horse	Not known	No multiplication in embryonated hen's eggs.	18, 24, 27, 28, 126
A. laidlawii PG8 (ATCC 23206) (NCTC 10116)	Avian, bovine, caprine, canine, equine, feline, murine, ovine, porcine, primates (including man), and plants?	Oral cavity, lymph nodes, serum, and respiratory and genital tracts of animals; contaminated tissue cultures (from animal serum sources and some dehydrated tissue culture components)	Not known	Not studied	16, 17, 20, 23 24, 26–41
A. modicum Squire (PG49) (NCTC 10134)	Bovine	Lung and lymph nodes	Not known	Not studied	20, 21, 26, 42
A. oculi 19L (ATCC 27350)	Caprine, equine	Eye of goats; nasopharynx of horse; contaminated tissue cultures	Not known	Not studied	9, 17, 22, 35

[a] ATCC = American Type Culture Collection; NCTC = National Collection of Type Cultures.

other organelles of locomotion have been demonstrated. Treatment of freshly prepared suspensions of a spiroplasma from insect hemolymph with 0.5% sodium deoxycholate results in destruction of the organisms and in the release of numerous striated fibrils.[58] The chemical nature of the fibrils and their possible role in motility are unknown at present. The helical morphology of spiroplasmas in plant, insect, and animal materials may be difficult to demonstrate[50] or is lost in preparative techniques for electron microscopy. Ultrathin sections of spiroplasmas in plant tissues infected with citrus stubborn or corn stunt diseases, examined retrospectively, show spherical forms or slightly curved filaments without clear evidence of a helical structure.[47,59-62] Thick sections (0.2 to 1.0 μm) show the typical helical forms more frequently.[47,53,62] Other studies have reported the importance of osmolarity of the fixatives used for plant and animal tissues and the deleterious effects of cacodylate buffer on the morphology of spiroplasmas.[47] Reproductive mechanisms have been incompletely studied, but there are some preliminary observations that binary fission occurs in some spiroplasmas.

Classification of Spiroplasmas

The general designation "spiroplasma" was first given[53] to helical, wall-free prokaryotes observed in corn plants infected with the disease agent.[63,64] At the time of this discovery, the organisms could not be continually cultured in an artificial medium. While this work on corn stunt was in progress, a mycoplasma-like agent was visualized[60,61,65] and eventually cultured[66-68] from citrus plants infected with stubborn disease. Preliminary characterization of one of the isolates[69] confirmed that this organism possessed the general properties of mycoplasmas, but it was also clear that further biochemical and serological comparisons were required for complete characterization and taxonomic consideration. These studies[47,70] revealed that the citrus stubborn agent was morphologically similar to the helical forms observed earlier in corn stunt disease. The complete characterization of the citrus stubborn agent, the first plant mycoplasma to be grown in culture, established this organism as a new and unique mycoplasma, for which a new genus and species designation (*Spiroplasma citri*) was proposed.[70] Investigations that excluded the presence of a modified bacterial cell wall or of constituents of cell wall in the organism[52] allowed the inclusion of this new organism within the class Mollicutes.[71] An additional proposal to establish a new family (Spiroplasmataceae) within the order Mycoplasmatales for the organism has also been made.[72]

Cultivation of corn stunt spiroplasmas in artificial media and documentation of the association of the cultured spiroplasmas to the plant disease were recently reported by two independent groups.[48,56] Since the morphology of corn stunt spiroplasmas was identical to that of *S. citri*, and since the two organisms had shown some serological cross reactions (discussed later), the respective authors wisely withheld a taxonomic decision, pending a more complete comparative characterization.

A third possible spiroplasma has been found in the hemolymph of natural populations of four closely related species of *Drosophila*. This agent, termed the sex ratio organism (SRO), is inherited maternally and is associated with the absence of males in the progeny of infected females.[73] The agent was first considered to be a spirochete,[73] but it was shown later to lack axial filaments, outer envelope, and a cell wall typical of spirochetes.[49,74] SRO spiroplasmas have not been cultured and are not known to be associated with plants or higher animals.[49,50]

The most recent spiroplasma described was previously termed the "suckling mouse cataract agent" (SMCA).[57] This agent was originally isolated from a pooled extract of rabbit ticks (*Haemaphysalis leporispalustris*) collected near Atlanta, Ga.[75,76] Following intracerebral inoculation, the agent grows to high titer in the eyes and brains of newborn mice and is associated with the production of cataract, uveitis, and chronic brain infection in newborn mice and rats.[77,78] SMCA can be grown to high titer in embryonated hen's eggs, in which it produces a lethal infection in 4 to 9 days.[75,76] Allantoic fluid from infected eggs contains numerous motile helical organisms that are morphologically indistinguishable from other spiroplasmas.[57] The occurrence of other motile helical organisms from ticks, possibly related to or identical to SMCA, has been noted.[79,128] SMCA spiroplasmas have been grown on artificial media, including both liquid and solid cultures, and the cultured spiroplasmas reproduce the expected lethal infection in embryonated eggs and induce cataracts in newborn rats.[80] The relationship of these organisms to *S. citri* and to other unclassified spiroplasmas will depend upon further comparative study of their biological and serological characteristics. These observations provide the first evidence that spiroplasmas possess pathological activities for vertebrate hosts. The ability of the organisms to multiply at 37°C (in contrast to other spiroplasmas, which have temperature optima between 29 and 32°C), combined with their persistence and multiplication in a variety of tissues of vertebrates and with the fact that SMCA spiroplasmas could not be grown on conventional media for mycoplasmas or other spiroplasmas upon initial recovery, has important implications in the possible role and occurrence of similar prokaryotes in human diseases of unknown etiology.

Biochemistry and Physiology of Spiroplasmas

The major biochemical and physiological properties of *S. citri* are given in Table 3. Since only a few strains of this organism have been examined, minor variations in the characteristics noted are to be expected. The lipid composition and lipid metabolism of *S. citri* is apparently very similar to most other sterol-requiring mycoplasmas.[124] Alterations in the lipid composition of the growth medium induced changes in the characteristic helical morphology of the cells, thus providing a useful model to study the effect of membrane lipid composition on the physical state of these organisms. Little information is available on the biochemical properties of other cultured (and unclassified) spiroplasmas.

One of the most important observations in the primary isolation and propagation of the first spiroplasma *(S. citri)* was that a variety of conventional mycoplasma media formulations were inadequate. The exact mechanism to explain why the addition of several carbohydrates (sucrose, fructose, and sorbitol) to conventional mycoplasma broth

Table 3
BIOCHEMICAL AND PHYSIOLOGICAL CHARACTERISTICS OF *SPIROPLASMA CITRI*[a]

Property	Characteristic
Glucose fermentation	+
Arginine hydrolysis	– (+)[b]
Urea hydrolysis	–
Optimal temperature for growth	32°C
Cholesterol requirement	+
Hemadsorption (guinea pig RBC's)	±
Hemolysis (guinea pig RBC's)	–
Tetrazolium reduction	±
Phosphatase activity	+
Aesculin fermentation	–
Film and spot reaction	–
G + C content of DNA (buoyant density), moles %	26.0–26.3
Genome size, daltons	10^9
Type strain	Morocco (R8-A2) (ATCC 27556)

[a] Data summarized principally from References 62 and 70.
[b] See Reference 125.

resulted in successful isolation of *S. citri* has not been determined. Speculation centers around the elevation of the osmolality of the medium to acceptable levels, or perhaps to levels similar to that of plant tissues. Cultivation of corn stunt spiroplasmas offered additional evidence that the specific nutritional requirements of spiroplasmas differed, either through variations exerted through their origin in different hosts or because the spiroplasmas themselves might represent separate and distinct organisms. Earlier attempts at cultivating the corn stunt organism,[81] though successful in maintaining prolonged viability of the agent, did not achieve continuous subculture. When successful isolation and continuous subculture were eventually achieved, the medium formulations varied widely. One study[56] used a medium based on insect tissuue culture media, and subsequent investigations[82] demonstrated the important influence of several organic acids, divalent cations, and osmolality on successful growth. However, a second group of investigators[48] used a very different medium to achieve similar ends, and eventually[83] they were able to cultivate the agent on a very simple medium supplemented with high levels of sucrose.

Successful isolations and growth of the SMCA spiroplasmas[80] were obtained from medium formulations used to assess various growth requirements of corn stunt spiroplasmas.[82] The outstanding advantage in all attempts to cultivate spiroplasmas is the availability of a microscopic technique for monitoring the adequacy of the medium under test. Dark-field microscopy provides good evidence of both an increase in the number of helical organisms and retention of their structural integrity in suitable media. Spiroplasmas maintained in suboptimal environments may not develop helical forms, or they may develop distorted helices.[50]

The frequency of recent isolations of new spiroplasmas suggests that others will be recovered in the future from a variety of plant and insect materials, or even from some unexpected sources. Observations on these newer spiroplasmas and more complete information on the unclassified spiroplasmas being cultured at this time will allow some generalizations to be made about the basic nutritional requirements of spiroplasmas.

Serological Characteristics of Spiroplasmas

The broad serological characteristics of spiroplasmas have not been fully defined. Cultured *S. citri* was examined by conventional mycoplasma serological tests, including growth inhibition and immunofluorescent procedures, as part of the requirements for classification of the organism.[70] In these tests, *S. citri* was serologically unrelated to *Mycoplasma* or *Acholeplasma* species. The availability of concentrated *S. citri* antigens did provide materials to develop preliminary information on the relationship of this organism to other spiroplasmas. Growth inhibition and ring precipitin tests with *S. citri* antigens and antisera to corn stunt-infected plant tissue showed considerable cross reactions, although antisera to healthy corn plants were devoid of serological activity to *S. citri.*[84] Similar relationships between *S. citri* and SRO spiroplasmas were noted in studies in which tests were performed with antisera to SRO-infected *Drosophila* hemolymph.[49] When cultured corn stunt spiroplasmas became available, additional comparisons among the three spiroplasmas were made. The consensus of the results of these tests indicated that some serological relationships exist between all three organisms, but that each spiroplasma possesses a number of distinct antigens.[48,49] These relationships were apparent not only with the classical mycoplasma serological procedures, including a modified metabolic inhibition test,[48] but with a microscopic deformation test.[49] SMCA spiroplasmas have been found to share some antigenic determinants with the three previously described spiroplasmas.[57] However, final evaluation of all these observations will depend upon further testing of antigens and antibody from filter-cloned spiroplasmas.

It seems evident at this point that spiroplasmas from widely different sources share a number of serological characters. Whether conventional serological tests usually employed for analysis of animal mycoplasmas are adquate to distinguish individual *Spiroplasma* species or whether these tests may only measure group antigens among various spiroplasmas has not been established. However, observations recorded to date suggest that other types of serological techniques should be applied to these organisms in the hope that better evaluation of their serological properties will be possible.

Habitat and Pathogenicity of Spiroplasmas

Spiroplasma citri has been cultured only from stubborn-diseased citrus leaves and seed coats,[66-68] and more recently from field-collected leafhoppers.[85,86] Demonstration of the association of *S. citri* with plant disease was established through inoculation of leafhoppers and the transmission of disease symptoms to white clover,[87-89] sweet orange,[62,89,90] and vinca.[86,89] The natural vector of *S. citri* is unknown at present, although several leafhopper species are suspected to play a direct role in transmission, and some natural carriers of the agent, but not transmitting vectors, have been found.[85,86] Experimental infection of insects with a variety of spiroplasmas has been reported, and the findings have been reviewed in several recent publications.[50,90-92]

Cultured corn stunt spiroplasmas have been shown to reproduce the disease in healthy plants through transmission by appropriate leafhopper vectors.[48,56] An important observation was made in conjunction with these studies[56] when it was noted that one of the cultured lines of corn stunt spiroplasmas, although retaining pathogenicity for the insect host, had lost the ability to induce disease in corn plants. Proper regard for possible attenuation of pathogencity in newly isolated spiroplasmas through extended in vitro cultivation appears to be a necessary precaution for other studies on plant and insect spiroplasmas. Vector relationships in corn stunt disease involves at least five leafhopper species, of which *Dalbulus eliminatus* and *D. maidis* are the most efficient.[93] Recent isolations and cultivation of corn stunt spiroplasmas from plants and insects, and the subsequent availability of quantitative cultural techniques, have not been fully utilized as yet in assessing vector relationships in this disease.

As noted earlier, SRO spiroplasmas occur only in females of at least four species of neotropical *Drosophila* and are associated with complete or near-complete elimination of the male sex from their progeny.[49] The inability to culture this organism in artificial medium and to separate possible mixed spiroplasma flora in *Drosophila* hemolymph has limited vector studies to artificial inoculation of infected hemolymph into natural and aberrant hosts.[50,91]

SMCA spiroplasmas apparently occur naturally in at least one tick species, as noted earlier. Pathogenicity of SMCA for mice, rats, and chick embryos has been confirmed with cultured spiroplasmas.[80] Techniques now available to grow SMCA spiroplasmas directly from ticks, and possibly from other hosts, should provide more specific information on distribution and epidemiology of this organism in a variety of hosts.

OTHER PLANT AND INSECT MYCOPLASMAS

In addition to plant diseases for which the etiological agent has been established as a spiroplasma, there is a very large group of plant diseases, the so-called "yellows-type," in which mycoplasma-like bodies have been observed by electron-microscopic techniques. Yellows disease is characterized by a variety of symptoms, but is most frequently associated with general stunting of plants, appearance of abnormal flowers, fruits, and seeds, and sometimes death of the plant (lethal yellowing). Mycoplasma-like bodies are essentially limited to the sieve tubes of host plants, but occur in a number of tissues in the insects that transmit them. A majority of these diseases are transmitted by leafhoppers. In 1967, investigations by Japanese workers (Doi, Ishiie, and associates – see

References 94, 97, and 101 for details of the early history) established the presence of pleomorphic bodies (80 to 800 nm in diameter) in sections of phloem tissue from mulberry dwarf and several other yellows-type diseases. These organisms were bounded only by a unit membrane and possessed ultrastructural elements resembling ribosomes and DNA-like strands, much like those observed in classical mycoplasmas. In addition, their studies involving treatment of plants with tetracycline antibiotics indicated that there was a direct association between the recovery of plants from the disease and the disappearance of mycoplasma-like bodies from phloem tissue. Recovery from disease was not permanent under the treatment schedules, and the observations correlated well with similar attempts to eradicate mycoplasmas from cell cultures with tetracyclines and several other antibiotics. Although chemotherapeutic experiments of this type provide important practical applications in plant protection, they do not permit firm conclusions about the nature of the etiological agent, since other microbial forms are susceptible to broad-spectrum antibiotics. However, reduction of microbial numbers following tetracycline treatment (and without alterations in microbial numbers from penicillin application) can suggest some role of a wall-free prokaryote in the etiology of the plant disease.

Since 1967, more than 70 plant diseases have been reported to be associated either with spiroplasmas or with mycoplasma-like bodies. Reports of successful isolation of classical sterol-requiring mycoplasmas and non-sterol-requiring acholeplasmas (especially *A. laidlawii*) have been offered in some of these diseases, but these reports have not been confirmed in other laboratories. It seems unlikely at this point that classical animal mycoplasmas might be involved in these diseases, since extensive attempts have been made to culture mycoplasmas from a variety of plants infected with yellows disease, using a number of media formulations that support the growth of animal mycoplasmas. It is possible, however, that these agents might represent classical mycoplasmas with different and unique growth requirements, which stem from plant or insect adaptation – a concept that cannot be summarily dismissed at this point.

More adequate review of this topic, along with extensive bibliographies, can be found in a number of recent publications.[94-103]

GENUS *THERMOPLASMA*

In 1970, several obligate thermophilic and acidophilic organisms were recovered from coal refuse piles that had undergone self-heating.[104] A detailed examination of two isolates (122-1B2 and 122-1B3) indicated that, although these organisms possessed many of the properties of mycoplasmas, their thermophilic and acidophilic characteristics were extremely unusual for a wall-free prokaryote. Morphologically, the organisms were predominantly small, pleomorphic spherical cells, although a few filamentous forms were observed. No cell wall was noted in sectioned cells examined by electron microscopy, and viable units passed through 450-nm filters. The two strains grew well in a basal salt medium containing yeast extract over a temperature range of 45 to 62°C (optimum at 59°C). Optimal growth also occurred between pH 1 and 2. The base composition (guanine + cytosine) of the DNA of the organisms was reported to be 25 moles % (see later comments). The organisms were designated *Thermoplasma acidophilum,* and a type strain (122-1B2) was deposited in a national reference collection (ATCC 25905).[104]

Subsequently, more than 100 other isolates of the organism were recovered from approximately 20 burning coal refuse piles; of these, 23 isolates were examined by a variety of biological and serological tests.[105] The organisms were shown to produce the classical "fried egg" colony on an agar medium prepared from the basal salt-yeast extract broth medium. The base morphology and nutritional requirements outlined above were confirmed for the newer isolates. At least five serogroups have been proposed, based on an immunofluorescent-adsorption test[105] and on a later study using immunodiffusion analysis of solubilized membrane preparations.[106] However, the serogrouping results between the two test procedures did not always agree.

Additional studies on the morphology of *T. acidophilum*, by scanning electron microscopy, verified the occurrence of pleomorphic spherical cells (with a size range from 0.5 to 1.9 μm).[107] The 122-1B3 isolate was reported to be more resistant to lysis by a variety of physical and chemical agents than most mycoplasmas.[108] The intracellular pH of *T. acidophilum* grown under high-temperature, low pH conditions was found to be close to neutral, indicating that the organism is capable of maintaining a normal intracellular pH when subjected to such a hostile environment.[109] Also, the occurrence of very long-chained isopranol ether lipids in *T. acidophilum* is thought to impart stability and maintenance of a proper liquid crystalline state of the lipid bilayer of the cell membrane during the acidic and high-temperature conditions during growth.[110,111] A lipopolysaccharide isolated from the organism has been found to be similar in its basic structure to lipopolysaccharides recovered from several other mycoplasmas.[112,113,123] The growth requirement for yeast extract, established in earlier studies,[105] was recently shown to be associated with a polypeptide.[114]

The DNA base composition of *T. acidophilum* as initially reported was subsequently reevaluated. It now appears that a G + C content of 46 moles % is more likely for both the 122-1B2 and the 122-1B3 strain of the organism,[115,116] as measured by buoyant density and by thin-layer chromatography. A genome size of approximately 1×10^9 daltons was also obtained in two independent studies.[115,116] The DNA of this organism is associated with a histone-like protein, which is thought to stabilize DNA against thermal denaturation.[117]

The thermoplasmas, therefore, appear to be a unique group of wall-free prokaryotes. Additional study of the various isolates and their biological and serological characteristics will be required before their more formal classification within the class Mollicutes can be considered.[71]

GENUS *ANAEROPLASMA*

A single isolate of a strictly anaerobic, filterable, and bacteriolytic microorganism was recovered from the bovine rumen. This strain had the ultrastructural, microscopic, and colonial morphology characteristics of mycoplasmas, was filterable through 450- and 220-nm filters, and was resistant to penicillin G. Since sterols could not be shown to be essential for growth, a taxonomic proposal was made to designate the isolate *Acholeplasma bactoclasticum*,[118] although only limited serological comparisons were made to previously established *Acholeplasma* species.

Recently, additional isolations of this and similar organisms were made from the bovine and ovine rumen, and a number of these strains were characterized by a variety of procedures.[119] On the basis of these tests, a proposal has been made to establish a new and separate genus *(Anaeroplasma)* for apparently two distinct organisms, represented by the earlier bovine rumen isolate (now designated *Anaeroplasma bactoclasticum*) and by a newly isolated organism (*Anaeroplasma abactoclasticum*) occurring in both the bovine and the ovine rumen.[120] Preliminary evidence would argue for the inclusion of these organisms within the class Mollicutes, but further information, especially serological comparisons to other mycoplasmas, will be needed to establish their taxonomic status with respect to order and family within this class.

A summary of the major biochemical and physiological distinctions of the two proposed anaeroplasmas is given in Table 4. Contrary to earlier findings,[118] *A. bactoclasticum* appears to have a sterol requirement, as do some strains of *A. abactoclasticum*. The two non-sterol-requiring strains of the latter species exhibited

Table 4
CHARACTERISTICS OF ANAEROPLASMAS[a]

Property	*Anaeroplasma bactoclasticum*	*Anaeroplasma abactoclasticum*
Range of temperatures enabling growth	30–47°C	30–47°C
Preferred atmosphere for growth	Strict anaerobe	Strict anaerobe
Cholesterol requirement for growth	+	+ and –[b]
Sensitivity to digitonin	+	+ and –
Fermentation of starch	+	+
Arginine hydrolysis	–	–
Urea hydrolysis	–	–
Bacteriolytic activity (lysis of *Escherichia coli* cells)	+	–
Inhibition by penicillin, 1000 IU/ml	–	–
Inhibition by thallium acetate, 2%	+	+
G + C content of DNA, moles %	32.5–33.7	29.3–29.5[b]
Type strain	JR	6-1
Habitat	Bovine rumen	Bovine and ovine rumen

[a] Data abstracted from Reference 119.
[b] Two strains of this proposed species do not require cholesterol and possess G + C ratios of 40.2 and 40.3 moles %.

exceedingly large variations in their G + C content of the DNA from that determined for the sterol-requiring strains. This suggests that additional biochemical and serological tests on these strains might indicate differences that would require further division of this particular species. The most unique properties of these organisms are their obligate anaerobic requirements and bacteriolytic activities. Growth is inhibited by oxygen, and therefore special equipment and techniques are required for isolation and propagation of these organisms. The bacteriolytic activity of *A. bactoclasticum* is directed toward autoclaved *Escherichia coli* cells added to the primary isolation medium for rumen bacteria. Lipid analysis of representative strains of the two *Anaeroplasma* species revealed the prevalence of plasmalogens (alk-1′-enyl glyceryl ethers)[121] and of lipopolysaccharides.[123] Plasmalogens have not been identified in other mycoplasmas, but the chemical nature of the lipopolysaccharides in anaeroplasmas appears to be similar to that of lipopolysaccharides recovered from other mycoplasmas.

REFERENCES

1. **Edward, D. G. ff.,** *Ann. N.Y. Acad. Sci.,* 143, 7–8, 1967.
2. **Tully, J. G. and Razin, S.,** *J. Bacteriol.,* 95, 1504–1512, 1968.
3. **Edward, D. F. ff. and Freundt, E. A.,** *J. Gen. Microbiol.,* 62, 1–2, 1970.
4. **Herring, P. K. and Pollack, J. D.,** *Int. J. Syst. Bacteriol.,* 24, 73–78, 1974.
5. **Pollack, J. D.,** *Int. J. Syst. Bacteriol.,* 25, 108–113, 1975.
6. **Neimark, H. C.,** *Ann. N.Y. Acad. Sci.,* 225, 14–21, 1973.
7. **Williams, C. O. and Wittler, R. G.,** *Int. J. Syst. Bacteriol.,* 21, 73–77, 1971.
8. **Razin, S. and Cleverdon, R. C.,** *J. Gen. Microbiol.,* 41, 409–415, 1965.
9. **Tully, J. G.,** *Ann. N.Y. Acad. Sci.,* 225, 74–93, 1973.
10. **Ernø, H. and Stipkovits, L.,** *Acta Microbiol. Acad. Sci. Hung.,* 20, 305–315, 1973.
11. **Stipkovits, L., Varga, L., and Schimmel, D.,** *Acta Vet. Acad. Sci. Hung.,* 23, 361–368, 1973.
12. **Cirillo, V. P. and Razin, S.,** *J. Bacteriol.,* 113, 212–217, 1973.
13. **Plackett, P., Smith, P. F., and Mayberry, W. R.,** *J. Bacteriol.,* 104, 798–807, 1970.
14. **Smith, P. F.,** in *Les Mycoplasmes,* Vol. 33, Bove, J. M. and Duplan, J. F., Eds., Colloques INSERM, Paris, 1974, pp. 63–67.
15. **Freundt, E. A.,** in *Les Mycoplasmes,* Vol. 33, Bove, J. M. and Duplan, J. F., Eds., Colloques INSERM, Paris, 1974, pp. 19–25.

16. **Kirchhoff, H.,** *Zentralbl. Veterinaermed. Reihe B,* 21, 207–210, 1974.
17. **Allam, N. M. and Lemcke, R. M.,** *J. Hyg.,* 74, 385–408, 1975.
18. **Switzer, W. P.,** in *The Mycoplasmatales and the L-Phase of Bacteria,* Hayflick, L., Ed., Appleton-Century-Crofts, New York, 1969, pp. 607–619.
19. **Jurmanova, K.,** *Zentralbl. Veterinaermed. Reihe B,* 22, 529–534, 1975.
20. **Leach, R. H.,** *J. Gen. Microbiol.,* 75, 135–153, 1973.
21. **Stipkovits, L.,** *Acta Vet. Acad. Sci. Hung.,* 23, 315–323, 1973.
22. **Al-Aubaidi, J. M., Dardiri, A. H., Muscoplatt, C. C., and McCauley, E. H.,** *Cornell Vet.,* 63, 117–129, 1973.
23. **Barile, M. F., Hopps, H. E., Grabowski, M., Riggs, D. B., and Del Giudice, R. A.,** *Ann. N.Y. Acad. Sci.,* 225, 251–264, 1973.
24. **Ogata, M., Watable, J., and Koshimizu, K.,** *Jpn. J. Vet. Sci.,* 36, 43–51, 1974.
25. **Stipkovits, L., Romvary, J., Nagy, Z., Bodon, L., and Varga, L.,** *J. Hyg.,* 72, 289–296, 1974.
26. **Stipkovits, L., Bodon, L., Romvary, J., and Varga, L.,** *Acta Microbiol. Acad. Sci. Hung.,* 22, 45–51, 1975.
27. **Gois, M., Cerny, M., Rozkosny, V., and Sovadina, M.,** *Zentralbl. Veterinaermed. Reihe B,* 16, 253–265, 1969.
28. **Taylor-Robinson, D. and Dinter, D.,** *J. Gen. Microbiol.,* 53, 221–229, 1968.
29. **Edward, D. G. ff.,** *J. Gen. Microbiol.,* 4, 4–15, 1950.
30. **Pan, I. J. and Ogata, M.,** *Jpn. J. Vet. Sci.,* 31, 83–93, 1969.
31. **Krauss, H. and Wandera, J. G.,** *J. Comp. Pathol.,* 80, 389–397, 1970.
32. **Madden, D. L., Hildebrandt, R. J., Moniff, G. R. G., London, W. T., Sever, J. L., and McCullough, N. B.,** *Lab. Anim. Care,* 20, 467–473, 1970.
33. **Tan, R. J. S. and Miles, J. A. R.,** *Br. Vet. J.,* 128, 87–90, 1972.
34. **Rosendal, S. and Laber, G.,** *Zentralbl. Bakteriol. Parasitenkd. Infektionskr. Hyg. Abt. 1 Orig.,* 225, 346–349, 1973.
35. **Allam, N. M., Powell, D. G., Andrews, B. E., and Lemcke, R. M.,** *Vet Rec.,* 93, 402, 1973.
36. **Barile, M. F. and Kern, J.,** *Proc. Soc. Exp. Biol. Med.,* 138, 432–437, 1971.
37. **Kirchhoff, H., Bisping, W., and Floer, W.,** *Berl. Muench. Tieraerztl. Wochenschr.,* 86, 401–403, 1973.
38. **Stipkovits, L., Schimmel, D., and Varga, L.,** *Acta Vet. Acad. Sci. Hung.,* 23, 307–313, 1973.
39. **Low, I. E.,** *Appl. Microbiol.,* 27, 1046–1052, 1974.
40. **Koshimizu, K. and Ogata, M.,** *Jpn. J. Vet. Sci.,* 36, 391–406, 1974.
41. **Hill, A.,** *Vet. Rec.,* 94, 385, 1974.
42. **Langer, P. H. and Carmichael, L. E.,** Proceedings, 67th Annual Meeting of the U.S. Livestock Sanitary Association, 1963, pp. 129–137.
43. **Harbourne, J. F., Hunter, D., and Leach, R. H.,** *Res. Vet. Sci.,* 6, 178–188, 1965.
44. **Gourlay, R. N. and Thomas, L. H.,** *Vet. Rec.,* 84, 416–417, 1969.
45. **Hoare, M. and Haig, D. A.,** *Vet. Rec.,* 76, 956–957, 1964.
46. **Hirth, R. S., Nielsen, S. W., and Tourtellotte, M. E.,** *Infect. Immun.,* 2, 101–104, 1970.
47. **Cole, R. M., Tully, J. G., Popkin, T. J., and Bove, J. M.,** *J. Bacteriol.,* 115, 367–386, 1973.
48. **Chen, T. A. and Liao, C. H.,** *Science,* 188, 1015–1017, 1975.
49. **Williamson, D. L. and Whitcomb, R. F.,** in *Les Mycoplasmes,* Vol. 33, Bove, J. M. and Duplan, J. F., Eds., Colloques INSERM, Paris, 1974, pp. 283–290.
50. **Whitcomb, R. F. and Williamson, D. L.,** *Ann. N.Y. Acad. Sci.,* 266, 260–275, 1975.
51. **Razin, S., Hasin, M., Ne'eman, Z., and Rottem, S.,** *J. Bacteriol.,* 116, 1421–1435, 1973.
52. **Bebear, C., Latrille, J., Fleck, J., Roy, B., and Bove, J. M.,** in *Les Mycoplasmes,* Vol. 33, Bove, J. M. and Duplan, J. F., Eds., Colloques INSERM, Paris, 1974, pp. 35–42.
53. **Davis, R. E. and Worley, J. F.,** *Phytopathology,* 63, 403–408, 1973.
54. **Davis, R. E.,** *Plant Dis. Rep.,* 58, 1109–1112, 1974.
55. **Fudl-Allah, A. A. and Calavan, E. C.,** *Phytopathology,* 64, 1309–1313, 1974.
56. **Williamson, D. L. and Whitcomb, R. F.,** *Science,* 188, 1018–1020, 1975.
57. **Tully, J. G., Whitcomb, R. F., Williamson, D. L., and Clark, H. F.,** *Nature,* 259, 117–120, 1976.
58. **Williamson, D. L.,** *J. Bacteriol.,* 117, 904–906, 1974.
59. **Granados, R. R., Maramorosch, K., and Shikata, E.,** *Proc. Natl. Acad. Sci. U.S.A.,* 60, 841–844, 1968.
60. **Igwegbe, E. C. K. and Calavan, E. C.,** *Phytopathology,* 60, 1525–1526, 1970.
61. **Zelcer, A., Bar-Joseph, M., and Loebenstein, G.,** *Isr. J. Agric. Res.,* 21, 137–142, 1971.
62. **Markham, P. G., Townsend, R., Bar-Joseph, M., Daniels, M. J., Plaskitt, A., and Meddins, B. M.,** *Ann. Appl. Biol.,* 78, 49–57, 1974.
63. **Davis, R. E., Worley, J. F., Whitcomb, R. F., Ishijima, T., and Steere, R. L.,** *Science,* 176, 521–523, 1972.

64. **Davis, R. E., Whitcomb, R. F., Chen, T. A., and Granados, R. R.,** in *Pathogenic Mycoplasmas* (Ciba Foundation Symposium), Elliott, K. and Birch, J., Eds., Elsevier, Amsterdam, 1973, pp. 205–225.
65. **Lafleche, D. and Bove, J. M.,** *Fruits,* 25, 455–465, 1970.
66. **Saglio, P., Lafleche, D., Bonissol, C., and Bove, J. M.,** *C. R. Acad. Sci. Ser. D,* 272, 1387–1390, 1971.
67. **Saglio, P., Lafleche, D., Bonissol, C., and Bove, J. M.,** *Physiol. Veg.,* 9, 569–582, 1971.
68. **Fudl-Allah, A. A., Calavan, E. C., and Igwegbe, E. C. K.,** *Phytopathology,* 62, 729–731, 1972.
69. **Saglio, P., Lafleche, D., L'Hospital, M., Dupont, G., and Bove, J. M.,** in *Pathogenic Mycoplasmas* (Ciba Foundation Symposium), Elliot, K. and Birch, J., Eds., Elsevier, Amsterdam, 1973, pp. 187–198.
70. **Saglio, P., L'Hospital, M., Lafleche, D., Dupont, G., Bove, J. M., Tully, J. G., and Freundt, E. A.,** *Int. J. Syst. Bacteriol.,* 23, 191–204, 1973.
71. ICSB Subcommittee on the Taxonomy of Mycoplasmatales, *Int. J. Syst. Bacteriol.* 25, 237–239, 1975.
72. **Skripal, I. G.,** *Mikrobiol. Zh. Akad. Nauk Ukr. SSR,* 36, 462–467, 1974.
73. **Poulson, D. F. and Sakaguchi, B.,** *Science,* 133, 1489–1490, 1961.
74. **Williamson, D. L.,** *Jpn. J. Genet.,* 44, Suppl. 1, 36–41, 1969.
75. **Clark, H. F.,** *J. Infect. Dis.,* 114, 476–487, 1964.
76. **Clark, H. F.,** *Prog. Med. Virol.,* 18, 307–322, 1974.
77. **Clark, H. F. and Karzon, D. T.,** *Proc. Soc. Exp. Biol. Med.,* 131, 693–696, 1969.
78. **Elizan, T. S., Fabiyi, A., and Clark, H. F.,** *Proc. Soc. Exp. Biol. Med.,* 139, 51–55, 1972.
79. **Pickens, E. G., Gerloff, R. K., and Burgdorfer, W.,** *J. Bacteriol.,* 95, 291–299, 1968.
80. **Tully, J. G., Whitcomb, R. F., Clark, H. F., and Williamson, D. L.,** *Science,* 195, 892–894, 1977.
81. **Chen, T. A. and Granados, R. R.,** *Science,* 167, 1633–1636, 1970.
82. **Jones, A., Whitcomb, R. F., Williamson, D. L., and Coan, M. E.,** *Phytopathology,* in press.
83. **Liao, C. H. and Chen, T. A.,** *Proc. Am. Phytopathol. Soc.,* 2, 100, 1975.
84. **Tully, J. G., Whitcomb, R. F., Bove, J. M., and Saglio, P.,** *Science,* 182, 827–829, 1973.
85. **Lee, I. M., Cartia, G., Calavan, E. C., and Kaloostian, G. H.,** *Calif. Agric.,* 27, 14–15, 1973.
86. **Kaloostian, G. H., Oldfield, G. N., Pierce, H. D., Calavan, E. C., Granett, A. L., Rana, G. L., and Gumpf, D. J.,** *Calif. Agric.,* 29, 14–15, 1975.
87. **Daniels, M. J., Markham, P. G., Meddins, B. M., Plaskitt, A. K., Townsend, R., and Bar-Joseph, M.,** *Nature,* 244, 523–524, 1973.
88. **Daniels, M. J. and Meddins, B. M.,** in *Les Mycoplasmes,* Vol. 33, Bove, J. M. and Duplan, J. F., Eds., Colloques INSERM, Paris, 1974, pp. 195–200.
89. **Markham, P. G. and Townsend, R.,** in *Les Mycoplasmes,* Vol. 33, Bove, J. M. and Duplan, J. F., Eds., Colloques INSERM, Paris, 1974, pp. 201–206.
90. **Rana, G. L., Kaloostian, G. H., Oldfield, G. N., Granett, A. L., Calavan, E. C., Pierce, H. D., Lee, I. M., and Gumpf, D. J.,** *Phytopathology,* 65, 1143–1145, 1974.
91. **Whitcomb, R. F., Williamson, D. L., Rosen, J., and Coan, M.,** in *Les Mycoplasmes,* Vol. 33, Bove, J. M. and Duplan, J. F., Eds., Colloques INSERM, Paris, 1974, pp. 275–282.
92. **Spaar, D., Kleinhempel, H., Muller, H. M., Stanarius, A., and Schimmel, D.,** in *Les Mycoplasmes,* Vol. 33, Bove, J. M. and Duplan, J. F., Eds., Colloques INSERM, Paris, 1974, pp. 207–214.
93. **Granados, R. R.,** in *Viruses, Vectors, and Vegetation,* Maramorosch, K., Ed., Interscience, New York, 1969, pp. 327–359.
94. **Whitcomb, R. F. and Davis, R. E.,** *Annu. Rev. Entomol.,* 15, 405–464, 1970.
95. **Maramorosch, K., Granados, R. R., and Hirumi, H.,** *Adv. Virus Res.,* 16, 135–193, 1970.
96. **Hull, R.,** *Rev. Plant Pathol.,* 50, 121–130, 1971.
97. **Davis, R. E. and Whitcomb, R. F.,** *Annu. Rev. Phytopathol.,* 9, 119–154, 1971.
98. **Hampton, R. O.,** *Annu. Rev. Plant Physiol.,* 23, 389–418, 1972.
99. **Whitcomb, R. F.,** *Proc. North Cent. Branch, Entomol. Soc. Am.,* 28, 38–60, 1973.
100. **Schneider, H.,** *Annu. Rev. Phytopathol.,* 11, 119–146, 1973.
101. **Maramorosch, K.,** *Annu. Rev. Microbiol.,* 28, 301–324, 1974.
102. **Bove, J. M. and Duplan, J. F. (Eds.),** *Les Mycoplasmes,* Vol. 33, Colloques INSERM, Paris, 1974.
103. **Borges, M. L. V.,** *Coleccao Natura, Nova Serie 2,* Sociedad Portuguesa de Ciencias Naturales, Lisboa, 1975, pp. 1–32.
104. **Darland, G., Brock, T. D., Samsonoff, W., and Conti, S. F.,** *Science,* 170, 1416–1418, 1970.
105. **Belly, R. T., Bohlool, B. B., and Brock, T. D.,** *Ann. N.Y. Acad. Sci.,* 225, 94–107, 1973.
106. **Bohlool, B. B. and Brock, T. D.,** *Infect. Immun.,* 10, 280–281, 1974.
107. **Mayberry-Carson, K. J., Roth, I. L., Harris, J. L., and Smith, P. F.,** *J. Bacteriol.,* 120, 1472–1475, 1974.

108. **Belly, R. T. and Brock, T. D.,** *J. Gen. Microbiol.,* 73, 465–469, 1972.
109. **Hsung, J. C. and Haug, A.,** *Biochim. Biophys. Acta,* 389, 477–482, 1975.
110. **Langworthy, T. A., Smith, P. F., and Mayberry, W. R.,** *J. Bacteriol.,* 112, 1193–1200, 1972.
111. **Smith, P. F., Langworthy, T. A., Mayberry, W. R., and Hougland, A. E.,** *J. Bacteriol.,* 116, 1019–1028, 1973.
112. **Mayberry-Carson, K. J., Roth, I. L., and Smith, P. F.,** *J. Bacteriol.,* 121, 700–703, 1975.
113. **Smith, P. F., Mayberry-Carson, K. J., Langworthy, T. A., Mayberry, W. R., and Sugiyama, T.,** in *Les Mycoplasmes,* Vol. 33, Bove, J. M. and Duplan, J. F., Eds., Colloques INSERM, Paris, 1974, pp. 63–67.
114. **Smith, P. F., Langworthy, T. A., and Smith, M. R.,** *J. Bacteriol.,* 124, 884–892, 1975.
115. **Christiansen, C., Freundt, E. A., and Black, F. T.,** *Int. J. Syst. Bacteriol.,* 25, 99–101, 1975.
116. **Searcy, D. G. and Doyle, E. K.,** *Int. J. Syst. Bacteriol.,* 25, 286–289, 1975.
117. **Searcy, D. G.,** *Biochim. Biophys. Acta,* 395, 535–547, 1975.
118. **Robinson, J. P. and Hungate, R. E.,** *Int. J. Syst. Bacteriol.,* 23, 171–181, 1973.
119. **Robinson, I. M., Allison, M. J., and Hartman, P. A.,** *Int. J. Syst. Bacteriol.,* 25, 173–181, 1975.
120. **Robinson, I. M. and Allison, M. J.,** *Int. J. Syst. Bacteriol.,* 25, 182–186, 1975.
121. **Langworthy, T. A., Mayberry, W. R., Smith, P. F., and Robinson, I. M.,** *J. Bacteriol.,* 122, 785–787, 1975.
122. **Razin, S. and Rottem, S.,** in *Biochemical Analysis of Membranes,* Maddy, A. H., Ed., Chapman and Hall, Ltd., London, 1976, pp. 3–26.
123. **Smith, P. F., Langworthy, T. A., and Mayberry, W. R.,** *J. Bacteriol.,* 125, 916–922, 1976.
124. **Freeman, B. A., Sissenstein, R., McManus, T. T., Woodward, J. E., Lee, I. M., and Mudd, J. B.,** *J. Bacteriol.,* 125, 946–954, 1976.
125. **Townsend, R.,** *J. Gen. Microbiol.,* 94, 417–420, 1976.
126. **Roberts, D. H. and Little, T. W. A.,** *Vet. Rec.,* 99, 13, 1976.
127. **Neimark, H.,** in *Les Mycoplasmes,* Vol. 33, Bove, J. M. and Duplan, J. F., Eds., Colloques INSERM, Paris, 1974, pp. 71–78.
128. **Brinton, L. P. and Burgdorfer, W.,** *Int. J. Syst. Bacteriol.,* 26, 554–560, 1976.

MAJOR CULTURE COLLECTIONS AND SOURCES

W. M. O'Leary

Very important tools for microbiologists are reliable sources of typical microbial species in pure culture which can be used as reference bases. There are about the world many culture collections and sources, which vary both in content and quality and can include a variety of microbe-related items such as bacteria, yeasts, fungi, viruses, algae, protozoa, and even tissue cultures. Some are general collections, while others specialize in human pathogens, plant parasites, etc. A complete listing of such sources would far exceed the constraints of this publication, but a listing of the major sources is given below. For a more inclusive listing, the interested reader should consult the excellent catalog of V. F. McGowan and V. B. Skerman (*World Directory of Collections and Cultures of Microorganisms,* 3rd edition, World Data Centre, University of Queensland, Australia, 1986).

Perhaps the world's premier collection of all sorts of microbial and related related materials is that of the American Type Culture Collection (12301 Parklawn Drive, Rockville, Maryland, 20852). This should usually be the first source to consult, since its holdings are vast and its quality is the best of the best.

In addition, there are other sources that can be contacted for cultures in other countries when necessary. These are listed here alphabetically by country and have been chosen as having both reliable and extensive collections. All, of course, can provide catalogs detailing their holdings.

Country	Address
Australia	Department of Microbiology University of Queensland St. Lucia, Brisbane Queensland 4067 Australia
Austria	Mikrobiologisches Institut Penzinger Strasse A 1140 Vienna Austria
Belgium	Laboratorium voor Microbiologie Faculte Wetenschappen Rijksuniversiteit Casinoplein 21 B-9000 Gent Belgium
Brazil	Instituto Zimotecnico Caixo Postal 56 Piraciaba Sao Paulo Brazil
Canada	Canadian Communicable Disease Center Ottawa 3, Ontario Canada
Denmark	WHO International Escherichia Center Statens Serum Institute Copenhagen S, DK-2300 Denmark
France	Centre de Collection de Types Microbiens 62 Boulevard Maréchal Vaillant 59800-Lille France
	Collection de l'Institut Pasteur Institut Pasteur 25 Rue du Dr. Roux 75015 Paris France

MAJOR CULTURE COLLECTIONS AND SOURCES (continued)

Country	Address	Country	Address
Germany (East)	Kulturensammlungen des Zentral Institutes für Microbiologie Deutsche Akademie der Wissenschaften Jena 6900 D.D.R.	Japan	Institute of Applied Microbiology University of Tokyo Tokyo Japan
Germany (West)	Robert Koch Institute Nordufer 20 1 Berlin 65 Bundesrepublic Deutschland	Spain	Coleccion Espanola de Culture Tipo Dept. de Microbiologica Faculdad de Ciencias Universidad de Bilbao Bilbao 10, Vizcaya Spain
	Deutsche Sammlung von Mikroorganismen Zentrale Griesebachstrasse 8 Gottingen 34 Bundesrepublik Deutschland	Sweden	Department of Microbiology Agricultural College Uppsala 7 Sweden
Great Britain	National Collection of Type Cultures Central Public Health Laboratory Colindale Avenue London, NW 9 England	Switzerland	International Centre for Type Cultures Rue de Bugnon Lausanne 17 Switzerland
Greece	Department of Microbiology National University of Athens Athens 609 Hellas		
Hungary	Hungarian National Collection of Medical Bacteria and Viruses National Institute of Public Health Gyali UT 2-6 Budapest IX Hungary	U.S.S.R.	Culture Collections of the U.S.S.R. Research Institute for Antibiotics Moscow M-105 U.S.S.R.
Italy	Instituto de Microbiologia Universita di Parma Ospedale Maggiore 90133 Parma Italy		U.S.S.R. All-Union Collection of Microorganisms Institute of Microbiology U.S.S.R. Academy of Sciences Moscow B-133 U.S.S.R.

YEASTS

J. D. Macmillan and H. J. Phaff

INTRODUCTION

Although the term yeast is used extensively in scientific literature, it does not represent a taxonomic designation that can be rigorously defined. Historically, the word itself originated from ancient words describing the visible changes occurring in fermenting liquids. During the 19th century, when the biological basis for the alcoholic fermentation became firmly established, the organisms responsible (such as *Saccharomyces cerevisiae*) were described as single hyaline round to oval budding cells capable of forming ascospores. After Pasteur related the fermentative process to the ability of yeasts to grow in the absence of air, it was assumed that all yeasts could grow anaerobically. As the years passed, other organisms were discovered that were similar – but not identical – in morphological and physiological properties, and the definition for yeast was expanded to include them. For example, although most yeast cells are hyaline, some – such as *Rhodotorula* – produce red or yellow carotenoid pigments. Budding, too, is a characteristic not common to all yeasts; species of the genus *Schizosaccharomyces* multiply exclusively by fission. Species of several yeast genera, such as *Saccharomycopsis* and *Rhodosporidium,* are capable of forming true mycelium. Even fermentation is not a universal characteristic since many species are incapable of anaerobic growth. Many yeasts are apparently unable to form ascospores or other kinds of sexually produced spores and are, therefore, assigned to the Fungi Imperfecti. Some budding yeasts produce spores that are forcibly discharged from the tips of sterigmata. Although these so-called ballistospores resemble basidiospores, they do not result from a sexual process. Other budding yeasts (*Rhodosporidium* and *Leucosporidium*) have been shown to produce sexual spores, known as sporidia; these yeasts have life cycles similar to the Ustilaginales of the Basidiomycetes.

With all these variations, it is little wonder that it has been difficult to state a precise definition of yeasts on the basis of general morphological and physiological considerations. According to Lodder,[6] "Yeasts may be defined as microorganisms in which the unicellular form is conspicuous and which belong to the fungi." This simple definition is perhaps the only one possible in view of the heterogeneous nature of this group of organisms. In spite of this, the world's leading yeast taxonomists apparently have had little difficulty in deciding what are yeasts. The most recent taxonomic study of 4300 strains led to a 1385-page tome[6] describing 39 genera and 349 species. The major part of the information presented here was obtained from this study. Characteristics of five new genera (*Aessosporon,*[14] *Filobasidium,*[15] *Phaffia,*[16] *Sympodiomyces,*[17] and *Wickerhamiella*[18]), described after the above-named volume was published, have also been included.

The taxonomy of yeasts, as with other groups of microorganisms, is in a continual state of flux. New genera and species are discovered and named. Frequently, yeasts classified earlier are shifted within subgroups, and sometimes new names are assigned. Obviously, such changes are necessary as a reflection of our ever-increasing knowledge about the organisms and their relationships. However, changes are slow to filter into general usage, particularly by scientists other than taxonomists. This has led to some confusing situations in the scientific literature that are difficult to unravel by those who do not have a special interest in taxonomy. For example, one recent change that may cause problems is the deletion of the name *Endomycopsis,* which is now called *Saccharomycopsis.* This change wouldn't cause much difficulty, except that there already existed a totally different group with the name *Saccharomycopsis;* this latter genus has been renamed *Cyniclomyces* (see the next chapter for the reasons for these changes). This new nomenclature has been employed throughout this chapter.

There are four groups of yeasts:

1. Ascomycetous yeasts – yeasts capable of forming ascospores in asci, considered to be primitive ascomycetes: 23 genera.
2. Basidiomycetous yeasts – yeasts having life cycles similar to those of the order Ustilaginales of the Basidiomycetes: 4 genera.
3. Ballistosporogenous yeasts – yeasts in the family Sporobolomycetaceae that forcibly discharge spores by the drop excretion mechanism: 3 genera. Morphologically, ballistospores resemble basidiospores, but they are generally considered to possess an asexual rather than a sexual means of reproduction.
4. Asporogenous yeasts – yeasts incapable of producing ascospores, ballistospores, or sporidia (some species produce asexual spores called endospores): 14 genera. Since sexual life cycles do not occur or have not been observed so far, these yeasts are members of the Fungi Imperfecti.

In addition to the above well-recognized categories, there are a number of microorganisms with yeast-like properties, which will also be included in this section.

The classification of yeasts is based on morphological, cultural, sexual, and physiological characteristics. Some of these properties are shown in Tables 1 to 6.

YEAST MORPHOLOGY

To a considerable extent, the morphology exhibited by a particular yeast is directly associated with the mechanism it employs for asexual reproduction. Various types of asexual or vegetative reproduction are illustrated in Figure 1. Although the majority of yeasts reproduce by budding, fission occurs in some, and in others there is a combination of the two processes. The following terms are used to describe various types of budding: multilateral – buds occurring at different sites on the surfaces of cells; bipolar – buds formed exclusively at the two opposite poles of a cell; monopolar – buds restricted to one pole only; bud fission – a so-called combination of budding and fission, in which a broadly based bud emerges from a yeast cell, usually polarly; later a cross wall forms by centripetal growth across the base of the bud.

Members of several genera, including *Saccharomycopsis* and *Nematospora,* produce true mycelium. The hyphae of some of these filamentous types disarticulate into cells called arthrospores. In some budding yeasts the buds do not detach themselves from one another; the chains of cells that result are reminiscent of filamentous fungal growth and are, therefore, termed pseudomycelia.

In addition to budding, members of the Sporobolomycetaceae characteristically reproduce by production of ballistospores. When agar plates containing colonies of these yeasts are inverted for a short time over fresh plates, spores are forcibly discharged onto the lower plates. Upon germination, these form new colonies that are "mirror images" of the original ones. Members of one genus, *Sterigmatomyces,* reproduce asexually by the formation of single conidia at the ends of sterigmata. These conidia, however, are not forcibly discharged. Conidia are also produced by members of the recently described genus *Sympodiomyces.* In this case, conidiophores and new terminal conidia can develop on the sides of conidia still attached to yeast cells. This process repeats itself, with sympodial formation of several successive conidia.

Table 1
MORPHOLOGICAL PROPERTIES OF ASCOMYCETOUS YEASTS

Genus	Number of species	Cell shape[a]	Budding[b]	Pseudomycelium[c]	Mycelium[c]	Pellicle[c]	Ascospore shape[a]	Ascospore number	Special morphological characteristics
Citeromyces	1	A, B	ML	–	–	–	A	1	Asci have thick walls; ascospore walls are warty.
Coccidiascus	1	A, H	ML		–		I	8	Spores are fusiform, arranged in a helix.
Cyniclomyces	1	C, K	ML	+	–	–	C, K	1–4	Double-walled spore.
Debaryomyces	11	A, B, L	ML	(±)	–	(M)	A, C, D	1–4	Bud and mother cell or two independent cells conjugate.
Dekkera	2	A, B, J, K, L	ML	+	–	(+)	E	1–4	Cells frequently ogival.
Hanseniaspora	5	M, C	BP	F	–	–	E, N, O	1–4	Apiculate cells.
Hansenula	25	A, B, H, K	ML	±	F	±	E, P, A, F	1–4	Colonies slimy, butyrous, or chalky dull.
Kluyveromyces	18	A, B, K, L	ML	M	–	(F)	Q, R, S, A, T	1–many	One species produces numerous ascospores.
Lipomyces	5	B, A	ML	–		–	B, U	1–16	"Active" buds may conjugate or directly form asci; spores are brown.
Lodderomyces	1	A, B, K	ML	+	–	–	S, T	1–2	Large, elongate ascospores.
Metschnikowia	5	A, B, V, K	ML	(M)	–	F	W	1–2	Elongate asci; chlamydospores in some species.

[a] A = spheroidal or globose; B = ellipsoidal; C = ovoidal; D = warty; E = hat-shaped; F = saturn-shaped; G = sickle-shaped; H = oblong; I = fusiform; J = ogival; K = cylindroidal; L = elongate; M = apiculate or lemon-shaped; N = helmet-shaped; O = walnut-shaped; P = hemispheroidal; Q = crescentiform; R = reniform; S = oblong with obtuse end; T = prolate-ellipsoidal; U = oblate-ellipsoidal or lenticular; V = pyriform; W = needle-shaped without appendage; X = spindle-shaped with appendage; Y = hemispheroidal with narrow ledge; Z = cap-shaped.

[b] BP = bipolar budding; ML = multilateral budding; BF = bud fission.

[c] + = all species possess the property; – = no species possess the property; ± = approximately half of species or strains possess the property; M = most species possess the property; F = few species possess the property; above symbols in ( ) = property primitive or weakly exhibited.

Table 1 (continued)
MORPHOLOGICAL PROPERTIES OF ASCOMYCETOUS YEASTS

Genus	Number of species	Cell shape[a]	Budding[b]	Pseudomycelium[c]	Mycelium[c]	Pellicle[c]	Ascospore shape[a]	Ascospore number	Special morphological characteristics
Nadsonia	3	M, C, L	BP, BF	–	–	(±)	A, D	1–2	Bud and mother cell conjugate.
Nematospora	1	Various (heteromorphic)	ML	M	+	–	X	8	Large ascus contains two bundles of four spindle-shaped spores.
Pachysolen	1	B, A	ML	±	–	–	Y	4	Ascus develops at tip of a long tube.
Pichia	35	A, B, H, K	ML	M	F	±	A, E, F, D	1–4	Colonies slimy, pasty, or chalky dull.
Saccharomyces	41	A, B, K, L	ML	(±)	–	–	A, T	1–4	Colonies pasty, semiglossy.
Saccharomycodes	1	M, L.	BP, BF	(±)	–	–	A	4	Large cells; spores conjugate in ascus; spores have a very narrow ledge.
Saccharomycopsis	10	Various	ML	M	+	(F)	A, E, F, G	1–4	Asci at hyphal tips or intercalary.
Schizosaccharomyces	4	A, K	–	–	±	–	A, C	4–8	Cells reproduce by fission.
Schwanniomyces	4	C, A, L, K	ML	–	–	–	O, D	1–2	Meiosis bud.
Wickerhamia	1	C, L, M	BP, BF	–	–	–	Z	1–16	Spores are shaped like sporting caps (usually one or two per ascus).
Wickerhamiella	1	A, B, C	ML	(±)	–	–	H, L	1	Cells may bud on a broad base; spores are small and rugose.
Wingea	1	A, B	ML	–	–	(±)	U	1–4	Lens-shaped spores.

Table 2
PHYSIOLOGICAL PROPERTIES OF ASCOMYCETOUS YEASTS

Genus	Number of species	Fermentation[a]	Nitrate utilization[a]	Growth without vitamins[a]	Acid production[a]	Cycloheximide resistance[a]	Growth at 37°C[a]	Growth on 50% glucose[a]	Growth in 10% NaCl[a]	Special physiological properties
Citeromyces	1	+	+	±	(+)	–	–	+	+	
Coccidiascus	1									Observed in drosophila; not cultivated.
Cyniclomyces	1	(+)		–	–		+	–		Inhabits digestive tract of rabbits; grows only between 30 and 40°C; requires CO_2 and organic nitrogen.
Debaryomyces	11	(M)	–	±			F	+		High salt tolerance.
Dekkera	2	+	±	–	+	+	+	–		Produces acetic acid.
Hanseniaspora	5	+	–	–	+		±	±		All species have an absolute requirement for inositol and pantothenate.
Hansenula	25	M	+	F	F		±		F	Some species produce phosphomannans or sphingolipids.
Kluyveromyces	18	+	–	–		M	±	F		Red (non-carotenoid) pigments may be formed in sporulating cultures.
Lipomyces	5	–	–	±	–		±	–		Cells produce lipids in high concentration.
Lodderomyces	1	(+)	–	±	–	+	+	(+)		Usually utilize higher paraffins.
Metschnikowia	5	M	–	–			F	M		Some species produce pulcherrimin.
Nadsonia	3	+	–	±	±		–	–		Maximum growth at less than 26°C.
Nematospora	1	+	–	–			+	–		Plant-parasitic; produces riboflavine.
Pachysolen	1	+	+	–			+		+	Produces a slimy extracellular phosphomannan.
Pichia	35	M	–	F		F	±	F		One species is common in olive brines; some species form phosphomannan.
Saccharomyces	41	+	–	±		F	±	F		All species ferment strongly.
Saccharomycodes	1	+	–	–	–		–	–		
Saccharomycopsis	10	(M)	F	F			(±)	(F)		Several species grow well on starch.
Schizosaccharomyces	4	+	–	–	±		+	M		No chitin is present in the cell wall.
Schwanniomyces	4	+	–	–	(±)		±	–		All species utilize soluble starch.

[a] + = all species possess the property; – = no species possess the property; ± = approximately half of species or strains possess the property; M = many species possess the property; F = few species possess the property; above symbols in ( ) = property slowly or weakly exhibited.

Table 2 (continued)
PHYSIOLOGICAL PROPERTIES OF ASCOMYCETOUS YEASTS

Genus	Number of species	Fermentation[a]	Nitrate utilization[a]	Growth without vitamins[a]	Acid production[a]	Cycloheximide resistance[a]	Growth at 37°C[a]	Growth on 50% glucose[a]	Growth in 10% NaCl[a]	Special physiological properties
Wickerhamia	1	+	–	–	+	+	–	+		Produces extracellular riboflavine.
Wickerhamiella	1	–	+	–			–	+		Uses very few carbon compounds.
Wingea	1	+	–	(±)		–	(±)	+		

Some yeast species, such as members of the genus *Metschnikowia,* form chlamydospores. They are thick-walled, nondeciduous, intercalary or terminal asexual spores, formed by rounding off a cell or cells. According to van der Walt,[10] the chlamydospores of *Candida albicans* may have a special role as a structure in which meiosis takes place (gonotocont) and are, therefore, not truly chlamydospores at all.

Some members of the genera *Trichosporon* and *Oosporidium* produce endospores in older cultures. These spores are vegetative cells, which are produced within the confines of other cells or hyphae.[2]

The morphology of vegetative yeast cells is sometimes very useful taxonomically, particularly when the cell shape is characteristic of an entire genus. For example, *Trigonopsis* has triangularly shaped cells; *Pityrosporum* is flask-shaped (i.e., because of monopolar budding); *Nadsonia, Hanseniaspora,* and *Saccharomycodes* are apiculate or lemon-shaped because of the build-up of scar material at the poles, due to bipolar budding; *Brettanomyces* and *Dekkera* are frequently ogival (i.e., pointed at one end). Other words describing cell shape, such as spheroidal, ellipsoidal, ovoidal, cylindrical, elongate, etc., are usually self-explanatory. Examples of typical shapes are shown in Figure 2. Reliance on cell shape or size for yeast identification can, however, be misleading, since considerable variation may occur within a single culture. As yeast cells age, they become scarred and misshapen from budding and, therefore, appear considerably different from younger cells.

CULTURAL CHARACTERISTICS

The cultural characteristics on solid or liquid media are sometimes sufficiently unique to be of taxonomic value. Distinctive growth on solid media, such as malt agar, may be a manifestation of hyphal or pseudohyphal growth or due to formation of carotenoids or pulcherrimin, etc. Growth in stationary liquid media results in the formation of a sediment, ring, islets, or pellicle – properties that are readily identifiable and of some value in species characterization.

SEXUAL CHARACTERISTICS

Ascomycetous Yeasts

Yeasts that produce ascospores are either homothallic or heterothallic. The term "homothallic" refers to yeasts (or fungi) in which sexual reproduction can take place with identical nuclei undergoing fusion. The term "heterothallic" refers to the opposite case, in which the fusing nuclei are not identical, because they originate from opposite mating types. Life cycles are further characterized on the basis of the ploidy of the vegetative reproductive stage, which is either haploid or diploid or a mixture of these two phases. Forms with higher ploidy have also been found to exist. In the life cycle of a

Table 3
MORPHOLOGICAL PROPERTIES OF ASPOROGENOUS YEASTS

Genus	Number of species	Cell shape	Budding[a]	Pseudomycelium[b]	Mycelium[b]	Arthrospores[b]	Endospores[b]	Conidia[b]	Chlamydospores[b]	Pigments[b]	Pellicle[b]	Special morphological characteristics
Brettanomyces	7	Ogival, ellipsoidal, spheroidal, elongate	ML	+	–	–	–	–	–	–	(±)	Ogival cells.
Candida	81	Spheroidal, ovoidal, cylindrical, elongate	ML	+	F	–	–	–	F	–	±	All produce pseudomycelium; may produce chlamydospores.
Cryptococcus	17	Spheroidal, ovoidal, elongate	ML	(F)	–	–	–	–	F	±	F	Cells of most species have capsules.
Kloeckera	4	Apiculate, ovoidal, elongate	BP	±	–	–	–	–	–	–	–	Apiculate cells.
Oosporidium	1	Various	ML	+		–	+	–	–	+	–	Produces asexual endospores; chains of cells.
Phaffia	1	Ellipsoidal	ML	(+)	–	–	–	–	+	+	+	Carotenoid pigment, mainly astaxanthin.
Pityrosporum	3	Flask-shaped, spheroidal, ellipsoidal	MP, BF	(F)	(F)	–	–	–	–	–	–	Flask-shaped cells.

[a] BP = bipolar budding; ML = multilateral budding; BF = bud fission.
[b] + = all species possess the property; – = no species possess the property; ± = approximately half of species or strains possess the property; M = most species possess the property; F = few species possess the property; above symbols in ( ) = property primitive or weakly exhibited.

Table 3 (continued)
MORPHOLOGICAL PROPERTIES OF ASPOROGENOUS YEASTS

Genus	Number of species	Cell shape	Budding[a]	Pseudomycelium[b]	Mycelium[b]	Arthrospores[b]	Endospores[b]	Conidia[b]	Chlamydospores[b]	Pigments[b]	Pellicle[b]	Special morphological characteristics
Rhodotorula	9	Spheroidal, ovoidal, elongate	ML	(F)	(F)	–	–	–	F	+	–	Carotenoid pigments include torulene and torularhodin.
Schizoblastosporion	1	Ellipsoidal, cylindrical, flask-shaped	BP, BF	±	–	–	–	–	–	–	–	Flask-shaped cells.
Sterigmatomyces	3	Spheroidal, ovoidal	–	–	–	–	–	+	–	–	+	Multiplies by cells on stalks (conidia).
Sympodiomyces	1	Various	–	–	(+)	–	–	+	–	–	–	Sympodial growth, forming successive terminal conidia.
Torulopsis	36	Spheroidal, ovoidal, elongate	ML	(F)	–	–	–	–	–	–	F	Few produce rudimentary pseudomycelium.
Trichosporon	8	Various	ML	+	+	+	F	–	–	–	±	Some species produce endospores (asexual).
Trigonopsis	1	Triangular, ellipsoidal		–	–	–	–	–	–	–	+	Triangular-shaped cells bud at apices.

Table 4
PHYSIOLOGICAL PROPERTIES OF ASPOROGENOUS YEASTS

Genus	Number of species	Fermentation[a]	Nitrate[a]	Starch synthesis[a]	Growth without vitamins[a]	Acid production[a]	Growth at 37°C[a]	Growth on 50% glucose[a]	Growth in 10% NaCl[a]	Assimilate inositol[a]	Gelatin liquefaction[a]	Special physiological properties
Brettanomyces	7	+	±		–	+	+	–		–		Produce acetic acid; resistant to cycloheximide; growth is slow.
Candida	81	M	±		(M)		M		±	F		
Cryptococcus	17	–	±	M	F	–	F	–		+	F	Inositol assimilated.
Kloeckera	4	+	–	–	–	±	–	±		–	(±)	Absolute requirement for inositol and pantothenate.
Oosporidium	1	–	+	(+)	–		–	–				
Phaffia	1	+	–	+	–	(+)	–	–	–	–		Carotenoid pigments; fermentation.
Pityrosporum	3	–					+					Requires lipids for growth.
Rhodotorula	9	–	±	–	F	–	±	–		–	F	Carotenoid pigments; no fermentation.
Schizoblastosporion	1	–	–		–		–	–		–		
Sterigmatomyces	2	–	±	–	–		–			–		
Sympodiomyces	1	–	–	–	–	–	–	–		(+)		
Torulopsis	36	M	±	–	(±)	F	M	±	±	–		
Trichosporon	8	(F)	F	F	±	–	M	–		±		
Trigonopsis	1	–	–	–	–		+	–		–		

[a] + = all species possess the property; – = no species possess the property; ± = approximately half of species or strains possess the property; M = many species possess the property; F = few species possess the property; above symbols in ( ) = property slowly or weakly exhibited.

Table 5
MORPHOLOGICAL PROPERTIES OF BASIDIOMYCETOUS AND BALLISTOSPOROGENOUS YEASTS

Genus	Number of species	Cell shape	Budding[a]	Pseudomycelium[b]	Mycelium[b]	Pellicle formation[b]	Teliospore formation[b]	Ballistospore formation[b]	Pigment formation[b]	Special morphological characteristics
				BASIDIOMYCETOUS YEASTS						
Leucosporidium	6	Ovoidal, elongate	B	M	+	–	+	–	–	Perfect form of certain *Candida*-like organisms.
Rhodosporidium	8	Spheroidal, ovoidal, elongate	B	±	+	–	+	–	+	Perfect form of *Rhodotorula.*
Aessosporon	2	Ovoidal, elongate, cylindrical	B	+	±	+	+	+	+	Asymmetrical ballistospores; perfect form of *Sporobolomyces.*
Filobasidium	2	Ovoidal, elongate	B	+	+	–	–	–	–	Forms *Tilletia*-like basidia with sessile basidiospores.
				BALLISTOSPOROGENOUS YEASTS						
Bullera	3	Spheroidal, ovoidal	B	–	–	–	–	+	±	Symmetrical ballistospores.
Sporidiobolus	2	Ovoidal, elongate, cylindrical	B	(±)	+	+	?	+	+	Asymmetrical ballistospores; brown, thick-walled chlamydospores.
Sporobolomyces	9	Ovoidal, elongate	B	±	±	±	–	+	M	Asymmetrical ballistospores.

[a] B = budding.
[b] + = all species possess the property; – = no species possess the property; ± = approximately half of species or strains possess the property; M = most species possess the property; above symbols in ( ) = property primitive or weakly exhibited.

typical haploid yeast, budding cells are predominantly haploid, and ascus formation occurs immediately following conjugation of the cells. The process of spore formation occurs so rapidly after conjugation that the shape of the ascus that results is reminiscent of the shape assumed by the two cells as they are engaged in the conjugation process. For example, in some haploid strains of *Saccharomyces* and *Schizosaccharomyces,* plasmogamy is initiated at the tips of tube-life outgrowths. Shortly after the union, karyogamy occurs, followed by meiosis, and spores are formed in asci that are characteristically elongated or dumbbell-shaped. Upon germination, the haploid spores produce haploid budding vegetative cells, and the cycle is complete. The life cycle of such a haploid yeast is shown in Figure 3a.

In diploid yeasts, conjugation followed by karyogamy takes place shortly after spore germination. Spore formation, however, is delayed. The diploid cells or zygotes may bud for many generations, producing additional diploid vegetative cells until conditions are

Table 6
PHYSIOLOGICAL PROPERTIES OF BASIDIOMYCETOUS AND BALLISTOSPOROGENOUS YEASTS

Genus	Number of species	Fermentation	Nitrate utilization[a]	Growth without vitamins[a]	Growth on 50% glucose[a]	Growth at 17°C[a]	Growth at 19°C[a]	Growth at 30°C[a]	Growth at 37°C[a]	Acid production[a]	Starch production[a]	Urea hydrolysis[a]	Gelatin liquefaction[a]
BASIDIOMYCETOUS YEASTS													
Leucosporidium	6	(±)	M	F	–	+	(±)	F	–	F	M	+	M
Rhodosporidium	8	–	+	±	–	+	+	+	(±)	–	–	+	±
Aessosporon	2	–	+	+	–	+	+	+	–	–	–	+	–
Filobasidium	2	±	±	–	–	+	+	+	(+)	(+)	+		–
BALLISTOSPOROGENOUS YEASTS													
Bullera	3	–	±	±	–	+	+	–	–	–	±	+	–
Sporidiobolus	2	–	+	+	–	+	+	±	±	–	–	+	(F)
Sporobolomyces	9	–	±	M	–	+	+	±	F	–	–	+	F

[a] + = all species possess the property; – = no species possess the property; ± = approximately half of species or strains possess the property; M = many species possess the property; F = few species possess the property; above symbols in ( ) = property slowly or weakly exhibited.

finally suitable for meiosis. Ascospores are formed within the cells, and the asci are usually about the same size and shape as the original diploid budding vegetative cells from which they were derived (Figure 4).

The life cycles described above are typical for haploid and diploid ascomycetous yeasts. Other life cycles are simply variations on these two themes. Three variations are known to occur among yeasts that spend their vegetative life as haploid cells (Figures 3a to c):

a. Conjugation tubes and dumbbell-shaped asci are formed, as previously described. Examples: *Schizosaccharomyces pombe* and *Saccharomyces elegans.*
b. A haploid cell produces a special bud, called a "meiosis" bud, which does not separate from the original cell. The nucleus of the mother cell divides mitotically, and both nuclei migrate into the daughter bud, where karyogamy and meiosis take place. The four nuclei migrate back into the original cell, and spore formation occurs. Usually, two of the four nuclei do not produce ascospores. (The life cycle depicted was for members of the genus *Schwanniomyces.* Similar life cycles may occur in species of *Hansenula, Pichia,* and some *Saccharomyces.*)
c. Two gametes fuse. (The case depicted is *Nadsonia,* where a mother cell and daughter cell conjugate.) The dikaryotic cell thus formed does not become the ascus. Instead,

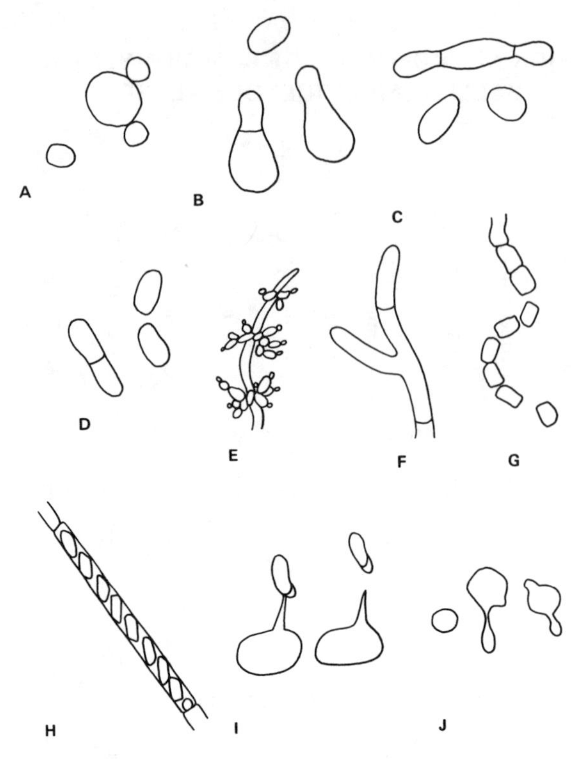

FIGURE 1. Kinds of asexual reproduction known to occur in yeasts. A. Multilateral budding. B. Monopolar budding. C. Bipolar budding, bud fission. D. Fission. E. Pseudomycelium. F. Mycelium. G. Arthrospores. H. Endospores. I. Ballistospores. J. Conidia.

FIGURE 2. Various shapes of yeast cells. From left to right: spheroidal; ovoidal; cylindrical; ogival; triangular; flask-shaped; apiculate. (From Phaff, H. J., Miller, M. W., and Mrak, E. M., *The Life of Yeasts,* Harvard University Press, Cambridge, Mass., 1966. With permission.)

karyogamy and ascospore formation occur in another specialized bud, which grows out of the opposite end of the dikaryotic structure. (Other examples for this type of life cycle can be found in *Saccharomycopsis* and in the yeast-like fungus *Eremascus.*)

There are four kinds of life cycles for yeasts that grow vegetatively as diploid cells (Figures 4a to d):

1. Two ascospores conjugate directly in the ascus, and the first bud from this zygote is a diploid. Example: *Saccharomycodes ludwigii.*
2. Ascospores may germinate and bud as haploid cells for a short time prior to conjugation. Example: strains of *Saccharomyces cerevisiae.*
3. Some spores germinate, bud for a while, and then one of the cells fuses with an ungerminated spore. Example: strains of *Saccharomyces cerevisiae.*

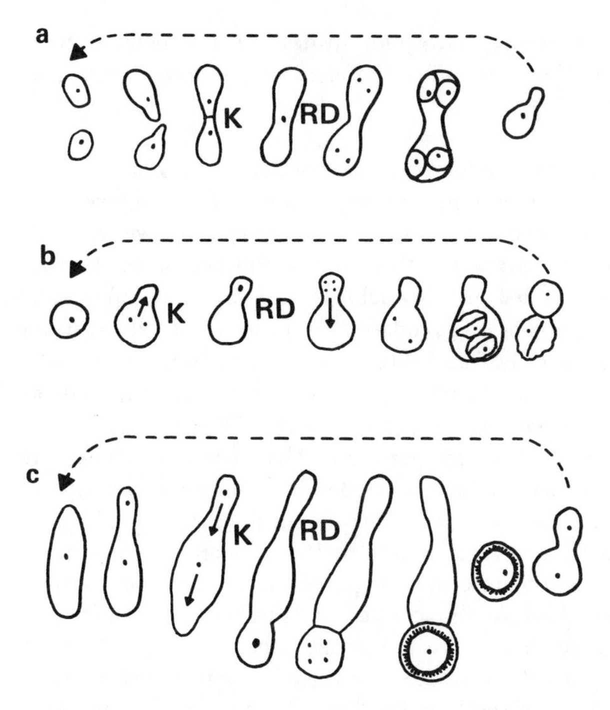

FIGURE 3. Life cycles of haploid yeasts. The heavier dots represent diploid nuclei; the lighter dots are haploid nuclei. K = karyogamy; RD = reduction division. (From Phaff, H. J., Miller, M. W., and Mrak, E. M., *The Life of Yeasts,* Harvard University Press, Cambridge, Mass., 1966. With permission.)

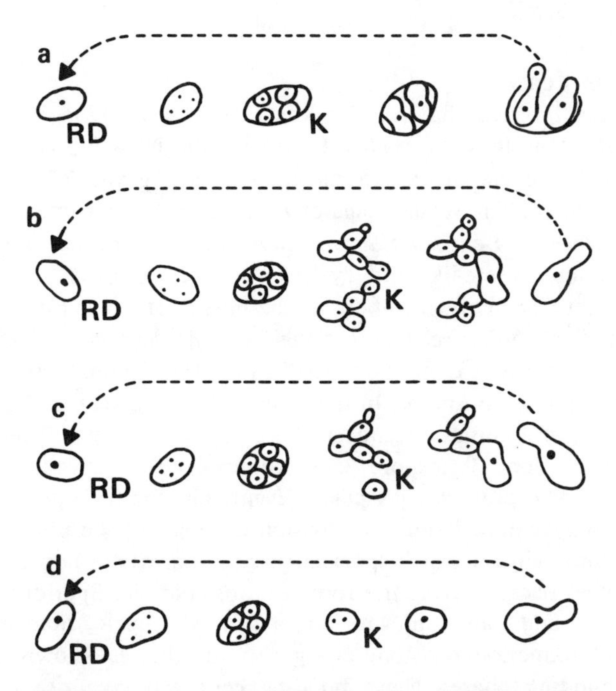

FIGURE 4. Life cycles of diploid yeasts. The heavier dots represent diploid nuclei; the lighter dots are haploid nuclei. K = karyogamy; RD = reduction division. (From Phaff, H. J., Miller, M. W., and Mrak, E. M., *The Life of Yeasts,* Harvard University Press, Cambridge, Mass., 1966. With permission.)

4. The nucleus in a swelling ascospore divides into two haploid nuclei that fuse prior to germination of the spore into a diploid cell (*Saccharomyces chevalieri* and some species of *Hanseniaspora*).

It is noteworthy that various yeasts cannot be categorized as strictly haploid or strictly diploid. In some yeast cultures, both haploid and diploid vegetative cells may exist together. For a more complete discussion of life cycles in yeasts, see References 4 and 7.

Ascomycetous yeasts are members of a primitive group known as the Hemiascomycetidae. Also included in this subclass are mycelial saprobes, such as *Dipodacus aggregatus,* which grows in tree exudates, and other yeast-like organisms, such as *Taphrina deformans,* which cause leaf curl disease in certain plants. Species of *Taphrina* reproduce by budding in culture, but form mycelium in the host plant. The ascomycetous yeasts have been differentiated from other members of the Hemiascomycetidae partly on the basis of the number of spores per ascus. Most yeasts produce 1 to 4 ascospores. By comparison, *Dipodacus* produces an indefinite number of ascospores (up to 100). This differentiation has not stood up taxonomically, since some species in the genera *Lipomyces* and *Kluyveromyces* can be multispored and produce more than 16 spores per ascus. The formation of more than 4 spores per ascus has been attributed to postmeiotic mitoses within the ascus. Ascomycetous yeasts usually do not produce the well-developed true mycelia that are characteristic of many other organisms in the Hemiascomycetidae.

The taxonomy used in Lodder[6] assigns ascomycetous yeasts to a single order, Endomycetales; three of the four families of this order contain yeast genera (Figure 5). Admittedly, this is not the final word, since considerable information is lacking for a natural phylogenetic classification of yeasts.

Considerable variation in the shapes of ascospores is encountered among different yeast species. Spore morphology is a fairly consistent property, which is useful in species identification. In some cases, all of the species in an entire genus have essentially the same spore shape. Typical shapes of spores are shown in Figure 6.

Basidiomycetous Yeasts

It has long been inferred that certain yeasts, such as *Rhodotorula* and *Sporobolomyces,* are related to the Basidiomycetes. It was not until 1967, however, when Banno[1] reported a sexual life cycle in certain members of the genus *Rhodotorula,* that this suspicion was confirmed. The sexual stage of *Rhodotorula* was named *Rhodosporidium.* Three additional genera – *Leucosporidium, Filobasidium* and *Aessosporon* – have been described with basidiomycete-like life cycles. Yeasts in the first two of these newly described genera were formerly members of the imperfect genus *Candida,* and those in the third genus of *Sporobolomyces,* before their sexual stages were elucidated. Members of this group of yeasts can be either homothallic or heterothallic. Characteristically, they form thick-walled diploid teliospores. In the heterothallic strains, haploid budding cells of compatible mating types conjugate and give rise to a dikaryotic mycelial phase, which exhibits clamp formation. Teliospores are produced either terminally or within the hyphal strands, and karyogamy takes place. Eventually the teliospores germinate, with formation of a promycelium. Reduction division occurs, and the promycelium becomes septate, forming four cells on which sporidia (basidiospores) are borne. Segregation into original mating types occurs during the formation of sporidia. Sporidia can reproduce as budding yeast cells and can conjugate and repeat the cycle. Sometimes teliospores germinate without reduction division, giving rise to what is known as a uninucleate self-sporulating budding phase. These budding cells can give rise to a uninucleate mycelium (usually without clamp connections), which produces teliospores, promycelia, and sporidia. In many cases the life cycles of the self-sporulating phase have not been completely worked out, so that the ploidy of these structures is unknown. In the case of

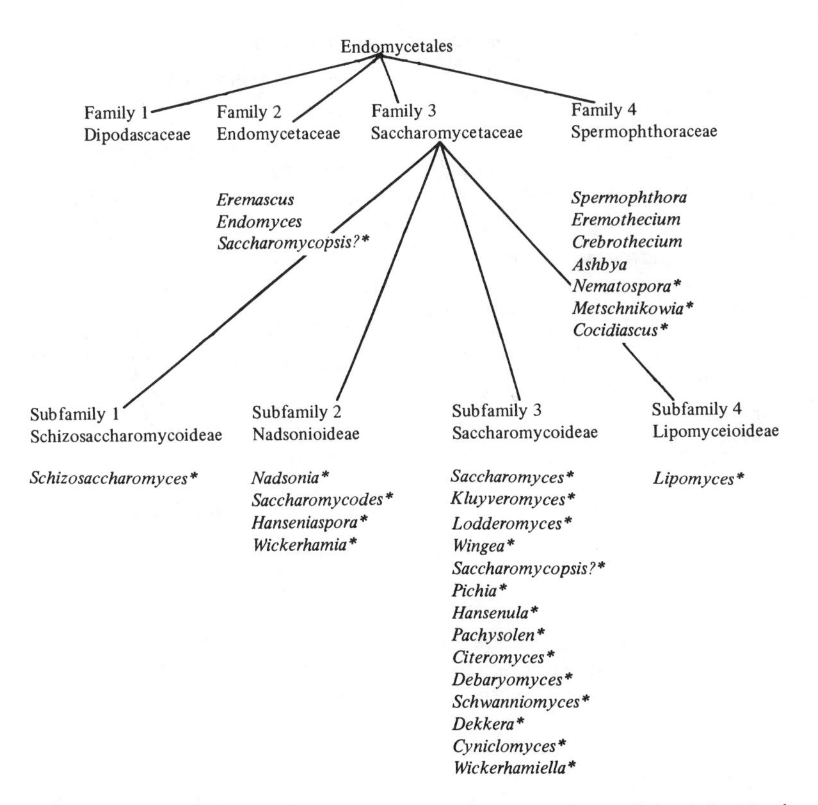

FIGURE 5. Classification of the ascomycetous yeasts according to Lodder.[6] Asterisked genera are yeasts. (From Lodder, J. (Ed.), *The Yeasts – A Taxonomic Study,* American Elsevier, New York, 1970. With permission.)

FIGURE 6. Shapes of various ascospores produced by yeasts. From top, left to right: spheroidal; ovoidal; reniform; crescent- or sickle-shaped; hat-shaped; helmet-shaped; spheroidal with warty surface; walnut-shaped; saturn-shaped; spheroidal with spiny surface; needle-shaped without appendage; spindle-shaped with appendage. (From Phaff, H. J., Miller, M. W., and Mrak, E. M., *The Life of Yeasts,* Harvard University Press, Cambridge, Mass., 1966. With permission.)

Rhodosporidium sphaerocarpum (Figure 7), sporidia from the self-sporulating phase are presumably haploid, since they mate with the appropriate haploid cells of the original mating types of this species.

Aessosporon salmonicolor formerly was classified as a ballistosporogenous yeast, *Sporobolomyces.* It was found to produce teliospores that germinated into a nonseptate promycelium, bearing two to four sporidia.[11] This homothallic yeast does not produce a dikaryotic mycelium phase. According to the postulated life cycle, ballistospores and buds are produced by both haploid- and diploid-phase cells.

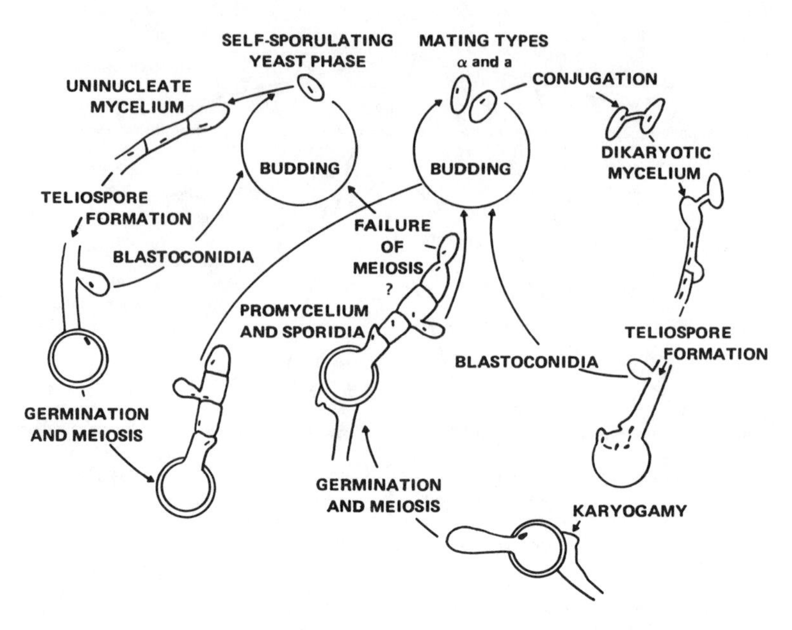

FIGURE 7. Life cycle of *Rhodosporidium sphaerocarpum.* (From Fell, J. W., Phaff, H. J., and Newell, S. Y., in *The Yeasts – A Taxonomic Study,* Lodder, J., Ed., American Elsevier, New York, 1970, pp. 801–814. With permission.)

Asporogenous Yeasts

"Asporogenous" is probably an inappropriate term for describing this group of yeasts, since certain strains produce arthrospores and other endospores or conidia. None, however, produce sexually derived spores, such as ascospores or sporidia, and none produce forcibly discharged spores (ballistospores). All genera reproduce either by budding or by fission, except two, *Sterigmatomyces* and *Sympodiomyces,* which are perhaps not yeasts, because they reproduce exclusively by formation of conidia.

Some genera are essentially the same as ascomycetous genera, except for the absence of formation of ascospores. For example, *Dekkera* is the perfect form of the asporogenous genus *Brettanomyces,* and *Hanseniaspora* is that of *Kloeckera.* It is natural that the asporogenous group of yeasts has become a catchall for those yeasts in which sexuality does not exist or simply has never been observed. As more information is obtained about specific strains, however, sexual processes sometimes are discovered, and the participating strains are removed from this group. (For example, sexually active strains of *Rhodotorula* are now called *Rhodosporidium.*)

Other forms of sexuality that do not include production of ascospores or sporidia have been suggested as occurring in certain members of this group. For example, van der Walt[12] reported that a kind of sexual cycle occurs in both *Candida albicans* and *Cryptococcus albidus.* Conjugation took place in certain homothallic strains. Zygotes can form chlamydospores, or they can give rise to a diploid generation that produces large, dormant, multispored cells. Ascospores are not produced, but the haploid generation probably is reestablished from the conidia found on these multinucleate cells. According to van der Walt, these multispored cells might be similar to basidia. Lodder,[6] however, suggests that this could be true for *Cryptococcus albidus,* but not for *Candida albicans,* which is an ascomycete-like yeast on the basis of DNA base composition, cell wall structure, and serological analysis.

Wickerham[13] has proposed that a sort of parasexual cycle is exhibited by certain asporogenous yeasts. He coined the term "protosexual" to describe species in which diploid vegetative cells give rise to haploid cells without the production of fruiting bodies or sexual spores. Unfortunately, the strains in which this type of sexuality was observed later were found to form ascospores; therefore, it is not known whether there are exclusively "protosexual" yeasts, as originally proposed by Wickerham.

PHYSIOLOGICAL CHARACTERISTICS

Although physiological characteristics generally are of limited value in strict delimitation of genera, they are frequently quite useful in narrowing down the field. For example, relatively few ascomycetous yeasts utilize nitrate as a sole source of nitrogen, and of these, *Hansenula* species are probably the most commonly encountered. As another example, until quite recently one could assume that the pigmented (carotenoid) yeasts, such as *Rhodotorula, Rhodospiridium* and *Sporobolomyces,* were all strictly oxidative and unable to ferment sugars. However, in 1972, certain pigmented yeasts, newly isolated from tree exudates, were able to ferment several sugars. On the basis of this fermentative ability and the fact that the pigment was mainly astaxanthin, a carotenoid not previously encountered in yeasts, a new genus named *Phaffia* was described.

A knowledge of physiological properties is essential in the identification of species. The most important properties are those related to fermentation and assimilation of carbon sources, nitrogen utilization, vitamin and temperature requirements, and ability to grow in high sugar or salt concentrations. Susceptibility to fungal antibiotics also has been employed. Recently, criteria for delimiting relationships among various yeasts have been based on immunological properties, analysis of base composition of deoxyribonucleic acid, and proton magnetic resonance spectra of cell wall components.

Many yeasts are capable of fermenting various carbon sources. There are probably no strictly anaerobic yeasts, and the same carbon sources that a yeast can ferment can also be assimilated oxidatively under appropriate conditions. The reverse is not true, in that ability to assimilate a certain carbon source does not mean that it necessarily can be fermented. All fermentative yeasts are able to ferment glucose, producing ethanol and CO_2. Generally, fermentation tests are conducted in tubes containing a basal medium (such as 0.5% yeast extract) and 2% sugar, and inverted vials (Durham tubes) are used for gas collection.

Carbon assimilation tests can be performed either in liquid or on solid medium, using a chemically defined basal medium, such as yeast nitrogen base (Difco), which contains ammonium sulfate as a nitrogen source and all vitamins, amino acids, and trace elements known to be required by yeasts. In the auxanographic method, an agar basal medium is seeded with a suspension of the yeast and plated. Small amounts of various carbon sources are then placed in specific locations on the dry agar surface. Plates are incubated for a few days, and growth occurs on those carbon sources that can be assimilated. Replicate plating methods have been designed for use when large numbers of yeast strains are to be tested.

Similar procedures have been employed for testing the ability to utilize various nitrogen sources. In this case, however, the basal medium lacks a nitrogen source and contains glucose as a carbon source (e.g., Yeast Carbon Base, Difco). Almost all yeasts are able to utilize peptone, asparagine, ammonium sulfate, and urea. The ability to utilize potassium nitrate, sodium nitrite, aliphatic amines, or amino acids is useful in the characterization of various yeast isolates.

A more complete description of these methods and descriptions of other physiological tests – such as splitting of arbutin, growth in vitamin-free media, growth in media of high osmolarity, growth at elevated temperatures, acid production, starch production, hydrolysis of urea, fat splitting, pigment formation, ester production, cycloheximide resistance, and gelatin liquefaction – have been published by van der Walt.[10]

YEAST-LIKE ORGANISMS

The literature abounds with descriptions of yeast-like organisms that yeast taxonomists have either refused to accept or that have at one time been accepted and later rejected. These include borderline genera, such as *Endomyces* (and its imperfect form *Geotrichum*), which would be similar to the yeast *Schizosaccharomyces* if it did not produce such extensive mycelium and gametangia, and related to *Saccharomycopsis* if it were also to reproduce by budding. Likewise, *Ashbya* might be a yeast if budding were not so "rare and atypical." The so-called "black yeasts" (*Pullularia, Aureobasidium*) produce mycelial phases and budding yeast phases, but because of the production of a black pigment, they are not accepted as yeasts. Other organisms are definitely not yeasts, yet they resemble them in colony morphology. For example, *Prototheca* is actually a colorless alga, probably related to *Chlorella;* the spheroidal to ellipsoidal cells of these organisms do not bud, but instead reproduce vegetatively by partitioning of the protoplasm into two or several irregularly shaped spore-like bodies (aplanospores). Other genera that can be confused with yeasts are *Taphrina, Eremothecium, Dipodascus,* and *Eremascus.* For brief descriptions of these genera and reasons for their exclusion from the yeasts, consult Lodder.[6]

YEASTS AND MAN

Throughout the last several thousand years, man has capitalized on the products formed by yeast cells. In addition to leavened bread, beer, wine, sake, and other forms of potable alcohol, yeasts have provided man with other products, such as glycerol, enzymes, coenzymes, and vitamins. Today there is considerable interest in employing yeast cells for upgrading waste materials into utilizable forms of single-cell protein. Yeasts have been convenient tools for biochemists, physiologists, geneticists, cell biologists, and other scientists. Indeed, the initiation of the fields of biochemistry and nutrition was based on the discovery of enzymes and vitamins in yeast cells.

The main genera of commercial significance are *Saccharomyces* and *Candida.* Frequently, species were named on the basis of the fermentation they were associated with – e.g., *Saccharomyces sake* and *Saccharomyces vini.* The morphological and physiological properties of many of these species are very similar, and separate names have not been justified from a taxonomic point of view. Special characteristics of certain yeast strains are extremely desirable for various industrial processes; therefore, strain differentiation may be more important than species classification.

A partial listing of yeast species either used or suitable for the manufacture of various products is shown in Table 7. Other commercial products isolated from *Saccharomyces cerevisiae* include alcohol dehydrogenase, hexokinase, L-lactate dehydrogenase, glucose-6-phosphate dehydrogenase, glyceraldehyde-3-phosphate dehydrogenase, inorganic pyrophosphatase, coenzyme A, oxidized and reduced diphosphopyridine nucleotides, and the mono-, di-, and triphosphates of adenosine, cytidine, guanosine, and uridine.[8]

Under certain circumstances, yeasts may cause infections of man and animals.[5] Treatment of patients with bactericidal antibiotics as well as the nutritional status of the infected individual is often of prime importance for the opportunistic invasion by yeasts. Table 8 is a listing of important pathogenic yeasts. Diseases range from superficial infections of cutaneous and mucosal sites to serious systemic diseases involving the viscera and circulatory fluids.

Yeasts have also been a detriment to man by causing food spoilage. Yeasts ordinarily do not compete well in mixed populations, and therefore cause spoilage under conditions that are adverse for the growth of other organisms. Bacteria usually outgrow yeasts at pH values in the neutral or slightly acidic range, but below pH 5 yeasts are readily able to compete. A listing of important food spoilage yeasts is shown in Table 9. An extensive discussion of food spoilage yeasts was published by Walker and Ayres.[9]

Table 7
INDUSTRIALLY IMPORTANT YEAST SPECIES[a]

Product	Yeast employed
Baker's yeast	*Saccharomyces cerevisiae*
Food yeasts	*Candida utilis* *Candida tropicalis* *Kluyveromyces fragilis* *Saccharomyces carlsbergensis* *Saccharomyces cerevisiae*
Protein (from hydrocarbons)	*Candida lipolytica* *Candida tropicalis* *Trichosporon japonicum*
Liquors and industrial alcohol	*Saccharomyces cerevisiae* (and others)
Lager beer	*Saccharomyces carlsbergensis*
Ale	*Saccharomyces cerevisiae*
Sakē	*Saccharomyces cerevisiae*
Wine	*Saccharomyces cerevisiae* *Saccharomyces fermentati* *Saccharomyces bayanus* Others
Yeast autolysates and extracts	*Saccharomyces cerevisiae* *Candida utilis*
Lipid	*Rhodotorula gracilis* (syn. *R. glutinis*) *Candida utilis* *Metchnikowia pulcherrima* *Metchnikowia reukaufii*
Invertase	*Saccharomyces cerevisiae* *Candida utilis*
Amylase	*Saccharomycopsis fibuligera* *Saccharomycopsis capsularis*
Lactase	*Kluyveromyces fragilis* *Kluyveromyces lactis* *Candida pseudotropicalis*
Uricase	*Candida utilis*
Polygalacturonase	*Kluyveromyces fragilis*
Ergosterol	*Saccharomyces cerevisiae* *Saccharomyces carlsbergensis* *Saccharomyces bayanus*
Ribonucleic acid	*Candida utilis* *Candida tropicalis*

[a] These data were compiled from appropriate sections of Reference 8.

Table 7 (continued)
INDUSTRIALLY IMPORTANT YEAST SPECIES[a]

Product	Yeast employed
Riboflavins	*Eremothecium ashbyi*
Lysine	*Candida utilis*
Cystine	*Rhodotorula gracilis*
Methionine	*Rhodotorula gracilis*

Table 8
YEASTS PATHOGENIC TO HUMANS[a]

Candida albicans	Candidiasis: superficial infections of various sites on the skin and in the digestive, genital, urinary, and respiratory tracts (for example, oral thrush, chronic paronychia, vaginitis, etc.), which can lead to deep-seated infections of individual viscera and the bloodstream.
Candida tropicalis	Associated with vulvovaginitis and with other forms of candidiasis; also free-living.
Candida stellatoidea	Associated with vulvovaginitis, also with thrush, and occasionally with other infections.
Candida pseudotropicalis	Frequently respiratory infections; rarely cutaneous lesions and septicaemia; also free-living. Imperfect form of *Kluyveromyces fragilis.*
Candida parapsilosis	Cutaneous, mucosal, and deep-seated lesions, especially on heart valves.
Candida guilliermondii	Implicated in infections of nails, heart, and blood; also free-living.
Candida krusei	Implicated in infections of vagina, mouth, heart, and blood; also occurs as a saprobe in the intestinal tract of certain warm-blooded animals or free-living.
Cryptococcus neoformans	Cryptococcosis: infections may occur in the skin, lungs, or other parts of the body; frequently the central nervous system is involved, causing a form of meningitis.
Torulopsis glabrata	Possible pathogen isolated from the urinary tract, skin, epidermal scales, nails, and genital secretions.
Torulopsis famata	Possible pathogen isolated from various skin lesions and genital secretions.
Pityrosporum orbiculare	Related to pityriasis versicolor (fine scales, lesions on skin).
Trichosporon cutaneum	Causes white piedra (soft white nodules on hair); also free-living in a large range of ecological niches.
Trichosporon capitatum	Possibly associated with respiratory complaints, such as bronchitis and asthma.

[a] Information in this table was obtained from Gentles and Touche.[5]

Table 9
TYPICAL FOOD SPOILAGE YEASTS[a]

Species	Source of isolation
Brettanomyces spp.	Beer, wine, cucumber brines
Candida catenulata	Frankfurters
Candida guilliermondii	Dates
Candida krusei	Fresh figs, dates, tomatoes, food brines, pickle brines
Candida lipolytica	Frankfurters, margarine
Candida pseudotropicalis	Cream, cheese, butter, sour milk
Candida vini	Wine, beer
Candida zeylanoides	Frankfurters, beef
Citeromyces matritensis	Condensed milk, fruit in syrup
Debaryomyces hansenii	Sausage, frankfurters, food brines, meat brines, sakē, lunch meats, tomato puree
Endomycopsis fibuligera	Spoiled starchy foods
Hanseniaspora uvarum	Tomatoes, fresh figs, grapes, etc.
Hanseniaspora valbyensis	Dates, fresh figs, fresh fruits
Hansenula subpelliculosa	Cucumber brines, foods of high sugar content
Kloeckera apiculata	Tomatoes, strawberries, fruit juices, fresh figs
Kluyveromyces fragilis	Fresh figs, yogurt and other dairy products
Kluyveromyces lactis	Cream, cheese, milk
Pichia kluyveri	Tomatoes
Pichia membranefaciens	Food brines, wine, olive brines
Rhodotorula spp.	Oysters, crabs, dairy products, olive brines
Saccharomyces bailii	Wine, apple juice, salad dressing, mayonnaise
Saccharomyces bisporus	Honey, maple sugar, cucumber brines, salted beans
Saccharomyces carlsbergensis	Figs
Saccharomyces chevalieri	Fruit juice, wine
Saccharomyces rouxii	Honey, maple syrup, citrus juice, sugar cane juice, dates, prunes, dried figs, jelly, strawberry juice, candied fruits, cucumber brines
Saccharomyces uvarum	Maple sugar, sugar cane juice, honey, sugar beet juice, fruit juice, lemonade

[a] Most information in this table was obtained from Walker and Ayres.[9] Many names of the yeasts listed there were changed here to conform to the taxonomy listed in Lodder,[6] from which additional examples of food spoilage yeasts were also obtained. Incorrect nomenclature may be responsible for unlikely habitats in a few instances.

Table 9 (continued)
TYPICAL FOOD SPOILAGE YEASTS[a]

Species	Source of isolation
Schizosaccharomyces octosporus	Prunes, figs, raisins
Schizosaccharomyces pombe	Sugar cane juice
Torulopsis lactis-condensi	Condensed milk, food brines
Torulopsis sphaerica	Cream, butter

REFERENCES

1. **Banno, I.,** Studies on sexuality of *Rhodotorula, J. Gen. Appl. Microbiol.,* 13, 167, 1967.
2. **do Carmo-Sousa, L.,** Proceedings of the Second International Symposium on Yeasts, Vydavatel'sto Slovenskej Akademie Vied, Bratislava, Czechoslavakia, 1966, p. 87.
3. **Fell, J. W., Phaff, H. J., and Newell, S. Y.,** Genus 2, *Rhodosporidium* Banno, in *The Yeasts – A Taxonomic Study,* Lodder, J., Ed., North Holland, Amsterdam, 1970, pp. 801–814.
4. **Fowler, R. R.,** Life cycles in yeasts, in *The Yeasts,* Vol. I, *The Biology of Yeasts,* Rose, A. H. and Harrison, J. S., Eds., Academic Press, London, England, and New York, 1969, pp. 461–471.
5. **Gentles, J. C. and Touche, C. J. L.,** Yeasts as human and animal pathogens, in *The Yeasts,* Vol. I, *The Biology of Yeasts,* Rose, A. H. and Harrison, J. S., Eds., Academic Press, London, 1969, pp. 107–182.
6. **Lodder, J., (Ed.),** *The Yeasts – A Taxonomic Study,* North Holland, Amsterdam, 1970.
7. **Phaff, H. J., Miller, M. W., and Mrak, E. M.,** *The Life of Yeasts,* Harvard University Press, Cambridge, Mass., 1966.
8. **Rose, A. H. and Harrison, J. S., (Eds.),** *The Yeasts,* Vol. III, *Yeast Technology,* Academic Press, London, 1970.
9. **Walker, H. W. and Ayres, J. C.,** Yeasts as spoilage organisms, in *The Yeasts,* Vol. III, Rose, A. H. and Harrison, J. S., Eds., Academic Press, London and New York, 1970, pp. 463–527.
10. **van der Walt, J. P.,** Criteria and methods used in classification, in *The Yeasts – A Taxonomic Study,* Lodder, J., Ed., North Holland, Amsterdam, 1970, pp. 34–113.
11. **van der Walt, J. P.,** The perfect and imperfect states of *Sporobolomyces salmonicolor, Antonie van Leeuwenhoek J. Microbiol. Serol.,* 36, 49, 1970.
12. **van der Walt, J. P.,** Sexually active strains of *Candida albicans* and *Cryptococcus albidus, Antonie van Leeuwenhoek J. Microbiol. Serol.,* 33, 246, 1967.
13. **Wickerham, L. J.,** A preliminary report on a perfect family of exclusively protosexual yeasts, *Mycologia,* 56, 253, 1964.
14. **van der Walt, J. P.,** The perfect and imperfect states of *Sporobolomyces salmonicolor, Antonie van Leeuwenhoek J. Microbiol. Serol.,* 36, 49, 1970.
15. **Rodrigues de Miranda,** *Filobasidium capsuligenum* nov. comb., *Antonie van Leeuwenhoek J. Microbiol. Serol.,* 38, 91, 1972.
16. **Miller, M. W., Yoncyama, M., and Soneda, M.,** *Phaffia,* a new yeast genus in the Deuteromycetes (Cryptococcales), *Int. J. Syst. Bacteriol.,* 26, 286, 1976.
17. **Fell, J. W. and Statzell, A. C.,** *Sympodiomyces* gen. w., a yeast-like organism from southern marine waters, *Antonie van Leeuwenhoek J. Microbiol. Serol.,* 37, 359, 1971.
18. **van der Walt, J. P. and Liebenberg, N. V. D. W.,** The yeast genus *Wickerhamiella* gen. nov. (Ascomycetes), *Antonie van Leeuwenhoek J. Microbiol. Serol.,* 39, 121, 1973.

INTRODUCTION TO THE SYSTEMATICS OF VIRUSES

K. Maramorosch

According to the rules of the International Committee on Nomenclature of Viruses (ICNV), laid down in 1966 during the Ninth International Congress for Microbiology, the taxonomic system for viruses does not classify them by the hosts they infect, but by such criteria as chemistry (DNA or RNA viruses), symmetry (helical or cubical), and other characteristics of the virions. Consequently, viruses should no longer be grouped as viruses of bacteria, fungi, algae, mycoplasma, higher plants, invertebrate and vertebrate animals, but as members of a single "kingdom" of viruses (Virales). Lwoff and Tournier[3] have presented a partial classification of all viruses according to the unified system of classification (Table 1). It utilized the nature of the genetic material, the symmetry of the capsid, the naked or enveloped nature of the nucleocapsid, the number of capsomeres for virions with cubical symmetry, and the diameter of the nucleocapsid for virions with helical symmetry.

The molecular weight of the nucleic acid, nature of the envelope, proportion of nucleotides, number of strands of nucleic acid, antigenicity of the viral proteins, and other characteristics are used for a more complete characterization of "families". Table 2 presents the characteristics of families according to Lwoff and Tournier.[3]

Viruses require ribosomes of host cells for their multiplication, and also in some cases the nuclear apparatus of the host cell, and these requirements represent the ultimate degree of parasitism.

Table 1
CLASSIFICATION OF VIRUSES

Nucleic acid	Capsid symmetry	Naked (N) or enveloped (E)	Helical diameter or number of capsomeres	Taxonomic designation
DNA	H	N	50Å	Inoviridae
		E	?	Poxviridae
			12	Microviridae
			32	Parvoviridae
	C	N	42	Densoviridae
			72	Papilloviridae
			252	Adenoviridae
			812	Iridoviridae
	B	E	162	Herpesviridae
		N Urovirales		Tailed bacteriophages
		N Rhabdovirales		
			90Å	Myxoviridae
			180Å	Paramyxoviridae
RNA	H	E Sagovirales		Stomatoviridae
			?	Thylaxoviridae
	C	N Gymnovirales	32	Napoviridae
			92	Reoviridae
			?	Blue tongue virus (sheep)
		E Togavirales	?	Encephaloviridae

Code: H = helical; C = cubic; B = binal.

From Lwoff, A. and Tournier, P., The "LHT System" of 1969, in *Comparative Virology*, Maramorosch, K. and Kurstak, E., Eds., Academic Press, New York, 1971. With permission.

Table 2
THE CHARACTERISTICS OF VARIOUS FAMILIES OF VIRUSES

				Virions with cubic symmetry			Virions with helical symmetry			
Viridae	Nucleic acid	Symmetry	Naked (N) or enveloped (E) nucleocapsid	Number of capsomers	Diameter (Å) of the nucleocapsid	Diameter (Å) of the envelope	Diameter and length (Å) of the nucleo-capsid	Dimensions (Å) of the enveloped virions	Mol wt (× 10^6) of nucleic acid	Number of nucleic acid strands
Ino-	D	H	N				5–6 × 760–850		1.7–3	1
Pox-	D	H?	E				?	2,500 × 1,600 3,000 × 2,300	160–240	2
Micro-	D	C	N	12	250				1.7	1
Parvo-	D	C	N	32	220				1.8	1
Denso-	D	C	N	42	200				160–240	1
Papilloma- (papova)	D	C	N	72	450–550				3–5	2
Adeno-	D	C	N	252	700				20–25	2
Irido–	D	C	N	812	1300				126	2
Herpes-	D	C	E	162	775	1500–2000			54–92	2
Uro–	D	BC	N							2
Rhabdo–	R	H	N				20 × 130 10 × 1,250			1
Myxo-	R	H	E				90 × ?	1,000	2–3	1
Paramyxo-	R	H	E				180 × ?	1,200	7.5	1
Stomato- (rhabdo)	R	H	E				180 × ?	1,750 × 680	6	1
Thylaxo-	R	H	E				?	10,000	10	1
Napo-	R	C	N	32	220–270				1.1–2	1
Reo-	R	C	N	92	700				10	2
Cyano-	R	C	N	32 or 42	540					2
Encephalo-	R	C	E	?	?		600–800		2–3	1

Code: D = DNA; R = RNA; H = helical; C = cubic; B = binal.

From Lwoff, A. and Tournier, P., Remarks on the classification of viruses, in *Comparative Virology*, Maramorosch, K. and Kurstak, E., Eds., Academic Press, New York, 1971 With permission.

The highly specific requirements for viral proliferation explain why many viruses have a fairly limited host range and why a few viruses appear to be limited to a single host. Years ago it was believed that certain viruses attack only vertebrates, invertebrates, higher plants, algae, fungi, or bacteria. Although this is generally the rule, the exceptions are fairly numerous. The host range of viruses may include a number of species within a family, order or class. Certain viruses infect vertebrate and invertebrate animals (arbo viruses). There are no reported instances in which the same virus would infect a higher plant and a vertebrate animal, although there are a few viruses that alternate between invertebrate animals and higher plants. When viruses multiply in plant and animal hosts, the invertebrate animals act not only as vectors of plant-pathogenic viruses, but also as reservoirs. Among the invertebrate animals that are known to transmit plant-pathogenic virues are insects, mites, and nematodes.[4] Certain plant viruses are transmitted by lower fungi, some are seed-borne, and several can be transmitted from plant to plant by parasitic higher plants.

Viruses seem to parasitize all forms of life. Viruses of ferns[5] and of mosses[1] have been reported, and in recent years viruses of mushrooms and of filamentous fungi have received increasing attention.[2] To date, all viruses of lower fungi have been found to contain RNA. There are reports of viruses that affect protozoa as well as mycoplasmas.

Since this handbook is not a treatise of systematic microbiology, but rather a handbook where people will be able to look up information about microorganisms, the viruses that are of primary interest to plant pathologists are described under the heading "Viruses of Plants", even though some infect insect vectors and even cause diseases of these invertebrate animals. Viruses that primarily cause diseases of insects are described under the heading "Viruses of Insects". Viruses of bacteria and mycoplasmas are listed as "Bacteriophages", and so on. This practical approach is based on the outmoded criterion of host affinity and disease, rather than on morphological and chemical characteristics of the virions.

At the Third International Congress of Virology in 1976, the International Committee on the Taxonomy of Viruses, ICTV, which now is a committee of the Section on Virology of IAMS, the International Association of Microbiological Societies, approved a number of new names for viral groups.[6] In addition to the newly designated families of Baculoviridae and Iridoviridae, the present chapter followed the suggestions of the Plant Virus Subcommittee of ICTV. This committee has been reluctant to designate various groups of plant viruses as families; the Latin names do not have the "family" status but are recognized merely as groups. We have also listed the reo-type and the rhabdoviruses that infect plants; a separate group of viruses, called by plant pathologists "mycoviruses" because they infect fungi, has also been added. Taxonomically, based on their morphology and chemical composition, viruses in the latter group fall into the families or groups of other viruses.

Finally, viroids have been listed here and described, although it is now apparent that they are not "naked viral nucleic acids", but a newly recognized group of very small pathogens. The reason for including them in the chapter on plant viruses is simply the fact that these agents have been studied primarily by plant virologists in the past.

Figures 1 and 2 present examples of viroids and invertebrate viruses.

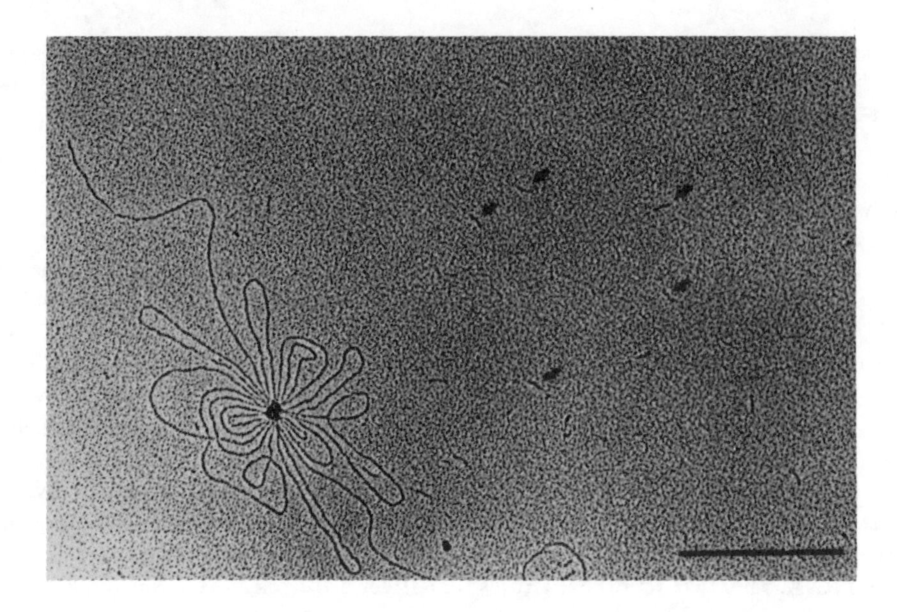

FIGURE 1. Electron micrograph of potato spindle tuber viroid mixed with double-stranded coliphage T7-DNA. Arrows point to several viroids. Magnification × 55,000; bar equals 0.5 nm. (Courtesy of T. Koller and J. M. Sogo, Swiss Federal Institute of Technology, Zurich.)

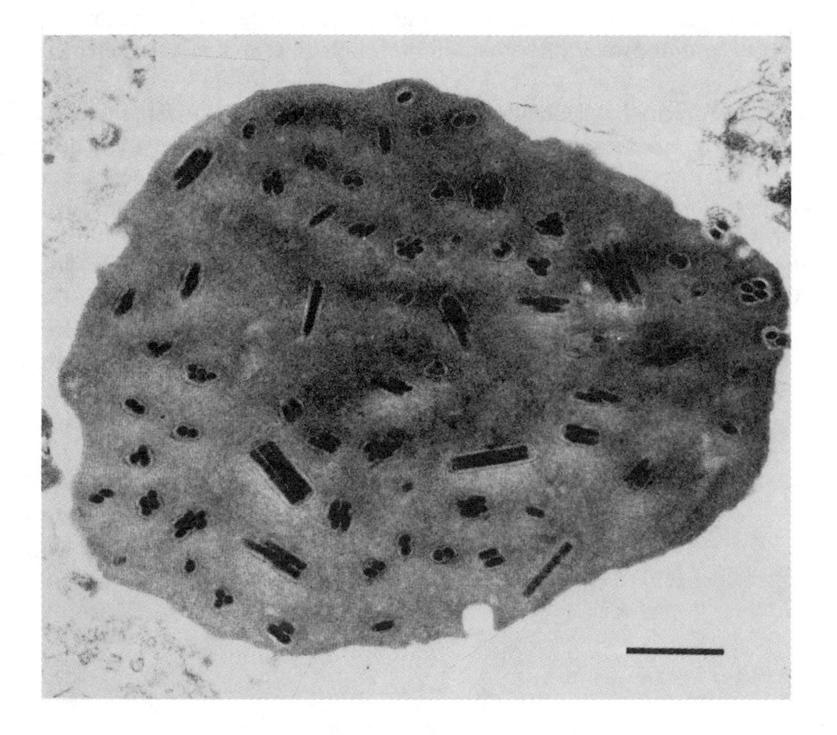

FIGURE 2. Electron micrograph of a cross section of a polyhedral inclusion body from a *Porthetria dispar* (gypsy moth) nuclear polyhedrosis virus (NPV, family Baculoviridae). The virus rods may occur singly or in bundles bounded by a single envelope within the proteinaceous inclusion body. Magnification × 42,750; bar equals 1 μm. (Original electron micrography by courtesy of Russell Riscoe, Waksman Institute of Microbiology, Rutgers University.)

REFERENCES

1. **Blattny, C., Pilous, Z., and Osvald, V.,** *Ochr. Rostl.,* 22, 136, 1949.
2. **Hollings, M. and Stone, O. M.,** *Annu. Rev. Phytopathol.,* 9, 93, 1971.
3. **Lwoff, A. and Tournier, P.,** in *Comparative Virology,* Maramorosch, K. and Kurstak, E., Eds., Academic Press, New York, 1971, p. 1.
4. **Maramorosch, K. (Ed.),** *Viruses, Vectors, and Vegetation,* Interscience, New York, 1969.
5. **Severin, H. H. P. and Tompkins, C. M.,** *Hilgardia,* 20, 81, 1950.
6. **Fenner, F.,** The classification and nomenclature of viruses: Summary of results of meetings of the International Committee on Taxonomy of Viruses in Madrid, September 1975, *J. Gen. Virol.,* 31, 463–470, 1976; *ASM News,* 42, 170, 1976.

PHAGE TYPING

S. S. Kasatiya and P. Nicolle

Isolation of a specific lytic agent was first reported by Twort.[1] Felix Hubert D'Hérelle coined the term "bacteriophagum",[2] and later "bacteriophage",[3] after having noticed that the agent was a living parasite of bacteria. He divided the bacterial species into two groups based on the lytic action of a bacteriophage.[4] As early as 1934, Marcuse[5] used five phages to divide *Salmonella typhi* into different phage types and reported that the strains isolated from the same patient at different time intervals belonged to the same phage type. Since then, many workers have isolated phages specific for different microbial species.

The microbial species are classified into various types according to their morphological, biological, and biochemical characters. According to the host range of more or less specific phages, biotypes or serotypes of a given species may further be subdivided into reasonable phage types. This subdivision is called phage typing. In an efficient phage-typing scheme, the microbial strains isolated from patients, carriers, contaminated food, and vectors during an outbreak all belong to the same phage type.

Phages may be virulent, temperate, or adapted. Phage-typing schemes employ either virulent or temperate (*Staphylococcus aureus*), adapted (*Salmonella typhi*), or a combination (*Salmonella paratyphi* B, *Salmonella typhimurium*) of these phages. Virulent phages are normally isolated from sewage, but they may be obtained from feces or from carrier microbial strains. They always lyse the microorganism they infect and produce clear lytic plaques that are easy to observe. Temperate phages, though very host-specific, are isolated from lysogenic bacteria, which carry them in intracellular form as prophage. These lysogenic microorganisms are potential producers of phage or phages as a stable heritable character; spontaneous lysis results from the transition of prophage to vegetative state under certain circumstances, but the mechanism is not fully known. However, the lysogenic cells are immune to lytic infection by the same or related phage or phages. Adapted phages were described by Craigie and Yen,[6] who, on the basis of the constant mutation rate, justified the presence of mutants in a phage population. By selectively propagating any one of these mutants on certain strains of *Salmonella typhi,* they obtained different new phage preparations that showed high affinity to the new host and suggested that the behavior of the adapted phage is conditioned by the strain on which it had last been propagated. The newly adapted phage is immunologically indistinguishable from its unadapted parent, but it shows a marked difference in host range. The authors proposed that these phage preparations, if used at the highest dilution giving confluent lysis with the homologous strain – known as routine test dilution (RTD) – and thus eliminating all but the dominant mutant, made it possible to identify *Salmonella typhi* strains similar to the ones on which they were last propagated.

For phage typing, the fresh culture to be examined is spread on the surface of a nutrient agar plate, and the phages are spotted in standard amounts, usually those of the RTD.[7] The lytic reactions are read after 6 to 18 hours incubation at 37°C. In order to avoid the risk of accidental contamination of the original phages with other bacteriophages, propagation and distribution of typing phages for *Salmonella typhi* and *Salmonella paratyphi* B is done by the International Centre for Phage Typing, Colindale, London, England.

THE IMPORTANCE OF PHAGE TYPING

Phage typing is a very useful epidemiological tool for tracing the origin of infection in epidemics by further subdividing the biotypes or serotypes of microorganisms. It establishes a relationship between the sick and the carriers in an epidemic or endemic focus due to the same phage type. It also indicates the distribution of various types of a pathogenic species throughout the world. It is very useful for investigating outbreaks of food poisoning by staphylococci, *Salmonella,* etc., and serves to establish a correlation between the strains isolated from the patients, contaminated food, and the source of infection. Species-specific phages play an important role in the taxonomy of bacterial species, such as phages of *Bacillus anthracis, Pasteurella pestis* and *Brucella,* which are employed in the identification of their specific hosts. Thus, they may also be employed in the purification of cultures. Phage typing can also be used for the identification of structural or antigenic characters of bacteria, due to the specific affinity of some bacteriophages to certain bacterial structures, e.g., smooth and rough strains, Vi antigen of *Salmonella typhi,* or flagella of *Escherichia coli.* Phage typing is employed to characterize bacterial strains in various fundamental and applied research programs.

There is considerable literature on the isolation and epidemiological applications of phage typing of *Salmonella typhi, Salmonella paratyphi* B, *Salmonella typhimurium, Staphylococcus aureus,* and some other bacteria, but Table 1 lists only the references that are of historical importance, that are landmarks in phage-typing schemes, or that are the latest publications covering the information on the phages or phage-typing schemes described earlier. The abbreviation (sp.) in the column giving the number of phage types indicates that the phages have been employed for species identification within a genus that shows that phage typing is feasible.

The most important phage-typing schemes, widely used throughout the world for their epidemiological value, have been reproduced in Tables 2 to 5. In Tables 2 to 4, the degrees of lysis are represented by various symbols; e.g., CL = confluent lysis, SCL = semiconfluent lysis, <SCL and <CL = intermediate degrees of lysis, OL = confluent "opaque" lysis, and ± to +++ = increasing numbers of discrete piaques. Different symbols are used for *Staphylococcus aureus* phage typing (Table 5); e.g., 5 = ++ reaction in the same dilution as on the propagating strain, 4 = ++ reaction in a dilution 10 to 10^2 times more concentrated than that giving ++ on the propagating strain, 3 = ++ reaction in a dilution 10^3 to 10^4 times more concentrated than that giving ++ on the propagating strain, 2 = ++ reaction in a dilution 10^5 to 10^6 times more concentrated than that giving ++ on the propagating strain, and 1 = very weak lysis. High-titer phages may "inhibit" the growth of many of the strains when used undiluted, but produce no discrete plaques when diluted. In some cases this inhibition may simulate confluent lysis, but generally it appears as a thinning of the growth in the drop area. Such reactions are recorded as 0.

Table 1
GUIDE TO LITERATURE ON PHAGE TYPING

Microorganism	Number of phages isolated or selected	Sources and types of phages	Number of phage types	Reference
PHAGE TYPING SCHEMES WITH INTERNATIONAL AGREEMENT				
Salmonella				
paratyphi B	12	Feces; adapted	48	8
	10	Feces; adapted	10	9
	6	Temperate, adapted	9	10
typhi	96	Adapted (various authors)	96	11
Staphylococcus				
aureus	27	Temperate, adapted	16	12
	18	Temperate, adapted	21	13

Table 1 (continued)
GUIDE TO LITERATURE ON PHAGE TYPING

Microorganism	Number of phages isolated or selected	Sources and types of phages	Number of phage types	Reference
PHAGE TYPING SCHEMES WITHOUT INTERNATIONAL AGREEMENT				
Aeromonas				
salmonicida	13	Sewage, water; temperate	13	14
Agrobacterium				
radiobacter	7	Sewage, soil; temperate	4	15
	19	Sewage, soil	4	16
Asticcacaulis	3	Sewage, pond water	2 (sp)	17
Bacillus				
pumilus	2	Not given	3	18
Brucella				
abortus	3	Soil; adapted	3 (sp)	19
melitensis	3	Temperate	4 (sp)	20
suis	15	Manure; temperate	3 (sp)	21
	1	Soil	3 (sp)	22
Caulobacter				
bacteroides	9			
crescentus	6			
fusiformis	1	Sewage, soil, pond water	6 (sp)	17
vibrioides	4			
Corynebacterium				
anaerobic	6	Temperate	11	23
diphtheriae	19	Temperate, adapted	9	24
	3	Temperate	4	25
	44	Temperate	35	26, 27
	8	Temperate, adapted	9	28
Escherichia				
coli				
026:B6		Feces; temperate	5	29
055:B5	28	Feces; temperate	9	29
0111:B4		Feces; temperate	11	29
0119:B14	10	Sewage	8	30
0124:K72(B17)	12	Temperate	11	31
0127:B8	9	Sewage	9	32
bovine	24	Sewage, feces	57	33
urinary	13	Sewage	109	34
Klebsiella				
rhinoscleromatis	15	Sewage, stools	12	35
gastrointestinal	12	Feces	13	36
respiratory	15	Sewage, stools	15	37
Listeria				
monocytogenes	5	Temperate	8	38
Mycobacterium sp.	6	Soil, sewage	5 (sp)	39
	6	Soil,stools; adapted	6 (sp)	40
tuberculosis	12	Soil; adapted (various authors)	3	41
	8	Soil; adapted (various authors)	2	42
	5	Soil; adapted (various authors)	3	43
	4	Soil	4	44
	3	Soil; adapted (various authors)	2	45
xenopi	3	Temperate	4	46
fast-growing	7	Soil; temperate	7 (sp)	47
	4	Soil; manure	3 (sp)	48
saprophytes	14	(Various authors)	9 (sp)	49
slow-growing	17	(Various authors)	3 (sp)	50

Table 1 (continued)
GUIDE TO LITERATURE ON PHAGE TYPING

Microorganism	Number of phages isolated or selected	Sources and types of phages	Number of phage types	Reference
Neisseria				
meningitidis	5	Body fluids,nasopharynx	18	51
Nocardia				
aster-oids	8	Temperate	2	52
pellegrini	13	Temperate	2	53
Propionibacterium				
acnes	6	Temperate	7	54
Proteus				
hauseri	20	Sewage; temperate	10	55
	15	Sewage, water; temperate	14	56
	12	Sewage, water; temperate	10	57
mirabilis				
OXK				
vulgaris	52	Sewage; temperate	5 (sp)	58
OX19				
OX2				
OXL				
mirabilis	15	Sewage	21	59
vulgaris				
morganii	7	Sewage	9	59
Pseudomonas				
aeruginosa	21	Water; temperate	6	60
	18	Sewage; temperate	>400	61
	13	Temperate	64	62
	12	Sewage; temperate	12	63
	13	ATCO phage; temperate	11	64
	24	(Various authors)	240	65
Rhizobium	28	Sewage, soil	31	66
trifolii	16	Sewage, soil	13	67
Ristella				
pseudo-insolita	9	Sewage; temperate	30	68
	6	Adapted	24	68
Salmonella				
adelaide	6	Temperate	6	69
blockley	3	Temperate	14	70
bovis-morbificans	10	Temperate	4	71
braenderup	3	Temperate	15	72
dublin	6	Temperate	11	73
enteritidis	6	Sewage, feces	8	74
gallinarum	2	Sewage, feces, dung water	2	75
heidelberg	8	Feces, intestinal contents	22	76
minnesota R forms	5	(Various authors)	7	77
oranienburg	14	Temperate	5	78
panama	8	Surface water; adapted	8	79
paratyphi A	4	Sewage; adapted	4	80
paratyphi B	6	Temperate, adapted; complementary to Felix and Callow, 1951	9	10
pullorum	4	Sewage, feces, dung water	4	75
newport	9	Temperate	35	81
thomp-son	7	Temperate	11	82
typhimurium	8	Sewage	13	83
	31	Temperate,adapted of Callow,1959	90	84
	29	Temperate, adapted	34	85
	12	Temperate, adapted	12	86

Table 1 (continued)
GUIDE TO LITERATURE ON PHAGE TYPING

Microorganism	Number of phages isolated or selected	Sources and types of phages	Number of phage types	Reference
	12	Sewage, feces; temperate, adapted	24	87
	6	Canal water; temperate, adapted	9	88
	32	Surface water; adapted	32	89
	6	Sewage; temperate	11	90
	20	Canal water; adapted	26	91
weltevreden	15	Temperate	15	92
Serratia	54	Surface water, adapted; extention of Schollens, 1962	88	93
marcescens	34	Sewage	71	94
Shigella				
boydii, different serotypes	20	Residual water, temperate, adapted	37	95
flexneri				
different serotypes	12	Temperate	40	96
different serotypes	19	Temperate	90	97
3a	14	Temperate	9	98
sonnei				
R forms, phase II	12	Sewage, feces, manure; temperate	68	99
	3	Temperate	3	100
	15	3 from feces, 12 from Hammarstrom,1949	38	101
S forms			12	102
R forms, phases I and II	10	Temperate	20	102
Staphylococcus				
epidermidis	18	Temperate,adapted	7	103
Streptococcus				
bovis				
durans				
faecalis	30	Monkey feces, water	30	104
var. *zymogenes*				
var. *liquefaciens*				
faecium				
cremoris	10	Cheese; temperate	12	105
faecalis	3	Sewage, monkey feces, nasopharyngeal washing; temperate	4	106
	7	Sewage,urogenital specimens	27	107
faecium	10	Sewage, urogenital specimens	22	107
hemolytic	4	Sludge, feces	8	108
lactis	12	Cheese; temperate	16	105
	24	(Various authors)	24	109
Vibrio				
cholerae	4	Stools,water; temperate	5	110
cholerae				
El-Tor	8	Stools, water; temperate	8	111
NAG (nonagglutinable)				
Xanthomonas				
malvacearum	14	Infected leaves	2	112
Yersinia				
enterocolitica	12	Temperate	12	113
	10	Temperate	5	114

Table 2
REACTIONS OF *SALMONELLA TYPHIMURIUM* TYPE STRAINS WITH ROUTINE TEST DILUTIONS OF THE NEW TYPING PHAGES

Type Strains		Phages in Routine Test Dilutions													
Old	New	1	2	3	4	5	6	7	8	9	10	11	12	13	14
1	1	CL	CL	CL	CL	CL	CL	CL			SCL	SCL	CL	CL	CL
1a	2	–	CL	CL	CL	CL	CL	–	–	±	CL	CL	CL	CL	CL
1a var. 1	3	–	+++	CL	CL	CL	CL	–	–	+	CL	SCL	–		CL
1b	4	–	++	–	CL	CL	CL	–	–	±	CL	SCL	+++	CL	+++
3	5	–	–	–	+++	CL	+++	–	–		–	++	–	–	+++
3a	6	–	++	–	++	±	CL	–	–	–	+++	–	–	–	
2d	7	–	–	–	–	–	–	CL	±		–	–	–	–	–
4	8	–	–	–	–	–	–	–	CL	CL	CL	CL			
	9	–	–	–	–	–	–	–	CL	CL	CL	CL	CL	CL	
2	10	–	–	–	–	–	–	–	–	±	CL	SCL	CL	CL	+++
	11	–	–	–	–	–	–	–	–	–	–	SCL	SCL	CL	
	12	–	–	–	–	–	–	–	–		–	–	CL	CL	
	12a	–	–	–	–	–	–	–	–	–	–	–	CL	CL	–
2a	13	–	–	–	–	–	–	–	–	–	–	SCL		CL	OL
	14	–	–	–	–	–	–	–	–	–	–	SCL	++	SCL	OL
2c	15	–	–	–	–	–	–	–	–	–			+++	SCL	–
	15a	–	–	–	–	–	–	–	–	–	–	++	CL	CL	–
	16	–	–	–	–	–	–	–	–	–		+++		–	OL
2b	17	–	–	–	–	–	–	–	–	–	–	–	–	–	–
	18	–	–	–	–	–	–	–	–	–	–	–	–	–	–
	19	–	–	–	–	–	–	–	–	–	–	–	–	–	–
	20	–	–	–	–	–	–	–	–	–	–	–	–	–	–
2d	20a	–	–	–	–	–	–	–	–	–	–	–	–	–	–
Untypable	21	–	–	–	–	–	–	–	–	–	–	–	–	–	–
Untypable	22	–	–	–	–	–	–	–	–	–	–	–	–	–	–
Untypable	23	–	–	–	–	–	–	–	–	–	–	–	–	–	–
Untypable	24	–	–	+++	–	–	–	–	–	–	–	–	–	–	CL
Untypable	25	–	–	–	–	–	–	–	–	–	–	–	–	–	–
Untypable	26	–	–	–	–	–	–	–	–	–	–	–	–	–	–
Untypable	27	–	–	–	–	–	–	–	–	–	–	–	–	–	–
2c	28	–	–	–	–	–	–	–	+++		–	–	+++	SCL	–
Untypable	29	–	–	–	–	–	–	–	–	–	–	–	–	–	–
Untypable	30	–	–	–	–	–	–	–	SCL		–	±	–	–	–
Untypable	31	–	–	–	–	–	–	–	–	–	–	–	–	–	–

Old	New	15	16	17	18	19	20	21	22	23	24	25	26	27	28	29
1	1	CL	CL	CL	CL	CL	CL	CL	CL	SCL	CL	SCL	CL	CL	±	CL
1*a*	2	CL	CL	CL	–	CL	CL	CL	CL	CL	CL	CL	CL	CL	++	CL
1*a* var. 1	3	CL	CL	CL	–	CL	CL	SCL	SCL	CL	CL	+++	SCL	CL	SCL	CL
1*b*	4	CL	CL	++	–	CL	SCL	±	CL	CL	–	CL	+++	CL	+++	CL
3	5	+++	+++	–	–	++	±	–	SCL	–	–	–	–	–	–	–
3*a*	6	–	+++	+++	–	–	–	+++	+++	+++	–	–	–	SCL	++	CL
2*d*	7	–	–	–	CL	±	SCL	–	–	–	–	–	–	–	–	–
4	8	–	+++	–	–	–	++	±	CL	CL	–	±	++	–	±	CL
	9	–	SCL	–	–	–	++	–	CL	CL	–	±	+++	–	–	CL
	10	–	SCL	++	–	–	++	++	CL	CL	–	±	++	++	–	CL
2	11	–	–	OL	CL	SCL	OL	–	–	–	–	–	–	–	–	–
	12	–	–	+++	–	–	–	–	–	–	–	–	–	–	–	–
	12*a*	–	–	–	CL	–	–	+++	–	–	–	–	–	OL	–	–
2*a*	13	–	++	–	+++	–	–	+++	–	–	–	–	OL	OL	++	OL
	14	–	–	–	–	–	–	–	–	–	–	–	+++	+++	–	OL
2*c*	15	SCL	–	OL	OL	OL	OL	–	–	–	–	–	–	++	+	–
	15*a*	OL	–	OL	–	OL	OL	–	–	–	–	–	–	++	++	–
	16	–	OL	+++	+++	–	–	–	±	–	–	SCL	+++	OL	±	OL
	17	+++	–	OL	+++	+++	SCL	–	–	–	–	–	–	+	++	–
2*b*	18	SCL	–	OL	CL	SCL	SCL	–	–	–	–	–	–	++	++	–
	19	SCL	–	–	–	SCL	SCL	–	–	–	–	–	–	+++	+++	–
	20	–	–	–	–	–	OL	–	–	–	–	–	–	–	–	–
2*d*	20*a*	–	–	–	CL	±	SCL	–	–	–	–	–	–	–	–	–
Untypable	21	–	–	–	–	–	–	OL	++	–	–	–	–	–	–	–
Untypable	22	–	–	–	–	–	–	±	OL	–	–	–	–	–	–	–
Untypable	23	–	–	–	–	–	–	–	–	OL	CL	+++	+++	–	–	++
Untypable	24	–	–	–	–	–	–	–	–	–	CL	–	SCL	–	–	–
Untypable	25	–	++	–	–	–	–	–	–	±	+++	OL	OL	–	–	+++
Untypable	26	–	–	–	–	–	–	–	–	–	+++	+++	OL	–	–	–
Untypable	27	–	–	–	–	–	–	+++	–	–	–	–	–	OL	–	–
2*c*	28	SCL	–	SCL	SCL	SCL	SCL	–	–	–	–	–	–	±	OL	–
Untypable	29	–	–	–	–	–	–	–	–	–	–	–	–	–	–	OL
Untypable	30	–	–	–	–	–	–	–	–	–	–	–	–	–	–	–
Untypable	31	–	–	–	–	–	–	–	–	–	–	SCL	+	–	–	+++

From Callow, B. R., *J. Hyg.*, 57, 352, 1959. With permission of Cambridge University Press.

Table 3
REACTIONS OF THE Vi-TYPE STRAINS OF *SALMONELLA TYPHI* WITH THE TYPING PHAGES IN ROUTINE TEST DILUTIONS[a]

Vi-Type Strains	Adapted Vi-Phage Preparations											
	A	B1	B2	B3	C1	C2	C3	C4	C5	C6	C7	C8
A	**CL**	CL	CL	CL	CL	CL	CL	CL	CL	CL	CL	CL
B1	±	**CL**	-	+++	-	-	+	+++	-	+	-	-
B2	+++	-	**CL**	++	+++	-	-	-	-	-	-	-
B3	-	-	-	**CL**	-	-	-	-	-	-	-	-
C1	+	+++	++	+++	**CL**	**CL**	**CL**	**CL**	**CL**	**CL**	**CL**	**CL**
C2	-	-	-	-	-	**CL**	-	-	-	-	-	-
C3	-	-	-	-	+	**CL**	**CL**	+	-	+	±	+
C4	-	-	-	-	-	-	-	**CL**	-	-	-	-
C5	-	-	-	-	-	-	-	-	**CL**	-	-	-
C6	-	-	-	-	-	-	-	-	-	**CL**	-	-
C7	-	-	-	-	+	-	-	+++	+	+++	**CL**	+++
C8	-	-	-	-	-	-	-	-	-	-	-	**CL**
C9	-	-	-	±	+++	-	-	-	-	-	-	-
D1	-	±	-	-	±	**CL**	-	-	-	-	-	-
D2	-	-	-	-	-	-	-	-	-	-	-	-
D4	-	-	-	-	-	-	-	-	-	-	-	-
D5	-	±	-	-	±	CL	-	-	-	-	-	-
D6	-	-	-	-	-	SCL	+++	-	-	-	-	-
D7	-	-	-	-	-	-	-	-	-	-	-	-
D8	-	-	-	-	-	-	-	-	-	-	-	-
D9	-	-	-	-	-	-	-	-	-	-	-	-
D10	-	-	-	-	-	-	-	-	-	-	-	-
D11	-	-	-	-	-	-	-	-	-	-	-	-
E1	-	+	-	-	±	-	-	-	-	-	-	-
E2	-	-	-	-	-	-	-	-	-	-	-	-
E3	-	-	-	-	-	-	-	-	-	-	-	-
E4	-	-	-	-	-	-	-	-	-	-	-	-
E5	-	-	-	-	-	-	-	-	-	-	-	-
E6	-	-	-	-	-	-	-	-	-	-	-	-
E7	-	-	-	-	-	-	-	-	-	-	-	-
E8	-	-	-	-	-	-	-	-	-	-	-	-
E9	-	-	-	-	-	+	-	-	-	-	-	-
E10	-	-	-	-	-	-	-	-	-	-	-	-
F1	-	-	-	-	-	-	-	-	-	-	-	-
F2	-	-	-	-	-	-	-	-	-	-	-	-
F3	-	-	-	-	-	-	-	-	-	-	-	-
F4	-	-	-	-	-	-	-	-	-	-	-	-
F5	-	-	-	-	-	-	-	-	-	-	-	-
G	-	-	-	-	±	-	-	-	-	-	-	-
H	-	+	+	+	+	-	+	+	+	-	-	-
J1	-	-	-	-	-	-	-	-	-	-	-	-
J2	-	-	-	-	-	-	-	-	-	-	-	-
J3	-	-	-	-	-	-	-	-	-	-	-	-
K1	-	-	-	-	-	-	-	-	-	-	-	-
K2	-	-	-	-	-	-	-	-	-	-	-	-
L1	-	-	-	-	-	-	-	-	-	-	-	-
L2	-	-	-	-	-	-	-	-	-	-	-	-
M1	-	-	-	-	-	-	-	-	-	-	-	-
M2	-	-	-	-	-	-	-	-	-	-	-	-
M3	-	-	-	-	-	-	-	-	-	-	-	-
N	-	-	-	-	-	-	-	-	-	-	-	-
O	-	-	-	+++	-	-	-	+++	-	-	-	-
T	-	+++	-	-	-	-	+++	++	-	-	-	-
25	-	-	-	-	-	-	-	-	-	-	-	-
26	-	-	-	-	-	-	-	-	-	-	-	CL
27	-	-	-	-	-	-	-	-	-	-	-	-
28	-	-	-	-	-	-	-	-	-	-	-	-
29	-	+	+++	+++	±	CL	CL	+	-	-	-	+
32	-	±	±	±	+	±	±	±	±	±	±	±
34	-	-	-	-	-	-	-	-	-	-	-	-
35	-	+++	-	-	-	-	-	-	-	-	-	-
36	-	-	-	-	-	-	-	-	-	-	-	-
37	-	-	-	-	-	-	-	-	-	-	-	-
38	-	-	-	-	-	-	-	-	-	-	-	-
39	-	±	-	-	-	-	-	-	-	-	-	-
40	-	-	-	-	-	-	-	-	-	-	-	-
41	-	+++	++	+++	+++	+++	+++	+++	+++	++	++	+++
42	-	++	-	SCL	++	-	+++	CL	-	SCL	SCL	CL
43	-	-	-	++	-	-	-	+++	-	-	-	-
44	-	-	-	-	-	-	-	-	-	-	-	-
45	-	-	-	-	-	-	-	CL	-	-	-	-
46	-	-	-	-	-	-	-	-	-	-	-	-

[a] Homologous and group reactions are shown in bold type.

Table 3 (continued)
REACTIONS OF THE Vi-TYPE STRAINS OF *SALMONELLA TYPHI* WITH THE TYPING PHAGES IN ROUTINE TEST DILUTIONS[a]

Vi-Type Strains	Adapted Vi-Phage Preparations											
	C9	D1	D2	D4	D5	D6	D7	D8	D9	D10	D11	E1
A	CL	CL	CL	CL	CL	CL	CL	CL	CL	CL	CL	CL
B1	+++	++	-	-	-	-	-	-	-	-	-	+
B2	+++	++	++	+++	+++	-	+++	++	+	+++	-	+++
B3	CL	+++	++	-	-	-	-	SCL	-	SCL	-	-
C1	**CL**	+++	-	+	++	++	+++	+	±	+	-	++
C2	+++	-	-	-	-	++	-	-	-	-	-	-
C3	+++	-	-	-	-	+++	-	-	-	-	-	-
C4	-	-	-	-	-	-	-	-	-	-	-	-
C5	-	-	-	-	-	-	-	-	-	-	-	-
C6	-	-	-	-	-	-	-	-	-	-	-	-
C7	±	-	-	-	-	-	-	-	-	-	-	-
C8	-	-	-	-	-	-	-	-	-	-	-	-
C9	**CL**	-	-	-	-	-	-	-	-	-	-	-
D1	++	**CL**	**CL**	**CL**	**CL**	**CL**	**CL**	**CL**	**CL**	**CL**	**CL**	-
D2	-	-	**CL**	±	±	-	-	-	-	-	-	-
D4	-	-	-	**CL**	-	-	++	+	+	+	-	-
D5	-	±	-	-	**CL**	**CL**	-	**CL**	**CL**	-	-	±
D6	-	+	-	-	-	**CL**	-	-	-	-	+++	-
D7	-	-	-	-	-	-	**CL**	-	-	**SCL**	-	-
D8	-	-	-	-	-	-	-	**CL**	-	**SCL**	-	-
D9	-	-	-	+++	-	-	**CL**	**SCL**	**CL**	**CL**	-	-
D10	-	-	-	-	-	-	-	-	-	**CL**	-	-
D11	-	-	-	-	-	-	-	-	-	-	**CL**	-
E1	±	-	-	-	-	-	-	-	-	-	-	**CL**
E2	-	-	-	-	-	-	-	-	-	-	-	-
E3	-	-	-	-	-	-	-	-	-	-	-	-
E4	-	-	-	-	-	-	-	-	-	-	-	-
E5	-	-	-	-	-	-	-	-	-	-	-	-
E6	-	-	-	-	-	-	-	-	-	-	-	-
E7	-	-	-	-	-	-	-	-	-	-	-	-
E8	-	-	-	-	-	-	-	-	-	-	-	-
E9	-	-	-	-	-	++	-	-	-	-	±	-
E10	-	-	-	-	-	-	-	-	-	-	-	-
F1	-	-	-	-	-	-	-	-	-	-	-	-
F2	-	-	-	-	-	-	-	-	-	-	-	-
F3	-	-	-	-	-	-	-	-	-	-	-	-
F4	-	-	-	-	-	-	-	-	-	-	-	-
F5	-	-	-	-	-	-	-	-	-	-	-	-
G	-	-	-	-	-	-	-	-	-	-	-	-
H	+	-	-	-	-	-	±	-	-	±	-	±
J1	-	-	-	-	-	-	-	-	-	-	-	-
J2	-	-	-	-	-	-	-	-	-	-	-	-
J3	-	-	-	-	-	-	-	-	-	-	-	-
K1	-	-	-	-	-	-	-	-	-	-	-	-
K2	-	-	-	-	-	-	-	-	-	-	-	-
L1	-	-	-	-	-	-	-	-	-	-	-	-
L2	-	-	-	-	-	-	-	-	-	-	-	-
M1	-	-	-	-	-	-	-	-	-	-	-	-
M2	-	-	-	-	-	-	-	-	-	-	-	-
M3	-	-	-	-	-	-	-	-	-	-	-	-
N	+++	-	-	-	-	-	-	+++	-	-	-	-
O	+++	-	-	-	-	-	-	-	-	-	-	-
T	-	+++	-	±	±	+++	±	-	-	-	-	-
25	-	-	-	-	-	-	-	-	-	-	-	-
26	-	-	-	-	-	-	-	-	-	-	-	-
27	-	-	-	-	-	-	-	-	-	-	-	-
28	-	-	-	-	-	-	-	-	-	-	-	-
29	+++	±	-	-	-	CL	-	-	-	-	CL	-
32	±	-	±	-	-	-	-	-	-	-	-	±
34	-	-	-	-	-	-	-	-	-	-	-	-
35	-	-	-	+	-	-	-	++	-	-	-	-
36	-	-	-	-	-	-	-	-	-	-	-	-
37	-	-	-	-	-	-	-	-	-	-	-	-
38	-	-	-	-	-	-	-	-	-	-	-	-
39	-	±	-	-	-	-	±	-	-	-	-	±
40	-	-	-	-	-	-	-	-	-	-	-	-
41	++	+++	+++	+++	+++	+++	+++	+++	+++	+	+	+++
42	CL	-	-	-	-	-	CL	-	-	-	CL	+++
43	+++	-	-	-	-	-	-	-	-	-	-	-
44	-	-	-	-	-	-	-	-	-	-	-	-
45	+++	-	-	-	-	-	-	-	-	-	-	-
46	-	-	-	-	-	-	-	-	-	-	-	-

[a] Homologous and group reactions are shown in bold type.

Table 3 (continued)
REACTIONS OF THE Vi-TYPE STRAINS OF *SALMONELLA TYPHI* WITH THE TYPING PHAGES IN ROUTINE TEST DILUTIONS[a]

Vi-Type	Adapted Vi-Phage Preparations											
Strains	E2	E3	E4	E5	E6	E7	E8	E9	E10	F1	F2	F3
A	CL	CL	CL	CL	CL	CL	CL	CL	CL	CL	CL	CL
B1	–	–	–	+	++	+++	++	–	+++	+++	+++	–
B2	–	+++	CL	+++	+++	+++	+++	–	SCL	+++	–	+++
B3	–	++	+++	–	–	–	–	–	+++	–	–	–
C1	–	±	++	±	±	+	±	–	++	++	+++	+
C2	–	–	–	–	–	–	–	+	–	–	–	–
C3	–	–	–	–	–	+++	–	+++	–	–	+++	–
C4	–	–	–	–	–	–	–	–	–	–	–	–
C5	–	–	–	–	–	–	–	–	–	–	–	–
C6	–	–	–	–	–	–	–	–	–	–	–	–
C7	–	–	–	–	–	–	–	–	–	–	–	–
C8	–	–	–	–	–	–	±	–	–	–	–	–
C9	–	–	–	–	–	–	–	–	–	–	–	–
D1	–	SCL	CL	–	–	±	–	CL	±	–	±	–
D2	–	–	+	–	–	–	–	–	–	–	–	–
D4	–	–	+	–	–	–	–	–	–	–	–	–
D5	–	SCL	++	–	–	–	–	CL	±	–	–	–
D6	–	–	–	–	–	+++	–	+++	–	–	+++	–
D7	–	–	–	–	–	–	–	–	–	–	–	–
D8	–	–	–	–	–	–	–	–	–	–	–	–
D9	–	–	–	–	–	–	–	–	–	–	–	–
D10	–	–	–	–	–	–	–	–	–	–	–	–
D11	–	–	–	–	–	–	–	–	–	–	–	–
E1	**CL**	**CL**	**CL**	**CL**	**CL**	**CL**	**CL**	**CL**	**CL**	±	–	–
E2	**CL**	–	–	–	–	–	–	–	–	–	–	–
E3	–	**CL**	**SCL**	–	–	–	–	**SCL**	–	–	–	–
E4	–	–	**CL**	–	–	–	–	–	–	–	–	–
E5	–	–	–	**CL**	–	–	–	–	–	–	–	–
E6	–	–	–	–	**CL**	–	–	–	++	–	–	–
E7	–	–	–	–	–	**CL**	–	**CL**	+++	–	–	–
E8	–	–	–	–	–	–	**CL**	–	–	–	–	–
E9	–	–	–	–	–	–	–	**CL**	–	–	–	–
E10	–	–	–	–	–	–	–	–	**CL**	–	–	–
F1	–	–	–	–	–	–	–	–	–	**CL**	**CL**	**CL**
F2	–	–	–	–	–	–	–	–	–	±	**CL**	–
F3	–	–	–	–	–	–	–	–	–	–	–	**CL**
F4	–	–	–	–	–	–	–	–	–	+++	++	–
F5	–	–	–	–	–	–	–	–	–	–	±	±
G	–	–	–	–	–	±	–	–	–	–	–	–
H	–	–	+	±	±	+	–	–	+	±	+	+
J1	–	–	–	–	–	–	–	–	–	–	–	–
J2	–	–	–	–	–	–	–	–	–	–	–	–
J3	–	–	–	–	–	–	–	–	–	–	–	–
K1	–	–	–	–	–	–	–	–	–	–	–	–
K2	–	–	–	–	–	–	–	–	–	–	–	–
L1	–	–	–	–	–	–	–	–	–	–	–	–
L2	–	–	–	–	–	–	–	–	–	–	–	–
M1	–	–	–	–	–	–	–	–	–	–	–	–
M2	–	–	–	–	–	–	–	–	–	–	–	–
M3	–	–	–	–	–	–	–	–	–	–	–	–
N	–	–	–	–	–	–	–	–	–	–	–	–
O	–	–	–	–	–	+++	–	–	++	–	–	–
T	–	–	–	–	–	+++	–	+++	CL	–	+++	–
25	–	–	–	–	–	–	–	–	–	–	–	–
26	–	–	–	–	–	–	CL	–	–	–	–	–
27	–	–	–	–	–	–	–	–	–	–	–	–
28	–	–	–	–	–	–	–	–	SCL	–	–	–
29	–	–	–	–	–	CL	–	CL	++	+	CL	–
32	–	–	±	CL	–	±	–	–	±	–	±	±
34	–	–	–	–	–	–	–	–	–	–	–	–
35	–	–	–	–	–	–	–	–	–	–	++	–
36	–	–	–	–	–	–	–	–	–	–	–	–
37	–	–	–	–	–	–	–	–	–	–	–	–
38	–	–	–	–	–	–	–	–	–	–	–	–
39	–	–	–	–	–	–	–	–	±	–	±	±
40	–	–	–	–	–	–	–	–	–	–	–	–
41	–	+	+++	+	+	+++	+	+++	++	++	+++	+++
42	–	–	–	++	–	CL	+++	–	+++	+++	SCL	CL
43	–	–	–	–	–	–	–	–	–	–	–	–
44	–	–	–	–	–	–	–	–	–	–	–	–
45	–	–	–	–	–	+++	–	–	–	–	–	–
46	–	–	–	–	–	–	–	–	–	–	–	–

[a] Homologous and group reactions are shown in bold type.

Table 3 (continued)
REACTIONS OF THE Vi-TYPE STRAINS OF *SALMONELLA TYPHI* WITH THE TYPING PHAGES IN ROUTINE TEST DILUTIONS[a]

Vi-Type Strains	Adapted Vi-Phage Preparations											
	F4	F5	G	H	J1	J2	J3	K1	K2	L1	L2	M1
A	CL	CL	CL	CL	CL	CL	CL	CL	CL	CL	CL	CL
B1	+++	+++	+++	+++	-	+	-	-	+++	±	-	+++
B2	+++	+++	+++	+++	SCL	-	SCL	++	-	+++	+++	+++
B3	-	-	-	-	SCL	+++	SCL	+++	±	SCL	SCL	CL
C1	-	+++	++	+	++	+++	+	-	++	++	-	++
C2	-	-	-	-	-	-	-	-	±	-	-	-
C3	-	-	-	-	±	±	-	-	++	±	-	±
C4	-	-	-	-	-	-	-	-	-	-	-	-
C5	-	-	-	-	-	-	-	-	-	-	-	-
C6	-	-	-	-	-	-	-	-	-	-	-	-
C7	-	-	-	-	-	++	+	-	±	-	-	++
C8	-	-	-	-	-	-	-	-	-	-	-	-
C9	-	-	-	-	±	-	-	-	-	-	-	±
D1	-	±	+	-	-	+++	-	-	SCL	-	-	-
D2	-	-	-	-	-	-	-	-	-	-	-	-
D4	-	-	-	-	-	-	-	-	±	-	-	-
D5	-	-	-	-	-	±	-	-	+++	-	-	-
D6	-	-	-	-	-	-	-	-	+++	-	-	+
D7	-	-	-	-	-	-	-	-	-	-	-	-
D8	-	-	-	-	-	-	-	-	-	-	-	-
D9	-	-	-	-	-	-	-	-	-	-	-	-
D10	-	-	-	-	-	-	-	-	-	-	-	-
D11	-	-	-	-	-	-	-	-	-	-	-	-
E1	+	±	-	-	-	-	-	-	-	-	-	-
E2	-	-	-	-	-	-	-	-	-	-	-	-
E3	-	-	-	-	-	-	-	-	-	-	-	-
E4	-	-	-	-	-	-	-	-	-	-	-	-
E5	-	-	-	-	-	-	-	-	-	-	-	-
E6	-	-	-	-	-	-	-	-	-	-	-	-
E7	-	-	-	-	-	-	-	-	±	-	-	-
E8	-	-	-	-	-	-	-	-	-	-	-	-
E9	-	-	-	-	-	-	-	-	-	-	-	-
E10	-	-	-	-	-	-	-	-	-	-	-	-
F1	**CL**	**CL**	-	-	-	-	-	-	-	-	-	-
F2	-	-	-	-	-	-	-	-	-	-	-	-
F3	-	-	-	-	-	-	-	-	-	-	-	-
F4	**CL**	**CL**	-	-	-	-	-	-	-	-	-	-
F5	-	**CL**	-	-	-	-	-	-	-	-	-	-
G	-	-	**CL**	-	-	-	-	-	-	-	-	-
H	±	++	+	CL	±	+	±	±	±	±	±	+
J1	-	-	-	-	**CL**	**CL**	**CL**	-	-	-	-	-
J2	-	-	-	-	-	**CL**	-	-	-	-	-	-
J3	-	-	-	-	-	-	**CL**	-	-	-	-	-
K1	-	-	-	-	-	-	-	**CL**	**CL**	-	-	-
K2	-	-	-	-	-	-	-	-	**CL**	-	-	-
L1	-	-	-	-	-	-	-	-	-	**CL**	**CL**	±
L2	-	-	-	-	-	-	-	-	-	**+++**	**CL**	-
M1	-	-	-	-	-	-	-	-	-	-	-	**CL**
M2	-	-	-	-	-	-	-	-	-	-	-	-
M3	-	-	-	-	-	-	-	-	-	-	-	-
N	+++	-	-	-	-	-	-	-	-	-	+++	-
O	-	-	-	-	-	-	-	-	++	-	±	++
T	-	CL	-	-	-	-	-	-	+++	-	-	-
25	-	-	-	-	-	-	-	-	-	-	-	-
26	-	-	-	-	-	-	-	-	-	-	-	-
27	-	-	-	-	-	-	-	-	-	-	-	-
28	-	-	-	-	-	-	-	-	-	-	-	-
29	-	-	+	±	±	-	-	-	CL	±	-	+++
32	-	±	±	±	-	±	-	-	-	-	-	-
34	-	-	-	-	-	-	-	-	-	-	-	-
35	-	-	-	-	-	-	-	-	-	-	-	-
36	-	-	-	-	-	-	-	-	-	-	-	-
37	-	-	-	-	-	-	-	-	-	-	-	-
38	-	-	-	-	-	-	-	-	-	-	-	-
39	-	±	-	-	-	±	-	-	-	-	-	-
40	-	-	-	-	-	-	-	-	-	-	-	-
41	+++	SCL	+++	+++	++	+++	+	+	++	+++	+	SCL
42	-	+++	+++	+++	-	CL	-	-	CL	+++	-	+++
43	SCL	-	-	-	-	-	-	-	-	-	-	+++
44	-	-	-	-	-	-	-	-	-	-	-	-
45	-	-	-	-	-	-	-	-	+++	±	-	±
46	-	-	-	-	-	-	-	-	-	-	-	-

[a] Homologous and group reactions are shown in bold type.

Table 3 (continued)
REACTIONS OF THE Vi-TYPE STRAINS OF *SALMONELLA TYPHI* WITH THE TYPING PHAGES IN ROUTINE TEST DILUTIONS[a]

Vi-Type Strains	Adapted Vi-Phage Preparations											
	M2	M3	N	O	T	25	26	27	28	29	32	34
A	CL	CL	CL	CL	CL	CL	CL	CL	CL	CL	CL	CL
B1	+++	++	—	++	++	+++	+++	++	SCL	+++	+	+
B2	+++	++	CL	+	SCL	SCL	CL	SCL	CL	—	+++	—
B3	CL	SCL	CL	CL	CL	SCL	CL	SCL	CL	±	SCL	+++
C1	++	++	±	—	++	±	+++	±	SCL	+	—	—
C2	—	±	—	—	—	—	—	—	—	—	—	—
C3	±	++	—	—	±	—	±	±	+	±	—	—
C4	—	—	—	—	—	—	—	—	—	—	—	—
C5	—	—	—	—	—	—	—	—	—	—	—	—
C6	—	—	—	—	—	—	—	—	—	—	—	—
C7	++	—	—	—	—	—	±	—	—	—	—	—
C8	—	—	—	—	—	—	+++	—	—	—	—	—
C9	+	±	—	—	±	—	+	—	++	—	—	—
D1	—	CL	+	—	—	—	—	—	±	—	±	—
D2	—	—	—	—	—	—	—	—	—	—	—	—
D4	—	++	—	—	—	—	—	—	—	—	—	—
D5	—	+++	—	—	—	—	—	—	—	—	—	—
D6	+	+++	—	—	—	—	—	—	—	++	—	—
D7	—	++	—	—	—	—	±	—	—	—	—	—
D8	—	++	—	—	—	—	—	—	—	—	—	—
D9	—	++	—	—	—	—	—	—	—	—	—	—
D10	—	—	—	—	—	—	—	—	—	—	—	—
D11	—	—	—	—	—	—	—	—	—	—	—	—
E1	±	±	—	—	—	—	—	—	—	—	—	—
E2	—	—	—	—	—	—	—	—	—	—	—	—
E3	—	—	—	—	—	—	—	—	—	—	—	—
E4	—	—	—	—	—	—	—	—	—	—	—	—
E5	—	—	—	—	—	—	—	—	—	—	—	—
E6	—	—	—	—	—	—	—	—	—	—	—	—
E7	—	±	—	—	—	—	—	—	—	—	—	—
E8	—	—	—	—	—	—	±	—	—	—	—	—
E9	—	±	—	—	—	—	—	—	—	—	—	—
E10	—	—	—	—	—	—	—	—	±	—	—	—
F1	—	—	—	—	—	—	—	—	—	—	—	—
F2	—	—	—	—	—	—	—	—	—	—	—	—
F3	—	—	—	—	—	—	—	—	—	—	—	—
F4	—	—	—	—	—	—	—	—	—	—	—	—
F5	—	—	—	—	—	—	—	—	—	—	—	—
G	—	—	—	—	—	—	—	—	—	—	—	—
H	+	±	±	—	±	±	+	—	+++	±	—	—
J1	—	—	—	—	—	—	—	—	—	—	—	—
J2	—	—	—	—	—	—	—	—	—	—	—	—
J3	—	—	—	—	—	—	—	—	—	—	—	—
K1	—	—	—	—	—	—	—	—	—	—	—	—
K2	—	—	—	—	—	—	—	—	—	—	—	—
L1	—	—	—	—	—	—	—	—	±	—	—	—
L2	—	—	—	—	—	—	±	—	—	—	—	—
M1	**CL**	**CL**	—	—	—	—	—	—	—	—	—	—
M2	**CL**	—	—	—	—	—	—	—	—	—	—	—
M3	—	**CL**	—	—	—	—	—	—	—	—	—	—
N	—	—	**CL**	—	±	—	±	—	—	—	—	—
O	+++	++	—	**CL**	—	—	—	—	+++	++	—	—
T	—	+++	++	—	**CL**	—	—	—	CL	SCL	—	—
25	—	—	—	—	—	**CL**	—	—	—	—	—	—
26	—	—	—	—	—	—	**CL**	—	—	—	—	—
27	—	—	—	—	—	—	—	**CL**	—	—	—	—
28	—	—	—	—	—	—	—	—	**CL**	—	—	—
29	+++	CL	—	±	—	—	±	—	+++	**CL**	—	±
32	±	±	—	—	—	—	—	—	±	±	**CL**	—
34	—	—	—	—	—	—	—	—	—	—	—	**CL**
35	±	—	—	—	—	—	—	—	—	—	—	—
36	—	—	—	—	—	—	—	—	—	—	—	—
37	—	—	—	—	—	—	—	—	—	—	—	—
38	—	—	—	—	—	—	—	—	—	—	—	—
39	—	—	—	—	—	—	±	—	±	—	—	—
40	—	—	—	—	—	—	—	—	—	—	—	—
41	SCL	SCL	++	+	+++	+	SCL	±	CL	++	+	+
42	SCL	—	—	+++	—	—	—	—	+++	+++	—	SCL
43	+++	—	—	—	—	—	—	—	—	—	—	—
44	—	—	—	—	—	—	—	—	—	—	—	—
45	+	++	—	—	SCL	—	—	—	+++	+++	—	—
46	—	—	+++	—	—	—	—	—	—	—	—	—

[a] Homologous and group reactions are shown in bold type.

Table 3 (continued)
REACTIONS OF THE Vi-TYPE STRAINS OF *SALMONELLA TYPHI* WITH THE TYPING PHAGES IN ROUTINE TEST DILUTIONS[a]

Vi-Type Strains	Adapted Vi-Phage Preparations 35	36	37	38	39	40	41	42	43	44	45	46
A	CL	CL	CL	CL	CL	CL	CL	CL	CL	CL	CL	CL
B1	++	SCL	—	+++	+++	CL	+++	+++	CL	CL	+++	CL
B2	+	—	SCL	CL	CL	—	CL	++	CL	—	SCL	—
B3	+++	—	SCL	SCL	CL	+	CL	SCL	CL	++	—	CL
C1	—	+++	—	SCL	+++	CL	SCL	±	++	++	+	±
C2	—	±	—	—	—	+++	—	—	—	—	—	—
C3	—	SCL	—	++	±	CL	++	±	++	±	—	—
C4	—	—	—	—	—	—	—	—	—	—	—	—
C5	—	—	—	—	—	—	—	—	—	—	—	—
C6	—	—	—	—	—	—	—	—	—	—	—	—
C7	—	±	—	—	—	—	—	±	±	+	—	—
C8	—	—	—	—	—	—	—	—	—	—	—	—
C9	—	—	—	+	+	—	+	—	+	—	—	—
D1	—	±	—	±	±	±	±	—	—	—	—	++
D2	—	—	—	—	—	—	—	—	—	—	—	—
D4	—	—	—	—	—	—	—	—	—	—	—	—
D5	—	—	—	±	—	+	—	—	—	—	—	+
D6	—	SCL	—	—	—	CL	—	+++	—	—	—	±
D7	—	—	—	—	—	—	—	—	—	—	—	—
D8	—	—	—	—	—	—	—	—	—	—	—	—
D9	—	—	—	—	—	—	—	—	—	—	—	+
D10	—	—	—	—	—	—	—	—	—	—	—	—
D11	—	—	—	—	—	—	—	—	—	—	—	—
E1	—	—	—	±	±	±	±	—	±	—	—	SCL
E2	—	—	—	—	—	—	—	—	—	—	—	+++
E3	—	—	—	—	—	—	—	—	—	—	—	—
E4	—	—	—	—	—	—	—	—	—	—	—	—
E5	—	—	—	—	—	—	—	—	—	—	—	—
E6	—	—	—	—	—	—	—	—	—	—	—	—
E7	—	±	—	—	—	+	—	—	—	—	—	±
E8	—	—	—	—	—	—	—	—	—	—	—	—
E9	—	±	—	—	—	++	—	—	—	—	—	—
E10	—	—	—	—	—	—	—	—	—	—	—	—
F1	—	—	—	±	±	±	—	—	—	—	—	—
F2	—	—	—	—	—	—	—	—	—	—	—	—
F3	—	—	—	—	—	—	—	—	—	—	—	—
F4	—	—	—	—	—	—	—	—	—	—	—	—
F5	—	—	—	—	—	—	—	—	—	—	—	—
G	—	±	—	±	—	—	—	—	±	—	—	±
H	±	++	±	+	+++	+++	+++	±	++	±	±	CL
J1	—	—	—	—	—	—	—	—	—	—	—	—
J2	—	—	—	—	—	—	—	—	—	—	—	—
J3	—	—	—	—	—	—	—	—	—	—	—	—
K1	—	—	—	—	—	—	—	—	—	—	—	—
K2	—	—	—	—	—	—	—	—	—	—	—	—
L1	—	±	—	±	—	—	±	—	±	—	—	±
L2	—	—	—	±	—	—	±	—	—	—	—	±
M1	—	—	—	—	—	—	—	—	—	—	—	—
M2	—	—	—	—	—	—	—	—	—	—	—	—
M3	—	—	—	—	—	—	—	—	—	—	—	—
N	—	—	—	—	—	—	—	—	CL	—	—	±
O	—	CL	—	++	—	CL	±	SCL	CL	+++	—	CL
T	+++	SCL	±	++	CL	CL	++	+++	±	—	—	+
25	—	—	—	—	—	—	—	—	—	—	—	—
26	—	—	—	—	—	—	—	—	—	—	—	—
27	—	—	—	—	—	—	—	—	—	—	—	CL
28	—	—	—	—	SCL	±	—	—	—	—	—	—
29	++	CL	—	++	+++	CL	±	CL	+++	+++	+++	±
32	—	—	—	+	±	+	±	—	±	—	—	+
34	+++	—	—	CL	—	—	—	—	—	—	—	CL
35	**CL**	—	—	CL	—	—	—	—	+	—	—	CL
36	—	**CL**	—	—	—	CL	—	—	—	—	—	—
37	—	—	**CL**	—	—	—	—	—	—	—	—	±
38	—	—	—	**CL**	—	—	—	—	—	—	—	—
39	—	—	—	±	**CL**	—	±	—	±	±	—	—
40	—	±	—	—	—	**CL**	—	—	—	—	—	—
41	+	SCL	+	CL	+++	CL	CL	±	CL	+++	+++	CL
42	SCL	CL	—	+++	+++	CL	—	**CL**	SCL	SCL	—	CL
43	—	—	—	—	—	—	—	—	**CL**	—	—	CL
44	—	—	—	—	—	—	—	—	—	**CL**	—	—
45	+++	SCL	—	++	±	SCL	—	+++	—	±	**CL**	±
46	—	—	—	SCL	—	—	—	—	—	—	—	**CL**

[a] Homologous and group reactions are shown in bold type.

Taken from Bernstein, A., and Wilson, E. M. J., *J. Gen. Microbiol.*, *32*, 350-351 (1963). Reproduced by permission of the Society for General Microbiology, Reading, Berkshire, England.

Table 4
REACTIONS OF TYPES, SUBTYPES, AND VARIATIONS OF *SALMONELLA PARATYPHI* B WITH TYPING PHAGES IN ROUTINE TEST DILUTIONS[a]

Type Strains	Typing Phages										
	1	2	3a	3b	Jersey	Beccles	Taunton	B.A.O.R.	Dundee	Battersea	Worksop	1010
1	CL	CL	+∓+	+∓	CL	+∓+	+∓+	—	+++	—	—	—
1 var. 1	CL	CL	+∓	+∓	CL	SCL	SCL	+∓	SCL	—	—	CL
1 var. 2	CL	CL	CL	CL	CL	CL	CL	OL	+++	SCL	++	CL
1 var. 3	CL	CL	CL	CL	CL	+∓	—	OL	SCL	+++	++	CL
1 var. 4	CL	CL	CL	CL	CL	+∓	+	OL	SCL	+++	++	—
1 var. 5	SCL	CL	SCL	—	SCL	—	—	+	—	OL	—	SCL
1 var. 6	+++ SCL	+++ SCL	—	—	CL	—	—	—	—	OL	—	SCL
1 var. 7	+++	+++	—	OL	—	+++ OL	SCL	—	SCL	—	++	CL
1 var. 8	OL	OL	+∓	—	OL	—	—	—	+++	—	—	CL
1 var. 9	+++	+++	—	OL	+∓	CL	CL	OL	SCL	—	+++	CL
1 var. 10	CL	CL	CL	—	CL	CL	OL	—	∓	+++	—	CL
1 var. 11	OL	CL	—	—	CL	—	—	OL	—	+++	—	CL
1 var. 12	CL	CL	CL	CL	CL	—	—	OL	—	OL	SCL	CL
2	—	CL	—	—	±	—	—	—	+++	—	—	—
2 var. 1	—	CL	—	—	—	SCL	CL	—	SCL	—	—	OL
3a	—	—	CL	CL	—	SCL	OL	OL	SCL	+∓	++	OL
3a var. 2	—	—	CL	CL	—	—	—	—	+++	—	++	SCL
3a var. 4	—	—	OL	+++	—	—	—	SCL	SCL	—	±	OL
3a var. 6	—	—	OL	OL	—	SCL	OL	OL	+++	—	++	—
3a var. 7	—	—	CL	CL	—	CL	SCL	OL	<SCL	CL	CL	CL
3aI	—	—	OL	—	—	—	—	—	+++	—	—	+++
3aI var. 1	—	—	OL	—	—	OL	OL	+±+	SCL	—	—	OL
3aI var. 4	—	—	OL	—	—	+∓	+∓	—	+++	—	—	OL
3b	—	—	—	OL	—	SCL	OL	OL	SCL	—	++	OL
3b var. 1	—	—	—	+++	—	SCL	OL	OL	—	—	—	OL
3b var. 2	—	—	—	OL	—	—	—	—	+++	—	++	+++
3b var. 3	—	—	—	OL	—	—	—	OL	OL	—	++	SCL
3b var. 6	—	±	—	OL	OL	SCL	OL	OL	+++	—	++	CL
3b var. 7	—	++	—	SCL	—	—	—	—	+++	SCL	<CL	CL
Jersey	—	±	—	—	CL	SCL	OL	—	+++	—	—	CL
Jersey var. 1	—	—	—	—	OL	—	—	—	+++	—	—	CL
Jersey var. 2	—	—	—	++	SCL	SCL	SCL	—	++	—	—	—
Beccles	—	±	—	—	—	SCL	OL	—	+++	—	—	OL
Beccles var. 1	—	—	—	++	—	SCL	OL	+++	+++	—	—	+++
Beccles var. 2	—	—	—	—	—	CL	CL	—	OL	—	—	—
Beccles var. 3	—	—	—	—	—	SCL	SCL	—	—	—	—	OL
Beccles var. 4	—	—	—	—	—	SCL	OL	—	—	+++	—	OL
Beccles var. 5	—	—	—	—	—	SCL	OL	—	—	—	—	—
Beccles var. 6	—	±	—	—	—	SCL	SCL	—	<OL	CL	—	CL
Scarborough	—	+	—	—	—	+∓+	+++ SCL	—	SCL	—	—	OL
Taunton	—	—	—	—	—	—	SCL	—	SCL	—	—	OL
Taunton var. 1	—	—	—	—	—	—	SCL	—	SCL	—	—	—
B.A.O.R.	—	—	—	—	—	—	—	OL	—	—	—	OL
Dundee	—	—	—	—	—	—	—	—	SCL	—	—	SCL
Dundee var. 1	—	±	—	—	—	—	—	—	+++	—	—	—
Dundee var. 2	—	—	—	—	—	—	—	—	SCL	<SCL <OL	—	CL
Battersea	—	—	—	—	—	— ±	— ±	—	—	OL	—	±
Worksop	—	—	—	+	—	—	—	—	—	±	OL	+++

[a] Where a range of reactions may be found with a particular phage, the highest expected reaction is shown beneath the lowest, thus: + ∓.

From Anderson, E. S., in *The World Problem of Salmonellosis*, Van Oye, E., Ed., W. Junk, The Hague, 1964, pp. 94—95. With permission.

Table 5
LYTIC SPECTRA OF *STAPHYLOCOCCUS AUREUS* PHAGES

Test Strain	Phages																										
	29	52	52A	79	80	3A	3B	3C	55	71	6	7	42E	47	53	54	75	77	42D	81	187	42B	47C	52B	69	73	78
29	5	0	0	0	0	—	—	—	—	—	—	—	—	—	—	—	—	—	—	—	—	—	—	—	3	—	—
52	0	5	4	0	4	—	—	—	—	—	—	—	—	—	—	—	—	—	—	—	—	—	0	3	—	—	—
52A/79	—	3	5	5	3	—	—	—	—	—	—	—	—	—	—	—	—	—	—	—	—	—	—	1	—	—	—
80	—	1	1	—	5	—	—	—	—	—	—	1	2	—	—	—	—	—	—	5	—	0	0	—	—	3	—
2009	3	5	—	—	—	—	—	—	—	—	—	—	—	—	—	—	—	—	—	—	—	0	0	4	3	—	—
3A	1	1	1	1	1	5	3	4	4	4	1	1	1	1	1	1	—	1	1	1	—	1	1	2	1	1	1
3B	—	—	—	—	—	3	5	5	5	5	—	—	—	—	—	—	1	—	—	—	—	—	1	—	—	—	—
71	1	—	—	—	—	—	—	5	5	5	—	—	—	—	—	—	1	—	—	—	—	—	—	—	—	—	—
8719	—	—	—	—	—	—	0	—	—	5	—	—	—	—	—	—	—	—	—	—	—	—	—	—	—	—	—
42C	2	0	0	0	0	3	3	4	0/2	4	—	0	3	2	3	3	2	0	—	3	—	2	—	—	—	—	—
42E	0	0	0	0	0	—	—	—	—	—	—	—	5	—	3	2	—	—	0/2	3	—	2	1	—	—	—	3
47	3	3	3	3	3	—	—	—	—	—	2	2	—	5	5	3	5	5	—	0	—	2	4	5	1	2	—
53	—	—	—	—	—	—	—	—	—	—	0	0	—	0	5	4	5	5	—	1	—	—	—	—	—	0	—
54	1	—	—	2	—	2	2	2	—	—	—	5	3	5	5	5	5	5	—	3	—	4	4	2	1	4	3
75	—	—	—	2	1	—	2	2	—	—	—	0	1	0	4	0	5	5	—	—	—	—	—	—	1	0	—
77	—	—	—	—	2	—	—	—	—	—	—	0	—	2	4	0	—	5	—	—	—	—	—	4	1	—	0

Notes:

1. Phages propagated on strains not included in the test set are also tested on their propagating strains, on which the reaction is by definition "5".
2. The few minor differences between this table and the corresponding tables previously published derive partly from further experience with the phages and partly from the results of a comparative test of the phages currently in use in five national laboratories.
3. The notation 0/2 is used for reactions that are variable and may appear as inhibition reactions on one occasion and as true lytic reactions on another.

From Blair, J. E. and Williams, R. E. O., *Bull. W.H.O.*, 24, 778, 1961. With permission.

REFERENCES

1. **Twort, F. W.,** *Lancet,* 2, 1241, 1915.
2. **D'Hérelle, F. H.,** *C. R. Acad. Sci.* (Paris), 165, 373, 1917.
3. **D'Hérelle, F. H.,** *Le bactériophage, son role dans l'immunité,* Masson, Paris, 1921.
4. **D'Hérelle, F. H.,** *Le bactériophage et son comportement,* Masson, Paris, 1926.
5. **Marcuse, R. J.,** *Pathol. Bacteriol.,* 38, 409, 1934.
6. **Craigie, J. and Yen, C. H.,** *Can. J. Public Health,* 29, 448, 1938.
7. **Adams, M. H.,** *Bacteriophages,* Interscience, New York, 1959.
8. **Anderson, E. S.,** in *The World Problem of Salmonellosis,* Van Oye, E., Ed., W. Junk, The Hague, and Humanities Press, New York, 1964, p. 92.
9. **Felix, A. and Callow, B. R.,** *Lancet,* 2, 10, 1951.
10. **Scholtens, R. T.,** *Antonie van Leeuwenhoek J. Microbiol. Serol.,* 25, 403, 1959.
11. International Committee for Enteric Phage-Typing, *J. Hyg.* (Cambridge), 1, 59, 1973.
12. **Blair, J. E. and Williams, R. E. O.,** *Bull. W.H.O.,* 24, 771, 1961.
13. **Wilson, G. S. and Atkinson, J. D.,** *Lancet,* 1, 647, 1945.
14. **Popoff, M. and Vieu, J. F.,** *C. R. Acad. Sci.* (Paris), 270, 2219, 1970.
15. **Conn, H. J., Bottcher, E. J., and Randall, C.,** *J. Bacteriol.,* 49, 359, 1945.
16. **Roslycky, E. B., Allen, O. N., and McCoy, E.,** *Can. J. Microbiol.,* 8, 71, 1962.
17. **Schmidt, J. M. and Stanier, R. Y.,** *J. Gen. Microbiol.,* 39, 95, 1965.
18. **Lovett, P. S.,** *Bacteriol. Proc.,* p. 207, 1971.
19. **Drimmelen, G. C.,** *Bull. W.H.O.,* 23, 127, 1960.
20. **Jacob, M. M.,** *Nature,* 219, 752, 1968.
21. **Parnas, J.,** *Pathol. Microbiol.* Suppl. 3, 1, 1963.
22. **Philippon, A.,** *Ann. Inst. Pasteur Paris,* 115, 367, 1968.
23. **Prévot, A. R. and Thouvenot, H.,** *Ann. Inst. Pasteur Paris,* 101, 966, 1961.
24. **Fahey, J. E.,** *Can. J. Public Health,* 43, 167, 1952.
25. **Rische, H. and Endemann, D.,** *Arch. Roum. Pathol. Exp. Microbiol.,* 21, 337, 1962.
26. **Saragea, A. and Maximesco, P.,** *Bull. W.H.O.* 35, 681, 1966.
27. **Maximescu, P., Saragea, A., and Drăgoi, T.,** *Arch. Roum. Path. Exp. Microbiol.,* 31, 357, 1972.
28. **Thibaut, J. and Fredecicq, P.,** *C. R. Soc. Biol.,* 150, 1039, 1956.
29. **Nicolle, P., LeMinor, S., Hamon, Y., LeMinor, L., and Brault, G.,** *Rev. Hyg. Med. Soc.,* 8, 523, 1960.
30. **Kasatiya, S. S.,** D. Sc. thesis, University, of Paris, 1963.
31. **Deak, Z.,** *Acta Microbiol. Acad. Sci. Hung.,* 12, 261, 1965.
32. **Ackermann, H.-W., Nicolle, P., LeMinor, S., and LeMinor, L.,** *Ann. Inst. Pasteur Paris,* 103, 523, 1963.
33. **Smith, H. W. and Crabb, W. E.,** *J. Gen. Microbiol.,* 15, 556, 1956.
34. **Parisi, J. T., Russell, J. C., and Merlo, R. J.,** *Appl. Microbiol.,* 17, 721, 1969.
35. **Przondo-Hessek, A., Slopek, S., and Miodonska, J.,** *Arch. Immunol. Ther. Exp.,* 16, 402, 1968.
36. **Milch, H. and Deak, S.,** *Acta Microbiol.,* 11, 250, 1965.
37. **Slopek, S., Przondo-Hessek, A., Milch, H., and Deak, S.,** *Arch. Immunol. Ther. Exp.,* 15, 589, 1967.
38. **Sword, C. P. and Pickett, M. J.,** *J. Gen. Microbiol.,* 25, 241, 1961.
39. **Buraczewsak, M., Manowska, W., and Rdultowska, H.,** *Med. Dosw. Mikrobiol.,* 18, 225, 1966.
40. **Redmond, W. B., Cater, J. C., and Ward, D. M.,** *Am. Rev. Respir. Dis.,* 87, 257, 1963.
41. **Rado, T. A., Bates, J. H., Engel, H. W. B., Mankiewicz, E., Murohashi, T., Mizuguchi, Y., and Sula, L.,** *Am. Rev. Respir. Dis.,* 111, 459, 1975.
42. **Baess, I.,** *Acta Pathol. Microbiol. Scand.,* 76, 464, 1969.
43. **Bates, J. and Mitchison, D. A.,** *Am. Rev. Respir. Dis.,* 100, 189, 1969.
44. **Froman, S., Will, D. W., and Bogen, E.,** *Am. J. Public Health,* 44, 1326, 1954.
45. **Tokunaga, T., Maruyama, Y., and Murohashi, T.,** *Am. Rev. Respir. Dis.,* 97, 469, 1968.
46. **Gunnels, J. J. and Bates, J. H.,** *Am. Rev. Respir. Dis.,* 105, 388, 1972.
47. **Juhasz, S. E. and Bönicke, R.,** *Can. J. Microbiol.,* 11, 235, 1965.
48. **Rodda, G. M. J.,** *Aust. J. Exp. Biol. Med. Sci.,* 42, 457, 1964.
49. **Tokunaga, T. and Murohashi, T.,** *Jpn. J. Med. Sci. Biol.,* 16, 21, 1963.
50. **Murohashi, T., Tokunaga, T., Mizuguchi, Y., and Maruyama, Y.,** *Am. Rev. Respir. Dis.,* 8, 664, 1963.
51. **Cary, S. G. and Hunter, D. E.,** *J. Virol.,* 1, 538, 1967.
52. **Pietkiewicz, D. and Andrzejewski, J.,** *Zentralbl. Bakteriol Parasitenkd. Infekhonskr. Hyg. Abt. 1 Orig. Reihe A,* 224, 376, 1973.

53. **Pietkiewicz, D., Andrzejewski, J., Manowska, W., and Bogunowicz, A.,** *Zentralbl. Bakteriol. Parasitenkd. Infektionskr. Hyg. Abt. 1 Orig.,*231, 214, 1975.
54. **Pulverer, G., Sorgo, W., and Ko, H. L.,** *Zentralbl. Bakteriol. Parasitenkd. Infektionskr. Hyg. Abt. 1 Orig.,* 225, 353, 1973.
55. **France, D. R. and Markham, N. R.,** *J. Clin. Pathol.* (London), 21, 97, 1968.
56. **Pavlatou, M., Hassikou-Kaklamani, E., and Zantioti, M.,** *Ann. Inst. Pasteur Paris,* 108, 402, 1965.
57. **Vieu, J. F.,** *Zentralbl. Bakteriol. Parasitenkd. Infektionskr. Abt. 1 Orig.,* 171, 612, 1958.
58. **Vieu, J. F. and Capponi, M.,** *Ann. Inst. Pasteur Paris,* 108, 103, 1965.
59. **Schmidt, W. C. and Jeffries, C. D.,** *Appl. Microbiol.,* 27, 47, 1974.
60. **Graber, C. D., Latta, R. L., Vogel, E. H., and Brame, R. E.,** *Ann. Inst. Pasteur Paris,* 37, 54, 1962.
61. **Lindberg, R. B., Latta, R. L., Brame, R. E., and Moncrief, J. A.,** *Bacteriol. Proc.,* p. 81, 1964.
62. **Meitert, E.,** *Arch. Roum. Pathol. Exp. Microbiol.,* 24, 439, 1965.
63. **Pavlatou, M. and Klakamani, E.,** *Ann. Inst. Pasteur Paris,* 101, 914, 1961.
64. **Postic, B. and Finland, M.,** *J. Clin. Invest.,* 40, 2064, 1961.
65. **Bergan, T.,** *Acta Pathol. Microbiol. Scand.,* 80, 177, 1972.
66. **Staniewski, R.,** *Can. J. Microbiol.,* 16, 1003, 1970.
67. **Staniewski, R., Jurzyk, I., and Lorkiewicz, Z.,** *Acta Microbiol. Pol.,* 5, 21, 1973.
68. **Prévot, A. R., Vieu, J. F., Thouvenot, H., and Brault, G.,** *Bull. Acad. Natl. Med. Paris,* 154, 681, 1970.
69. **Atkinson, N. and Klauss, C.,** *Aust. J. Exp. Biol. Med. Sci.,* 33, 375, 1955.
70. **Sechter, I. and Gerichter, C. B.,** *Ann. Inst. Pasteur Paris,* 116, 190, 1969.
71. **Atkinson, N., Geytenbeek, H., Swann, M. C., and Wollaston, J. M.,** *Aust. J. Exp. Biol.,* 30, 333, 1952.
72. **Sechter, I. and Gerichter, C. B.,** *Appl. Microbiol.,* 16, 1708, 1968.
73. **Smith, H. W.,** *J. Gen. Microbiol.,* 5, 919, 1951.
74. **Lilleengen, K.,** *Acta Pathol. Microbiol. Scand.,* 27, 625, 1950.
75. **Lilleengen, K.,** *Acta Pathol. Microbiol. Scand.,* 30, 194, 1952.
76. **Ibrahim, A. E.,** *Appl. Microbiol.,* 18, 748, 1969.
77. **Schmidt, G. and Lüderitz, O.,** *Zentralbl. Bakteriol. Parasitenkd. Infektionskr. Hyg. Abt. 1 Orig.,* 210, 381, 1969.
78. **Bordini, A.,** *C. R. Acad. Sci.,* 270, 567, 1970.
79. **Guinée, P. A. M. and Scholtens, R. T.,** *Antonie van Leeuwenhoek J. Microbiol. Serol.,* 33, 25, 1967.
80. **Banker, D. D.,** *Nature,* 175, 309, 1955.
81. **Petrow, S., Kasatiya, S. S., Pelletier, J., Ackerman, H.-W., and Peloquin, J.,** *Ann. Microbiol.,* 125, 433, 1974.
82. **Smith, H. W.,** *J. Gen. Microbiol.,* 5, 472, 1951.
83. **Gershman, M.,** *Appl. Microbiol.,* 23, 831, 1972.
84. **Anderson, E. S.,** in *Health Congress Papers,* R. S. H. Health Congress, Torquay, Devonshire, England, 1960, p. 96.
85. **Callow, B. R.,** *J. Hyg.,* 57, 346, 1959.
86. **Felix, A.,** *J. Gen. Microbiol.,* 14, 208, 1956.
87. **Lilleengen, K.,** *Acta Pathol. Microbiol. Scand. Suppl.,* 77, 1948.
88. **Popovici, M., Nestoresco, N., Szégli, L., Bercovici, C., Iosub, C., and Besleaga, V.,** *Arch. Roum. Pathol. Exp. Microbiol.,* 21, 359, 1962.
89. **Scholtens, R. T.,** *Antonie van Leeuwenhoek J. Microbiol. Serol.,* 28, 373, 1962.
90. **Sechter, I. and Gerichter, C. B.,** *Ann. Inst. Pasteur Paris,* 113, 399, 1967.
91. **Wilson, V. R., Hermann, G. J., and Balows, A.,** *Appl. Microbiol.,* 21, 774, 1971.
92. **Garg, D. N. and Singh, I. P.,** *Antonie van Leeuwenhoek J. Microbiol. Serol.,* 39, 41, 1973.
93. **Guinée, P. A. M., VanLeeuwen, W. J., and Pruys, D.,** *Zentralbl. Bakteriol. Parasitenkd. Infektionskr. Abt. 1 Orig.,* 226, 194, 1974.
94. **Hamilton, R. L. and Brown, W. J.,** *Appl. Microbiol.,* 24, 899, 1972.
95. **Bercovici, C., Iosub, C., Besleaga, V., and Popa, S.,** *J. Hyg. Epidemiol. Microbiol. Immunol.,* 16, 282, 1972.
96. **Slopek, S. and Mulczyk, M.,** *Arch. Immunol. Ther. Exp.,* 8, 417, 1960.
97. **László, V. G., Milch, H., and Hajnal, A.,** *Acta Microbiol. Acad. Sci. Hung.,* 20, 135, 1973.
98. **Ogawa, T., Inagaki, Y., Takamatsu, M., Yamamoto, T., and Yoshikane, M.,** *Nagoya Med. J.,* 10, 119, 1964.
99. **Hammarström, E.,** *Acta Med. Scand. Suppl.,* 223, 1949.
100. **Rische, H.,** *Arch. Immunol. Ther. Exp.,* 16, 392, 1968.

101. **Slopek, S., Krukowska, A., and Mulczyk, M.,** *Arch. Immunol. Ther. Exp.*, 16, 519, 1968.
102. **Gromkova, R. and Trifonova, A.,** *Zentralbl. Bakteriol. Parasitenkd. Infektionskr. Hyg. Abt. 1 Orig.*, 204, 212, 1967.
103. **Van Boven, C. P. A.,** *Antonie van Leeuwenhoek J. Microbiol. Serol.*, 35, 232, 1969.
104. **Baldovin, A. C., Balteanu, E., Mihalco, F., Beloui, I., and Pleceas, P.,** *Arch. Roum. Pathol. Exp. Microbiol.*, 21, 385, 1962.
105. **Hunter, G. J. E.,** *J. Hyg.*, 44, 264, 1946.
106. **Ciuca, M., Baldovin, A. C., Mihalco, F., Beloiu, I., and Caffé, I.,** *Arch. Roum. Pathol. Exp. Microbiol.*, 18, 519, 1959.
107. **Caprioli, T., Zacour, F., and Kasatiya, S. S.,** *J. Clin. Microbiol.*, 2, 311, 1975.
108. **Evans, A. C.,** *U.S. Public Health Rep.*, 49, 1386, 1934.
109. **Wilkowske, H. H., Nelson, F. E., and Parmelee, C. E.,** *Iowa Agricultural Station Research Project No. 652*, 1954.
110. **Mukerjee, S.,** *Bull. W.H.O.*, 28, 337, 1963.
111. **Nicolle, P., Gallut, J., Ducrest, P., and Quiniou, J.,** *Rev. Hyg. Med. Soc.*, 10, 91, 1962.
112. **Hayward, A. C.,** *J. Gen. Microbiol.*, 35, 287, 1964.
113. **Nicolle, P., Mollaret, H., and Brault, J.,** *Bull. Acad. Natl. Med.*, 156, 712, 1972.
114. **Niléhn, B. and Ericsson, H.,** *Acta Pathol. Microbiol. Scand.*, 75, 177, 1969.

STERILIZATION, DISINFECTION, AND ANTISEPSIS

E. R. L. Gaughran and P. R. Borick

STERILIZATION

Sterilization is the process by which living organisms are removed or killed to the extent that they are no longer detectable in standard culture media in which they have previously been found to proliferate. The methods most commonly employed and the conditions required to effect sterilization are listed below in Tables 1 to 7.

STERILITY

Sterility is the freedom from living microorganisms achieved by the application of a process in which living organisms are removed or killed to the extent that they are no longer detectable in standard culture media in which they have previously been found to proliferate. Although sterility is in theory an absolute term, in practice it must be regarded as relative at best. Sterility is a probability phenomenon. Assurance of sterility depends upon the verification and reproducibility of the sterilization process. The sterilization process (other than filtration) involves the application of a set of conditions proven to be effective in killing expected loads of contaminating microorganisms (Table 8). From this concept developed the idea of challenging the process with large numbers of microorganisms especially resistant to a particular method of sterilization. These organisms, placed on carriers or product samples and commonly referred to as biological indicators, are incorporated in each batch of product to be sterilized.[15,21] In this way sterility can be assessed with a far greater degree of assurance than by testing product samples. This practice is generally employed by industry and hospitals.

Spores of the following organisms applied to a carrier or product sample are recommended for use as biological indicators in the various sterilization processes. In order to be acceptable as biological indicators, the designated resistance under specific parameters of exposure must be met (see Table 8).

The regulatory agencies of most countries permit a number of alternative procedures for certifying sterility. Tables 9, 10, and 11 give guidelines for sterility testing.[40] Details of the sterility tests may be found in References 26 and 40, information on culture media is given in References 1, 11, 12, and 39.

DISINFECTION AND ANTISEPSIS

Disinfection is the treatment of inanimate objects with a chemical in order to destroy pathogenic microorganisms other than spores. Some disinfectants fall short of this broad spectrum of activity; others go beyond it, due to their ability to kill bacterial spores.

Antisepsis is the application of a chemical agent to living tissues in order to control growth of microorganisms either by killing them or by preventing their growth.

Since many interrelated conditions (e.g., time, temperature, presence of organic matter, etc.) influence the action of disinfectants and antiseptics, Tables 12 and 13 attempt to indicate in general terms the antimicrobial activity usually attributed to these agents.

Table 1
DRY HEAT

150—160°C (302—320°F)	> 3 hr
160—170°C (320—338°F)	2—3 hr
170—180°C (338—356°F)	1—2 hr

Table 2
MOIST HEAT

(Steam under Pressure)

121°C (250°F)	15—30 min
132°C (270°F)	3—5 min
140°C (284°F)	1—2 min

Table 3
BOILING LIQUIDS

Boiling point of liquids — e.g., cumene (isopropylbenzene): 152°C (306°F)

Indirect	1 hr
Direct	2 min

Table 4
FILTRATION

TYPES OF FILTERS[a]

- Membrane filters[b]
  - Acrylonitryl polyvinylchloride copolymer
  - Cellulose acetate[c]
  - Cellulose nitrate
  - Cellulose triacetate
  - Mixture of cellulose acetate and nitrate
  - Nylon[c]
  - Polycarbonate
  - Polytetrafluoroethylene
  - Polyvinyl chloride[c]
  - Regenerated cellulose
  - Silver
- Asbestos-cellulose pads
- Unglazed porcelain candles
- Diatomaceous earth candles
- Sintered glass

[a] All filters available for use with positive or negative pressure, except diatomaceous earth candles.
[b] Available as disks or cartridges, except silver, which is available in disk form only.
[c] Not autoclavable.

Table 5
RADIATION

Source	Commonly used dose	Time to deliver dose
Isotopes	2.5 Mrad	Hours (dependent upon curies in the source)
Cobalt 60		
Cesium 137		
Electron Accelerators	2.5 Mrad	< 1 second
Electrostatic (Van de Graaff)		
Electromagnetic (Linac)		
Direct current		
Pulsed transformer		

Table 6
GASEOUS CHEMICALS

Gas employed	Temperature	Time	Concentration of chemosterilant	Relative humidity
Ethylene oxide, 100%	> 25°C (77°F)	> 3 hr	400—1200 mg/l	40—70%
Ethylene oxide diluted with dichlorodifluoromethane[a]	> 25°C (77°F)	> 3 hr	400—1200 mg/l	40—70%
Ethylene oxide diluted with carbon dioxide[a]	> 25°C (77°F)	> 3 hr	400—1200 mg/l	40—70%
Propylene oxide, 100%	> 25°C (77°F)	> 3 hr	800—2000 mg/l	30—60%
Propylene oxide diluted with dichlorodifluoromethane	> 25°C (77°F)	> 3 hr	800—2000 mg/l	30—60%
Propylene oxide diluted with carbon dioxide	> 25°C (77°F)	> 3 hr	800—2000 mg/l	30—60%
Formaldehyde	> 18°C (64°F)	24 hr	5—15 mg/l	Saturated
β-Propiolactone	> 18°C (64°F)	> 3 hr	5—10 mg/l	Saturated

[a] Advantage: less flammable or explosive than 100% ETO.

Table 7
LIQUID CHEMICALS

Solution for sporicidal action	Time
1% formaldehyde (water or alcohol)	18—24 hr
8% formaldehyde (70% ethanol or isopropanol)[a]	3—10 hr
2% aqueous glutaraldehyde[a]	3—10 hr
1% peracids	3—10 hr
2.5 *N* hydrochloric acid	1—3 hr
2.5 *N* sodium hydroxide	6—10 hr
1% chlorine (10,000 ppm)	3—10 hr
1% iodine	3—10 hr

[a] Approved by EPA for liquid chemical sterilization.

Table 8
STERILIZATION PROCESS

		Parameters			Performance	
Indicator[a] organism (as spores)	Process	Temperature (°C)	Relative humidity (%)	Sterilant concentration (mg/l)	Survival time (min)	Kill time (min)
Bacillus stearothermophilus	Steam	121 ± 0.5	100		5	5
Bacillus subtilis var. *niger* or a nontoxigenic strain of *Clostridium tetani*	Dry heat	165 ± 5			30	120
Bacillus subtilis var. *niger* or *Bacillus pumilus*	Ethylene oxide	25 ± 1	60 ± 20	300 ± 30	60	360
		54 ± 1	60 ± 20	600 ± 60	15	120
		54 ± 1	60 ± 20	1200 ± 120	5	30
					Absorbed dose (Mrad)	
Bacillus pumilus	Radiation				0.5	1.0

[a] Other organisms showing an order of resistance to the sterilization process comparable to or greater than that suggested may be used.

Table 9
GUIDELINES FOR STERILITY TESTING OF PRODUCTS STERILIZED BY FILTRATION

Sample	Number of test units[a]	Medium	Incubation temperature (°C)	Minimum incubation time (days)
Unit containers of product	30	Fluid thioglycollate	30—35	14[b]
	and			
	30	Soybean-casein digest	20—25	14[b]

[a] If the volume in the unit container is sufficient to permit testing in both media, the same test unit may be used to inoculate both media.

[b] If the membrane filtration test is employed, 7 days incubation is acceptable.

Table 10
GUIDELINES FOR STERILITY TESTING OF PRODUCTS STERILIZED BY OTHER METHODS[a]

	Sample	Number of test units	Medium	Incubation temperature (°C)	Minimum incubation time (days)
1.	Product in final container	20	Fluid thioglycollate	30—35	14
		and			
		20	Soybean-casein digest	20—25	14
2.	Product previously inoculated with indicator organism	10	Optimal[b]	Optimal[b]	Optimal[b]
3.	Carrier inoculated with indicator organism	10	Optimal[b]	Optimal[b]	Optimal[b]
	and				
	Product in final container	10	Fluid thioglycollate	30—35	10
		and			
		10	Soybean-casein digest	20—25	10

[a] In some countries, in the case of materials treated by gamma radiation (from isotope sources) and alpha radiation (electron beam), sterility may also be certified and the product released on the basis of dosimetry, without resorting to sterility testing or the use of biological indicators.
[b] Medium, temperature, and time of incubation are selected to be optimal for the indicator organism.

Table 11
GUIDELINES FOR STERILITY TESTING OF PRODUCTS STERILIZED IN THE FINAL CONTAINER BY STEAM UNDER PRESSURE

	Sample	Number of test units	Medium	Incubation temperature (°C)	Minimum incubation time (days)
1.	Product in final container	10	Fluid thioglycollate	30—35	7
		and			
		10	Soybean-casein digest	20—25	7
2.	Product previously inoculated with indicator organism	10	Optimal[a]	Optimal[a]	Optimal[a]
3.	Carrier inoculated with indicator organism	10	Optimal[a]	Optimal[a]	Optimal[a]
	and				
	Product in final container	5	Fluid thioglycollate	30—35	7
		and			
		5	Soybean-casein digest	20—25	7

[a] Medium, temperature, and time of incubation are selected to be optimal for the indicator organism.

Table 12
EFFECTS OF VARIOUS DISINFECTANTS ON MICROORGANISMS

+ = active; − = inactive

Class	Example	Concentration, %	Vegative bacteria	Tubercle bacillus	Bacterial spores	Higher fungi	Virus
Alcohols, aliphatic	Ethyl	70	+	+	−	+	+
	Isopropyl	70—90	+	+	−	+	−
Aldehydes	Formaldehyde	1—8	+	+	+	+	+
	Glutaraldehyde	2	+	+	+	+	+
Bisguanidines	Chlorhexidine	0.1—1.0	+	−	−	+	−
β-Propiolactone	β-Propiolactone	1	+		+	+	+
Cresols	Coal tar disinfectants	1—5	+	−	−	+	−
Epoxides	Ethylene oxide	1	+	+	+	+	+
	Propylene oxide	1	+	+	+	+	+
Halogens	Organic and inorganic halogen-releasing compounds	0.005—0.02 available halogen	+	−	−	+	+
Organotin compounds	Tri-*n*-butyltin salts	0.01—0.02	+	−	−	+	−
Peracids	Peracetic acid	1	+	+	+	+	+
Phenols	Phenol, phenyl phenol, and halogenated derivatives	0.5—5.0	+	+	−	+	+[a]
Surfactants							
Anionic	Alkyl aryl sulfonates[b]	0.1—0.25	+	−	−	−	−
Cationic	Quaternary ammonium salts[c]	0.02	+	−	−	+	−
Amphoteric	Alkyl di(aminoethyl)glycine[c]	1—5	+	−	−	+	−

[a] Phenol only.
[b] Acid pH (< 3).
[c] Alkaline pH.

Table 13
EFFECTS OF VARIOUS ANTISEPTICS ON MICROORGANISMS[a]

+ = active; − = inactive

Class	Example	Application	Concentration, %	Gram-positive bacteria	Gram-negative bacteria	Fungi
Acridines	Aminoacridine, HCl	Wounds	0.1	+	−	+
Alcohols	Ethyl	Intact skin	70	+	+	+
	Isopropyl	Intact skin	70—90	+	+	+
Bisguanidines	Chlorhexidine	Intact skin	0.5	+	+	+
		Wounds	0.05	+	−	−
Bisphenols	Hexachlorophene	Intact skin	2—3	+	−	+
		Wounds	0.5	+	−	−
Carbanilides	Trichlorocarbanilide	Intact skin	2	+	−	+
Heavy silver salts	Silver nitrate	Burns	0.5	+	+	+
Iodine compounds	I_2-NaI or KI, aqueous	Wounds	2—5	+	+	+
	I_2-NaI or KI in 50% alcohol	Intact skin	2	+	+	+
	Iodophors	Intact skin and wounds	1	+	+	+
Mercurials, organic	Sodium (ethylmercurithio) salicylate	Intact skin: tincture	0.1	+	−	−
		Wounds: aqueous	0.1	+	−	−
Nitrofurans	Nitrofurazone	Wounds	0.2	+	+	−
Oxyquinolines	Chloroquinaldol	Wounds	3	+	−	+
Peroxides	Hydrogen peroxide	Wounds	1.5—3.0	+	−	−
Pyrithiones	Zinc pyridinethione	Intact skin	0.1—0.5	+	+	+
Salicylanilides	Chloro- and bromosalicylanilides	Intact skin	2	+	−	+
Sulfa compounds	Mafenide HCl	Wounds	5	+	+	−

Table 13 (continued)
EFFECTS OF VARIOUS ANTISEPTICS ON MICROORGANISMS[a]

+ = active; − = inactive

Class	Example	Application	Concentration, %	Gram-positive bacteria	Gram-negative bacteria	Fungi
Surfactants						
Cationic	Quaternary ammonium salts	Intact skin	0.1—0.5	+	+	+
		Wounds	0.1—0.3	+	−	−
Amphoteric	Alkyl and acyl amino acids	Intact skin	1	+	+	−
Thiocarbamates	Tolnaftate	Wounds	1	−	−	+
Triphenylmethane dyes	Crystal violet	Wounds	0.5	+	−	+
Xylenols	2,4-Dichloro-*sym-meta*-xylenol	Intact skin	2	+	−	−

[a] See Reference 13 for a report of the FDA Advisory Review Panel on over-the-counter (OTC) topical antimicrobial drug products.

REFERENCES

1. *BBL Manual of Products and Laboratory Procedures,* 5th ed., BioQuest, Division of Becton, Dickinson & Co., Cockeysville, Md., 1968.
2. **Bernarde, M. A. (Ed.),** *Disinfection,* Marcel Dekker, New York, 1970.
3. **Block, S. S.,** *Disinfection, Sterilization and Preservation,* Lea & Febiger, Philadelphia, 1977.
4. **Borick, P. M.,** Antimicrobial agents as liquid chemosterilizers, *Biotechnol. Bioeng.,* 7, 435, 1965.
5. **Borick, P. M.,** Chemical sterilizers (chemosterilizers), *Adv. Appl. Microbiol.,* 10, 291, 1968.
6. **Borick, P. M. (Ed.),** *Chemical Sterilization,* Dowden, Hutchinson & Ross, Stroudsburg, Pa., 1973.
7. **Borick, P. M. and Borick, J. A.,** Sterility testing of pharmaceuticals, cosmetics and medical devices, in *Quality Control in the Pharmaceutical Industry,* Cooper, M. S., Ed., Academic Press, New York, 1972, pp. 1–38.
8. **Bruch, C. W.,** Gaseous sterilization, *Annu. Rev. Microbiol.,* 15, 245, 1961.
9. **Corum, C. J. (Ed.),** *Federal Regulations and Practical Control Microbiology for Disinfectants, Drugs and Cosmetics* (Special Publication No. 4), Society for Industrial Microbiology, Washington, D.C., 1969.
10. **Davis, J. G.,** Chemical sterilization, *Prog. Ind. Microbiol.,* 8, 141, 1968.
11. *Difco Manual,* 9th ed., Difco Laboratories, Detroit, 1953.
12. *Difco Supplementary Literature,* Difco Laboratories, Detroit, 1972.
13. *Fed. Regist.* (U.S.), 39(179), 33102–33141, 1974; 42(63), 17642–17681, 1977; 43(4), 1210–1249, 1978.
    Multiscience Press, Montreal, 1974, 1978.
14. **Gaughran, E. R. L. and Goudie, A. J. (Eds.),** *Sterilization by Ionizing Radiation,* Vols. I and II, Multiscience Press, Montreal, 1974, 1977.
15. **Gaughran, E. R. L. and Kereluk, K. (Eds.),** *Sterilization of Medical Products,* Johnson & Johnson, New Brunswick, N.J., 1977.
16. **Goldsmith, M.,** Ionizing radiation and the sterilization of medical products, in *Proceedings of the First International Symposium,* Taylor and Francis, London, 1964.
17. **Gucklhorn, I. R.,** Antimicrobials in cosmetics, *Manuf. Chem. Aerosol News,* 40(6), 25, 40(7), 38, 40(8), 71, 40(9), 33, 40(10), 33, 40(11), 35, 40(12), 43, 1969; 41(1), 42, 41(2), 30, 41(3), 26, 41(4), 34, 41(6), 44, 41(7), 51, 41(8), 28, 41(9), 82, 41(10), 49, 41(11), 48, 41(12), 50, 1970; 42(1), 34, 42(2), 33, 1971.
18. **Hedgecock, L. W.,** *Antimicrobial Agents,* Lea & Febiger, Philadelphia, 1967.
19. **Horn, H.,** *Biologische Prüfung und Leistungskriterien der Sterilisation,* VEB Verlag Volk und Gesundheit, Berlin, 1968.
20. **Hugo, W. B. (Ed.),** *Inhibition and Destruction of the Microbial Cell,* Academic Press, New York, 1971.
21. **Kereluk, K.,** Quality control in sterilization procedures: Biological indicators, in *Quality Control in Microbiology,* Prier, J. E., Bartola, J. T., and Friedman, H., Eds., University Park Press, Baltimore, 1975, pp. 25–39.
22. **Kereluk, K. and Lloyd, R. S.,** Ethylene oxide sterilization, a current review of principles and practices, *J. Hosp. Res.,* 7, 7, 1969.
23. **Lawrence, C. A. and Block, S. S. (Eds.)** *Disinfection, Sterilization, and Preservation,* Lea & Febiger, Philadelphia, 1968.
24. *Manual on Radiation Sterilization of Medical and Biological Materials,* (Technical Report Series No. 149), International Atomic Energy Agency, Vienna, 1973.
25. **McCulloch, E. C.,** *Disinfection and Sterilization,* 2nd ed. Lea & Febiger, Philadelphia, 1945.
26. American Pharmaceutical Association, *National Formulary XIV,* Mack, Easton, Pa., 1975.
27. **Parisi, A. N. and Borick, P. M.,** Pharmaceutical sterility testing, *Contam. Control Biomed. Environ.,* 8, 31, 1969.
28. **Perkins, J. J.,** *Principles and Methods of Sterilization in Health Sciences,* 2nd ed., Charles C Thomas, Springfield, Ill., 1969.
29. **Phillips, G. B. and Miller, W. S. (Eds.),** *Industrial Sterilization,* Duke University Press, Durham, N.C., 1973.
30. **U.S.P.,** *Proceedings of the U.S.P. Conference on Radiation Sterilization, October 1972, Washington, D.C.,* Jacob Slonim, Tel-Aviv, 1974.
31. *Radiosterilization of Medical Products and Recommended Code of Practice* (Proceedings Series), International Atomic Energy Agency, Vienna, 1967.
32. *Radiosterilization of Medical Products, 1974* (Proceedings Series), International Atomic Energy Agency, Vienna, 1975.
33. *Radiosterilization of Medical Products, Pharmaceutical and Bioproducts* (Technical Reports Series, No. 72), International Atomic Energy Agency, Vienna, 1967.

34. The Pharmaceutical Society of Great Britain, *Recent Development in the Sterilization of Surgical Materials* (Symposium held at the University of London, April 11–13, 1961), Smith and Nephew, London, 1961.
35. **Richards, J. W.,** *Introduction to Industrial Sterilization,* Academic Press, London, 1968.
36. **Rubbo, S. D. and Gardner, J. F.,** *A Review of Sterilization and Disinfection as Applied to Medical, Industrial and Laboratory Practice,* Lloyd-Duke, London, 1965.
37. **Stellmacher, W., Scholz, K., and Preissler, K.,** *Desinfektion,* Gustav Fischer, Jena, Germany, 1970.
38. **Sykes, G.,** *Disinfection and Sterilization,* 2nd ed., E. and F. N. Spon, London, 1965.
39. *The Oxoid Manual,* 3rd ed., Oxoid, London, 1971.
40. *The Pharmacopeia of the United States of America, XIX,* U.S.P., Bethesda, Md., 1975.
41. **Wallhäusser, K. H. and Schmidt, H.,** *Sterilisation, Desinfektion, Konservierung, Chemotherapie; Verfahren, Wirkstoffe, Prufüngmethoden,* Georg Thieme, Stuttgart, 1967.

ENUMERATION OF MICROORGANISMS

J. J. Gavin and D. P. Cummings

The following pages present general information regarding the methodology used to *estimate* the number of microorganisms in a given sample. The scope is limited to those methods used for the enumeration of cells; methods for the measurement of cellular mass are not considered in this section. Specific procedural details have not been included, because these will vary according to the interest of the investigator. For example, if one is interested in *Mycoplasma,* the number of cells can be estimated by direct observation of the sample or by turbidimetric measurement; however, prior to estimating their numbers it is necessary to concentrate these organisms rather than to dilute them, as is done with other types of cells. In applications that concern legal regulations, specific details of the procedures to be used are promulgated both in official and in nonofficial but expert manuals. Further, the investigator must decide the degree of importance that is attached to the enumeration, since no *absolute* value is obtainable. The value of the numbers is directly related to the skill and precision of the individual worker, his knowledge of his particular needs, and a not-too-rigid interpretation of his data. In many research situations it is advisable to determine cell mass rather than to enumerate the organisms.

The problem is most complex when working with mixed cell populations, such as those occurring in samples of natural origin, foods, soils, raw materials used in pharmaceutical preparations, water, air, and other sources. No universal culture medium is available that will provide optimal growth conditions for all the microorganisms that could be present. Other factors are the specific interaction between various organisms in competition for nutrients and the elaboration of stimulatory and/or inhibitory metabolic by-products by certain species, which change the environmental conditions so that they favor the growth of some types of cells and suppress the development of others. Relative numbers are also of importance, because dilution to obtain a countable range could eliminate a large number of organisms if a single organism were predominant. Each dilution in a tenfold dilution series eliminates 90% of the organisms contained in the previous dilution. Even with pure cultures there is no assurance that observed counts correlate with actual numbers. Viable counts in any nutrient-agar medium are only accurate to the extent of dispersion of the microorganisms throughout the diluting medium. Single colonies will develop from a single cell; but single colonies may also develop from clumps of 10, 100, or 1000 cells, a factor that cannot be determined.

Physical methods, such as particle-counting and turbidimetric measurement, do not differentiate between viable and nonviable cells. In addition, absorbance measurements may be more nearly related to total cellular mass than to the number of cells.[1] Environmental changes that take place during the growth of various organisms in pure culture may lead to aggregation simulating lysis. Turbidity measurements, in such cases, would underestimate the actual count.

Mycelial cultures are difficult to evaluate by turbidimetric methods. The hyphae clump, and the culture becomes thick with aggregates. Although it is possible to dilute such cultures to obtain absorbance values, the error is considerable. It is impossible to obtain more than a general impression of the quantity of these and other organisms that tend to grow in clusters by any enumeration method. Modifications of standard procedures can be applied in specific instances if enumeration is desired, but measurement of cellular mass would be more reliable.

Direct enumeration by microscopic examination may produce ambiguous results. The best precision is obtained when viral particles in pure culture are counted. The virions are easily recognizable in the electron microscope; however, as with the counting of other cells by this method, one cannot distinguish between viable (infectious) and nonviable (noninfectious) particles.

Counting of plaque or focal lesions approaches the ideal enumeration method for virions, because each plaque arises from a single viral particle. This is qualified by the fact that the absolute efficiency of infection (i.e., the total number of virions present in a sample as determined by electron microscope count relative to the total number of infectious particles) varies widely among different viruses – and even for the same virus counted in different host cell systems – due not only to the presence of noninfectious particles, but also to the failure of potential infectious particles to initiate an infection in a specific cell under appropriate conditions for multiplication. Plaque overlapping is a further source of error.

Despite the shortcomings of enumerative methods, it is possible to obtain valuable information from such procedures if the investigator (a) has an understanding of the nature of the sample to be tested, (b) gives attention and care to the preparation of the sample, (c) utilizes the proper methodology in context with the nature of the sample and the type(s) of microorganism(s) to be enumerated, (d) includes calibration standards when required and/or possible, (e) provides sufficient replicates to obtain statistically precise – if not accurate in the absolute sense – results, and (f) realizes that numerical results are only indicative.

Standard references dealing, in part, with procedures for the enumeration of microorganisms include the following:

1. *Federal Register* Promulgates regulations enforced by the Food and Drug Administration, the Division of Biological Standards (NIH), and the Department of Agriculture.
2. *National Formulary XIV* American Pharmaceutical Association, Washington, D.C.
3. *NASA Standards for the Microbiological Examination of Space Hardware* NHB 5340.1A, October 1968. U.S. Government Printing Office, Washington, D.C.
4. *Official Methods of Analysis of the Association of Official Agricultural Chemists* Association of Official Agricultural Chemists, Washington, D.C.
5. *Recommended Methods for the Microbiological Examination of Foods,* 2nd Edition American Public Health Association, Inc., Washington, D.C.
6. *Standard Methods for the Examination of Water and Waste Water,* 13th Edition American Public Health Association, Inc., Washington, D.C.
7. *Standard Methods for the Examination of Dairy Products,* 13th Edition American Public Health Association, Inc., Washington, D.C.
8. *Recommended Procedures for the Examination of Sea Water and Shellfish,* 4th Edition American Public Health Association, Inc., Washington, D.C.
9. *The United States Pharmacopeia,* 19th Revision U.S.P., Rockville, Md.

An excellent general reference is the series *Methods in Microbiology* (Norris, J. R. and Ribbons, D. W., Eds., Academic Press, New York, 1969.) Volumes 1, 3A, and 3B contain information that is applicable to the enumeration of microorganisms. A further reference containing pertinent information is *Isolation Methods for Microbiologists* (Shopton, D. A. and Gould, G. W., Eds., Academic Press, London, 1969).

MOST-PROBABLE-NUMBER (MPN) ESTIMATES

Theory

On the basis that microorganisms are uniformly distributed in liquid medium, it may be presumed that repeated samples of equal volumes from a single source will contain, *on the average,* the same number of microorganisms. This average is the most probable number (MPN). If a number of samples of varying volumes (for example, 5 samples of

10.0 ml each, 5 samples of 1.0 ml each, and 5 samples of 0.1 ml each) are inoculated into individual test tubes of nutrient medium and observed for growth after a suitable incubation period, it is possible to calculate the most probable number of organisms per unit volume, usually 100 ml, of the original samples.

Tables 1 to 4 indicate the estimated number of bacteria of the coliform group present in 100 ml of water according to various combinations of positive and negative results in the quantities used for the test. They are basically the tables originally computed by McCrady,[2] with certain amendments due to more precise calculations by Swaroop;[3] a few values have also been added to the tables from other sources, corresponding to further combinations of positive and negative results likely to occur in practice. Swaroop[4] has tabulated limits within which the real density of coliform organisms is likely to fall; his paper should be consulted by those anxious to know the precision of these estimates.

Basic Equipment Required

Sufficient glassware of adequate size to contain the selected volume(s) of sample plus medium; suitable environmental conditions for the desired application.

Sources of Error

Preparation of nonliquid samples, both in the extraction of microorganisms and in the even distribution of the material in the diluent used.

Applications

Because this method is amenable for the testing of a variety of samples, there are many applications for it. Although relatively inaccurate, it permits detection of very low concentrations of microorganisms.

TURBIDIMETRIC MEASUREMENT

Theory

The development of turbidity in suitable liquid nutrient medium inoculated with a given microorganism is a function of growth, reflecting increases in both mass and cell number. For enumeration purposes, changes in turbidity can be correlated with changes in cell numbers. Standard curves can be constructed, which – within an appropriate concentration range – may be used to estimate the microbial population from observed turbidity values.

Basic Equipment Required

A suitable instrument for the measurement of turbidity; numerous filter photometers, spectrophotometers, and direct-reading turbidimeters (nephelometers).

Optional Equipment and/or Accessories

Optically matched tubes; calibration standards; recording instruments; specialized equipment.

Sources of Error

INSTRUMENTAL

1. Nonlinearity of response of the light-sensitive devices and associated measuring circuits
2. Variation in the intensity of the light source
3. Stray light
4. Light scatter

Table 1
ONE TUBE OF 50 ml AND FIVE TUBES OF 10 ml

50-ml tubes positive	10-ml tubes positive	MPN per 100 ml
0	0	0
0	1	1
0	2	2
0	3	4
0	4	5
0	5	7
1	0	2
1	1	3
1	2	6
1	3	9
1	4	16
1	5	18+

From *The Bacteriological Examination of Water Supplies* Reports on Public Health and Medical Subjects, No. 71, 1957. Reproduced by permission of the Controller of Her Majesty's Stationery Office, London.

Table 2
ONE TUBE OF 50 ml, FIVE TUBES OF 10 ml, AND FIVE TUBES OF 1 ml

50-ml tubes positive	10-ml tubes positive	1-ml tubes positive	MPN per 100 ml	50-ml tubes positive	10-ml tubes positive	1-ml tubes positive	MPN per 100 ml
0	0	0	0	1	2	1	7
0	0	1	1	1	2	2	10
0	0	2	2	1	2	3	12
0	1	0	1	1	3	0	8
0	1	1	2	1	3	1	11
0	1	2	3	1	3	2	14
0	2	0	2	1	3	3	18
0	2	1	3	1	3	4	20
0	2	2	4	1	4	0	13
0	3	0	3	1	4	1	17
0	3	1	5	1	4	2	20
0	4	0	5	1	4	3	30
1	0	0	1	1	4	4	35
1	0	1	3	1	4	5	40
1	0	2	4	1	5	0	25
1	0	3	6	1	5	1	35
1	1	0	3	1	5	2	50
1	1	1	5	1	5	3	90
1	1	2	7	1	5	4	160
1	1	3	9	1	5	5	180+
1	2	0	5				

From *The Bacteriological Examination of Water Supplies,* Reports on Public Health and Medical Subjects, No. 71, 1957. Reproduced by permission of the Controller of Her Majesty's Stationery Office, London.

Table 3
FIVE TUBES OF 10 ml, FIVE TUBES OF 1 ml, AND FIVE TUBES OF 0.1 ml

10-ml tubes positive	1-ml tubes positive	0.1-ml tubes positive	MPN per 100 ml	10-ml tubes positive	1-ml tubes positive	0.1-ml tubes positive	MPN per 100 ml
0	0	0	0	3	2	2	20
0	0	1	2	3	3	0	17
0	0	2	4	3	3	1	20
0	1	0	2	3	4	0	20
0	1	1	4	3	4	1	25
0	1	2	6	3	5	0	25
0	2	0	4	4	0	0	13
0	2	1	6	4	0	1	17
0	3	0	6	4	0	2	20
1	0	0	2	4	0	3	25
1	0	1	4	4	1	0	17
1	0	2	6	4	1	1	20
1	0	3	8	4	1	2	25
1	1	0	4	4	2	0	20
1	1	1	6	4	2	1	25
1	1	2	8	4	2	2	30
1	2	0	6	4	3	0	25
1	2	1	8	4	3	1	35
1	2	2	10	4	3	2	40
1	3	0	8	4	4	0	35
1	3	1	10	4	4	1	40
1	4	0	11	4	4	2	45
2	0	0	5	4	5	0	40
2	0	1	7	4	5	1	50
2	0	2	9	4	5	2	55
2	0	3	12	5	0	0	25
2	1	0	7	5	0	1	30
2	1	1	9	5	0	2	45
2	1	2	12	5	0	3	60
2	2	0	9	5	0	4	75
2	2	1	12	5	1	0	35
2	2	2	14	5	1	1	45
2	3	0	12	5	1	2	65
2	3	1	14	5	1	3	85
2	4	0	15	5	1	4	115
3	0	0	8	5	2	0	50
3	0	1	11	5	2	1	70
3	0	2	13	5	2	2	95
3	1	0	11	5	2	3	120
3	1	1	14	5	2	4	150
3	1	2	17	5	2	5	175
3	1	3	20	5	3	0	80
3	2	0	14	5	3	1	110
3	2	1	17	5	3	2	140

Table 3 (continued)
FIVE TUBES OF 10 ml, FIVE TUBES OF 1 ml, AND FIVE TUBES OF 0.1 ml

10-ml tubes positive	1-ml tubes positive	0.1-ml tubes positive	MPN per 100 ml	10-ml tubes positive	1-ml tubes positive	0.1-ml tubes positive	MPN per 100 ml
5	3	3	175	5	4	5	425
5	3	4	200	5	5	0	250
5	3	5	250	5	5	1	350
5	4	0	130	5	5	2	550
5	4	1	170	5	5	3	900
5	4	2	225	5	5	4	1600
5	4	3	275	5	5	5	1800+
5	4	4	350				

From *The Bacteriological Examination of Water Supplies* (Reports on Public Health and Medical Subjects, No. 71), 1957. Reproduced by permission of the Controller of Her Majesty's Stationery Office, London.

5. Temperature rise in the measuring photocell
6. Dust, scratches, and imperfections in the optical system
7. Clarity and color of the suspending medium

MICROBIAL

1. Cell shape
2. Cell aggregation (clumping)
3. Cell settlement with uneven distribution of particles
4. Cell lysis
5. Viable/nonviable cell ratio

Applications

Studies of growth and metabolism of microorganisms; evaluation of stimulating and inhibiting effects of various compounds; osmotic effects; microbiological assay; control of commercial production of the products of microbial biosynthesis.

PARTICLE-COUNTING

Theory

Cells suspended in an electrolyte that passes through an electrical field of standard resistance within a small aperture will alter this resistance as they pass through the aperture. If a constant voltage is maintained, a cell passing through will cause a transient decrease in current as the resistance changes; if a constant current is maintained, a transient increase in voltage occurs. Such changes are amplified and recorded electronically, providing a count of the number of cells flowing through the opening. By relating the count obtained to either a metered volume or to a fixed time period for flow under a constant pressure head, an estimate of the total number of cells in a sample may be obtained.

Basic Equipment Required

Commercially available counting equipment.

Sources of Error

1. Foreign particles or bubbles
2. Ratio of signal to electronic noise background
3. Elevated background counting rates when the sensitivity of detection is increased for counting low concentrations of cells
4. Partial blocking of the aperture
5. High cell concentrations, which increase the probability of the coincident passage of two or more cells through the aperture

Table 4
THREE TUBES EACH INOCULATED WITH 10 ml, 1 ml, AND 0.1 ml OF SAMPLE

10-ml tubes positive	1-ml tubes positive	0.1-ml tubes positive	MPN per 100 ml	10-ml tubes positive	1-ml tubes positive	0.1-ml tubes positive	MPN per 100 ml
0	0	1	3	2	0	1	14
0	0	2	6	2	0	3	20
0	0	3	9	2	0	3	26
0	1	0	3	2	1	0	15
0	1	1	6	2	1	1	20
0	1	2	9	2	1	2	27
0	1	3	12	2	1	3	34
0	2	0	6	2	2	0	21
0	2	1	9	2	2	1	28
0	2	2	12	2	2	2	35
0	2	3	16	2	2	3	42
0	3	0	9	2	3	0	29
0	3	1	13	2	3	1	36
0	3	2	16	2	3	2	44
0	3	3	19	2	3	3	53
1	0	0	4	3	0	0	23
1	0	1	7	3	0	1	39
1	0	2	11	3	0	2	64
1	0	3	15	3	0	3	95
1	1	0	7	3	1	0	43
1	1	1	11	3	1	1	75
1	1	2	15	3	1	2	120
1	1	3	19	3	1	3	160
1	2	0	11	3	2	0	93
1	2	1	15	3	2	1	150
1	2	2	20	3	2	2	210
1	2	3	24	3	2	3	290
1	3	0	16	3	3	0	240
1	3	1	20	3	3	1	460
1	3	2	24	3	3	2	1100
1	3	3	29	3	3	3	1100+
2	0	0	9				

From Jacobs, M. B. and Gerstein, M. J., *Handbook of Microbiology,* D. Van Nostrand, New York, 1960.

Applications

All types of cell enumeration.

FOCAL-LESION DETERMINATION

Theory

Infection of a suitable host with a specific virus can produce a focal-lesion response that may be quantitated if – within an appropriate range of dilutions of virus – the average number of lesions is a linear function of the quantity of virus. The method is a true measure of the virus because the lesion is at the site of the activity of an individual viral particle. The relationship between the lesion count and virus dilution is consistent with a Poisson distribution.

Basic Equipment Required

Dependent upon the test system employed: focal lesions produced on the skin of whole animals, on the chorioallantoic membrane of chicken embryos, or in cell monolayers; focal lesions produced in agar plate cultures of bacterial viruses; focal lesions produced on the leaves of a plant rubbed with a mixture of virus and abrasive.

Sources of Error

1. Nonspecific inhibition of the response by unknown factors
2. Dilution errors
3. Counting errors
4. Overlapping of lesions
5. Size of the lesions and the number countable on the surface area
6. Insufficient replication to provide statistical significance

Applications

The enumeration of plant, animal, and bacterial viruses.

QUANTAL MEASUREMENT

Theory

The concentration of viruses may be calculated by statistical methods if an all-or-none (mortality) response results from infection. By use of a dilution series, the endpoint of activity is considered to be the highest dilution of the virus at which there were 50% or more positive responses. The exact endpoint is determined – by interpolation from the cumulative frequencies of positive and negative response observed at various dilutions – to be that dilution at which there are 50% positive and 50% negative responses. The reciprocal of the dilution yields an estimate of the number of viral units per inoculum volume of the undiluted sample and is expressed in multiples of the 50% endpoint.

Basic Equipment Required

Dependent upon the test system employed: groups of animals, chick embryos, or tubes of cell cultures.

Sources of Error

1. Dilution errors
2. Insufficient replication to provide statistical significance

Applications
Titration of animal viruses.

VIABLE-CELL COUNT

Theory
A single microorganism will, if viable, give rise to a visible colony under appropriate conditions for growth. Thus, if a microbial population is dispersed in a suitable diluent and added to a medium that allows for the fixation of individual cells at single, discrete points, the number of cells in a sample can be calculated by counting the number of colonies that develop after incubation and multiplying the number so obtained by the dilution factor.

Basic Equipment Required
Petri dishes; glassware for dilutions; calibrated pipettes; suitable environmental conditions for the desired application (incubators).

Optional Equipment and/or Accessories
Commercial colony counter, either with magnification alone or with magnification combined with an electronic counting probe; membrane filters; wire or glass spreaders; mechanical mixers, dip slides for specific applications; tubes for the roll-tube method; Pasteur pipettes for the drop-count method.

Sources of Error

1. Factors that influence the development of inoculated cells into viable colonies
2. Preparation of nonliquid samples, both in the extraction of microorganisms from the sample and in the even distribution of the material in the diluent used
3. Pipetting in the preparation of the dilution series
4. Cell aggregation (clumping)
5. Adhesion of cells to the spreaders.

Applications
All situations where an estimate of the viable number of microorganisms is required.

DIRECT MICROSCOPIC EXAMINATION

Theory
The number of cells in a given sample may be determined by microscopic examination of a portion of the material, counting the number of cells observed. By use of an aliquot of known volume or by inclusion of a counting standard, the number of cells in the original sample can be calculated. The precision and accuracy of this method are related to the number of cells per field and to the number of fields counted.

Basic Equipment Required
Standard light microscope; some type of support for the sample, slides, counting chamber, and membrane filter; biological stains for certain applications; internal counting standard, such as latex particles or India ink, if required; electron microscope for virus determinations.

Optional Equipment and/or Accessories
Phase-contrast microscope; fluorescent microscope; commercial counting chambers; capillary tubes; slide cultures.

Sources of Error

1. The precision of the individual performing the count
2. Differences in the thickness of the layer of the cell suspension
3. Cell aggregation (clumps or chains)
4. Staining techniques, if used, may cause error in either direction; loss of cells while preparing slides, or counting precipitated stain as cells.
5. Surface/volume ratio of the container in which the cell suspension is prepared or stored
6. Preparation of samples other than liquid
7. Use of vital stains when the ratio of viable to dead cells is determined

Applications

Milk and water examination; tissue cultures; clinical microbiology; virology.

REFERENCES

1. **Koch, A. L.,** *Biochem. Biophys. Acta,* 51, 429, 1961.
2. **McCrady, M. H.,** *Can. J. Public Health,* 9, 201, 1918.
3. **Swaroop, S.,** *Indian J. Med. Res.,* 26, 353, 1938.
4. **Swaroop, S.,** *Indian J. Med. Res.,* 39, 107, 1951.

STAINS FOR LIGHT MICROSCOPY

H. A. Lechevalier and F. J. Roisen

General information about the chemistry of dyes used as biological stains, their nomenclature, and methods of testing can be found in Reference 1. The following is a selection of staining methods that may be of assistance to microbiologists.

STAINING SOLUTIONS

Aceto Carmine – Heat an excess of powdered carmine in 45% acetic acid to boiling. Cool, then filter.[2]

Aceto Carmine, Billing's Modification – Add 1 g of carmine to 100 ml of boiling 45% acetic acid. Boil for about 2 minutes, cool, then filter. To half of this mixture add a few drops of a solution of ferric hydroxide in 45% acetic acid until the liquid becomes bluish red, but without visible precipitate, then add the untreated portion of the aceto-carmine mixture.[2]

Alcoholic Safranin – Dissolve 0.25 g of safranin O in 10 ml of 95% ethyl alcohol. Mix with 100 ml of distilled water.[2]

Ammonium Oxalate Crystal Violet – Dissolve 2 g of crystal violet in 20 ml of 95% ethyl alcohol. Mix with 80 ml of 1% (w/v) aqueous ammonium oxalate.[2]

Aqueous Nigrosin – Add 6 to 8 g of nigrosin to 100 ml of distilled water and dissolve by placing the mixture for 30 minutes in a boiling water bath. Replace any water lost by evaporation, then add 0.5 ml of formalin. Filter twice through a double thickness of filter paper.[2]

Aqueous Safranin – Prepare a 0.5% (w/v) solution of safranin in water.[2]

Carbol Rose Bengal – Dissolve 1 g of rose bengal in 100 ml of 5% (w/v) aqueous phenol. It is sometimes well to add 0.01 to 0.03 g of calcium chloride ($CaCl_2$) to the solution.[2]

Cotton Blue – Prepare 0.1% (w/v) cotton blue solution in lactic acid (U.S.P. grade). This may be further diluted with water.[3]

Crystal Violet in Dilute Alcohol – Dissolve 2 g of crystal violet in 20 ml of 95% ethyl alcohol. Mix with 80 ml of distilled water.[2]

Loeffler's Alkaline Methylene Blue – Dissolve 0.3 g of methylene blue in 30 ml of 95% ethyl alcohol. Mix with 100 ml of 0.01% (w/v) aqueous potassium hydroxide (KOH) solution.[2]

Melzer's Solution – Dissolve 1.5 g of potassium iodide (KI) and 0.5 g of iodine in 20 ml of water, then add 22 g of chloral hydrate.[3]

Methylene Blue in Dilute Alcohol – Dissolve 0.3 g of methylene blue in 30 ml of 95% ethyl alcohol. Mix with 100 ml of distilled water.[2]

Schiff Reagent – Dissolve 0.5 g of basic fuchsin by pouring over it 100 ml of boiling distilled water. Cool to 50°C. Filter the solution, then add 10 ml of 1*N* hydrochloric acid and 0.5 g of anhydrous potassium metabisulfite to the filtrate. Allow the solution to stand in the dark overnight. The solution should become colorless or pale straw-colored. If the solution is not completely decolorized, add 0.25 to 0.50 g of charcoal, shake thoroughly, and filter immediately. This solution will keep for several weeks in a tightly stoppered bottle.

Sudan Black B – Dissolve 0.5 g of Sudan black B in 100 ml of ethylene glycol.

Ziehl's Carbolfuchsin – Dissolve 0.3 g of basic fuchsin in 10 ml of 95% ethyl alcohol. Mix with 100 ml of 5% aqueous phenol.[3]

GENERAL STAINS FOR BACTERIA

Loeffler's alkaline methylene blue, methylene blue in dilute alcohol, ammonium oxalate crystal violet, alcoholic safranin, and aqueous safranin solutions are all good stains for bacterial preparations. Bacterial endospores will show as unstained elements.

NEGATIVE STAINING

India ink is a good, easy-to-use substance for negative staining in wet mounts. The ink, or a dilution thereof, is mixed with the microbial suspension, giving a black to sepia background against which the unstained microbial elements stand out clearly. Unfortunately, India ink is usually contaminated with bacteria and is, therefore, not very suitable for their study.

Aqueous nigrosin solution can be used as follows. With the help of a wire loop, stir a very small sample of a bacterial (or other) colony into a loopful of nigrosin on a coverslip. Place a second coverslip with a small drop of nigrosin (without bacteria) over the first one. The second slip should be turned through an angle of 45 degrees relative to the first one. Let the two drops flow together; then, gripping the corners of upper and lower slip, slide the two swiftly and gently apart (in a sideways movement). Let the two films so produced dry in air. Place the best one, face down, on a slide. Tack the corners down with wax.[5]

GRAM STAIN FOR BACTERIA: HUCKER'S MODIFICATION[2]

Staining Schedule

1. Stain bacterial smears for 1 minute with ammonium oxalate crystal violet.
2. Wash in tap water.
3. Immerse for 1 minute in an iodine solution prepared by adding 1 g of iodine and 2 g of potassium iodide (KI) to 300 ml of distilled water.
4. Wash in tap water; blot dry.
5. Decolorize for 30 seconds, with gentle agitation, in 95% alcohol; blot dry.
6. Counterstain for 10 seconds with alcoholic safranin solution.
7. Wash in tap water.
8. Dry, then examine the smears.

Results

1. Gram-positive organisms: blue
2. Gram-negative organisms: red

ACID-FAST STAIN FOR BACTERIA[2]

Staining Schedule

1. Prepare smears and fix them on a flat surface over boiling water.
2. Stain for 3 to 5 minutes with Ziehl's carbolfuchsin, applying heat to permit gentle steaming.
3. Rinse in tap water.
4. Decolorize in 95% alcohol containing 3% by volume of concentrated hydrochloric acid (HCl) until only a suggestion of pink remains.
5. Wash in tap water.

6. Counterstain with saturated aqueous methylene blue or Loeffler's methylene blue.
7. Wash in tap water.
8. Dry, then examine the smears.

Results

1. Acid-fast bacteria: red
2. Other organisms: blue

FAT-STAINING

Spread bacteria in a loop of distilled water on a grease-free coverslip. Let the film dry, then invert it over a drop of Sudan black B. Cytoplasm will be colorless, and fat droplets will be dark.[5]

OBSERVATION MEDIA FOR FUNGI

Fungi can, of course, be observed in water mounts, but clearing can be achieved by using 5 to 20% potassium hydroxide (KOH) or lactophenol.

Lactophenol of Amann[6] – Dissolve 1 g of crystalline phenol in a mixture of 1 g lactic acid, 2 g glycerol, and 1 g distilled water. Keep in a dark bottle.

GENERAL STAINS FOR FUNGI

Fungi can be stained for general observation by some of the stains used for bacteria. Methylene blue and cotton blue solutions are especially useful. Concentrated cotton blue may be diluted in lactophenol.

Melzer's solution will give reddish to black reactions with various parts of mycelia. In general, it is a detector of starch, glycogen, and related substances, giving the so-called amyloid (blue to black) reaction.

PERIODIC ACID-SCHIFF STAIN FOR FUNGI IN ANIMAL TISSUES[4]

Staining Schedule

1. Fix the specimen to be examined.
2. Dehydrate, embed, and section it.
3. Deparaffinize the sections, then dehydrate them with absolute alcohol.
4. Wash in distilled water.
5. Immerse in 1% periodic acid for 5 minutes.
6. Wash in running tap water for 15 minutes.
7. Stain in Schiff reagent for 10 to 15 minutes.
8. Transfer directly to two changes of either of the following solutions, allowing 5 minutes for each change:

(1)	10% potassium metabisulfite	5 ml
	1*N* hydrochloric acid	5 ml
	Distilled water	100 ml
(2)	Thionyl chloride	5 ml
	Distilled water	100 ml

9. Wash in running tap water for 10 minutes.
10. Counterstain with light green.
11. Dehydrate, clear, and mount.

Results

1. Fungal elements: red

TOLUIDINE BLUE-SAFRANIN STAIN FOR FUNGI IN PLANT TISSUES[7]

Staining Schedule

1. Flood sections of plant tissue with 1% (w/v) toluidine blue in a 1% (w/v) solution of borax in distilled water; place on a hot plate at 60 to 85°C until steam or a few bubbles appear.
2. After washing in tap water, flood with 0.25*N* solution of HCl in 50% (v/v) ethanol.
3. Decolorize in a 0.2*N* solution of NaOH in 50% (v/v) ethanol.
4. After washing in tap water, counterstain with 1% (w/v) safranin in water.
5. After washing in water, the sections may be dehydrated in alcohol and mounted as desired.

Results

1. Fungal hyphae and spores: deep red
2. Plant tissues: blue or greenish blue

REFERENCES

1. **Lillie, R. D. (Ed.),** *H. J. Conn's Biological Stains, A Handbook on the Nature and Uses of the Dyes Employed in the Biological Laboratory,* 8th ed., Williams & Wilkins, Baltimore, 1969.
2. **Conn, H. J., Darrow, M. A., and Emmel, V. M.,** *Staining Procedures Used by the Biological Stain Commission,* 2nd ed., Williams & Wilkins, Baltimore, 1960.
3. **Kühner, R. and Romagnesi, H.,** *Flore Analytique des Champignons Supérieurs,* Masson, Paris, 1953.
4. **Kligman, A. M. and Mescon, H.,** *J. Bacteriology,* 60, 415, 1950.
5. **Robinow, C.,** personal communication.
6. **Segretain, G., Drouhet, E., and Mariat, F.,** *Diagnostic de Laboratoire en Mycologie Médicale,* Editions de la Tourelle, St. Mandé, France, 1958.
7. **Pomerleau, R.,** *Can. J. Bot.,* 48, 2043, 1970.

PRESERVATION OF MICROORGANISMS

S. P. Lapage, K. F. Redway, and R. Rudge

INTRODUCTION

Preservation may be by subculture (including in living animals and in cell lines), by reduced metabolism, or by suspended animation (or nearly so). Emphasis in recent years has been on preservation by drying, by freeze-drying (lyophilization), and by freezing, including storage in liquid nitrogen. General accounts[1,2] give references to earlier work, and practical techniques[3] are described.

METHODS

Serial Subculture

This is the least satisfactory method, due to the inherent difficulties of mislabeling, contamination, loss of cultures, and changes in the cultures. However, it is simple and may be necessary for organisms that do not withstand freezing or drying. Stored on agar media in sealed tubes, many species will survive for years without subculture,[4] although others may require periodic transfer. Storage under mineral oil[5] is widely used for fungi; although messy, it prevents dehydration and prolongs survival.

Storage at Low Temperatures*

Storage at 4°C prolongs survival of many organisms, although sudden cooling, even above freezing temperatures, can lead to death due to "cold shock".[6,7] Storage at temperatures between 0°C and -30°C, especially without protective additives, is not recommended, because concentration of electrolytes above the eutectic point (common in this temperature range, although also occurring at lower temperatures) may damage the microorganisms.[8,9]

Storage in Distilled Water

Successful storage of *Pseudomonas* species[10,11] and of fungi[12] in distilled water at room temperature for several years has been reported. Coliform organisms have not survived under similar conditions.[13] Storage in saline has also proved successful for some bacteria.[14] These methods do not appear to have been widely used, however.

Drying

Sterile Soil or Sand

These have been used for the preservation of spore-bearing organisms.[15-17] Suspensions of the organisms are added to sterile soil or sand and are dried at room temperature. Kieselguhr and silica gel[18-21] have also been used.

Paper disks

Bacteria freeze-dried on filter paper disks and sealed in air between two layers of plastic film have survived short-term storage.[22]

Gelatin disks

These are prepared by mixing bacterial suspensions with melted gelatin, allowing the mixtures to set as drops and then drying them in a desiccator.[23] The method is simple and satisfactory for many microorganisms.

*See also "Freezing".

L-Drying

With this method, suspensions of the organisms are dried under vacuum from the liquid state without freezing taking place. The method may thus prove successful for organisms that do not withstand freeze-drying. There are several techniques for L-drying:

1. By using small volumes of suspension spread over a large surface area, the material dries by rapid evaporation before freezing can occur. Predried plugs of freeze-dried suspensions of peptone, starch or dextran have been used;[24] small volumes of suspensions of the organisms are dropped on to the plugs and then subjected to a vacuum.
2. Freezing can be prevented by restricting the water vapor flow from the material being dried, either by inserting cotton plugs into the ampoules[25] or by controlling the vacuum by means of a valve.[25,26]
3. Immersion of the ampoules in a water bath can maintain sufficient heat input to the suspensions to prevent them from freezing under vacuum.[27,28]

Freezing[8,29]

The process of injury to cells during freezing is not fully understood.[30] More resistant organisms – e.g., most bacteria, yeasts, and many fungi – are readily frozen. More sensitive organisms may require protective agents, such as glycerol[31] or dimethyl sulfoxide (DMSO),[30,32] to prevent injury due to intracellular ice formation caused by rapid freezing[33–35] or due to exposure to electrolyte concentrations during slow freezing.[36] Protective agents have been reviewed.[37]

There is conflicting evidence concerning the optimal rate of cooling, whether slow[38,39] or rapid,[40,41] and a two-step method (slow to around -20°C, then rapid) may be preferable.[3,42] However, whichever method is used, the rate of rewarming should generally be rapid.[30] Other factors – e.g., culture age, pH,[43,44] and cell concentration[44,45] – can also affect survival. The frozen cultures should preferably be stored below -30°C; -70°C is commonly used. Storage at -196°C is said to reduce the metabolism to virtually nil,[46] and liquid nitrogen systems have proved successful in the preservation of a wide range of microorganisms (see Separate Taxa).

Freeze-drying[47–49]

This widely used method is that of choice for microorganisms that will withstand the process. The ampoules produced are easily transported, and special storage apparatus is not essential, although viability may be prolonged if the ampoules are stored at 4°C, -30°C, or even lower, in preference to room temperature.[50]

The machinery and techniques[3,51,52] used for drying differ, but the process essentially consists of the removal of water by sublimation from suspensions either prefrozen or frozen by evaporative freezing under vacuum, as in centrifugal freeze-drying. Predrying cultural conditions may affect the viability of some organisms when freeze-dried – e.g., growth medium,[53] pH,[54] and culture age,[54–56] as well as the concentration of the suspension used.[57] Many media have been used for the suspension of organisms during freeze-drying (see Separate Taxa). The medium should contain the following:[58] a substance capable of maintaining the residual moisture content at the optimal level – e.g., carbohydrates such as glucose,[59] sucrose, or inositol (not all carbohydrates are protective; some are deleterious[60–62]); a compound that contains amino groups – e.g., glutamate, which neutralize carbonyl groups, said to be toxic;[63–65] minimal electrolytes; and a substance that forms a "cake" when dried – e.g., serum, dextran, or PVP – for additional protection[66] and to aid both sublimation and reconstitution. The addition of sodium glutamate to the medium will also improve stability at high storage temperatures.[54,67]

The residual water content in freeze-dried preparations is an important factor in their longevity,[63,68–74] and determination of the moisture content may be made by several methods.[75–78] Most freeze-drying processes involve a primary stage (often using a refrigerated vapor trap), in which the bulk of the water is rapidly removed, followed by a longer secondary stage (using a desiccant) to reduce the residual moisture content to around 1%.[79] P_2O_5 is commonly used as a desiccant, although molecular sieves may also be used.[80,81] Freeze-dried microorganisms are sensitive to oxygen[82,83] and to humidity, and therefore are commonly stored under high vacuum.[84] Various gases, usually nitrogen, may also be used,[85,86] although there can be problems with checking and sealing gas-filled ampoules.[87]

Sordelli's Method

This is a simple method of drying. An inner tube containing a small volume of the microorganisms emulsified in horse serum is placed in an outer tube containing P_2O_5, and a vacuum is applied for a few minutes; the outer tube is then sealed under the vacuum. The method has been described, and survival of many microorganisms for years has been obtained.[88–90]

Recovery

The percentage of recovery of dried microorganisms may be affected by the conditions of recovery; for example, for some organisms the temperature of recovery,[91–94] the recovery medium,[93–96] and even the rate of rehydration[97] and volume of recovery medium used[93] may be of importance. The rate of warming of frozen microorganisms[29,30] and the recovery medium[98,99] can also affect the percentage of recovery.

Viability

Both freezing and freeze-drying may cause some loss in viability, often extensive, with losses being up to 1000-fold compared to the original viable count. However, this is acceptable if the aim of preservation is to maintain viable cultures, provided that sufficient organisms are still easily recoverable. If methods of preservation are designed to produce products with a given number of viable organisms, such as vaccines, different techniques may be required and greater accuracy may be needed. The viabilities of a wide range of organisms have been given.[100,101] Acceleration of death of freeze-dried microorganisms by storage at elevated temperatures has been used to predict the stabilities at normal storage temperatures.[50,102–104]

Changes after Freezing, Drying, and Freeze-drying[105]

After drying and freeze-drying, changes have been observed in colonial appearance,[106,107] DNA,[107–110] enzyme systems,[111] pathogenicity[112,113] and antigenic structure.[114] Other studies however, have found no changes.[115–117] Damage to enzyme systems after freezing has been reported,[118–120] but in some other studies, as with pathogenicity,[121,122] no changes have been found.

In practice, after revival from the dried or frozen state, it is advisable to subculture the strain several times; this may restore characteristics lost during processing.[113]

SEPARATE TAXA

Algae*

Microalgae are usually maintained by periodic transfer of illuminated cultures. When

* Microalgae only.

the temperature is lowered (for example, to 10–15°C) and the light intensity decreased, some species need only annual transfer; others require subculture more frequently.[123] Storage in dry sand has proved successful for *Nostoc,*[124] and some algae can be maintained in soil.[125] *Tolpothrix tenuis* and *Calothrix brevissima* have been preserved for at least 2 years by both freeze-drying and drying on a special volcanic earth.[126] Some algae can be freeze-dried,[127-130] and members of the Chlorophyta, Chrysophyta, and Cyanophyta have survived 4 to 10 months by this method.[131] *Euglena gracilis* and other unicellular green algae have been preserved in liquid nitrogen for up to 3 years by use of cryoprotective agents.[132,133]

Bacteria[26,134-137,*]

Most bacteria can be readily maintained by periodic transfer on a variety of media[137,138] and by storage under oil.[139,140] Longevity may be increased by storage at low temperatures. However, many species can be successfully preserved by freeze-drying[26,134,137,141-143] or in liquid nitrogen.[101,144] Liquid nitrogen may be required for storage of organisms that do not readily withstand freeze-drying (for example, some autotrophic bacteria), although L-drying (q.v.) may also be used for some of these organisms.

Widely used media for suspension with freeze-drying are 7.5% glucose horse serum,[26] 7.5% glucose in a mixture of 1/3 nutrient broth and 2/3 horse serum (*Mist desiccans*),[145] and skim milk;[146,147] 5% inositol serum has been found superior for preservation of some bacteria.[62] Other media have been reviewed.[134,135] Special techniques may be useful for particular species, such as sterile soil for *Bacillus* species,[16] or porcelain beads.[148]

Gelatin disks (q.v.) or Sordelli's method (q.v.) are useful methods for a laboratory without specialized equipment. Paper disks (q.v.) may be valuable for easy distribution and short-term storage, but their value in long-term storage is untested.

Many species have survived in the freeze-dried state for up to 20 years or more.[26,100,137,143,149-152]

Bacteriophages

Bacteriophages can often be stored in broth at 4°C,[153-155] which has been found more satisfactory than freeze-drying for long-term storage.[155] Freeze-drying[153-160] and L-drying[25,161] have been used successfully for some phages, as has freezing at various temperatures.[154] However, storage in liquid nitrogen appears to be the most satisfactory method for long-term preservation.[156,162-164]

Cell Lines

Primary cells, diploid cells, and continuous cell lines can be successfully preserved in liquid nitrogen[165] with the addition of DMSO or glycerol. Most cell lines can be readily stored at -70°C.

Fungi[15,166,167,**]

Many fungi and yeasts withstand freeze-drying.[168-177] Suitable suspending media are serum,[117,178] sugars, and skim milk. Simple desiccation in a suitable medium can also be used for some species.[89,179]

Fungi can be readily preserved in liquid nitrogen,[177,180-190] which in conjunction with protective agents has proved successful for groups of fungi that do not withstand freeze-drying – e.g., *Entomophorales,* aquatic phycomycetes, strictly mycelial forms, and species with extra large or delicate spores.[166]

* Also see Mycoplasmas and Spirochaetes.

** Also see Yeasts.

Storage under mineral oil is a common technique.[5,191,192] Periodic transfer on various media is also used.[166,193] Special methods, such as spray-drying[194] and preservation in sterile soil,[15,195,196] anhydrous silica gel,[18-20] and dried-up agar cultures[15] have been successfully employed. Storage at around $-20°C$ has also proved successful.[197-201]

Freeze-dried fungi and yeasts have survived many years.[152,173-177] Many species have survived storage in liquid nitrogen for up to 8 years.[177,187]

Mycoplasmas

Many species withstand freeze-drying and storage for years,[202-206] although initial survival may be poor. Many protective compounds used for the freeze-drying of bacteria and viruses have no effect on the survival of mycoplasmas,[204] although sucrose has been successfully used.[205,207] Preservation by freezing at various temperatures and storage in liquid nitrogen have also proved successful with species difficult to freeze-dry, and the methods have been compared.[204,205] Otherwise, mycoplasmas are fully maintained by subculture and preserved on blocks of agar at approximately $-30°C$, although the more delicate species – e.g., *M. pneumoniae* – survive better at $-70°C$.

Protozoa

Preservation of protozoa has been carried out by subculture,[208] often because there is no alternative method. The temperature of storage may be reduced – e.g., to 10–15°C.

Freeze-drying has been used for a few protozoa: amoebae,[209,210] *Stentor* and *Frontonia* species,[209] slime molds,[210] and *Mastigina* species.[211] Successful preservation of *Strigomonas oncopelti* for up to 5 years has been obtained by L-drying.[212-214]

However, storage at low temperatures, using a cryoprotective agent such as glycerol or DMSO, is a more common method. Species of the following genera have been preserved in this way: *Crithidia,*[215] *Entamoeba,*[216] *Leishmania,*[217,218] *Paramecium,*[219] *Plasmodium,*[217,220-223] *Tetrahymena,*[219] *Toxoplasma,*[224] *Trichomonas,*[217,225-228] and *Trypanosoma.*[215,217,229-231] The cultures are usually stored at approximately $-70°C$ with the use of solid carbon dioxide.

Storage in liquid nitrogen has been successfully employed for the preservation of coccidia,[232,233] species of *Crithidia,*[234] *Entamoeba,*[234,235] *Paramecium,*[219] *Tetrahymena,*[236,237] *Toxoplasma,*[238] *Trichomonas,*[234,235] and *Trypanosoma.*[234,235]

Survival times at low temperatures vary considerably with the species and storage temperature, but some protozoa remain viable for months[226] and even years.[234]

Spirochaetes

These organisms have often proved difficult to preserve, except by subculture, which in the case of some *Treponema* species requires living animals. Stored in polysorbate medium at 24°C, some *Leptospira* species have survived for up to 4 years.[239] L-drying has been used for the successful preservation of *Leptospira*[214,240-243] and the Reiter treponeme.[214,244] Some species of *Leptospira, Borrelia* and *Treponema* have been freeze-dried.[245,246] *Borrelia anserina* has been successfully frozen,[247] as have *Treponema pallidum*[248] and various species of *Leptospira,*[249] including in liquid nitrogen.[250-252]

Viruses*[154,253-255]

In general, plant viruses are easier to preserve than animal viruses. Long-term storage of some plant viruses can be achieved by drying infected plant tissue[254] or by freeze-drying.[254,256,257] Some labile plant viruses have been preserved frozen at various temperatures[254,258] and by storage in liquid nitrogen.[254,259]

* Also see Bacteriophages.

Animal viruses are often preserved for short periods of time by refrigeration (approximately 4°C) in normal saline, or in glycerol solutions for transport.[255] Freeze-drying[255,260-265] with various suspending media has also been used for the preservation of some animal viruses: measles virus,[261,266-268] influenza virus,[73,74,269,270] polio virus,[261,271,272] vaccinia virus,[273-277] rubella virus,[278] foot-and-mouth disease virus,[279-282] various herpes viruses,[283-285] and the scrapie agent.[286] Many other viruses are held in lyophilized form by the American Type Culture Collection.[287] Freezing at various temperatures with protective substances can also be used for preserving animal viruses.[255,260-266,288-291] The animal viruses vary considerably in their resistance to freezing, freeze-drying, and storage.[263,265] Some viruses need continuous subculture in living cells.

Rickettsiaceae and Chlamydiaceae, though probably more related to bacteria than to viruses, are handled by viral techniques, and most species can be satisfactorily freeze-dried.[287]

Yeasts[89,170,292]

Methods used for fungi are satisfactory for yeasts. The mineral oil technique[293] has been used. Most yeasts freeze-dry readily[294-302] and have also been preserved in liquid nitrogen.[303] Freeze-dried yeasts survive well,[152,304,305] but attention may have to be paid to the reconstitution medium if the original fermentative powers are to be retained.[306]

REFERENCES

1. **Martin, S. M.,** Conservation of microorganisms, *Annu. Rev. Microbiol.,* 18, 1, 1964.
2. **Clark, W. A. and Loegering, W. Q.,** Functions and maintenance of a type-culture collection, *Annu. Rev. Phytopathol.,* 5, 319, 1967.
3. **Muggleton, P. W.,** The preservation of cultures, *Prog. Ind. Microbiol.,* 4, 191, 1963.
4. **Antheunisse, J.,** Preservation of microorganisms, *Antonie van Leeuwenhoek J. Microbiol. Serol.,* 38, 617, 1972.
5. **Buell, C. B. and Weston, W. H.,** Application of the mineral oil conservation method to maintaining collections of fungus cultures, *Am. J. Bot.,* 34, 555, 1947.
6. **Strange, R. E. and Dark, F. A.,** Effect of chilling on *Aerobacter aerogenes* in aqueous suspension, *J. Gen. Microbiol.,* 29, 719, 1962.
7. **Farrell, J. and Rose, A. H.,** Temperature effects on microorganisms, in *Thermobiology,* Rose, A. H., Ed., Academic Press, London, 1967, p. 147.
8. **Meryman, H. T. (Ed).,** Review of biological freezing, in *Cryobiology,* Academic Press, London, 1966, p. 1.
9. **Speck, M. L. and Cowman, R. A.,** Preservation of lactic streptococci at low temperatures, in *Proceedings of the First International Conference on Culture Collections,* Iizuka, H. and Hasegawa, T., Eds., University of Tokyo Press, Tokyo, 1970, p. 241.
10. **Berger, L. R.,** Proposed basis for the storage of viable bacterial clones in distilled water, in *Proceedings of the First International Conference on Culture Collections,* Iizuka, H. and Hasegawa, T., Eds., University of Tokyo Press, Tokyo, 1970, p. 265.
11. **DeVay, J. E. and Schnathorst, W. C.,** Single-cell isolation and preservation of bacterial cultures, *Nature London,* 199, 775, 1963.
12. **Castellani, A.,** Maintenance and cultivation of the common pathogenic fungi of man in sterile distilled water: Further researches, *J. Trop. Med. Hyg.,* 70, 181, 1967.
13. **Strange, R. E., Dark, F. A., and Ness, A. G.,** The survival of stationary phase *Aerobacter aerogenes* stored in aqueous suspension, *J. Gen. Microbiol.,* 25, 61, 1961.
14. **Chance, H. L.,** Salt – a preservative for bacterial cultures, *J. Bacteriol.,* 85, 719, 1963.
15. **Fennell, D. I.,** Conservation of fungus cultures, *Bot. Rev.,* 26, 79, 1960.
16. **Gordon, R. E. and Rynearson, T. K.,** Maintenance of strains of *Bacillus* species, in *Culture Collections: Perspectives and Problems,* Martin, S. M., Ed., University of Toronto Press, Toronto, 1963, p. 118.
17. **Raper, K. B.,** General methods for preserving cultures, in *Culture Collections: Perspectives and Problems,* Martin, S. M., Ed., University of Toronto Press, Toronto, 1963, p. 81.

18. **Perkins, D. D.**, Preservation of *Neurospora* stock cultures with anhydrous silica gel, *Can. J. Microbiol.*, 8, 591, 1962.
19. **Ogata, W. N.**, Preservation of *Neurospora* stock cultures with anhydrous silica gel, *Neurospora News Lett.*, 1, 13, 1962.
20. **Bell, J. V. and Hamalle, R. J.**, Viability and pathogenicity of entomogenous fungi after prolonged storage on silica gel at -20°C, *Can. J. Microbiol.*, 20, 639, 1974.
21. **Grivell, A. R. and Jackson, J. F.**, Microbial culture preservation with silica gel, *J. Gen. Microbiol.*, 58, 423, 1969.
22. **Coe, A. W. and Clark, S. P.**, Short-term preservation of cultures for transmission by post, *Mon. Bull. Minist. Health Public Health Lab. Serv.*, 25, 97, 1966.
23. **Lord Stamp**, The preservation of bacteria by drying, *J. Gen. Microbiol.*, 1, 251, 1947.
24. **Annear, D. I.**, The preservation of bacteria by drying in peptone plugs, *J. Hyg.*, 54, 487, 1956.
25. **Iijima, T. and Sakane, T.**, A method for preservation of bacteria and bacteriophages by drying *in vacuo*, *Cryobiology*, 10, 379, 1973.
26. **Lapage, S. P., Shelton, J. E., Mitchell, T. G., and MacKenzie, A. R.**, Culture collections and the preservation of bacteria, in *Methods in Microbiology*, Vol. 3A, Norris, J. R. and Ribbons, D. W., Eds., Academic Press, London, 1970, p. 135.
27. **Annear, D. I.**, Observations on drying bacteria from the frozen and from the liquid state, *Aust. J. Exp. Biol. Med. Sci.*, 36, 211, 1958.
28. **Annear, D. I.**, Preservation of microorganisms by drying from the liquid state, in *Proceedings of the First International Conference on Culture Collections*, Iizuka, H. and Hasegawa, T., Eds., University of Tokyo Press, Tokyo, 1970, p. 273.
29. **Smith, A. U.**, Effect of freezing and thawing on microorganisms, in *Biological Effects of Freezing and Supercooling* (Monographs of the Physiological Society, No. 9), Edward Arnold, London, 1961, p. 74.
30. **Mazur, P.**, Physical and chemical basis of injury in single-celled microorganisms subjected to freezing and thawing, in *Cryobiology*, Meryman, H. T., Ed., Academic Press, London, 1966, p. 213.
31. **Polge, C., Smith, A. U., and Parkes, A. S.**, Revival of spermatozoa after vitrification and dehydration at low temperatures, *Nature London*, 164, 666, 1949.
32. **Lovelock, J. E. and Bishop, M. W. H.**, Prevention of freezing damage to living cells by dimethyl sulphoxide, *Nature London*, 183, 1394, 1959.
33. **Nei, T., Araki, T., and Matsuaka, T.**, Freezing injury of aerated and non-aerated cultures of *Escherichia coli*, in *Freezing and Drying of Microorganisms*, Nei, T., Ed., University of Tokyo Press, Tokyo, 1969, p. 3.
34. **Postgate, J. R. and Hunter, J. R.**, The survival of frozen bacteria, *J. Gen. Microbiol.*, 26, 367, 1961.
35. **Nash, T., Postgate, J. R., and Hunter, J. R.**, Similar effects of various neutral solutes on the survival of *Aerobacter aerogenes* and of red blood cells after freezing and thawing, *Nature London*, 199, 1113, 1963.
36. **Lovelock, J. E.**, The mechanism of the protective action of glycerol against haemolysis by freezing and thawing, *Biochim. Biophys. Acta*, 11, 28, 1953.
37. **Vos, O. and Kaalen, M. C. A. C.**, Prevention of freezing damage to proliferating cells in tissue culture – a quantitative study of a number of agents, *Cryobiology*, 1, 249, 1965.
38. **Mazur, P.**, Physical and temporal factors involved in the death of yeast at subzero temperatures, *Biophys. J.*, 1, 247, 1961.
39. **Mazur, P., Rhian, M. A., and Mahlandt, B. G.**, Survival of *Pasteurella tularensis* in gelatin-saline after cooling and warming at subzero temperatures, *Arch. Biochem. Biophys.*, 71, 31, 1957.
40. **Levy, L.**, The effect of several rates of freezing and thawing on the viability of *Mycobacterium leprae*, *Cryobiology*, 6, 42, 1969.
41. **Calcott, P. H. and MacLeod, R. A.**, Survival of *Escherichia coli* from freeze-thaw damage: A theoretical and practical study, *Can. J. Microbiol.*, 20, 671, 1974.
42. **Polge, C. and Soltys, M. A.**, Protective action of some neutral solutes during the freezing of bull spermatozoa and trypanosomes, in *Recent Research in Freezing and Drying*, Parkes, A. S. and Smith, A. U., Eds., Blackwell Scientific, Oxford, 1960, p. 87.
43. **Lamprech, E. D. and Foster, E. M.**, The survival of starter organisms in concentrated suspensions, *J. Appl. Bacteriol.*, 26, 359, 1963.
44. **Johannsen, E.**, Influence of various factors on the survival of *Lactobacillus leichmannii* during freezing and thawing, *J. Appl. Bacteriol.*, 35, 415, 1972.
45. **Major, C. P., McDougal, J. D., and Harrison, A. P., Jr.**, The effect of the initial cell concentration upon survival of bacteria at –22°C, *J. Bacteriol.*, 69, 244, 1955.
46. **Meryman, H. T.**, Mechanics of freezing in living cells and tissues, *Science*, 124, 515, 1956.

47. **Meryman, H. T. (Ed.),** Freeze-drying, in *Cryobiology,* Academic Press, London, 1966, p. 609.
48. **Parkes, A. S. and Smith, A. U. (Eds.),** *Recent Research in Freezing and Drying,* Blackwell Scientific, Oxford, 1960.
49. **Rey, L. (Ed.),** *Aspects Theoriques et Industriels de la Lyophilisation,* Hermann, Paris, 1964.
50. **Greiff, D. and Rightsel, W. A.,** An accelerated storage test for predicting the stability of suspensions of measles virus dried by sublimation *in vacuo, J. Immunol.,* 94, 395, 1965.
51. **Rowe, T. W. G.,** The theory and practice of freeze-drying, *Ann. N.Y. Acad. Sci.,* 85, 679, 1960.
52. **Rowe, T. W. G.,** Machinery and methods in freeze-drying, *Cryobiology,* 8, 153, 1971.
53. **Ungar, J.,** Recent improvements in BCG vaccine, *Postgrad. Med. J.,* 40, 86, 1964.
54. **Miller, R. and Goodner, K.,** Studies on the stability of lyophilized BCG vaccine, *Yale J. Biol. Med.,* 25, 262, 1953.
55. **Fisher, P. J.,** The effects of freeze-drying on the viability of *Chromobacterium lividum, J. Appl. Bacteriol.,* 26, 502, 1963.
56. **Amarger, N., Jacquemetton, M., and Blond, G.,** Influence de l'âge de la culture sur la survie de *Rhizobium melitoti* à la lyophilisation et à la conservation après lyophilisation, *Arch. Mikrobiol.,* 81, 361, 1972.
57. **Record, B. R. and Taylor, R.,** Some factors influencing the survival of *Bacterium coli* on freeze-drying, *J. Gen. Microbiol.,* 9, 475, 1953.
58. **Greaves, R. I. N.,** Fundamental aspects of freeze-drying bacteria and living cells, in *Aspects Théoriques et Industriels de la Lyophilisation,* Rey, L., Ed., Hermann, Paris, 1964, p. 407.
59. **Ungar, J., Farmer, P., and Muggleton, P. W.,** Freeze-dried BCG vaccine, *Br. Med. J.,* 2, 568, 1956.
60. **Morichi, T.,** Nature and action of protective solutes in freeze-drying of bacteria, in *Proceedings of the First International Conference on Culture Collections,* Iizuka, H. and Hasegawa, T., Eds., University of Tokyo Press, Tokyo, 1970, p. 351.
61. **Marshall, B. J. and Scott, W. J.,** The effect of some solutes on preservation of dried bacteria during storage *in vacuo,* in *Proceedings of the First International Conference on Culture Collections,* Iizuka, H. and Hasegawa, T., Eds., University of Tokyo Press, Tokyo, 1970, p. 363.
62. **Redway, K. F. and Lapage, S. P.,** Effect of carbohydrates and related compounds on the long-term preservation of freeze-dried bacteria, *Cryobiology,* 11, 73, 1974.
63. **Scott, W. J.,** A mechanism causing death during storage of dried microorganisms, in *Recent Research in Freezing and Drying,* Parkes, A. S. and Smith, A. U., Eds., Blackwell Scientific, Oxford, 1960, p. 188.
64. **Marshall, B. J., Coote, G. G., and Scott, W. J.,** Some factors affecting the viability of dried bacteria during storage *in vacuo, Appl. Microbiol.,* 27, 648, 1974.
65. **Takano, M. and Terui, G.,** Correlation of dehydration with death of microbial cells in the secondary stage of freeze-drying, in *Freezing and Drying of Microorganisms,* Nei, T., Ed., University of Tokyo Press, Tokyo, 1969, p. 131.
66. **Record, B. R., Taylor, R., and Miller, D. S.,** The survival of *Escherichia coli* on drying and rehydration, *J. Gen. Microbiol.,* 28, 585, 1962.
67. **Cho, C. and Obayashi, Y.,** Effect of adjuvant on preservability of dried BCG vaccine at 37°C, *Bull. W.H.O.,* 14, 657, 1956.
68. **Greiff, D.,** Protein structure and freeze-drying: The effects of residual moisture and gases, *Cryobiology,* 8, 145, 1971.
69. **Scott, W. J.,** The effects of residual water on the survival of dried bacteria during storage, *J. Gen. Microbiol.,* 19, 624, 1958.
70. **Rey, L. (Ed),** L'humidité résiduelle des produits lyophilisés. Nature – Origine et méthodes d'étude, in *Aspects Théoriques et Industriels de la Lyophilisation,* Hermann, Paris, 1964, p. 199.
71. **Nei, T., Araki, T., and Souza, H.,** Studies of the effect of drying conditions on residual moisture content and cell viability in the freeze-drying of microorganisms, *Cryobiology,* 2, 68, 1965.
72. **Nei, T., Souza, H., and Araki, T.,** Effect of residual moisture content on the survival of freeze-dried bacteria during storage under various conditions, *Cryobiology,* 2, 276, 1966.
73. **Greiff, D. and Rightsel, W. A.,** Stability of suspensions of influenza virus dried to different contents of residual moisture by sublimation *in vacuo, Appl Microbiol.,* 16, 835, 1968.
74. **Greiff, D.,** Stabilities of suspensions of influenza virus dried by sublimation of ice *in vacuo* to different contents of residual moisture and sealed under different gases, *Appl. Microbiol.,* 20, 935, 1970.
75. **Fabián, J. and Sindelář, L.,** Determination of residual water in freeze-dried bacterial cultures by Fischer's methods, *Folia Microbiol. Prague,* 7, 247, 1962.
76. **Heckly, R. J.,** Rapid and precise measurement of moisture in biological materials, *Science,* 122, 760, 1955.

77. **Baker, P. R. W.**, The microdetermination of residual moisture in freeze-dried biological materials, *J. Hyg.*, 53, 426, 1955.
78. **Robinson, L. C.**, A gas-chromatographic method of measuring residual water in freeze-dried smallpox vaccine, *Bull. W.H.O.*, 47, 7, 1972.
79. **Robson, E. M. and Rowe, T. W. G.**, The physics of secondary drying, in *Recent Research in Freezing and Drying*, Parkes, A. S. and Smith, A. U., Eds., Blackwell Scientific, Oxford, 1960, p. 146.
80. **Barrer, R. M.**, Some aspects of molecular sieve science and technology, *Chem. Ind.*, 1968, p. 1203.
81. **Robson, E. M.**, The vacuum use of molecular sieves and other desiccants, *Vaccum*, 11, 10, 1961.
82. **Lion, M. B. and Bergmann, E. D.**, The effect of oxygen on freeze-dried *Escherichia coli*, *J. Gen. Microbiol.*, 24, 191, 1961.
83. **Israeli, E., Giberman, E., and Kohn, A.**, Membrane malfunctions in freeze-dried *Escherichia coli*, *Cryobiology*, 11, 473, 1974.
84. **Christian, R. T. and Stockton, J. J.**, The influence of sealing pressure on survival of *Serratia marcescens* and *Micrococcus pyogenes* var. *aureus* desiccated from the frozen state, *Appl. Microbiol.*, 4, 88, 1956.
85. **Greiff, D. and Rightsel, W.**, Stabilities of dried suspensions of influenza virus sealed in a vacuum or under different gases, *Appl. Microbiol.*, 17, 830, 1969.
86. **Marshall, B. J., Coote, G. G., and Scott, W. J.**, Effects of various gases on the survival of dried bacteria during storage, *Appl. Microbiol.*, 26, 206, 1973.
87. **Greiff, D., Melton, H., and Rowe, T. W. G.**, On the sealing of gas-filled glass ampoules, *Cryobiology*, 12, 1, 1975.
88. **Rhodes, M.** Viability of dried bacterial cultures, *J. Gen. Microbiol.*, 4, 450, 1950.
89. **Rhodes, M.**, Preservation of yeasts and fungi by desiccation, *Trans. Br. Mycol. Soc.*, 33, 35, 1950.
90. **Soriano, S.**, Sordelli's method for preservation of microbial cultures by desiccation in vacuum, in *Proceedings of the First International Conference on Culture Collections*, Iizuka, H. and Hasegawa, T., Eds., University of Tokyo Press, Tokyo, 1970, p. 269.
91. **Speck, M. L. and Myers, R. P.**, The viability of dried skim-milk cultures of *Lactobacillus bulgaricus* as affected by the temperature of reconstitution, *J. Bacteriol.*, 52, 657, 1946.
92. **Wasserman, A. E. and Hopkins, W. J.**, Studies in the recovery of viable cells of freeze-dried *Serratia marcescens*, *Appl. Microbiol.*, 5, 295, 1957.
93. **Leach, R. H. and Scott, W. J.**, The influence of rehydration on the viability of dried microorganisms, *J. Gen. Microbiol.*, 21, 295, 1959.
94. **Ray, B., Jezeski, J. J., and Busta, F. F.**, Effect of rehydration on recovery, repair and growth of injured freeze-dried *Salmonella anatum*, *Appl. Microbiol.*, 22, 184, 1971.
95. **Baird-Parker, A. C. and Davenport, E.**, The effect of recovery medium on the isolation of *Staphylococcus aureus* after heat treatment and after the storage of frozen or dried cells, *J. Appl. Bacteriol.*, 28, 390, 1965.
96. **Weiler, W. A. and Hartsell, S. E.**, Diluent composition and the recovery of *Escherichia coli*, *Appl. Microbiol.*, 18, 956, 1969.
97. **Cox, C. S., Harris, W. J., and Lee, J.**, Viability and electron microscope studies of phages T3 and T7 subjected to freeze-drying, freeze-thawing and aerosolization, *J. Gen. Microbiol.*, 81, 207, 1974.
98. **Postgate, J. R. and Hunter, J. R.**, Metabolic injury in frozen bacteria, *J. Appl. Bacteriol.*, 26, 405, 1963.
99. **Ray, B. and Speck, M. L.**, Repair of injury induced by freezing *Escherichia coli* as influenced by recovery medium, *Appl. Microbiol.*, 24, 258, 1972.
100. **Rhoades, H. E.**, Effects of 20 years' storage on lyophilized cultures of bacteria, molds, viruses and yeasts, *Am. J. Vet. Res.*, 31, 1867, 1970.
101. **Clark, W. A.**, The American Type Culture Collection: Experiences in freezing and freeze-drying microorganisms, viruses, and cell lines, in *Proceedings of the First International Conference on Culture Collections*, Iizuka, H. and Hasegawa, T., Eds., University of Tokyo Press, Tokyo, 1970, p. 309.
102. **Damjanović, V. and Radulovia, D.**, Predicting the stability of freeze-dried *Lactobacillus bifidus* by the accelerated storage test, *Cryobiology*, 5, 101, 1968.
103. **Damjanović, V.**, Kinetics of thermal death and prediction of the stabilities of freeze-dried streptomycin-dependent live *Shigella* vaccines, *J. Biol. Stand.*, 2, 297, 1974.
104. **Mitic, S., Otenhajamer, I., and Damjanović, V.**, Predicting the stabilities of freeze-dried suspensions of *Lactobacillus acidophilus* by the accelerated storage test, *Cryobiology*, 11, 116, 1974.

105. **Servin-Massieu, M.**, Effects of freeze-drying and sporulation on microbial variation, *Curr. Top. Microbiol. Immunol.*, 54, 119, 1971.
106. **Subramaniam, M. K. and Prahalado-Rao, P. L.**, Lyophilization and mutation in yeast, *Experientia*, 7, 98, 1951.
107. **Servin-Massieu, M. and Cruz-Camarillo, R.**, Variants of *Serratia marcescens* induced by freeze-drying, *Appl. Microbiol.*, 18, 689, 1969.
108. **Webb, S. J.**, Mutation of bacterial cells by controlled desiccation, *Nature London*, 213, 1137, 1967.
109. **Webb, S. J.**, Some effects of dehydration on the genetics of microorganisms, in *Freezing and Drying of Microorganisms*, Nei, T., Ed., University of Tokyo Press, Tokyo, 1969, p. 153.
110. **Hieda, K. and Ito, T.**, Induction of genetic change by drying in yeast, in *Freeze-Drying of Biological Materials* (Proceedings of C-1 Symposium, Sapporo), International Institute of Refrigeration, Paris, 1973, p. 71.
111. **Vaschenko, L. N.**, The effect of nutritional conditions on the growth of bacterial cells intact and injured in the process of lyophilization, *Zh. Mikrobiol. Epidemiol. Immunobiol.*, 46, 107, 1969.
112. **Priestley, F. W.**, Freeze-drying of the organism of contagious bovine pleuropneumonia, *J. Comp. Pathol. Ther.*, 62, 125, 1952.
113. **Heckly, R. J., Anderson, W. W., and Rockenmacher, M.**, Lyophilization of *Pasteurella pestis*, *Appl. Microbiol.*, 6, 255, 1958.
114. **Lambin, S., German, A., and Sigrist, W.**, Influence de la lyophilisation sur le constituants et le pouvoir antigéniques d' *E. typhosa* et de *S. paratyphi B*, *C. R. Soc. Biol.* (Paris), 152, 1650, 1958.
115. **Jennens, M. G.**, The effect of desiccation on antigenic structure, *J. Gen. Microbiol.*, 10, 127, 1954.
116. **Sharpe, M. E. and Wheater, D. M.**, The physiological and serological characters of freeze-dried lactobacilli, *J. Gen. Microbiol.*, 12, 513, 1955.
117. **Mehrotra, B. S., Tandon, G. D., Maurya, J. N., Chopra, B. K., and Prasad, R.**, Preservation of industrial cultures by lyophilization, in *Proceedings of the First International Conference on Culture Collections*, Iizuka, H. and Hasegawa, T., Eds., University of Tokyo Press, Tokyo, 1970, p. 319.
118. **MacLeod, R. A., Smith, L. D. H., and Gelinas, R.**, Metabolic injury to bacteria. I. Effect of freezing and storage on the requirements of *Aerobacter aerogenes* and *Escherichia coli* for growth, *Can. J. Microbiol.*, 12, 61, 1966.
119. **Moss, C. W. and Speck, M. L.**, Identification of nutritional components in trypticase responsible for recovery of *Escherichia coli* injured by freezing, *J. Bacteriol.*, 91, 1098, 1966.
120. **Speck, M. L. and Cowman, R. A.**, Metabolic injury to bacteria resulting from freezing, in *Freezing and Drying of Microorganisms*, Nei, T., Ed., University of Tokyo Press, Tokyo, 1969, p. 39.
121. **Sorrells, K. M., Speck, M. L., and Warren, J. A.**, Pathogenicity of *Salmonella gallinarum* after metabolic injury by freezing, *Appl. Microbiol.*, 19, 39, 1970.
122. **Davies, G., Hebert, N., and Casey, A.**, The preservation of *Brucella abortus* (strain 544) in liquid nitrogen and its virulence when subsequently used as a challenge, *J. Biol. Stand.*, 1, 165, 1973.
123. **Starr, R. C.**, Culture collections of algae, in *Culture Collections: Perspectives and Problems*, Martin, S. M., Ed., University of Toronto Press, Toronto, 1963, p. 136.
124. **Venkataraman, G. S.**, A method of preserving blue-green algae for seeding purposes, *J. Gen. Appl. Microbiol.*, 7, 96, 1961.
125. **Dietz, A.**, General discussion, in *Culture Collections: Perspectives and Problems*, Martin, S. M., Ed., University of Toronto Press, Toronto, 1963, p. 116.
126. **Watanabe, A.**, Some devices for preserving blue-green algae in viable state, *J. Gen. Appl. Microbiol.*, 5, 153, 1959.
127. **Holm-Hansen, O.**, Effect of varying residual moisture content on the viability of lyophilized algae, *Nature London*, 198, 1014, 1963.
128. **Holm-Hansen, O.**, Viability of lyophilized algae, *Can. J. Bot.*, 42, 127, 1964.
129. **Holm-Hansen, O.**, Factors affecting the viability of lyophilized algae, *Cryobiology*, 4, 17, 1967.
130. **Takano, M., Sado, J., Ogawa, T., and Terui, G.**, Freezing and freeze-drying of *Spirulina platensis*, *Cryobiology*, 10, 440, 1973.
131. **Daily, W. A. and McGuire, J. M.**, Preservation of some algal cultures by lyophilization, *Butler Univ. Bot. Stud.*, 11, 139, 1953–54.
132. **Hwang, S. W. and Horneland, W.**, Survival of algal cultures after freezing by controlled and uncontrolled cooling, *Cryobiology*, 1, 305, 1965.
133. **Hwang, S. W.**, Problems in superlow temperature preservation of microorganisms, in *Freezing and Drying of Microorganisms*, Nei, T., Ed., University of Tokyo Press, Tokyo, 1969, p. 169.

134. **Heckly, R. J.**, Preservation of bacteria by lyophilization, in *Advances in Applied Microbiology,* Vol. 3, Umbreit, W. W., Ed., Academic Press, New York, 1961, p. 1.
135. **Fry, R. M.**, Freezing and drying of bacteria, in *Cryobiology,* Meryman, H. T., Ed., Academic Press, London, 1966, p. 665.
136. **Sturdza, S. A.**, Dessiccation directe et lyophilisation en bactériologie (revue générale), *Arch. Roum. Pathol. Exp. Microbiol.,* 30, 25, 1971.
137. **Lapage, S. P. and Redway, K. F.**, *Preservation of Bacteria with Notes on other Microorganisms* (Public Health Laboratory Service Monograph No. 7) Willis, A. T. and Collins, C. H., Eds., Her Majesty's Stationery Office, London, 1974.
138. **Lapage, S. P., Shelton, J. E., and Mitchell, T. G.**, Media for the maintenance and preservation of bacteria, in *Methods in Microbiology,* Vol. 3A, Norris, J. R. and Ribbons, D. W., Eds., Academic Press, London, 1970, p. 1.
139. **Hartsell, S. E.**, The preservation of bacterial cultures under paraffin oil, *Appl Microbiol.,* 1, 36, 1953.
140. **Hartsell, S. E.**, Maintenance of cultures under paraffin oil, *Appl. Microbiol.,* 4, 350, 1956.
141. **Floodgate, G. D. and Hayes, P. R.**, The preservation of marine bacteria, *J. Appl. Bacteriol.,* 24, 87, 1961.
142. **Lelliott, R. A.**, The preservation of plant-pathogenic bacteria, *J. Appl Bacteriol.,* 28, 181, 1965.
143. **Sourek, J.**, Long-term preservation by freeze-drying of pathogenic bacteria of the Czechoslovak National Collection of Type Cultures, *Int. J. Syst. Bacteriol.,* 24, 358, 1974.
144. **Jarvis, J. D., Wynne, C. D., and Telfer, E. R.**, Storage of bacteria in liquid nitrogen, *J. Med. Lab. Technol.,* 24, 312, 1967.
145. **Fry, R. M. and Greaves, R. I. N.**, The survival of bacteria during and after drying, *J. Hyg.,* 49, 220, 1951.
146. **Hornibrook, J. W.**, A useful menstruum for drying organisms and viruses, *J. Lab. Clin. Med.,* 35, 788, 1950.
147. **Sinha, R. N., Dudani, A. T., and Ranganathan, B.**, Protective effect of fortified skim milk as suspending medium for freeze-drying of different lactic acid bacteria, *J. Food Sci.,* 39, 641, 1974.
148. **Hunt, G. A., Gourevitch, A., and Lein, J.**, Preservation of cultures by drying on porcelain beads, *J. Bacteriol.,* 76, 453, 1958.
149. **Steel, K. J. and Ross, H. E.**, Survival of freeze-dried bacterial cultures, *J. Appl. Bacteriol.,* 26, 370, 1963.
150. **Harrison, A. P. and Pelczar, M. J.**, Damage and survival of bacteria during freeze-drying and during storage over a ten-year period, *J. Gen. Microbiol.,* 30, 395, 1963.
151. **Aktan, M.**, A laboratory investigation on the percentage of viable cells in cultures of bacteria 12 years after lyophilisation, *Inf. Bull. Int. Cent. Inf. Distrib. Type Cult.,* 2(4), 29, 1967–68.
152. **Davis, R. J.**, Viability and behaviour of lyophilized cultures after storage for 21 years, *J. Bacteriol.,* 85, 486, 1963.
153. **Wahl, R.**, Discussion II, in *Culture Collections: Perspectives and Problems,* Martin, S. M., Ed., University of Toronto Press, Toronto, 1963, p. 159.
154. **Williams, R. E. O. and Asheshov, E. A.**, Preservation of viruses and bacteriophage, in *Culture Collections: Perspectives and Problems,* Martin, S. M., Ed., University of Toronto Press, Toronto, 1963, p. 147.
155. **Clark, W. A.**, Comparison of several methods for preserving bacteriophages, *Appl. Microbiol.,* 10, 466, 1962.
156. **Clark, W. A. and Geary, D.**, The collection of bacteriophages at the American Type Culture Collection, in *Freezing and Drying of Microorganisms,* Nei., T., Ed., University of Tokyo Press, Tokyo, 1969, p. 179.
157. **Zierdt, C. H.**, Preservation of staphylococcal bacteriophage by means of lyophilization, *Am. J. Clin. Pathol.,* 31, 326, 1959.
158. **Prouty, C. C.**, Storage of the bacteriophage of the lactic acid streptococci in the desiccated state with observations on longevity, *Appl. Microbiol.,* 1, 250, 1953.
159. **Davies, J. D. and Kelly, M. J.**, The preservation of bacteriophage H1 of *Corynebacterium ulcerans* U 103 by freeze-drying, *J. Hyg.,* 67, 573, 1969.
160. **Carne, H. R. and Greaves, R. I. N.**, Preservation of corynebacteriophages by freeze-drying, *J. Hyg.,* 72, 467, 1974.
161. **Annear, D. I.**, The preservation of bacteriophage by drying, *J. Appl. Bacteriol.,* 20, 21, 1957.
162. **Clark, W. A., Horneland, W., and Klein, A. G.**, Attempts to freeze some bacteriophages to ultralow temperatures, *Appl. Microbiol.,* 10, 463, 1962.
163. **Meyle, J. S. and Kempf, J. E.**, Preservation of T_2 bacteriophage with liquid nitrogen, *Appl. Microbiol.,* 12, 400, 1964.

164. **Clark, W. A. and Klein, A.**, The stability of bacteriophages in long-term storage at liquid nitrogen temperatures, *Cryobiology,* 3, 68, 1966.
165. **Stevenson, R. E.**, Preservation of cultured cell lines at low temperatures, in *Aspects Théoriques et Industriels de la Lyophilisation,* Rey, L., Ed., Hermann, Paris, 1964, p. 279.
166. **Onions, A. H. S.**, Preservation of fungi, in *Methods in Microbiology,* Vol. 4, Booth, C., Ed., Academic Press, London, 1971, p. 113.
167. **Codner, R. C.**, Preservation of fungal cultures and the control of mycophagous mites, in *Safety in Microbiology,* Shapton, D. A. and Board, R. G., Eds., Academic Press, London, 1972, p. 213.
168. **Raper, K. B. and Alexander, D. F.**, Preservation of molds by the lyophil process, *Mycologia,* 37, 499, 1945.
169. **Sharp, E. L. and Smith, F. G.**, Preservation of *Puccinia* urediospores by lyophilization, *Phytopathology,* 42, 263, 1952.
170. **Haynes, W. C., Wickerham, L. J., and Hesseltine, C. W.**, Maintenance of cultures of industrially important microorganisms, *Appl. Microbiol.,* 3, 361, 1955.
171. **Staffeldt, E. E.**, Observations on lyophil preservation and storage of *Pythium* species, *Phytopathology,* 51, 259, 1961.
172. **Boyd, I. and Bullock, K.**, Freeze-dried preparations of *Penicillium spinulosum, J. Pharm. Pharmacol.,* 18, Suppl. 28S, 1966.
173. **Mehrotra, B. S. and Hesseltine, C. W.**, Further evaluation of the lyophil process for the preservation of aspergilli and penicillia, *Appl. Microbiol.,* 6, 179, 1958.
174. **Hesseltine, C. W., Bradle, B. J., and Benjamin, C. R.**, Further investigations on the preservation of molds, *Mycologia,* 52, 762, 1960.
175. **Ellis, J. J. and Roberson, J. A.**, Viability of fungus cultures preserved by lyophilization, *Mycologia,* 60, 399, 1968.
176. **Bosmans, J.**, Ten years lyophilization of pathogenic fungi, *Mycopathol. Mycol. Appl.,* 53, 13, 1974.
177. **Butterfield, W., Jong, S. C., and Alexander, M. T.**, Preservation of living fungi pathogenic for man and animals, *Can. J. Microbiol.,* 20, 1665, 1974.
178. **Fennell, D. I., Raper, K. B., and Flickinger, M. H.**, Further investigations on the preservation of mold cultures, *Mycologia,* 42, 135, 1950.
179. **Goldie-Smith, E. K.**, Maintenance of stock cultures of aquatic fungi, *J. Elisha Mitchell Sci. Soc.,* 72, 158, 1956.
180. **Hwang, S. W.**, Effects of ultralow temperatures on the viability of selected fungus strains, *Mycologia,* 52, 527, 1960.
181. **Hwang, S. W.**, Long-term preservation of fungus cultures with liquid nitrogen refrigeration, *Appl. Microbiol.,* 14, 784, 1966.
182. **Hwang, S. W.**, Investigation of ultralow temperature for fungal cultures. I. An evaluation of liquid-nitrogen storage for preservation of selected fungal cultures, *Mycologia,* 60, 613, 1968.
183. **Hwang, S. W. and Howells, A.**, Investigation of ultralow temperature for fungal cultures. II. Cryoprotection afforded by glycerol and dimethyl sulfoxide to 8 selected fungal cultures, *Mycologia,* 60, 622, 1968.
184. **Hwang, S. W.**, Longevity of fungal cultures in liquid nitrogen refrigeration, in *Proceedings of the First International Conference on Culture Collections,* Iizuka, H. and Hasegawa, T., Eds., University of Tokyo Press, Tokyo, 1970, p. 251.
185. **Davis, E. E.**, Preservation of myxomycetes, *Mycologia,* 57, 986, 1965.
186. **Loegering, W. Q., McKinney, H. H., Harmon, D. L., and Clark, W. A.**, A long-term experiment for preservation of urediospores of *Puccinia graminis tritici* in liquid nitrogen, *Plant Dis. Rep.,* 45, 384, 1961.
187. **Loegering, W. Q., Harmon, D. L., and Clark, W. A.**, Storage of urediospores of *Puccinia graminis tritici* in liquid nitrogen, *Plant Dis. Rep.,* 50, 502, 1966.
188. **Wellman, A. M. and Walden, D. B.**, Qualitative and quantitative estimates of viability for some fungi after periods of storage in liquid nitrogen, *Can. J. Microbiol.,* 10, 585, 1964.
189. **Wellman, A. M.**, Growth of some fungi before and after cryogenic storage, in *Proceedings of the First International Conference on Culture Collections,* Iizuka H. and Hasegawa, T., Eds., University of Tokyo Press, Tokyo, 1970, p. 255.
190. **Bugbee, W. M. and Kernkamp, M. F.**, Storage of pycniospores of *Puccinia graminis secalis* in liquid nitrogen, *Plant Dis. Rep.,* 50, 576, 1966.
191. **Dade, H. A.**, in *Herbarium I.M.I. Handbook,* Commonwealth Mycological Institute, Kew, Surrey, England, 1960, p. 40.
192. **Little, G. N. and Gordon, M. A.**, Survival of fungus cultures maintained under mineral oil for twelve years, *Mycologia,* 59, 733, 1967.
193. **Booth, C. (Ed.)**, Fungal culture media, in *Methods in Microbiology,* Vol. 4, Academic Press, London, 1971, p. 49.

194. **Mazur, P. and Weston, W. H.**, The effects of spray-drying on the viability of fungus spores, *J. Bacteriol.*, 71, 257, 1956.
195. **Bakerspigel, A.**, Soil as a storage medium for fungi, *Mycologia*, 45, 596, 1953.
196. **Bakerspigel, A.**, A further report on the soil storage of fungi, *Mycologia*, 46, 680, 1954.
197. **Carmichael, J. W.**, Viability of mold cultures stored at –20°C, *Mycologia*, 54, 432, 1962.
198. **Meyer, E.**, The preservation of dermatophytes at subfreezing temperatures, *Mycologia*, 47, 664, 1955.
199. **Kramer, C. L. and Mix, A. J.**, Deep-freeze storage of fungus cultures, *Trans. Kans. Acad. Sci.*, 60, 58, 1957.
200. **O'Brien, M. J. and Webb, R. E.**, Preservation of conidia of *Albugo occidentalis* and *Peronospora effusa*, obligate parasites of spinach, *Plant Dis. Rep.*, 42, 1312, 1958.
201. **Wester, R. E., Drechsler, C., and Jorgensen, H.**, Effect of freezing on viability of the lima bean downy mildew fungus (*Phytophthora phaseoli* Thaxt), *Plant Dis. Rep.*, 42, 413, 1958.
202. **Kelton, W. H.**, Storage of mycoplasma strains, *J. Bacteriol.*, 87, 588, 1964.
203. **Tully, J. G. and Ruchman, I.**, Recovery, identification, and neurotoxicity of Sabin's Type A and C mouse mycoplasmas (PPLO) from lyophilized cultures, *Proc. Soc. Exp. Biol. Med.*, 115, 554, 1964.
204. **Addey, J. P., Taylor-Robinson, D., and Dimic, M.**, Viability of mycoplasmas after storage in frozen or lyophilised states, *J. Med. Microbiol.*, 3, 137, 1970.
205. **Norman, M. C., Franck, E. B., and Choate, R. V.**, Preservation of mycoplasma strains by freezing in liquid nitrogen and by lyophilization with sucrose, *Appl. Microbiol.*, 20, 69, 1970.
206. **Doğuer, M.**, Studies on the viability of the lyophilized *Mycoplasma capri* strains, *Etlik. Vet. Bakteriol. Enst. Derg.*, 4, 75, 1972.
207. **Yugi, H., Suzuki, M., Sato, S., and Ozaki, Y.**, Freeze-drying of mycoplasma, *Cryobiology*, 10, 464, 1973.
208. **Taylor, A. E. R. and Baker, J. R.**, Cultivation of protozoa, in *The Cultivation of Parasites in Vitro*, Blackwell Scientific, Oxford, 1968, p. 3.
209. **Hjelm, K. K. and Møller, K. M.**, A freeze-drying procedure for protozoa, *C. R. Trav. Lab. Carlsberg*, 33, 301, 1963.
210. **Raper, K. B.**, General discussion, in *Culture Collections: Perspectives and Problems*, Martin, S. M., Ed., University of Toronto Press, Toronto, 1963, p. 145.
211. **Wickerham, L. J. and Page, F. C.**, Cultivation and lyophilization of *Mastigina* sp., *J. Protozool.*, 17, 518, 1970.
212. **Annear, D. I.**, Preservation of *Strigomonas oncopelti* in the dried state, *Nature London*, 178, 413, 1956.
213. **Annear, D. I.**, Recovery of *Strigomonas oncopelti* after drying from the liquid state, *Aust. J. Exp. Biol. Med. Sci.*, 39, 295, 1961.
214. **Annear, D. I.**, Recoveries of *Strigomonas oncopelti*, leptospirae and Reiter's treponeme from desiccates after storage for 5 years, *Aust. J. Exp. Biol. Med. Sci.*, 43, 683, 1965.
215. **O'Connell, K. M., Hutner, S. H., Fromentin, H., Frank, O., and Baker, H.**, Cryoprotectants for *Crithidia fasciculata* stored at –20°C, with notes on *Trypanosoma gambiense* and *T. conorhini*, *J. Protozool.*, 15, 719, 1968.
216. **Fulton, J. D. and Smith, A. U.**, Preservation of *Entamoeba histolytica* at –79°C in the presence of glycerol, *Ann. Trop. Med. Parasitol.*, 47, 240, 1953.
217. **Weinman, D. and McAllister, J.**, Prolonged storage of human-pathogenic protozoa with conservation of virulence: Observations on the storage of helminths and leptospiras, *Am. J. Hyg.*, 45, 102, 1947.
218. **Most, H., Alger, N., and Yoeli, M.**, Preservation of *Leishmania donovani* by low-temperature freezing, *Nature London*, 201, 735, 1964.
219. **Wang, G. T. and Marquardt, W. C.**, Survival of *Tetrahymena pyriformis* and *Paramecium aurelia* following freezing, *J. Protozool.*, 13, 123, 1966.
220. **Manwell, R. D.**, The low-temperature freezing of malaria parasites, *Am. J. Trop. Med.*, 23, 123, 1943.
221. **Manwell, R. D. and Edgett, R.**, The relative importance of certain factors in the low-temperature preservation of malaria parasites, *Am. J. Trop. Med.*, 23, 551, 1943.
222. **Wolfson, F.**, Effect of preservation by freezing upon the virulence of *Plasmodium* for ducks, *Am. J. Hyg.*, 42, 155, 1945.
223. **Saunders, G. M. and Scott, V.**, Preservation of *Plasmodium vivax* by freezing, *Science*, 106, 300, 1947.
224. **Eyles, D. E., Coleman, N., and Cavanaugh, D. J.**, Preservation of *Toxoplasma gondii* by freezing, *J. Parasitol.*, 42, 408, 1956.
225. **McEntegart, M. G.**, The maintenance of stock strains of trichomonads by freezing, *J. Hyg.*, 52, 545, 1954.

226. **McEntegart, M. G.,** Prolonged survival of *Trichomonas vaginalis* at −79°C, *Nature London,* 183, 270, 1959.
227. **Levine, N. D., Mizell, M., and Houlahan, D. A.,** Factors affecting the protective action of glycerol on *Trichomonas foetus* at freezing temperatures, *Exp. Parasitol.,* 7, 236, 1958.
228. **Levine, N. D., Andersen, F. L., Losch, M. B., Notzold, R. A., and Mehra, K. N.,** Survival of *Trichomonas foetus* stored at −28 and −95°C after freezing in the presence of glycerol, *J. Protozool.,* 9, 347, 1962.
229. **Polge, C. and Soltys, M. A.,** Preservation of trypanosomes in the frozen state, *Trans. R. Soc. Trop. Med. Hyg.,* 51, 519, 1957.
230. **Cunningham, M. P. and Harley, J. M. B.,** Preservation of living metacylic forms of the *Trypanosoma brucei* subgroup, *Nature London,* 194, 1186, 1962.
231. **Polge, C.,** Freezing and freeze-drying in parasitology, in *Aspects Théoriques et Industriels de la Lyophilisation,* Rey, L., Ed., Hermann, Paris, 1964, p. 417.
232. **Doran, D. J. and Vetterling, J. M.,** Preservation of coccidial sporozoites by freezing, *Nature London,* 217, 1262, 1968.
233. **Norton, C. C. and Joyner, L. P.,** The freeze preservation of coccidia, *Res. Vet. Sci.,* 9, 598, 1968.
234. **Diamond, L. S.,** Freeze-preservation of protozoa, *Cryobiology,* 1, 95, 1964.
235. **Diamond, L. S., Meryman, H. T., and Kafig, E.,** Preservation of parasitic protozoa in liquid nitrogen, in *Culture Collections: Perspectives and Problems,* Martin, S. M., Ed., University of Toronto Press, Toronto, 1963, p. 189.
236. **Hwang, S. W., Davis, E. E., and Alexander, M. T.,** Freezing and viability of *Tetrahymena pyriformis* in dimethylsulfoxide, *Science,* 144, 64, 1964.
237. **Simon, E. M. and Hwang, S. W.,** *Tetrahymena:* Effect of freezing and subsequent thawing on breeding performance, *Science,* 155, 694, 1967.
238. **Paine, G. D. and Meyer, R. C.,** *Toxoplasma gondii* propagation in cell cultures and preservation at liquid nitrogen temperatures, *Cryobiology,* 5, 270, 1969.
239. **Ellinghausen, H. C., Jr.,** Growth temperatures, virulence, survival and nutrition of leptospiras, *J. Med. Microbiol.,* 6, 487, 1973.
240. **Annear, D. I.,** Preservation of leptospirae by drying, *J. Pathol. Bacteriol.,* 72, 322, 1956.
241. **Annear, D. I.,** Observations on the preservation by drying of leptospirae and some other bacteria, *Aust. J. Exp. Biol. Med. Sci.,* 36, 1, 1958.
242. **Annear, D. I.,** The preservation of leptospires by drying from the liquid state, *J. Gen. Microbiol.,* 27, 341, 1962.
243. **Annear, D. I.,** Recovery of leptospires after dry storage for ten years, *Int. J. Syst. Bacteriol.,* 24, 399, 1974.
244. **Annear, D. I.,** Preservation of the Reiter treponeme by drying from the liquid state, *J. Bacteriol.,* 83, 932, 1962.
245. **Otsuka, S. and Manako, K.,** Studies on the preservation of leptospirae by freeze-drying, *Jpn. J. Microbiol.,* 5, 141, 1961.
246. **Hanson, A. W. and Cannefax, G. R.,** Recovery of *Treponema* and *Borrelia* after lyophilization, *J. Bacteriol.,* 88, 811, 1964.
247. **Hart, L.,** Freeze preservation of *Borrelia anserina, Aust. Vet. J.,* 46, 455, 1970.
248. **Hardy, P. H. and Nell, E. E.,** Maintenance of virulence and motility of *Treponema pallidum* in the frozen state, *Bacteriol. Proc.,* p. 109, 1971.
249. **Schubert, J. H. and Sulzer, C. R.,** Maintenance of *Leptospira* by storage at −75°C, *Health Lab. Sci.,* 10, 96, 1973.
250. **Torney, H. L. and Bordt, D. E.,** Viability quantitation of leptospires after rapid and controlled rate freezing, *Cryobiology,* 5, 352, 1969.
251. **Stalheim, O. H. V.,** Viable, avirulent *Leptospira interrogans* serotype *pomona* vaccine: Preservation in liquid nitrogen, *Appl. Microbiol.,* 22, 726, 1971.
252. **Alexander, A. D., Lessel, E. F., Evans, L. B., Franck, E., and Green, S. S.,** Preservation of leptospiras by liquid nitrogen refrigeration, *Int. J. Syst. Bacteriol.,* 22, 165, 1972.
253. **Harris, R. J. C. (Ed.),** The preservation of viruses, in *Biological Applications of Freezing and Drying,* Academic Press, New York, 1954, p. 201.
254. **McKinney, H. H. and Silber, G.,** Methods of preservation and storage of plant viruses, in *Methods in Virology,* Vol. 4, Maramorosch, K. and Koprowski, H., Eds., Academic Press, New York, 1968, p. 491.
255. **Ward, T. G.,** Methods of storage and preservation of animal viruses, in *Methods in Virology,* Vol. 4, Maramorosch, K. and Koprowski, H., Eds., Academic Press, New York, 1968, p. 481.
256. **Hollings, M. and Lelliott, R. A.,** Preservation of some plant viruses by freeze-drying, *Plant Pathol.,* 9, 63, 1960.

257. **Worley, J. F. and Schneider, I. R.,** Long-term storage of purified southern bean mosaic virus freeze-dried in the presence of lactose, *Phytopathology,* 56, 1327, 1966.
258. **Best, R. J.,** On maintaining the infectivity of a labile plant virus by storage at −69°, *Virology,* 14, 440, 1961.
259. **McKinney, H. H., Greeley, L. W., and Clark, W. A.,** Preservation of plant viruses in liquid nitrogen, *Plant Dis. Rep.,* 45, 755, 1961.
260. **Greiff, D.,** The effect of freezing, low-temperature storage and drying by vacuum sublimation on the activities of viruses and cellular particulates, in *Recent Research in Freezing and Drying,* Parkes, A. S. and Smith, A. U., Eds., Blackwell Scientific, Oxford, 1960, p. 167.
261. **Greiff, D. and Rightsel, W.,** The effects on the activities of viruses of rates and temperatures of freezing, cyclic freezing and thawing, storage at low temperatures, or drying by sublimation *in vacuo,* in *Aspects Théoriques et Industriels de la Lyophilisation,* Rey, L., Ed., Hermann, Paris, 1964, p. 369.
262. **Greiff, D. and Rightsel, W.,** Freezing and freeze-drying of viruses, in *Cryobiology,* Meryman, H. T., Ed., Academic Press, London, 1966, p. 697.
263. **Greiff, D. and Rightsel, W.,** Recent research in freezing, drying and kinetics of thermal degradation of viruses, in *Lyophilisation: Récherches et Applications Nouvelles,* Rey, L., Ed., Hermann, Paris, 1966, p. 103.
264. **Greiff, D. and Rightsel. W. A..** Stabilities of suspensions of viruses after freezing or drying by vacuum sublimation and storage, *Cryobiology,* 3, 432, 1967.
265. **Rightsel, W. A. and Greiff, D.,** Freezing and freeze-drying of viruses, *Cryobiology,* 3, 423, 1967.
266. **Greiff, D., Rightsel, W. A., and Schuler, E. E.,** Effects of freezing, storage at low temperatures, and drying by sublimation *in vacuo* on the activities of measles virus, *Nature London,* 202, 624, 1964.
267. **Damjanović, V. and Klašnja, A.,** Freeze-drying of the attenuated measles virus (Belgrade's strain), *Mikrobiologija* (Beograd), 5, 87, 1968.
268. **Mareš, I., Kittnar, E., Srbová, H., and Časny, J.,** The lyophilization of measles virus, *J. Hyg. Epidemiol. Microbiol. Immunol.,* 13, 279, 1969.
269. **Annear, D. I. and Beswick, T. S. L.,** A note on the preservation of influenza virus, *J. Hyg.,* 54, 509, 1956.
270. **Beardmore, W. B., Clark, T. D., and Jones, K. V.,** Preservation of influenza virus infectivity by lyophilization, *Appl. Microbiol.,* 16, 362, 1968.
271. **Tyrrell, D. A. J. and Ridgwell, B.,** Freeze-drying of certain viruses, *Nature London,* 206, 115, 1965.
272. **Portocală, R., Samuel, I., and Popescu, M.,** Effect of lyophilization on picornaviruses, *Arch. Gesamte Virusforsch.,* 28, 97, 1969.
273. **Collier, L. H.,** The preservation of vaccinia virus by freeze-drying, in *Freezing and Drying,* Harris, R. J. C., Ed., Institute of Biology, London, 1951, p. 133.
274. **Collier, L. H.,** The development of a stable smallpox vaccine, *J. Hyg.,* 53, 76, 1955.
275. **Slonim, D. and Kittner, E.,** Remarks on lyophilization and stability of lyophilized vaccinia virus, *Cesk. Epidemiol. Mikrobiol. Imunol.,* 16, 325, 1967.
276. **Suzuki, M.,** Studies on freeze-drying of vaccinia virus: Effect of suspending media on infectivity titers, *Jpn. J. Vet. Res.,* 16, 87, 1968.
277. **Suzuki, M.,** Effect of suspending media on freeze-drying and preservation of vaccinia virus, *J. Hyg.,* 68, 29, 1970.
278. **Hekker, A. C., Smith, L., and Huisman, P.,** Stabilizer for lyophilization of rubella virus, *Arch. Gesamte Virusforsch.,* 29, 257, 1970.
279. **Verge, J., Goret, P., and Merteax, C.,** Nouvelle note sur la conservation de quelques ultravirus par la dessiccation, *Ann. Inst. Pasteur,* 72, 499, 1946.
280. **Fellowes, O. N.,** Freeze-drying of foot-and-mouth disease virus and storage stability of the infectivity of dried virus at 4°C, *Appl. Microbiol.,* 13, 496, 1965.
281. **Ramyar, H. and Traub, E.,** Lyophilizing foot-and-mouth disease virus at low drying temperature, *Am. J. Vet. Res.,* 28, 1605, 1967.
282. **Fellowes, O. N.,** Comparison of cryobiological and freeze-drying characteristics of foot-and-mouth disease virus and of vesicular stomatitis virus, *Cryobiology,* 4, 223, 1968.
283. **Dundarov, S., Trendafilova, P., and Andonov, P.,** Lyophilization of herpes simplex virus, *Ann. Res. Inst. Epidemiol. Microbiol.* (Sofia), 14, 167, 1969.
284. **Calnek, B. W., Hitchner, S. B., and Adldinger, H. K.,** Lyophilization of cell-free Marek's disease herpes virus and a herpes virus from turkeys, *Appl. Microbiol.,* 20, 723, 1970.
285. **Jung, M. and Krech, U.,** Lyophilization of human cytomegaloviruses, *Pathol. Microbiol.,* 37, 47, 1971.
286. **Pattison, I. H., Jones, K. M., and Kimberlin, R. H.,** Observations on a freeze-dried preparation containing the scrapies agent, *Res. Vet. Sci.,* 10, 214, 1969.

287. *Catalogue of Viruses, Rickettsiae, Chylamydiae,* 4th ed., American Type Culture Collection, Rockville, Md., 1971.
288. **Olitsky, P. K., Casals, J., Walker, D. L., Ginsberg, H. S., and Horsfall, F. L.,** Preservation of viruses in a mechanical refrigerator at −25°C, *J. Lab. Clin. Med.,* 34, 1023, 1949.
289. **Allen, E. G., Kaneda, B., Girardi, A. J., Scott, T. F. M., and Sigel, M. M.,** Preservation of viruses of the psittacosis-lymphogranuloma venereum group and herpes simplex under various conditions of storage, *J. Bacteriol.,* 63, 369, 1952.
290. **Melnick, J. L.,** Preservation of viruses by freezing, *Fed. Proc. Suppl.,* 15, S-280, 1965.
291. **Dobrowolska, H. and Kańtoch, M.,** Survival of poliomyelitis virus in a frozen state, *Med. Dośw. Mikrobiol.,* 21, 305, 1969.
292. **Beech, F. W. and Davenport, R. R.,** Isolation, purification and maintenance of yeasts, in *Methods in Microbiology,* Vol. 4, Booth, C., Ed., Academic Press, London, 1971, p. 153.
293. **Henry, B. S.,** The viability of yeast cultures preserved under mineral oil, *J. Bacteriol.,* 54, 264, 1947.
294. **Wickerham, L. J. and Andreasen, A. A.,** The lyophil process: Its use in the preservation of yeasts, *Wallerstein Lab. Commun.,* 5, 165, 1942.
295. **Guibert, L. and Bréchot, P.,** La lyophilisation des levures, *Ann. Inst. Pasteur,* 88, 750, 1955.
296. **Kirsop, B.,** Maintenance of yeasts by freeze-drying, *J. Inst. Brew. London,* 61, 466, 1955.
297. **Atkin, L., Moses, W., and Gray, P. P.,** The preservation of yeast cultures by lyophilization, *J. Bacteriol.,* 57, 575, 1949.
298. **Bréchot, P., Guibert, L., and Croson, M.,** La lyophilisation des levures, *Ann. Inst. Pasteur,* 95, 62, 1958.
299. **Brady, B. L.,** Some observations on the freeze-drying of yeasts, in *Recent Research in Freezing and Drying,* Parkes, A. S. and Smith, A. U., Eds., Blackwell Scientific, Oxford, 1960, p. 243.
300. **Wynants, J.,** Preservation of yeast cultures by lyophilization, *J. Inst. Brew. London,* 68, 350, 1962.
301. **Rose, D.,** Some factors influencing the survival of freeze-dried yeast cultures, *J. Appl. Bacteriol.,* 33, 228, 1970.
302. **Annear, D. I.,** Preservation of yeasts by drying, *Aust. J. Exp. Biol. Med. Sci.,* 41, 575, 1963.
303. **Tsuji, K.,** Liquid nitrogen preservation of *Saccharomyces carlsbergensis* and its use in a rapid biological assay of vitamin B_6 (pyridoxine), *Appl. Microbiol.,* 14, 456, 1966.
304. **Wickerham, L. J. and Flickinger, M. H.,** Viability of yeasts preserved two years by the lyophil process, *Brew. Dig.,* 21, 55, 65, 1946.
305. **Burns, M. E.,** Survival of lyophilized yeasts, *Sabouraudia,* 1, 203, 1962.
306. **Scheda, R. and Yarrow, D.,** The instability of physiological properties used as criteria in the taxonomy of yeasts, *Arch. Mikrobiol.,* 55, 209, 1966.

ENRICHMENT CULTURE

S. Aaronson

Enrichment (elective) culture is the creation of a special environment in vitro for the growth and multiplication of microorganisms. This special environment permits the selection of the desired microorganism or microbial group from a mixture of microorganisms with which it is likely to be found in its natural environment. The special environment may function in one of the following ways: (1) by permitting the desired microorganism to outgrow all the others in the medium and thus make it the most common and therefore easy to isolate; (2) by inhibiting the growth of all microorganisms except the desired species or group; (3) by permitting the desired microorganism to multiply or perhaps metabolize in such a conspicuous way that it can be distinguished from other microorganisms.

The enrichment culture was first exploited in the late 19th century by several of the most famous microbiologists of that century, including M. W. Beijerinck and S. Winogradsky. Through their brilliant use of this technique and through the work of their 20th century colleagues we have learned a great deal about the role of microorganisms in the natural cycles – carbon, iron, nitrogen, oxygen, and sulfur; this technique has also been exploited to isolate microorganisms capable of specific metabolic reactions for the study of metabolic pathways or for the experimental or industrial conversion of one molecule to another, more desirable molecule.

Enrichment culture may be performed in liquid media by successive incubation in the same selective medium until the population is predominantly the desired organism, or by successive incubation in increasing concentrations of the selective chemical or physical environment. Another method is that of continuous-liquid culture, in which a selective medium enhances the growth of a favored microorganism while washing away its products and the less-favored species (see Specific References 1, 2, and 3 for details). Enrichment cultures may also be performed on media solidifed with agar (0.5 to 2.0%), gelatin (1 to 2%), or silica gel (see Specific Reference 4 for details). Solid media have the advantage of permitting the selection of the desired microorganism in an isolated and probably purified state. It is common to use liquid media for enrichment and solid media for the isolation of the desired microorganism.

The medium to be used for a specific organism must be tailored to that organism. However, all media and enrichments should include the following:

1. **Energy**
   Energy may be supplied solely as organic molecules for heterotrophs (see below), or as reduced inorganic molecules or ions (H_2, CO, H_2S, NH_4^+, NO_2^-, N, Fe^{++}, Mn^{++}, etc.) for chemoautotrophs or as artificial or sunlight for photosynthetics. Photosynthetic bacteria do better with light rich in the red part of the spectrum (tungsten lamp); on the other hand, light rich in the blue part of the spectrum tends to select blue-green algae (fluorescent lamp). Most photosynthetic eukaryotes do best with light relatively rich in the red end of the spectrum.
2. **Source of carbon**
   Autotrophic organisms may be satisfied by bubbling carbon dioxide through their medium or by supplying carbonate. Most microorganisms are heterotrophic, and their carbon and energy requirements may be met by any one or by a combination of the organic supplements listed in Table 1 or the compounds listed in Table 2. Many heterotrophs, as well as some algae, require exogenous organic nutrients they cannot synthesize; these may include one or more of the compounds listed in Table 3. A good

rule to remember here is that the more organic material the medium contains, the more likely you are to select for heterotrophic "weeds", which quickly utilize the available organic compounds and swamp the autotrophs or more fastidious heterotrophs.

3. **Source of trace and major elements**
   All microorganisms need a source of sulfer (usually sulfate), nitrogen (usually nitrate, ammonium), and phosphorus (phosphate), as well as the trace elements. Examples of trace element solutions for marine and fresh-water microorganisms are shown in Table 4.
4. **pH, temperature, and oxygen tension**
   These must be varied to suit the needs and limits of the desired microorganism. Marine microorganisms tend to prefer a somewhat alkaline pH and lower temperatures (10 to 20°C; fresh-water microorganisms show wider environmental preferences, depending on their natural source.

Specific groups of microorganisms may be excluded from growth in appropriate media by the use of the antimicrobial agents shown in Table 5. The inhibition of fast-growing microorganisms allows slower-growing or more fastidious microorganisms to appear under circumstances where they might normally be overgrown or suppressed. One of the simplest demonstrations of an enrichment culture is the Winogradsky column (Figure 1), which permits the enrichment for different photosynthetic microorganisms on the basis of their need for a reducing (anaerobic) environment. The bottom of a transparent tube (glass or plastic) that contains wet bits of paper mixed with several grams of calcium sulfate and calcium carbonate as well as some mud is gradually filled with fine mud, preferably marine, from which pebbles, twigs, leaves, and other debris have been

Table 1
ORGANIC SUPPLEMENTS

		Major constituents					
					Lipids		
Supplement	Concentration, g/l	Amino acids	Nucleic acid derivatives	Water-soluble vitamins	Fatty acids	Steroids	Mineral
Soybean lecithin	0.1				+	+	
TEM[a]	0.1				+	+	
Tween 40	0.1				+	+	
Tween 60	0.1				+	+	
Tween 80	0.1				+	+	
Gelatin hydrolysate[b]	5.0	+					
Trypticase[c]	5.0	+		+			
Hycase[d]	5.0	+		+			
Protease peptone	5.0	+		+			
Liver extract	1.0	+	+	+			
Soil extract[e]	5.0 ml	+		+			

[a] TEM = diacetyl tartaric ester of tallow glycerides.
[b] Except L-methionine and L-tryptophan.
[c] Enzymatic digest of casein.
[d] Acid digest of casein.
[e] See Specific Reference 4 for preparation.

Table 2
SOME CARBON AND ENERGY SOURCES[a,b]

CARBOHYDRATES, SUGARS, AND SUGAR ALCOHOLS	
Arabinose	Maltose
Cellulose	Mannitol, adonitol
Dextrin	Melibiose
Fructose	Raffinose
Galactose	Rhamnose
Glucose	Ribose
Glycerol, dulcitol, sorbitol	Starch
Glycogen	Trehalose
Inulin	Xylose
Lactose	
TCA CYCLE ACIDS	
Acetic acid	α-Ketoglutaric acid
Aconitic acid	Malic acid
Citric acid	Oxaloacetic acid
Fumaric acid	Succinic acid
COMPOUNDS LEADING TO THE TCA CYCLE	
Amino Acids	**Fatty Acids**
Alanine	Butyric acid
Arginine	Glyoxylic acid
Aspartic acid, asparagine	Lactic acid
Glutamic acid, glutamine	Propionic acid
Glycine	Pyruvic acid
Serine	

[a] Use at 1 to 10 g/l.
[b] Any one of these molecules provided with required trace or major elements may be used to isolate microorganisms capable of using the chosen molecule as the sole source of carbon and energy. Other organic molecules, such as hydrocarbons, aromatic compounds, etc., may also be used.

removed. Avoid air bubbles. The column of mud is allowed to settle until there is a clear liquid zone of approximately 1/4" on top. Excess water is poured off. The tube is tightly covered with a transparent plastic cover (e.g., Saran) to minimize evaporation, and then it is incubated for 2 to 4 weeks in sunlight or tungsten bulb at room temperature. Do not allow the column of mud to dry or to become too warm. Add distilled water as needed. After a time, areas of the mud surface will become rust-colored, red, or dark green, indicating the appearance of specific types of photosynthetic bacteria. The liquid area on top of the mud column will appear pigmented and contain a variety of aerobic algae. Some of the algae may be isolated on the selected agar media in Table 6; some photosynthetic bacteria may be isolated in the anaerobic media shown in Table 7.

Procedures for the isolation of a variety of other special microbial groups are outlined in Tables 8 and 9. Table 8 describes media for sulfur bacteria. Table 9 describes media for the isolation of bacteria and/or fungi that hydrolyze a variety of organic substrates.

Table 3
EXOGENOUS ORGANIC NUTRIENTS AND THEIR RECOMMENDED AMOUNTS[a]

Organic nutrient	Amount, mg/l	Present in simple vitamin mixture	Organic nutrient	Amount, mg/l	Present in simple vitamin mixture
		WATER-SOLUBLE VITAMINS			
p-Aminobenzoic acid	0.1	+	Nicotinic acid	1.0	+
Betaine	1.0	–	Putrescine 2HCl	0.2	–
Biotin	0.01	+	Pyridoxal ethylacetal	0.1	+
Calcium pantothenate	1.0	+	Pyridoxamine 2HCl	0.2	+
DL-Carnitine HCl	0.4	–	Pyridoxine	0.2	–
Choline H_2 ·citrate	16.0	+	Riboflavine	0.1	+
Cystamine 2HCl	0.1	–	Sodium riboflavine $PO_4 \cdot 2H_2O$	0.1	+
Ferulic acid	0.1	–	Spermidine $PO_4 \cdot 3H_2O$	0.1	–
Folic acid	0.5	+	Thiamine NO_3	0.4	+
p-Hydroxybenzoic acid	0.1	–	DL-Thioctic acid	0.05	+
Inositol	10.0	+	Vitamin B_{12}	0.004	+
		PURINES AND PYRIMIDINES			
Adenine	5.0		Sodium guanylate H_2O	20.0	
Adenosine	10.0		Sodium inosinate	40.0	
Adenylic acid (2′, 3′, or 5′)	20.0		Thymidine	10.0	
Cytidylic acid	20.0		Thymine	5.0	
Deoxyguanosine	10.0		Uracil	5.0	
Guanine	4.0		Uridine	10.0	
Guanosine H_2O	10.0		Xanthine	5.0	
Hypoxanthine	4.0		Xanthosine	10.0	
Orotic acid	4.0				
		AMINO ACIDS			
DL-Alanine	0.4		DL-Methionine	0.06	
L-Arginine (free base)	0.3		DL-Phenylalanine	0.04	
DL-Aspartic acid	0.5		L-Proline	0.04	
L-Glutamic acid	1.0		DL-Serine	0.1	
Glycine	0.5		DL-Threonine	0.1	
L-Histidine	0.2		DL-Tryptophan	0.05	
DL-Isoleucine	0.05		L-Tyrosine	0.04	
DL-Leucine	0.05		DL-Valine	0.05	
DL-Lysine HCl	0.45				

[a] Any one of these molecules provided with required trace or major elements may be used to isolate microorganisms capable of using the chosen molecule as the sole source of carbon and energy. Other organic molecules, such as hydrocarbons, aromatic compounds, etc., may also be used.

Table 4
MAJOR AND MINOR ELEMENTS FOR SELECTED MICROORGANISMS[a]

	Fresh-water			Marine		
	1	2	3	4	5	6
Compounds for Major Elements, g/l						
NH_4Cl	0.5		0.5			
$(NH_4)_2SO_4$		0.5				
KCl		0.03		0.3		
NaCl				24.0	23.0	25.0
$MgCl_2 \cdot 6H_2O$				3.0	11.0	
$MgSO_4 \cdot 7H_2O$	1.0	0.3	2.0	0.3		5.0
Na_2SO_4					4.0	
$MgCO_3$	0.4					
$NaHCO_3$					2.0	
KNO_3			0.1	0.1		
$Ca(NO_3)_2 \cdot 4H_2O$					0.1	
$CaCO_3$	0.05		0.2	0.2		0.25
K_2HPO_4					0.02	0.1
KH_2PO_4	0.3	Footnote b	Footnote b	0.02		
Trace Elements, mg/l						
Fe	2.0	1.0	4.0	3.0	0.5	2.0
Zn	1.0	0.01	10.0	4.0	0.3	20.0
Mn	0.5	0.09	8.0	10.0	0.1	20.0
Cu	0.08	0.0003	1.0	0.003	0.1	
Co	0.1	0.003	1.0	0.03	Footnote c	Footnote c
B	0.1		4.0	2.0	0.1	
Mo	0.05	0.002	0.8	0.5	0.1	
V	0.01					
Ca		20.0	10.0			
Sr				6.5		
Al				0.25		
Rb				0.1		
Li				0.05		
I				0.025		
Br				32.5		
pH	5.0	5.5	7.0	7.5	8.0	8.0–8.2

[a] Code: 1. *Ochromonas danica*;[5] 2. *Phaeus pyrum*;[6] 3. *Micrococcus sodonensis*;[7] 4. *Gyrodinium* sp.;[8] 5. *Stichococcus* sp.;[8] 6. *Labyrinthula* sp.[8]

[b] Supplied as Na_2 glycerophosphate $5H_2O$, 5.0 g.

[c] Supplied as vitamin B_{12}, 0.004 mg.

Table 5
ANTIBIOTICS INHIBITING SPECIFIC MICROBIAL GROUPS

Antibiotic	Concentration to be used, mg/l
Antiprokaryotic	
Bacitracin	50
Chloramphenicol	0.5–75
D-Cycloserine	1–400
Penicillin G	0.2–100
Streptomycin	0.1–20
Tetracyclines	0.1–50
Antieukaryotic	
Amphotericin B	0.1–20
Candicidin	0.1–20
Cycloheximide (actidione)	0.2–100
Filipin	0.1–20
Griseofulvin	1–20
Nystatin (mycostatin)	0.1–20

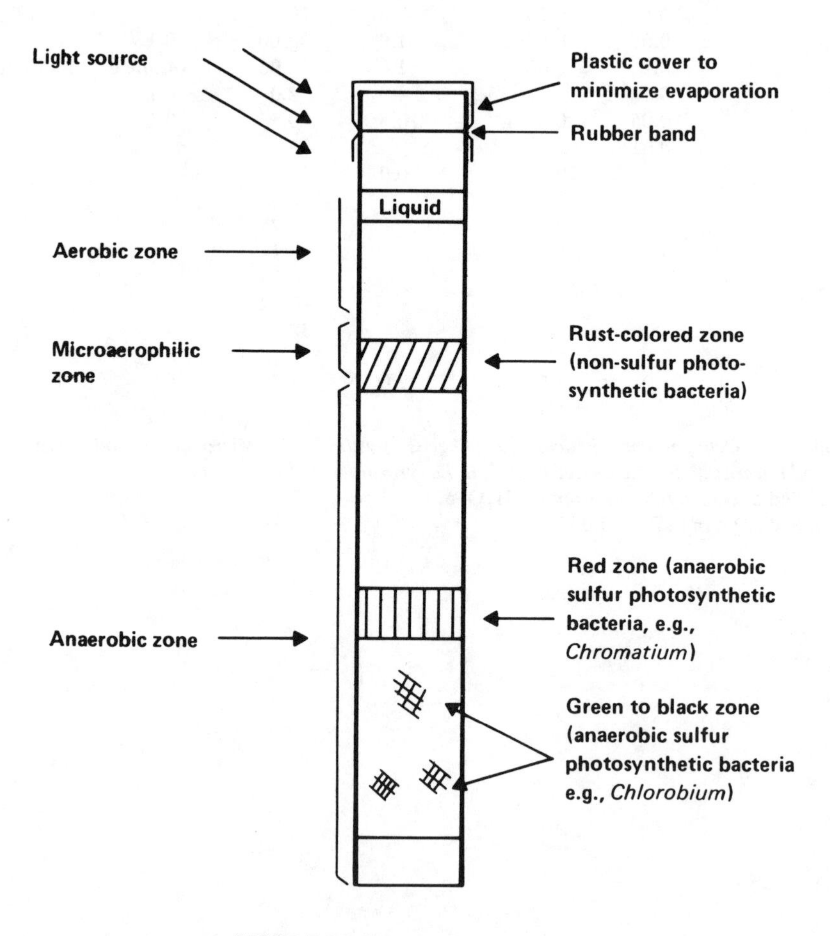

FIGURE 1. Winogradsky column.

Table 6
SOME MEDIA FOR THE SELECTION OF SPECIFIC MICROBIAL GROUPS[a]

Additions, g/l[b]	Blue-green algae	Blue-green algae	Blue-green algae	Acid-living phytoflagellates[c]	Desmids	Diatoms[d]
$Ca(NO_3)_2 \cdot 4H_2O$	0.04		0.025			0.1
Citric acid	0.003					
Ferric ammonium citrate	0.003	0.5	1.0			
Ferric chloride $6H_2O$			0.001	0.001		
K_2HPO_4	0.01	0.1	1.0	0.2	0.01	0.02
KNO_3		5.0	0.05		0.1	
$MgSO_4 \cdot 7H_2O$	0.025	0.05	0.25	0.1	0.01	
Na_2CO_3	0.02		1.5[e]			
$NaSiO_3 \cdot 9H_2O$	0.025					0.05
NH_4NO_3			1.0	1.0		
Soil extract					50.0 ml	
Tryptone						1.0
B_{12}						1.0 μg
Fe, as $Fe(NH_4)_2(SO_4)_2 \cdot 6H_2O$						0.5 mg
Zn, as $ZnSO_4 \cdot 7H_2O$						0.3 mg
B, as H_3BO_3						0.1 mg
Co, as $CoSO_4 \cdot 7H_2O$						0.1 mg
Cu, as $CuSO_4 \cdot 5H_2O$						0.1 mg
Mn, as $MnSO_4 \cdot H_2O$						0.1 mg
Mo, as $(NH_4)_6Mo_7O_{24} \cdot 4H_2O$						0.1 mg
Agar	15	15	1–15		7.5	10
pH	8–9.5	6–8	6–8		5–8	–
Reference	9	10	11	12	10	13

[a] Use fresh- or salt-water mud or water as the inoculum, at room temperature or at the temperature of the source of inoculum.
[b] Unless indicated as mg, ml, or μg.
[c] $CaCO_3$ may be added to make the medium alkaline.
[d] Use filtered sea water.
[e] Add aseptically.

Table 7
MEDIA FOR THE ENRICHMENT AND ISOLATION OF SOME ANAEROBIC OR MICROAEROPHILIC PHOTOSYNTHETIC BACTERIA[14,a]

Additions, g/l medium	Anaerobic sulfide oxidizers: *Chlorobium limicola*	*Chlorobium thiosulfatophilum*	*Chromatium*	*Thiopedia*	*Rhodopseudomonas Rhodospirillum*
$(NH_4)_2SO_4$	1	1	1	1	
$NaHCO_3$[b]	2	2	2	2	2
$NaS \cdot 9H_2O$[b]	1	0.2	0.2	0.2	
$Na_2S_2O_2 \cdot 5H_2O$[b]		1	1	1	
Sodium malate			1	1	1
Yeast extract (or hydrolyzate)					1
Ethanol (or other alcohol)					20 ml
pH	7–7.2[c]	7–7.2[c]	8–8.4[c]	8–8.4[c]	7–7.5
Color of growth	Green	Green	Red	Red	Rust

[a] Fill the container (a 50-ml bottle or smaller tube) with medium to the top. Add agar (15 g/l) for solid media. All the above media contain the following ingredients (in g/l, unless otherwise indicated): KH_2PO_4, 1; NH_4Cl, 1; $MgCl_2 \cdot 6H_2O$, 0.5; NaCl (for marine forms only), 10; vitamin B_{12}, 1.0 μg; 1.0 ml of each of the following trace metals: $FeCl_3 \cdot 6H_2O$, 1.5 mg; $MnCl_2 \cdot 6H_2O$, 5mg; $CuCl_2 \cdot 2H_2O$, 1 mg; $NiCl_4 \cdot 6H_2O$, 2mg; Na_2MoO_4; $Co(NO_3)_2 \cdot 6H_2O$,0.05; $CuSO \cdot 5H_2O$, 0.005.

[b] Sterilize by filtration and add separately just before use to yield the above concentrations.

[c] Adjust the pH with H_3PO_4.

Table 8
MEDIA FOR THE ENRICHMENT AND ISOLATION OF SOME SULFUR BACTERIA[14]

	Sulfur (thiosulfate) oxidizers				
	Thiobacillus (Thiobacterium)				
Additions, g/l	*thiooxidans*	*thioparus*	*denitrificans*[a]	*ferrooxidans*	Sulfate reducers
Sulfur	10	10			
(or $Na_2S_2O_3 \cdot 5H_2O$)[b]	5	5	5		
$(NH_4)_2SO_4$	2–4	2–4			0.15
KH_2PO_4	2–4	2–4	2[c]	0.05	0.5
$CaCl_2$	0.25	0.25			
$MgSO_4 \cdot 7H_2O$	0.5	0.5		0.5	2
$FeSO_4 \cdot 7H_2O$	0.01	0.01	0.01[c]	1[c]	0.5
NH_4Cl			0.5		
$MgCl_2 \cdot 6H_2O$			0.5		
KNO_3			2		
$NaHCO_3$			1[c]		
KCl				0.05	
$Ca(NO_3)_2$				0.01	
$CaSO_4$					1
Yeast extract					1
Sodium lactate					3.5
Ascorbic acid					1
Thioglycollic acid					1
Agar (or silica gel)	30[d]	10–20	15	30[d]	15
pH	5	7.5	7	3.5	7–7.5

[a] Anaerobically at 30°C.
[b] Sterilize by steaming if sulfur is the substrate.
[c] Sterilize separately and then add it to the medium.
[d] Sterilize the agar separately as a 15% solution and then add it to the medium.

Table 9
ENRICHMENT AND ISOLATION FOR BACTERIA AND/OR FUNGI THAT HYDROLYZE ORGANIC SUBSTRATES WITH VISIBLE CLEARING OR PIT FORMATION AROUND THE COLONIES ON SOLID MEDIUM

Addition	Grams per liter[a]									
Sodium alginate	25									
Agar		15								
Cellulose powder			10							
Chitin				1						
Soluble starch					5					
Skim milk (casein)						20				
Lipid (olive oil, tributyrin, etc.)							50			
Chondroitin sulfate								0.4[b]		
Sodium hyaluronidate									0.4[b]	
Pectin										5
$CaCl_2 \cdot 2H_2O$										0.2
K_2HPO_4	0.3		0.5							
KNO_3		1								
$FeCl_3 \cdot 6H_2O$										0.01
NaCl	30									
$MgSO_4 \cdot 7H_2O$			0.5							
NH_4Cl			0.5							
Trypticase	2.5									
Soil extract (see Specific Reference 4)			0.5 ml							
Yeast extract				1						1
Brain-heart infusion broth								1 liter	1 liter	
Bovine albumin, fraction V								10	10	
Agar	15	15	15	15			15	10	10	15
Nutrient agar					15	15				
pH	7		7	7.5–7.8	7	7	7	6.8	6.8	7
Source of inoculum	Soil, salt water	Soil, salt water	Soil	Soil, salt water	Soil	Soil	Soil	Humans, animals		Plants, soil
Reference	15	16	4	17	4	4	18	19	19	20
Special notes	1	2	3	3, 4	3	3	3	3, 6	3, 6	3, 7

Instructions:

1. Preparation of the agar top layer: sodium alginate, NaCl, K_2HPO_4. Preparation of the bottom layer: K_2HPO_4, trypticase, agar.
2. Look for pits around the colonies.
3. Look for a clear halo.
4. Seal the plates to prevent evaporation and incubate them for 4 to 40 days in the dark at room temperature.
5. After incubation, flood the plates with Gram's iodine (iodine, 1 g; KI, 2 g; water to 100 ml).
6. After incubation, flood the plates with 2*N* acetic acid for 10 minutes. Nondegraded substrate precipitates with the albumin, leaving a clear zone around colonies that produce enzyme.
7. After incubation, flood the plates with 1% aqueous solution of hexadecyltrimethylammonium bromide (J. T. Baker Chemical Co.); areas of pectolysis usually appear within 15 minutes.

[a] Unless otherwise indicated.
[b] Sterilize by filtration and add aseptically.

REFERENCES

General References

1. **Aaronson, S.,** *Experimental Microbial Ecology,* Academic Press, New York, 1970.
2. **Alexander, M.,** *Microbial Ecology,* John Wiley & Sons, 1970.
3. **Schlegel, H. G. (Ed.),** *Anreicherungskultur und Mutantenauslese,* G. Fischer, Stuttgart, 1965.
4. **Stein, J. R. (Ed.),** *Handbook of Phycological Methods,* Cambridge University Press, Cambridge, England, 1973.

Specific References

1. **Jannasch, H. W.,** *Arch. Mikrobiol.,* 45, 323, 1963.
2. **Jannasch, H. W.,** *Arch. Mikrobiol.,* 59, 165, 1967.
3. **Jannasch, H. W.,** *Limnol. Oceanogr.,* 12, 264, 1967.
4. **Aaronson, S.,** *Experimental Microbial Ecology,* Academic Press, New York, 1970.
5. **Aaronson, S. and Scher, S.,** *J. Protozool.,* 7, 156, 1960.
6. **Provasoli, L. and Pintner, I. J.,** in *The Ecology of Algae,* Tryon, C. A. and Hartman, R. T., Eds., University of Pittsburgh Press, Pittsburgh, 1960, p. 84.
7. **Aaronson, S.,** *J. Bacteriol.,* 69, 67, 1955.
8. **Provasoli, L., McLaughlin, J. J., and Droop, M. R.,** *Arch. Mikrobiol.,* 25, 392, 1957.
9. **Gerloff, G. C., Fitzgerald, G. P., and Skoog, F.,** *Am. J. Bot.,* 37, 216, 1950.
10. **Starr, R. C.,** *Am. J. Bot.,* 51, 1013, 1964.
11. **Kratz, W. A. and Myers, J.,** *Am. J. Bot.,* 42, 282, 1955.
12. **Pringsheim, E. G.,** in *Manual of Phycology,* Smith, G. M., Ed., Chronica Botanica, Waltham, Mass., 1951, p. 347.
13. **Lewin, J. C. and Lewin, R. A.,** *Can. J. Microbiol.,* 6, 127, 1960.
14. **Postgate, J. R.,** *Lab. Pract.,* 15, 1239, 1966.
15. **Yaphe, W.,** *Nature,* 196, 1120, 1962.
16. **Stanier, R. Y.,** *J. Bacteriol.,* 42, 527, 1941.
17. **Lear, E. W., Jr.,** in *Symposium on Marine Microbiology,* Oppenheimer, C. H., Ed., Charles C Thomas, Springfield, Ill., 1963, p. 594.
18. **Rhodes, M. E.,** *J. Gen. Microbiol.,* 21, 221, 1959.
19. **Smith, R. F. and Willett, N. P.,** *Appl. Microbiol.,* 16, 1434, 1968.
20. **Hankin, L., Zucker, M., and Sand, D. C.,** *Appl. Microbiol.,* 22, 205, 1971.

BACTERIAL CELL WALL STRUCTURE

C. S. Cummins

INTRODUCTION

The bacterial cell wall is not composed of a single substance; it is a complex of polymers that can be separated in various ways in the laboratory, although it is likely that they are bound together covalently in the wall of the living cell. The study of bacterial wall structure essentially started in 1951, when Salton and Horne,[1] reported chemical studies on cell wall fractions that could be seen to be pure by electron microscopy. Since that classical study, a great deal of work has been done, and the subject has been reviewed at fairly frequent intervals.[2-8] Reviews on specialized aspects of the subject (e.g., biosynthesis, structure of O-antigens, etc.) are referred to in the appropriate section below.

PREPARATION OF CELL WALL FRACTIONS FOR ANALYSIS

Detailed discussions of methods for cell disruption can be found in References 1 and 9 to 11, as well as in Volume III. The matter is considered here briefly because of the considerable influence that methods of preparation have on the final product. An outline of the essential features of the three commonly used procedures is given below. An extensive review of methods of preparation and chemical analysis of cell walls, with much practical detail, can be found in Reference 261.

Ultrasonic disruption — Ultrasonic vibrations at 20 kHz or higher, produced in either a magnetostrictive or a piezoelectric oscillator, are applied to the bacterial suspension, usually by a probe connected to the oscillator. The bacteria are disrupted by the great pressure changes produced by the alternate development and collapse of gas cavities in the liquid. Considerable heating of the fluid takes place, and some form of cooling is necessary.

Shaking with small glass beads — This method was introduced by Dawson in 1949.[12] It was used by Salton from the start of his classic investigations into cell wall composition[1] and has remained a standard procedure. For bacteria, glass beads (Ballotini) 0.1 to 0.12 mm in diameter are most satisfactory, although beads up to 0.2 mm have been used. The original device was the Mickle tissue disintegrator,[13] which shook a mixture of glass beads and bacteria at 50 Hz. The principal disadvantage of this apparatus is the rather small amount of material that can be treated at one time; other machines — such as the Braun shaker,[14] the Bühler "cell mill",[15] or the Nossal apparatus[16] — will disintegrate 1 to 2 g dry weight of bacteria in a total shaking time of about 5 minutes, although the whole process may take longer because of the intervals for cooling. After shaking, the glass beads are removed by filtration through a coarse sintered glass filter.

Extrusion at high pressure — If a bacterial suspension is placed under high pressure, at 20,000 psi or more, and then extruded through a narrow orifice by slightly opening a fine ball or needle valve, the combination of release of pressure and shear at the orifice will disrupt the organisms. In other methods, a paste of bacteria frozen at −20°C or below is forced through a narrow orifice by the impact of a fly press on a closely fitting piston. In this case the ice crystals in the frozen paste probably have an abrasive effect. Examples of commercial machines are the Aminco® French pressure cell,[17] the Sorvall®-Ribi refrigerated cell fractionator,[18] and the Hughes press.[19]

Of these three methods, the best for the preparation of cell wall fractions is probably shaking with small glass beads. The major disadvantage of ultrasonic methods is that the cell wall fragments first produced are soon disintegrated still further to almost colloidal dimensions, so that by the time the majority of the cells have been ruptured, a very wide range of sizes of wall fragments has been produced. This makes the clean recovery of cell walls in the centrifuge very difficult. Extrusion at high pressure works well with Gram-negative rods, but is much less efficient for Gram-positive organisms. Gram-positive cocci are particularly resistant to rupture by this method.

Preparation of Cell Wall Fractions from the Crude Disintegrate

After disintegration, an essential preliminary step in purification is centrifugation to separate unbroken cells from the crude cell wall fractions. Alternate low-speed (ca 3,000 × G) and high-speed (ca 30,000 × G) centrifugation in distilled water or buffer is a standard procedure. However, it is often possible after a single high-speed centrifugation to get a clear separation in the deposit into a lower, more opaque layer of unbroken cells and an upper, more translucent layer of cell walls. The upper cell wall layer can then usually be removed by gentle washing with a wash bottle or Pasteur pipette, and a relatively pure cell wall preparation is achieved in a single step.

Subsequent treatment of the crude cell wall fractions will depend on the purpose for which the walls are being prepared. If it is desired to recover cell wall fragments in as nearly native a state as possible, centrifugation in a sucrose or other density gradient is the method of choice.[20-22] If, on the other hand, it is desired to prepare peptidoglycan for analysis, treatment with proteolytic enzymes — such as trysin or Pronase — will be followed by extraction with formamide at 165°C, or with trichloroacetic acid or phenol, depending on the nature of the organism. Cell walls are very liable to autolysis because of the lytic enzymes liberated at the time of cell disruption, and it is usual to inactivate these by a short treatment at 100°C (10 to 15 minutes) or by a wash in 0.1% formalin. Alternatively, autolysis may be prevented if the cultures to be disintegrated are killed by 2% formalin before being washed and shaken with glass beads. However, this may make the organisms somewhat more difficult to disintegrate.

Estimation of Purity of Cell Wall Fragments

Electron microscopy is the ideal method for determining whether or not cell wall fragments are free from contamination with other cell elements. Undisrupted cells, ribosomes, and other cytoplasmic particles are almost all more electron-dense than the wall, and examination of shadowed or negatively stained preparations will show contamination with membrane fragments. In the absence of facilities for electron microscopy, it is possible to assess the degree of contamination in other ways. For example, ribose or deoxyribose should be absent from cell wall hydrolysates, and there should be no UV absorption in the region between 260 and 270 nm. In many Gram-positive organisms, treatment with proteolytic enzymes will remove all sources of amino acids, except the peptide of peptidoglycan, so that the finding of a very simple amino acid pattern in hydrolysates — e.g., alanine, glutamic acid, and *meso*-diaminopimelic acid, or alanine, glutamic acid, and lysine — is evidence of purity, because almost all contaminating material will contain substantial amounts of nonpeptidoglycan amino acids, such as leucine and isoleucine. This test, of course, cannot be applied to cell walls of Gram-negative organisms, since the protein fraction of the O-antigen is unaffected by proteolytic enzymes.

Preservation of Cell Wall Samples

Probably the most convenient method of preservation is freeze-drying. In most cases, especially if proteolytic enzymes have been used in their preparation, cell wall

Table 1
GENERAL DIFFERENCES IN CELL WALL COMPOSITIONS BETWEEN GRAM-POSITIVE AND GRAM-NEGATIVE ORGANISMS

Gram-positives	Gram-negatives
Wall thick, about 200 Å; usually structureless	Wall thinner, about 100—150 Å; double membrane structure usually obvious
Amino sugar content high: 10—30%	Amino sugar content low: 1—10%
Lipid content generally low: 0—2% (except in mycobacteria, nocardias, and corynebacteria)	Lipid content high: 10—20%
3 or 4 major amino acids	14 to 18 amino acids (i.e., the full range found in protein, plus diamino pimelic acid)

fractions after lyophilization form a very light, flaky white material. It may need brief sonication or shaking with beads to redisperse the dry cell wall fragments in saline or water. Suspensions of cell wall fragments in distilled water or dilute buffer appear to be stable for years in the refrigerator, provided that autolytic enzymes have been inactivated during their preparation. An inhibitor such as 0.1% sodium azide should be present to prevent attack by bacterial contaminants, many of which will actively digest cell wall fragments.

DIFFERENCES BETWEEN CELL WALLS FROM GRAM-POSITIVE AND GRAM-NEGATIVE BACTERIA: NATURE OF POLYMERS PRESENT

There are certain broad chemical and structural differences between the cell walls of Gram-positive and Gram-negative bacteria. The major physical and chemical differences are set out in Table 1, and the structural differences as shown in electron microscope profiles of thin sections of walls are illustrated diagrammatically in Figure 1. These broad differences in structure are reflected in the different kinds of polymers found in cell walls of Gram-positive and of Gram-negative organisms, and the polymers are detailed in Table 2. The most important of them is undoubtedly peptidoglycan (mucopeptide, murein), since it is responsible for the structural integrity of the cell in the face of its internal osmotic pressure.

Peptidoglycan is present in almost all kinds of bacteria. The only exceptions found so far are the extreme halophiles, where the osmotic pressure of the environment presumably makes it unnecessary, and some kinds of methane bacteria.[262] It is not yet known what material replaces peptidoglycan in the latter group. Peptidoglycan is also present in *Spirillum, Spirochaeta, Rickettsia* and in the blue-green algae, but not in viruses, yeasts, fungi, higher plants, or animals. It is, therefore, confined to the cell wall of prokaryotic cells.

Peptidoglycan forms a large proportion of the wall in Gram-positive organisms, where it may make up 50% or more of the wall's dry weight. In Gram-negatives, on the other hand, it forms usually less than 10% of the wall, and sometimes as little as 1%. These differences are largely responsible for the higher proportion of amino sugars in the walls of Gram-positives.

In the electron microscope, the difference in appearance of cross sections of the wall in Gram-positives and in Gram-negatives is generally rather obvious (Figure 1).The

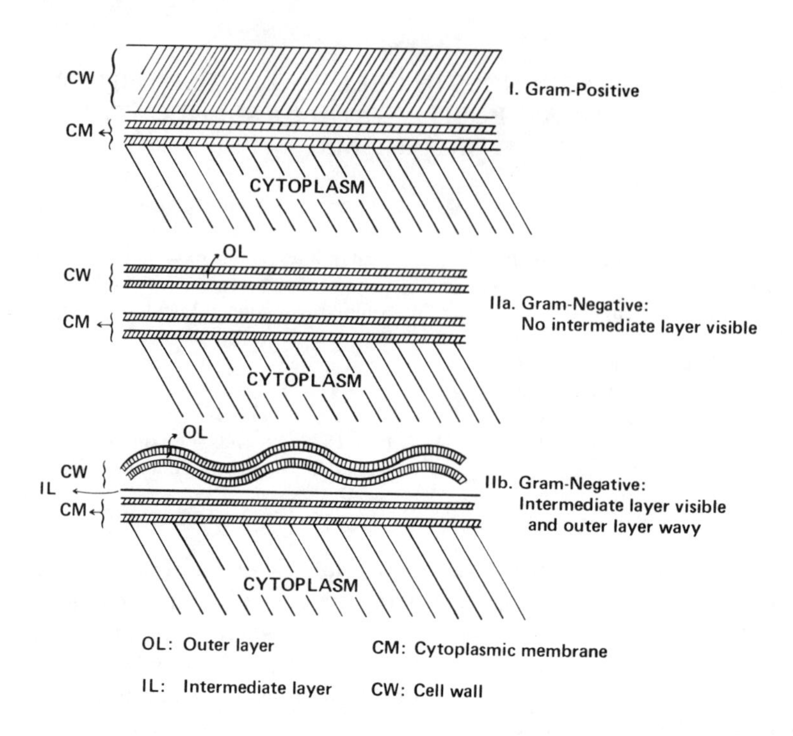

FIGURE 1. Electron microscope profiles of bacterial cell walls.

Table 2
PRINCIPAL POLYMERS FOUND IN CELL WALLS OF GRAM-POSITIVE AND GRAM-NEGATIVE BACTERIA

GRAM-POSITIVES	
Peptidoglycan	
Polysaccharides	
Teichoic acids	
Teichuronic acids	
Proteins	
Glycolipids Mycosides	mycobacteria, corynebacteria, nocardias
GRAM-NEGATIVES	
Peptidoglycan	
O-antigen complex	
Lipoprotein	

typical Gram-positive wall shows a thick (200 to 500 Å) featureless structure, inside which is the double-track (dark-light-dark) appearance of the cytoplasmic membrane. The typical Gram-negative wall, however, is generally thinner (100 to 150 Å) and shows two double-track membranes separated by a space in which an intermediate layer may or may not be visible (Figure 1, IIb); the inner of the two double-track structures is the cytoplasmic membrane, and the intermediate layer is almost certainly the peptidoglycan. These "typical" appearances may vary considerably in different organisms and with different methods of fixation.[23-26]

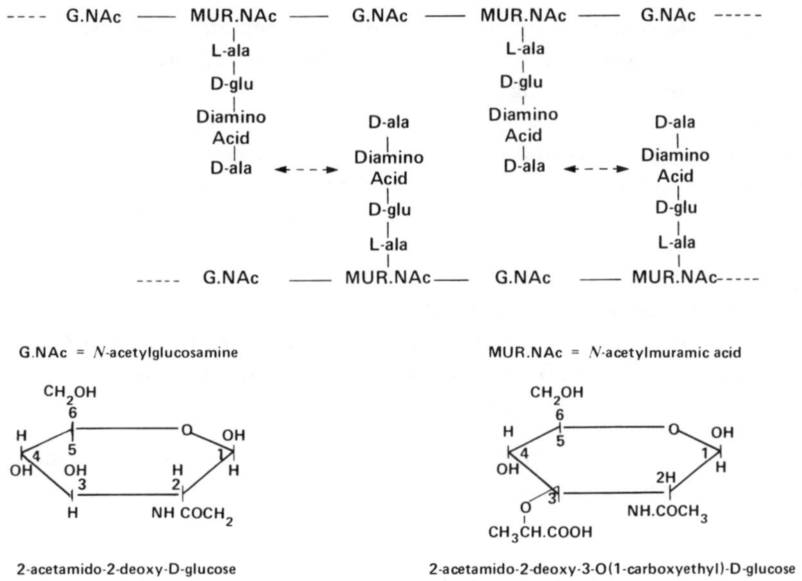

---- G.NAc —— MUR.NAc —— G.NAc —— MUR.NAc —— G.NAc -----

L-ala, D-glu, Diamino Acid, D-ala ←--→ D-ala, Diamino Acid, D-glu, L-ala

----- G.NAc —— MUR.NAc—— G.NAc —— MUR.NAc-----

G.NAc = *N*-acetylglucosamine

MUR.NAc = *N*-acetylmuramic acid

2-acetamido-2-deoxy-D-glucose

2-acetamido-2-deoxy-3-O(1-carboxyethyl)-D-glucose

FIGURE 2. A general structure for cell wall peptidoglycan.

Because of the rather complex structure of the outer layers of Gram-negative organisms, there is not the same clear distinction between "cell wall" and "cytoplasmic membrane" as is possible with Gram-positives. From the practical point of view, preparation of "cell walls" from Gram-negative bacteria made by standard methods — e.g., shaking with glass beads or disruption in a French pressure cell — will contain portions of all the layers visible in EM profiles. Salton[259] has proposed that the general term *cell envelope* rather than *cell wall* be used to describe collectively the outer membrane system in Gram-negative cells.

The different parts of the envelope complex are usually referred to as the *outer membrane* (= outer double-track layer), the *cytoplasmic membrane* (= inner double-track layer), and between them the *periplasmic space,* in which lies the *peptidoglycan layer.* It seems probable, in fact, that in many Gram-negatives peptidoglycan is covalently bound to a lipoprotein as a *peptidoglycan-lipoprotein layer,* which is probably connected via the lipoprotein to the inner part of the outer membrane. Methods have been devised for separating these various layers from each other (e.g., Osborn et al.[258]), and these techniques have made it clear that the outer membrane contains the lipopolysaccharide O-antigen complex and that the inner membrane behaves functionally as a cytoplasmic membrane.

For recent work on the nature and function of Gram-negative cell envelopes, see reviews by Costerton et al.[257] and by Braun and Handtke.[256]

PEPTIDOGLYCAN

Peptidoglycan is a polymer with a backbone of amino sugar chains, which are cross-linked through tetrapeptide side chains. An idealized diagram of the structure is given in Figure 2. The backbone of amino sugar is composed of alternate residues of N-acetylglucosamine and N-acetylmuramic acid, linked by β-1 > 4 glycosidic bonds, and the tetrapeptide is attached to the −COOH group of the lactic acid side chain on C-3 of muramic acid. Since muramic acid is a substituted glucosamine, the carbohydrate skeleton of peptidoglycan resembles that of chitin.

The tetrapeptide is almost always composed of four amino acids in the following order, starting at muramic acid: L-alanine, D-glutamic acid, L-diamino acid, D-alanine. One of several diamino acids may be present, as described below. Since the lactic acid side chain of muramic acid is in the D-configuration, this gives the configurational sequence "D-L-D-L-D" down the peptide chain. This sequence is a constant feature of peptidoglycan structure. The D-glutamic acid residue is linked in the chain by the γ-COOH, and the α-COOH is frequently amidated. For a review of peptidoglycan structure see Reference 27.

Variations in Peptidoglycan Structure

In theory, variation may occur in any of the three major elements of peptidoglycan structure — i.e., in the amino sugar backbone, in the tetrapeptide, and in the linking amino acids. However, some kinds of variations are much more common than others.

Variation in the backbone structure — No major variations in the amino sugar backbone have been reported, except that some aerobic actinomycetes have N-glycolyl muramic acid instead of N-acetyl muramic acid.[28]

Variations in tetrapeptide — Any one of four diamino acids may be present: L-lysine, diaminopimelic acid (DAP), L-ornithine, or L-diaminobutyric acid. Rarely, no diamino acid is present, as in the homoserine-containing tetrapeptide of some plant-pathogenic corynebacteria. In the case of DAP, the *meso-* or L-isomers may be present, but the end of the molecule involved in the tetrapeptide has the L-configuration.[27] The D-isomer of DAP has also been found in the cell walls of some organisms (e.g., in *Bacillus megaterium* and in some aerobic actinomycetes,[34,45] but always in association with other DAP isomers, and its position in the structure is not certain. It is thought not to occur in the tetrapeptide, since this would violate the "D-L-D-L-D" rule. In some actinomycetes the 3-hydroxyderivative of DAP is found.[120] Most variations in tetrapeptide structure concern the diamino acid, but in rare instances the L-alanine is replaced by glycine or L-serine (see Table 4). For a review of the distribution of diamino acids in cell walls, see Reference 121. The diamino acids that have been found in different genera of bacteria are listed in Table 3.

Variations in the linking amino acids — In some cases — e.g., with *meso*-DAP — the link between tetrapeptide chains is a direct peptide bond between one $-NH_2$ group of the diamino acid and the $-COOH$ group of D-ala on another tetrapeptide chain. More often, however, the linkage involves a single additional amino acid or a short peptide composed of 2 to 5 residues. In unusual types of peptidoglycan structure, the link may involve the α-COOH of glutamic acid. Table 4 gives the principal variations in linking amino acids that detailed structural studies have so far revealed (for a review, see Reference 255).

Recognition of Peptidoglycan Components in Acid Hydrolysates of Whole Cells

Two components of peptidoglycan — diaminopimelic acid and muramic acid — are so distinctive that their presence in hydrolysates of whole cells may be taken as evidence of their occurrence in the peptidoglycan of that particular strain, or more generally, as evidence that peptidoglycan is present in a particular group of organisms.

Table 3
DISTRIBUTION OF DIAMINOPIMELIC ACID (DAP) AND OTHER DIAMINO ACIDS IN BACTERIA AND OTHER MICROORGANISMS

Organism	DAP			Other diamino acids				References
	meso/DD	LL	3-Hydroxy	L-Lysine	L-Ornithine	L-DABA	L-Homoserine[a]	
GRAM-POSITIVE BACTERIA								
Aerococcus								
A. viridans				+				29[b]
Aerobic actinomycetes								
Streptomyces spp.		+						
Micromonospora spp., *Actinoplanes* spp., *Streptosporangium* spp.	+	Variable, generally a small amount	+ to −					30,[b] 31,[b] 32,[b] 33,[b] 34[b]
Nocardia	+							
Actinomyces, microaerophilic								
A. israelii, A. viscosus, A. naeslundii				Lysine or ornithine (structure not known in detail)				35, 36
A. bovis				+				
Arachnia								
A. propionica		+						36[b], 37[b]
Arthrobacter								
A. albidus and other strains	+							38,[b] 39[b]
A. simplex and other strains		+						40, 41[b]
A. globiformis and other strains				+				42
Bacillus								
Most species	+							43,[b] 44, 45, 46,[b] 47,[b] 48—52, 53[b]

Table 3 (continued)
DISTRIBUTION OF DIAMINOPIMELIC ACID (DAP) AND OTHER DIAMINO ACIDS IN BACTERIA AND OTHER MICROORGANISMS

Organism	DAP			Other diamino acids				References
	meso/DD	LL	3-Hydroxy	L-Lysine	L-Ornithine	L-DABA	L-Homoserine[a]	
B. sphaericus, B. pasteurii				+				47,[b] 53[b]
Bacterionema								
B. matruchottii	+							54
Bifidobacterium								
B. infantis and other species				+				53
B. bifidum and other species					+			
Brevibacterium								
ca. 50% of strains (15 named species)	+							
B. acetylicum				+				41[b]
B. helvolum (1 strain)						+		
B. albidum (and some other strains)							+	
Butyribacterium								
B. rettgeri				+				55
Cellulomonas								
6 species (1 strain each)				+				38
Clostridium								
Most species	+							
C. perfringens, C. pectinovorum		+						56
C. paraputrificum, C. tertium, C. innocuum				+				
GRAM-POSITIVE BACTERIA (continued)								
Micrococcus								
Most species				+				29

M. varians	+				
M. radiodurans				+	66, 67
Mycobacterium					
All species examined	+				68,[b] 69[b]
Planococcus			+		29
Pediococcus					
P. cerevisiae			+		53,[b] 70
Peptostreptococcus					71, 72[b]
Propionibacterium					
Most species		+			37[b]
P. shermanii, P. freudenreichii	+				
Rothia					
R. dentocariosa			+		73
Sarcina					
S. ventricule, S. maxima		+			29[b]
Spororsarcina					
S. ureae			+		29[b]
Straphylococcus					
All species examined			+		74, 75, 76[b]
Streptococcus					
All species examined			+		57, 77,[b] 78,[b] 79, 80
GRAM-NEGATIVE BACTERIA					
Gram-negative Bacilli					
Aerobacter	+				81,[b,c] 82; 83
Agrobacterium	+				83
Brucella	+				81,[b,c] 84,[b] 85
Citrobacter	+				86
Erwinia	+				87
Escherichia	+				48,[b] 86
Ferrobacillus	+				88
Haemophilus	+				81[b,c]
Klebsiella	+				86
Proteus	+				86, 89
Pseudomonas	+				86, 89, 90,[b] 91[b]

Table 3 (continued)
DISTRIBUTION OF DIAMINOPIMELIC ACID (DAP) AND OTHER DIAMINO ACIDS IN BACTERIA AND OTHER MICROORGANISMS

	DAP			Other diamino acids				
Organism	*meso*/DD	LL	3-Hydroxy	L-Lysine	L-Ornithine	L-DABA	L-Homoserine[a]	References
Salmonella	+							89, 92, 93
Serratia	+							86, 89
Thiobacillus	+							94
Gram-negative Cocci								
Neisseria, Veillonella	+							95
Vibrios								
V. metschnikovi	+							92
V. fetus (Campylobacter fetus)	+							96
V. comma	+							81[b,c]
OTHER ORGANISMS								
Blue-green algae	+							81,[b,c] 97, 98, 99
Myxobacteria	+							81,[b,c] 100
Caryophanon				+				101
Spirillum	+							102,[b] 103, 104, 105
Spirochaeta	Traces?			Probably lysine or ornithine				106, 107
Corynebacterium								
C. diphtheriae (and most human and animal pathogenic coryneforms)	+							57
C. pyogenes, C. haemolyticum				+				58
C. acnes (Propionibacterium acnes) and related organisms	(+)[d]	+						37[b]
C. poinsettiae and some other plant pathogens							+	59

C. insidiosum, C. tritici, C. sepedonicum				+	59
Diplococcus					
D. pneumoniae		+			60
Erysipelothrix					
E. rhusiopathiae		+			61
Gaffkya					
G. homari		+			29[b]
Lactobacillus					
Most species		+			
L. plantarum, L. inulinus	+				53[b]
L. cellobiosus			+		
Leuconostoc					
All species examined		+			53, 62
Listeria					
L. monocytogenes	+				61,[b] 63, 64
Microbacterium					
M. flavum, M. thermosphactum	+				
M. lacticum		+			41[b], 65[b]
M. liquefaciens				+	

[a] Homoserine is not a diamino acid, but where it occurs it occupies the position in the tetrapeptide structure normally occupied by a diamino acid.
[b] Isomer of DAP definitely identified, or lysine, ornithine, etc., definitely identified as L-isomer.
[c] Isomer of DAP identified in whole-cell hydrolysate only.
[d] Occasional strains of *P. acnes* and *P. avidum* have the *meso*-isomer of DAP[37]

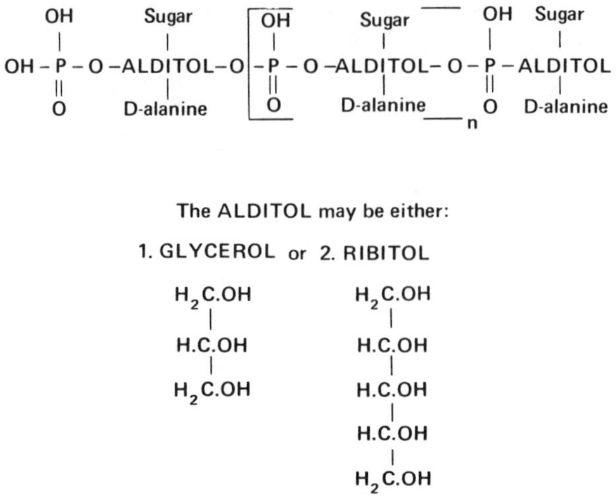

```
  OH        Sugar      ┌OH        Sugar   ─┐  OH   Sugar
  |          |         │|          |       │  |     |
OH-P-O-ALDITOL-O│-P-O-ALDITOL-O-│-P-ALDITOL
  ||         |         │||         |       │  ||    |
  O      D-alanine     └O      D-alanine  ─┘n O  D-alanine
```

The ALDITOL may be either:

1. GLYCEROL or 2. RIBITOL

```
H2C.OH        H2C.OH
  |             |
H.C.OH        H.C.OH
  |             |
H2C.OH        H.C.OH
                |
              H.C.OH
                |
              H2C.OH
```

FIGURE 3. General structure of teichoic acids. (From Archibald, A. R. and Baddiley, J., *Adv. Carbohydr. Chem*, 21, 323, 1966. With permission.)

DIAMINOPIMELIC ACID (DAP)

DAP occurs as an intermediate in the more common of the two pathways of lysine biosynthesis,[260] and as such it is found in small quantities in the cytoplasm of bacteria and other prokaryotic organisms, in green algae, and in plants. As a structural amino acid, however, it is found only in peptidoglycan.

In Gram-positive bacteria, the cell wall may constitute 15 to 20% or more of the dry weight of the cell, and 50% or more of the wall may be peptidoglycan. Under these circumstances, DAP, if present, is likely to be a major constituent of acid hydrolysates of whole cells and can readily be detected by paper chromatography. Occasional strains of lactobacilli with defective lysine metabolism may produce large amounts of free DAP, although the diamino acid of their peptidoglycan is lysine.[122] In Gram-negative bacteria, however, the wall forms a smaller proportion of the cell (5 to 10%), and peptidoglycan may form less than 1% of the wall, so that it may be difficult to detect DAP in crude whole cell hydrolysates. (For recognition of DAP and its isomers and derivatives by paper chromatography, see References 45, 81, 120, and 123.)

MURAMIC ACID

This is a more reliable indicator substance for peptidoglycan than DAP, because it appears to be an invariable component of the backbone glycan chain and is not known to occur elsewhere. The presence of muramic acid in whole-cell hydrolysates has been used to demonstrate the occurrence of peptidoglycan in *Leptospira* and *Borrelia*[124] and in *Rickettsia.*[125]

OTHER CELL WALL POLYMERS

In most cases detailed knowledge about other cell wall polymers is confined to a few selected groups of organisms, since they have not been investigated as thoroughly or as widely as peptidoglycan. In the following sections, polymers occurring in Gram-positive and in Gram-negative organisms are discussed separately.

Gram-positive Organisms

TEICHOIC ACIDS AND OTHER ACIDIC POLYSACCHARIDES

The teichoic acids are complex polymers of polyols and phosphate, originally described by Baddiley and his co-workers.[126,127] Most commonly they have the general structure illustrated in Figure 3, which shows a backbone of repeating units of glycerol or ribitol phosphate with associated residues of sugars or amino sugars and D-alanine. However, teichoic acid from *Staphylococcus lactis* has alternate residues of N-acetylglucosamine and glycerol in the backbone structure,[128] and the C-substance from pneumococcal cell walls is a teichoic acid containing choline phosphate.[129]

Although originally named because of their association with cell walls (*teichos* is the Greek word for wall), teichoic acids are also widely distributed in Gram-positive organisms in association with the cytoplasmic membrane, in which case they are often referred to as intracellular or cytoplasmic teichoic acids, although they probably accumulate between the wall and the membrane.

In the membrane, teichoic acids normally occur in the form of lipoteichoic acids,[263] in which the teichoic acid is linked to a glycolipid through the terminal glycerol phosphate residue of the teichoic acid chain. This is a rather labile linkage, which is broken by cold trichloracetic acid at 4°C, so that the usual extraction procedure (see below) gives teichoic acid devoid of lipid. For the methods of extraction of the complete teichoic acid complex, see Reference 263.

Cell wall teichoic acids appear to be especially prominent in certain genera, such as *Staphylococcus, Micrococcus, Bacillus,* and *Lactobacillus,* where they may form up to 50% of the wall. They may also be important cell wall antigens (see below). The presence of appreciable amounts of phosphorus (i.e., 2 to 5%) in cell walls may be taken as presumptive evidence of the presence of teichoic acids, since in their absence the phosphorus content of bacterial walls is low (i.e., 0.2 to 0.5%).

Acidic polysaccharides containing uronic acids, but no phosphate, have been isolated from some organisms. For example, Janczura et al.[130] isolated from *Bacillus licheniformis* a polysaccharide consisting of equal parts of glucuronic acid and N-acetyglucosamine, which they called teichuronic acid; a similar substance, composed of glucose and aminomannuronic acid, was found in the walls of *Micrococcus lysodeikticus* by Perkins.[131] Since uronic and (especially) aminouronic acids are acid-labile, it is possible that they occur rather widely in bacterial walls, but have remained unrecognized because they are destroyed by the usual methods of hydrolysis.

There appears to be a close relationship between synthesis of teichoic and teichuronic acids, because in *Bacillus subtilis* teichuronic acid replaces teichoic acid if the organisms are deprived of phosphate.[132] However, NaCl concentration is also important.[133,253,254]

Extraction of Teichoic Acid from Cell Walls

Teichoic acids are generally prepared from cell walls by extraction with dilute trichloroacetic acid (TCA, 5 to 10% aqueous) at 0 to 5°C. The teichoic acid is precipitated from the extract with ethanol (2 to 5 volumes). The extraction may need to be long-continued (2 to 3 days) and repeated several times to obtain the bulk of the material. The process can be followed by estimating the phosphorus content of the residual cell walls. Extraction is much faster at higher temperatures — i.e., at 37°C, or even at 80 to 90°C — but this is usually undesirable because other polysaccharides, if present, will also be readily extracted, and because degradation of teichoic acid may occur. Cell wall teichoic acids may also be obtained by extraction of intact bacterial cells, but they will then be contaminated by nucleic acids and by intracellular teichoic acids, from which it may be difficult to purify them.

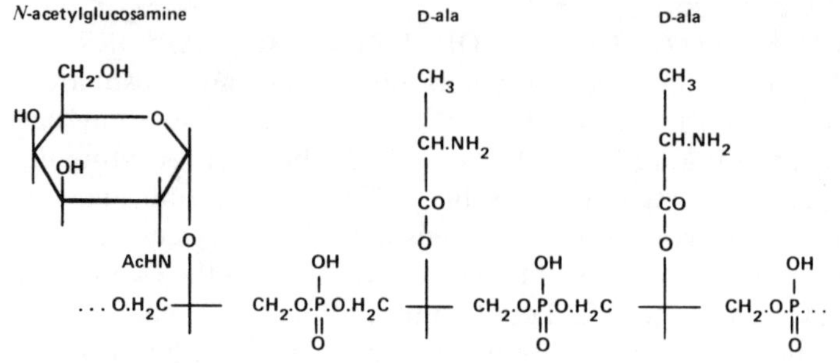

FIGURE 4. Repeating unit of the glycerol teichoic acid in cell walls of *Staphylococcus albus*. (Reprinted by permission from Ellwood, D. C., Kelleman, M. V., and Baddiley, J., *Biochem. J.*, 86, 213, 1963. Copyright© 1963, The Biochemical Society, London.)

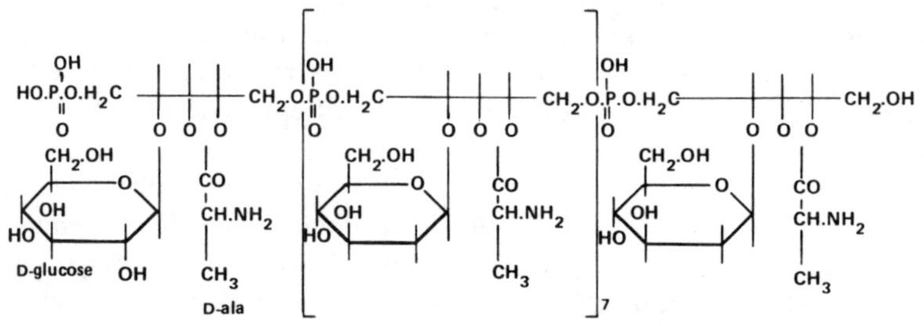

FIGURE 5. Ribitol teichoic acid from cell walls of *Bacillus subtilis*. (Reprinted by permission from Armstrong, J. J., Baddiley, J., and Buchanan, J. G., *Biochem. J.*, 80, 254, 1961. Copyright© 1963, The Biochemical Society, London.)

Details of TCA extraction procedures as applied to walls of *Lactobacillus buchneri, Lactobacillus arabinosus, Staphylococcus albus,* and *Bacillus subtilis* may be found in References 134, 135, 136, and 137. Other extraction procedures, using dilute alkali or dilute aqueous N,N-dimethylhydrazine, will be found in References 138 and 139.

The detailed structures of a glycerol teichoic acid from *Staphylococcus albus* and a ribitol teichoic acid from *Bacillus subtilis* are shown in Figures 4 and 5, respectively.[136,140]

Teichoic Acids as Antigens

Teichoic acids are good antigens and have been used as the basis for defining antigenic groups in several genera. The determinants involved are usually the sugars or amino sugars attached to the alditols. (See Figure 3.) For example, in *Staphylococcus aureus*[141] the walls contain ribitol teichoic acid that has N-acetylglucosamine attached to the ribitol by either an α- or a β-linkage, so that two serologically distinct teichoic acids may occur in the same organism. On the other hand, *Staphylococcus epidermidis* strains have glycerol teichoic acids, which may have α- or β-linked glucose or α-linked glucosamine.[142,143] As a further example of the use of teichoic acid antigens in classification, Table 5 shows the chemical nature and anatomical locaton of the group antigen in various serological groups of lactobacilli.[144] It may be noted that not all lactobacilli contain teichoic acids in the cell wall.

Table 4
PEPTIDOGLYCAN LINKAGES ASSOCIATED WITH DIFFERENT DIAMINO ACIDS[a]

Diamino acid	Linking amino acids or peptides: Organism	Linkage	Ref.
L-Lysine	*Staphylococcus*	−(gly)$_5$− −(gly)$_2$−ser− (gly)$_2$− −L-ala−(gly)$_4$−	29,76
	Micrococcus *Sarcina*	−L-ala− −D-ala−L-lys−D-glu−L-ala− gly −gly−L-glu−(L-ala)$_3$−	29, 108—111
	Aerococcus *Gaffkya*	Direct peptide bond between L-lys and D-ala	29
	Planococcus	−D-glu −	111
	Pediococcus	−D-asp−	29,53,70
	Leuconostoc	−L-ser−L-ala)$_2$− −L-ser−L-ala−L-ala−L-ser−	53,112
	Lactobacillus	−D-asp−L-ser−(L-ala)$_2$−(L-ala)^2−	53,112—115
	Streptococcus	−L-asp−thr−L-ala−(L-ala)$_3$−	77,78,114,116
	Bifidobacterium	−gly−	53,117,118
meso-DAP	*Bacillus, Arthrobacter, Corynebacterium, Clostridium,* etc. Gram-negative bacteria	Direct peptide bond between *meso*-DAP and D-ala	See Table 3 for genera having *meso*-DAP
L-DAP	*Propionibacterium, Streptomyces, Clostridium*	−(gly)$_n$− (n usually 1 to 4)	119
L-Ornithine	*Bifidobacterium*	−L-ser− L-ala−L-thr−L-ala− −L-lys−D-asp− −L-lys−L-ser−(L-ala)$_2$− −L-lys−(L-ala)$_2$−	53
L-DABA	*Corynebacterium tritici, C. insidiosum, C. michiganense*[b]	−D-DABA−	41,59
L-Homoserine	*Corynebacterium betae, C. flaccumfaciens, C. poinsettiae*	−D-orn−	41,115
	Some strains in *Arthrobacter, Microbacterium,* and *Brevibacterium*[c]	−gly−D-orn− −(gly)$_2$−L-lys−	41

[a] The material in the table has been arranged broadly in terms of the linkages occurring in different genera, but it must be noted that (1) the same linkage may occur in several different genera, and (2) organisms classified in a single genus may have different diamino acids and types of linkage.

[b] The peptidoglycan in these organisms is unusual because L-ala is replaced by gly as the amino acid attached to muramic acid, and the cross-linkage is through the α-COOH of glutamic acid instead of through the diamino acid (L-DABA).

[c] Peptidoglycans with homoserine are unusual, because gly is the amino acid joined to muramic acid, because the cross-link is through the α-COOH of glutamic acid, and because homoserine is not a diamino acid; D-glu in these strains may be replaced by 3-threo-hydroxyglutamic acid.

Table 5
LOCATION AND CHEMICAL NATURE OF GROUP ANTIGENS IN DIFFERENT LACTOBACILLI

Serological group	Species included in group	Location of antigen in cell	Chemical nature of antigen	Sugar	Alkali hydrolysis product
A	*L. helveticus*	Membrane	GTA[a]	Glucose (trace)	
	L. jugurti			Ribose (trace)	Glucosyl glycerol
B	*L. casei*	Wall	Polysaccharide	Rhamnose	
C	*L. casei*	Wall	Polysaccharide	Glucose	
D	*L. plantarum*	Wall	RTA[b]	Glucose	Glucosyl ribitol
E	*L. lactis*	Wall	GTA[a]	Glucose	Glucosyl glycerol
	L. bulgaricus				
	L. brevis				
	L. buchneri				
F	*L. fermenti*	Membrane	GTA[b]		
G	*L. salivarius*	?	?		

[a] Glycerol teichoic acid.
[b] Ribitol teichoic acid.

From Sharpe, E., *Int. J. Syst. Bacteriol.*, 24, 509, 1970. Reproduced by permission of the International Association of Microbiological Societies.

NEUTRAL POLYSACCHARIDES

Apart from acidic polysaccharides such as teichoic and teichuronic acids, many Gram-positive organisms have cell wall polysaccharides composed of neutral sugars, usually hexoses and pentoses, and less commonly amino sugars. The polysaccharides of Gram-negative organisms will not be dealt with here, since they are an essential part of the O-antigen complex, which is considered in this article under the heading "Lipopolysaccharide-Protein Complexes: O-Antigens, Endotoxins."

The cell wall polysaccharides of streptococci have been most thoroughly investigated, probably because of their economic importance and because of the pioneer work of Lancefield in developing an antigenic scheme for these organisms.[145,146] Qualitatively, most streptococcal polysaccharides consist of rhamnose and some combination of glucose, galactose, glucosamine, and galactosomine: mannose is found in some groups. However, a number of strains in Lancefield's Groups K, M, and O have polysaccharides without rhamnose.

The results of qualitative analysis of the sugars in cell wall polysaccharides of more than 200 strains of streptococci can be found in papers by Slade and Slamp[147] and by Colman and Williams.[79] The quantitative composition of polysaccharides from strains in several Lancefield groups is given in Table 6.

In other groups of Gram-positive organisms, knowledge of cell wall polysaccharide composition is fragmentary and largely confined to reports of the sugar components released on hydrolysis. A list of the sugars found in some genera of Gram-positive organisms other than streptococci is given in Table 7. More than one polysaccharide may be identified in cell wall extracts. For example, Michel and Krause[179] found two polysaccharides in formamide extracts of Group F streptococci, and two antigenically distinct carbohydrates were found in strains of *Lactobacillus casei* by Knox.[180] Since in some cases teichoic acids and neutral polysaccharides may be found together in walls of the same organism, some of the sugars reported may come from teichoic acid.

Most methods of extraction of polysaccharides from cell walls were originally devised to get material from whole cells into solution for immunological testing. The

Table 6
PERCENT COMPOSITION OF CELL WALL POLYSACCHARIDES IN STREPTOCOCCI[a]

Lancefield serological group	Rhamnose[b]	Glucose[b]	Galactose[b]	Mannose	N-Acetyl-glucosamine	N-Acetyl-galactosamine	Method of preparation	Reference
A	39				24.3		TCA[c]	148
	42—49				23—28		Enzyme[d]	149
B	50.2		8.9		12.3		Formamide	150, 151
	50.5		11.0		11.4		Formamide	150, 151
C	43				3.9	35.1	Formamide	152
E	44.2	22.0			2.2		TCA[c]	148
	36.4	19.4			2.3		TCA[c]	148
F	48.2	20.6	3		1	27	Formamide	152
	17.3	13	13	20.5	18.9	17.3	Formamide	152
	55.3	4.5	14.6	5.9	4.5	11.9	Formamide	152
	36.4	11.7	24.3	12.1	0.8	14.6	Formamide	152
G	40.7		23.7			20.6	Formamide	150, 153
	38.7		20.1			17.8	Formamide	150, 153
	36.8		21.2			16.2	Formamide	150, 153
	34.8		16.0			18.3	TCA[c]	148
T	22.6	7.0	26.3		15.9	2.8	TCA[c]	148

[a] During hydrolysis of the carbohydrates to liberate individual components, some sugar is destroyed; hence the figures for percentage composition, based on weight before hydrolysis, normally total 75 to 80% rather than 100%. The figures for Group F streptococci, however, were calculated on the basis of 100% recovery (see Reference 152).

[b] Wherever the isomers have been determined, the sugars have been found to be L-rhamnose, D-glucose, and D-galactose.

[c] Trichloroacetic acid.

[d] Muralytic enzyme.

Table 7
SUGARS FOUND IN CELL WALL POLYSACCHARIDES FROM GRAM-POSITIVE BACTERIA

Genus or group	Sugars present[a]	References
Aerobic actinomycetes	Very variable; at least four patterns recognized: (1) arabinose, galactose, (2) madurose (3-O-methyl-D-galactose),[b] (3) no sugars, (4) arabinose, xylose[b]	154,155
	Teichoic acids may also occur in these strains.	127
Actinomyces	*A. israelii* — (1) galactose, (2) galactose and rhamnose	156
	A. bovis, A. naeslundii — rhamnose, glucose, fucose, 2-deoxytalose	157,158
Arachnia	*A. propionica* — (1) galactose, (2) glactose and glucose	37,159
Arthrobacter	Some combination of rhamnose, glucose, galactose, and mannose; also, rarely, fucose, unknown sugars	38,160
Bacillus	Glucose and/or galactose appear to be present in the walls of most strains, but are probably from teichoic acids, as in the following:	
	B. subtilis	137
	B. licheniformis	130
	B. coagulans	43
	B. stearothermophilus	43
	However, a polysaccharide composed of glucose, glucosamine, and galactosamine was obtained from *B. thuringiensis* var. *thuringiensis*, and one composed of galactose and glucosamine was obtained from *B. anthracis*.	49, 161
Bacterionema	*Bacterionema matruchottii* — arabinose, galactose, glucose	54, 162
Bifidobacterium	Some combination of rhamnose, galactose, and glucose; considerable variation; some strains may also have teichoic acids	163—165
Clostridium	Usually either (1) glucose, (2) glucose, galactose, or (3) rhamnose, glucose, galactose, mannose	556, 166, 167
Corynebacterium	Human and animal pathogens:	
	C. diphtheriae and most others — arabinose, galactose (may also be mannose and glucose)	168, 169
	C. pyogenes — rhamnose, glucose	58
	Plant pathogens:	170—172
	C. poinsettiae, C. betae — rhamnose, mannose, galactose, fucose	
	C. tritici, C. insidosum, C. sepedonicum — xylose, glucose, mannose	
Erysipelothrix	Galactose, glucose, mannose	61
Leuconostoc	Rhamnose, glucose (plus a glycerol teichoic acid)	62
Lactobacillus	*L. casei* — rhamnose, galactose, glucose	144, 169, 173
	Other lactobacilli — some combination of glucose, galactose, and mannose, probably in association with teichoic acids	
Listeria	Rhamnose, galactose, glucose, mannose; some strains are also reported to have fucose and xylose.	25, 41
Micrococcus	Cell walls of most strains probably have teichoic or teichuronic acids, but some have polysaccharides — e.g., *M. roseus*, which has a polysaccharide containing glucose, galactose, and mannose.	75, 174
Mycobacterium	Arabinose, galactose, glucose; the polysaccharides of mycobacteria occur in combination with lipid.	175
Peptostreptococcus	Some combination of rhamnose, glucose, mannose	71
Propionibacterium	*P. freudenreichii, P. shermanii* — rhamnose, galactose, mannose	35, 157, 176
	Other propionibacteria, including *P. acnes* and related organisms — some combination of glucose, galactose, and mannose	
Rothia	*Rothia dentocariosa (Nocardia salivae)* — (1) galactose, (2) galactose, glucose, ribose, fructose	73, 177, 178
Staphylococcus	Teichoic acids	75, 127, 174
Streptococcus	Some combination of rhamnose, galactose, glucose, mannose, glucosamine, and galactosamine	79, 147 (See also Table 6)

[a] Amino sugars are not generally included in this survey; by analogy with the results for streptococci, it is likely that they are present in a considerable number of cell wall polysaccharides.

[b] These sugars have been found in whole-cell hydrolysates, but it is not yet certain that they are from cell wall polysaccharides.

Table 8
EXTRACTION OF POLYSACCHARIDES FROM GRAM-POSITIVE BACTERIA

Organism	Method	Reference
Streptococcus	*N*/20 HCl at 100°C, 15 min	184
	Formamide at 160—165°C	181
	Cell wall lytic enzymes from *Streptomyces* spp.	182
	Cell wall lytic enzymes from bacteriophage	185
	Activation of autolytic enzyme systems	186
	5% trichloracetic acid at 90°C	148
Actinomyces	(1) Formamide at 160—165°C	156
	(2) 5% trichloracetic acid at 55°C for 15 min	156
Listeria monocytogenes	Formamide at 145—150°C, 20 min	64
Rothia dentocariosa	10% perchloric acid in the cold, followed by column chromatography	178
Propionibacterium	10% trichloracetic acid at 56°C for 30 min	265

classic method is that used by Lancefield for streptococci, which involved heating with *N*/20 HCl (pH 2.0) at 100°C for 15 minutes and then cooling and neutralizing the extract. Fuller,[181] in 1938, introduced extraction with neutral formamide at 160 to 165°C, and Maxted[182] introduced the use of muralytic enzymes from *Streptomyces*. References to the use of these and other methods are given in Table 8; for a general discussion of the action of hot formamide on cell walls, see Reference 183.

The material extracted by these methods usually behaves antigenically as a haptene in that it reacts strongly in precipitin tests against suitable antisera, but it is not itself immunogenic. To prepare antisera against cell wall polysaccharides, animals are usually immunized with suspension of whole cells, but suspensions of crude or purified cell wall fragments can also be used.

In many organisms the wall polysaccharides are the main polysaccharide elements in the cell, and it may reasonably be assumed that the sugars found in whole-cell hydrolysates are from the cell wall. Obviously, the presence of capsular material or of intracellular polysaccharides is a source of error to be guarded against. Patterns of sugar components in whole-cell hydrolysates may be of considerable value in identification, especially if unusual sugars, such as arabinose, are present.[187]

CELL WALL LIPIDS

In most cases the walls of Gram-positive bacteria are characterized by a low lipid content (< 5%). However, mycobacteria, nocardias, and most human- and animal-pathogenic corynebacteria are exceptional in having a high content of cell wall lipids, which may be more than 50% of the dry weight of the wall in some mycobacteria. Many of these lipids are characterized by the presence of mycolic acids, which are α-substituted, β-hydroxy long-chain acids with the following general formula:

$$\begin{array}{l} R-CH-CH-COOH \\ \quad\;\; | \qquad \backslash \\ \quad\; OH \qquad R_2 \end{array}$$

In mycobacteria, mycolic acids with high molecular weights usually occur with numbers of carbon atoms from C_{79} to C_{85}, whereas acids with lower molecular weights are found in nocardias (C_{48} to C_{58}) and in corynebacteria (C_{32} to C_{36}).

In the cell wall, mycolic acids are found in combination in the form of glycolipids. Thus they may be esterified with the arabino-galactan polysaccharide to form *WaxD,* or with trehalose to form Cord factor (trehalose-6′-6-dimycolate). The cell walls of mycobacteria and related organisms may also contain other lipids in the form of mycosides or lipoproteins, but these have been little studied so far.

The lipids of the mycobacterial wall, especially the mycolic acids, are associated with acid-fastness, and this property is lost if the cells are thoroughly delipidated. However, acid-fastness is also dependent on the integrity of the wall and cell membrane, since only intact cells are acid-fast. For detailed information on mycobacterial cell wall lipids, see References 175 and 188.

The cell wall lipids of Gram-negative bacteria are associated with the endotoxin complex and are considered in the section on Gram-negative organisms.

For general information on bacterial lipids, see References 189 to 191.

PROTEIN COMPONENTS

Except in one or two genera, the protein components of the cell walls of Gram-positive organisms have been little investigated. As in the case of wall polysaccharides, the most detailed information on cell wall proteins is available for streptococci, in which numerous acid-soluble protein antigens have been recognized (M-, T-, and R-antigens). Immunological and electron microscope studies agree in locating these antigens in the superficial layers of the cell wall;[192,193] they are not necessary for the integrity of the wall, since strains that lack them appear to grow as well as those in which they are present, and they can be removed by proteolytic enzymes without affecting the other wall components. In the case of a Group A, Type 12 streptococcus acid-extractable protein forms about 12% of the wall.[194]

In *Staphylococcus aureus,* the ability of a suspension of the organism to absorb agglutinins from antisera is completely abolished by treatment with Pronase,[195] suggesting that the antigens concerned are protein. A protein (antigen A) that was isolated and purified from the walls of some strains was shown to be a surface component responsible for agglutination.[196]

There is immunological evidence for the presence of a protein in the cell walls of *Corynebacterium diphtheriae*[197] and a trypsin-soluble protein was found in the wall of *Bacillus licheniformis.*[198] In fact, it seems likely that cell wall proteins are present quite widely in Gram-positive bacteia, but it must be admitted that as yet little concrete evidence supports this statement.

Gram-negative Organisms

LIPOPOLYSACCHARIDE-PROTEIN COMPLEXES: O-ANTIGENS, ENDOTOXINS

These substances form a major part of the cell wall of Gram-negative bacteria. In their most complete form they are macromolecular complexes of lipid, polysaccharide, and protein, with a molecular weight of at least several million. The name endotoxin was given because the toxic principle was regarded as being firmly bound to the cell substance, although in fact variable amounts occur free in young cultures, especially in some strains (see Reference 199). The polysaccharide part of the complex carries antigenic groupings that determine the serological reactions of agglutinating suspensions of the organisms (O-antigens) unless the cell surface is covered by some other material, such as a capsule, or unless flagellar antigens interfere.

tracted, but contamination with nucleic acid and other cell components is avoided if cell walls are used. The original classic method is that of Boivin and Mesrobeanu,[236] in which acetone-dried cells are extracted with dilute trichloracetic acid (0.25*M* to 0.5*M*) at 4°C. This gives an antigenic, endotoxic, lipoprotein-polysaccharide complex. Another classic method, described by Westphal et al.,[237] uses 45% aqueous phenol at 68°C and gives a lipopolysaccharide fraction devoid of protein. A method that is especially useful for the more hydrophobic complexes, such as R-antigens, is that of Galanos et al.[264] The state of aggregation and other properties of the product obtained will thus vary considerably according to the method of extraction used; the type of complex extracted by different methods is given in Table 10.

The lipid, protein, and polysaccharide elements of the O-antigen complex may be separated by various mild chemical treatments. For example, hydrolysis in 0.2*N* acetic acid at 100°C will liberate lipid, polysaccharide, and protein components. For a thorough discussion of the composition and properties of O-antigens, see Reference 203 and various articles on bacterial endotoxins in Volume 4 of *Microbial Toxins*.[242]

Table 9
SUGAR CONSTITUENTS OF LIPOPOLYSACCHARIDES

Structure	Configuration		Trivial name	Occurrence	References
Hexoses	Glucose	D		Frequent	
	Galactose	D		Frequent	
	Mannose	D		Frequent	
6-Deoxyhexoses	Galacto-	L	Fucose	Frequent	
	Manno-	L	Rhamnose	Frequent	
	Manno-	D	Rhamnose	*Xanthomonas campestris*	212
	Talo-	L		*Escherichia coli* 045, 066, 084, 088	213
3,6-Dideoxyhexoses	Gluco-	D	Paratose	*Salmonella* group A	214
				Pasteurella pseudotuberculosis I, III	215
	Galacto-	L	Abequose	*Salmonella* groups B, C2, C3	214
				Citrobacter 4.5	216
				P. pseudotuberculosis II	215
	Galacto-	L	Colitose	*Salmonella* groups 35, 50	214
				E. coli 055, 0111	216
				Arizona 9, 20	216
	Manno-	D	Tyvelose	*Salmonella* group D	214
				P. pseudotuberculosis IV	215
	Manno-	L	Ascarylose	*P. pseudotuberculosis* V	215
Pentoses	Ribose	D		*Salmonella* groups 28, 52, 56	214, 217
	Ribose			*E. coli* 0114	213
	Xylose			*Citrobacter freundii* 08	218
				Chromobacterium violaceum Lewitus MWB	215
Heptoses					
L-Glycero-	Manno-	D		Frequent	
D-Glycero-	Manno-	D		*C. violaceum* NCTC 7917	215
				Serratia marcescens	219
				Proteus mirabilis	220
				Salmonella, E. coli, and others	208
				Brucella melitensis	221
D-Glycero-	Galacto-	D		*C. violaceum* Brown, Birch, BN	215
2-Amino-2-deoxyhexoses	Glu-	D	Glucosamine	Frequent	
	Galacto-	D	Galactosamine	Frequent	
	Manno-	D	Mannosamine	*Salmonella* groups 17, 42	222
				E. coli K235, 031	223, 225
				Arizona 15	

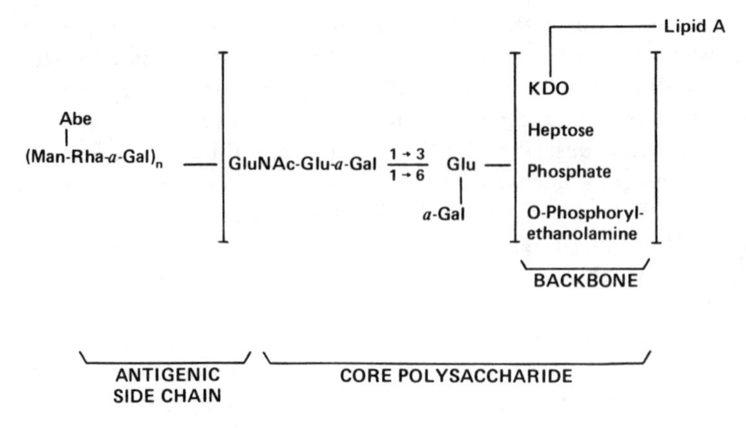

Abe = abequose = 3:6-dideoxy D-galactose.

KDO = 2-keto-3-deoxyoctanic acid

FIGURE 6. Schematic structure of O-antigen in *Salmonella typhimurium.* (From Weiner, I. M., Higuchi, T., Osborn, M. J., and Horecker, B. L., *Ann. N.Y. Acad. Sci.,* 133, 391, 1966. With permission.)

One major characteristic shared by all high-molecular-weight O-antigen complexes is that they are toxic and pyrogenic in man and animals. An active preparation from, for example, *Escherichia coli* will cause a febrile reaction in a rabbit at a dosage level of 0.1 μg/kg or less. Other effects of endotoxin injection are the Schwartzmann phenomenon (hemorrhage at the site of intradermal injection in a sensitized animal), hemorrhagic necrosis of tumors, enhanced antigenicity of proteins, and increased nonspecific resistance to infection. For a general review of the properties of endotoxin, see Reference 200.

Variation In Lipopolysaccharide Structure

It is not proposed to do more than summarize briefly the results of the extensive work on the polysaccharides of Gram-negative O-antigens that has been done over the past 20 years. Several detailed reviews are available by Westphal, Lüderitz, Simmonds, Nikaido, Kauffmann, and others, which cover the subject both adequately and authoritatively, since the authors of the reviews have themselves been the principal investigators in the field. (See References 201 to 206.)

The general structure of the lipopolysaccharide portion of a typical O-antigen (from *Salmonella typhimurium*) is shown in Figure 6.[207] Apart from laboratory-induced mutants, the structure of the core polysaccharide appears to be uniform over a rather wide range of Gram-negative organisms. However, heptose may be absent in some groups — e.g., *Xanthomonas* and *Bacteroides* (see References 208 to 211). In contrast to the relative stability of the core, the polysaccharide side chains are very variable in composition, and these variations are responsible for the large number of serologically distinct O-antigens that exist. More than thirty different sugar components have been found in different combinations in the lipopolysaccharides from Gram-negative organisms; a list of the components so far identified is given in Table 9, which is taken from Reference 206. Details of the chemical properties of some of the more unusual of these constitutents can be found in Reference 235.

Extraction of the O-Antigen Complex

Several different extraction methods can be used to obtain O-antigen complexes from Gram-negative bacteria.[238-241] For convenience, intact cells are frequently ex-

Table 9 (continued)
SUGAR CONSTITUENTS OF LIPOPOLYSACCHARIDES

Structure	Configuration		Trivial name	Occurrence	References
2-Amino-2-deoxy-D-glyceroheptose	Ido- or gulo-	D		Blue-green alga *Anacystis nidulans*	224
2-Amino-2,6-dideoxyhexoses	Gluco-	D	Quinovosamine	Unidentified strain No. COC21	225
	Gluco-		Quinovosamine	*Salmonella* groups 41, 58	222
				Arizona 1, 33	222
				P. vulgaris	222
	Galacto-	D	Fucosamine	*C. violaceum* NCTC 7917, Frazier's Hill	226
	Galacto-	L	Fucosamine	*Salmonella* group 48	
				Citrobacter freundii 5	
				Arizona 5, 29	227
	Galacto-	D, L	Fucosamine	*Pseudomonas aeruginosa*	227
	Galacto-		Fucosamine	*E. coli* 04, 012, 015, 016, 025, 026, 029, 045, 057	213
	Manno-	L	Rhamnosamine	*E. coli* 03	229
3-Amino-3,6-dideoxyhexoses	Gluco-	D		*C. freundii* 8090	230
	Gluco-			*Salmonella* groups 28, 39	231
				C. freundii 896	231
				E. coli 05, 065, 070, 071, 0114	213
	Galacto-	D		*Xanthomonas campestris*	212
				E. coli 02, 074	213
	Galacto-			*Salmonella* group 55	231
				Arizona 24	231
				B. melitensis	232
4-Amino-4,6-dideoxyhexoses	Gluco-	D	Viosamine	*Chromobacterium violaceum* NCTC 7917	233
				E. coli 07	234
	Galacto-	D	Thomosamine	*E. coli* 010	234

Taken from Lüderitz, O., Westphal, O., Staub, A. M., and Nikaido, H., in *Microbial Toxins,* Vol. 4, Weinbaum, G., Kadis, S., and Ajl, S., Eds., Academic Press, New York, 1971. With permission.

Lipid Components

These are divided into lipid A (firmly bound) and lipid B (loosely bound). Lipid A is obtained by mild acid hydrolysis of lipopolysaccharide; polysaccharide is liberated at the same time. Lipid B is obtained from the complete antigen complex by extraction with neutral fat solvents.

It seems probable that lipid A has a backbone structure composed of N-β-hydroxymyristoylglucosamine phosphate, in which the available −OH groups are esterified with fatty acids.[243,244] However, other hydroxylated fatty acids may occur — e.g., in *Pseudomonas aeruginosa*.[245] Lipid B, on the other hand, is a phospholipid of the cephalin type, with mainly palmitic and oleic acids.

Protein Components

Compared to the lipid and polysaccharide components of endotoxin, little is known about the protein, which has been obtained in two different forms; conjugated protein, and simple amphoteric protein. The conjugated protein is released, together with lipid B and polysaccharide, when complete endotoxin is heated with 1% acetic acid at 100°C. Since lipid A is not liberated separately, it is assumed that conjugated protein is protein + lipid A. Simple protein is released when either conjugated protein or complete endotoxin is dissociated with phenol. Since lipid A is found associated with

Table 10
TYPE OF O-ANTIGEN COMPLEX EXTRACTED FROM GRAM-NEGATIVE BACTERIA BY DIFFERENT METHODS

Method	Type of complex extracted	Antigenicity (rabbits)	Endotoxic activity	Molecular weight	Ability to sensitize RBC
TCA[242] Diethylene glycol[244] 50% aqueous pyridine[245] 50% aqueous glycol[245] EDTA at alkaline pH[246]	Lipid-polysaccharide protein	+ + +	+ + +	One to several millions	+
Phenol/water[247] Aqueous ether[248]	Lipopolysaccharide (lipid)	+	+ + +	One to several millions	±
Dilute alkali (e.g., 0.25 *N* NaOH at 56°)[205]	Lipopolysaccharide (deacylated lipid A)	Only if fixed to RBC	±	200,000	+ + +
Dilute acid (e.g., 0.1 *N* acetic acid at 90°)[205]	Polysaccharide	Only if coupled to protein	–	20,000—30,000	–

Modified from Table 2 of Luderitz, O., Staub, A. M., and Westphal, O., *Bacteriol. Rev.*, 30, 192, 1966. Reproduced by permission of the American Society for Microbiology.

polysaccharide when the complete endotoxin is dissociated in this way, it is assumed that the simple protein does not contain lipid A.

There is some evidence that the protein part of the O-antigen or endotoxic complex may be associated with bacteroicine activity.[246-248]

SMOOTH AND ROUGH FORMS: S→ R VARIATION

Originally, S → R variation referred to the change in colony type and stability in saline corresponding to the loss of antigens that characterized the smooth type. In terms of the structure shown in Figure 6, S-organisms have both the side chain and the core, whereas R-organisms have the core polysaccharide only, and the S → R variation consists of loss in ability to synthesize the side chain. However, numerous intermediate mutants are known, in which some or all of the sugars of both the side chain and the core are missing. Since "S" and "R" have now come to have rather specialized meanings in immunochemical terms, it is probably better to describe the morphological appearance or stability of cultures by using the words "smooth" and "rough" written in full.

BACTERIAL CELL WALL SYNTHESIS

The basic synthetic and energy-yielding mechanisms of the cell are inside the cytoplasmic membrane, whereas the cell wall is outside it. The process of cell wall synthesis therefore essentially requires that the building blocks for the wall be passed through the membrane at some stage. It has now become evident that this transfer is performed by carrier lipids and that some of the intermediate stages of the synthesis of several cell wall polymers are carried out while the growing polymers are attached to the cytoplasmic membrane.

The carrier lipid in peptidoglycan synthesis has been identified as a phosphorylated C_{55} polyisoprenoid alcohol; similar or identical substances act as carriers in the synthesis of O-antigens and teichoic acids.

An outline of the probable course of peptidoglycan synthesis is given in Figure 7. The uridine-diphosphomuramyl-pentapeptide units are built up by the stepwise addition of amino acids, each step being catalyzed by a specific soluble enzyme. The dipeptide D-alanyl-D-alanine is added last as a single unit. This part of the process therefore differs from protein synthesis in that it is extraribosomal and t-RNA is not involved. In the formation of the linking peptides, however, which occurs after the UDP-muramyl-pentapeptide units have become linked to the carrier lipid, at least some of the amino acids (e.g., glycine) are incorporated via t-RNA, although others are not (e.g., D-aspartic acid). For detailed discussions of peptidoglycan synthesis, see Reference 8.

Effects of Antibiotics on Cell Wall Synthesis

A number of antibiotics are inhibitory or lethal to bacterial cells because they affect some aspect of peptidoglycan synthesis. In the first stage (in the cytoplasm), D-cycloserine interferes with the formation of the D-alanyl-D-alanine dipeptide, which is added to the developing peptide chain to form the pentapeptide. In the second stage (cell membrane), vancomycin and ristocetin become attached to the peptidoglycan subunits and prevent their subsequent incorporation into the cell wall. Bacitracin also acts at this stage, since it inhibits the dephosphorylation of the lipid carrier so that the lipid cannot reenter the cycle. In the third stage, penicillins and cephalosporins are specific inhibitors of the cross-linking reaction, in which the transpeptidase removes the terminal D-alanine residue and forms the peptide bond that links two chains. For the effects of antibiotics on wall synthesis, see References 8 and 249 to 252.

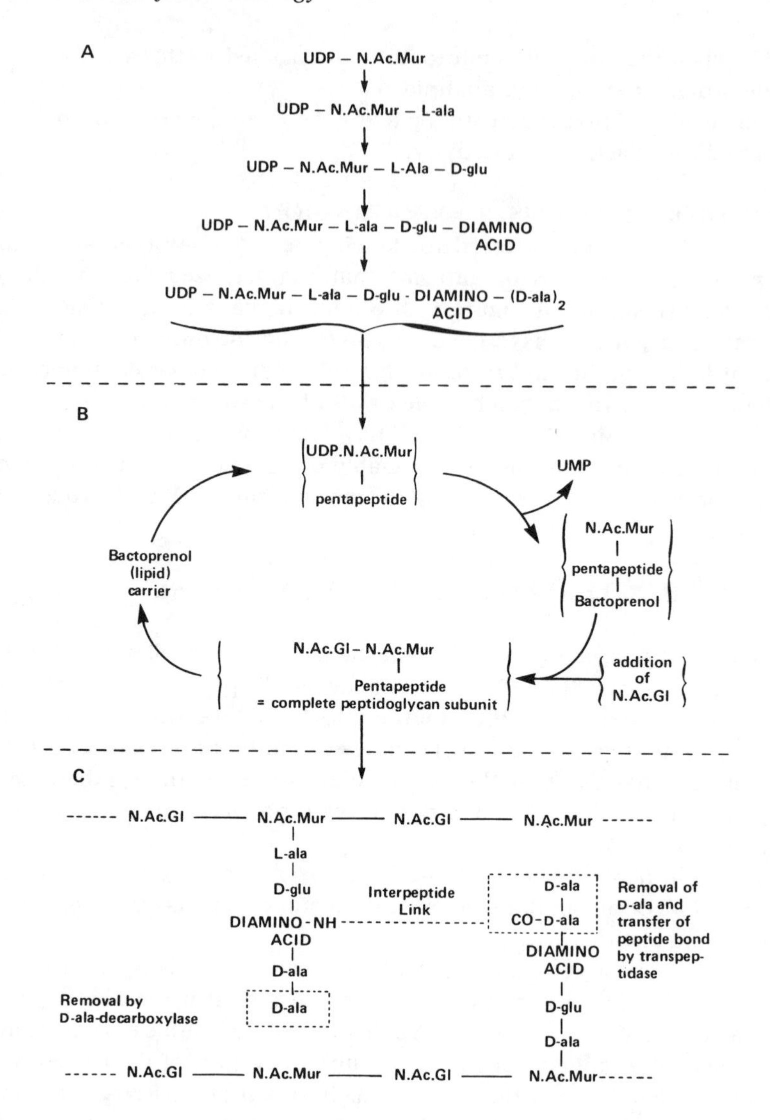

FIGURE 7. Outline of peptidoglycan synthesis. (A) Cytoplasm: soluble enzymes and building blocks. (B) Cytoplasmic membrane: attachment to lipid carrier and transport through membrane. (C) Cell wall: complete unit incorporated and cross-linked in wall.

REFERENCES

1. **Salton, M. R. J. and Horne, R. W.,** *Biochim. Biophys. Acta,* 7, 177, 1951.
2. **Cummins, C. S.,** *Int. Rev. Cytol.,* 5, 25, 1957.
3. **Perkins, H. R.,** *Bacteriol. Rev.,* 27, 18, 1963.
4. **Salton, M. R. J.,** *The Bacterial Cell Wall,* American Elsevier, New York, 1964.
5. **Martin, H. H.,** *Annu. Rev. Biochem.,* 35, 457, 1966.
6. **Ghuysen, J. M.,** *Bacteriol. Rev.,* 32, 425, 1968.
7. **Rogers, H. J. and Perkins, H. R.,** *Cell Walls and Membranes,* E. and F. N. Spon, London, 1968.

8. **Osborn, M. J.**, *Annu. Rev. Biochem.*, 38, 501, 1969.
9. **Hugo, W. B.**, *Bacteriol. Rev.*, 18, 87, 1954.
10. **Edebo, L.**, *Disintegration of Microorganisms*, Almqvist and Wiksell, Uppsala, Sweden, 1961.
11. **Ribi, E. and Milner, K. C.**, in *Methods in Immunology and Immunochemistry*, Vol. 1, Williams, C. A. and Chase, M. W., Eds, Academic Press, New York, 1967, p. 13.
12. **Dawson, I. M.**, *Symp. Soc. Gen. Microbiol.*, No. 1, 119, 1949.
13. **Mickle, H.**, *J. R. Microsc. Soc.*, 68, 10, 1948.
14. **Braun, B.**, Melsungen, Germany; *Bronwill Scientific*, Rochester, N.Y.
15. **Bühler, E.**, Tübingen, Germany; *RHO Scientific*, Commack, N.Y.
16. **Nossal, P. M.**, *Aust. J. Exp. Biol. Med. Sci.*, 31, 583, 1953.
17. **American Instrument Co.**, Silver Spring, Md.
18. **Ivan Sorvall, Inc.**, Norwalk, Conn.
19. **Colab Laboratories, Inc.**, Chicago Heights, Ill.
20. **Roberson, B. S. and Schwab, J. H.**, *Biochim. Biophys. Acta*, 44, 436, 1960.
21. **Yoshida, A., Hedén, G. C., Cedergren, B., and Edebo, L.**, *J. Biochem. Microbiol. Technol. Eng.*, 3, 151, 1961.
22. **von Hofsten, B. and Baird, G. D.**, *Biotechnol. Bioeng.*, 4, 403, 1962.
23. **Nermuth, M. V. and Murray, R. G. E.**, *J. Bacteriol.*, 93, 1949, 1967.
24. **Glauert, A. M. and Thornley, M. J.**, *Annu. Rev. Microbiol.*, 23, 159, 1969.
25. **Buckmire, F. L. A.**, *Int. J. Syst. Bacteriol.*, 20, 345, 1970.
26. **Freer, J. H. and Salton, M. R. J.**, in *Microbial Toxins*, Vol. 4, Weinbaum, G., Kadis, S., and Ajl, S., Eds., Academic Press, New York, 1971, p. 67.
27. **Tipper, D. J.**, *Int. J. Syst. Bacteriol*, 20, 361, 1970.
28. **Azuma, I., Thomas, D. W., Adam, A., Ghuysen, J. M., Bonaly, R., Petit, J. F., and Lederer, E.**, *Biochim. Biophys. Acta*, 208, 444, 1970.
29. **Kandler,O., Schleifer, K. H., Niebler, E., Nakel, M., Zahradnik, H., and Reid, M.**, *Publ. Fac. Sci. Univ. Brno*, 47, 143, 1970.
30. **Becker, B., Lechevalier, M. P., and Lechevalier, H. A.**, *Appl. Microbiol.*, 13, 236, 1965.
31. **Lechevalier, H. A., Lechevalier, M. P., and Becker, B.**, *Int. J. Syst. Bacteriol.*, 16, 151, 1966.
32. **Szaniszlo, P. J. and Gooder, H.**, *J. Bacteriol.*, 94, 2037, 1967.
33. **Yamaguchi, T.**, *J. Bacteriol.*, 89, 444, 1965.
34. **Hoare, D. S. and Work, E.**, *Biochem. J.*, 65, 441, 1957.
35. **Cummins, C. S.**, *Nature*, 206, 1272, 1965.
36. **DeWeese, M. S., Gerencser, M. A., and Slack, J. M.**, *Appl. Microbiol.*, 16, 1713, 1968.
37. **Johnson, J. L. and Cummins, C. S.**, *J. Bacteriol.*, 109, 1047, 1972.
38. **Keddie, R. M., Leask, B. G. S., and Grainger, J. M.**, *J. Appl. Bacteriol.*, 29, 17, 1966.
39. **Krulwich, T. A., Ensign, J. C., Tipper, D. J., and Strominger, J. L.**, *J. Bacteriol.*, 94, 741, 1967.
40. **Gillespie, D. C.**, *Can. J. Microbiol.*, 9, 515, 1963.
41. **Fiedler, F., Schleifer, K. H., Cziharz, B., Interschick, E., and Kandler, O.**, *Publ. Fac. Sci. Univ. Brno*, 47, 111, 1970.
42. **Cummins, C. S. and Harris, H.**, *Nature*, 184, 831, 1959.
43. **Forrester, I. T. and Wicken, A. J.**, *J. Gen. Microbiol.*, 42, 147, 1966.
44. **Boylen, C. W. and Ensign, J. C.**, *J. Bacteriol.*, 96, 421, 1968.
45. **Bricas, E., Ghuysen, J. M., and Dezelée, P.**, *Biochemistry*, 6, 2598, 1967.
46. **Hughes, R. C.**, *Biochem. J.*, 106, 41, 1968.
47. **Hungerer, K. D. and Tipper, D. J.**, *Biochemistry*, 8, 3577, 1969.
48. **Van Heijenoort, J., Elbaz, L., Dezelée, P., Petit, J. F., Bricas, E., and Ghuysen, J. M.**, *Biochemistry*, 8, 207, 1969.
49. **Kingan, S. L. and Ensign, J. C.**, *J. Bacteriol.*, 95, 724, 1968.
50. **Ratney, R. S.**, *Biochim. Biophys. Acta*, 101, 1, 1965.
51. **Reynolds, P. E.**, *Biochim. Biophys. Acta*, 237, 239, 1971.
52. **Sutow, A. B. and Welker, N. E.**, *J. Bacteriol.*, 93, 1452, 1967.
53. **Kandler, O.**, *Int. J. Syst. Bacteriol.*, 20, 491, 1970.
54. **Baboolal, R.**, *J. Gen. Microbiol.*, 58, 217, 1969.
55. **Guinand, M., Ghuysen, J. M., Schleifer, K. H., and Kandler, O.**, *Biochemistry*, 8, 200, 1969.
56. **Cummins, C. S. and Johnson, J. L.**, *J. Gen. Microbiol.*, 67, 33, 1971.
57. **Cummins, C. S. and Harris, H.**, *J. Gen. Microbiol.*, 14, 583, 1956.
58. **Barksdale, W. L., Li, K., Cummins, C. S., and Harris, H.**, *J. Gen. Microbiol.*, 16, 749, 1957.
59. **Perkins, H. R.**, *Biochem. J.*, 121, 417, 1971.
60. **Mosser, J. C. and Tomasz, A.**, *J. Biol. Chem.*, 245, 287, 1970.
61. **Mann, S.**, *Zentralbl. Bakteriol. Parasitenkd. Infektionskr. Hyg. Abt. 1 Orig.*, 209, 510, 1969.

62. Harney, S. J., Simopoulos, N. D., and Ikawa, M., *J. Bacteriol.*, 93, 273, 1967.
63. Keeler, R. F. and Gray M. L., *J. Bacteriol.*, 80, 683, 1960.
64. Ullman, W. W. and Cameron, J. A., *J. Bacteriol.*, 98, 486, 1969.
65. Robinson, K., *J. Appl. Bacteriol.*, 29, 616, 1966.
66. Work, E. and Griffiths, H., *J. Bacteriol.*, 95, 641, 1968.
67. Work, E., *Nature*, 201, 1107, 1964.
68. Cummins, C. S. and Harris, H., *J. Gen. Microbiol.*, 18, 173, 1958.
69. Acharya, P. V. N. and Goldman, D. W., *J. Bacteriol.*, 102, 733, 1970.
70. White, P. J., *J. Gen. Microbiol.*, 50, 107, 1968.
71. Bahn, A. N., Kung, P. C. Y., and Hayashi, J. A., *J. Bacteriol.*, 91, 1672, 1966.
72. Schleifer, K. H. and Kandler, O., *Arch. Mikrobiol.*, 61, 292, 1968.
73. Davis, G. H. G. and Freer, J. H., *J. Gen. Microbiol.*, 23, 163, 1960.
74. Cummins, C. S., and Harris, H., *Int. Bull. Bacteriol. Nomencl. Taxon.*, 6, 111, 1956.
75. Baird-Parker, A. C., *J. Gen. Microbiol.*, 38, 363, 1965.
76. Tipper, D. J. and Berman, M. F., *Biochemistry*, 8, 2183, 1969.
77. Schleifer, K. H. and Kandler, O., *Arch. Mikrobiol.*, 57, 335, 1967.
78. Schleifer, K. H. and Kandler, O., *Arch. Mikrobiol.*, 57, 365, 1967.
79. Colman, G. and Williams, R. E. O., *J. Gen. Microbiol.*, 41, 375, 1965.
80. Slade, H. D. and Slamp, W. C., *J. Bacteriol.*, 109, 691, 1972.
81. Hoare, D. S. and Work, E., *Biochem. J.*, 61, 562, 1955.
82. Jusic, D., Roy, C., and Watson, R. W., *Can. J. Biochem.*, 42, 1553, 1964.
83. Manasse, R. J. and Corpe. W. A., *Can. J. Microbiol.*, 13, 1591, 1967.
84. LaCave, C. and Roux, J., *C. R. Acad. Sci.*, 260, 1514, 1965.
85. Mardarowitz, C., *Z. Naturforsch, Teil B*, 21, 1006, 1966.
86. Mandelstam, J., *Biochem. J.*, 84, 294, 1962.
87. Grula, E. A., Smith, G. L., and Grula, M. M., *Can. J. Microbiol.*, 11, 605, 1965.
88. Wang, W. S. and Lundgren, D. G., *J. Bacteriol.*, 95, 1851, 1968.
89. Braun, V., Rehn, K., and Wolff, H., *Biochemistry*, 9, 5041, 1970.
90. Clarkson, C. E. and Meadow, P. M., *J. Gen. Microbiol.*, 66, 161, 1971.
91. Forsberg, C. W., Rayman, M. K., Costerton, J. W., and MacLeod, R. A., *J. Bacteriol.*, 109, 895, 1972.
92. Salton, M. R. J. and Shafa, F., *Nature*, 181, 1321, 1958.
93. Colobert, L. and Creach, O., *Ann. Inst. Pasteur*, 99, 672, 1960.
94. Crum, E. H. and Siehr, D. J., *J. Bacteriol.*, 94, 2069, 1967.
95. Graham, R. K. and May, J. W., *J. Gen. Microbiol.*, 41, 243, 1965.
96. Winter, A. J., Katz, W., and Martin, H. H., *Biochim. Biophys. Acta*, 244, 58, 1971.
97. Drews, G. and Meyer, H., *Arch. Mikrobiol.*, 48, 259, 1964.
98. Frank, H., Lefort, M., and Martin, H. H., *Z. Naturforsch. Teil B*, 17, 262, 1962.
99. Höcht, H., Martin, H. H., and Kandler, O., *Z. Pflanzenphysiol.*, 53, 39, 1965.
100. Verma, J. P. and Martin, H. H., *Arch. Mikrobiol.*, 59, 355, 1967.
101. Becker, B., Worzel, E. M., and Nelson, J. H., *Nature*, 213, 300, 1967.
102. Kolenbrander, P. E. and Ensign, J. C., *J. Bacteriol.* 95, 201, 1968.
103. Martin, H. H. and Frank, H., *Zentralbl. Bakteriol. Parasitenkd. Infektionskr. Hyg. Abt. 1 Orig.*, 184, 306, 1962.
104. Newton, J.W., *Biochim. Biophys. Acta*, 165, 534, 1968.
105. Preusser, H. J., *Arch. Mikrobiol.*, 68, 150, 1969.
106. Tinelli, R. and Pillot, J., *C. R. Acad. Sci.*, 263, 739, 1966.
107. Cummins, C. S. and Smibert, R. M., unpublished data.
108. Campbell, J. N., Leyh-Bouille, M., and Ghuysen, J. M., *Biochemistry*, 8, 193, 1969.
109. Ghuysen, J. M., Bricas, E., Lache, M., and Leyh-Bouille, M., *Biochemistry*, 7, 1450, 1968.
110. Niebler, E., Schleifer, K. H., and Kandler, O., *Biochem. Biophys. Res. Commun.*, 34, 560, 1969.
111. Schleifer, K. H. and Kandler, O., *Biochem. Biophys. Res. Commun.*, 28, 965, 1967.
112. Kandler, O., Plapp, R., and Holzapfel, W., *Biochim. Biophys. Acta*, 147, 252, 1967.
113. Hungerer, K. D., Fleck, J., and Tipper, D. J., *Biochemistry*, 8, 3567, 1969.
114. Plapp, R., Schleifer, K. H., and Kandler, O., *Folia Microbiol.*, 12, 205, 1967.
115. Plapp, R. and Kandler, O., *Arch. Mikrobiol.*, 58, 305, 1967.
116. Kandler, O., Schleifer, K. H., and Dandl, R., *J. Bacteriol.*, 96, 1935, 1968.
117. Koch, D., Schleifer, K. H., and Kandler, O., *Z. Naturforsch. Teil B*, 25, 1294, 1970.
118. Kandler, O., Koch, D., and Schleifer, K. H., *Arch. Mikrobiol.*, 61, 181, 1968.
119. Schleifer, K. H., Plapp, R., and Kandler, O., *FEBS Lett.*, 1, 287, 1968.
120. Perkins, H. R., *Nature*, 208, 872, 1965.
121. Work, E., *Int. J. Syst. Bacteriol.*, 20, 425, 1970.

122. **Bottazzi, V., Weiss, N., and Kandler, O.,** *Arch. Mikrobiol.,* 58, 35, 1967.
123. **Rhuland, L. E., Work, E., Denman, R. F., and Hoare, D. S.,** *J. Am. Chem. Soc.,* 77, 4844, 1955.
124. **Ginger, C. D.,** *Nature,* 199, 159, 1963.
125. **Perkins, H. R. and Allison, A. C.,** *J. Gen. Microbiol.,* 30, 496, 1963.
126. **Baddiley, J.,** *J. R. Inst. Chem.,* 86, 366, 1962.
127. **Archibald, A. R. and Baddiley, J.,** *Adv. Carbohydr. Chem.,* 21, 323, 1966.
128. **Archibald, A. R., Baddiley, J., and Button, D.,** *Biochem. J.,* 110, 543, 1968.
129. **Brundish, D. E. and Baddiley, J.,** *Biochem. J.,* 110, 573, 1968.
130. **Janczura, E., Perkins, H. R., and Rogers, H. J.,** *Biochem. J.,* 80, 82, 1961.
131. **Perkins, H. R.,** *Biochem. J.,* 86, 475, 1963.
132. **Ellwood, D. C. and Tempest, D. W.,** *Biochem. J.,* 111, 1, 1969.
133. **Ellwood, D. C.,** *Biochem. J.,* 121, 349, 1971.
134. **Shaw, N. and Baddily, J.,** *Biochem. J.,* 93, 317, 1964.
135. **Archibald, A. R., Baddiley, J., and Buchanan, J. G.,** *Biochem. J.,* 81, 124, 1961.
136. **Ellwood, D. C., Kelleman, M. V., and Baddiley, J.,** *Biochem. J.,* 86, 213, 1963.
137. **Armstrong, J. J., Baddiley, J., and Buchanan, J. G.,** *Biochem. J.,* 76, 610, 1960.
138. **Archibald, A. R., Coapes, H. E., and Stafford, G. H.,** *Biochem. J.,* 113, 899, 1969.
139. **Hughes, R. C.,** *Biochem. J.,* 117, 431, 1970.
140. **Armstrong, J. J., Baddiley, J., and Buchanan, J. G.,** *Biochem. J.,* 80, 254, 1961.
141. **Davison, A. L., Baddiley, J., Hofstad, T., Losengard, N., and Oeding, P.,** *Nature,* 202, 872, 1964.
142. **Davison, A. L. and Baddiley, J.,** *Nature,* 202, 874, 1964.
143. **Oeding, P., Mykelstad, B., and Davison, A. L.,** *Acta Pathol. Microbiol. Scand.,* 69, 458, 1967.
144. **Sharpe, E.,** *Int. J. Syst. Bacteriol.,* 24, 509, 1970.
145. **Lancefield, R. C.,** *Harvey Lect.,* 36, 251, 1941.
146. **McCarty, M.,** *Harvey Lect.,* 65, 73, 1970.
147. **Slade, H. D. and Slamp, W. C.,** *J. Bacteriol.,* 84, 345, 1962.
148. **Slade, H. D.,** *J. Bacteriol.,* 90, 667, 1965.
149. **McCarty, M. and Lancefield, R. C.,** *J. Exp. Med.,* 102, 11, 1955.
150. **Curtis, S. N. and Krause, R. M.,** *J. Exp. Med.,* 120, 629, 1964.
151. **Krause, R. M. and McCarty, M.,** *J. Exp. Med.,* 115, 49, 1962.
152. **Willers, J. M. N., Michel, M. F., Sysma, M. J., and Winkler, K. C.,** *J. Gen. Microbiol.,* 36, 95, 1964.
153. **Curtis, S. N. and Krause, R. M.,** *J. Exp. Med.,* 119, 997, 1964.
154. **Lechevalier, M. P. and Lechevalier, H.,** *Int. J. Syst. Bacteriol.,* 20, 435, 1970.
155. **Lechevalier, M. P. and Gerber, N. N.,** *Carbohydr. Res.,* 13, 451, 1970.
156. **Cummins, C. S.,** in *The Actinomycetales,* Prauser, H., Ed., Gustav Fisher, Jena, Germany, 1970, p. 29.
157. **Cummins, C. S. and Harris, H.,** *J. Gen. Microbiol.,* 18, 173, 1958.
158. **Pine, L. and Boone, C. J.,** *J. Bacteriol.,* 94, 875, 1967.
159. **Pine, L. and Georg, L. K.,** *Int. J Syst. Bacteriol.,* 19, 267, 1969.
160. **Cummins, C. S. and Harris, H.,** *Nature,* 184, 831, 1959.
161. **Smith, H., Strange, R. E., and Zwartouw, H. T.,** *Nature,* 178, 865, 1965.
162. **Cummins, C. S. and Harris, H.,** unpublished data.
163. **Cummins, C. S., Glendenning, O. M., and Harris, H.,** *Nature,* 180, 337, 1957.
164. **Cummins, C. S.,** unpublished data.
165. **Veerkamp, J. H., Lambert, R., and Saito, Y.,** *Arch. Biochem. Biophys.,* 112, 120, 1965.
166. **Cato, E. P., Cummins, C. S., and Smith, L. DS.,** *Int. J. Syst. Bacteriol.,* 20, 305, 1970.
167. **Haythornthwaite, S. U.,** Ph.D. thesis, University of London, 1968.
168. **Holdsworth, E. S.,** *Biochim. Biophys. Acta,* 8, 110, 1952.
169. **Cummins, C. S. and Harris, H.,** *J. Gen. Microbiol.,* 14, 583, 1956.
170. **Perkins, H. R.,** *Biochem. J.,* 97, 3C, 1965.
171. **Perkins, H. R.,** *Biochem. J.,* 102, 29C, 1967.
172. **Perkins, H. R.,** *Int. J. Syst. Bacteriol.,* 20, 379, 1970.
173. **Glastonbury, J. and Knox, K.W.,** *J. Gen. Microbiol.,* 31, 73, 1963.
174. **Baird-Parker, A. C.,** *Int. J. Syst. Bacteriol.,* 20, 483, 1970.
175. **Lederer, E.,** *Pure Appl. Chem.,* 25, 135, 1971.
176. **Allsop, J., and Work, E.,** *Biochem. J.,* 87, 512, 1963.
177. **Georg, L. K. and Brown, J. M.,** *Int. J. Syst. Bacteriol.,* 17, 79, 1967.
178. **Hammond, B. F.,** *J. Bacteriol.,* 103, 634, 1970.
179. **Michel, M. F. and Krause, R. M.,** *J. Exp. Med.,* 125, 1075, 1967.
180. **Knox, K. W.,** *J. Gen. Microbiol.,* 31, 59, 1963.
181. **Fuller, A. T.,** *Br. J. Exp. Pathol.,* 19, 130, 1938.

182. **Maxted, W. R.,** *Lancet*, 2, 255, 1948.
183. **Perkins, H. R.,** *Biochem. J.*, 95, 876, 1965.
184. **Lancefield, R. C.,** *J. Exp. Med.*, 57, 571, 1933.
185. **Krause, R. M.,** *J. Exp. Med.*, 108, 803, 1958.
186. **Bleiweiss, A. S., Young, F. E., and Krause, R. M.,** *J. Bacteriol.*, 94, 1381, 1967.
187. **Becker, B., Lechevalier, M. P., Gordon, R. E., and Lechevalier, H. A.,** *Appl. Microbiol.*, 12, 421, 1964.
188. **Goren, M. B.,** *Bacteriol. Rev.*, 36, 33, 1972.
189. **Carter, H. E., Johnson, P., and Weber, E. J.,** *Annu. Rev. Biochem.*, 34, 109, 1965.
190. **Asselineau, J.,** *The Bacterial Lipids*, Holden-Day, San Francisco, 1967.
191. **O'Leary, W. M.,** *The Chemistry and Metabolism of Microbial Lipids*, World, Cleveland, 1967.
192. **Hahn, J. J. and Cole, R. M.,** *J. Exp. Med.*, 118, 659, 1963.
193. **Swanson, J., Hsu, K. C., and Gotschlich E. C.,** *J. Exp. Med.*, 130, 1063, 1969.
194. **Lange, C. F., Lee, R., and Merdinger, E.,** *J. Bacteriol.*, 100, 1277, 1969.
195. **Pillot, J., Rouyer, M., and Orta, B.,** *Ann. Inst. Pasteur*, 88, 662, 1955.
196. **Yoshida, A., Mudd, S., and Lenhart, N. A.,** *J. Immunol.*, 91, 777, 1963.
197. **Cummins, C. S.,** *Br. J. Exp. Pathol.*, 35, 166, 1956.
198. **Hughes, R. C.,** *Biochem. J.*, 96, 100, 1965.
199. **Work, E., Knox, K. W., and Vesk, M.,** *Ann. N.Y. Acad. Sci.*, 133, 438, 1966.
200. **Milner, K. C., Rudbach, J. A., and Ribi, E.,** in *Microbial Toxins*, Vol. 4, Weinbaum, G., Kadis, S., and Ajl, S., Eds., Academic Press, New York, 1971, p. 1.
201. **Kauffmann, F.,** *Die Bakteriologie der Salmonella Species*, Munksgaard, Copenhagen, 1961.
202. **Lüderitz, O., Staub, A. M., and Westphal, O.,** *Bacteriol. Rev.*, 30, 192, 1966.
203. **Lüderitz, O., Jann, K., and Wheat, R.,** in *Comprehensive Biochemistry*, Vol. 26A, Florkin, M. and Stotz, E. H., Eds., American Elsevier, New York, 1968, p. 105.
204. **Nikaido, H.,** *Int. J. Syst. Bacteriol.*, 20, 383, 1970.
205. **Simmons, D. A. R.,** *Bacteriol. Rev.*, 35, 117, 1971.
206. **Lüderitz, O., Westphal, O., Staub, A. M., and Nikaido, H.,** in *Microbial Toxins*, Vol. 4, Weinbaum, G., Kadis, S., and Ajl, S., Eds., Academic Press, New York, 1971, p. 145.
207. **Weiner, I. M., Higuchi, T., Osborn, M. J., and Horecker, B. L.,** *Ann. N.Y. Acad. Sci.*, 133, 391, 1966.
208. **Adams, G. A., Quadling, C., and Perry, M. B.,** *Can. J. Microbiol.*, 13, 1605, 1967.
209. **Volk, W. A.,** *J. Bacteriol.*, 91, 39, 1966.
210. **Volk, W. A.,** *J. Bacteriol.*, 95, 980, 1968.
211. **Hofstad, T. and Kristoffersen, T.,** *Acta Pathol. Microbiol. Scand. Sect. B*, 79, 12, 1971.
212. **Hickman, G. and Ashwell, J.,** *J. Biol. Chem.*, 241, 1424, 1966.
213. **Orskov, F., Orskov, I., Jann, B., Jann, K., Müller-Seitz, E., and Westphal, O.,** *Acta Pathol. Microbiol. Scand.*, 71, 339, 1967.
214. **Kauffman, F., Lüderitz, O., Stierlin, H., and Westphal, O.,** *Zentralbl. Bakteriol. Parasitenkd. Infektionskr. Hyg. Abt. 1 Orig.*, 178, 442, 1960.
215. **Davies, D. A. L.,** *Adv. Carbohydr. Chem.*, 15, 271, 1960.
216. **Westphal, O., Kauffmann, F., Lüderitz, O., and Stierlin, H.,** *Zentralbl. Bakteriol. Parasitenkd. Infektionskr. Hyg. Abt. 1 Orig.*, 179, 336, 1960.
217. **Kauffmann, F., Jann, B., Kruger, L., Lüderitz, O., and Westphal, O.,** *Zentralbl. Bakteriol. Parasitenkd. Infektionskr. Hyg. Abt. 1 Orig.*, 186, 509, 1962.
218. **Fromme, I., Lüderitz, O., and Westphal, O.,** *Z. Naturforsch. Teil B*, 9, 303, 1954.
219. **Adams, G. A. and Young, R.,** *Can. J. Biochem.*, 43, 1499, 1965.
220. **Bagdian, G., Droge, W., Kotelko, K., Lüderitz, O., Westphal, O., Yamakawa, T., and Ueta, N.,** *Biochem. Z.*, 344, 197, 1966.
221. **Lacave, C., Asselineau, J., Serre, A., and Roux, J.,** *Eur. J. Biochem.*, 9, 189, 1969.
222. **Lüderitz, O., Gmeiner, J., Kickhofen, B., Mayer, H., Westphal, O., and Wheat, R. W.,** *J. Bacteriol.*, 95, 490, 1968.
223. **Rude, E. and Goebel, W. F.,** *J. Exp. Med.*, 116, 73, 1962.
224. **Weise, G., Drews, G., Jann, B., and Jann, K.,** *Arch. Mikrobiol.*, 71, 89, 1970.
225. **Smith, E. J.,** *Biochem. Biophys. Res. Commun.*, 15, 593, 1964.
226. **Crumpton, M. J. and Davies, D. A. L.,** *Biochem. J.*, 70, 729, 1958.
227. **Barry, G. T.,** *Bull. Soc. Chim. Biol.*, 47, 529, 1964.
228. **Suzuki, N.,** *Biochim. Biophys. Acta*, 177, 371, 1969.
229. **Jann, B. and Jann, Kr.,** *Eur. J. Biochem.*, 5, 173, 168.
230. **Raff, R. A. and Wheat, R. W.,** *J. Biol. Chem.*, 242, 4610, 1967.
231. **Lüderitz, O., Rusehmann, E., Westphal, O., Raff, R., and Wheat, R.,** *J. Bacteriol.*, 93, 1681, 1967.

232. **Lacave, C.**, Thesis, University of Toulouse, France, 1969.
233. **Stevens, C. L., Blumbergs, P., Daniher, F. A., Wheat, R. W., Kukomoto, A., and Rollins, E.**, *J. Am. Chem. Soc.*, 85, 3061, 1963.
234. **Jann, B. and Jann, K.**, *Eur. J. Biochem.*, 2, 26, 1967.
235. **Ashwell, G. and Hickman, J.**, in *Microbial Toxins*, Vol. 4, Weinbaum, G., Kadis, S., and Ajl, S., Eds., Academic Press, New York, 1971, p. 235.
236. **Boivin, A. and Mesrobeanu, L.**, *Rev. Immunol.*, 1, 553, 1935.
237. **Westphal, O., Luderitz, O., and Bister, F.**, *Z. Naturforsch. Teil B*, 7, 148, 1952.
238. **Morgan, W. T. J.**, *Biochem. J.*, 31, 2003, 1937.
239. **Goebel, W. F., Binkley, F., and Perlman, E.**, *J. Exp. Med.*, 81, 315, 1945.
240. **Westphal, O. and Jann, K.**, in *Methods in Carbohydrate Chemistry*, Vol. 5, Whistler, R. L., Ed., Academic Press, New York, 1965, p. 83.
241. **Ribi, E., Anacker, R. L., Fukushi, K., Haskins, W. T., Landy, M., and Milner, K. C.**, in *Bacterial Endotoxins*, Landy, M. and Braun, W., Eds., Rutgers University Press, New Brunswick, N.J., 1964, p. 16.
242. **Weinbaum, G., Kadis, S., and Ajl, S., (Eds.)** *Microbial Toxins*, Vol. 4, Academic Press, New York, 1971.
243. **Nowotny, A.**, *J. Am. Chem. Soc.*, 83, 501, 1961.
244. **Burton, A. J. and Carter, H. E.**, *Biochemistry*, 3, 411, 1964.
245. **Roberts, N. A., Gray, G. W., and Wilkinson, S. G.**, *Biochim. Biophys. Acta*, 135, 1068, 1967.
246. **Goebel, W. F. and Barry, G. T.**, *J. Exp. Med.*, 107, 185, 1958.
247. **Homma, J. Y. and Suzuki, N.**, *Ann. N.Y. Acad. Sci.*, 133, 508, 1966.
248. **Mesrobeanu, L.**, *Ann. N.Y. Acad. Sci.*, 133, 685, 1966.
249. **Ghuysen, J. M., Strominger, J. L., and Tipper, D. J.**, in *Comprehensive Biochemistry*, Vol. 26A, Florkin, M. and Stoltz, E. H., Eds., American Elsevier, New York, 1968, p. 53.
250. **Perkins, H. R.**, *Biochem. J.*, 111, 195, 1969.
251. **Siewert, G. and Strominger, J. L.**, *Proc. Natl. Acad. Sci. U.S.A.*, 57, 767, 1967.
252. **Rothfield, L. and Romeo, D.**, *Bacteriol. Rev.*, 35, 14, 1971.
253. **Reaveley, D. A. and Burge, R. E.**, in *Advances in Microbial Physiology*, Vol. 7, Rose, A. H. and Tempest, D. W., Eds., Academic Press, New York, 1972, p. 1.
254. **Ellwood, D. C. and Tempest, D. W.**, in *Advances in Microbial Physiology*, Vol. 7, Rose, A. H. and Tempest, D. W., Eds., Academic Press, New York, 1972, p. 83.
255. **Schliefer, K. H. and Kandler, O.**, *Bacteriol. Rev.*, 36, 407, 1972.
256. **Braun, V. and Handtke, K.**, *Annu. Rev. Biochem.*, 43, 89, 1974.
257. **Costerton, J. W., Ingram, J. M., and Cheng, K. J.**, *Bacteriol. Rev.*, 38, 87, 1974.
258. **Osborn, M. J., Gander, J. E., Parisi, E., and Carson, J.**, *J. Biol. Chem.*, 247, 3962, 1972.
259. **Salton, M. R. J.**, *Annu. Rev. Microbiol.*, 21, 417, 1967.
260. **Vogel, H. J.**, in *Proceedings, 5th International Congress of Biochemistry, Vol. III*, Pergamon Press, New York, 1963, p. 341.
261. **Work, E.**, in *Methods in Microbiology*, Vol. 5A, Norris, J. R. and Ribbons, D. W., Eds., Academic Press, New York, 1971, p. 361.
262. **Kandler, O. and Hippe, H.**, *Arch. Microbiol.*, 113, 57, 1977.
263. **Wicken, A. J. and Knox, K. W.**, *Science*, 187, 1161, 1975.
264. **Galanos, C., Luderitz, O., and Westphal, O.**, *Eur. J. Biochem.*, 9, 245, 1969.
265. **Cummins, C. S.**, *J. Clin. Microbiol.*, 2, 104, 1975.

FUNGAL CELL WALL COMPOSITION

Dr. S. Bartnicki-Garcia and E. Lippman

MONOSACCHARIDES

Unless otherwise specified by a footnote, values are in % wall dry weight. Whenever feasible, values were recalculated from available data and expressed as anhydrosugar. Multiple and range values were averaged.

Abbreviations

GlcNAc	=	N-acetylglucosamine	Man	=	mannose	Fuc	=	fucose
GlcN	=	glucosamine	Gal	=	galactose	Xyl	=	xylose
GalN	=	galactosamine	GlcUA	=	glucuronic acid	Ara	=	arabinose
Glc	=	glucose	Rha	=	rhamnose	Rib	=	ribose

AMINO ACIDS AND PROTEIN

Amino acid composition is expressed in mole % of the sum total of amino acids. Values in parentheses pertain to glycoprotein fractions isolated from cell walls. All other values correspond to unfractionated cell walls. Whenever necessary, values were recalculated from published data.

Protein content of isolated cell walls is expressed in % wall dry weight.

Table 1
MONOSACCHARIDE COMPOSITION OF FUNGAL CELL WALLS

Fungus	Form	Hexosamine	GlcNAc	GlcN	GalN	Neutral sugar	Glc	Man	Gal	GlcUA	Rha	Fuc	Xyl	Ara	Rib	Ref.
ACRASIALES																
Polysphondylium																
pallidum	Microcyst			0	0		55									142
MYXOMYCETES																
Physarum																
flavicomum	Microcyst				32.1	10.4	+		+						+	65
	Microslerotium				54.9	4.0	+		+							65
polycephalum	Spherule				88.4											94
	Spore				81.4											94
TRICHOMYCETES																
Amoebidium																
parasiticum	Thallus				30				10				3			146
Smittium																
culisetae	Mycelium			35			13	5.5	4							128
OOMYCETES																
Achlya																
flagellata	Mycelium		2.0[aa]	2.4		81										49
orion	Mycelium		1.0[aa]	3.1		72										49
pseudoradiosa	Mycelium		1.0[aa]	3.8		78										49
Apodachyla																
sp.	Mycelium		22.8[bb]				67.1									82
brachynema	Mycelium			3.2			87	tr							0	132

Atkinsiella																
dubia	Mycelium			1.8	0.03		78.4									9
Dictyuchus																
monosporus	Mycelium		2.5[aa]	1.8		75										49
sterile	Mycelium			2.5			82	tr							0	132
Phytophthora																
cinnamomi	Mycelium			0.3	tr		88	0.6		0					tr	16
heveae	Mycelium			2.3			90	+	+		+				+	107
megasperma var. *sojae*	Mycelium	<1				~90	(95)[h]	(3)[h]								11
	Oospore/ oogonium			0.43		77.9	75.9[x]	1.9[x]	tr							83
palmivora	Mycelium			0.6			90.1									143
	Cyst			0.1			92.9									143
	Sporangium			0.2			93.5									143
parasitica	Mycelium			0.3	tr		86	0.7		0					tr	16
Pythium																
sp.	Mycelium		0.5[aa]	1.2		73										49
	Mycelium			1.2			79	tr							0	132
butleri	Mycelium			1.3			81	+	+		+				0	107
debaryanum	Mycelium			0.5			82.4	+							0	43
ultimum	Mycelium		0.5[aa]	1.3		79										49
Saprolegnia																
diclina	Mycelium			1.8		92										49
	Mycelium			0.89	72.5	68.6	0.4	2.85	1.33[cc]		0.3	0.1	0.2	0.05		33
ferax	Mycelium			2.7			84	tr							0	132
	Mycelium			1.7			93	tr	tr		+				+	107
Sapromyces																
elongatus	Mycelium			tr			89.1									111
Schizochytrium																
aggregatum[dd]	Thallus						25.5	0.4[x]		~25[ee,x]						45
Thraustochytrium																
sp.[dd]	Thallus						34	<0.2[x]		~22[ee,x]			11			45
CHYTRIDIOMYCETES																
Allomyces																
arbuscula		45	+													
	Cyst			30.7		10.9[gg]	1.8[x]	0.9[x]	3.0[x]		1.8[x]	+	+	0.6[x]		79
	Cyst + Rhizoid			36.4		13.0[gg]	6.4[x]	0.9[x]	3.5[x]		1.0[x]	+	+	0.5[x]		79
	Cyst + Rhizoid + Hypha			35.0		17.4[gg]	12.0[x]	0.6[x]	1.7[x]		1.9[x]	+	+	0.3[x]		79

Table 1 (continued)
MONOSACCHARIDE COMPOSITION OF FUNGAL CELL WALLS

Fungus	Form	Hexosamine	GlcNAc	GlcN	GalN	Neutral sugar	Glc	Man	Gal	GlcUA	Rha	Fuc	Xyl	Ara	Rib	Ref.
				35.0		17.4[xx]	12.0[x]	0.6[x]	1.7[x]		1.9[x]	+	+	0.3[x]		79
macrogynus	Mycelium		58				16									9
(gametophyte)	Mycelium	38	+			12	+	+	+							58
(sporophyte)	Mycelium	33	+			11.6	+	+	+							58
neomoniliformis	Mycelium	39	+			12.4	+	+	+							58
Blastocladiella																
britannica	Thallus		+			10.0	+	+	+							58
emersonii	Thallus	39	+				+	+	+							58
Phlyctochytrium																
articum	Sporangium/ rhizoid			45			+	+	tr							60
Rhizophydium																
patellarium	Sporangium/ rhizoid			24			+	+	tr							60
ZYGOMYCETES																
Basidiobolus																
ranarum	Mycelium			11.49			+		+	3.88		+				66
Blakeslea																
trispora	Mycelium (White)			5		42.6	+	tr	+				+			38
	Mycelium (Yellow)			6.7		56.7	+	tr	+				+			38
Choanephora																
cucurbitarum	Mycelium			28.4				0.3	3.2	2.4		0.8				81
Conidiobolus																
stromoideus	Mycelium			22.76			20.02			2.65						66
Entomophthora																
coronata	Mycelium			10.81			7.94	7.44		2.63						66
exitialis	Mycelium			14.13			12.93	8.40		4.05						66
thaxteriana	Mycelium			16.75			9.97	5.55		5.30						66
virulenta	Mycelium			19.41			11.38	4.61		5.16						66

Muca-													
erectus	Mycelium	19.2											151
javanicus	Mycelium	24.3											151
mucedo	Mycelium	32				7		+	+	12	+		46
pusillus	Mycelium	21.2											151
rouxii	Mycelium		9.4	32.7			0	1.6	1.6	11.8	3.8		17, 20
	Yeast		8.4	27.9			0	8.9	1.1	12.2	3.2	+	17, 20
	Sporangio-sphore		18.0	20.6			tr	0.9	0.8	25.0	2.1		19
	Sporangio-spore		2.1	9.5			42.6	4.8	0	1.9?	0		18, 20
Mycotypha													
poitrasii	Mycelium			51.6	0	5.8	0.06[x]	1.0[x]	1.05[x]		3.2[x]		40
	Intermediate			44.5	0	7.0	0.19[x]	3.6[x]	1.18[x]		2.0[x]		40
	Yeast			46.6	0	14.7	0.5[x]	11.1[x]	0.6[x]		2.5[x]		40
Phycomyces													
blakesleeanus	Sporangio-spore	9.8[h]					52.6[h]	11.8[h]	0.7[h]	22.6[h]	0.4[h]	2.1[h]	149
	Mycelium	45.6[h]					9.8[h]	2.3[h]	6.2[h]	24.3[h]	0.4[h]	11.4[h]	149
	Sporangio-phore	89.9[h]					0.9[h]	1.0[h]	0.6[h]	6.4[h]	0.6[h]	0.6[h]	149
Rhizopus													
nigricans	Mycelium	53						+	+	+	+		99
Zygorhynchus													
vuilleminii	Mycelium		+	31.5			0	tr	5.1	16.0	6.8		13
HEMIASCOMYCETES													
Hansenula													
saturnus	Yeast						45	19.3					150
Hanseniaspora													
uvarum	Yeast		0.05				30.4	34.9					97
Nadsonia													
elongata	Yeast			0.39			50[x]	34.1[x]					53
Pichia													
farinosa	Yeast						54	42.4					150

Table 1 (continued)
MONOSACCHARIDE COMPOSITION OF FUNGAL CELL WALLS

Fungus	Form	Hexosamine	GlcNAc	GlcN	GalN	Neutral sugar	Glc	Man	Gal	GlcUA	Rha	Fuc	Xyl	Ara	Rib	Ref.
Saccharomyces																
carlsbergensis	Yeast						43	38.1								150
cerevisiae																
baker's	Yeast			1.4			29	31								78, 105
	Yeast			<1.0			∿34	∿34								119
	Yeast			2.9			∿42	∿42								32
	Yeast			2.7		84.4	+	+								56
	Yeast (small)						39.9[e]	40.6[e]								74
	Yeast (large)						40.8[e]	42.2[e]								74
	Yeast			1.5			49	40								21
brewer's	Yeast			0.7			47.9	31.6								114
	Abnormal[d]			4.2			47.9	15.4								114
	Yeast		0.85			80	∿58	∿22								54
	Yeast[a]			1.2			47	44								96
	Yeast[b]			1.1			46	43								96
	Yeast						∿44[e]	∿42[e]								36
Saccharomycopsis																
guttulata	Yeast			1.7	72		∿48	∿24								32
	Filament			2.3	74		∿50	∿25								32
Schizosaccharomyces																
pombe	Yeast			0			91.3	5.6	3.2							34
	Yeast						64[e]	10.6[e]	0.7[e]							47
	Abnormal[c]						64.6[e]	4.5[e]	3.6[e]							47
EUASCOMYCETES																
Aspergillus																
sp.	Mycelium		16.5			48	36.7	4.3	4.8							121
carbonarius	Mycelium			6.38												151
clavatus	Mycelium		23				48									44

fumigatus	Mycelium			7.23								151
nidulans	Mycelium	19.1				57.6	52.9[x]	2.0[x]	2.6[x]			159
wild type	Mycelium		+		+	60.5	+	+	+	1.9		33
albino mutant	Mycelium	25.1	2.3	12.0	10.8	82.8	28.9	2.8	3.8	3.5		33
wild type at 30°C	Mycelium			16.9	0.28	38	32	+	+		tr	75
wild type at 41°C	Mycelium			16.7	0.28	27	18.8	+	+		tr	75
Ts6 mutant at 30°C	Mycelium			21.9	0.8	42.5	29.0	+	+		tr	75
Ts6 mutant at 41°C	Mycelium			1.9	0.05	41.5	20.7	+	+		tr	75
wild type	Mycelium	23				48.3						141
pcnb-2[k]	Mycelium	18.7				46.3						141
wild type	Mycelium[xx]	30.9				41.7						141
pcnb-2[k]	Mycelium[xx]	24.4				49.6						141
niger	Mycelium			12			52	1.0	4		1.0	80
	Mycelium	11	+		+	78[x]	63[x]	2.53[x]	11.29[x]		tr	69
	Mycelium	14.9			0		45.54[x]	4.19[x]	14.19[x]			41
	Conidium outer wall	2.2			0		0.94[x]	0.79[x]	0.61[x]			41
	Conidium inner wall	6.5			0		11.36[x]	9.20[x]	9.58[x]			41
oryzae	Mycelium			48		54	+	+	+			67
	Conidium			20		27	+	+	+			57
phoenicis	Mycelium		20.9[x]		+	67.7[x]	58.5[x]	+	+			24
	Conidium		32.1		+		51.3[x]	+	+			24
	Mycelium		11.7			72	55[x]	12.7[x]	4.5[x]			109
wentii	Mycelium			8.88								151
Blastomyces *dermatitidis*	Yeast		44[x]				42.5[x]	0	0			73
	Yeast	37					36.2	tr	tr			72
	Mycelium		13				44[x]	8.8[x]	4.4[x]			73
	Mycelium	22.8					44[x]	8.8[x]	4.4[x]			72
Ceratocystis *ulmi*	Mycelium	3.4				56	43.5	+	1.8			63
	Synnema	4.5				67	54.5	+	0.3			63
Chaetomium *globosum*	Mycelium			19.5[x]	tr	50.5[x]	29.4[x]	9.2[x]	11.9[x]			90
Geotrichum *lactis*	Yeast			6.1		60	26.8	27.8	5.4			14
Helminthosporium *spiciferum*	Mixedy	14.0	+			41.4	+	+	+			23

Table 1 (continued)
MONOSACCHARIDE COMPOSITION OF FUNGAL CELL WALLS

Fungus	Form	Hexosamine	GlcNAc	GlcN	GalN	Neutral sugar	Glc	Man	Gal	GlcUA	Rha	Fuc	Xyl	Ara	Rib	Ref.
Histoplasma																
farciminosum	Mycelium	32.0				29.6	+	+	+							125
	Yeast	28.0				58.0	+	+	+							125
Oospora																
suaveolens	Yeast															
	16 hr old			4.6		65	35.6	23.7	7.9							14
	72 hr old			4.2		70	29.4	30.3	6.2							14
Cordyceps																
militaris	Mycelium	11		6.55			46	9	7							91
	Unicells[pp]	5.5					52	17.5	5.7							92
	Unicells[qq]	6.8					60	9.6	6.1							92
Epidermophyton																
floccosum	Mycelium															
	(albino)					48.3	4.3[h]	1[h]								103
	Mycelium															103
	(pigmented)			23.5		30.5	2.7[h]	1	sh							103
Fusarium																
sp.	Mycelium									0.5						59
solani f. *phaseoli*	Mycelium		47				14	+	+	+						135
sulphureum	Mycelium		39.0				12.0	7.0	3.0	5.0						15
	Chlamydospore		19.0				10.7	5.4	4.0	9.5						129
	Macroconidium		12.5				24.7	6.2	6.7	14.5						129
Geotrichum																
candidum	Mycelium	14.7				58.6	28.5	12.6	14.7							134
Histoplasma																
capsulatum	Yeast		25				21.1	1.2								50
	Mycelium		4				5.5	5.1								50
Hypomyces																
chlorinus	Mycelium[rr]	19.2	27			49	10[h]	9[h]	5.5[h]		tr			tr		144
	Mycelium[ss]	26.7	28.9			41	10[h]	8[h]	3[h]		tr			tr		144

Microsporum										
audouinii	Mycelium		22		45.5	1.4[h]	1[h]			103
canis	Mycelium	26.6		0.3[x]	48.3	37.5	11.4	tr		131
	Mycelium		24.5		45.1	1.4[h]	1[h]			103
cookei	Mycelium		19.7		39.3	1.4[h]	1[h]			103
gypseum	Mycelium	31.2		0.25[x]	45.9	36.6	10.3	tr		131
	Mycelium		22.1		33.1	1.4[h]	1[h]			103
Morchella	Mycelium				17.2	7.9[x]	4.6[x]		4.6[x]	122
Neurospora										
crassa	Mycelium[c]	11.9				48.6				112
SYR 9-7a	Mycelium[d]		7.1			56.4				48
Perkins	Mycelium[d]		6.4			56.8				48
	Mycelium[d]	8.0			14[e]	48[f]				89
	Conidium[d]	7.4			30[e]	40[f]				48
St. Lawrence	Mycelium[d]		6.9			58.2				48
B 6	Mycelium[g]		11.3			50.1				48
B28	Mycelium[g]		12.4			51.9				48
colonial-1	Mycelium[g]		17.7			49.7				48
melon-1	Mycelium[g]		12.3			51.2				48
	Mycelium[g,hh]		9.4			43.4				48
B132	Mycelium[i]		6.7			56.6				48
B110	Mycelium[i]		6.7			54.4				48
B4	Mycelium[i]		8.6			58.3				48
B54	Mycelium[j]		6.8			62.4				48
RL-3-8-A	Mycelium[d]	12.3[x]		1.5		47.5	+		+	88
RL	Mycelium[d]	10.0			16[e]	50[f]				88
	Conidium[d]	9.0			28[e]	35[f]				88
STL 74A	Mycelium[d]	5.1		0.27	75.4	71.4				84
os-4-NM201o	Mycelium[l]	2.6		0.06	81.3	78.9				84
os-3-S2	Mycelium[l]	4.9		0.15	79.6	58.2				84
os-5-C24	Mycelium[l]	4.7		0.27	72.8	59.5				84
cut-A49	Mycelium[l]	10.6		0.25	70.4	60.2				84
os-1-B135	Mycelium[l]	7.9		0.37	82.4	58.9				84
STA-4	Mycelium[d]		8.2			51.3				87
cnb-1	Mycelium[kk]		7.2			57.0				87
Lindegren A	Mycelium[rr]		3.43	1.49		21.37		7.84		42
	Mycelium/ conidiophore[ss]		0.62	10.46		15.42		5.33		42
FGSC-305	Vesicular[rr]		1.35	3.96		17.91		8.26		42
	Mycelium/ conidiophore[ss]		1.06	9.09		10.46		3.09		42

Table 1 (continued)
MONOSACCHARIDE COMPOSITION OF FUNGAL CELL WALLS

Fungus	Form	Hexosamine	GlcNAc	GlcN	GalN	Neutral sugar	Glc	Man	Gal	GlcUA	Rha	Fuc	Xyl	Ara	Rib	Ref.
FGSC-45	Mycelium[u]			1.83	2.57		25.09		7.93							42
	Mycelium[u]			1.16	11.32		8.72		4.72							42
Penicillium																
album	Mycelium									0.2						59
chrysogenum	Mycelium			19.4			38.9	4.4	12.0		1.9	2.1				62
	Mycelium		>42			52	40	8	4							147
	Mycelium			18.0			33.9	2.65	7.1		0.9		0			118
	Conidium			11.4			26.3	2.6	19.4		0.3		0			118
digitatum	Mycelium			5.7			45.4	+	3.8		tr		tr			61
italicum	Mycelium			9.0			51.6	+	3.8		tr		tr			61
notatum	Mycelium			18.5	tr		43	1	7		0		0			4
	Mycelium	9.8		5.6[x]	1.2[x]	58.6	27.1	6.2	8.6		0					93
	Germ./conidium[u]	3.2		1.4[x]	0	58.5	17.9	6.1	7.3		0					93
	Germ./conidium[w]	5.3		4.5[x]	0.5[x]	56.8	24.5	5.7	8.2		0					93
	Conidium	1.0		0.1[x]	0	60.9	13.3	8.8	22.2		tr					93
patulum	Mycelium			12.3	0.5											5
roqueforti	Mycelium			13	<0.6		41	1	14							6
rubrum	Mycelium	15.4			tr	52.0	44.6	0.6	5.4							148
stoloniferum																
ATCC 14586	Mycelium[w]			18.5[x]	20.1[x]											30
	Mycelium[v]			15.2[x]	0.8[x]											30
ATCC 10111	Mycelium			25.6[x]	0.7[x]											30
CMI 31200	Mycelium			25.5[x]	0.8[x]											30
CMI 91960	Mycelium			25.1[x]	1.1[x]											30
CMI 92219	Mycelium			18.2[x]	0.45[x]											30
Trichophyton																
ferrugineum	Mycelium			23.5		33.5	1.4[h]	1[h]								103
fluviomuniense	Mycelium			22.4		44	1.6[h]	1[h]								103
gloriae	Mycelium			17.5		46.5	3.8[h]	1[h]								103
megninii	Mycelium			18.6		45	3.6[h]	1[h]								103
mentagrophytes	Mycelium			30.4			36.2	11.7	tr							104
	Mycelium		28.1		0.4[x]	55.5	45.9	7.8								131
	Microconidium		16.0			56.1	3[h]	1[h]								156

var. *asteroides*	Mycelium		26.2		43.3	2.3[h]	1[h]						103
var. *interdigitale*	Mycelium		26.2		44.8	2.3[h]	1[h]						103
rubrum	Mycelium		21.3		32.9	2.3[h]	1[h]						103
tonsurans	Mycelium		16.3		51.4	3.8[h]	1[h]						103
violaceum	Mycelium		27.2		38.1	1.6[h]	1[h]						103
					LOCULOASCOMYCETES								
Cochliobolus													
miyabeanus	Mycelium		38	+	51.7	+	+	+	+	+	+	+	102
Helminthosporium													
sativum	Mycelium		8.6	8.3		37.0	2.0	2.0					7
Leptosphaeria													
albopunctata	Mycelium[o]		4.8	<0.1		51.6	20.4	4.4					138
allorgei	Mycelium[k]		11.35	<0.1		45.9	18.0	8.0					138
discors	Mycelium[o]		20.6	<0.1		39.8	18.9	4.1					138
nitschkei	Mycelium[k]		14.5	<0.1		32.8	22.2	6.8					138
orae-maris	Mycelium[o]		10.75	<0.1		40.3	12.2	10.4					138
robusta	Mycelium[k]		17.4	<0.1		32.6	16.3	9.6					138
Venturia													
inaequalis	Mycelium	7.3				+	+						68
					HETEROBASIDIOMYCETES								
Puccinia													
graminis v. *tritici*	Uredospore	+	7.2		68.4	1.35	37.42	1.32					71
	Germinating uredospore	14.2	2.0		55.0	11.47	10.69	1.97					71
Rhodotorula													
aurantiaca	Yeast[nn]		3.4		74.7								27
	Yeast[oo]		3.5		76.0								26
glutinis	Yeast[nn]		2.5		68.4								27
	Yeast[oo]		9.2		70.7								27
	Yeast		6			20	38	13		6			8
mucilaginosa	Yeast[nn]		2.4		75.6								27
	Yeast[oo]		3.9		80.0								27

Table 1 (continued)
MONOSACCHARIDE COMPOSITION OF FUNGAL CELL WALLS

Fungus	Form	Hexosamine	GlcNAc	GlcN	GalN	Neutral sugar	Glc	Man	Gal	GlcUA	Rha	Fuc	Xyl	Ara	Rib	Ref.
Tremella																
mesenterica	Yeast			2.39		70.3	41.4	9.6	0.3	0	1.75	0.85	16.1	0.25		35
	Yeast			3.7		80	47	6.3	tr	2.1			7.1			116
	Conjugation tube			5.4		74	53	5.8	tr	1.5			4.9			116
Uromyces																
phaseoli	Uredospore			6[z]			6[z]	36[z]								145
	Germinated uredospore hull			7[z]			7[z]	33[z]								145
	Germ tube			16[z]			28[z]	16[z]								116

HOMOBASIDIOMYCETES

Fungus	Form	Hexosamine	GlcNAc	GlcN	GalN	Neutral sugar	Glc	Man	Gal	GlcUA	Rha	Fuc	Xyl	Ara	Rib	Ref.
Agaricus																
bisporus	Mycelium			43.2		39.9	40.7	tr					tr			95
Rhizoctonia																
solani	Mycelium		+	+		31.6	22.4	+	+	+						113
	Mycelium					38.0	+	+		+			+			70
	Sclerotium					33.4	+	+		+			+			70
Schizophyllum																
commune	Mycelium		12.5				67.6	3.4					0.2			133
	Mycelium		5.0				81.4						+			154
	Primordium		3.1				86.8						+			154
699	Mycelium[d]	8.7					54.1									152
1019	Mycelium[d]	8.2					67.8									152
699D	Mycelium[p]	8.8					56.1									152
1737T	Mycelium[p]	7.6					63.2									152
3532	Mycelium[q]	12.6					45.2									152
3535	Mycelium[s]	16.1					41.2									152
Sclerotium																
rolfsii	Mycelium			3.5		67.5	+	+								39
	Sclerotium			1.6		39.5	+	+								39
	Mycelium			61.0[x]	+		18.3[x]	+	+							24

	Mycelium					66.0	+	+						70
	Sclerotium					81.7	+	+						70
tuliparum	Mycelium					50.0	+	+		+		+			70
	Sclerotium					34.3	+	+		+		+			70
DEUTEROMYCETES														
Alternaria														
sp.	Mycelium									0.7					59
Aureobasidium														
(Pullularia)														
pullulans	Yeast			1.9			54	9.3	4.2	1.8	0.9				29
	Filament			2.7			70	7.1	3.8	tr	1.9				29
	Yeast	1					13	15	9						28
	Chlamydospore	1					36	8	7						28
Candida														
albicans	Yeast[r]			1.5[x]		42.8[x]	25.4[x]	17.5[x]							37
	Yeast[s]			1.7[x]		44.7[x]	29.5[x]	15.2[x]							37
	Mycelium[t]			6.5[x]		47.4[x]	30.4[x]	17.0[x]							37
	Mycelium[pp]			6.5		41 .15[k]	23.2[k]	18.0[k]							37
krusei	Yeast	1.3				41.2	+	tr	0						124
utilis	Yeast					0.35	78	48	30						106
	Tubular[v]			16		50	48	1.5							106
Cladosporium														
carrionii			7.6				16.6	13.7							137
Colletotrichum														
lagenarium	Conidium		12.5				10.4	7.4	7.7		7.5		0.6	0.9	55
lindemuthianum	Mycelium						91[h]	4[h]	5[h]						1
Cryptococcus														
albidus	Yeast		+				∼75								12
terreus	Yeast		+				∼75								12
Dactylium														
dendroides	Mycelium									2					59
Dendryphiella														
vinosa			∼2.5		0		21.5	+	+	0					25,117
Drechslera														
sorokiniana														
wild type	Yeast		9.6			56.2	+	+	+						2
albino	Yeast		9.4			58.1	+	+	+						2

Table 1 (continued)
MONOSACCHARIDE COMPOSITION OF FUNGAL CELL WALLS

Fungus	Form	Hexosamine	GlcNAc	GlcN	GalN	Neutral sugar	Glc	Man	Gal	GlcUA	Rha	Fuc	Xyl	Ara	Rib	Ref.
Epidermophyton floccosum	Mycelium		29.7			0.43[v]	56.6	45.8	6.7							131
	Mycelium			30.9		46.3	35.2	5.3	tr							108
	Mycelium[ww]			17[ww]		11[ww]		10[ww]	tr[ww]							77
Fusicoccum amygdali	Mycelium			4.5		79	69	2.8	3.8		1.3		1.0	0.5		31
Geotrichum lactis	Yeast			6.1		60	26.8	27.8	5.4							14
Helminthosporium spiciferum	Mixedy	14.0	+			41.4	+	+	+							23
Histoplasma farciminosum	Mycelium	32.0				29.6	+	+	+							125
	Yeast	28.0				58.0	+	+	+							125
Oospora suaveolens	Yeast															
	16 hr old			4.6		65	35.6	23.7	7.9							14
	72 hr old			4.2		70	29.4	30.3	6.2							14
Paracoccidioides brasiliensis																
7193	Mycelium	11.7				38.0	38	+	+							73
	Yeast	40.2				37.2	37	0	0							73
IVIC Pb9	Yeast[ii]	30.0				41.5	+	+	tr							126
IVIC Pb9H	Yeast[ii]	41.0				37.7	+	+	tr							127
IVIC Pb140	Yeast	44.1				36.5	+	+	tr							126
Phialophora pedrosoi	Mycelium		5.7				27.7	9.2								137
yerrucosa	Mycelium		5.9				30.9	8.3								137
Piricularia oryzae	Mycelium		+	12		71	62	4	0.5							101

Pithomyces												
chartarum	Mycelium			9.5		39	+	+	+			123
	Spore	3.6		+		45	+	+	+			136
Pityrosporum												
ovale												
MRL-3074	Yeast											
	2 days old			18			48	0	0			140
	9 days old			15			57	0	0			140
	14 days old			12.5			52	0	0			140
ATCC 14521	Yeast											130
	2 days old			8			40	0	0			
	9 days old			10			47	0	0			140
Sclerotinia												
fructigena	Mycelium					43	11			2	2	70
sclerotiorum	Mycelium				62.0	+	+					70
	Sclerotium				85.0	+	+					70
Sepedonium												
sp.	Mycelium	27.6	24.5	3.1		18.2	3.8	4.4				51
chrysospermum	Mycelium	18.8	17.9	0.9		20.3	2.5	2.1				51
Sporothrix												
schenkii	Mycelium	7.0					24.86[x]	12.67[x]	tr		7.13[x]	115
	Yeast	7.0					30.4[x]	21.96[x]	tr		9.58[x]	115
	Conidium	8.3					28.0[x]	11.79[x]	tr		7.99[x]	115
Torulopsis												
aeris				2.3		85	70.2	12.3	0			14
pintolopesii	Yeast[uu]					41	22.5	185				3
	Yeast[vv]					45	26.5	18.5				3
Trichoderma												
viride	Mycelium											
	18 hr old	15	12			40	30	4	6			22
	24 hr old	20	22			30	25	1	4			22
	Aberrant[tt]	2	2.5			58	55	1	2			22
Trigonopsis												
variabilis	Triangular	1.9				80.6	+	+				130
	Ellipsoidal	1.9				91.0	+					130
Verticillium												
albo-atrum	Yeast		8.8	2.6			49	6.2	8.2	1.3		153

Table 1 (continued)
MONOSACCHARIDE COMPOSITION OF FUNGAL CELL WALLS

[a] Nonflocculent.
[b] Flocculent
[c] Caused by isomytilitol.
[d] Wild type.
[e] Alkali-soluble fraction only.
[f] Alkali-insoluble -1,3-glucan fractions only.
[g] Sorbose-type semicolonial.
[h] Relative proportion.
[i] Semicolonial.
[j] Colonial.
[k] Terrestrial fungus.
[l] Osmotic mutant.
[m] Virus infected.
[n] Virus free.
[o] Marine fungus.
[p] Thin mutants.
[q] Puff mutants.
[r] In starch medium at 30°C.
[s] In glucose medium at 37°C.
[t] In starch medium at 40°C.
[u] In ox serum at 37°C.
[v] Regenerating protoplast.
[w] Calculated from published data.
[x] Aerial organs (conidia + conidiophores + mycelium).
[y] % ^{14}C from ^{14}C-labeled walls.
[aa] After enzymatic hydrolysis.
[bb] If all GlcN residues were acetylated.
[cc] Uronic acid identity not established.
[dd] Formerly classified as Oomycetes; present status unresolved (See Ref. 41).
[ee] Mostly L-galactose.
[ff] After several years.
[gg] After 2*N* trifuoroacetic acid hydrolysis.
[hh] Grown on sorbose.
[ii] Caused by inositol deficiency.
[jj] After passage through hamster.
[kk] Pentachloronitrobenzene resistant mutant.
[ll] "Swollen" stage.
[mm] Germ tube stage.
[nn] Grown in glucose-yeast extract medium.
[oo] Grown in sucrose-ammonium sulfate medium.
[pp] Grown in Czapek-Dox medium.
[qq] Grown in Moyer's medium.
[rr] Grown on sucrose medium.
[ss] Grown on acetate medium.
[tt] Aberrant tubes from regenerating protoplasts.
[uu] Choline-less strain grown on choline.
[vv] Choline-less strain grown on methionine.
[ww] Outermost wall layer only.
[xx] Grown with pentachloronitrobenzene.
[yy] Grown with biotin.
[zz] Grown without biotin.

Table 2
AMINO ACID COMPOSITION AND PROTEIN CONTENT OF CELL WALLS OF FUNGI

Fungus	Form	Protein	Ala	Arg	Asp	Cys	Glu	Gly	His	Ile	Leu	Lys	Met	Phe	Pro	Hyp	Ser	Thr	Tyr	Val	Others	Ref.
									MYXOMYCETES													
Physarum																						
flavicomun	Microcyst	21.2	9	2	10	1	8	8	4	5	7	16	tr	3	5		6	7	2	6		65
	Microsclerotium	15.4	9	2	10	1	9	8	4	4	8	16	tr	3	4		6	6	2	6		65
polycephalum	Spherule	2	12.2	2.2	7.8	0	9	11.3	1.3	6.3	8.9	3.5	0	4.7	6.8		6.9	7.4	2.2	9.4		94
	Spore	2	9.0	4.3	11.6	0	13.4	12.5	2.1	4.8	6.9	6.0	0	3.3	6.4		5.7	5.9	1.3	6.9		94
									OOMYCETES													
Apodachlya	Mycelium	6.4	11.7	3.1	9.5	0.4	7.9	12.8	1.9	5.5	8.0	7.8	0.3	2.6	4.9	0.5	0.68	5.9	1.8	6.9	1.8[f]	82
Atkinsiella																						
dubia	Mycelium	13.7	6.1	1.3	8.7	2.5	7.3	5.1	2.9	1.5	1.9	5.5	0.2	1.2	4.1	20.4	3.6	17.9	3.0	3.8		9
Phytophthora																						
cinnamoni	Mycelium	3.5	9.0	4.2	9.0	1.1	11.1	4.5	1.1	1.8	4.5	4.7	1.1	2.1	5.5	5.0	9.8	14.8	2.6	5.0		16
megasperma var. *sojae*	Oospore/oogon-ium	11.45	10.14	9.95	9.35		8.15	20.88	0.48	1.41	4.22	3.15	0.74	1.79	5.61	3.66	7.37	7.73	2.54	2.83		83
Saprolegnia																						
diclina	Mycelium	1.33	10.73	4.39	8.62		9.10	7.64	3.04	3.60	6.61	7.85	1.97	2.86	6.65	4.38	4.49	8.89	2.61	5.77	0.81[h]	35
Sapromyces																						
elongatus	Mycelium	3.7	4.3	3.5	23.6		8.8	5.0	3.0	2.6	5.3	9.1	2.8	3.6	3.3	2.5	7.4	6.1	3.1	4.6		111
Schizochytrium																						
aggregatum	Thallus	36.5	5.2	3.1	8.9	2.7	28.4	5.4	1.2	3.3	4.9	6.2	5.5	2.8	3.2		1.7	3.2	2.0	5.3	g	45
									ZYGOMYCETES													
Blakeslea																						
trispora	Mycelium(white)	12.9	6.88	4.57	16.48	4.09	5.06[i]	2.55	4.56	5.51[k]		5.76			9.56	9.63		11.16		10.69	4.50[i]	38
	Mycelium(yellow)	13.4	51.3	3.41	19.72	5.21	4.03[i]	3.10	5.78	4.08[k]		5.77			11.41	10.76		8.02		11.42	2.37[i]	38
Mucor																						
rouxii	Mycelium		11.49	2.5	10.96		9.44	17.24	2.2	4.76	6.55	8.29			2.21	4.7		5.49	6.42	tr	7.74	52
	Yeast		12.74	4.04	12.69		7.38	11.34	3.42	3.2	4.52	16.37		0.41	1.64	4.93		6.45	4.87	0.95	5.06	52
Mycotypha																						
poitrasii	Mycelium	8.6	20.20		8.29		6.46	15.83		2.33	5.44	7.78			5.34			7.59	7.83	3.47	9.42	40
	Intermediate	3.8	23.86		9.54		6.78	15.61		3.74	2.91	5.89			4.29			9.07	8.24	2.26	7.82	40
	Yeast	6.4	22.71		8.83		7.51	16.92		8.23	5.02	5.96			5.13			6.94	7.50	tr	10.24	40
									HEMIASCOMYCETES													
Hensenula																						
wingei	Yeast[b]	(2.5)	(14.50)	(0.95)	(6.32)		(7.34)	(5.55)	(0.61)	(3.05)	(3.31)	(1.69)	(0.17)	(1.46)	(2.08)		(27.57)	(11.2)	(2.67)	(7.71)		158
Nadsonia																						
elongata	Yeast[a]	8.9	10.3	1.0	9.7	3.0	11.9	2.9	1.45	4.35	4.9	4.2	0.4	3.0	4.0		10.1	17.2	2.75	7.7	0.9[h]	53
Saccharomyces																						
cerevisiae	Yeast[c]		(9.0)	(2.6)	(28.0)	(1.5)	(9.3)	(5.5)		(4.3)	(4.4)	(4.3)		(2.7)	(8.5)		(4.8)	(6.25)	(1.5)	(7.0)		76
	Yeast[a]	15.9	11.2	2.1	7.0	0.3	7.5	5.9	2.4	5.3	6.4	5.9	0.3	2.8	4.9		14.6	14.0	1.9	7.4		114
	Abnormal[a,d]	11.1	12.0	1.0	8.4	0.2	9.8	6.5	0.9	4.2	3.9	4.2		2.3	5.8		15.8	17.3	1.5	6.2		114
G1406	Yeast[d]		(13.91)	(0)	(12.44)		(12.44)	(6.80)	(0.71)	(2.49)	(1.47)	(1.47)		(0.31)	(6.22)		(15.99)	(16.88)	(0.62)	(8.26)		36
G1406	Yeast		12.10	1.70	10.21		10.59	5.86	1.13	4.73	4.92	4.73		2.84	5.29		12.10	13.23	1.89	7.18	1.51[h]	36
NF1040	Yeast	10.14	9.75	1.46	12.37		9.00	6.52	1.24	4.05	4.84	5.81	0.19	1.50	5.25		12.75	16.50	2.25	6.52		86

Table 2 (continued)
AMINO ACID COMPOSITION AND PROTEIN CONTENT OF CELL WALLS OF FUNGI

Fungus	Form	Protein	Ala	Arg	Asp	Cys	Glu	Gly	His	Ile	Leu	Lys	Met	Phe	Pro	Hyp	Ser	Thr	Tyr	Val	Others	Ref.
NF1004	Yeast	16.7	10.69	0.88	9.16		8.78	6.11	0.38	3.82	4.20	3.63	0.08	1.53	6.68		17.18	17.56	2.44	6.87		86
F1005	Yeast	13.5	8.12	1.68	9.92		12.89	6.28	1.03	2.26	5.51	4.83	0.32	1.42	3.22		10.95	19.97	2.06	9.89		86
F1026	Yeast	12.5	11.91	1.17	13.13		8.89	9.29	1.25	3.63	4.56	4.32	tr	1.90	3.76		12.92	14.54	2.42	6.10		86
Schizosaccharomyces pombe	Yeast		8.2	4.9	9.9	0.3	8.9	9.5	2.1	6.2	8.5	6.8	1.55	4.3	4.2		7.8	6.2	2.9	7.6		47
	Abnormal[d]		8.1	5.2	9.8	0.45	8.75	9.5	2.4	6.6	8.8	7.0	1.7	4.75	4.15		6.5	5.2	3.0	7.9		47
										EUASCOMYCETES												
Aspergillus niger	Mycelium	1.15	9.3	2.25	8.95	3.0	5.95	11.0	tr	2.0	2.6	4.5	2.3	3.1	6.5		19.3	11.0	2.9	4.8		98
	Mycelium	6.6	13.92	tr	6.46		6.57	16.79	tr	3.17	7.95	7.28		11.13	tr		12.28	7.52		6.92		41
	Conidium outer wall	63.4	10.52	tr	14.52		8.52	22.52	0.79	1.08	6.92	3.78		7.29	tr		11.09	5.60	0.89	6.49		41
	Conidium inner wall	20.0	12.50	tr	10.91		9.69	21.64	0.76	2.53	3.71	4.02	tr	5.94	tr		9.09	9.08	1.30	8.02		41
Blastomyces dermatitidis	Mycelium	26.8	7.3	2.1	6.1		9.6	22.3	1.4	5.1	5.75	4.2	0.6	2.8	12.6		6.3	8.5		5.1		73
	Yeast	7.1	6.8	2.9	15.7		10.3	9.9	11.2	2.7	2.3	13.2	1.2	2.5	5.6		6.6	5.60		3.9		73
	Yeast[n]		(9.74)	(2.05)	(9.59)	(0.35)	(9.38)	(16.33)	(0.98)	(3.09)	(5.05)	(3.49)	(0.68)	(0.31)	(12.43)		(9.72)	(7.28)	(3.38)	(6.15)		120
	Yeast[m]		(7.65)	(2.22)	(11.63)	(0.72)	(8.15)	(11.45)	(2.45)	(2.60)	(5.19)	(3.97)		(2.06)	(19.36)		(8.18)	(5.74)	(5.68)	(2.91)		120
	Mycelium[m]		(10.98)	(2.49)	(9.80)	(0.46)	(9.62)	(15.45)	(1.17)	(4.83)	(6.36)	(2.72)	(0.96)	(1.31)	(10.70)		(7.51)	(6.08)	(2.76)	(6.80)		120
Ceratocystis ulmi	Mycelium	7.92	10.92	4.95	9.55		9.48	7.88	1.42	6.02	8.19	5.22	0.44	3.58	4.50		7.65	9.24	2.91			63
	Synnema	5.02	11.53	3.47	8. 36		7.63	7.93	1.23	5.34	7.07	4.51	1.37	2.42	5.17		8.17	12.13	2.61	11.06		63
Chaetomium globosum	Mycelium	4.96	8.5	3.8	10.4	1.7	12.2	9.8	tr	2.6	6.0	1.0	0.7	2.65	9.9		9.2	14.0	2.2	5.4		98
	Mycelium		8.78	3.82	7.98	1.0	14.37	8.46	2.23	3.82	6.53	5.40	0.79	3.19	8.46		7.98	8.62	2.23	6.0		90
Cordyceps militaris	Mycelium	6.3	9.9	4.3	9.7		8.9	8.1	1.8	3.7	6.4	5.4	1.4	3.1	8.2		9.9	11.0	2.3	5.7		91
	Unicell[r]	8.2	9.98	3.71	10.54	1.26	7.82	7.80	1.71	2.29	5.01	7.83	1.05	2.44	9.50		9.85	11.68	1.90	5.45	0—17[b]	92
	Unicell[s]	9.7	10.29	4.93	10.65	0.83	9.14	7.94	1.68	2.63	5.18	6.66	1.12	1.79	9.44		10.23	10.49	2.08	4.93		92
Fusarium sulphureum	Mycelium	7.3	11.8	4.2	10.3	0.3	9.1	10.2	2.8	2.8	5.8	2.8	15	4.6			6.7	6.4	1.3	5.7		15
	Chlamydospores	28.5	7.9	2.0	10.6	4.1	12.1	9.1	1.8	3.5	5.4	9.2	1.3	2.8	7.4		6.0	7.8	2.7	6.4		129
	Macroconidium	21.6	6.3	1.5	9.2	3.2	9.7	8.8	1.6	3.2	2.3	7.1	1.2	2.0	11.4		9.8	12.6	3.6	6.3		129
Histoplasma capsulatum	Mycelium	10	7.9	2.2	7.9		13.5	19.1	2.2	3.3	4.5	6.7	tr	1.1	tr		11.2	12.3	tr	7.9	tr[c]	50
	Yeast	5	7.9	7.9	2.6		13.15	15.8	2.6	7.9	10.5	10.5	tr	7.9	tr		5.3	5.3	2.6			50
Hypomyces chlorinus	Mycelium[t]	6.48	13.52	3.78	9.0		6.81	11.20	tr	2.72	5.13	8.34	tr	2.42	7.79		10.93	10.70	2.40	5.68		144
	Mycelium[u]	7.35	12.81	2.99	8.85		7.46	10.0	tr	2.41	2.55	7.30	tr	3.16	10.24		11.39	13.78	3.08	4.55		144
Microsporum canis	Mycelium	6.8	4.4	tr	5.8	tr	7.5	8.5	1.4	2.0	19.4	3.7	2.4	7.8	4.1		7.5	8.8	4.8	5.8	6.1[k]	131

gypseum	Mycelium	8.0	6.9	tr	6.3		7.2	10.8	3.0	2.1	4.5	2.4	2.4	8.1	9.6	10.8	12.3	6.0	3.6	3.6[t]	131
	Spore[v]		(6.5)	(0.2)	(11.0)	(tr)	(8.5)	(10.2)	(0.3)	(3.1)	(15.6)	(1.2)	(6.6)	(7.2)	(6.9)	(7.2)	(5.2)	(2.2)	(4.5)		110
	Spore[w]		(7.2)	(0.4)	(5.7)	(n.d.)	(11.8)	(11.0)	(4.4)	(1.2)	(2.3)	(3.9)	(tr)	(0.9)	(19.8)	(9.7)	(15.3)	(1.0)	(5.0)		110
	Mycelium[v]		(4.2)	(1.4)	(4.6)	(n.d.)	(4.5)	(7.4)	(1.6)	(1.9)	(23.9)	(9.7)	(1.6)	(9.5)	(4.7)	(4.6)	(4.5)	(1.5)	(2.8)		110
	Mycelium[w]		(9)	(0.2)	(5.6)	(n.d.)	(9.1)	(11.4)	(7.2)	(0.5)	(2.0)	(3.0)	(0.4)	(0.3)	(16.0)	(10.7)	(13.4)	(0.6)	(7.8)		110
Neurospora																					
crassa	Mycelium	6.0	13.9	3.5	8.5		6.2	13.7	1.2	3.9	6.2	5.6		2.5	8.1	9.75	8.7	1.0	7.0		84
	Mycelium[x]	8.63	(10.85)	(3.67)	(12.43)	(0.71)	(10.24)	(13.36)	(2.26)	(3.26)	(6.85)	(6.43)	(1.21)	(2.25)	(0.72)	(8.64)	(8.18)	(2.91)	(5.82)		155
Penicillium																					
expansum	Conidium	7.7	7.9	3.9	9.9	10.2	8.9	8.0	4.8	3.1	6.0	2.4	1.0	2.4	6.0	8.7	8.4	2.4	5.7		57
notatum	Mycelium	8.9	10.7	2.6	9.9		9.1	11.4	1.7	3.1	7.5	3.4		5.7	9.4	12.8	9.7	2.9			4
	Mycelium		11.19	tr	9.84	[aa]	9.26	10.44	tr	3.45	6.18	2.59	[aa]	3.68	9.49	12.40	12.65	2.33	6.50		93
	Conidium		10.3	2.25	10.46	[aa]	11.58	10.98	2.55	3.71	6.76	4.47	0.65	4.67	8.39	7.84	6.34	2.40	6.63		93
	Germ. conidium[y]		10.93	0.67	10.79	[aa]	12.41	10.89	0.99	4.16	7.93	2.05	[aa]	4.40	9.16	9.01	7.07	2.64	6.87		93
	Germ. conidium[z]		10.90	1.43	9.74	[aa]	10.04	11.26	1.94	3.96	7.42	3.16	[aa]	3.96	7.86	11.92	7.83	2.56	6.29		93
roqueforti	Mycelium	9.7	11.4	2.4	8.6		8.6	10.1	0.4	2.4	6.2	3.4		5.8	9.1	14.9	11.4	4.3			6
rubrum	Mycelium	7.2	9.73	4.11	8.21		8.45	8.53	2.19	5.46	10.82	6.19	1.09	4.49	6.66	5.02	7.59	2.36			148
Trichophyton																					
mentagrophytes	Mycelium	7.1	6.3	tr	4.8	3.3	6.3	9.2	1.7	2.1	14.6	4.0	tr	7.1	5.4	7.9	8.8	6.4	6.9	5.2[t]	131
	Mycelium	7.8	6.8	6.0	11.6		8.3	6.1	0.9	7.1	14.9	6.8		5.1	tr	7.2	5.6	6.5	7.0		104
	Microconidium	22.6	6.4	1.9	12.2	3	9.5	6.1	tr	3.2	6.0	4.3	0.7	2.6	3.8	5.6	5.4	1.5	5.8		156
	Microconidium[bb]	(82.5)	(6.97)	(1.11)	(15.67)	(1.90)	(8.26)	(9.02)	(1.26)	(3.23)	(7.40)	(7.86)	(0.14)	(2.69)	(3.38)	(3.99)	(5.10)	(1.29)	(7.47)		64
											LOCULOASCOMYCETES										
Helminthosporium																					
sativum	Mycelium	18	11.2	2.0	9.9	1.4	10.3	8.9	0.7	4.1	7.5	4.6	0.7	2.0	9.0	9.0	7.1	3.0	7.6		7
Leptoshaeria																					
albopunctata	Mycelium	4.8	12.6	0.7	5.3		9.45	10.1	2.7	1.5	2.4	5.7	1.3	1.3	17.8	12.3	10.0	2.8	3.9		138
allorgei	Mycelium	9.5	11.0	3.1	10.4		10.6	10.0	1.6	2.7	5.1	4.8	0	2.1	11.8	10.7	8.9	2.2	4.9		138
discors	Mycelium	8.65	11.2	2.0	8.0		8.85	10.9	1.4	2.4	3.8	4.7	0.6	2.7	12.9	10.8	10.7	2.9	6.0		138
nitschkei	Mycelium	12.8	10.6	2.2	8.0		9.0	10.6	1.9	2.2	3.6	5.8	0.4	1.8	12.9	11.1	10.7	3.5	5.6		138
orae-maris	Mycelium	10.9	10.0	3.0	7.0		10.6	10.5	1.75	2.2	5.3	4.5	0	2.9	12.4	9.4	12.3	3.3	4.8		138
robusta	Mycelium	11.2	10.2	2.55	8.6		10.1	9.55	2.2	2.6	4.7	5.0	0.65	2.3	13.1	8.3	11.7	3.2	5.2		138
Venturia																					
inaequalis	Mycelium	2.7	9.9	3.5	7.3	0.8	9.3	10.4	2.7	5.2	8.3	7.0	2.3	4.2	5.4	6.9	5.0	3.0	6.0	0.6[t]	68
											HETEROBASIDIOMYCETES										
Puccinia																					
graminis	Germ tube	12.5	10.99		9.55		8.38	16.80		5.74	7.18		0.95	3.93	9.57	7.88	5.77	5.30	7.96		71
	Spore	1.2	6.60		22.83		15.32	25.50		3.74	5.23		0.66	2.97	+	8.39	4.11	1.62	5.02		71
Rhodotorula																					
aurantiaca	Yeast[hh]	∼5	11.59		7.25		7.25	21.74	3.62	4.35	7.97	7.25		4.35		8.70	5.80	4.35	5.80		26
	Yeast[ii]	∼5	13.16	3.29	8.55		10.53	19.08	2.63	3.29	3.95	7.90		2.63		13.16	5.26	1.97	4.61		26
glutinis	Yeast[hh]	∼5	14.65	1.95	11.72		9.77	17.58	1.95	2.34	5.86	1.95		3.91		13.67	5.86	4.88	3.91		26
	Yeast[ii]	∼5	9.45		7.09		11.81	31.50	8.66		7.87	9.45				14.17					26
mucilaginosa	Yeast[hh]	∼5	13.74	3.05	13.99		11.45	16.29	1.78	2.55	6.36	2.55		4.33		10.19	5.85	2.80	5.09		26
	Yeast[ii]	∼5	13.21	4.53	9.81		10.19	12.45	3.02	2.64	12.45	4.91		3.77		10.19	5.66	2.64	4.53		26

Table 2 (continued)
AMINO ACID COMPOSITION AND PROTEIN CONTENT OF CELL WALLS OF FUNGI

Fungus	Form	Protein	Ala	Arg	Asp	Cys	Glu	Gly	His	Ile	Leu	Lys	Met	Phe	Pro	Hyp	Ser	Thr	Tyr	Val	Others	Ref.
Tremella mesenterica	Yeast	1.6	11.6	4.62	8.63		8.59	9.18	tr	3.50	7.01	12.58	0.77	3.48	5.99		9.84	5.79	2.54	5.89		116
	Conjugation tube	1.8	9.53	4.33	9.92		9.62	8.80	1.82	3.60	7.20	9.69	0.63	4.00	6.56		10.77	7.13	1.56	4.83		116
	Yeast	9.1	9.77	5.29	9.81		13.83	8.01	2.06	4.98	8.59	5.95	2.32	3.49	5.89	0.78	4.51	5.48	2.90	6.23	0.13[h]	35
Uromyces Phaseoli	Uredospore hull	8	11.81	6.04	10.99		8.24	11.81		4.39	7.69	6.87	tr	3.85	tr.		11.54	7.14	2.75	6.87		145
	Germ tube		15.66	5.79	7.12		6.08	21.68	0.92	3.00	5.87	4.57	0.16	2.89	tr		11.23	7.01	2.17	5.85		145
HOMOBASIDIOMYCETES																						
Agaricus bisporus	Fruiting body	9.6	8.08	4.73	9.53	0.76	11.03	10.28	2.78	7.48	7.58	7.16	1.23	3.25	6.14		5.85	6.26	1.59	6.26		139
DEUTEROMYCETES																						
Candida albicans	Yeast		9.61	2.53	10.21	1.12	9.24	9.69	3.20	2.61	8.05	4.84	tr	2.98	4.77		14.46	8.27	2.61	5.81		100
	Mycelium	15.7	10.7	5.3	6.5		9.1	7.4	0	5.9	7.5	0.8	2.2	p	3.7	3.1[f]	5.6	7.4	3.6	9.4		37
	Mycelium[cc]		(10.9)	(1.0)	(5.3)	(1.3)	(30.9)	(5.7)	(2.6)	(4.1)	(10.4)	(9.8)	(1.4)	(4.3)	(3.0)		(0.3)	(0.5)	(2.9)	(5.6)		157
	Mycelium[c]		(13.3)	(1.5)	(16.0)		(10.6)	(1.3)	(1.2)	(10.3)	(12.3)	(1.6)	(tr)	(5.2)	(8.2)		(1.8)	(1.9)	(1.6)	(13.3)		157
	Yeast[cc]		(9.1)	(1.2)	(12.3)	(0.6)	(11.8)	(9.0)	(2.8)	(5.3)	(11.0)	(13.7)	(1.4)	(4.9)	(4.4)		(0.7)	(0.8)	(3.3)	(7.8)		157
	Yeast[c]		(15.4)	(tr)	(15.9)		(11.7)	(2.2)	(tr)	(13.6)	(14.6)	(1.7)		(7.7)	(tr)		(1.7)	(1.8)	(tr)	(13.2)		157
	Yeast	25.6	8.9	0	6.5		6.55	6.9	11.2	5.3	6.3	2.6	7.9	p	6.3		5.6	6.1	5.1	5.1		37
Cladosporium carrionii	Mycelium	42.4	9.5	7.46	10.21		8.01	8.85	2.53	3.92	7.60	8.90	4.05	3.66	5.5		6.90	4.82	2.67	5.42		137
wernickii	Yeast[f]		(12.8)	(0.7)	(6.8)		(8.6)	(6.8)	(0.5)	(3.2)	(3.9)	(1.7)	(tr)	(2.1)	(5.4)		(16.3)	(26.4)	(tr)	(4.71)		85
Dendryphiella vinosa	Mycelium	4	8.33	5.47	7.55		11.16	10.53	5.35	4.87	6.92	4.47	4.18	5.06	6.92		7.23	1.89	3.62	6.45		25
Epidermophyton floccosum	Mycelium	7.4	6.8	tr	5.4	2.2	6.8	10.2	2.2	2.2	5.4	4.1	tr	5.85	10.0		9.5	10.2	5.6	4.9		131
	Mycelium[dd]		(10.1)	(0.9)	(10.5)		(13.3)	(17.0)		(4.7)	(9.6)	(2.4)		(5.4)			(8.0)	(9.4)	(4.0)	(4.1)		77
	Mycelium	10.3	9.4	4.4	9.4	6.6	8.7	10.0	2.2	4.7	7.6	5.6		3.3	8.7		18.5	18.2	1.7			108
Fusicoccum amygdali	Mycelium	5	9.99	14.31	10.16		7.98	9.01	2.75	3.80	6.51	2.68	+	5.82	+		9.48	5.98	3.93	7.60		31
Paracoccidioides brasiliensis	Mycelium	37	8.9	3.6	9.3		11.8	14.8	1.0	3.9	5.3	2.8	2.0	3.1	11.7		8.1	7.6		6.0		73
	Yeast	11.2	8.35	2.3	8.9		9.9	11.8	10.8	2.7	4.7	9.4	2.1	3.5	6.4		7.6	8.0		3.5		73
Phialophora pedrosoi	Mycelium	29.5	6.79	4.22	9.42		8.52	8.06	1.95	2.64	4.94	5.62	11.89	2.88	6.38		9.05	8.35	2.63	6.65		137
verrucosa	Mycelium	10.8	4.34	11.10	4.3		11.91	10.07	3.1	0.26	7.60	7.76	1.61	0.42	6.27		8.17	8.36	2.66	5.57		137
Piricularia oryzae	Mycelium	4.6	9.5	2.1	11.4	0.8	7.0	7.6	0.8	2.7	4.6	5.3	0.8	3.6	9.5		5.3	8.6	1.9	5.8		101
Sporothrix schenkii	Mycelium	21.7	13.1	4.6	10.3		10.9	7.1	1.2	3.7	8.2	4.5	tr	3.7	8.0		8.2	7.9	2.4	6.0		115
	Yeast	14.4	15.2	3.5	9.7		8.4	7.8	1.6	3.5	6.1	3.6	tr	2.9	7.6		11.7	10.2	2.4	5.6		115

	Conidium	24.2	12.3	4.3	9.7		10.4	8.9	1.7	3.9	7.6	4.7	tr	3.7	7.0		8.8	8.4	2.5	5.9		115
Torulopsis pintolopesii	Yeast[ee]		6.81	3.69	11.43	0.82	9.90	6.72	2.02	6.70	7.74	7.44	0.94	3.68	7.79		6.48	6.18	2.81	8.85		3
	Yeast[ff]		6.94	3.72	10.68	0.87	10.13	6.75	2.09	6.56	7.72	7.32	0.94	3.61	8.06		6.76	5.99	2.87	9.30		3
Verticillium albo-atrum	Yeast	12.5	7.9	0.9	8.2	1.9	8.5	10.0	2.2	2.7	2.9	4.8	tr	1.9	13.2		8.6	14.0	3.8	8.4		153

[a] Excluding NH_3.
[b] Tryptophan.
[c] Composition of a glucan-protein fraction.
[d] Caused by isomytilitol.
[e] Ornithine.
[f] Hydroxylysine.
[g] Two unidentified amino acids, not included.
[h] Composition of a mannan protein (Y-5 mannan).
[i] Three unidentified components, amounting to 2.8%.
[j] Mixture of 3- and 4-hydroxyproline.
[k] Unidentified amino acid, amounting to 4.3%.
[l] Composition of a galactomannan-protein fraction.
[m] Composition of an alkali-soluble, ammonium sulfate precipitable fraction.
[n] Composition of an alkali-soluble, ammonium sulfate soluble fraction.
[o] Unresolved.
[p] Caused by inositol deficiency.
[q] Grown on Czapek-Dox medium.
[r] Grown on Moyer's medium.
[s] On biotin-deficient medium.
[t] On biotin-rich medium.
[u] Composition of a water-insoluble glycoprotein.
[v] Compositoion of a water-soluble glycoprotein.
[w] Composition of the alkali-soluble fraction.
[x] Swollen state.
[y] With germ tube.
[aa] Not measured because of interference with other unidentified substances.
[bb] Rodlet layer.
[cc] Composition of a glucomannan protein fraction.
[dd] Outermost wall layer.
[ee] Choline-less strain grown on methionine.
[ff] Choline-less strain grown on choline.
[gg] Composition of a phosphomannan protein.
[hh] Grown in glucose-yeast extract medium.
[ii] Grown in sucrose-ammonium sulfate medium.
[jj] Plus threonine.
[kk] Plus leucine.
[ll] Plus methionine.

REFERENCES

1. Albersheim, P. and Valent, B. S., *Plant Physiol.*, 53, 684, 1974.
2. Al-Rikabi, K. H. and Bonaly, R., *Can. J. Bot.*, 23, 1508, 1977.
3. Angluster, J. and Travassos, L. R., *Arch. Mikrobiol.*, 83, 303, 1972.
4. Applegarth, D. A., *Arch. Biochem. Biophys.*, 120, 471, 1967.
5. Applegarth, D. A. and Bozoian, G., *J. Bacteriol.*, 94, 1787, 1967.
6. Applegarth, D. A. and Bozoian, G., *Can. J. Microbiol.*, 14, 489, 1968.
7. Applegarth, D. A. and Bozoian, G., *Arch. Biochem. Biophys.*, 135, 285, 1969.
8. Arai, M. and Murao, S., *Agric. Biol. Chem.*, 41, 617, 1977.
9. Aronson, J. M. and Fuller, M. S., *Arch. Microbiol.*, 68, 295, 1969.
10. Aronson, J. M. and Machlis, L., *Am. J. Bot.*, 46, 292, 1959.
11. Ayers, A. R., Ebel, J., Valent, B., and Albersheim, P., *Plant Physiol.*, 57, 760, 1976.
12. Bacon, J. S. D., Jones, D., Farmer, V. C., and Webley, D. M., *Biochim. Biophys. Acta*, 158, 313, 1968.
13. Ballesta, J. P. G. and Alexander, M., *J. Bacteriol.*, 106, 938, 1971.
14. Ballesta, J. P. G. and Villanueva, J. R., *Trans. Br. Mycol. Soc.*, 56, 403, 1971.
15. Barran, L. R., Schneider, E. F., Wood, P. J., Madhosingh, C., and Miller, R. W., *Biochim. Biophys. Acta*, 392, 148, 1975.
16. Bartnicki-Garcia, S., *J. Gen. Microbiol.*, 42, 57, 1966.
17. Bartnicki-Garcia, S. and Nickerson, W. J., *Biochim. Biophys. Acta*, 58, 102, 1962.
18. Bartnicki-Garcia, S. and Reyes, E., *Arch. Biochem. Biophys.*, 108, 125, 1964.
19. Bartnicki-Garcia, S. and Reyes, E., *Biochim. Biophys. Acta*, 165, 32, 1968.
20. Bartnicki-Garcia, S. and Reyes, E., *Biochim. Biophys. Acta*, 170, 54, 1968.
21. Bauer, H., Horisberger, M., Bush, D. A., and Sigarlakie, E., *Arch. Microbiol.*, 85, 202, 1972.
22. Benitez, T., Villa, T. G., and Garcia Acha, I., *Arch. Microbiol.*, 105, 277, 1975.
23. Berthe, M. C. and Bonaly, R., *Can. J. Microbiol.*, 22, 929, 1976.
24. Bloomfield, B. J. and Alexander, M., *J. Bacteriol.*, 93, 1276, 1967.
25. Bonaly, R., *Carbohydr. Res.*, 24, 355, 1972.
26. Bonaly, R., Moulki, H., Touimi-Benjelloun, A., and Pierfitte, M., *Biochim. Biophys. Acta*, 244, 484, 1971.
27. Bonaly, R. and Reisinger, O., *C. R. Acad. Sci.*, 272, 2309, 1971.
28. Brown, R. G., Hanic, L. A., and Hsiao, M., *Can. J. Microbiol.*, 19, 163, 1973.
29. Brown, R. G. and Nickerson, W. J., Unpublished results and *Bacteriol. Proc.*, p. 26, 1965.
30. Buck, K. W., Chain, E. B., and Darbyshire, J. E., *Nature (London)*, 223, 1273, 1969.
31. Buck, K. W., and Obaidah, M. A., *Biochem. J.*, 125, 461, 1971.
32. Buecher, E. J. and Phaff, H. J., in *Proc. 2nd Int. Symp. Yeast Protoplasts*, p. 165, 1968.
33. Bull, A. T., *J. Gen. Microbiol.*, 63, 75, 1970.
34. Bush, D. A., Horisberger, M., Horman, I., and Wursch, P., *J. Gen. Microbiol.*, 81, 199, 1974.
35. Cameron, D. S. and Taylor, I. E. P., *Biochim. Biophys. Acta*, 444, 212, 1976.
36. Cawley, T. N., Harrington, M. G., and Letters, R., *Biochem. J.*, 129, 711, 1972.
37. Chattaway, F. W., Holmes, M. R., and Barlow, A. J. E., *J. Gen. Microbiol.*, 51, 367, 1968.
38. Chenoudo, M. S., *J. Gen. Appl. Microbiol.*, 18, 143, 1972.
39. Chet, I., Henis, Y., and Mitchell, R., *Can. J. Microbiol.*, 13, 137, 1967.
40. Cole, G. T., Sekiya, T., Kasai, R., Yokoyama, T., and Nozawa, Y., *Exper. Mycol.*, 3, 132, 1979.
41. Cole, G. T., Sekiya, T., Kasai, R., Yokoyama, T., and Nozawa, Y., *Exper. Mycol.*, 3, 132, 1979.
42. Coniordos, N. and Turian, G., *Ann. Microbiol. (Paris)*, 124(A), 5, 1973.
43. Cooper, B. A. and Aronson, J. M., *Mycologia*, 59, 658, 1967.
44. Corina, D. L. and Munday, K. A., *J. Gen. Microbiol.*, 65, 253, 1971.
45. Darley, W. M., Porter, D., and Fuller, M. S., *Arch. Mikrobiol.*, 90, 89, 1973.
46. Datema, R., Van Den Ende, H., and Wessels, J. G. H., *Eur. J. Biochem.*, 80, 611, 1977.
47. Deshusses, J., Berthoud, S., and Posternak, T., *Biochim. Biophys. Acta*, 176, 803, 1969.
48. De Terra, N. and Tatum, E. L., *Am. J. Bot.*, 50, 669, 1963.
49. Dietrich, S. M. C., *Biochim. Biophys. Acta*, 313, 95, 1973.
50. Domer, J. E., Hamilton, J. G., and Harkin, J. C., *J. Bacteriol.*, 94, 466, 1967.
51. Domer, J. E. and Harmon, R. D., *Sabouraudia*, 10, 56, 1972.
52. Dow, J. M. and Rubery, P. H., *J. Gen. Microbiol.*, 99, 29, 1977.
53. Dyke, K. G. H., *Biochim. Biophys. Acta*, 82, 374, 1964.
54. Eddy, A. A., *Proc. R. Soc. Ser. B*, 149, 425, 1958.
55. Esquerre-Tugays, M. T. and Touze, A., *Phytochemistry*, 10, 821, 1971.
56. Falcone, G. and Nickerson, W. J., *Science*, 124, 272, 1956.
57. Fisher, D. J. and Richmond, D. V., *J. Gen. Microbiol.*, 64, 205, 1970.

58. **Fultz, S. A.,** Ph.D. thesis, *Diss. Abstr.* p. 2433, 1965.
59. **Gancedo, J. M., Gancedo, C., and Asensio, C.,** *Biochem. Z.*, 346, 328, 1966.
60. **Gerhart, S. and Barr, D. J. S.,** *Mycologia*, 663, 107, 1974.
61. **Grisaro, V., Sharon, N., and Barkai-Golan, R.,** *J. Gen. Microbiol.*, 51, 145, 1968.
62. **Hamilton, P. B. and Knight, S. G.,** *Arch. Biochem. Biophys.*, 99, 282, 1962.
63. **Harris, J. L. and Taber, W. A.,** *Can. J. Bot.*, 51, 1147, 1973.
64. **Hashimoto, T., Wu-Yuan, C. D., and Blumenthal, H. J.,** *J. Bacteriol.*, 127, 1543, 1976.
65. **Henney, H. R., Jr. and Chu, P.,** *Exp. Mycol.*, 1, 83, 1977.
66. **Hoddinott, J. and Olsen, O. A.,** *Can. J. Bot.*, 50, 1675, 1972.
67. **Horikoshi, K. and Iida, S.,** *Biochim. Biophys. Acta*, 83, 197, 1964.
68. **Jaworski, E. G. and Wang, L. C.,** *Phytopathology*, 55, 401, 1965.
69. **Johnston, I. R.,** *Biochem. J.*, 96, 651, 1965.
70. **Jones, D., Farmer, V. C., Bacon, J. S. D., and Wilson, M. J.,** *Trans. Br. Mycol. Soc.*, 59, 11, 1972.
71. **Joppien, S., Burger, A. and Reisener, H. J.,** *Arch. Mikrobiol.*, 82, 337, 1972.
72. **Kanetsuna, F. and Carbonell, L. M.,** *J. Bacteriol.*, 106, 946, 1971.
73. **Kanetsuna, F., Carbonell, L. M., Moreno, R. E., and Rodriguez, J.,** *J. Bacteriol.*, 97, 1036, 1969.
74. **Katohda, S., Abe, N., Matsui, M., and Hayashibe, M.,** *Plant Cell Physiol.*, 17, 909, 1976.
75. **Katz, D. and Rosenberger, R. F.,** *Biochim. Biophys. Acta*, 208, 452, 1970.
76. **Kessler, G. and Nickerson, W. J.,** *J. Biol. Chem.*, 234, 2281, 1959.
77. **Kitajima, Y. and Nozawa, Y.,** *Biochim. Biophys. Acta*, 394, 558, 1975.
78. **Korn, E. D. and Northcote, D. H.,** *Biochem. J.*, 75, 12, 1960.
79. **Kroh, M., Knuiman, B., Kirby, E. G., and Sassen, M. M. A.,** *Arch. Microbiol.*, 113, 73, 1977.
80. **Leopold, J. and Seichertova, O.,** *Folia Microbiol. (Prague)*, 12, 345, 1967.
81. **Letourneau, D. R., Deven, J. M., and Manocha, M. S.,** *Can. J. Microbiol.*, 22, 486, 1976.
82. **Lin, C. C., Sicher, R. C. Jr., and Aronson, J. M.,** *Arch. Microbiol.*, 108, 85, 1976.
83. **Lippman, E., Erwin, D. C., and Bartnicki-Garcia, S.,** *J. Gen. Microbiol.*, 80, 131, 1974.
84. **Livingston, L. R.,** *J. Bacteriol.*, 99, 85, 1969.
85. **Lloyd, K. O.,** *Biochemistry*, 9, 3446, 1970.
86. **Lyons, T. P. and Hough, J. S.,** *J. Inst. Brew., London*, 77, 300, 1971.
87. **Macris, B. and Georgopoulos, S. G.,** *Zeitschrift Allg. Mikrobiol.*, 13, 415, 1973.
88. **Mahadevan, P. R. and Tatum, E. L.,** *J. Bacteriol.*, 90, 1073, 1965.
89. **Mahadevan, P. R. and Mahadkar, U. R.,** *Indian J. Exp. Biol.*, 8, 207, 1970.
90. **Maret, R.,** *Arch. Mikrobiol.*, 8, 68, 1972.
91. **Marks, D. B., Keller, B. J., and Guarino, A.J.,** *Biochim. Biophys. Acta*, 183, 58, 1969.
92. **Marks, D. B., Keller, B. J., and Guarino, A. J.,** *J. Gen. Microbiol.*, 69, 253, 1971.
93. **Martin, J. F., Nicolas, G., and Villanueva, J. R.,** *Can. J. Microbiol.*, 19, 789, 1973.
94. **McCormick, J. J., Blomquist, J. C., and Rusch, H. P.,** *J. Bacteriol.*, 104, 1119, 1970.
95. **Michalenko, G. O., Hohl, H. R., and Rast, D.,** *J. Gen. Microbiol.*, 92, 251, 1976.
96. **Mill, P. J.,** *J. Gen. Microbiol.*, 44, 329, 1966.
97. **Miller, M. W. and Phaff, H. J.,** *Antonie van Leeuwenhoek J. Microbiol. Serol.*, 24, 225, 1958.
98. **Mitchell, A.D. and Taylor, I. E. P.,** *J. Gen. Microbiol.*, 59, 103, 1969.
99. **Miyazaki, T. and Irino, T.,** *Chem. Pharm. Bull. (Tokyo)*, 19, 2545, 1971.
100. **Morilhat, J. P., Leleu, J. B., and Bonaly, R.,** *Mycopath.*, 61, 49, 1977.
101. **Nakajima, T., Tamari, K., Matsuda, K., Tanaka, H., and Ogasawara, N.,** *Agric. Biol. Chem.*, 34, 553, 1970.
102. **Nanba, H. and Kuroda, H.,** *Chem.. Pharm. Bull.*, 19, 252, 1971.
103. **Noguchi, T., Banno, Y., Watanabe, T., Nozawa, Y., and Ito, Y.,** *Mycopath.*, 55, 71, 1975.
104. **Noguchi, T., Kitazima, Y., Nozawa, Y., and Ito, Y.,** *Arch. Biochem., Biophys.*, 146, 506, 1971.
105. **Northcote, D. H. and Horne, R. W.,** *Biochem. J.*, 51, 232, 1952.
106. **Novaes-Ledieu, M. and Garcia-Mendoza, C.,** *J. Gen. Microbiol.*, 61, 335, 1970.
107. **Novaes-Ledieu, M., Jimenez-Martinez, A., and Villanueva, J. R.,** *J. Gen. Microbiol.*, 47, 237, 1967.
108. **Nozawa, Y., Kitajima, Y., and Ito, Y.,** *Biochim. Biophys. Acta*, 307, 92, 1973.
109. **Oliver, G., Pesce de Ruiz Holgado, A. A., Pons, J. J., and Manca, M. C.,** *Sep. Arch. Bioquim. Quim. Farm.*, 17, 91, 1971.
110. **Page, W. J. and Stock, J. J.,** *J. Bacter.*, 119, 44, 1974.
111. **Pao, V. M. and Aronson, J. M.,** *Mycologia*, 62, 531, 1970.
112. **Potgieter, H. J. and Alexander M.,** *Can. J. Microbiol.*, 11, 122, 1965.
113. **Potgieter, H. J. and Alexander, M.,** *J. Bacteriol.*, 91, 1526, 1966.
114. **Power, D. M. and Challinor, S. W.,** *J. Gen. Microbiol.*, 55, 169, 1969.
115. **Previato, J. O., Gorin, P. A. J., and Travassos, L. R.,** *Exper. Mycol.*, 3, 83, 1979.
116. **Reid, I. D. and Bartnicki-Garcia, S.,** *J. Gen. Microbiol.*, 96, 35, 1976.
117. **Reisinger, O. and Bonaly, R.,** *C. R. Acad. Sci.*, 274, 50, 1972.

118. **Rizza, V. and Kornfeld, J. M.,** *J. Gen. Microbiol.*, 58, 307, 1969.
119. **Roelofsen, P. A.,** *Biochim. Biophys. Acta*, 10, 477, 1953.
120. **Roy, I. and Landau, J. W.,** *Can. J. Microbiol.*, 18, 473, 1972.
121. **Ruiz-Herrera, J.,** *Arch. Biochem. Biophys.*, 122, 118, 1967.
122. **Ruiz-Herrera, J. and Osorio, E.,** *Antonie van Leeuwenhoek J. Microbiol. Serol.*, 40, 57, 1974.
123. **Russell, D. W., Sturgeon, R. J., and Ward, V.,** *J. Gen. Microbiol.*, 36, 289, 1964.
124. **Rzueidlo, L., Stachow, A., Nowakowska, A., and Kubica, J.,** *Bull. Acad. Pol. Sci. Ser. Sci. Biol.*, 6, 15, 1958.
125. **San-Blas, G. and Carbonell, L. M.,** *J. Bacteriol.*, 119, 602, 1974.
126. **San-Blas, F., San-Blas, G., and Cova, L. J.,** *J. Gen. Microbiol.*, 93, 209, 1976.
127. **San-Blas, G., San-Blas, F., and Serrano, L. E.,** *Infect. Immun.*, 15, 343, 1977.
128. **Sangar, V. K. and Dugan, P. R.,** *Mycologia*, 65, 421, 1973
129. **Schneider, E. F., Barran, L. R., Wood, P. J., and Siddiqui, I. R.,** *Can. J. Microbiol.*, 23, 763, 1977.
130. **Sentheshanmuganathan, S. and Nickerson, W. J.,** *J. Gen. Microbiol.*, 27, 451, 1962.
131. **Shah, V. K. and Knight, S. G.,** *Arch. Biochem. Biophys.*, 127, 229, 1968.
132. **Sietsma, J. H., Eveleigh, D. E., and Haskins, R. H.,** *Biochim. Biophys. Acta*, 184, 306, 1969.
133. **Sietsma, J. H. and Wessels, J. G. H.,** *Biochim. Biophys. Acta*, 496, 225, 1977.
134. **Sietsma, J. H. and Wouters, J. T. M.,** *Arch. Mikrobiol.*, 79, 263, 1971.
135. **Skujins, J. J., Potgieter, H. J., and Alexander, M.,** *Arch. Biochem. Biophys.*, 11, 358, 1965.
136. **Sturgeon, R. J.,** *Nature (London)*, 209, 204, 1966.
137. **Szaniszlo, P. J., Cooper, B. H., and Voges, H. S.,** *Sabouraudia*, 10, 94, 1972.
138. **Szaniszlo, L. and Mitchell, R.,** *J. Bacteriol.*, 106, 640, 1971.
139. **Temeriusz, A.,** *Rocz. Chem. Ann. Soc. Chim. Polonorum*, 49, 1803, 1975.
140. **Thompson, E. and Colvin, J. R.,** *Can. J. Bacteriol.*, 16, 263, 1970.
141. **Threlfall, R. J.,** *J. Gen. Microbiol.*, 71, 173, 1972.
142. **Toama, M. A. and Raper, K. B.,** *J. Bacteriol.*, 94, 1150, 1967.
143. **Tokunaga, J. and Bartnicki-Garcia, S.,** *Arch. Microbiol.*, 79, 293, 1971.
144. **Touze-Soulet, J. M. and Dargent, R.,** *Can. J. Bot.*, 55, 227, 1977.
145. **Trocha, P., Daly, J. M., and Longenbach, R. J.,** *Plant Physiol.*, 53, 519, 1974.
146. **Trotter, M. J. and Whisler, H. C.,** *Can. J. Bot.*, 43, 869, 1965.
147. **Troy, F. A. and Koffler, H.,** *J. Biol. Chem.*, 244, 5563, 1969.
148. **Unger, P. D. and Hayes, A. W.,** *J. Gen. Microbiol.*, 91, 201, 1975.
149. **Van Laere, A. J., Carlier, A. R., and Van Assche, J. A.,** *Arch. Microbiol.*, 112, 303, 1977.
150. **Wai, N.,** *Bull. Inst. Chem. Acad. Sinica*, 7, 41, 1970.
151. **Wai, N.,** *Bull. Inst. Chem. Acad. Sinica*, 7, 1, 1970.
152. **Wang, M. C. and Bartnicki-Garcia, S.,** *J. Gen. Microbiol.*, 64, 41, 1970.
153. **Wang, C.-S., Schwalb, M. N., and Miles, P. G.,** *Can. J. Microbiol.*, 14, 809, 1968.
154. **Wessels, J. G. H.,** *Wentia*, 13, 1, 1965.
155. **Wrathall, C. R. and Tatum, E. L.,** *J. Gen. Microbiol.*, 78, 139, 1973.
156. **Wu-Yuan, C. D. and Hashimoto, T.,** *J. Bacteriol.*, 129, 1584, 1977.
157. **Yamaguchi, H.,** *J. Gen. Appl. Microbiol.*, 20, 217, 1974.
158. **Yen, P. H. and Ballou, C. E.,** *Biochemistry*, 13, 2420, 1974.
159. **Zonneveld, B. J. M.,** *Biochim. Biophys. Acta*, 249, 506, 1971.

BACTERIAL ENDOTOXIN

R. J. Elin and S. M. Wolff

Bacterial endotoxins are biologically active compounds present in the cell wall of Gram-negative bacteria. Endotoxin-like materials have been found in other organisms; however, this review will be concerned only with the endotoxins of Gram-negative bacteria. Using chemical methods of extraction and purification, bacterial endotoxins can be obtained as relatively pure chemical complexes that consist mainly of lipids and polysaccharides and contain a small amount of amino acids. The chemical name, lipopolysaccharides (LPS), is used synonymously with bacterial endotoxins.

The chemical and biological properties of lipopolysaccharides have been under investigation for more than a century. These investigations have been hampered by two intrinsic problems: (1) different bacterial strains produce chemically dissimilar LPS, which may modify their biological properties, and (2) different extraction and purification methods produce a heterogeneity of LPS preparations. Although endotoxins have a myriad of biological activities that affect all systems of the body, no clear-cut molecular-biological basis for these activities is known.

CELLULAR LOCALIZATION

The term "endotoxin" (LPS) was initially used because it was thought that the toxic substances were present within the bacterial cell and could only be liberated by disruption or digestion of the cell. This is in contrast to classical toxins (exotoxins), which are usually proteins and are excreted by the bacterial cells. It is now known that lipopolysaccharides are located in the outer layer of the cell wall.[299] Electron microscopy of Gram-negative bacteria shows three distinct layers to the cell wall (Figure 1): (1) an inner layer, which is the cytoplasmic membrane; (2) a middle layer, which is composed of a rigid peptidoglycan, and (3) an outer layer containing LPS, phospholipid, and protein.[14, 350, 351] It is postulated that a large portion of the LPS molecule is exposed on the cell surface, since a carbohydrate moiety of the LPS molecule is responsible for the somatic O-antigenic determinants of the bacteria.[350]

The attachment of the LPS to the cell wall is uncertain. It has been shown that LPS are synthesized by the cytoplasmic membrane. The molecules move to the surface of the bacterial cell wall, possibly because of their hydrophilic properties. At the cell wall surface, LPS are probably linked to protein granules that are covalently bound to the peptidoglycan layer. The treatment of Gram-negative bacterial cells with EDTA[224] or with inhibitors of protein synthesis[212, 334] results in the liberation of chemical complexes composed of LPS, phospholipids, and proteins. It is, therefore, postulated that LPS are found in the outer layer of the cell wall of Gram-negative bacteria, complexed to phospholipids and proteins.

CHEMICAL PROPERTIES

Methods of Extraction and Purification

As previously indicated, LPS preparations vary with the methods of extraction and purification. This had led to major difficulties in comparing the work of different investigators who used different methods to obtain LPS. Several methods of extraction and purification of LPS are listed in Table 1.

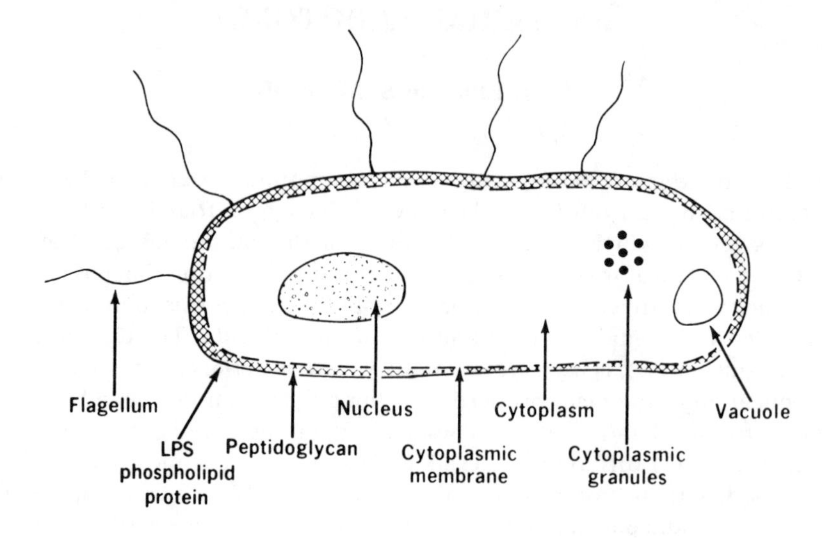

FIGURE 1. Diagrammatic structure of a Gram-negative bacterium, showing the location of LPS in the outer layer of the cell wall.

Table 1
EXTRACTION AND PURIFICATION METHOD FOR LPS

Extraction	Purification	Reference
0.25 *M* trichloroacetic acid at 4°C	Dialysis	37
Diethylene glycol	Alcohol or acetone precipitation	270
2% phenol and autolysis	Filtration + $(NH_4)_2SO_4$ precipitation	264
95% phenol	Precipitation with alcohol	306
2.5 *M* urea	Acetone and alcohol precipitation	397
45% phenol	Precipitation with alcohol	233
Water and ethyl ether	Precipitation with alcohol	325
Detergent (Cetavlon®)	Precipitation with alcohol	298
Dimethyl sulfoxide	Precipitation with acidic acetone	2
Phenol, chloroform, and petroleum ether	Precipitation with water	121
Ethylenediaminetetraacetic acid and tris(hydroxymethyl)aminomethane	Membrane partition chromatography	340

Structure of Endotoxins (LPS)

Considerable work has been done on the structure of LPS, primarily using *Salmonella* mutants. On the basis of these studies, it is possible to divide the LPS molecule into three parts: (1) the lipid moiety, (2) the R-core, and (3) the O-polysaccharide (Figure 2 and Table 2). Chemical analyses of LPS isolated from several different strains of Gram-negative bacteria have shown a consistency of certain organic molecules for the three parts of the LPS molecule (Figures 2 and 3).

Lipid Moiety

Initial studies of the structure of the lipid moiety of the LPS molecule were conducted by Boivin and co-workers.[37] They found that acid hydrolysis of LPS resulted in a phosphorus-containing lipid precipitate, designated as "fraction A" (later called

Table 2
CHEMICAL STRUCTURE OF LPS

Chemical unit	Investigations	References
Lipid Moiety	Structure	3, 4, 212, 312
	Biosynthesis	44, 188, 284, 386
	Binding to cell wall	131
R-Core	Structure	83—85, 131, 132, 149, 153, 169, 293
	Biosynthesis	163, 284, 302
O-Polysaccharide	Structure	14, 87, 132, 153, 170, 171, 231, 232, 298
	Biosynthesis	72, 163, 236, 284, 287, 288, 303—305, 328
	Binding to R-core	285, 286
	Review	232, 233, 236, 328
Protein		409
	General reviews	198, 302, 321, 334, 380, 409, 418

Basic Structure	O-polysaccharide O-polysaccharide O-polysaccharide	R-Core	Lipid Moiety
Function	Antigenic Specificity	Link between polysaccharide and lipid moieties	Biological Activity (Toxicity)
Chemical Composition	1) Hexoses 2) Hexosamines 3) Deoxyhexosamines 4) 6-Deoxyhexoses 5) 3,6-Dideoxyhexoses	1) Hexoses 2) Hexosamine 3) Heptose 4) Octose (acid) 5) O-phosphorylethanolamine 6) Phosphate	1) Fatty acids (C_{10} to C_{22}) 2) Glucosamine 3) Ethanolamine 4) Phosphate
Specific Chemical Components	1) D-Galactose D-Glucose 2) D-Galactosamine D-Glucosamine 3) L and D-Fucosamine D-Viosamine 4) L-Fucose L-Rhamnose 5) Abequose Colitose Paratose	1) Glucose Galactose 2) N-acetylglucosamine 3) L-glycero-D-mannoheptose 4) 2-Keto-3-deoxyoctonate (KDO)	1) 3-hydroxymyristic acid Myristic acid Lauric acid Palmitic acid

FIGURE 2. Function and chemical composition of the three parts of the LPS molecule. The numbers of the specific chemical components are correlated with the numbers in the chemical composition above.

"lipid A"), and a soluble "fraction B", which contained polysaccharides. The term "lipid A" has persisted through the years and refers to all chemical compounds that become insoluble after acid hydrolysis of LPS. Lipid A, therefore, is a mixture of several chemical compounds rather than one chemically defined molecule.

The chemical structure of the lipid moiety of LPS has been evaluated by careful dissection and analysis of the several components of the lipid A precipitate. These studies show that the lipid component of LPS is composed entirely of carboxylic acids (fatty acids), which have a range of chain length from at least C_{10} to C_{22}.[44] Ten differ-

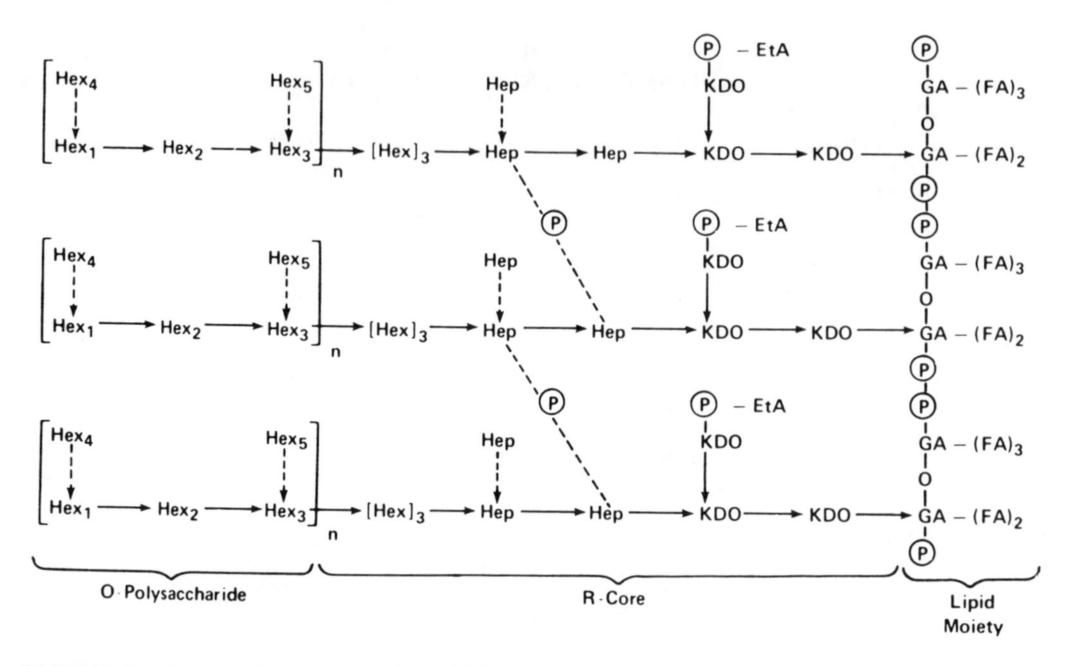

FIGURE 3. A general structure of a LPS molecule based upon the work of many different investigators.[83,85,169,232,233,284,302] Intermittent occurrence of a structure is indicated by the broken lines. Hex = hexose; Hep = heptose; P = phosphate; KDO 13 = 2-keto-3-deoxyoctonate; EtA = ethanolamine; GA = glucosamine; FA = fatty acid.

ent carboxylic acids were found in a mutant of *Salmonella typhimurium;* the major component was β-hydroxymyristic acid, which appears to be unique to the LPS molecule.[232]

One of the basic structural components of the lipid A precipitate is a diglucosamine unit that contains fatty acids attached to amino and hydroxyl groups (Figure 4). Although there is no unanimity of opinion, it appears that the glucosamine units are bound together by 1,6-glycosidic linkages and 1,4-phosphodiester bridges.[3,4,44,131,321]

R-Core

The R-core is the middle portion of the LPS molecule, linking the lipid moiety to the O-polysaccharide moiety. Investigation of the structure of the R-core has been greatly facilitated by the group of mutants termed R or rough forms, which are devoid of O-polysaccharide. These mutants are easily identified by their rough colony morphology, in contrast to the S or smooth forms, which contain the O-polysaccharide moiety.

Chemical analyses of the R-core have shown five basal sugars, phosphate, and O-phosphorylethanolamine.[169,233,284] For *Salmonella*, the five sugars are the following: the amino sugar, N-acetylglucosamine; the hexoses glucose and galactose; the heptose, L-glycero-O-manno heptose; and the octose 2-keto-3-deoxyoctonate (KDO). These basal polysaccharides seem to be common constituents in all *Salmonellae.*[169,302] Thus far, KDO has been found only with LPS (Figure 5).

The lipid moiety of the LPS molecule is covalently bound to the R-core by attachment of KDO (Figures 5 and 6). The O-phosphorylethanolamine is linked to a different KDO unit by its phosphate group. The outermost KDO unit serves as a link to the outer polysaccharide portion of the R-core (Figure 6). Molecular-weight determinations suggest that there is an average of three R-core units, per LPS molecule, which are probably linked by phosphodiester bridges.[418]

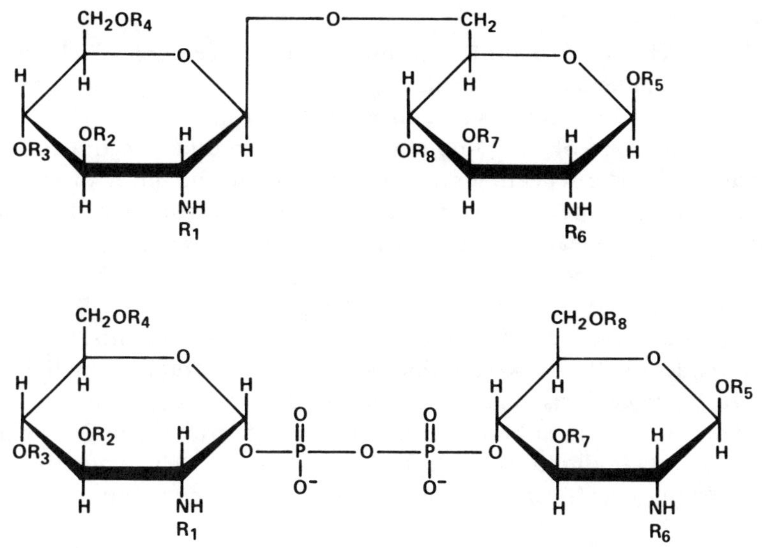

FIGURE 4. Structure of the diglucosamine unit of the lipid moiety of LPS.[3,4,44,131] The β-1,6-glycosidic linkage is shown above, and the 1,4-phosphodiester bridge is shown below. The R positions are the sites of attachment for carboxylic acid, KDO, a glycosidic linkage, or a phosphodiester bridge.

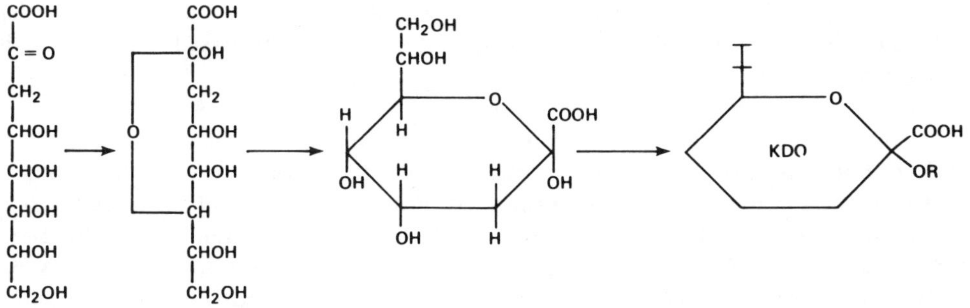

FIGURE 5. The structure of 2-keto-3-deoxyoctonate (KDO). This eight-carbon α-ketonic acid is present in the R-core of the LPS molecule as a six-membered ring.

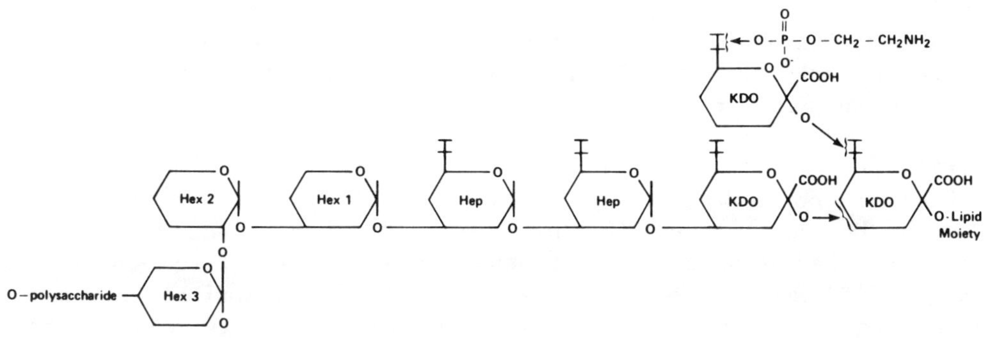

FIGURE 6. General structure of the R-core of the LPS molecules.[169,233,284,302] KDO = 2-keto-3-deoxyoctonate; Hep = heptose; Hex = hexose.

O-Polysaccharide

The O-polysaccharide structure determines the O-antigenic specificity of the bacteria and their LPS. Members of a particular serogroup have at least one O-antigenic determinant in common with members of a different serogroup.

Biochemical analyses of the O-polysaccharide have established a regular sequence of the sugars.[170,171] The repeating sequences may either be linear or may contain one or more side branches of the component sugars (see Figure 3). The length and the composition of these repeating units vary from strain to strain.[328]

Amino Acids

All LPS preparations contain amino acids in varying amounts, depending on the method of extraction. The lowest amino acid content is obtained by the phenol-water extraction or by proteolysis. The importance of amino acids to the structure and biological activity of LPS is unknown. Furthermore, it is uncertain whether the amino acids are covalently bound to the LPS molecule or are merely contaminants. It is possible that the amino acids serve to bind the LPS to the peptidoglycan layer of the cell wall.[409]

ASSAY METHODS FOR LPS

Several methods have been devised to assay for LPS in tissues and body fluids (Table 3). Most of these procedures are based upon their toxicity in vivo, — i.e., the ability of LPS to produce fever, abortion, dermal necrosis, or death. Recently, in vitro methods of assay for LPS have been introduced, which appear to offer greater precision and sensitivity. The *Limulus* method for the detection of LPS, based upon the in vitro reaction between LPS and a lysate prepared from amoebocytes, the only circulating blood cell of the *Limulus* crab, seems to have great potential for the future because of its simplicity and sensitivity.[227,228,382] However, since the biological activities of each LPS are not a uniform gravimetric property, but vary, it may be virtually impossible for any method to measure LPS quantitatively with any meaningful accuracy.[414]

The *Limulus* test has been used clinically to detect endotoxin in blood,[90,100,109,227,243,381] urine,[196] and cerebrospinal fluid.[277] In addition, this test has been found to be useful for routine pyrogen screening of short-lived radiopharmaceuticals and parenteral solutions.[89,197,245] However, the specificity of the test to detect only endotoxin has been questioned.[93,408]

BIOLOGICAL PROPERTIES OF LPS

The biological and pathological effects of LPS in humans and in experimental animals are multiple and diverse. These biological changes are briefly summarized in the following paragraphs and in Table 4.

Fever

The fever produced by LPS may occur by direct LPS action on the thermoregulatory center in the brain or, more likely, by indirect LPS action on blood leukocytes, on reticuloendothelial cells, or perhaps on other tissues, to liberate an endogenous pyrogen.[79,330,412] The evidence for the direct action of LPS on the brain is based upon studies in which the injection of LPS into the central nervous system resulted in the abrupt onset of fever after 3 minutes.[22,276,394] After an intravenous injection of LPS, there is a lag period of 10 to 20 minutes in the rabbit, and of 1 to 2 hours in man, before the temperature elevation begins. The length of the lag period is dose-dependent. During

Table 3
ASSAY METHODS FOR ENDOTOXIN

Method	End point of assay	Sensitivity, μg/mℓ	References
Chick embryo lethality	Lethality	0.01	267, 367
Intradermal injection in rabbits	Dermal inflammation	0.15	222
Fever in rabbits	0.5°C rise in temperature	10^{-3}	204, 267, 414
Actinomycin-treated mice	Lethality	1	82, 311
Adrenalectomized mice	Lethality	0.5	54
Hepatic enzyme induction	Inhibition of enzyme induction	1	30
Macrophage migration	Inhibition of migration	0.1	168
Rabbit epinephrine skin test	Dermal inflammation	1	248, 249
Hypoferremia	Decrease in serum iron concentration	0.1	12
Limulus lysate	Gelation	10^{-6}	66, 226—228, 320, 329, 382, 423

Table 4
BIOLOGICAL PROPERTIES OF LPS

LPS reaction	Type of study	References
Fever	Mechanism	22, 45, 79, 99, 276, 292, 354, 394, 420
	Relationship to endogenous pyrogen	10, 11, 79, 80, 152, 330
	Tolerance	15, 79, 273, 309, 398
	Measurement	70, 204, 414, 417
	Review	8, 9, 20, 21, 371, 412
Effect on blood	Leukocytes	64, 129, 239, 272, 354, 416
	Leukocytosis	35, 156, 239
	Leukopenia	81, 272
	Leukopoiesis	156, 317, 370
	Leukotaxis	42, 117, 347
	Leukotoxic effects	129, 300, 347, 407
	Other leukocyte changes	117, 133, 300, 347, 407
	Erythrocytes	63, 118, 119
	Platelets	77, 78, 187, 251, 374
	Coagulation	68, 158, 227, 253
	Mechanism	97, 424
	Disseminated intravascular coagulation (DIC)	69, 97, 116, 125, 158, 271, 424
	Complement	32, 104, 110, 122, 127, 215, 260, 261, 275
Effect on the endocrine system	Pituitary	49, 115, 208, 216, 235, 268, 383, 402, 403
	Adrenal	98, 208, 259, 268, 383, 403, 420
	Reproduction-abortion	52, 254, 326, 361
Effect on metabolism	Carbohydrates	31, 154, 193, 352, 353
	Lipids	123, 124, 183, 225, 227
	Proteins and enzymes	29, 269, 356, 364, 372, 422
	Minerals	
	Iron	12, 92, 199—201
	Zinc	308
Effect on the reticuloendothelial system (RES)	Description	6, 17, 19, 135, 144, 165, 392
	Mechanism	136, 137, 166, 167
	Morphologic changes	164, 167

Table 4 (continued)
BIOLOGICAL PROPERTIES OF LPS

LPS reaction	Type of study	References
	Distribution of injected LPS	38, 55, 151, 172, 338
Effect on the vascular system	Shock	
	Mechanism	18, 173, 215, 331, 391, 396
	Therapy	28, 103, 265, 315, 366, 387, 399
	Experimental models	23, 217
	Primate	50, 51, 176, 178, 252, 282
	Dog	43, 375, 393
	Rat	120
	Biochemical mediators	50, 175, 191, 214, 258, 331
	Catecholamines	139, 375, 421
	Histamine	74, 140, 178, 185, 281, 312
	Kinins	48, 95, 209, 282, 364
	Serotonin	74, 75
	Steroids	120, 181, 265, 387
	Hemodynamic changes	34, 41, 50, 57, 58, 128, 182, 217, 230
	Heart	33, 43, 60, 174, 176, 177, 373, 388
	Capillaries	59, 159, 160, 250, 251
	Lung	43, 159, 160
	Liver	34, 102, 180, 258, 290
	Other organs	160, 179, 190, 359
	Pathological changes	175, 251, 252
	Reviews	61, 128, 175, 181, 230, 427
Immunological phenomena	Immunogenicity of LPS	56, 150, 221, 233, 264, 266, 270, 295
	Description	332, 333, 341, 406, 413
	Effect on endotoxicity	206, 207, 256, 318, 327, 384
	Immune response	111, 255, 280, 310
	Humoral	105, 106, 150, 155, 219
	Cellular	88, 105, 108, 130, 147, 219
	Adjuvant effect	46, 111, 112, 134, 255, 404
	Immunosuppression	108, 111, 112, 134, 255, 404
	Thymus	46, 126, 335
	Lymphocytes	126, 155, 255
	Reviews	194, 278
Detoxification	Biological	70, 203, 240, 362, 365, 392
	Plasma	53, 211, 220, 343, 344, 348, 349, 425, 426
	Leukocytes	53, 102
	Reticuloendothelial system	102, 362, 365, 392
	Immunochemical	206, 207, 297, 318
	Chemical	241, 242, 279, 291, 294, 295, 322—324
	Alkali	63, 192, 240, 279, 289
	Acids	161, 324
	Enzymes	205, 345
	Reduction of particles size	161, 244, 301, 323, 324
	Physical	342
Endotoxin tolerance	Description	15—17, 309, 398
	Mechanism	17, 47, 54, 134, 141, 144—146, 148, 273, 309, 339, 415
	Clinical Correlation	1, 142, 143, 246, 339

Table 4 (Continued)
BIOLOGICAL PROPERTIES OF LPS

LPS reaction	Type of study	References
Nonspecific resistance to infection	Description	24, 39, 65
	Bacteria	26, 86, 247, 263, 336
	Fungi	91, 210, 213, 419
	Parasites	234, 360
	Viruses	314, 346, 395
	Mechanism	65, 91, 272, 297, 337, 405
	Interferon	184, 314, 346, 379
	Comparison with other compounds	25, 96, 401
	Reviews	39, 65, 337, 355
Shwartzman phenomenon	Description	
	Localized	357, 358
	Generalized	186, 223, 363, 390
	Mechanism	67, 68, 138, 186, 223, 238, 253, 274, 377, 389
	Clinical correlation	68, 237, 274, 319, 363, 376, 385
Pathology	Cytotoxicity	27, 40, 76, 113, 164, 167
	Lymphocytes	5, 255
	Liver	5, 76, 101, 145, 180, 183, 229, 378
	Other organs	114, 251, 255
	Effect on tumor growth	162, 189
Protection against irradiation	Description	157, 257, 369
	Mechanism	119, 368, 370
General reviews	Books	198, 202, 218, 400, 411
	Journals	296

this period, LPS is rapidly removed from the blood by phagocytic cells. These cells are activated by LPS to release a fever-producing substance, which has been named "endogenous pyrogen".[417] The activation process requires new protein and RNA synthesis.[292] It has been recently discovered that at least two human phagocytic blood cells — i.e., granulocytes and monocytes — produce chemically distinct pyrogens.[80] These low-molecular-weight proteins, which are heat-labile, then act on the anterior hypothalamus to produce fever.[9] The fever is produced by alteration of vasomoter control of either heat redistribution or heat production.[45] Man is the most sensitive of all animals studied to the pyrogenic effects of endotoxin on a weight-for-weight basis.[410]

Effect on Blood

Leukocytes

The intravenous administration of LPS to humans and to experimental animals produces transient leukopenia followed by marked granulocytosis. Depth and length of the leukopenia are dose-dependent.[354] The leukopenia has been shown to result predominantly from a decrease in granulocytes, due to an increased margination of circulating granulocytes to the endothelium of capillaries.[272,416] The granulocytosis seen after LPS stimulation is dependent on an adequate bone marrow reserve of granulocytes and on a normal granulocyte-releasing mechanism.[239] The release of granulocytes may be controlled by a humoral factor, which is increased by LPS injection[351] These findings have been used clinically for the estimation of bone marrow granulocyte reserves.[36,71,107] In addition, LPS injection increase the production of a colony-stimulating factor, which may regulate granulocytopoiesis.[317] The development of tolerance to the effects of endotoxin has been shown for both the granulocyte-releasing and the colony-stimulating factor in mice,[316] but not in man.[416] In vitro studies have shown that LPS in the presence of complement are ingested by, or adhere to, the human granulocyte and induce a series of metabolic changes resembling those associated with phagocytosis.[64]

Erythrocytes

Studies in mice have shown that a single injection of LPS results in a significant shift in erythropoiesis from the bone marrow to the spleen.[118] The leukocytic hyperplasia of the bone marrow following LPS administration may cause this shift in erythropoiesis.[119]

Platelets

Platelet aggregation in vivo following LPS injection has been demonstrated in rats and in rabbits.[251,377] After aggregation, degranulation of platelets occurs, which results in an increased activity of platelet factor 3.[187] Electron microscopy investigations have shown that LPS adheres to platelets, causing them to aggregate, fuse, and fragment.[374] A platelet—endotoxin interaction seems to be necessary for the final detoxification of intravascular endotoxin by the RES.[73]

Disseminated Intravascular Coagulation (DIC)

One of the major complications of Gram-negative septicemia in man is the frequent occurrence of DIC.[424] The DIC is considered to be secondary to the LPS of the bacteria, since there is experimental evidence that LPS may initiate DIC by two different pathways. The first is the interaction between platelets and LPS liberates platelet factor 3, which in the presence of calcium and factor V acts on activated factor X to form blood thromboplastin, which in turn transforms prothrombin to thrombin.[187] Secondly, LPS can activate directly factor XII (Hageman factor), the initial component of the intrinsic clotting system.[271] The activation occurs by the formation of a complex between the lipid moiety of LPS and factor XII.[271] The importance of the coagulation sequence in LPS-induced DIC and subsequent mortality is unclear,[116,125] since in patients with Gram-negative bacteremia and DIC correction of the DIC with heparin did not decrease the mortality rate.[69]

Complement

LPS has been shown to activate the complement system via the alternate pathway, as demonstrated by consumption of the terminal complement components (C′3—C′9) and by sparing of the early complement components (C′1, C′4, and C′2).[110,260] However, recent studies suggest that LPS is capable of activating both the classical and the alternate complement pathway.[104] The biologically active products generated due to the activation of the complement system by LPS effect changes in the clotting system, in vascular permeability, in smooth-muscle reactivity, and in neutrophil chemotaxis.[260] Indeed, several of the biological changes induced by endotoxin may be mediated by activated components of the complement system.

Effect on the Endocrine System

LPS have a selective effect on the hypothalamic-pituitary system, causing the release of adrenocorticotropic hormone (ACTH) and of growth hormone (GH), but having no influence on thyrotropin (TSH) and on luteinizing-hormone (LH) release.[49,115,216] The site of LPS action on the hypothalamic-pituitary axis has not been settled. There is evidence that LPS act upon or through the neuronal elements of the hypothalamus to cause the release of ACTH by the pituitary gland.[268] On the other hand, studies show LPS can by-pass the medial hypothalamus and act directly upon the pituitary gland to release ACTH.[235,402] Also, there is a circadian rhythm in the pituitary response after LPS injection.[383]

A significant elevation of plasma cortisol can be demonstrated 2 hours after LPS injection.[98,208,259] Fever and the associated nonspecific stress do not appear to be the stimuli for the release of ACTH by LPS.[208,420]

Effects on Metabolism

Carbohydrates

The injection of LPS into experimental animals produces initial hyperglycemia, which is followed by a prompt decrease in blood sugar to hypoglycemic levels and concomitant decrease in liver glycogen.[31,193] In the mouse, the hypoglycemia has been shown to be of sufficient magnitude to cause death.[352] The mechanism of the hypoglycemia appears to be inhibition of the synthesis of glucose from noncarbohydrate sources.[353]

Lipids

The injection of LPS into rabbits produces hyperlipidemia, which is characterized by an early transitory rise in plasma free fatty acids and, at 24 hours, by an elevation of serum cholesterol, serum phospholipids, and plasma triglycerides.[183] Patients with Gram-negative infections have been shown to have a marked elevation of their plasma triglycerides or free fatty acids.[123]

Proteins and Enyzmes

Protein synthesis by the liver is stimulated by LPS injection.[356] However, the influence of LPS on specific enzyme activities in the liver indicates a selective effect, as increases and decreases have been demonstrated.[269,372] LPS injection causes an increase in several serum enzymes (lactic dehydrogenase, isocitric dehydrogenase, transaminases, and creatine phosphokinase) secondary to tissue damage.[364] Also, rats given repeated injections of endotoxin develop amyloidosis.[13]

Minerals

Hypoferremia proportional to the dose of LPS has been demonstrated in man and in rats, with the maximum decrease occurring 8 to 10 hours after injection.[92,199] The total iron-binding capacity of the serum also decreases following LPS administration.[200] Endogenous pyrogen liberated by LPS may in part mediate the hypoferremia, since endogenous pyrogen from rabbit polymorphonuclear leukocytes lowers the serum iron concentration of endotoxin-tolerant rats.[201] In addition, serum zinc concentrations decrease significantly in a dose-dependent response after LPS injection in the rat.[308]

Effect on the Reticuloendothelial System (RES)

The RES was first implicated in the host response to LPS by Beeson in 1947.[17] Although a variety of studies have supported this basic hypothesis, they also have indicated that the relationship between host RES activity and LPS toxicity is complex.[19,392] In general, LPS are potent stimulators of RES activity, as evidenced by enhanced blood clearance of colloidal carbon.[136,166] Such alterations in the activity of the RES have, in turn, been related to many of the biological responses of LPS, such as the enhancement of nonspecific resistance,[136] the adjuvant effects of LPS,[136] the development of endotoxin tolerance,[17] and biological detoxification of these substances. The importance of the RES in LPS metabolism is made evident by the fact that the major portion of an injected dose of LPS can be traced to the Kupffer cells of the liver and to the spleen.[38]

Effect on the Vascular System

The complex interaction of LPS with the cardiovascular system and the production of shock have been intensively investigated since the syndrome of endotoxin shock in man was first described in 1951.[396] The slow progress in the understanding of the pathogenesis and of the hemodynamic and biochemical events of endotoxin shock is indicated by the still appallingly high mortality rate of this condition in man.

The syndrome of endotoxin shock begins with Gram-negative septicemia and resultant endotoxemia. The LPS reacts with leukocytes, platelets, complement system, and other serum proteins to increase blood levels of proteolytic enzymes and of certain vasoactive substances, such as histamine, kinins, and serotonin.[175] This results in pooling of the blood, primarily in the pulmonary and splanchnic vasculature.[230] Cardiac output is diminished by the direct myocardial depressant action of LPS[60] and by a decrease in venous return secondary to the pooling of the blood. The decrease in cardiac output initiates a circular series of events, beginning with a drop in blood pressure and peripheral perfusion. Endogenous catecholamines are liberated to compensate for the falling blood pressure; however, the arteriolar-venular constriction produced by the catecholamines leads to tissue acidosis and anoxia.[331] This initiates capillary congestion and dilatation, which completes the circle by accentuating the decreased venous return and cardiac output. In time, the cellular hypoxia results in lysosomal disruption, leading to cell lysis, propagation of tissue injury, and death.[61]

A concomitant feature of this syndrome is disseminated intravascular coagulation (DIC). Whether DIC is a cause or an effect of endotoxin shock is unknown, but the localized or generalized hemorrhagic diathesis due to the consumption of clotting factors certainly intensifies the shock condition.

Current therapy for endotoxin shock attempts to disrupt the circle of events mentioned above by restoring and maintaining the integrity of the microcirculation. This treatment regimen begins with massive fluid administration and plasma expanders. Although of unproven benefit, β-adrenegic stimulators, α-adrenergic blockage, and large doses of corticosteroids are frequently administered.[62,366] In most cases the DIC can be reversed with small doses of heparin.[103,315] In man, the most effective therapy is the use of specific antibiotic treatment of the underlying infection.

Immunological Phenomena

In 1927, White[406] established that the somatic polysaccharides of LPS on the surface of Gram-negative bacteria confer on these organisms their characteristic serologic specificities. This was followed in the early 1930s by the pioneer work of Boivin and his colleagues,[37] who extracted and isolated these LPS and showed that the same complex carried both antigenic and toxic attributes. It has not been clearly established that the repeating oligosaccharide units of the O-polysaccharide portion of the LPS molecule represent, in part or completely, the antigenic determinant group.[233] The grouping of *Salmonella* by the Kauffmann-White schema depends on the O-polysaccharide antigen. In general, the O-polysaccharide of LPS is a potent antigen that can elicit antibody formation in animals in submicrogram quantities.[341]

The immune response to LPS in all species studied is quite characteristic.[333] At 3 to 4 days after intravenous injection of LPS, circulating antibodies to that LPS are measurable, reaching their maxmum concentration at 6 to 8 days.[332] Although the initial antibody response is predominantly IgM, IgG, and IgA, antibodies are present within a few days.[332]

Another interesting property of LPS is mitogenicity. The existence of two main lymphocyte populations — thymus-derived (T) and thymus-independent (B) — is well established. The mitogenic response to LPS appears to be carried out by B cells only.[126] This response is not mediated by immunologic mechanisms.[307]

Extensive studies have unequivocally documented the adjuvant effect of LPS. Adjuvancy has been shown by increase of titers of circulating antibodies, increase in the number of antibodies, increase in the number of antibody-producing cells, accelerated antigen eliminations, shortening of the lag period of the immune response, and/or longer persistence of circulating antibodies.[278] This adjuvant effect has been demonstrated in several animal species, including mouse,[111] rat,[310] rabbit,[169] guinea pig,[278] and chicken.[278] On the other hand, LPS can act as an immunosuppressant by manipulation of the dose and time of injection relative to giving the antigen.[255]

Detoxification

Several methods are known by which the toxic effects of LPS can be altered. These range from biological observations through chemical and physiochemical changes in the LPS molecule. The detoxifying properties of human serum for LPS have been studied in the greatest detail.[220] However, there is experimental evidence for the following two mechanisms of detoxification: (1) enzymatic degradation of the LPS molecule into nontoxic fragments,[220,349,426] and (2) reversible complex formation between LPS and serum proteins, which effects loss of toxicity.[343,344] In addition, acid and alkaline hydrolysis and reduction of LPS particle size result in detoxification.[296]

Endotoxin Tolerance

Since the classical experiments of Beeson[15] in the rabbit, it has been known that administration of LPS results in a progressive increase in the resistance to some of its biological effects. This phenomenon, termed endotoxin tolerance, can be readily induced by single or repetitive doses. Tolerance to the pyrogenic and lethal properties of LPS has been most widely studied.

The precise mechanism underlying the development of endotoxin tolerance is unknown. Early experiments suggested that enhanced RES activity effected more rapid clearance of subsequent LPS injections, thereby producing tolerance.[17] However, other studies have shown that endotoxin tolerance persists following RES blockade,[141,414] that tolerance develops in man without a change in RES activity,[144,339] that passive transfer of tolerance readily occurs,[146,147] and that stimulation of the RES with zymosan or BCG (bacille Calmette Guérin) fails to induce a state of tolerance.[309] Another study has suggested that a 19S γ-globulin fraction of serum is responsible for tolerance.[398] Still other investigations have failed to demonstrate a specific antibody as the cause of endotoxin tolerance.[273,415] Recent evidence indicates that hepatic mechanisms dominate the second phase of the biphasic febrile response to LPS and that the tolerant state may be an inability of the Kupffer cells of the liver to produce or release endogenous pyrogen.[145] The relevance of endotoxin tolerance to human disease is equivocal.[1,142,143,246,339]

Nonspecific Resistance to Infection

LPS administration can alter resistance to infection by bacteria,[26,86,247,263,336] fungi,[91,210,213,419] viruses,[314,346,395] and parasites[234,360] in several animal species. This change in susceptibility to infection is characterized by a transient increased susceptibility (so-called "negative phase") followed by a prolonged increase in nonspecific resistance to infection. The mechanism of the increased resistance to infection secondary to LPS injection is unclear; however, enhanced macrophage and reticuloendothelial-system activity,[6] granulocytosis,[65] and changes in iron metabolism[91,94] may be factors.

Shwartzman Phenomenon

A local tissue reactivity was described in 1928 by Shwartzman,[357] who observed that, if a single injection of a culture filtrate of *Salmonella typhosa* was made into the skin of a rabbit and the same material was injected intravenously 24 hours later, a severe hemorrhagic, necrotic lesion developed at the site of the original skin injection. The reaction was nonspecific, since the filtrate used for the preparatory skin injection showed the same reaction after intravenous injection of the filtrate from another organism. The component of the culture filtrate that produced the localized Shwartzman reaction was later shown to be LPS.

The classical generalized Shwartzman reaction (GSR) requires two properly spaced (usually 24 hours apart) intravenous injections of LPS into a rabbit. This results in bilateral renal cortical necrosis, due to fibrin deposition secondary to diffuse intravascular clotting. A precipitous drop in the leukocytes and platelets is essential for the reaction.[186] An attempt to define the mechanism of this phenomenon has implicated granulocyte products as the clot-promoting factor;[186] other studies implicate stimulation of α-adrenergic receptor sites in the microcirculation as the cause of the formation of fibrin thrombi.[223,238] Heparin effectively blocks the GSR,[186] but the mechanism is unclear.[274] Since there are striking similarities between the pathologic abnormalities of the GSR in rabbits and certain clinical conditions, it has been suggested that certain human diseases are correlates of the animal model.[237,319,376,385]

Pathology

The injection of a single dose of LPS causes profound depletion of the lymphoid elements of the spleen and lymph nodes, which persists for several days.[5,255] If an antigen is given at the time of the LPS injection, regeneration of the lymphoid tissue is evident after 24 hours.[255]

In the liver, LPS causes vesiculation of hepatocytes with progressive necrosis in the centrilobular zone and congestion of the central veins and sinusoids.[229] In addition, there is progressive increase in the number, size, and variety of lysosomes, depletion of glycogen, swelling of the Kupffer cells, and fibrin deposition in the sinusoids.[76] In the lung there is hemorrhage into the alveolar wall with clumping of polymorphonuclear leukocytes in the pulmonary capillaries.[76] Examination of the myocardium by electron microscopy shows marked distortion of the intercalated discs.[76]

REFERENCES

1. **Abernathy, R. S. and Spink, W. W.**, Studies with *Brucella* endotoxin in humans: The significance of susceptibility to endotoxin in the pathogenesis of brucellosis, *J. Clin. Invest.*, 37, 219, 1958.
2. **Adams, G. A.**, Extraction of lipopolysaccharides from Gram-negative bacteria with dimethyl sulfoxide, *Can. J. Biochem.*, 45, 422, 1967.
3. **Adams, G. A., and Singh, P. P.**, The chemical constitution of lipid A from *Serratia marcescens*, *Can. J. Biochem.*, 48, 55, 1970.
4. **Adams, G. A. and Singh, P. P.**, Structural features of lipid A preparations isolated from *Escherichia coli* and *Shigella flexneri*, *Biochim. Biophys. Acta*, 202, 553, 1970.
5. **André-Schwartz, J., Rubenstein, H. S., and Coons, A. H.**, Electron microscopy of cellular responses following immunization with endotoxin, *Am. J. Pathol.*, 53, 331, 1968.

6. **Arredondo, M. I. and Kampschmidt, R. F.,** Effect of endotoxin on phagocytic activity of the reticuloendothelial system of the rat, *Proc. Soc. Exp. Biol. Med.*, 112, 78, 1963.
7. **Athens, J. W., Haab, O. P., Raab, S. O., Mauer, A. M., Ashenbrucker, H., Cartwright, G. E., and Wintrobe, M. M.,** Leukokinetic studies. IV. The total blood, circulating and marginal granulocyte pools and the granulocyte turnover rate in normal subjects, *J. Clin. Invest.*, 40, 989, 1961.
8. **Atkins, E. A.,** Pathogenesis of fever, *Physiol. Rev.*, 40, 580, 1960.
9. **Atkins, E. A. and Bodel, P.,** Fever, *N. Engl. J. Med.*, 286, 27, 1972.
10. **Atkins, E. A. and Wood, W. B., Jr.,** Studies on the pathogenesis of fever. I. The presence of transferable pyrogen in the blood stream following injection of typhoid vaccine, *J. Exp. Med.*, 101, 519, 1955.
11. **Atkins, E. A. and Wood, W. B., Jr.,** Studies on the pathogenesis of fever. II. Identification of an endogenous pyrogen in the blood stream following the injection of typhoid vaccine, *J. Exp. Med.*, 102, 499, 1955.
12. **Baker, P. J., and Wilson, J. B.,** Hypoferremia in mice and its application to the bioassay of endotoxin, *J. Bacteriol.* 90, 903, 1965.
13. **Barth, W. F., Willerson, J. T., Asofsky, R., Sheagren, J. H., and Wolff, S. M.,** Experimental murine amyloid. III. Amyloidosis induced with endotoxins, *Arthritis Rheum.*, 12, 615, 1969.
14. **Bayer, M. and Anderson, T. F.,** The surface structure of *Escherichia coli, Proc. Natl. Acad. Sci. U.S.A.*, 54, 1592, 1965.
15. **Beeson, P. B.,** Development of tolerance to typhoid bacterial pyrogen and its abolition by reticuloendothelial blockade, *Proc. Soc. Exp. Biol. Med.*, 61, 248, 1946.
16. **Beeson, P. B.,** Tolerance to bacterial pyrogens, I. Factors influencing its development, *J. Exp. Med.*, 86, 29, 1947.
17. **Beeson, P. B.,** Tolerance to bacterial pyrogens. II. Role of the reticuloendothelial system, *J. Exp. Med.*, 86, 39, 1947.
18. **Bell, M. L., Herman, A. H., Smith, E. E., Egdahl, R. H., and Rutenburg, A. M.,** Role of lysosomal instability in the development of refractory shock, *Surgery*, 70, 341, 1971.
19. **Benacerraf, B. and Sebestyen, M. M.,** Effect of bacterial endotoxins on the reticuloendothelial system, *Fed. Proc.*, 16, 860, 1957.
20. **Bennett, I. L., Jr.,** Pathogenesis of fever, *Bull. N.Y. Acad. Med.*, 37, 440, 1961.
21. **Bennett, I. L., Jr. and Beeson, P. B.,** Pathogenesis of fever, *Med. Sci. Law*, 29, 365, 1950.
22. **Bennett, I. L., Jr., Petersdorf, R. G., and Keene, W. R.,** Pathogenesis of fever: Evidence for direct cerebral action of bacterial endotoxins, *Trans. Assoc. Am. Physicians*, 70, 64, 1957.
23. **Berczi, I., Bertok, L., and Bereznay, T.,** Comparative studies on the toxicity of *Escherichia coli* lipopolysaccharide endotoxin in various animal species, *Can. J. Microbiol.*, 12, 1070, 1966.
24. **Berger, F. M.,** The effect of endotoxin on resistance to infection and disease, *Adv. Pharmacol.*, 5, 19, 1967.
25. **Berger, F. M., Fukui, G. M., Gustafson, R. H., and Rosselet, J. P.,** Studies on the mechanism of protodyne-induced protection against microbial infections, *Proc. Soc. Exp. Biol. Med.*, 138, 391, 1971.
26. **Berger, F. M., Fukui, G. M., Ludwig, B. J., and Rosselet, J. P.,** Increased host resistance to infection elicited by lipopolysaccharides from *brucella abortus, Proc. Soc. Exp. Biol. Med.*, 131, 1376, 1969.
27. **Bergman, S. and Nilsson, S. B.,** Effect of endotoxin on embryonal chick fibroblasts cultured in monolayer, *Acta Pathol. Microbiol. Scand.*, 59, 161, 1963.
28. **Berk, J. L., Hagen, J. F., Beyer, W. H., Gerber, M. J., and Dochat, G. R.,** The treatment of endotoxin shock by beta adrenergic blockade, *Ann. Surg.* 169, 74, 1969.
29. **Berry, L. J. and Smythe, D. S.,** Effects of bacterial endotoxins on metabolism. VII. Enzyme induction and cortisone protection, *J. Exp. Med.*, 120, 721, 1964.
30. **Berry, L. J., Smythe, D. S. and Colwell, L. S.,** Inhibition of hepatic enzyme induction as a sensitive assay for endotoxin, *J. Bacteriol.*, 96, 1191, 1968.
31. **Berry, L. J., Smythe, D. S., and Young, L. G.,** Effects of bacterial endotoxin on metabolism. I. Carbohydrate depletion and the protective role of cortisone, *J. Exp. Med.*, 110, 389, 1959.
32. **Bladen, H. A., Gewurz, H., and Mergenhagen, S. E.,** Interactions of the complement system with the surface and endotoxin lipopolysaccharide of *Veillonella alcalescens J. Exp. Med.*, 125, 767, 1967.
33. **Blattberg, B. and Levy, M. N.,** Nature of bradycardia evoked by bacterial endotoxin, *Am. J. Physiol.*, 216, 249, 1969.
34. **Blattberg, B. and Levy, M. N.,** Early hepatic and extraheptic pooling in response to endotoxin, *Am. J. Physiol.*, 219, 460, 1970.
35. **Boggs, D. R.,** Chervenick, P. A., Marsh, J. C., Cartwright, G. E., and Wintrobe, M. M., Neutrophil-releasing activity in plasma of dogs injected with endotoxin, *J. Lab. Clin. Med.* , 72, 177, 1968.

36. **Boggs, D. R., Marsh, J. C., Chervenick, P. A., Cartwright, G. E., and Wintrobe, M. M.,** Neutrophil-releasing activity in plasma of normal human subjects injected with endotoxin, *Proc. Soc. Exp. Biol. Med.,* 127, 689, 1968.
37. **Boivin, A., Mesrobeanu, I., and Mesrobeanu, L.,** Technique pour la préparation des polysaccharides microbiens spécifiques, *C.R. Soc. Biol.* 113, 490, 1933.
38. **Braude, A. I.,** Absorption, distribution and elimination of endotoxins and their derivatives, in *Bacterial Endotoxins,* Landy, M. and Braun, W., Eds., Rutgers University Press, New Brunswick, N.J., 1964, p. 98.
39. **Braude, A. I. and Siemienski, J.,** The influence of endotoxin on resistance to infection, *Bull. N.Y. Acad. Med.,* 37, 448, 1961.
40. **Braun, W. and Kessel, R. W. I.,** Cytotoxicity of endotoxins in relation to their effects on host resistance, in *Bacterial Endotoxins,* Landy, M. and Braun, W., Eds., Rutgers University Press, New Brunswick, N.J., 1964, p. 397.
41. **Brockman, S. K., Thomas, C. S., and Vasko, J. S.,** The effect of *Escherichia coli* endotoxin on the circulation, *Surg. Gynecol. Obstet. Int. Abstr. Surg.,* 125, 763, 1967.
42. **Bryant, R. E., Des Prez, R. M., and Rogers, D. E.,** Studies on human leukocyte mobility. II. Effects of bacterial endotoxin on leukocyte migration, adhesiveness, and aggregation, *Yale J. Biol. Med.,* 40, 192, 1968.
43. **Burch, G. E., Giles, T. D., Quiroz, A. C., and Shen, Y.,** Influence of *E. coli* endotoxin on the pulmonary veins, circulatory system and work and tone of the heart of intact dog, *Cardiologia,* 53, 77, 1968.
44. **Burton, A. J. and Carter, H. E.,** Purification and characterization of the lipid A component of the lipopolysaccharides from *Escherichia coli, Biochemistry,* 3, 411, 1964.
45. **Buskirk, E. R., Thompson, R. H., Rubenstein, M., and Wolff, S. M.,** Heat exchange in men and women following intravenous injection of endotoxin, *J. Appl. Physiol.,* 19, 907, 1964.
46. **Campbell, P. A., Rowlands, D. T., Jr., Harrington, M. J., and Kind, P. D.,** The adjuvant action of endotoxin in thymectomized mice, *J. Immunol.,* 96, 849, 1966.
47. **Carey, F. J., Braude, A. I., and Zalesky, M.,** Studies with radioactive endotoxin. III. The effect of tolerance on the distribution of radioactivity after intravenous injection of *Escherichia coli* endotoxin labeled with Cr^{51}, *J. Clin. Invest.,* 37, 441, 1958.
48. **Carretero, O. A., Nasjletti, A., and Fasciolo, J. C.,** Kinins and kininogen in endotoxin shock, *Experientia,* 26, 63, 1970.
49. **Carroll, B. J., Pearson, M. J., and Martin, F. I. R.,** Evaluation of three acute tests of hypothalamic-pituitary-adrenal function, *Metabolism,* 18, 476, 1969.
50. **Cavanagh, D. and Rao, P. S.,** Endotoxin shock in the subhuman primate. I. Hemodynamic and biochemical changes, *Arch. Surg.,* 99, 107, 1969.
51. **Cavanagh, D., Rao, S. P., Sutt, D. M., Bhagat, B. D., and Bachmann, F.,** Pathophysiology of endotoxin shock in the primate, *Am. J. Obstet. Gynecol.,* 108, 705, 1970.
52. **Chedid, L., Boyer, F., and Parant, M.,** Etude de l'action abortive des endotoxines injectées a la souris gravide normale, castrée ou hypophysectomisée, *Ann. Inst. Pasteur Paris,* 102, 77, 1962.
53. **Chedid, L., Parant, M., Boyer, F., and Skarnes, R. C.,** Nonspecific host responses in tolerance to the lethal effect of endotoxins, in *Bacterial Endotoxins,* Landy, M. and Braun, W., Eds., Rutgers University Press, New Brunswick, N.J., 1964, p. 500.
54. **Chedid, L., Lamensans, A., and Prixova, J.,** Comparison of the effects of an antibacterial leukocytic extract and the serum endotoxin-detoxifying component of lipopolysaccharides extracted from rough and smooth *Salmonella, J. Infect. Dis.,* 121, 634, 1970.
55. **Chedid, L., Skarnes, R. C., and Parant, M.** Characterization of a ^{51}Cr-labeled endotoxin and its identification in plasma and urine after parenteral administration, *J. Exp. Med.,* 117, 561, 1963.
56. **Chernokhvostova, E., Luxemburg, K. I., Starshinova, V., Andreeva, N., and German, G.,** Study on the production of IgG, IgA, and IgM antibodies to somatic antigens of *Salmonella typhi* in humans, *Clin. Exp. Immunol.,* 4, 407, 1969.
57. **Chien, S., Chang, C., Dellenback, J., Usami, S., and Gregsen, M. I.,** Hemodynamic changes in endotoxin shock, *Am. J. Physiol.,* 210, 1401, 1966.
58. **Chien, S., Dellenback, J., Usami, S., Treitel, K., Chang, C., and Gregsen, M. I.,** Blood volume and its distribution in endotoxin shock, *Am. J. Physiol.,* 210, 1411, 1966.
59. **Chien, S., Sinclair, D. G., Dellenback, R. J., Chang, C., Peric, B., Usami, S., and Gregsen, M. I.,** Effect of endotoxin on capillary permeability to macromolecules, *Am. J. Physiol.,* 207, 518, 1964.
60. **Cho, Y. W.,** Direct cardiac action of *E. coli* endotoxin, *Proc. Soc. Exp. Biol. Med.,* 141, 705, 1972.
61. **Christy, J. H.,** Pathophysiology of Gram-negative shock, *Am. Heart J.,* 81, 694, 1971.
62. **Christy, J. H.,** Treatment of Gram-negative shock, *Am. J. Med.,* 50, 77, 1971.
63. **Ciznar, I. and Shands, J. W., Jr.,** Effect of alkali-treated lipopolysaccharide on erythrocyte membrane stability, *Infect. Immun.,* 4, 362, 1971.

64. **Cline, M. J., Melmon, K. L., David, W. C., and Williams, H. E.,** Mechanism of endotoxin interaction with human leukocytes, *Br. J. Haematol.*, 15, 539, 1968.
65. **Cluff, L. E.,** Effects of endotoxins on susceptibility to infections, *J. Infect. Dis.*, 122, 205, 1970.
66. **Cooper, J. F., Levin, J., and Wagner, H. N., Jr.,** Quantitative comparison of in vitro and in vivo methods for the detection of endotoxin, *J. Lab. Clin. Med.*, 78, 138, 1971.
67. **Corrigan, J. J., Jr.,** Effect of anticoagulating and non-anticoagulating concentrations of heparin on the generalized Shwartzman reaction, *Thromb. Diath. Haemorrh.*, 24, 136, 1970.
68. **Corrigan, J. J., Jr., Abildgaard, C. F., Vanderheiden, J. F., and Schulman, I.,** Quantitative aspects of blood coagulation in the generalized Shwartzman reaction. I. Effects of variation of preparation and provocative dose of *E. coli* endotoxin, *Pediatr. Res.*, 1, 39, 1967.
69. **Corrigan, J. J., Jr. and Jordan, C. M.,** Heparin therapy in septicemia with disseminated intravascular coagulation, *N. Engl. J. Med.*, 283, 778, 1970.
70. **Cundy, K. R. and Nowotny, A.,** Quantitative comparison of toxicity parameters of bacterial endotoxins, *Proc. Soc. Exp. Biol. Med.*, 127, 999, 1968.
71. **Dale, D. C. and Wolff, S. M.,** Studies of the neutropenia of acute malaria, *Blood*, 41, 197, 1973.
72. **Dankert, M. A., Wright, A., Kelley, W. S., and Robbins, P. W.,** Isolation, purification and properties of the lipid-linked intermediates of O-antigen biosynthesis, *Arch. Biochem. Biophys.*, 116, 425, 1966.
73. **Das, J., Schwartz, A. A., and Folkman, J.,** Clearance of endotoxin by platelets: Role in increasing the accuracy of the *Limulus* gelatin test and in combating experimental endotoxemia, *Surgery*, 74, 235, 1973.
74. **Davis, R. B., Bailey, W. L., and Hanson, N. P.,** Modification of serotinin and histamine release after *E. coli* endotoxin administration, *Am. J. Physiol.*, 205, 560, 1963.
75. **Davis, R. B., Meeker, W. R., Jr., and Bailey, W. L.,** Serotonin release by bacterial endotoxin, *Proc. Soc. Exp. Biol. Med.*, 108, 774, 1961.
76. **Depalma, R. G., Coil, J., David, J. H., and Holden, W. D.,** Cellular and ultrastructural changes in endotoxemia: A light and electron microscopic study, *Surgery*, 62, 505, 1967.
77. **Des Prez, R. M.,** Effects of bacterial endotoxin on rabbit platelets. III. Comparison of platelet injury induced by thrombin and by endotoxin, *J. Exp. Med.*, 120, 305, 1964.
78. **Des Prez, R. M. and Bryant, R. E.,** Effects of bacterial endotoxin on rabbit platelets. IV. The divalent ion requirements of endotoxin induced and immunologically induced platelet injury, *J. Exp. Med.*, 124, 971, 1966.
79. **Dinarello, C. A., Bodel, P. T., and Atkins, E.,** The role of the liver in the production of fever and in pyrogenic tolerance, *Trans. Assoc. Am. Physicians*, 81, 334, 1968.
80. **Dinarello, C. A., Goldin, N. P., and Wolff, S. M.,** Demonstration and characterization of two distinct human leukocytic pyrogens, *J. Exp. Med.*, 139, 1369, 1974.
81. **Donald, W. D., Winkler, C. H., and Hare, K.,** Investigations on the mechanisms of the leukopenic response to *Shigella* endotoxin, *J. Clin. Invest.*, 37, 1100, 1958.
82. **Dowling, J. N., and Feldman, H. A.,** Quantitative biological assay of bacterial endotoxins, *Proc. Soc. Exp. Biol. Med.*, 134, 861, 1970.
83. **Dröge, W., Lehmann, V., Lüderitz, O., and Westphal, O.,** Structural investigations on KDO region of lipopolysaccharides, *Eur. J. Biochem.*, 14, 175, 1970.
84. **Dröge, W., Lüderitz, O., and Westphal, O.,** Biochemical studies on lipopolysaccharides of *Salmonella* R mutants. III. The linkage of the heptose units, *Eur. J. Biochem.*, 4, 126, 1968.
85. **Dröge, W., Ruschmann, E., Lüderitz, O., and Westphal, O.,** Biochemical studies on lipopolysaccharides of *Salmonella* R mutants. IV. Phosphate groups linked to heptose units and their absence in some R lipopolysaccharides, *Eur. J. Biochem.*, 4, 134, 1968.
86. **Dubos, R. J., Schaedler, R. W., and Bohome, D.,** Effects of bacterial endotoxins on susceptibility to infection with Gram-positive and acid-fast bacteria, *Fed. Proc.*, 16, 856, 1957.
87. **Edstrom, R. D. and Heath, E. C.,** The biosynthesis of cell wall lipopolysaccharide in *Escherichia coli*. VI. Enzymatic transfer of galactose, glucose, N-acetyl-glucosamine and colitose into the polymer, *J. Biol. Chem.*, 242, 3581, 1967.
88. **Elekes, E., Meretey, K., and Kocsar, L.,** The action of aluminium hydroxide or endotoxin on natural sheep haemolysin producing cells in rats, *Pathol. Microbiol.*, 32, 345, 1968.
89. **Elin, R. J., Lundberg, W. B., and Schmidt, P. J.,** Evaluation of bacterial contamination in blood processing, *Transfusion*, 15, 260, 1975.
90. **Elin, R. J., Robinson, R. A., Levine, A. S., and Wolff, S. M.,** Lack of clinical usefulness of the *Limulus* test in the diagnosis of endotoxemia, *N. Engl. J. Med.*, 293, 521, 1975.
91. **Elin, R. J. and Wolff, S. M.,** Iron and endotoxin induced nonspecific resistance to infection, *Fed. Proc.*, 31, 802, 1972.
92. **Elin, R. J. and Wolff, S. M.,** Effect of fever on serum iron in man, *Clin. Res.*, 21, 598, 1973.

93. **Elin, R. J. and Wolff, S. M.**, Nonspecificity of the *Limulus* amebocyte lysate test: Positive reactions with polynucleotides and proteins, *J. Infect. Dis.*, 128, 349, 1973.
94. **Elin, R. J. and Wolff, S. M.**, The role of iron in nonspecific resistance to infection induced by endotoxin, *J. Immunol.*, 112, 737, 1974.
95. **Erdos, E. G. and Miwa, I.**, Effect of endotoxin shock on the plasma kallikrein-kinin system of the rabbit, *Fed. Proc.*, 27, 92, 1968.
96. **Erlondson, A. L., Jr.**, The induction of nonspecific resistance in mice by papain, *J. Infect. Dis.*, 116, 297, 1966.
97. **Evensen, S. A., Jeremic, M., and Hjort, P. F.**, The effect of endotoxin on factor VII in rats: in vivo and in vitro observations, *Scand. J. Clin. Lab. Invest.*, 18, 509, 1966.
98. **Farmer, T. A., Jr., Hill, S. R., Jr., Pittman, J. A., Jr., and Herod, J. W., Jr.**, The plasma 17-hydroxycorticosteroid response to corticotrophin, SU-4885, and lipopolysaccharide pyrogen, *J. Clin. Endocrinol. Metab.*, 21, 433, 1961.
99. **Fekety, R. F.**, Heat balance and reactivity to endotoxin, *Am. J. Physiol.*, 204, 719, 1963.
100. **Feldman, S. and Pearson, T. A.**, The *Limulus* test and Gram-negative bacillary sepsis, *Am. J. Dis. Child.*, 128, 173, 1974.
101. **Filkins, J. P.**, Hepatic vascular response to endotoxin, *Proc. Soc. Exp. Biol. Med.*, 131, 1235, 1969.
102. **Filkins, J. P.**, Comparison of endotoxin detoxification by leukocytes and macrophages, *Proc. Soc. Exp. Biol. Med.*, 137, 1396, 1971.
103. **Filkins, J. P. and DiLuzio, N. R.**, Heparin protection in endotoxin shock, *Am. J. Physiol.*, 214, 1074, 1968.
104. **Fine, D. P.**, Activation of the classic and alternate complement pathways by endotoxin, *J. Immunol.*, 112, 763, 1974.
105. **Finger, H., Beneke, G., and Fresenius, H.**, Cellular kinetics of 19S and 7S hemolysin production in mice under the influence of bacterial endotoxins, *Pathol. Microbiol.*, 35, 324, 1970.
106. **Finger, H., Emmerling, P., and Busse, M.**, Increased priming for the secondary response in mice to sheep erythrocytes by bacterial endotoxins, *Int. Arch. Allergy Appl. Immunol.*, 38, 598, 1970.
107. **Fink, M. E. and Calabresi, P.**, The granulocyte response to an endotoxin (pyrexal) as a measure of functional marrow reserve in cancer chemotherapy, *Ann. Intern. Med.*, 57, 732, 1962.
108. **Floersheim, G. L.**, Suppression of cellular immunity by Gram-negative bacteria, *Antibiot. Chemother.*, 15, 407, 1969.
109. **Fossard, D. P., Kakkar, V. V., and Elsey, P. A.**, Assessment of *Limulus* test for detecting endotoxaemia, *Br. Med. J.*, 2, 465, 1974.
110. **Frank, M. M., May, J. E., and Kane, M. A.**, Contributions of the classical and alternate complement pathways to the biological effects of endotoxin, *J. Infect. Dis.*, 128 (Suppl.), 176, (1973).
111. **Franzl, R. E. and McMaster, P. D.**, The primary immune response in mice. I. The enhancement and suppression of hemolysin production by a bacterial endotoxin, *J. Exp. Med.* , 127, 1087, 1968.
112. **Freedman, H. H., Fox, A. E., and Schwartz, B. S.**, Antibody formation at various times after previous treatment of mice with endotoxin, *Proc. Soc. Exp. Biol. Med.*, 125, 583, 1967.
113. **Fritz, H. and Nordenfelt, E.**, Endotoxin-induced cytotoxicity of rabbit serum, *Acta Pathol. Microbiol. Scand.*, 75, 631, 1969.
114. **Frohlich, E. D.**, Effect of *Salmonella typhosa* endotoxin on perfused dog spleen, *Proc. Soc. Exp. Biol. Med.*, 113, 559, 1963.
115. **Frohman, L. A., Horton, E. S., and Lebovitz, H. E.**, Growth hormone releasing action of a *Pseudomonas* endotoxin (piromen.), *Metabolism*, 16, 57, 1967.
116. **From, A. H. L., Spink, W. W., Knight, D., and Gewurz, H.**, Significance of intravascular coagulation in canine endotoxin shock, *Infect. Immun.*, 11, 1010, 1975.
117. **Fruhman, G. J.**, Mobilization of neutrophils into peritoneal fluid following intraperitoneal injection of bacterial endotoxins, *Proc. Soc. Exp. Biol. Med.*, 102, 423, 1959.
118. **Fruhman, G. J.**, Bacterial endotoxin: effects on erythropoiesis, *Blood*, 27, 363, 1966.
119. **Fruhman, G. J.**, Endotoxin-induced shunting of erythropoiesis in mice, *Am. J. Physiol.*, 212, 1095, 1967.
120. **Fukuda, T. and Hata, N.**, Mechanisms of endotoxin shock in rats and the anti-endotoxic effect of glucocorticoids and endotoxin-conditioning, *Jpn. J. Physiol.*, 19, 509, 1969.
121. **Galanos, C., Lüderitz, O., and Westphal, O., A new method for the extraction of R lipopolysac-**charides, *Eur. J. Biochem.* , 9, 245, 1969.
122. **Galanos, C., Rietschel, E. T., Lüderitz, O., and Westphal, O.**, Interaction of lipopolysaccharides and lipid A with complement, *Eur. J. Biochem.*, 19, 143, 1971.
123. **Gallin, J. I., Kaye, D., and O'Leary, W. M.**, Serum lipids in infection, *N. Engl. J. Med.*, 281, 1081, 1969.
124. **Gallin, J. I., O'Leary, W. M., and Kaye, D.**, Serum concentrations of lipids in rabbits infected with *Escherichia coli* and *Staphylococcus aureus*, *Proc. Soc. Exp. Biol. Med.*, 133, 309, 1970.

125. **Garner, R. and Evensen, S. A.,** Endotoxin-induced intravascular coagulation and shock in dogs: The role of factor VII, *Br. J. Haematol.*, 27, 655, 1974.
126. **Gery, I., Krüger, J., and Spiesel, S. Z.,** Stimulation of B-lymphocytes by endotoxin. Reactions of thymus-deprived mice and karyotypic analysis of dividing cells in mice bearing T_6 T_6 thymus grafts, *J. Immunol.*, 108, 1088, 1972.
127. **Gewurz, H., Mergenhagen, S. E., Nowotny, A., and Philips, J. K.,** Interactions of the complement system with native and chemically modified endotoxins, *J. Bacteriol.*, 95, 397, 1968.
128. **Gilbert, R. P.,** Mechanisms of the hemodynamic effects of endotoxin, *Physiol. Rev.*, 40, 245, 1960.
129. **Gimber, P. E. and Rafter, G. W.,** The interaction of *Escherichia coli* endotoxin with leukocytes, *Arch. Biochem. Biophys*, 135, 14, 1969.
130. **Gingold, J. L. and Freedman, H. H.,** Inhibition by endotoxin of the migration of peritoneal exudate cells from endotoxin-sensitive mice, *Proc. Soc. Exp. Biol. Med.*, 128, 599, 1968.
131. **Gmeiner, J., Lüderitz, O., and Westphal, O.,** Biochemical studies on lipopolysaccharides of *Salmonella* R mutants. VI. Investigations on the structure of the lipid A component, *Eur. J. Biochem.*, 7, 370, 1969.
132. **Gmeiner, J., Simon, M., and Lüderitz, O.,** The linkage of phosphate groups and of 2-keto-3-deoxyoctonate to the lipid A component in a *Salmonella minnesota* lipopolysaccharide, *Eur. J. Biochem.*, 21, 355, 1971.
133. **Goldstein, I. M., Wünschmann, B., Astrup, T., and Henderson, E. S.,** Effects of bacterial endotoxin on the fibrinolytic activity of normal human leukocytes, *Blood*, 37, 447, 1971.
134. **Golub, E. S., and Weigle, W. O.,** Studies on the induction of immunologic unresponsiveness. I. Effects of endotoxin and phytohemagglutinin, *J. Immunol.*, 98, 1241, 1967.
135. **Golub, S., Groschel, D. H. M., and Nowotny, A.,** RES uptake of endotoxin in mice, *Bacteriol. Proc.*, p. 79, 1967.
136. **Golub, S., Groschel, D. H. M., and Nowotny, A.,** Factors which affect the reticuloendothelial system uptake of bacterial endotoxins, *J. Reticuloendothel. Soc.*, 5, 324, 1968.
137. **Golub, S., Groschel, D. H. M., and Nowotny, A.,** Studies on the opsonization of a bacterial endotoxoid, *J. Reticuloendothel. Soc.*, 7, 518, 1970.
138. **Good, R. A., and Thomas, L.,** Studies on the generalized Shwartzman reaction. IV. Prevention of the local and generalized Shwartzman reactions with heparin, *J. Exp. Med.*, 97, 871, 1953.
139. **Gourzis, J. T., Hollenberg, M W., and Nickerson, M.,** Involvement of adrenergic factors in the effects of bacterial endotoxin, *J. Exp. Med.*, 114, 593, 1961.
140. **Greisman, S. E.,** Activation of histamine-releasing factor in normal rat plasma by *E. coli* endotoxin, *Proc. Soc. Exp. Biol. Med.*, 103, 628, 1960.
141. **Greisman, S. E., Carozza, F. A., Jr., and Hill, J. D.,** Mechanisms of endotoxin tolerance. I. Relationship between tolerance and reticuloendothelial system phagocytic activity in the rabbit, *J. Exp. Med.*, 117, 663, 1963.
142. **Greisman, S. E., Hornick, R. B., Carozza, F. A., Jr., and Woodward, T. E.,** The role of endotoxin during typhoid fever and tularemia in man. I. Acquisition of tolerance to endotoxin, *J. Clin. Invest.*, 42, 1064, 1963.
143. **Greisman, S. E., Hornick, R. B., and Woodward, T. E.,** The role of endotoxin during typhoid fever and tularemia in man. III. Hyperreactivity to endotoxin during infection, *J. Clin. Invest.*, 43, 1747, 1964.
144. **Greisman, S. E., Wagner, H. N., Iio, M., and Hornick, R. B.,** Mechanisms of endotoxin tolerance. Relationship between endotoxin tolerance and reticuloendothelial system phagocytic activity in man, *J. Exp. Med.*, 119, 241, 1964.
145. **Greisman, S. E. and Woodward, C. L.,** Mechanisms of endotoxin tolerance. VII. The role of the liver, *J. Immunol.*, 105, 1468, 1970.
146. **Greisman, S. E. and Woodward, W. E.,** Mechanisms of endotoxin tolerance. III. The refractory state during continuous intravenous infusions of endotoxin, *J. Exp. Med.*, 121, 911, 1965.
147. **Greisman, S. E. and Young, E. J.,** Mechanisms of endotoxin tolerance. VI. Transfer of the "anamnestic" tolerance response with primed spleen cells, *J. Immunol.*, 103, 1237, 1969.
148. **Greisman, S. E., Young, E. J., and Carozza, F. A., Jr.,** Mechanisms of endotoxin tolerance V. Specificity of the early and late phases of pyrogenic tolerance, *J. Immunol.*, 103, 1223, 1969.
149. **Grollman, A. P. and Osborn, M. J.,** O-Phosphorylethanolamine: A component of lipopolysaccharide in certain gram-negative bacteria, *Biochemistry*, 3, 1571, 1964.
150. **Gupta, J. D. and Reed, C. E.,** The direct reaction between *Salmonella enteritidis* endotoxin and antibody — measurement of 7S and 19S antibody in normal, tolerant, and immune sera, *J. Immunol.*, 98, 1093, 1967.
151. **Gupta, J. D. and Reed, C. E.,** Distribution and degradation of sublethal doses of I^{125} labeled endotoxin from *Salmonella enteritidis* in mice, *Proc. Soc. Exp. Biol. Med.*, 131, 481, 1969.

152. **Hahn, H., Char, D. C., Postel, W. B., and Wood, W. B.**, Studies on the pathogenesis of fever. XV. The production of endogenous pyrogen by peritoneal macrophages, *J. Exp. Med.*, 126, 385, 1967.
153. **Hammerling, G., Lüderitz, O., and Westphal, O.**, Structural investigations of the core polysaccharide of *Salmonella typhimurium* mode of attachment to the O-specific chains, *Eur. J. Biochem.*, 15, 48, 1970.
154. **Hamosh, M. and Shapiro, B.**, The mechanism of glycogenolytic action of endotoxin, *Br. J. Exp. Pathol.*, 41, 372, 1960.
155. **Han, S. S., Johnson, A. G., and Han, I. H.**, The antibody response in the rat. I. A histometric study of the spleen following a single injection of bovine gamma globulin with and without endotoxin, *J. Infect. Dis.*, 115, 149, 1965.
156. **Handler, E. S., Varsa, E. E., and Gordon, A. S.**, Mechanisms of leukocyte production and release. V. Studies on the leukotosis-inducing factor in the plasma of rats treated with typhoid-paratyphoid vaccine, *J. Lab. Clin. Med.*, 67, 398, 1966.
157. **Harcourt, K. F., Robertson, R. D., and Fletcher, W. S.**, Effect of pretreatment with endotoxin on the response of skin to irradiation in the rabbit, *Cancer*, 21, 812, 1968.
158. **Hardaway, R. M. and Johnson, D.**, Clotting mechanism in endotoxin shock, *Arch. Intern. Med.*, 112, 775, 1963.
159. **Harrison, L. H., Beller, J. J., Hinshaw, L. B., Coalson, J. J., and Greenfield, L. J.**, Effects of endotoxin on pulmonary capillary permeability, ultra-structure and surfactant, *Surg. Gynecol. Obstet.*, 129, 723, 1969.
160. **Harrison, L. H., Jr., Hinshaw, L. B., Coalson, J. J., and Greenfield, L. J.**, Effects of *E. coli* septic shock on pulmonary hemodynamics and capillary permeability, *J. Thorac. Cardiovasc. Surg.*, 61, 795. 1971.
161. **Haskins, W. T., Landy, M., Milner, K. C., and Ribi, E.**, Biological properties of parent endotoxins and lipoid fractions with a kinetic study of acid-hydrolyzed endotoxin, *J. Exp. Med.*, 114, 665, 1961.
162. **Havas, H. F., Groesbeck, M. E., and Donnelly, A. J.**, Mixed bacterial toxins in the treatment of tumors. I. Methods of preparation and effects on normal and sarcoma 37 bearing mice, *Cancer Res.*, 18, 141, 1958.
163. **Heath, E. C., Mayer, R. M., Edstrom, R. D., and Beaudreau, C. A.**, Structure and biosynthesis of the cell wall lipopolysaccharide of *Escherichia coli, Ann. N.Y. Acad. Sci.*, 133, 315, 1971.
164. **Heilman, D. H.**, Cellular aspects of the action of endotoxin: The role of the macrophage, in *Bacterial Endotoxins,* Landy, M. and Braun, W., Eds., Rutgers University Press, New Brunswick, N.J., 1964, p. 610.
165. **Heilman, D. H.**, In vitro studies on changes in the reticuloendothelial system of rabbits after an injection of endotoxin, *J. Reticuloendothel. Soc.*, 2, 89, 1965.
166. **Heilman, D. H.**, Effect of dosage on endotoxin-induced changes in the reticuloendothelial system of rabbits, *J. Reticuloendothel. Soc.*, 2, 273, 1965.
Int. Arch. Allergy Appl. Immunol., 26, 63, 1965.
168. **Heilman, D. H. and Bast, R. C., Jr.**, In vitro assay of endotoxin by the inhibition of macrophage migration, *J. Bacteriol.*, 93, 15, 1967.
169. **Hellerqvist, C. G. and Lindberg, A. A.**, Structural studies on the common core polysaccharide of the cell wall lipopolysaccharide from *Salmonella typhimurium, carbohydr. Res.*, 16, 39, 1971.
170. **Hellerqvist, C. G., Lindberg, B., Svensson, S., Holme, T., and Lindberg, A. A.**, Structural studies on the O-specific side-chains of the cell wall lipopolysaccharide from *Salmonella typhimurium* 395 MS, *Carbohydr. Res.*, 8, 43, 1968.
171. **Hellerqvist, C. G., Lindberg, B., Svensson, S., Holme, T., and Lindberg, A. A.**, Structural studies on the O-specific side-chains of the cell wall lipopolysaccharide from *Salmonella typhimurium* LT2, *Carbohydr. Res.*, 9, 237, 1969.
172. **Herring, W. B., Herion, J. C., Walker, R. I., and Palmer, J. G.**, Distribution and clearance of circulating endotoxin, *J. Clin. Invest.*, 42, 79, 1963.
173. **Hildebrand, G. J., Ng, J., Seys, Y., and Madin, S. H.**, Differential between pathogenic mechanisms of early and late phase of endotoxin shock, *Am. J. Physiol.*, 210, 1451, 1966.
174. **Hinshaw, L. B., Archer, L. T., Greenfield, L. J., and Guenter, C. A.**, Effects of endotoxin on myocardial hemodynamics, performance, and metabolism, *Am. J. Physiol.*, 221, 504, 1971.
175. **Hinshaw, L. B., Brake, C. M., and Emerson, T. E., Jr.**, Biochemical and pathologic alterations in endotoxin shock, in *Shock and Hypotension: Pathogenesis and Treatment,* Mills, L. J. and Moyer, J. H., Eds., Grune and Stratton, New York, 1965, p. 431.
176. **Hinshaw, L. B., Emerson, T. E., Jr., and Reins, D. A.**, Cardiovascular responses of the primate in endotoxin shock, *Am. J. Physiol.*, 210, 335, 1966.

177. **Hinshaw, L. B., Greenfield, L. J., Archer, L. T., and Guenter, C. A.,** Effects of endotoxin on myocardial hemodynamics, performance, and metabolism during beta adrenergic blockade, *Proc. Soc. Exp. Biol. Med.,* 137, 1217, 1971.
178. **Hinshaw, L. B., Jordan, M. M., and Vick, J. A.,** Histamine release and endotoxin shock in the primate, *J. Clin. Invest.* , 40, 1631, 1961.
179. **Hinshaw, L. B. and Nelson, D. L.,** Venous response of intestine to endotoxin, *Am. J. Physiol.,* 203, 870, 1962.
180. **Hinshaw, L. B., Reins, D. A., and Hill, R. J.,** Response of isolated liver to endotoxin, *Can. J. Physiol. Pharmacol.,* 44, 529, 1966.
181. **Hinshaw, L. B., Solomon, L. A., Freeny, P. C., and Reins, D. A.,** Endotoxin shock, *Arch. Surg.,* 94, 61, 1967.
182. **Hinshaw, L. B., Vick, J. A., Jordan, M. M., and Wittmers, L. E.,** Vascular changes associated with development of irreversible endotoxin shock, *Am. J. Pathol.,* 202, 103, 1962.
183. **Hirsch, R. L., MacKay, D. G., and Travers, A. I.,** Hyperlipidemia, fatty liver, and bromsulfophthalein retention in rabbits injected intravenousy with bacterial endotoxins, *J. Lipid Res.,* 5, 563, 1964.
184. **Ho, M.,** Interferon-like viral inhibitor in rabbits after intravenous administration of endotoxin, *Science,* 146, 1472, 1964.
185. **Hook, W. A., Snyderman, R., and Mergenhagen, S. E.,** Histamine-releasing factor generated by the interaction of endotoxin with hamster serum, *Infect. Immun.,* 2, 462, 1970.
186. **Horn, R. G. and Collins, R. D.,** Studies on the pathogenesis of the generalized Shwartzman reaction. The role of granulocytes, *Lab. Invest.,* 18, 101, 1968.
187. **Horowitz, H. I., Des Prez, R. M., and Hook, E. W.,** Effect on bacterial endotoxin on rabbit platelets factor 3 activity in vitro and in vivo, *J. Exp. Med.,* 116, 619, 1962.
188. **Humphreys, G. O. and Meadow, P. M.,** The biosynthesis of lipid A in *Pseudomonas, J. Gen. Microbiol.,* 68, 5, 1971.
189. **Ikawa, M., Koepfli, J. B., Mudd, S. G., and Niemann, C.,** An agent from *E. coli* causing hemorrhage and regression of an experimental mouse tumor. I. Isolation and properties, *J. Natl. Cancer Inst.,* 13, 157, 1952.
190. **Jacobson, E. D., Dooley, E. S., Scott, J. B., and Frohlich, E. D.,** Effects of endotoxin on the hemodynamics of the stomach, *J. Clin. Invest.,* 42, 391, 1963.
191. **Jacobson, E. D., Mehlman, B., and Kalas, J. P.,** Vasoactive mediators as the "trigger-mechanism" of endotoxin shock, *J. Clin. Invest.,* 43, 1000, 1964.
192. **Jarvis, F. G., Mesenko, M. T., Martin, D. G., and Perrine, T. D.,** Physicochemical properties of the V_i antigen before and after mild alkaline hydrolysis, *J. Bacteriol.,* 94, 1406, 1967.
193. **Jeffries, C. D.,** Liver carbohydrate levels in mice treated with endotoxin, cortisone, and elipten, *Proc. Soc. Exp. Biol. Med.,* 132, 540, 1969.
194. **Johnson, A. G.,** The adjuvant action of bacterial endotoxins on the primary antibody response, in *Bacterial Endotoxins,* Landy, M. and Braun, W., Eds., Rutgers University Press, New Brunswick, N.J., 1964, p. 252.
195. **Johnson, A. G., Gaines, S., and Landy, M.,** Studies on the O-antigen of *Salmonella typhosa.* V. Enhancement of antibody response to protein antigens by the purified lipopolysaccharide, *J. Exp. Med.,* 103, 225, 1956.
bacteriuria by use of the *Limulus* endotoxin assay, *Appl. Microbiol.,* 26, 38, 1973.
197. **Jorgensen, J. H. and Smith, R. F.,** Rapid detection of contaminated intravenous fluids using the *Limulus* in vitro endotoxin assay, *Appl. Microbiol.,* 26, 521, 1973.
198. **Kadis, S., Weinbaum, G., and Ajl, S. J.,** (Eds.), *Bacterial endotoxins, in Microbial Toxins,* Vol. 5, Academic Press, New York, 1971.
199. **Kampschmidt, R. F. and Schultz, G. A.,** Hypoferremia in rats following injection of bacterial endotoxin, *Proc. Soc. Exp. Biol. Med.,* 106, 870, 1961.
200. **Kampschmidt, R. F. and Upchurch, H. F.,** Effect of endotoxin upon total iron-binding capacity of the serum, *Proc. Soc. Exp. Biol. Med.,* 116, 420, 1964.
201. **Kampschmidt, R. F. and Upchurch, H. F.,** A comparison of the effects of rabbit endogenous pyrogen on the body temperature of the rabbit and lowering of plasma iron in the rat, *Proc. Soc. Exp. Biol. Med.,* 133, 128, 1970.
202. **Kass, E. H. and Wolff, S. M. (Eds.),** *Bacterial Lipopolysaccharides,* University of Chicago Press, Chicago, 1973.
203. **Keene, W. R.,** Detoxification of bacterial endotoxin by soluble tissue extracts, *J. Lab. Clin. Med.,* 60, 433, 1962.
204. **Keene, W. R., Silberman, H. S., and Landy, M.,** Observations on the pyrogenic response and its application to the bioassay of endotoxin, *J. Clin. Invest.,* 40, 295, 1961.

205. **Kim, Y. B and Watson, D. W.**, Inactivation of Gram-negative bacterial edotoxins by papain, *Proc. Soc. Exp. Biol. Med.*, 115, 140, 1964.
206. **Kim, Y. B. and Watson, D. W.**, Modification of host responses to bacterial endotoxins. II. Passive transfer of immunity to bacterial endotoxin with fractions containing 19S antibodies, *J. Exp. Med.*, 121, 751, 1965.
207. **Kim, Y. B. and Watson, D. W.**, Role of antibodies in reaction to Gram-negative bacterial endotoxins, *Ann. N.Y. Acad. Sci.*, 133, 727, 1966.
208. **Kimball, H. R., Lipsett, M. B., Odell, W. D., and Wolff, S. M.**, Comparison of the effect of the pyrogens, etiocholanolone and bacterial endotoxin on plasma cortisol and growth hormone in man, *J. Clin. Endocrinol. Metab.*, 28, 337, 1968.
209. **Kimball, H. R., Melmon, K. L., and Wolff, S. M.**, Endotoxin-induced kinin production in man, *Proc. Soc. Exp. Biol. Med.*, 139, 1078, 1972.
210. **Kimball, H. R., Williams, T. W., and Wolff, S. M.**, Effect of bacterial endotoxin on experimental fungal infections, *J. Immunol.*, 100, 24, 1968.
211. **Kimball, H. R. and Wolff, S. M.**, Febrile responses of rabbits to bacterial endotoxin following incubation in homologous serum and plasma, *Proc. Soc. Exp. Biol. Med.*, 124, 269, 1967.
212. **Knox, K. W., Cullen, J., and Work, E.**, An extracellular lipopolysaccharide-phospholipid-protein complex produced by *Escherichia coli* grown under lysine-limiting conditions, *Biochem. J.*, 103, 192, 1967.
213. **Kobayashi, H., Yasuhira, K., and Uesaka, I.**, Effect of *Escherichia coli* and its endotoxin on the resistance of mice to experimental cryptococcal infection, *Jpn. J. Microbiol.*, 13, 223, 1969.
214. **Kobold, E. E., Lovell, R., Katz, W., and Thal, A. P.**, Chemical mediators released by endotoxin, *Surg. Gynecol. Obstet.*, 118, 807, 1964.
215. **Kohler, P. F. and Spink, W. W.**, Complement in endotoxin shock: Effect of decomplementation by aggregated gamma globulin, *Proc. Soc. Exp. Biol. Med.*, 117, 207, 1964.
216. **Kohler, P. O., O'Malley, B. W., Rayford, P. L., Lipsett, M. B., and Odell, W. D.**, Effect of pyrogen on blood levels of pituitary tropic hormones: Observations of the usefulness of the growth hormone response in the detection of pituitary disease, *J. Clin. Endocrinol. Metab.*, 27, 219, 1967.
217. **Kuida, H., Gilbert, R. P., Hinshaw, L. B., Brunson, J. G., and Visscher, M. B.**, Species differences in effect of Gram-negative endotoxin on circulation, *Am. J. Physiol.*, 200, 1197, 1961.
218. **Landy, M. and Braun, W., (Eds.)**, *Bacterial Endotoxins*, Rutgers University Press, New Brunswick, N.J., 1964.
219. **Landy, M., Sanderson, R. P., and Jackson, A. L.**, Humoral and cellular aspects of the immune response to the somatic antigen of *Salmonella enteritidis*, *J. Exp. Med.*, 122, 483, 1965.
220. **Landy, M., Trapani, R. J., and Rosen, F. S.**, Inactivation of endotoxin by a humoral component. VI. Two separate systems required for viable and killed *Salmonella typhosa*, *J. Clin. Invest.*, 39, 352, 1960.
221. **Landy, M. and Weidanz, W. P.**, Natural antibodies against Gram-negative bacteria, in *Bacterial Endotoxins*, Landy, M. and Braun, W., Eds., Rutgers University Press, New Brunswick, N.J., 1964, p. 275.
222. **Larson, C., Ribi, E., Milner, K., and Lieberman, J.**, A method for titrating endotoxic activity in the skin of rabbits, *J. Exp. Med.*, 111, 1, 1960.
223. **Latour, J. G., Prejean, J. B., and Margaretten, W.**, Corticosteroids and the generalized Shwartzman reaction. Mechanisms of sensitization in the rabbit, *Am. J. Pathol.*, 65, 189, 1971.
224. **Leive, L., Shovlin, V. K., and Mergenhagen, S. E.**, Physical, chemical, and immunological properties of lipopolysaccharide released from *Escherichia coli* by ethylenediaminetetraacetate, *J. Biol. Chem.*, 243, 6384, 1968.
225. **Lequire, V. S., Hutcherson, J. D., Hamilton, R. L., and Gray, M. E.**, Effects of bacterial endotoxin on lipid metabolism, *J. Exp. Med.*, 110, 293, 1959.
226. **Levin, J. and Bang, F. B.**, Clottable protein in *Limulus:* Its localization and kinetics of its coagulation by endotoxin, *Thromb. Diath. Haemorrh.*, 19, 186, 1968.
227. **Levin, J. and Bang, F. B.**, Clottable protein in *Limulus*. Its localization and kinetics of its coagulation by endotoxin, *Thromb. Diath. Haemorrh.*, 19, 186, 1968.
228. **Levin, J., Poore, E., Young, N. S., Margolis, S., Zauber, N. P., Townes, A. S., and Bell, W. R.**, Gram-negative sepsis detection of endotoxemia with the *Limulus* test with studies of associated changes in blood coagulation, serum lipids, and complement, *Ann. Intern. Med.*, 76, 1, 1972.
229. **Levy, E., Path, F. C., and Ruebner, B. H.**, Hepatic changes produced by a single dose of endotoxin in the mouse, *Am. J. Pathol.*, 51, 269, 1967.
230. **Lillehei, R. C., Longerbeam, J. K., Bloch, J. H., and Manax, W. G.**, Hemodynamic changes in endotoxin shock in *Shock and Hypotension: Pathogenesis and Treatment*, Mills, L. J. and Moyer, J. H., Eds., Grune and Stratton, New York, 1965, p. 442.

231. **Lüderitz, O., Galanos, C., Risse, H. J., Ruschmann, E., Schlecht, S., Schmidt, G., Schulte-Holthausen, H., Wheat, R. and Westphal, O.,** Structural relationships of *Salmonella* O and R antigens, *Ann. N.Y. Acad. Sci.*, 133, 349, 1966.
232. **Lüderitz, O., Jann, K., and Wheat, R.,** Somatic and capsular antigens of Gram-negative bacteria, in *Comprehensive Biochemistry*, Florkin, M. and Stotz, E. H., Eds., Elsevier, Amsterdam, 1968, p. 105.
233. **Lüderitz, O., Staub, A. M., and Westphal, O.,** Immunochemistry of O and R antigens of *Salmonella* and related enterobacteriaceae, *Bacteriol. Rev.*, 30, 192, 1966.
234. **MacGregor, R. R., Sheagren, J. N., and Wolff, S. M.,** Endotoxin-induced modification of *Plasmodium berghei* infection in mice, *J. Immunol.*, 102, 131, 1969.
235. **MaKara, G. B., Stark, E., and Meszaros, T.,** Corticotrophin release induced by *E. coli* endotoxin after removal of the medial hypothalamus, *Endocrinology*, 88, 412, 1971.
236. **Mäkelä, P. H. and Stocker, B. A. D.,** Genetics of polysaccharide synthesis, *Annu. Rev. Genet.*, 3, 291, 1969.
237. **Margaretten, W.,** Some clinical correlations of the generalized Shwartzman reaction, *Rev. Can. Biol.*, 25, 69, 1966.
238. **Margaretten, W. and McKay, D. G.,** The effect of leukocyte antiserum on the generalized Shwartzman reaction, *Am. J. Pathol.*, 57, 299, 1969.
239. **Marsh, J. C. and Perry, S.,** The granulocyte response to endotoxin in patients with hematologic disorders, *Blood*, 23, 581, 1964.
240. **Martin, A. R.,** The toxicity for mice of certain fractions isolated from *Bact. aertrycke, Br. J. Exp. Pathol.*, 15, 137, 1934.
241. **Martin, W. J. and Marcus, S.,** Detoxified bacterial endotoxins. I. Preparation and biological properties of an acetylated crude endotoxin from *Salmonella typhimurium, J. Bacteriol.*, 91, 1453, 1966.
242. **Martin, W. J. and Marcus, S.,** Detoxified bacterial endotoxins. II. Preparation and biological properties of chemically modified crude endotoxins from *Salmonella typhimurium, J. Bacteriol.*, 91, 1750, 1966.
243. **Martinez-G., L. A., Quintiliani, R., and Tilton, R. C.,** Clinical experience on the detection of endotoxemia with the *Limulus* test, *J. Infect. Dis.*, 127, 102, 1973.
244. **Marx, A., Musetescu, M., Sendrea, M., and Mihalca, M.,** Relationship between particle size and biological activity of *Salmonella typhimurium* endotoxin, *Zentralbl. Bakteriol. Parasitenkd. Infektionskr. Hyg. Abt. 1, Orig.*, 207, 313, 1968.
245. **McAuley, R. J., Ice, R. D., and Curtis, E. G.,** The *Limulus* test for in vitro pyrogen detection, *Am. J. Hosp. Pharm.*, 31, 668, 1974.
246. **McCabe, W. R.,** Endotoxin tolerance. II. Its occurrence in patients with pyelonephritis, *J. Clin. Invest.*, 42, 618, 1963.
247. **McCabe, W. R.,** Immunization with R mutants of *S. minnesota*. I. Protection against challenge with heterologous Gram-negative bacilli, *J. Immunol.*, 108, 601, 1972.
248. **McGill, M. W., Porter, P. J., and Kass, E. H.,** The use of a bioassay for endotoxin in clinical infections, *J. Infect. Dis.*, 121, 103, 1970.
249. **McGill, M. W., Porter, P. J., Vivaldi, E., and Kass, E. H.,** Use of a bioassay for endotoxin in clinical infections and in experimental vasomotor collapse, in *Antimicrobial Agents and Chemotherapy — 1967*, Hobby, G. L., Ed., American Society for Microbiology, Ann Arbor, Mich. 1968, p.132.
250. **McGrath, B. J. and Stewart, G. J.,** The effects of endotoxin on vascular endothelium, *J. Exp. Med.*, 129, 833, 1969.
251. **McKay, D. G., Margaretten, W., and Csavossy, I.,** An electron microscope study of the effects of bacterial endotoxin on the blood-vascular system, *Lab. Invest.*, 15, 815, 1966.
252. **McKay, D. G., Margaretten, W., and Csavossy, I.,** An electron microscope study of endotoxin shock in rhesus monkeys, *Surg. Gynecol. Obstet.*, 125, 825, 1967.
253. **McKay, D. G. and Müller-Berghaus, G.,** Hageman factor (HF) and the generalized Shwartzman reaction (GSR). *Fed. Proc.*, 27, 436, 1968.
254. **McKay, D. G. and Wong, T. C.,** The effect of bacterial endotoxin on the placenta of the rat, *Am. J. Pathol.*, 42, 357, 1963.
255. **McMaster, P. D. and Franzl. R. E.,** The primary immune response in mice. II. Cellular responses of lymphoid tissue accompanying the enhancement or complete suppression of antibody formation by a bacterial endotoxin, *J. Exp. Med.*, 121, 1109, 1968.
256. **Medearis, D. N., Camitta, B. M., and Heath, E. C.,** Cell wall composition and virulence in *Escherichia coli, J. Exp. Med.*, 128, 399, 1968.
257. **Mefferd, R. B., Henkel, D. T., and Loefer, J. B.,** Effect of piromen on survival of irradiated mice, *Proc. Soc. Exp. Biol. Med.*, 83, 54, 1953.
258. **Mela, L., Bacalzo, L. V., Jr., and Miller, L. D.,** Defective oxidative metabolism of rat liver mitochondria in hemorrhagic and endotoxin shock, *Am. J. Physiol.*, 220, 571, 1971.

259. **Melby, J. C., Egdahl, R. H., and Spink, W. W.,** Secretion and metabolism of cortisol after injection of endotoxin, *J. Lab. Clin. Med.*, 56, 50, 1960.
260. **Mergenhagen, S. E., Gewurz, H., Bladen, H. A., Nowotny, A., Kasai, N., and Lüderitz, O.,** Interactions of the complement system with endotoxins from *Salmonella minnesota* mutant deficient in O-polysaccharide and heptose, *J. Immunol.*, 100, 227, 1968.
261. **Mergenhagen, S. E., Snyderman, R., Gewurz, H., and Shin, H. S.,** Significance of complement to the mechanism of action of endotoxin, *Curr. Top. Microbiol. Immunol.*, 50, 37, 1969
262. **Meritt, K. and Johnson, A. G.,** Studies on the adjuvant action of bacterial endotoxins on antibody formation. V. The influence of endotoxin and 5-fluoro-2-deoxyuridine on the primary antibody response of the Balb mouse to a purified protein antigen, *J. Immunol.*, 91, 266, 1963.
263. **Michael, J. G. and Massell, B. F.,** Factors involved in the induction of non-specific resistance to streptococcal infection in mice by endotoxin, *J. Exp. Med.*, 116, 101, 1962.
264. **Miles, A. A. and Pirie, N. W.,** The properties of antigenic preparations from *Brucella melitensis.* I. Chemical and physical properties of bacterial fractions, *Br. J. Exp. Pathol.*, 20, 83, 1939.
265. **Mills, L. C.,** Corticosteroids in endotoxin shock, *Proc. Soc. Exp. Biol. Med.*, 138, 507, 1971.
266. **Milner, K. C., Anacker, R. L., Fukushi, K., Haskins, W. T., Landy, M., Malmgren, B., and Ribi, E.,** Symposium on relationship of structure of microorganisms to their immunological properties. III. Structure and biological properties of surface antigens from Gram-negative bacteria, *Bacteriol. Rev.*, 27, 352, 1963.
267. **Milner, K. C. and Finkelstein, R. A.,** Bioassay of endotoxin: Correlation between pyrogenicity for rabbits and lethality for chick embryos, *J. Infect. Dis.*, 116, 529, 1966.
268. **Moberg, G. P.,** Site of action of endotoxins on hypothalamic-pituitary-adrenal axis, *Am. J. Physiol.*, 220, 397, 1971.
269. **Moon, R. J., and Berry, L. J.,** Role of tryptophan pyrrolase in endotoxin poisoning, *J. Bacteriol.*, 95, 1247, 1968.
270. **Morgan, W. T. J.,** Studies in immunochemistry. II. The isolation and properties of a specific antigenic substance from *B. dysenteriae* (Shiga), *Biochem. J.*, 31, 2003, 1937.
271. **Morrison, D. C. and Cochrane, C. G.,** Direct evidence for Hageman Factor (factor XII) activation by bacterial lipopolysaccharides (endotoxins), *J. Exp. Med.*, 140, 797, 1974.
272. **Mulholland, J. H. and Cluff, L. E.,** The effect of endotoxin upon susceptibility to infection: The role of the granulocyte, in *Bacterial Endotoxins*, Landy, M. and Braun, W., Eds., Rutgers University Press, New Brunswick, N.J., 1964, p. 211.
273. **Mulholland, J. H., Wolff, S. M., Jackson, A. L., and Landy, M.,** Quantitative studies of febrile tolerance and levels of specific antibody evoked by bacterial endotoxin, *J. Clin. Invest.*, 44, 920, 1965.
274. **Müller-Berghaus, G. and Schneberger, R.,** Hageman factor activation in the generalized Shwartzman reaction induced by endotoxin, *Br. J. Haematol.*, 21, 513, 1971.
275. **Muschel, L. H., Schmoker, K., and Webb, P. M.,** Anticomplementary action of endotoxin, *Proc. Soc. Exp. Biol. Med.*, 117, 639, 1964.
276. **Myers, R. D., Rudy, T. A., and Yaksh, T. L.,** Fever produced by endotoxin injected into the hypothalamus of the monkey and its antagonism by salicylate, *J. Physiol.*, 243, 167, 1974.
277. **Nachum, R., Lipsey, A., and Siegel, S. E.,** Rapid detection of Gram-negative bacterial meningitis by the *Limulus* lysate test, *N. Engl. J. Med.*, 289, 931, 1973.
278. **Neter, E.,** Endotoxins and the immune response, *Curr. Top. Microbiol. Immunol.*, 47, 82, 1969.
279. **Neter, E., Westphal, O., Lüderitz, O., Gorzynski, E. A., and Eichenberger, E.,** Studies on enterobacterial lipopolysaccharides: Effect of heat and chemicals on erythrocyte modifying antigenic, toxic and pyrogenic properties, *J. Immunol.*, 76, 377, 1956.
280. **Neter, E., Whang, H. Y., Lüderitz, O., Gorzynski, E. A., and Westphal, O.,** Immunological priming without production of circulating antibodies conditioned by endotoxin and its lipoid A component, *Nature London*, 212, 420, 1966.
281. **Netzer, W. and Vogt, W.,** Anaphylatoxin bildung durch pyrogenes Lipopolysaccharid, *Naunyn Schmiedebergs Arch. Exp. Pathol. Pharmacol.*, 248, 261, 1964.
282. **Nies, A. S., Forsyth, R. P., Williams, H. E., and Melmon, K. L.,** Contribution of kinins to endotoxin shock in unanesthetized rhesus monkeys, *Circ. Res.*, 22, 155, 1968.
283. **Nikaido, H.,** Studies on the biosynthesis of cell wall polysaccharide in mutant strains of *Salmonella*, I. *Proc. Natl. Acad. Sci. U.S.A.*, 48, 1337, 1962.
284. **Nikaido, H.,** Biosynthesis of cell wall lipopolysaccharide in Gram-negative enteric bacteria, *Adv. Enzymol.*, 31, 77, 1969.
285. **Nikaido, H.,** Structure of cell wall lipopolysaccharide from *Salmonella typhimurium*, I. Linkage between O side chains and R-core, *J. Biol. Chem.*, 244, 2835, 1969.
286. **Nikaido, H.,** Structure of cell wall lipopolysaccharide from *Salmonella typhimurium*. Further studies on the linkage between O side chains and R-core, *Eur. J. Biochem.*, 15, 57, 1970.

287. **Nikaido, H. and Nikaido, K.,** Biosynthesis of cell wall polysaccharide in mutant strains of *Salmonella*. IV. Synthesis of S-specific side chains, *Biochem. Biophys. Res. Commun.*, 19, 322, 1965.
288. **Nikaido, H., Nikaido, K., and Mäkelä, P. H.,** Genetic determination of enzymes synthesizing O-specific sugars of *Salmonella* lipopolysaccharides, *J. Bacteriol.*, 91, 1126, 1966.
289. **Niwa, M., Milner, K. C., Ribi, E., and Rudbach, J. A.,** Alteration of physical, chemical, and biological properties of endotoxin by treatment with mild alkali, *J. Bacteriol.*, 97, 1069, 1969.
290. **Nolan, J. P. and O'Connell, C. J.,** Vascular response in the isolated rat liver. I. Endotoxin, direct effects, *J. Exp. Med.*, 122, 1063, 1965.
291. **Noll, H. and Braude, A. I.,** Preparation and biological properties of a chemically modified *Escherichia coli* endotoxin of high immunogenic potency and low toxicity, *J. Clin. Invest.*, 40, 1935, 1961.
292. **Nordlund, J. J., Root, R. K., and Wolff, S. M.,** Studies on the origin of human leukocytic pyrogen, *J. Exp. Med.*, 131, 727, 1970.
293. **Nowotny, A.,** Chemical structure of a phosphomucopeptide and its occurrence in some strains of *Salmonella, J. Am. Chem. Soc.*, 83, 501, 1961.
294. **Nowotny, A.,** Chemical detoxification of bacterial endotoxins, in *Bacterial Endotoxins*, Landy, M. and Braun, W., Eds., Rutgers University Press, New Brunswick, N.J., 1964, p. 29.
295. **Nowotny, A.,** Immunogeniticy of toxic detoxified endotoxin preparations, *Proc. Soc. Exp. Biol. Med.*, 127, 745, 1968.
296. **Nowotny, A.,** Molecular aspects of endotoxin reactions, *Bacteriol. Rev.*, 33, 72, 1969.
297. **Nowotny, A., Radvany, R., and Neale, N. L.,** Neutralization of toxic bacterial O-antigens with O-antibodies while maintaining their stimulus on non-specific resistance, *Life Sci.*, 4, 1107, 1965.
298. **Nowotny, A. M., Thomas, S., Duron, S., and Nowotny, A.,** Relation of structure in bacterial O-antigens. I. Isolation methods, *J. Bacteriol.*, 85, 418, 1963.
299. **Ogura, M.,** High resolution electron microscopy on the surface structure of *Escherichia coli, J. Ultrastruct. Res.*, 8, 251, 1963.
300. **Oppenheim, J. J. and Perry, S.,** Effects of endotoxins on cultured leukocytes, *Proc. Soc. Exp. Biol. Med.*, 118, 1014, 1965.
301. **Oroszlan, S. I. and Mora, P. T.,** Dissociation and reconstitution of an endotoxin, *Biochem. Biophys. Res. Commun.*, 12, 345, 1963.
302. **Osborn, M. J.,** Structure and biosynthesis of the bacterial wall, *Annu. Rev. Biochem.*, 38, 501, 1969.
303. **Osborn, M. J., Rosen, S. M., Rothfield, L., Zeleznick, L. D., and Horecker, B. L.,** Lipopolysaccharide of the Gram-negative cell wall: Biosynthesis of a complex heteropolysaccharide occurs by successive addition of specific sugar residues, *Science*, 145, 783, 1964.
304. **Osborn, M. J. and Weiner, I. M.,** Biosynthesis of a bacterial lipopolysaccharide. VI. Mechanism of incorporation of abequose into the O-antigen of *Salmonella typhimurium, J. Biol. Chem.*, 243, 2631, 1968.
305. **Osborn, M. J. and Yuan Tze-Yeuen, R.,** Biosynthesis of bacterial lipopolysaccharide. VII. Enzymatic formation of the first intermediate in biosynthesis of the O-antigen of *Salmonella typhimurium, J. Biol. Chem.*, 243, 5145, 1968.
306. **Palmer, J. W. and Gerlough, T. D.,** Scientific apparatus and laboratory methods: A simple method for preparing antigenic substances from the typhoid bacillus, *Science*, 92, 155, 1940.
307. **Peavy, D. L., Shands, J. W., Jr., Adler, W. H., and Smith, R. T.,** Mitogenicity of bacterial endotoxins: Characterization of the mitogenic principle, *J. Immunol.*, 111, 352, 1973.
308. **Pekarek, R. S. and Beisel, W. R.,** Effect of endotoxin on serum zinc concentrations in the rat, *Appl. Microbiol.*, 18, 482, 1969.
309. **Petersdorf, R. G. and Shulman, J. A.,** The role of tolerance in the action of bacterial endotoxins, in *Bacterial Endotoxins*, Landy, M. and Braun, W., Eds., Rutgers University Press, New Brunswick, N.J., 1964, p. 487.
310. **Pierce, W. C.,** The effects of endotoxin on the immune response in the rat. III. Elimination of I^{125}-labeled bovine γ-globulin from the circulation of rats, *Lab. Invest.*, 17, 380, 1967.
311. **Pieroni, R. E., Broderick, E. J., Bundeally, A., and Levine, L.,** A simple method for the quantitation of submicrogram amounts of bacterial endotoxin, *Proc. Soc. Exp. Biol. Med.*, 133, 790, 1969.
312. **Pieroni, R. E., Broderick, E. J., and Levine, L.,** Endotoxin-induced hypersensitivity to histamine in mice. I. Contrasting effects of bacterial lipopolysaccharides and the classical histamine-sensitizing factor of *Bordetella pertussis, J. Bacteriol.*, 91, 2169, 1966.
313. **Porter, P. J., Spievack, A. R., and Kass, E. H.,** Endotoxin-like activity of serum from patients with severe localized infections, *N. Engl. J. Med.*, 271, 445, 1964.
314. **Postic, B., DeAngelis, C., Breinig, M. K., and Ho, M.,** Effect of temperature on the induction of interferons by endotoxin and virus, *J. Bacteriol.*, 91, 1277, 1966.

315. **Priano, L. L., Wilson, R. D., and Traber, D. L.,** Lack of significant protection afforded by heparin during endotoxin shock, *Am. J. Physiol.*, 200, 901, 1971.
316. **Quesenberry, P., Halperin, J., Ryan, M., and Stohlman, F., Jr.,** Tolerance to the granulocyte-releasing and colony-stimulating factor elevating effects of endotoxin, *Blood*, 45, 789, 1975.
317. **Quesenberry, P., Morley, A., Stohlman, F., Richard, K., Howard, D., and Smith, M.,** Effect of endotoxin on granulopoiesis and colony-stimulating factor, *N. Engl. J. Med.*, 286, 227, 1972.
318. **Radvany, R., Neale, N. L., and Nowotny, A.,** Relation of structure to function in bacterial O-antigens. VI. Neutralization of endotoxic O-antigens by homologous O-antibody, *Ann. N.Y. Acad. Sci.*, 133, 763, 1966.
319. **Rapaport, S. I., Tatter, D., Coeur-Barron, N., and Hjort, P. F.,** *Pseudomonas* septicemia with intravascular clotting leading to the generalized Shwartzman reaction, *N. Engl. J. Med.*, 271, 80, 1964.
320. **Reinhold, R. B. and Fine, J.,** A technique for quantitative measurement of endotoxin in human plasma, *Proc. Soc. Exp. Biol. Med.*, 137, 334, 1971.
321. **Rietschel, E. Th.,** Chemical structure and biological activity of endotoxins (lipopolysaccharides) and lipid A, *Arch. Pharmacol.*, 287, 73, 1975.
322. **Rietschel, E. T., Galanos, C., Tanaka, A., Ruschmann, E., Lüderitz, O., and Westphal, O.,** Biological activities of chemically modified endotoxins, *Eur. J. Biochem.*, 22, 218, 1971.
323. **Ribi, E., Anacker, L., Brown, R., Haskins, W. T., Malmgren, B., Milner, K. C., and Rudbach, J. A.,** Reaction of endotoxin and surfactants. I. Physical and biological properties of endotoxins treated with sodium desoxycholate, *J. Bacteriol.*, 92, 1493, 1966.
324. **Ribi, E., Haskins, W. T., Milner, K. C., Anacker, R. L., Ritter, D. B., Goode, G., Trapani, R. J., and Landy, M.,** Physicochemical changes in endotoxin associated with loss of biological potency, *J. Bacteriol.*, 84, 803, 1962.
325. **Ribi, E., Milner, K. C., and Perrine, T. D.,** Endotoxic and antigenic fractions from the cell wall of *Salmonella enteritidis*. Methods for separation and some biologic activities, *J. Immunol.*, 82, 75, 1959.
326. **Rieder, R. F. and Thomas, L.,** Studies of the mechanisms involved in the production of abortion by endotoxin, *J. Immunol.*, 84, 189, 1960.
327. **Roantree, R. J.,** *Salmonella* O-antigens and virulence, *Annu. Rev. Microbiol.*, 21, 443, 1967.
328. **Robbins, P. W. and Wright, A.,** Biosynthesis of O-antigens, in *Microbial Toxins*, Vol. 4, Weinbaum, G., Kadis, S., and Ajl, S. J., Eds., Academic Press, New York, 1971, p. 351.
329. **Rojas-Corona, R. R., Skarnes, R., Tamakuma, S., and Fine, J.,** The *Limulus* coagulation test for endotoxin. A comparison with other assay methods, *Proc. Soc. Exp. Biol. Med.*, 132, 599, 1969.
330. **Root, R. K., Nordlund, J. J., and Wolff, S. M.,** Factors affecting the quantitative production and assay of human leukocytic pyrogen, *J. Lab. Clin. Med.*, 75, 679, 1970.
331. **Rosenberg, J. C. and Rush, B. F.,** Lethal endotoxin shock: Oxygen deficit, lactic acid levels, and other metabolic changes, *J. Am. Med. Assoc.*, 196, 767, 1966.
332. **Rossen, R. D., Wolff, S. M., and Butler, W. T.,** The antibody response in nasal washings and serum to *S. typhosa* endotoxin administered intravenously, *J. Immunol.*, 99, 246, 1967.
333. **Rossen, R. D., Wolff, S. M., Butler, W. T., and Vannier, W. E.,** The identification of low molecular weight bentonite flocculating antibodies in the serum of the rabbit, monkey and man, *J. Immunol.*, 98, 764, 1967.
334. **Rothfield, L. and Pearlman-Kothencz, M.,** Synthesis and assembly of bacterial membrane components: A lipopolysaccharide-phospholipid-protein complex excreted by living bacteria, *J. Mol. Biol.*, 44, 477, 1969.
335. **Rowlands, D. T., Jr., Claman, H. N., and Kind, P. D.,** The effect of endotoxin on the thymus of young mice, *Am. J. Pathol.*, 46, 165, 1965.
336. **Rowley, D.,** Stimulation of natural immunity to *Escherichia coli* infections, *Lancet*, 1, 232, 1955.
337. **Rowley, D.,** Endotoxin-induced changes in susceptibility to infection, in *Bacterial Endotoxins*, Landy, M., and Braun, W., Eds., Rutgers University Press, New Brunswick, N.J., 1964, p. 359.
338. **Rowley, D., Howard, J. G., and Jenkin, C. R.,** The fate of P^{32}-labeled bacterial lipopolysaccharide in laboratory animals, *Lancet*, 1, 366, 1956.
339. **Rubenstein, M., Mulholland, J. H., Jeffery, G. M., and Wolff, S. M.,** Malaria-induced endotoxin tolerance, *Proc. Soc. Exp. Biol. Med.*, 118, 283, 1965.
340. **Rubio, N. and Lopez, R.,** Purification of *Pseudomonas aeruginosa* endotoxin by membrane partition chromatography, *Appl. Microbiol.*, 23, 211, 1972.
341. **Rudbach, J. A.,** Molecular immunogenicity of bacterial lipopolysaccharide antigens: Establishing a quantitative system, *J. Immunol.*, 106, 993, 1971.
342. **Rudbach, J. A., Anacker, R. L., Haskins, W. T., Johnson, A. G., Milner, K. C., and Ribi, E.,** Physical aspects of reversible inactivation of endotoxin, *Ann. N.Y. Acad. Sci.*, 133, 629, 1966.

343. **Rudbach, J. A. and Johnson, A. G.**, Restoration of endotoxin activity following alteration by plasma, *Nature London*, 202, 811, 1964.
344. **Rudbach, J. A. and Johnson, A. G.**, Alteration and restoration of endotoxin activity after complexing with plasma proteins, *J. Bacteriol.*, 92, 892, 1966.
345. **Rudbach, J. A., Ribi, E., and Milner, K. C.**, Reaction of papain-treated endotoxin, *Proc. Soc. Exp. Biol. Med.*, 119, 115, 1965.
346. **Sauter, C. and Gifford, G. E.**, Interferon-like inhibitor and lysosomal enzyme induced in mice injected with endotoxin, *Nature London*, 212, 626, 1966.
347. *Schmidt, G., Eichenberger, E., and Westphal, O.*, Die Wirkung der Lipoid- und Polysaccharide-Komponente endotoxischer Lipopolysaccharide Gram-negativer Bakterien auf die Leukozyten-Kultur, *Experientia*, 14, 289, 1958.
348. **Schultz, D. R. and Becker, E. L.**, The alteration of endotoxin by postheparin plasma and its purified fractions. I. Comparison of the ability of guinea pig postheparin and normal plasma to detoxify endotoxin, *J. Immunol.*, 98, 473, 1967.
349. **Schultz, D. R. and Becker, E. L.**, The Alteration of Endotoxin by Postheparin Plasma and Its Purified Fractions. II. Relationship of the endotoxin detoxifying activity of euglobulin from postheparin plasma to lipoprotein lipase, *J. Immunol.*, 98, 482, 1967.
350. **Shands, J. W.**, Localization of somatic antigen on Gram-negative bacteria by electron microscopy, *J. Bacteriol.*, 90, 266, 1965.
351. **Shands, J. W.**, Localization of somatic antigens on Gram-negative bacteria using ferritin antibody conjugates, *Ann. N.Y. Acad. Sci.*, 133, 292, 1966.
352. **Shands, J. W., Jr., Miller, V., and Martin, H.**, The hypoglycemic activity of endotoxin. I. Occurrence in animals hyperreactive to endotoxin, *Proc. Soc. Exp. Biol. Med.*, 130, 413, 1969.
353. **Shands, J. W., Jr., Miller, V., Martin, H., and Senterfitt, V.**, Hypoglycemic activity of endotoxin. II. Mechanism of the phenomenon in BCG-infected mice, *J. Bacteriol.*, 98, 494, 1969.
354. **Sheagren, J. N., Wolff, S. M., and Shulman, N. R.**, Febrile and hematologic responses of rhesus monkeys to bacterial endotoxin, *Am. J. Physiol.*, 212, 884, 1967.
355. **Shilo, M.**, Non-specific resistance to infections, *Annu. Rev. Microbiol.*, 13, 255, 1959.
356. **Shtasel, T. F. and Berry, L. J.**, Effect of endotoxin and cortisone on synthesis of ribonucleic acid and protein in livers of mice, *J. Bacteriol.*, 97, 1018, 1969.
357. **Shwartzman, G.**, Studies of *Bacillus typhosus* toxic substances. I. The phenomenon of local skin reactivity to *B. typhosus* culture filtrate, *J. Exp. Med.*, 48, 247, 1928.
358. **Shwartzman, G., and Michailovsky, N.**, Phenomenon of local skin reactivity to bacterial filtrates in the treatment of mouse sarcoma 180, *Proc. Soc. Exp. Biol. Med.*, 29, 737, 1931.
359. **Silk, M. R.**, The effect of endotoxin on renal hemodynamics, *Arch. Surg.*, 93, 531, 1966.
360. **Singer, I., Kimble, E. T., III, and Ritts, R. E., Jr.**, Alterations of the host-parasite relationship by administration of endotoxin to mice with infections of trypanosomes, *J. Infect. Dis.*, 114, 243, 1964.
361. **Skarnes, R. C. and Harper, M. J. K.**, Relationship between endotoxin-induced abortion and the synthesis of prostaglandin F, *Prostaglandins*, 1, 191, 1972.
362. **Skarnes, R., Rutenburg, S., and Fine, J.**, Fractionation of an esterase from calf spleen implicated in the detoxification of bacterial endotoxin, *Proc. Soc. Exp. Biol. Med.*, 128, 75, 1968.
363. **Skjorten, F.**, Bilateral renal cortical necrosis and the generalized Shwartzman reaction. I. Review of literature and report of seven cases, *Acta Pathol. Microbiol. Scand.*, 61, 394, 1964.
364. Sleeman, H. K., Lamborn, P. B., Diggs, J. W., and Emery, C. E., Effects of endotoxin and histamine on serum enzyme activity, *Proc. Soc. Exp. Biol. Med.*, 138, 536, 1971.
365. **Smith, E. E., Rutenburg, S. H., Rutenburg, A. M., and Fine, J.**, Detoxification of endotoxin by splenic extracts, *Proc. Soc. Exp. Biol. Med.*, 113, 781, 1963.
366. **Smith, L. L., Muller, W., and Hinshaw, L. B.**, The management of experimental endotoxin shock, *Arch. Surg.*, 89, 630, 1964.
367. **Smith, R. T. and Thomas, L.**, The lethal effect of endotoxins in the chick embryo, *J. Exp. Med.*, 104, 217, 1956.
368. **Smith, W. W., Alderman, I. M., and Cornfield, J.**, Granulocyte release by endotoxin in normal and irradiated mice, *Am. J. Physiol.*, 201, 396, 1961.
369. **Smith, W. W., Alderman, I. M., and Gillespie, R. E.**, Increased survival in irradiated animals treated with bacterial endotoxins, *Am. J. Physiol.*, 191, 124, 1957.
370. **Smith, W. W., Marston, R. A., and Cornfield, J.**, Patterns of hemopoietic recovery in irradiated mice, *Blood*, 14, 737, 1959.
371. **Snell, E. S. and Atkins, E. A.**, The mechanisms of fever, in *The Biological Basis of Medicine*, Vol. 2, Bittar, E. E., Ed., Academic Press, London, 1968, p. 397.
372. **Snyder, I. S., Deters, M., and Ingle, J.**, Effect of endotoxin on pyruvate kinase activity in mouse liver, *Infect. Immun.*, 4, 138, 1971.

373. **Solie, R. T. and Downing, S. E.,** Effects of *E. coli* endotoxemia on ventricular performance, *Am. J. Physiol.*, 211, 307, 1966.
374. **Spielvogel, A. R.,** An ultrastructural study of the mechanisms of platelet—endotoxin interaction, *J. Exp. Med.*, 126, 235, 1967.
375. **Spink, W. W., Reddin, J., Zak, S. J., Peterson, M., Starzecki, B., and Seljeskog, E.,** Correlation of plasma catecholamine levels with hemodynamic changes in canine endotoxin shock, *J. Clin. Invest.*, 45, 78, 1966.
376. **Starzl, T. E., Lerner, R. A., Dixon, F. J., Groth, C. G., Brettschneider, L., and Terasaki, P. I.,** Shwartzman reaction after human renal homotransplantation, *N. Engl. J. Med.*, 278, 642, 1968.
377. **Stetson, C. A., Jr.,** Studies on the mechanisms of the Shwartzman phenomenon, *J. Exp. Med.*, 93, 489, 1951.
378. **Stewart, G. J.,** Effect of endotoxin on the ultrastructure of liver and blood cells of hamsters, *Br. J. Exp. Pathol.*, 51, 114, 1970.
379. **Stinebring, W. R. and Youngner, J. S.,** Patterns of interferon appearance in mice injected with bacteria or bacterial endotoxin, *Nature London*, 204, 712, 1964.
380. **Stocker, B. A. D. and Mäkelä, P. H.,** Genetic aspects of biosynthesis and structure of *Salmonella* lipopolysaccharide, in *Microbial Toxins*, Vol. 4, Weinbaum, G., Kadis, S., and Ajl, S. J., Eds., Academic Press, New York, 1971, p. 369.
381. **Stumacher, R J., Kovnat, M. J., and McCabe, W. R.,** Limitations of the usefulness of the *Limulus* assay for endotoxin, *N. Engl. J. Med.*, 288, 1261, 1973.
382. **Sullivan, J. D., Jr. and Watson, S. W.,** Factors affecting the sensitivity of *Limulus* lysate, *Appl. Microbiol.*, 28, 1023, 1974.
383. **Takebe, K., Setaishi, C., Hirama, M., Yamamoto, M., and Horiuchi, Y.,** Effects of a bacterial pyrogen on the pituitary-adrenal axis at various times in 24 hours, *J. Clin. Endocrinol. Metab.*, 26, 437, 1966.
384. **Tate, W. J., III, Douglas, H., Braude, A. I., and Wells, W. W.,** Protection against lethality of *E. coli* endotoxin with "O" antiserum, *Ann. N.Y. Acad. Sci.*, 133, 746, 1966.
385. **Taub, R. N., Rodriguez-Erdmann, F., and Dameshek, W.,** Intravascular coagulation, the Shwartzman reaction and the pathogenesis of T.T.P., *Blood*, 24, 775, 1964.
386. **Taylor, S. S. and Heath, E. C.,** The incorporation of β-hydroxy fatty acids into phospholipid of *Escherichia coli* B., *J. Biol. Chem.*, 244, 6605, 1969.
387. **Thomas, C. S., Jr. and Brockman, S. K.,** The role of adrenal corticosteroid therapy in *Escherichia coli* endotoxin shock, *Surg. Gynecol. Obstet.*, 126, 61, 1968.
388. **Thomas, L.,** Possible new mechanism of tissue damage in the experimental cardiovascular effects of endotoxin, *Am. Heart J.*, 52, 507, 1956.
389. **Thomas, L. and Good, R. A.,** Bilateral cortical necrosis of kidneys in cortisone-treated rabbits following injection of bacterial toxins, *Proc. Soc. Exp. Biol. Med.*, 76, 604, 1951.
390. **Thomas, L. and Good, R. A.,** Studies on the generalized Shwartzman reaction. I. General observations concerning the phenomenon, *J. Exp. Med.*, 96, 605, 1952.
391. **Thomas, L. and Zweifach, B. W.,** Mechanisms in the production of tissue damage and shock by endotoxin, *Trans. Assoc. Am. Physicians*, 70, 54, 1957.
392. **Trejo, R. A. and DiLuzio, N. D.,** Influence of reticuloendothelial system (RES) functional modification on endotoxin detoxification by liver and spleen, *J. Reticuloendothel. Soc.*, 10, 515, 1971.
393. **Tsagaris, T. J., Gani, M., and Lange, R. L.,** Central blood volume during endotoxin shock in dogs, *Am. J. Physiol.*, 212, 498, 1967.
394. **Villablanca, J. and Myers, R. D.,** Fever produced by microinjections of typhoid vaccine into hypothalamus of cats, *Am. J. Physiol.*, 208, 703, 1965.
395. **Wagner, R. R., Snyder, R. M., Hook, E. W., and Luttrell, C. N.,** Effect of bacterial endotoxin on resistance of mice to viral encephalitides including comparative studies of the interference phenomenon, *J. Immunol.*, 83, 87, 1959.
396. **Waisbren, B. A.,** Bactermia due to Gram-negative bacilli other than *Salmonella*, *Arch. Intern. Med.*, 88, 467, 1951.
397. **Walker, J.,** A method for the isolation of toxic and immunizing fractions from bacteria of the *Salmonella* group, *Biochem. J.*, 34, 325, 1940.
398. **Watson, D. W. and Kim, Y. B.,** Modification of Host Responses to bacterial endotoxins I. Specificity of pyrogenic tolerance and the role of hypersensitivity in pyrogenicity, lethality, and skin reactivity, *J. Exp. Med.*, 118, 425, 1963.
399. **Weil, M. H., Shubin, H., Udhoji, V. N., and Rossoff, L.,** Effects of vasopressor agents and corticosteroid hormones in endotoxin shock, in *Shock and Hypotension: Pathogenesis and Treatment*, Mills, L. J. and Moyer, J. H., Eds., Grune and Stratton, New York, 1965, p. 470.
400. **Weinbaum, G., Kadis, S., and Ajl, S. J., Eds.,** Bacterial endotoxins, in *Microbial Toxins*, Vol. 4, Academic Press, New York, 1971.

401. **Weinstein, M. J., Waitz, J. A., and Came, P. E.**, Induction of resistance to bacterial infections of mice with poly I poly C, *Nature London*, 226, 170, 1970.
402. **Wexler, B.**, Effects of a bacterial polysaccharide (piromen) on the pituitary-adrenal axis: modification of ACTH release, *Metabolism*, 12, 49, 1963.
403. **Wexler, B. C., Dolgin, A. E., and Tryczynski, E. W.**, Effects of a bacterial polysaccharide (piromen) on the pituitary-adrenal axis: Adrenal ascorbic acid, cholesterol, and histologic alterations, *Endocrinology*, 61, 300, 1965.
404. **Whang, H. Y. and Neter, E.**, Immunosuppression by endotoxin and its lipoid A component, *Proc. Soc. Exp. Biol. Med.*, 124, 919, 1967.
405. **Whitby, J. L., Michael, J. G., Woods, M. W., and Landy, M.**, Symposium on bacterial endotoxins. II. Possible mechanism whereby endotoxins evoke increased nonspecific resistance to infection, *Bacteriol. Rev.*, 25, 437, 1961.
406. **White, P. B.**, On the relation of the alcohol-soluble constituents of bacteria to their spontaneous agglutination, *J. Pathol. Bacteriol.*, 30, 113, 1927.
407. **Wiener, E., Beck, A., and Shilo, M.**, Effect of bacterial lipopolysaccharides on mouse peritoneal leukocytes, *Lab. Invest.*, 14, 475, 1965.
408. **Wildfeuer, A., Heymer, B., Schleifer, K. H., and Haferkamp, O.**, Investigations on the specificity of the *Limulus* test for the detection of endotoxin, *Appl. Microbiol.*, 28, 867, 1974.
409. **Wober, W. and Alaupovic, P.**, Studies on the protein moiety of endotoxin from Gram-negative bacteria: Characterization of the moiety isolated by phenol treatment of endotoxin from *Serratia marcescens* 08 and *Escherichia coli* 0141:K85 (B), *Eur. J. Biochem.*, 19, 340, 1971.
410. **Wolff, S. M.**, Biological effects of bacterial endotoxins in man, *J. Infect. Dis.*, 128 (Suppl.), 259, 1973.
411. **Wolff, S. M. and Bennett, J. V.**, Editorial: Gram-negative-rod bacteremia, *N. Engl. J. Med.*, 291, 733, 1974.
412. **Wolff, S. M. and Dinarello, C. A.**, Pathogenesis of fever, *N. Engl. J. Med.*, in press.
413. **Wolff, S. M., Mulholland, J. H., and Rubenstein, M.**, Suppression of the immune response to bacterial endotoxins, in *Bacterial Endotoxins*, Landy, M. and Braun, W., Eds., Rutgers University Press, New Brunswick, N.J., 1964, p. 319.
414. **Wolff, S. M., Mulholland, J. H., and Ward, S. B.**, Quantitative aspects of the pyrogenic response of rabbits to endotoxin, *J. Lab. Clin. Med.*, 65, 268, 1965.
415. **Wolff, S. M., Mulholland, J. H., Ward, S. B., Rubenstein, M., and Mott, P. D.**, Effect of 6-mercaptopurine on endotoxin tolerance, *J. Clin. Invest.*, 44, 1402, 1965.
416. **Wolff, S. M., Rubenstein, M., Mulholland, J. H., and Alling, D. W.**, Comparison of hematologic and febrile response to endotoxin, *Blood*, 26, 190, 1965.
417. **Wood, W. B., Jr.**, Studies on the cause of fever, *N. Engl. J. Med.*, 258, 1023, 1958.
418. **Wright, A. and Kanegasaki, S.**, Molecular aspects of lipopolysaccharides, *Physiol. Rev.*, 51, 748, 1971.
419. **Wright, L. J., Kimball, H. R., and Wolff, S. M.**, Alterations in host responses to experimental *Candida albicans* infections by bacterial endotoxin, *J. Immunol.*, 103, 1276, 1969.
420. **Wright, L. J., Lipsett, M. B., Ross, G. T., and Wolff, S. M.**, Effects of dexamethasone and aspirin on the responses to endotoxin in man, *J. Clin. Endocrinol. Metab.*, 34, 13, 1972.
421. **Wright, R. C. and Winkelmann, R. K.**, The epinephrine response of isolated rabbit vascular strips after in vivo and in vitro endotoxin exposure, *Angiology*, 22, 495, 1971.
422. **Yen-Watson, B. and Kushner, I.**, Rabbit CRP response to endotoxin administration: Dose—response relationship and kinetics, *Proc. Soc. Exp. Biol. Med.*, 146, 1132, 1974.
423. **Yin, E. T., Galanos, C., Kinsky, S., Bradshaw, R. A., Wessler, S., Luderitz, O., and Sarmiento, M. E.**, Picogram-sensitive assay for endotoxin: gelation of *Limulus polyphemus* blood cell lysate induced by purified lipopolysaccharides and lipid A from Gram-negative bacteria, *Biochim. Biophys. Acta*, 261, 284, 1972.
424. **Yoshikawa, T., Tanaka, K. R., and Guze, L. B.**, Infection and disseminated intravascular coagulation, *Medicine Baltimore*, 50, 237, 1971.
425. **Yoshioka, M. and Johnson, A. B.**, Characteristics of endotoxin-altering fractions derived from normal human serum, *J. Immunol.*, 89, 326, 1962.
426. **Yoshioka, M. and Konno, S.**, Characteristics of endotoxin-altering fractions derived from normal serum. III. Isolation and properties of horse serum α_2-macroglobulin, *Infect. Immun.*, 1, 431, 1970.
427. **Zweifach, B. W.**, Vascular effects of bacterial endotoxin, in *Bacterial Endotoxins*, Landy, M. and Braun, W., Eds., Rutgers University Press, New Brunswick, N.J., 1964, p. 110.

LIPOPOLYSACCHARIDES OF GRAM-NEGATIVE BACTERIA*

Table 1
SUGAR COMPOSITION IN BACTERIA BELONGING TO ENTEROBACTERIACEAE

Chemo-type[a]	Sugar composition[b]						Isolated from					
	Glucosamine	KDO	L-Glycero-D-mannoheptose	Galactose	Glucose	Other sugars	*E. coli* O-type	*Salmonella* O-group	*Arizona* O-group	*Shigella*	*Citrobacter* O-group	Others
I	+	+	+	+	+		24, 28, 30, 42, 56, 64, 82, 83, 85, 118, 141 (1)[c]	V, X, Y (2)	8, 19, 26, 29 (2, 3)	*S. boydii* 3, 7, 8, 15 (4)	9, 10, 13, 14, 29, 33 (3, 5)	*Klebsiella* 1, 2, 6, 8, 9 (6); *Yersinia pestis* (7); *Proteus morganii* (8)
II	+	+	+	+	+	GalN	21–23, 27, 33, 37, 46, 61, 76, 81, 87 (1)	L, P, 51 (2)	16 (2, 3)	*S. boydii* 16 (4); *S. dysenteriae* 3, 6 (9)	5, 12, 16, 17, 29, 30, 42 (5)	
III	+	+	+	+	+	Man	8, 9, 40, 58, 73, 78 (1)	C_1, C_4, H (2)	30 (2, 3)	*S. dysenteriae* 5, 7 (9)	21, 48 (5)	*Klebsiella* 3 (6); *P. rettgeri* (8)
IV	+	+	+	+	+	GalN, Man	6 (1)	K, R (2)			2, 28 (5)	
V	+	+	+	+	+	Fuc	41, 52 (1)	W (2)		*S. dysenteriae* 5 (9)	31 (5)	
VI	+	+	+	+	+	GalN, Fuc	80, 86, 90, 127, 128 (1)	G, N, U (2)	21, 25 (2, 3)		6 (5)	

Note: Numbers in parentheses indicate references.

[a] Chemotypes I through XLIII are those found in *E. coli, Shigella,* or *Salmonella.* CC-, CY-, and CK- chemotypes are those found in *Citrobacter, Yersinia,* and *Klebsiella.*
[b] Sugars present either in trace amounts or in only some of the strains examined are shown in parentheses. Systematic names of uncommon sugars are shown in brackets.
[c] *E. coli* O14, which had been thought to belong to this chemotype, turned out to be an encapsulated R strain (1a).

*From Nikaido, H., in *Handbook of Biochemistry and Molecular Biology*, 3rd ed., *Lipids, Carbohydrates, Steroids,* Fasman, G. D., Ed., CRC Press, Boca Raton, Fla., 1975.

Table 1 (continued)
SUGAR COMPOSITION
IN BACTERIA BELONGING TO ENTEROBACTERIACEAE

Chemo-type[a]	Sugar composition[b]						Isolated from					
	Glucosamine	KDO	L-glycero-D-mannoheptose	Galactose	Glucose	Other sugars	*E. coli* O-type	*Salmonella* O-group	*Arizona* O-group	*Shigella*	*Citrobacter* O-group	Others
VII	+	+	+	+	+	Rha	1, 13, 18, 19, 31, 35, 39, 50, 53, 54, 60, 69, 99, 100, 102 119, 129 (1, 10)[d]	59 (2)	6 (2, 3)	*S. flexneri* (11); *S. boydii* 1, 2, 4, 9, 10, 11 14(4)		*Klebsiella* 12 (6)
VIII	+	+	+	+	+	GalN, Rha	48, 49, 51, 117 (1)	53, 57 (2)			1, 7, 12, 15, 18 (5)	
IX	+	+	+	+	+	GalN, Rib		56 (2)				
X	+	+	+	+	+	Colitose[3,6-dideoxy-L-*xylo*-hexose]	111 (1)	O (2)	9a, c, 20 (2, 3, 3a)			
XI	+	+	+	+	+	GalN, colitose	55 (1)	Z (2)				*Y. pseudotuberculosis* VI (13)
XII	+	+	+	+	+	GalN, Man, Fuc	11, 43, 125 (1)	I (2)				*Y. pseudotuberculosis* VB (13)
XIII	+	+	+	+	+	Man, Rha	34, 68, 75, 79 (1)	E, F, 54 (2)	17 (2, 3)	*S. boydii* 5, 12 (4)	3,8 (5)	
XIV	+	+	+	+	+	Man, Rha, abequose[3,6-dideoxy-D-*xylo*-hexose]		B,C_2, C_3 (2)			22, 38 (5, 14)	
XV	+	+	+	+	+	Man, Rha, paratose[3,6-dideoxy-D-*ribo*-hexose]		A (2)				

[d]*E. coli* K12 (an R strain) belongs to this chemotype (12).

XVI	+	+	+	+	+	Man, Rha, tyvelose[3,6-dideoxy-D-*arabino*-hexose]		D_1, D_2 (2)			
XVII	+	+	+		+	Man	44, 59, 77 (1)				
XVIII	+	+	+	+	+	Man, Fuc	126 (1)				
XIX	+	+	+		+	Man, Rha	17 (1)				
XX	+	+	+	+	+	FucN	12, 15, 29, 57 (1)				
XXI	+	+	+	+	+	FucN, Rha	4, 6, 25, 26 (1)				
XXII	+	+	+	+	+	FucN, 6-deoxytalose	45 (1)				
XXIII	+	+	+	+	+	2-Amino-2,6-dideoxymannose	3 (1)				
XXIV	+	+	+	+	+	Fuc, Rha	36 (1)				
XXV	+	+	+	+	+	Rib		52 (2)	15 (2, 3)		*Klebsiella* 4, 11 (6)
XXVI	+	+	+	+	+	Man, 6-deoxytalose	66, 88 (1)				
XXVII	+	+	+	+	+	Fuc, 6-deoxytalose	84 (1)				
XXVIII	+	+	+	+	+	3-Amino-3,6-dideoxy-D-galactose	74 (1)				
XXIX	+	+	+	+	+	Rha, 3-amino-3,6-dideoxy-D-galactose	2 (1)				
XXX	+	+	+	+	+	GalN, 3-amino-3,6-dideoxy-D-glucose	5, 65 (1)				
XXXI	+	+	+	+	+	GalN, Fuc, 3-amino-3,6-dideoxy-D-glucose	70 (1)				
XXXII	+	+	+	+	+	GalN, Rha, 3-amino-3,6-dideoxy-D-glucose	71 (1)	M (2, 15)		20 (3, 5, 15)	
XXXIII	+	+	+	+	+	Rib, 3-amino-3,6-dideoxy-D-glucose	114 (1)				
XXXIV	+	+	+	+	+	ManN		J (2, 16)			
XXXV	+	+	+	+	+	ManN, Rha		T (2, 16)			
XXXVI	+	+	+	+	+	Quinovosamine[2-amino-2,6-dideoxy-D-glucose]		58 (2, 16)	1, 33 (3, 16)		
XXXVII	+	+	+	+	+	Man, quinovosamine		S (2, 16)			

Table 1 (continued)
SUGAR COMPOSITION
IN BACTERIA BELONGING TO ENTEROBACTERIACEAE

Chemo-type[a]	Sugar composition[b]						Isolated from					
	Glucosamine	KDO	L-glycero-D-mannoheptose	Galactose	Glucose	Other sugars	*E. coli* O-type	*Salmonella* O-group	*Arizona* O-group	*Shigella*	*Citrobacter* O-group	Others
XXXVIII	+	+	+	+	+	GalN, 3-amino-3,6-dideoxy-D galactose		55 (2, 15)	24 (3, 15)			
XXXIX	+	+	+	+	+	GalN, Rib, 3-amino-3,6-dideoxy-D-glucose		M (2, 15)				
XL	+	+	+	+	+	GalN, Man, Fuc, 3-amino-3,6-dideoxy-D-glucose		Q (2, 15)				
XLI	+	+	+	+	+	Rha, 4-amino-4,6-dideoxy-D-galactose	10 (16a)					
XLII	+	+	+	+	+	Rha, 4-amino-4,6-dideoxy-D-glucose	7 (16a)					
XLIII	+	+	+	+	+	GalN, 2-aminohexuronic acid				*S. sonnei* I (17)		
CC-A	+	+	+	+	+	GalN, Fuc, Rha					26 (5)	
CC-B	+	+	+	+	+	GalN, FucN, Man					11 (5)	
CC-C	+	+	+	+	+	GalN, 3-amino-3,6-dideoxy-D-glucose, 6-deoxytalsoe					19 (5)	
CC-D	+	+	+	+	+	GalN, 3-amino-3,6-dideoxy-D-galactose, 6-deoxytalose					36, 41 (5)	
CC-E	+	+	+	+	+	GalN, Xyl, Rha					8 (5)	
CC-F	+	+	+	+	+	Man, Xyl, Rha					25, 32 (5)	
CC-G	+	+	+	+	+	GalN, 4-deoxy-*arabino*-hexose					4, 27, 36 (5)	
CC-H	+	+	+	+	+	GalN, Man, 4-deoxy-*arabino*-hexose					23 (5)	
CC-K	+	+	+	+	+	Fuc, Rha, unknown sugar					35 (5)	
CC-L	+	+	+	+	+	Rha, 3-amino-3,6-dideoxyglucose					*C. freundii* strain 8090	

							(18)	
CY-A	+	+	+	+	+	Paratose, 6-deoxy-D-*manno*-heptose		*Y. pseudotuberculosis* IA (13)
CY-B	+	+	+	+	+	Abequose, 6-deoxy-D-*manno*-heptose		*Y. pseudotuberculosis* IIA (13)
CY-C	+	+	+	+	+	Tyvelose, 6-deoxy-D-*manno*-heptose		*Y. pseudotuberculosis* IVB (13)
CY-D	+	+	+	(+)	+	GalN, Man, Fuc, paratose		*Y. pseudotuberculosis* III (13)
CY-E	+	+	+	+	+	GalN, Man, Fuc, abequose		*Y. pseudotuberculosis* IIB (13)
CY-F	+	+	+	(+)	+	Man, Fuc, paratose		*Y. pseudotuberculosis* IB (13)
CY-G	+	+	+		+	GalN, Man, Fuc, tyvelose		*Y. pseudotuberculosis* IVA (13)
CY-H	+	+	+		+	GalN, Man, Fuc, ascarylose-[3,6-dideoxy-L-*arabino*-hexose]		*Y. pseudotuberculosis* VA (13)
CK-A	+	+	+	+	+	Rib, Rha		*Klebsiella* 7 (6)
CK-B	+	+	+	+	+	Rib, Rha, 3-O-Me-Rha		*Klebsiella* 10 (6)
CK-C	+	+	+	+	+	Man, 3-O-Me-Man		*Klebsiella* 5 (6)
CO-A	+	+	+	+	+	GalN, quinovosamine		*Proteus vulgaris* (16)
CO-B	+	+	+	(+)	+	GalN, (GlcUA), GalUA, D-*glycero*-D-*manno*-heptose		*Proteus mirabilis* (19)

Table 2
SUGAR COMPOSITION IN BACTERIA NOT BELONGING TO ENTEROBACTERIACEAE

Chemotype	Sugar composition						Found in
	Glucosamine	KDO	L-glycero-D-manno-heptose	Galactose	Glucose	Other sugars	
CP-A	+	+	+	+	+	GalN, Rha	*Pseudomonas alcaligenes*[19a]
CP-B	+	+	+		+	GalN, Rha, Rib, quinovosamine	*P. syncyanea*[20]
CP-C	+	+	+		+	GalN, Rha, FucN	*P. aeruginosa*[21]
CP-D	+	+	+			GalN, Rha, quinovosamine	*P. stutzeri*[20]
CP-E		+	+		+	Man	*P. diminuta*[20]
CP-F	+		+	+	+	(GalN), (FucN)	*P. rubescens*[20]
CP-G	+	+		+	(+)		*P. pavonaceae*[20]
CX-A	+	+			+	Man, Rha, GalUA, Xyl	*Xanthomonas*[22]
CX-B	+	+			+	Man, Fuc, Rha, GalUA	*Xanthomonas*[22]
CX-C	+	+			+	Man, Rha, GalUA[a]	*Xanthomonas*[22]
CR-A	+	+			+	Fuc, GlcUA	*Rhizobium trifolii*[b,24]
CR-B	+	+	+		+	Fuc, GlcUA	*Rhizobium trifolii*[b,24]
CN-A	+	+	+		+	GalN, Rha	*Neisseria perflava*[26]
CN-B	+	+			+	GalN	*N. sicca*[27]
CN-C	+			+	+	GalN	*Branhamella catarrhalis*[28] *Moraxella duplex*[28] *Acinetobacter calcoaceticus*[28]
CI-A	+		+		+	Quinovosamine, fructose	*Vibrio cholerae*[29-31]
CS-A	+	?		+	+	GalN, Rha, Ketose?	*Treponema pallidum*[32]
CM-A	+	+			+	Man, Rib, Ara, (3-O-Me-Xyl)	*Cytophaga*[33]
CM-B	+	+		+	+	(GalN), Man, (Rha), Rib, Ara, (3-O-Me-Xyl)	*Cytophaga*[33]
CM-C	+	+		+	+	Man, Rib, (Xyl), Ara, (3-O-Me-Xyl)	*Sporocytophaga*[33]
CM-D	+	+		+	+	Man, Rha, Rib, (Xyl), Ara, (3-O-Me-Xyl)	*Polyangium*[33] *Stigmatella*[33]
CM-E	+	+		+	(+)	GalN, (Man), (Rha), Rib, (Xyl), Ara, (3-O-Me-Xyl)	*Myxococcus*[33,34] *Sorangium*[33]
CA-A	+	+		+	+	Man, Fuc, Rha, 2-amino-2-deoxyheptose	*Anacystis nidulans*[35]

CA-B	+			+	+	Man, Rha, 3-O-Me-Rha	*Anabaena variabilis*[35],[a]
CH-A	+	+		+	+	Rha, 3-O-Me-Rha, neuraminic acid	*Rhodopseudomonas capsulata*[35],[b]
CH-B	+	+		+	+	Quinovosamine, Man, 2,3-diamino-dideoxy-hexose, 3-O-Me-Man, 3-O-Me-Xyl, GalNUA	*R. viridis*[36]
CH-C	+	+	+	+		Quanovosamine, Man, 2,3-diamino-dideoxy-hexose, 4-O-Me-Xyl, unidentified amino sugar	*R. palustris*[37]
CH-D	+	+	+	+		Quinovosamine, Man, 2,3-diamino-dideoxy-hexose, 6-deoxytalose, Xyl, 3-O-Me-6-deoxytalos	*R. palustris*[37]
CH-E	+	+	+	+	+	Quinovosamine, Man, 2,3-diamino-dideoxy-hexose, GalN, Rha, Xyl, 6-O-Me-GlcN	*R. palustris*[37]
CV-A	+	+	+	+		GalN	*Veillonella*[38]
CV-B	+	+	+		+	GalN	*Veillonella*[38]
CV-C	+	+	+		+	GalN, Rha	*Veillonella*[38]
CV-D	+	+	+	+	+	GalN	*Veillonella*[38]

[a] One of the strains also produces a "phenol-soluble" lipopolysaccharide, which contains GlcN, Glc, Rha, GalUA, 3-amino-3,6-dideoxy-D-galactose, but not KDO or heptose.[23]

[b] In another investigation on 22 strains of *Rhizobium* and *Agrobacterium* species, Glc and Rha were found in all strains, whereas GlcN, Man, Fuc, Gal, and 4-O-methylglucuronic acid were present in many strains. A few strains also contained Xyl or Ara. The presence of heptose or KDO was not examined.[25]

Table 3
SUGAR SEQUENCES

General Structure of Lipopolysaccharide

```
(Repeating unit)n  →  (Core oligosaccharide) → (Lipid A)
-----------------     --------------------------------------
 "O side chain"                  "R core"
```

Organism	Structure[a]	Reference

O SIDE CHAIN REPEATING UNIT

Salmonella paratyphi A (O-antigen: 2, 12) — Reference 39

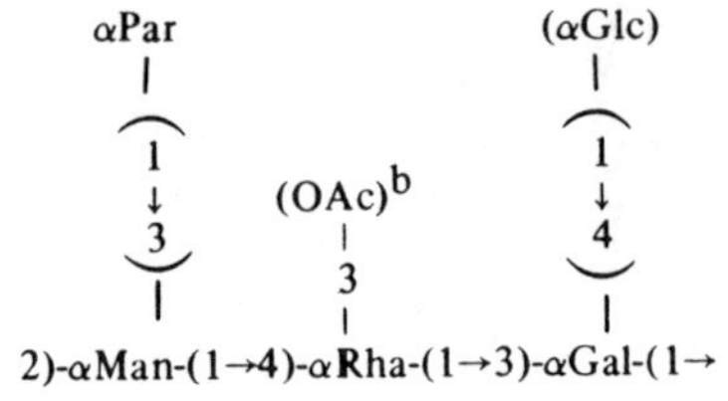

```
    αPar                     (αGlc)
     |                         |
     1                         1
     ↓       (OAc)[b]          ↓
     3          |              4
     |          3              |
     |          |              |
2)-αMan-(1→4)-αRha-(1→3)-αGal-(1→
```

S. typhimurium (O-antigen: 4, 5, 12) — Reference 40

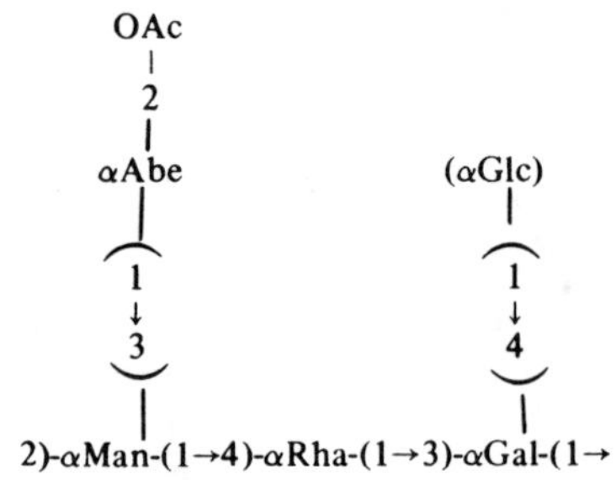

```
     OAc
      |
      2
      |
    αAbe                     (αGlc)
      |                        |
      1                        1
      ↓                        ↓
      3                        4
      |                        |
2)-αMan-(1→4)-αRha-(1→3)-αGal-(1→
```

S. bredeney (O-antigen: 1, 4, 12) — Reference 41

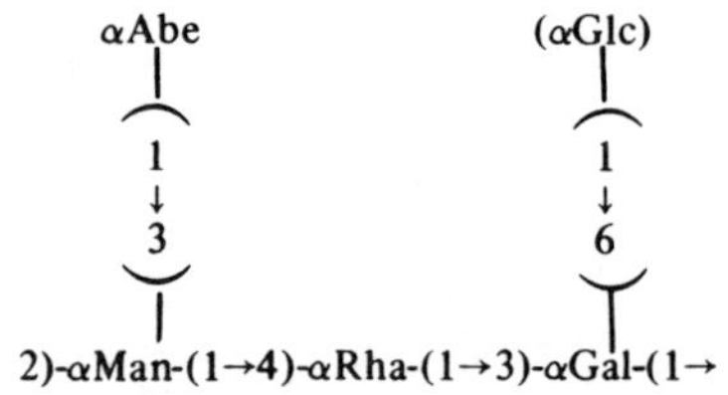

```
    αAbe                     (αGlc)
      |                        |
      1                        1
      ↓                        ↓
      3                        6
      |                        |
2)-αMan-(1→4)-αRha-(1→3)-αGal-(1→
```

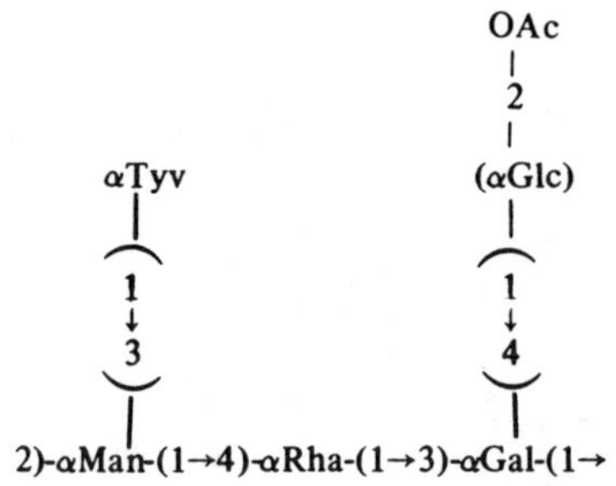

S. typhi (O-antigen: 9, 12) — Reference 42

```
                               OAc
                                |
                                2
                                |
    αTyv                     (αGlc)
      |                        |
      1                        1
      ↓                        ↓
      3                        4
      |                        |
2)-αMan-(1→4)-αRha-(1→3)-αGal-(1→
```

[a]Except for L-rhamnose (Rha), L-fucose (Fuc), and colitose, sugar residues in this table presumably belong to the D-series.
[b]Substitutions which are not always present are shown in parentheses.

Table 3 (continued)
SUGAR SEQUENCES

Organism	Structure[a]	Reference
	O SIDE CHAIN REPEATING UNIT (continued)	
S. strasbourg (O-antigen: (9), 46)	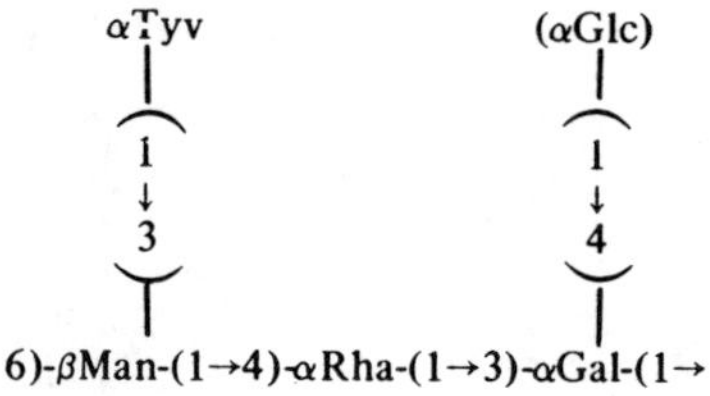 αTyv (αGlc) 1→3 1→4 6)-βMan-(1→4)-αRha-(1→3)-αGal-(1→	43
S. anatum and *S. muenster* (O-antigen: 3, 10)	OAc or (Glc) | 6)-βMan-(1→4)-αRha-(1→3)-αGal-(1→	44, 45
S. newington (O-antigen: 3, 15)	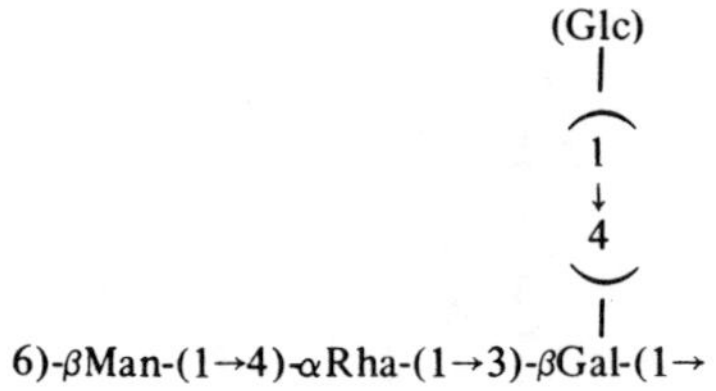 (Glc) 1→4 6)-βMan-(1→4)-αRha-(1→3)-βGal-(1→	44, 46
S. minneapolis (O-antigen: 3, (15), 34)	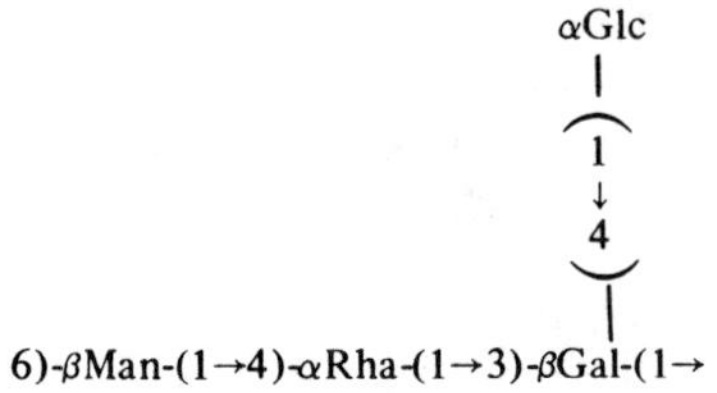 αGlc 1→4 6)-βMan-(1→4)-αRha-(1→3)-βGal-(1→	44
S. senftenberg (O-antigen: 1, 3, 19)	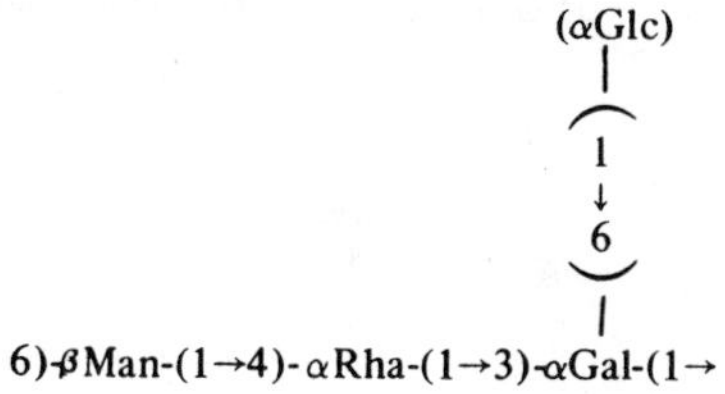 (αGlc) 1→6 6)-βMan-(1→4)- αRha-(1→3)-αGal-(1→	47, 48
S. cholerae suis (O-antigen: 6_2, 7)	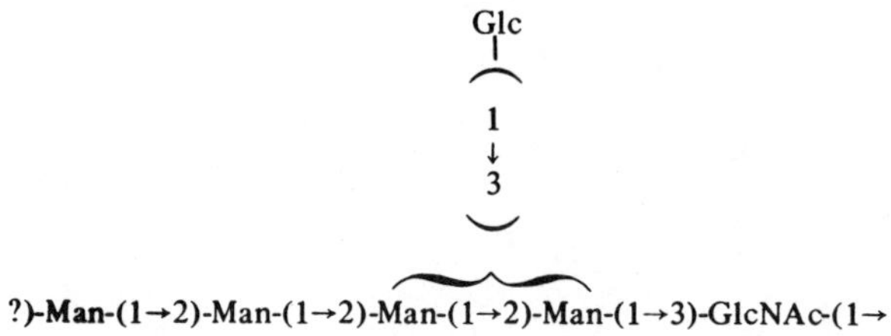 Glc 1→3 ?)-**Man**-(1→2)-Man-(1→2)-Man-(1→2)-Man-(1→3)-GlcNAc-(1→	49, 50

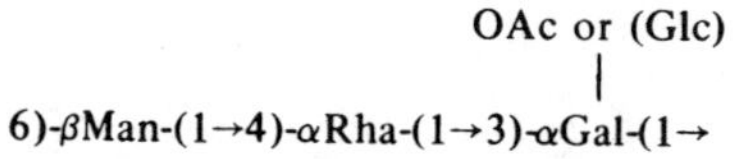

Table 3 (continued)
SUGAR SEQUENCES

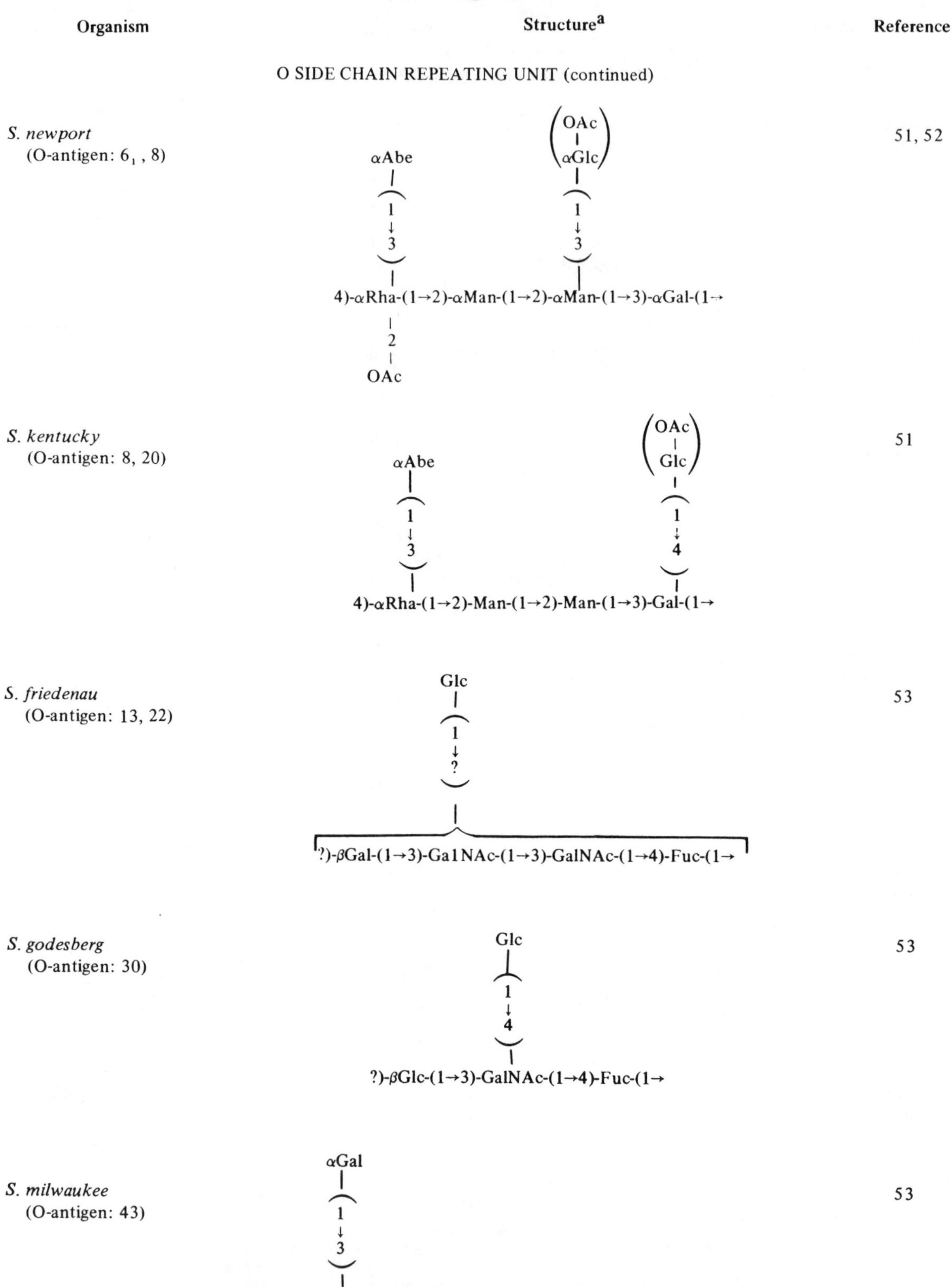

Organism	Structure[a]	Reference
	O SIDE CHAIN REPEATING UNIT (continued)	
S. newport (O-antigen: 6_1, 8)	αAbe (1→3) on αRha; (OAc–αGlc) (1→3) on second αMan; OAc on 2 of αRha 4)-αRha-(1→2)-αMan-(1→2)-αMan-(1→3)-αGal-(1→	51, 52
S. kentucky (O-antigen: 8, 20)	αAbe (1→3) on αRha; (OAc–Glc) (1→4) on Gal 4)-αRha-(1→2)-Man-(1→2)-Man-(1→3)-Gal-(1→	51
S. friedenau (O-antigen: 13, 22)	Glc (1→?) ?)-βGal-(1→3)-GalNAc-(1→3)-GalNAc-(1→4)-Fuc-(1→	53
S. godesberg (O-antigen: 30)	Glc (1→4) on GalNAc ?)-βGlc-(1→3)-GalNAc-(1→4)-Fuc-(1→	53
S. milwaukee (O-antigen: 43)	αGal (1→3) on βGal ?)-βGal-(1→3)-GalNAc-(1→3)-GlcNAc-(1→4)-Fuc-(1→	53

Table 3 (continued)
SUGAR SEQUENCES

Organism	Structure[a]	Reference
	O SIDE CHAIN REPEATING UNIT (continued)	
S. minnesota (O-antigen: 21)	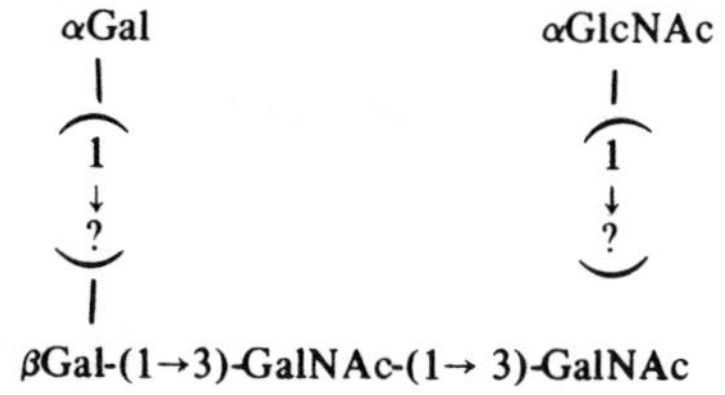 αGal αGlcNAc 1 1 ↓ ↓ ? ? βGal-(1→3)-GalNAc-(1→ 3)-GalNAc	54
Salmonella T_1-form (T_1-antigen instead of O-antigen)	**2) βRibf-(1→2)-βRibf-(1→2)-βRibf-(1→** **and** **6) βGalf-(1→3)-βGalf-(1→3)-βGal-(1→**	55
Salmonella T_2-form (T_2-antigen instead of O-antigen)	X-3 or 4)-GlcNAc (1→[c]	56
Escherichia coli O8	3)-αMan-(1→2)-αMan-(1→2)-αMan-(1→	57
E. coli O86	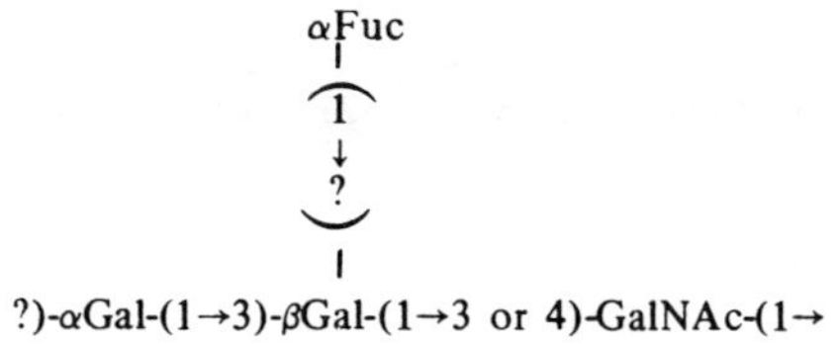 αFuc 1 ↓ ? ?)-αGal-(1→3)-βGal-(1→3 or 4)-GalNAc-(1→	58
E. coli O100	Glycerol O⁻-P=O O ?)-GlcNAc-(1→?)-Gal-(1→?)-Rha-(1→?)-Rha-(1→ or O⁻ Glycerol-O-P-O-Rha O 1 ↓ ? ?)-GlcNAc-(1→?)-Gal-(1→?)-Rha-(1→	59
E. coli O111	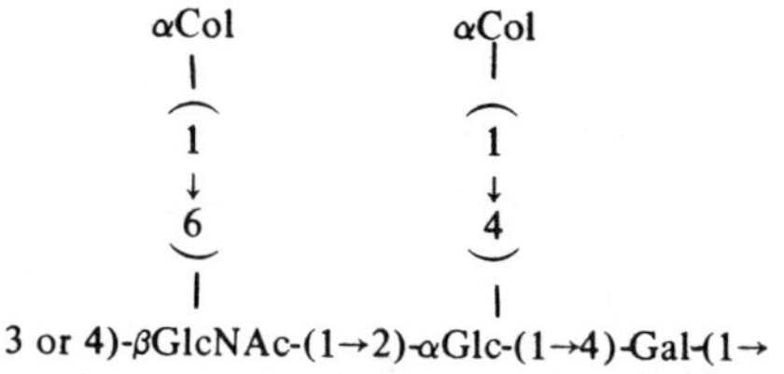 αCol αCol 1 1 ↓ ↓ 6 4 3 or 4)-βGlcNAc-(1→2)-αGlc-(1→4)-Gal-(1→	60

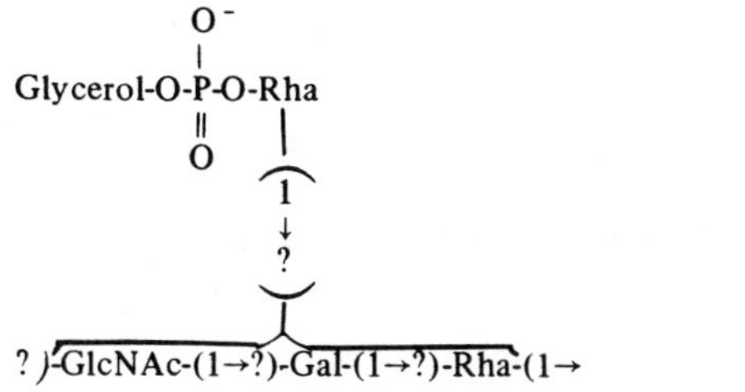

[c]X: unknown substitution. The entire side chain consists of this single, substituted *N*-acetylglucosamine residue; this structure strictly is not a *repeating* unit.

Table 3 (continued)
SUGAR SEQUENCES

Organism	Structure[a]	Reference
	O SIDE CHAIN REPEATING UNIT (continued)	
Shigella flexneri Y	3)-GlcNAc-(1→2)-Rha-(1→2)-Rha-(1→3)-Rha-(1→ or 3)-GlcNAc-(1→2)-Rha-(1→3)-Rha-(1→2)-Rha-(1→	61
Sh. flexneri 2a[d]	αGlc 1 ↓ 4 3)-GlcNAc-(1→2)-Rha-(1→2)-Rha-(1→3)-Rha-(1→	61, 62
Sh. flexneri 3a[d]	αGlc 1 ↓ 3 3)-GlcNAc-(1→2)-Rha-(1→2)-Rha-(1→3)-Rha-(1→	61, 62
Sh. flexneri 4a[d]	αGlc 1 ↓ 6 3)-GlcNAc-(1→2)-Rha-(1→2)-Rha-(1→3)-Rha-(1→	61, 62
Citrobacter O22 (strain 1026)	αAbe 1 ↓ 3 2)-αMan-(1→4)-Rha-(1→3)-αGal-(1→	62a
Citrobacter O22 (strain 86/57)	αAbe 1 ↓ 3 6)-αMan-(1→4)-αRha-(1→3)-αGal-(1→	62a

[d]These side chains correspond to glucosylated forms of *Sh. flexneri* Y O side chain. Here only the structures corresponding to the first of the two possible structures of *Sh. flexneri* Y are shown. Alternative structures, generated by the glucosylation of the second structure for *Sh. flexneri* Y chain, are equally possible.

Table 3 (continued)
SUGAR SEQUENCES

Organism	Structure[a]	Reference
	O SIDE CHAIN REPEATING UNIT (continued)	
Citrobacter O396	Glc Abe-2-OAc 1 1 ↓ ↓ 3 3 ?)-Man-(1→2)-Man-(1→2)-Man-(1→2)-Man-(1→3)-GlcNAc-(1→	62a
Yersinia pseudotuberculosis type IA (DDH[e] = paratose); type IIA (DDH = abequose); and type IVB (DDH = tyvelose)	αDDH 1 ↓ 3 α-6-deoxy-Hep 1 ↓ 4 3)-Gal(1→	13
Y. pseudotuberculosis type IB (DDH = paratose); and type IIB (DDH = abequose)	αDDH 1 ↓ 3 2)-αMan-(1→3)-αFuc-(1→	13
Y. pseudotuberculosis type III (DDH = paratose; and type IVA (DDH = tyvelose)	αDDH 1 ↓ 4 2)-αMan-(1→3)-αFuc-(1→	13
Y. pseudotuberculosis type VA (DDH = ascarylose); and type VB (DDH = none)	DDH 1 ↓ 3 3)-Man-(1→4)-Fuc-(1→	13

[e] DDH = 3,6-dideoxyhexose

Table 3 (continued)
SUGAR SEQUENCES

Organism	Structure[a]	Reference
	CORE OLIGOSACCHARIDE	
Salmonella typhimurium and *S. minnesota*	αGlcNAc (1→2), (Glc) (1→6), αGal (1→6) [(O chain)-(1→4)-αGlc-(1→2)-αGal-(1→3)-αGlc-(1→	63–71
E. coli strain B	Hep (1→6); Ⓟ-Ⓟ-CH_2CH_2-$NH_3^{\oplus}$ (6) Glc-(1→3)-Glc-(1→3)-Hep-(1→3)-Hep-(1→5)-KDO-(2→ Ⓟ KDO-7-Ⓟ-CH_2CH_2-$NH_3^{\oplus}$ (2→4) →7 or 8)-KDO	71a

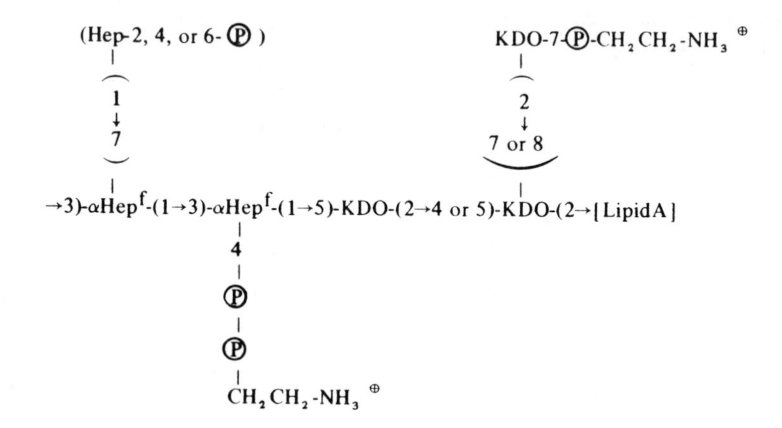

[f]The anomeric configuration of these heptosyl residues can be expressed as L-*glycero*-α-D-*manno*pyranosyl (British-American Rules of Carbohydrate Nomenclature) or β-L-*glycero*-D-*manno*pyranosyl (IUPAC-IUB Rules).

Table 3 (continued)
SUGAR SEQUENCES

Organism	Structure[a]	Reference

CORE OLIGOSACCHARIDE (continued)

Escherichia coli O100 ("R2-type core")[g] — Reference 72

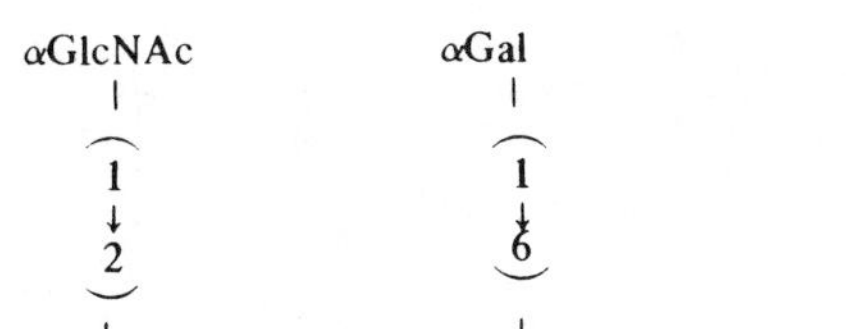

[(O chain)-(1]→4)-αGlc-(1→2)-αGlc-(1→3)-Glc-(1→

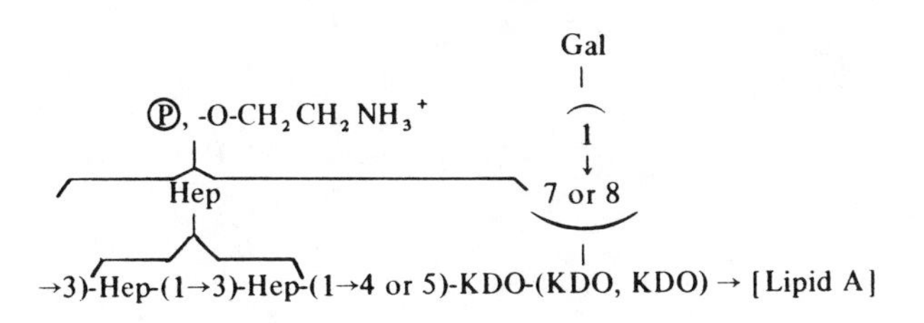

Shigella flexneri 4b ("R3-type core")[h,i] — Reference 75

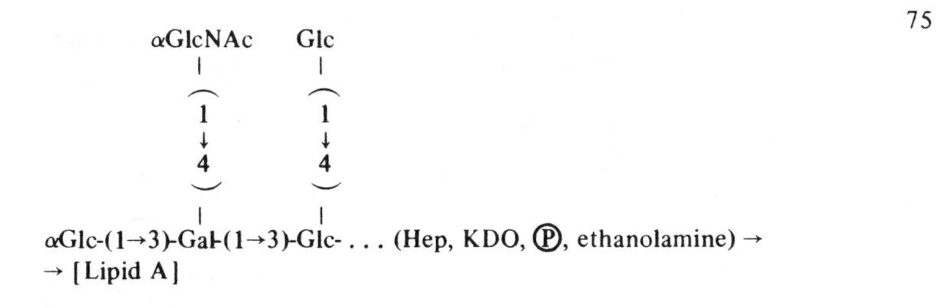

[g] *E. coli* O8:K42 produces a core of very similar structure.[73,74]

[h] *E. coli* O111:K58 and *Citrobacter* O10, 9b produce a core of very similar structure.[74,76]

[i] Immunological data suggest the presence of at least two other types of core structure in the *E. coli-Shigella* group. One, "R1", is found in *E. coli* O8:K27, *Sh. boydii* 3 and 6, and *Sh. sonnei,* whereas the other, "R4", is found in *E. coli* O14:K7 as well as in *Sh. dysenteriae* type 4.[73,74,76]

Table 3 (continued)
SUGAR SEQUENCES

Organism	Structure[a]	Reference
	CORE OLIGOSACCHARIDE (continued)	
E. coli 0111[j]	Glc-(1→4)-GlcN-(1→6 or 7)-Hep; . . . . Gal-(1→?)-Glc-(1→3)-Hep-(1→3)-Hep-(1→5)-KDO	77
	Glc-(1→6 or 7)-Hep; . . . . Gal-(1→?)-Glc-(1→4)-Glc-(1→3)-Hep-(1→3)-Hep-(1→5)-KDO. . . .	
	Glc-(1→6 or 7)-Hep; . . . . Glc-(1→4)-GlcN-(1→3)-Hep-(1→3)-Hep-(1→5)-KDO.	
	LIPID A	
Salmonella minnesota	(O chain)–(Core)-(2→3)-; [Ⓟ-4-βGlcN-(1→6)[k]-βGlcN-1-Ⓟ]$_n$; GlcN 6–R_2, N–R_1; GlcN 4–R_2, 3–R_2, N–R_1	78–81

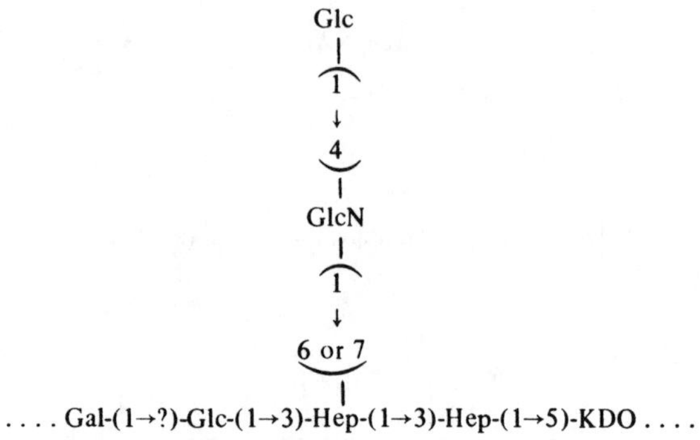

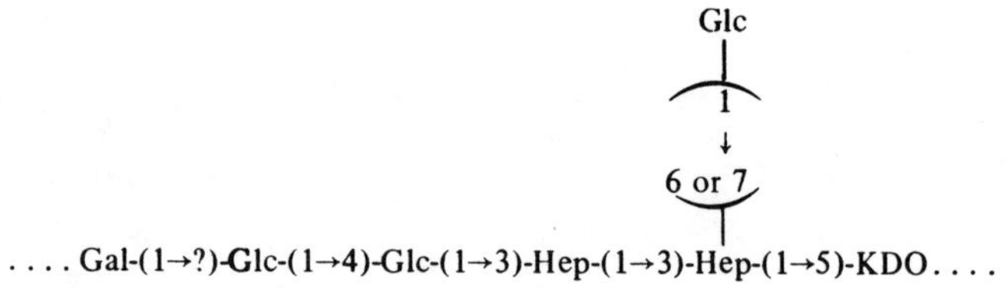

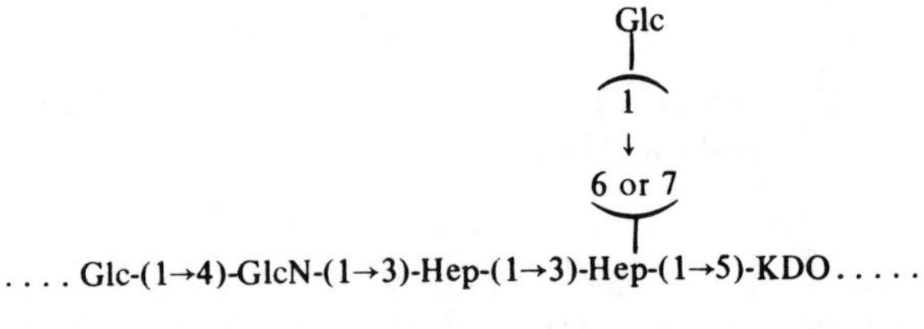

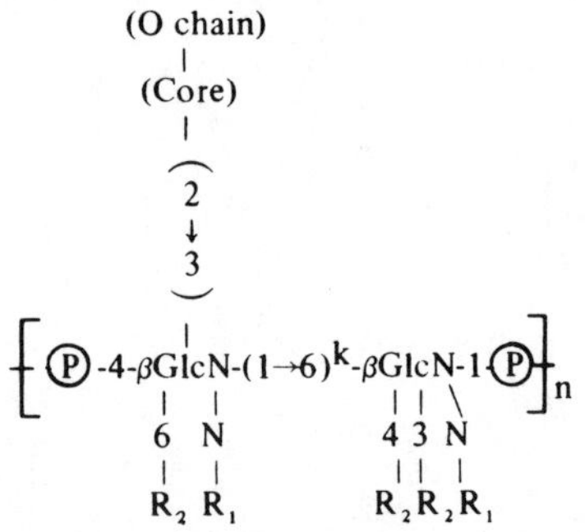

[j]The three structures shown were all found in a single strain. This strain lacks UDP-galactose 4-epimerase, and the core produced is incomplete in its structure.

Table 3 (continued)
SUGAR SEQUENCES

Organism	Structure[a]	Reference
	$R_1{}^l$ = 3-hydroxytetradecanoyl $[CH_3\text{-}(CH_2)_{10}\text{-}CHOH\text{-}CH_2\text{-}C(=O)\text{-}]$	
	$R_2{}^m$ = dodecanoyl $[CH_3\text{-}(CH_2)_{10}\text{-}C(=O)\text{-}]$,	
	hexadecanoyl $[CH_3\text{-}(CH_2)_{14}\text{-}C(=O)\text{-}]$, and	
	3-*O*-tetradecanoyl-3-hydroxytetradecanoyl $[CH_3\text{-}(CH_2)_{10}\text{-}CH(O\text{-}C(=O)\text{-}(CH_2)_{12}\text{-}CH_3)\text{-}CH_2\text{-}CO\text{-}]$	

[k] βGlcN-1(1→6)-βGlcN is the backbone of lipid A also in *Serratia marcescens*,[82,83] *Pseudomonas aeruginosa*,[84] *P. alcaligenes*,[84] and *Selenomonas ruminantium*.[85] In contrast, *E coli* O86 and *Shigella flexneri* are reported to produce lipid A with β- (1→4)-linked glucosamine disaccharides.[86] In lipid A of *Rhodopseudomonas* glucosamine is absent and a 2,3-diamino-hexose forms the backbone.[37]

[l] 3-Hydroxytetradecanoic acid also occurs as the main amide-bound acid in the lipid A of *E. coli*,[87,88] *Proteus mirabilis*,[89] *Aerobacter aerogenes*,[90] *Bordetella pertussis*,[91] *Rhodopseudomonas palustris*,[35] *Acinetobacter calcoaceticus*,[20] *Moraxella duplex*,[20] and *Neisseria perflava*.[27] Other hydroxy acids, however, occur in this position in some other organisms. Examples are: 3-hydroxydodecanoic acid in *Pseudomonas aeruginosa*,[92,93] *P. alcaligenes*,[12] *P. syncyanea*,[25] *P. diminuta*,[25] *P. pavonacea*,[25] and *4zotobacter agilis*;[94] 3-hydroxyhexadecanoic acid in *Branhamella catarrhalis*[20] and *Rhizobium trifolii*;[95] e-hydroxytridecanoic and 3-hydroxypentadecanoic acids in *Veillonella*[28] and *Selenomonas ruminantium*;[85] 3-hydroxy-iso-pentadecanoic and heptadecanoic acids in *Myxococcus, Polyangium, Flexibacter,* and *Cytophaga*;[34] and 3-hydroxy-*iso*-fridecanoic acid in *P. rubescens*.[25] The lipid A of *Brucella* does not contain any hydroxy fatty acids.[96,97]

[m] Other fatty acids occur in lipid A of other bacteria. These include *iso*-pentadecanoic and *iso*-heptadecanoic acids in myxobacteria,[34] octadecanoic acid in *Anacystis*[29] and *Brucella*,[96,97] docosanoic acid in Anacystis,[31] and 2-hydroxy-dodecanoic acid in *P. aeruginosa*[92,93] and *P. syncyanea*.[25]

Compiled by Hiroshi Nikaido, who acknowledges the advice and help received from Otto Lüderitz, K. Jann, and H. Mayer. Earlier results have been tabulated in References 98 and 99.

REFERENCES

1. **Orskov, Orskov, Jann, Jann, Muller-Seitz, and Westphal,** *Acta Pathol. Microbiol. Scand.*, 71, 339, 1967.

1a. **Schmidt, Jann, and Jann,** *Eur. J. Biochem.*, 42, 303, 1974.

2. **Kauffmann, Luderitz, Stierlin, and Westphal,** *Zentralbl. Bakteriol. Parasitenkd. Infektionskr. Hyg. Abt. 1 Orig.*, 178, 442, 1960; **Kauffmann, Jann, Kruger, Luderitz, and Westphal,** *Zentralbl. Bakteriol. Parasitenkd. Infektionskr. Hyg. Abt. 1 Orig.*, 186, 509, 1962.

3. **Westphal, Kauffmann, Luderitz and Stierlin,** *Zentralbl. Bakteriol. Parasitenkd. Infektionskr. Hyg. Abt. 1 Orig.*, 179, 336, 1960.

3a. **Schwarzmuller, Mayer, and Westphal,** *Abstr. Joint Meet. Eur. Soc. Immunol.*, Strasbourg, 1974.

4. **Seltmann and Hofmann,** *Zentralbl. Bakteriol. Parasitenkd. Infektionskr. Hyg. Abt. 1 Orig.* 199, 497, 1966; **Seltmann,** *Arch. Immunol. Ther. Exp.*, 16, 367, 1968.

5. **Keleti, Luderitz, Mlynarcik, and Sedlak,** *Eur. J. Biochem.*, 20, 237, (1971),; **Keleti, Mayer, Fromme, and Luderitz,** *Eur. J. Biochem.*, 16, 284, 1970.

6. **Nimmich and Korten,** *Pathol. Microbiol.* 36, 179, 1970.

7. **Hartley, Adams, and Tornabene,** *J. Bacteriol.*, 118, 848, 1974.

8. **Checcacci, Nava, Garofano, and Bo,** *G. Microbiol.*, 5, 87, 1958; **Luderitz,** unpublished data, 1969.

9. **Dimitriev, Backinowsky, Lvov, Kochetkov, and Hofman,** *Eur. J. Biochem.*, 40, 355, 1973.

10. **Lopes and Innis,** *Can. J. Microbiol.*, 16, 1117, 1970.

11. **Simmons,** *Biochem. J.*, 84, 353, 1962; 98, 903, 1966.

12. Rapin and Mayer, *Experientia,* 29, 756, 1973.
13. Samuelson, Lindberg, and Brubaker, *J. Bacteriol.,* 117, 1010, 1974.
14. Yuan and Horecker, *J. Bacteriol.,* 95, 2242, 1968.
15. Luderitz, Ruschmann, Westphal, Raff, and Wheat, *J. Bacteriol.,* 93, 1681, 1967.
16. Luderitz, Gmeiner, Kickhofen, Mayer, Westphal, and Wheat, *J. Bacteriol.,* 95, 490, 1968.
16a. Jann and Jann, *Eur. J. Biochem.* 2, 26, 1967.
17. Romanowska and Mulczyk, *Biochim. Biophys. Acta,* 136, 312, 1967; Romanowska and Reinhold *Eur. J. Biochem.,* 36, 160, 1973.
18. Raff and Wheat, *J. Biol. Chem.,* 242, 4610, 1967; *J. Bacteriol.,* 95, 2035, 1968.
19. Kotelko, Luderitz, and Westphal, *Biochem. Z.,* 343, 227, 1965; Bagdian, Droge, Kotelko, Luderitz, Westphal, Yamakawa, and Ueta *Biochem. Z.,* 344, 197, 1966; Sidorczyk and Kotelko *Arch. Immunol. Ther. Exp.,* 21, 829, 1973; Kotelko, Fromme, and Sidorczyk, *Bull. Acad. Pol. Sci.,* 23, 249, 1975.
19a. Key, Gray, and Wilkinson, *Biochem. J.,* 120, 559, 1970.
20. Wilkinson, Galbraith, and Lightfoot, *Eur. J. Biochem.,* 33, 158, 1973.
21. Chester, Gray, and Wilkinson, *Biochem. J.,* 126, 395, 1972.
22. Volk, *J. Bacteriol.,* 91, 39, 1966; 95, 980, 1968.
23. Hickman and Ashwell, *J. Biol. Chem.,* 241, 1424, 1966.
24. Humphrey and Vincent, *J. Gen. Microbiol.,* 59, 411, 1969.
25. Graham and O'Brien, *Antonie van Leeuwenhoek J. Microbiol. Serol.,* 34, 326, 1965.
26. Adams, Kates, Shaw, and Yaguchi, *Can. J. Biochem.,* 46, 1175, 1968.
27. Adams, *Can. J. Biochem.,* 49, 243, 1971.
28. Adams, Tornabene, and Yaguchi, *Can. J. Microbiol.,* 15, 365, 1969; Adams, Quadling, Yaguchi, and Tornabene, *Can. J. Microbiol.,* 16, 1, 1970.
29. Jackson and Redmond, *FEBS Lett.,* 13, 117, 1971.
30. Redmond, Korsch, and Jackson, *Aust. J. Exp. Biol. Med. Sci.,* 51, 229, 1973.
31. Jann, Jann, and Beyaert, *Eur. J. Biochem.,* 37, 531, 1973.
32. Nell and Hardy, *Immunochemistry,* 3, 233, 1966.
33. Sutherland and Smith, *J. Gen. Microbiol.,* 74, 259, 1973.
34. Rosenfelder, Luderitz, and Westphal, *Eur. J. Biochem.,* 44, 411, 1974.
35. Weise, Drews, Jann, and Jann, *Arch. Mikrobiol.,* 71, 89, 1970.
35a. Weckesser, Katz, Drews, Mayer, and Fromme, *J. Bacteriol.,* 120, 672, 1974.
35b. Weckesser, Drews, and Fromme, *J. Bacteriol.,* 109, 1106, 1972; Weckesser, Mayer, and Drews, *Eur. J. Biochem.,* 16, 158, 1970.
36. Weckesser, Rosenfelder, Mayer, and Luderitz, *Eur. J. Biochem.,* 24, 112, 1971; Weckesser, Mayer, and Fromme, *Biochem. J.,* 135, 293, 1973; Weckesser, Drews, Roppel, Mayer and Fromme, *Arch. Mikrobiol.,* 101, 233, 1974; Roppel, Mayer, and Weckesser, *Carbohydr. Res.,* 40, 31, 1975.
37. Weckesser, Drews, Fromme, and Mayer, *Arch. Mikrobiol.,* 92, 123, 1973; Mayer, Framberg, and Weckesser, *Eur. J. Biochem.,* 44, 181, 1974.
38. Hewett, Knox, and Bishop, *Eur. J. Biochem.,* 19, 169, 1971.
39. Hellerqvist, Lindberg, Samuelsson, and Lindberg, *Acta Chem. Scand.,* 25, 955, 1971.
40. Hellerqvist, Lindberg, Svensson, Holme, and Lindberg, *Carbohydr. Res.,* 8, 43, 1968; 9, 237, 1969.
41. Hellerqvist, Larm, Lindberg, Holme, and Lindberg, *Acta Chem. Scand.,* 23, 2217, 1969.
42. Hellerqvist, Lindberg, Svensson, Holme, and Lindberg, *Acta Chem. Scand.,* 23, 1588, 1969.
43. Hellerqvist, Lindberg, Pilotti, and Lindberg, *Acta Chem. Scand.,* 24, 1168, 1970.
44. Robbins and Uchida, *Biochemistry,* 1, 323, 1962.
45. Hellerqvist, Lindberg, Lonngrenn, and Lindberg, *Carbohydr. Res.,* 16, 289, 1971.
46. Hellerqvist, Lindberg, Lonngrenn, and Lindberg, *Acta Chem. Scand.,* 25, 939, 1971.
47. Staub and Girard, *Bull. Soc. Chim. Biol.,* 47, 1245, 1965.
48. Hellerqvist, Lindberg, Pilotti, and Lindberg, *Carbohydr. Res.,* 16, 297, 1971.
49. Fuller and Staub, *Eur. J. Biochem.,* 4, 286, 1968.
50. Hellerqvist, Lindberg, Svensson, Lindberg, and Holme, personal communication, 1968.
51. Hellerqvist, Lindberg, Svensson, Holme, and Lindberg, *Carbohydr. Res.,* 14, 17, 1970.
52. Hellerqvist, Lindberg, Lonngrenn, and Lindberg, *Acta Chem. Scand.,* 25, 601, 1971.
53. Simmons, Luderitz, and Westphal, *Biochem. J.,* 97, 807, 815 and 820, 1965.
54. Luderitz, Galanos, Risse, Ruschmann, Schlecht, Schmidt, Schulte-Holthausen, Wheat, Westphal, and Schlosshardt, *Ann. N.Y. Acad. Sci.,* 133, 349, 1966.
55. Berst, Luderitz, and Westphal, *Eur. J. Biochem.,* 18, 361, 1971.
56. Bruneteau, Volk, Singh, and Luderitz, *Eur. J. Biochem.,* 43, 501, 1974.
57. Reske and Jann, *Eur. J. Biochem.,* 31, 320, 1972.
58. Springer, Wang, Nichols, and Shear, *Ann. N.Y. Acad. Sci.,* 133, 566, 1966.

59. **Jann, Jann, Shmidt, Orskov, and Orskov,** *Eur. J. Biochem.*, 15, 29, 1970.
60. **Edstrom and Heath,** *J. Biol. Chem.*, 242, 4125, 1967.
61. **Lindberg, Lonngrenn, Ruden, and Simmons,** *Eur. J. Biochem.*, 32, 15, 1973.
62. **Simmons,** *Eur. J. Biochem.*, 11, 554, 1969.
62a. **Jann,** personal communication.
63. **Osborn, Rosen, Rothfield, Zeleznick, and Horecker,** *Science*, 145, 783, 1964.
64. **Sutherland, Luderitz, and Westphal,** *Biochem. J.*, 96, 439, 1965.
65. **Nikaido,** *J. Biol. Chem.*, 244, 2835, 1969; *Eur. J. Biochem.*, 15, 57, 1970.
66. **Hammerling, Luderitz, and Westphal,** *Eur. J. Biochem.*, 15, 48, 1970.
67. **Hellerqvist and Lindberg,** *Carbohydr. Res.*, 16, 39, 1971.
68. **Droge, Luderitz, and Westphal,** *Eur. J. Biochem.*, 4, 126, 1968.
69. **Droge, Lehmann, Luderitz, and Westphal,** *Eur. J. Biochem.*, 14, 175, 1970.
70. **Lehmann, Luderitz, and Westphal,** *Eur. J. Biochem.*, 21, 339, 1971.
71. **Hammerling, Lehmann, and Luderitz,** *Eur. J. Biochem.*, 38, 453, 1973.
71a. **Prehm, Jann, Stirm, and Jann,** *Eur. J. Biochem.*, in press.
72. **Hammerling, Luderitz, Westphal, and Makela,** *Eur. J. Biochem.*, 22, 331, 1971.
73. **Schmidt, Jann, and Jann,** *Eur. J. Biochem.*, 10, 501, 1969.
74. **Schmidt, Fromme, and Mayer,** *Eur. J. Biochem.*, 14, 357, 1970.
75. **Johnston, Johnston, and Simmons,** *Biochem. J.*, 105, 79, 1967.
76. **Mayer and Schmidt,** *Zentralbl. Bakteriol. Parasitenkd. Infektionskr. Hyg. Abt. 1 Orig. A.*, 224, 345, 1973.
77. **Fuller, Wu, Wilkinson, and Heath,** *J. Biol. Chem.*, 248, 7938, 1973.
78. **Gmeiner, Luderitz, and Westphal,** *Eur. J. Biochem.*, 7, 370, 1969.
79. **Gmeiner, Simon, and Westphal,** *Eur. J. Biochem.*, 21, 355, 1971.
80. **Rietschel, Gottert, Luderitz, and Westphal,** *Eur. J. Biochem.*, 28, 166, 1973.
81. **Luderitz, Galanos, Lehmann, Nurminen, Rietschel, Rosenfelder, Simon, and Westphal,** *J. Infect. Dis.*, 128, S17, 1973.
82. **Adams and Singh,** *Can. J. Biochem.*, 48, 55, 1970.
83. **Bundle and Shaw,** *Carbohydr. Res.*, 21, 211, 1972.
84. **Drewry, Lomax, Gray, and Wilkinson,** *Biochem. J.*, 133, 563, 1973.
85. **Kamio, Kim, and Takahashi,** *J. Biochem. Tokyo*, 70, 189, 1971.
86. **Adams and Singh,** *Biochim. Biophys. Acta*, 202, 553, 1970.
87. **Ikawa, Koepfli, Mudd, and Niemann,** *J. Am. Chem. Soc.*, 75, 1035, 1953.
88. **Burton and Carter,** *Biochemistry*, 3, 411, 1964.
89. **Nesbitt and Lennarz,** *J. Bacteriol.*, 89, 1020, 1965.
90. **Gallin and O'Leary,** *J. Bacteriol.*, 96, 660, 1968.
91. **Kasai,** *Ann. N.Y. Acad. Sci.*, 133, 486, 1966.
92. **Fensom and Gray,** *Biochem. J.*, 114, 185, 1969.
93. **Hancock, Humphreys, and Meadow,** *Biochim. Biophys. Acta*, 202, 389, 1970.
94. **Kaneshiro and Marr,** *Biochim. Biophys. Acta*, 70, 271, 1963.
95. **Russa and Lorkiewicz,** *J. Bacteriol.*, 119, 771, 1974.
96. **Lacave, Asselineau, Serre, and Roux,** *Eur. J. Biochem.*, 9, 189, 1969.
97. **Berger, Fukui, Ludwig, and Rosselet,** *Proc. Soc. Exp. Biol. Med.*, 131, 1376, 1969.
98. **Luderitz, Jann, and Wheat,** in *Comparative Biochemistry*, Vol. 26, Florkin and Stotz, Eds., Elsevier, Amsterdam, 1968, p. 105.
99. **Luderitz, Westphal, Staub, and Nikaido,** in *Microbial Toxins*, Vol. IV, Weinbaum, Kadis, and Ajl, Eds., Academic Press, New York, 1971, p. 145.

LIPIDS IN BACTERIAL TAXONOMY*

M. P. Lechevalier

INTRODUCTION

Fifty years ago, the pioneering work of Anderson on mycobacterial lipids[16,267] gave the initial impetus to what was to be an explosive development of interest in the study of bacterial lipids in general. Early attempts at analyzing the rapidly accumulating data from the point of view of the role of lipids in bacterial taxonomy were made by Asselineau,[15,16] by O'Leary,[266,267] and by Kates.[152] As the mountain of data continued to grow, the reviews became more specialized: phosphatidyl glycerols and aminoacyl phosphatidyl glycerols,[218] lipids of bacterial membranes,[219] phospholipids,[126,278] glycolipids of mycobacteria and their relatives,[210] bacterial glycolipids, excluding mycobacteria and closely allied forms,[324] mycobacterial lipids,[111] bacterial plasmalogens,[104] bacterial glycophospholipids,[328] lipids of Gram-negative bacteria,[308] glycosyl glycerides,[321] lipids of mycobacteria grown in vivo,[142] and sugar-containing lipids.[326] Some of the more general recent reviews dealing with lipids in bacterial taxonomy are those of Lennarz,[213] Goldfine,[103] Váczi,[385] Shaw,[325] and O'Leary.[268-270] A certain number of reviews have also appeared in Japanese: Kojima,[175] Azuma,[19] Matsumoto and Miwa,[230] Muramatsu,[259] Uchida,[375] and Suto and Ryozo.[343]

Which Lipids?

For the purpose of this review, lipids are defined as fatty acids, fatty alcohols, hydrocarbons, or compounds containing these substances that are soluble in organic solvents. This eliminates the lower fatty acids and alcohols with less than six carbons, which, as products of fermentation of carbohydrates, are widely used in the taxonomy of anaerobic bacteria as well as of certain aerobes. It also excludes such fatty acid-containing substances as lipopolysaccharides, lipoproteins, and lipoteichoic acids that are water-soluble. A great deal is known about the structure and location of the lipopolysaccharides commonly found in Gram-negative bacteria; however, their usefulness in taxonomy has not been well established. [62,81,217,313] Likewise, lipoproteins have been studied largely from a functional or structural point of view; little work has been done on their relation to taxonomy.[369] In contrast, lipoteichoic acids were used in the serological classification of "group D" streptococci and "group F" lactobacilli before their chemical structure and cellular location were known. These typical glycerol- or ribitol-phosphate polymers are substituted with various glycosyl and D-alanyl ester groups and are covalently linked to a glycolipid of the plasma membrane. They are found exclusively in Gram-positive bacteria.[406]

This leaves us with the fatty acids having six or more carbons, and with glycerides, fatty alcohols, glyco-, sulfo-, and peptidolipids, phospholipids, waxes, and hydrocarbons. As we shall see in the following pages, some of these are of potential value as taxonomically important characters.

Which Bacteria?

The term "bacteria" will be restricted here to all free-living, cell wall-producing prokaryotic organisms that contain two types of nucleic acids and, if photosynthetic, do not produce oxygen. This excludes the Mycoplasmatales, the Bdellovibriales, and the blue-green algae (Cyanobacteria).

* Reproduced with slight modifications from *Crit. Rev. Microbiol.*, 5(2), 109, 1977.

For this review, actinomycetes are classified according to a modified version of the scheme of Cross and Goodfellow;[60] all other bacteria are grouped according to the eighth edition of *Bergey's Manual of Determinative Bacteriology*.[47b] Where the nomenclature has changed, taxa are referred to under the new epithet, with the older one in parentheses.

Why Not More Detailed Quantitative Data?

Lipid data on the same strain, grown and analyzed under the same conditions in two different laboratories, or even in the same laboratory, may vary appreciably from a quantitative point of view (see Effect of Environmental Conditions). Thus, the most important facts to consider are the types of lipids that occur in a given organism and which of these represent the major types. Actual percentages will not be of much use.

For most bacteria the total lipid content is quite low, usually less than 5% by dry weight. Some exceptions to this rule include the mycobacteria and certain allied forms, such as the corynebacteria of the *diphtheriae* type and the nocardiae, which commonly have as much as 20% lipids by dry weight. Higher values for these genera have been reported; however, such data are frequently based on extracts of extremely old cells, where many of the normal nonlipidic constituents have been autolyzed. Certain other actinomycetes that are less closely related to the mycobacteria, such as members of the genus *Chainia*, contain very large amounts of lipids (35 to 50%).[207] Bacteria producing the polymer poly-β-hydroxybutyrate also have a high lipid content.

FATTY ACIDS

The largest amount of data on bacterial lipids accumulated to date concerns bacterial fatty acid composition. In most cases, certain types of fatty acids predominate in a given taxon. For example, the major fatty acids of members of the genus *Bacillus* are almost exclusively of the branched-chain type. Thus, an atypical, nonsporulating *Bacillus* sp. might be partially characterized by means of its fatty acid composition.

Bacteria contain the straight-chain (normal) fatty acids common in higher plants and animals (Table 1). Although the more common monounsaturated acids with a double bond between carbons 9 and 10 (oleic series) do occur in bacteria, their monounsaturated acids are usually different from those of higher forms, having a double bond between carbons 7 and 8, counting from the methyl end (vaccenic series). However, the fatty acids that make bacteria unique are the β-hydroxy, cyclopropane, and branched-chain fatty acids, which are not common elsewhere. Furthermore, bacteria do not contain the polyunsaturated fatty acids, the molecules with acetylenic linkages, or the sterols that are so common in other living forms.

In some bacteria, fatty acids occur free or in the form of glycerides (Table 1); however, most of the fatty acids of bacteria are linked to larger molecules, primarily phospholipids and glycolipids or lipoproteins, lipopolysaccharides, and lipoteichoic acids. Fatty acids, as well as phospholipids, have also been reported to be released into the growth medium,[2,42,47a] but these may be products of cellular decomposition. A summary of recent data on bacterial fatty acids is given in Tables 2 to 5.

Glycerides

The bacteria and actinomycetes that have been reported to contain tri- (or di- or mono-) glycerides include members of the genera *Acinetobacter*,[225] *Arthrobacter*,[182] *Bacteroides*,[189] *Bacillus*,[47c] *Campylobacter (Vibrio)*,[368] *Chainia*,[207] *Clostridium*,[123] *Corynebacterium*,[193] *Enterobacter*,[370] *Mycobacterium*,[33,55] *Nocardia*,[200,384] *Pseudomonas*,[285] *Rhizobium*,[86] *Staphylococcus*,[310] and *Streptomyces*.[159,200] According

Table 1
FATTY ACIDS AND GLYCERIDES

Name	General structure	Abbreviation[a]
Normal (straight-chain, saturated)	$CH_3(CH_2)_XCOOH$	n-$C_{n:0}$
Monounsaturated (vaccenic series)[b]	$CH_3(CH_2)_5-CH=CH-(CH_2)_XCOOH$	$C_{n:1}$
Monounsaturated (oleic series)	$CH_3(CH_2)_7-CH=CH-(CH_2)_XCOOH$	$C_{n:1}$
Diunsaturated (animals, plants)	$CH_3(CH_2)_X-CH=CH-CH_2-CH=CH-(CH_2)_XCOOH$	$C_{n:2}$
Diunsaturated [*Aerococcus viridans (Pediococcus homari)*][378]	$CH_3(CH_2)_X-CH=CH-CH=CH-(CH_2)_XCOOH$	$C_{n:2}$
Polyunsaturated (*Mycobacterium phlei*)[14]	$CH_3(CH_2)_X-(CH=CH-CH_2-CH_2)_YCOOH$ x = 12, 14; y = 5, 6, or 4	
Branched (*iso*)	$CH_3-CH(CH_3)-(CH_2)_XCOOH$	*iso* $C_{n:0}$
Branched (*anteiso*)	$CH_3-CH_2-CH(CH_3)-(CH_2)_XCOOH$	
Tuberculostearic acid	$CH_3(CH_2)_7-CH(CH_3)-(CH_2)_8COOH$	*anteiso*-$C_{n:0}$
Cyclopropane (Δ)	$CH_3(CH_2)_XCH-CH-(CH_2)_XCOOH$ (the two CH carbons bridged by CH_2)	10-Me-$C_{16:0}$
Hydroxy (α)	$CH_3(CH_2)_X-CH(OH)-COOH$	Δ-$C_{n:0}$
Hydroxy (β)	$CH_3(CH_2)_X-CH(OH)-CH_2-COOH$	2-OH-$C_{n:0}$
Mycolic acid (β-OH, α-branched) (*Mycobacterium*) R = $C_{43\text{-}61}H_x$ n = 19, 21, 23	$R-CH(OH)-CH((CH_2)_nCH_3)-COOH$	3-OH-$C_{n:0}$ Eumycolate

Table 1 (continued)
FATTY ACIDS AND GLYCERIDES

Name	General structure	Abbreviation[a]
Mycolic acid (β-OH, α-branched) (*Nocardia*) $R = C_{18-52}H_x$ n = 9, 11, 13, 15	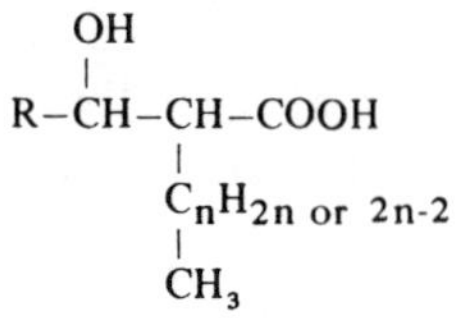 $R-CH(OH)-CH(C_nH_{2n \text{ or } 2n-2}-CH_3)-COOH$	Nocardomycolate
Mycolic acid (β-OH, α-branched) (*Corynebacterium*) $R = C_{14-18}H_x$ m = 2, 4, 6 n = 2	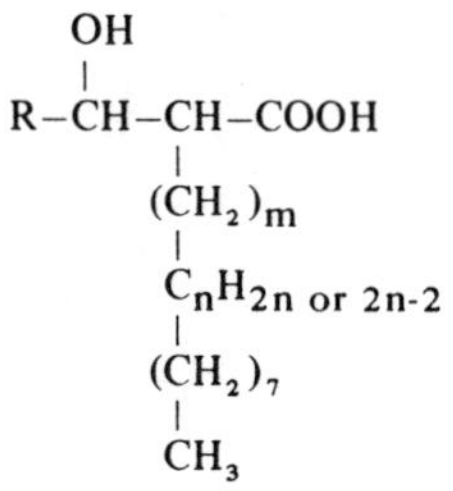 $R-CH(OH)-CH((CH_2)_m-C_nH_{2n \text{ or } 2n-2}-(CH_2)_7-CH_3)-COOH$	Corynomycolate
Glycerides: Triglyceride R = R′ = R″ = fatty acid; Diglyceride R = R′ = fatty acid[c], R″ = H; Monoglyceride R = fatty acid[c], R′ = R″ = H	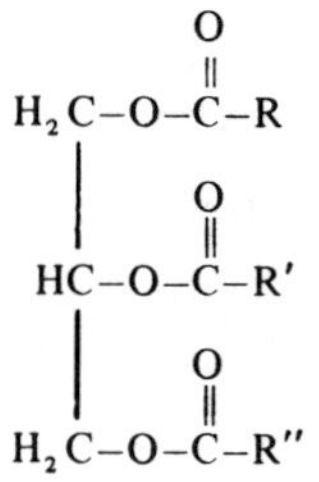 $H_2C-O-C(=O)-R$ / $HC-O-C(=O)-R'$ / $H_2C-O-C(=O)-R''$	

Note: 0, 1, 2, 3, or 4 = number of unsaturations; 10-Me = 10-methyl (methyl group on carbon 10); 2-OH = 2-hydroxy (hydroxyl group on carbon 2).

[a] n = number of carbons in the main chain.

[b] Position of unsaturation of any given acid is symbolized by Δ. Δ^9 would mean that the unsaturation is between carbon 9 and carbon 10, counting from the carboxyl end:

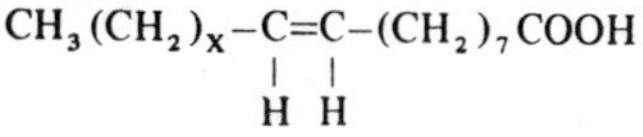

$$CH_3(CH_2)_x-CH=CH-(CH_2)_7COOH$$

[c] Positions of the fatty acids are not absolute.

to O'Leary,[267] this character, since it is not widespread among bacteria, may be a potentially useful taxonomic marker.

Polyunsaturated Fatty Acids

Until recently, polyunsaturated fatty acids (PFA) were unknown in bacteria. The few reports in the literature have generally been dismissed as mistakes or as the result of cellular uptake of polyunsaturated fatty acids present in the medium.[213,267,271] However, even those PFA that were recently described from *Mycobacterium phlei*[14] and from *Nocardia kirovani*[383] are quite unlike the usual types found in animals and plants in that they have chain lengths of 35 to 45 carbons. The diunsaturated fatty acids from *Aerococcus viridans (Pediococcus homari)* have double bonds that are conjugated,[378] unlike those found in animals and plants, which are not (Table 1).

Table 2
FATTY ACIDS OF GRAM-NEGATIVE BACTERIA[a]

Organism	Types of lipids analyzed	S	U	Br	C	OH	UK	Reference
BUDDING OR APPENDAGED BACTERIA[a]								
Caulobacter								
crescentus	F, B	b	b	b	b			58
Hyphomicrobium sp.	F	b	b		b			18
GLIDING BACTERIA								
Cystobacteriaceae								
Stigmatella								
aurantiaca	F, B	b	b	b				397
Myxococcaceae								
Myxococcus								
xanthus	F, B	b	b	b				397
PHOTOTROPHIC BACTERIA								
Chlorobiaceae								
Chlorobium								
limicola	F, B	b	b		c			162
limicola forma sp. *thiosulfatophilum*	F, B	b	b		b			162
phaeobacteriodes	F, B	b	b		c			162
vibrioforme forma sp. *thiosulfatophilum*	F, B	b	b		c			162
Chloroflexus								
aurantiacus[d]	F, B	b	b					162
Pelodictyon								
luteolum	F, B	b	b		b			162
Rhodospirillaceae								
Rhodopseudomonas								
viridis	P	b	b					300
Rhodospirillum								
rubrum	P	b	b					46b

Table 2 (continued)
FATTY ACIDS OF GRAM-NEGATIVE BACTERIA[a]

Organism	Types of lipids analyzed	S	U	Br	C	OH	UK	Reference
CHEMOLITHOTROPHIC BACTERIA								
Nitrobacteriaceae								
Nitrobacter								
winogradskyi (agilis)	F, B	b	b	e				38
	F	b	b		b			18
winogradskyi	F	b	b		b			18
Nitrococcus								
mobilis	F, B	b	b					38
Nitrosolobus								
multiformis	F, B	b	b					38
Nitrosomonas								
europaea	F, B	b	b					38
sp.	F, B	b	b					38
Nitrospina								
gracilis	F, B	b	b	c				38
SULFUR BACTERIA								
Thiobacillus								
concretivorus	F, B	b	b					3
denitrificans	F, B	b	b				c	3
ferrooxidans	F, B	b	b					3
intermedius	F, B	b	b					3
neopolitanus	F, B	b	b				b	3
novellus	F, B	b	b		b			215
	F, B	b	b				b	3
thiocyanooxidans	F, B	b	b				b	3
thiooxidans	F, B	b	b				b	3
	F, B	b	b		b			214
thioparus	F, B	b	b				b	3

SPIROCHETES						
Leptospira						
australis[f]	F, B	b	b			179
autumnalis[f]	F, B	b	b			179
biflexa[f]	F, B	b	b			179
canicola[f]	F, B	b	b			179
interrogans (icterohaemorrhag-iae)	F, B	b	b			179
patoc	F, B	b	b			179
Spirochaeta						
aurantia	F	b	b	b		216[a]
litoralis	F	b		b		216[a]
stenostrepta	F	b		b		137
	B	b		b		137
	F	b		b		216a
zuelzerae	F, B	b	b	b		233
	F	b	b	b		216a
Treponema						
denticola	F	b	b			216a
pallidum	F, B	b	b			233
phagedenis	F	b	b			216a
refringens	F	b	b			216a
scoliodontum	F	b	b			216a
vincentii	F	b	b			216a
SPIRAL AND CURVED BACTERIA						
Spirillaceae						
Campylobacter						
fetus						
(Serotype I)	F	b	b	e	b	368
(Serotype V)	F	b	b	e		368
Spirillum						
linum	F	b	b	e		271
AEROBIC RODS AND COCCI						
Azotobacteriaceae						
Azotobacter						
vinelandii	F	b			b	226

Table 2 (continued)
FATTY ACIDS OF GRAM-NEGATIVE BACTERIA[a]

Organism	Types of lipids analyzed	S	U	Br	C	OH	UK	Reference
Pseudomonadaceae								
Pseudomonas								
acidovorans	F, B	b	b			b		256
aeruginosa	F	b	b					35
	F, B	b	b		c	b		256
	F, P	b	b	e				11a
	F	b	b	e				271
alcaligenes	F, B	b	b	b	b	b		65
	F, B	b	b			b		256
cepacia (multivorans)	F, B	b	b		b	b	c	319
	F, B	b	b		b	b		256
diminuta	F, B	b	b		b	b		140
	F, B	b	b			b	b	254
	B	b	b			b	b	412
fluorescens	F	b	b		c			61
	P	b	b		b			61
fragi	F	b	b					235
	B	b	b				b	235
kingii[d]	F, B	b	b		b	b		256
maltophila	F, B	b	b	b		b	b	255
	F, B	b	b	b		b	b	254, 256
	F	b	b	b				255
pavonacea	F	b	b					412
	B	b	c			b		412

perfectomarinus	F	b	b	e				271
pseudoalkaligenes	F, B	b	b	b	b	b		65
putida	F, B	b	b			b		256
putrefaciens[f]	F, B	b	b	b				65
rubescens[d]	F	b	b	b				412
	B	b	b	b		b	b	412
	P	b	b	b				409
	OL	b	b	b		b		409
stutzeri	F	b	b				b	412
	B	b	b			b		412
	F, B	b	b			b		256
syncyanea	F	b	b					412
	B	b	b			b		412
testosteroni	F, B	b	b			b		256
vesicularis	F, B	b	b			b		140
Rhizobiaceae								
Rhizobium sp.	F	b	b	b	b		g	97, 98
	F	b	b	b	b			98
TAXA OF UNKNOWN AFFILIATION								
Alcaligenes								
aquamarinus	F	b	b	e				271
denitrificans	F, B	b	b		b	b		65
	P	b	b		b		b	356
eutrophus	F	b	b		b		b	356
faecalis	F	b	b		b	b		99
	B	b	b		b	b		99
	F, B	b	b		b	b		65
odorans	F, B	b	b		b	b		65
sp.	P	b	c		b			352
Bordetella								
pertussis	F	b	b					353
Brucella								
melitensis	F	b	b		b		b	353

Table 2 (continued)
FATTY ACIDS OF GRAM-NEGATIVE BACTERIA[a]

Organism	Types of lipids analyzed	S	U	Br	C	OH	UK	Reference
Thermus								
aquaticus	F	c		b				280
	F	b		b				130
	F, B	b	b	b				306, 307
flavis[d]	F			b				280
sp.	F	b		b				130
FACULTATIVELY ANAEROBIC RODS								
Enterobacteriaceae								
Citrobacter								
freundii	F	b	b		b			220
Enterobacter								
aerogenes	F	b	b	c			c	370
Escherichia								
coli	F	b	b		b			77
	B	b	b		b			77
	F	b	b		b			35
	F, B?	b	b		b	b		310
	F	b	b		b			34
	P	b	b		b			59
	P	b	b	c	b			310
	P	b	b		b			77
Klebsiella								
pneumoniae(aerogenes)	F, B	b	b		b			77
	P	b	b		b			77
	F, B	b	b	b?			b	74
Proteus								
inconstans (Providencia alcali-faciens)	F	b	b		b			220

inconstans (Providencia stuarti)	F	b	b		b			220
vulgaris	F, B	b	b					339
sp.	F	b	b		b			220
Salmonella								
arizonae	F	b	b		b			220
typhimurium	F	b	b		b			362
	N	b	b		b			362
Serratia								
marcescens	F	b	b					48, 339
	F	b	b		b			46a
	F, B	b	b				b	74
Yersinia								
pestis	F	b	b		b	c	c	363
pseudotuberculosis	F, B	b	b				e	297
	F	b	b		b	c	c	363
Vibrionaceae								
Vibrio								
algosus	F	b	b					271
cholerae	F	b	b	c				271
fischeri	F	b	b	e				271
fischeri (marinus)	F	b	b	e				271
marinofulvus[f]	F	b	b	e				271
parahaemolyticus	F	b	b	e				271
parahaemolyticus (alginolyticus)	F	b	b	e				271
spp.	F	b	b	c				271
TAXA OF UNCERTAIN AFFILIATION								
Flavobacterium								
thermophilum	F	b		b				280
Haemophilus								
aphrophilus	F, B	b	b				e	297
Pasteurella								
haemolytica	F, B	b	b				e	297
multocida	F, B	b	b				e	297

Table 2 (continued)
FATTY ACIDS OF GRAM-NEGATIVE BACTERIA[a]

Organism	Types of lipids analyzed	S	U	Br	C	OH	UK	Reference
ANAEROBES								
Bacteroidaceae								
Bacteroides								
corrodens	F, B	b	b					297
fragilis var. *fragilis*	F, B	b	e				b	297
	P	b		b			b	92
fragilis var. *thetaiotamicron*	P	b		b			b	92
	P	b	b	b		b		340
fragilis var. *vulgatus*	P	b		b			b	92
melanogenicus	F, B	b		b?			b	297
putredinis	P	b		b			b	92
ruminicola	N	b		b				189
	P	b		b				189
	Sp	b		b			b	189
Fusobacterium								
mortiferum	G, P	b	b	b				92
necrophorum	G, P	b	b	b				92
varium	G, P	b	b	b				92
TAXA OF UNCERTAIN AFFILIATION								
Desulfovibrio								
gigas	OL	b	b	b				224
COCCI AND COCCOBACILLI								
Neisseriaceae								
Acinetobacter								
calcoaceticus (Micrococcus cerificans)	F	b	b					101

sp.	F	b	b				225
	F	b	b				284
Branhamella							
catarrhalis	F	b	b				101
	F, B	b	b		b	b	195
	F, B	b	b			b	74
	F, B	b	b		b	c	132
Moraxella							
bovis	F, B	b	b		b	b	132
kingae[f]	F, B	b	b		b	e	132
lacunata	F, B	b	b		b	b	132
nonliquefaciens	F, B	b	b		b	b	132
osloensis	F, B	b	b		b	e	132
phenylpyrouvica	F, B	b	b		b	v	132
urethralis[f]	F, B	b	b		b	b	132
Neisseria							
canis[f]	F, B	b	b		b		195
caviae[f]	F, B	b	b		b		195
	F, B	b	b		b	b	132
cinerea[f]	F, B	b	b		b	e	132
	F, B	b	b		b	b	195
cuniculi[f]	F, B	b	b		b	b	195
denitrificans[f]	F, B	b	b		b	b	195
elongata[f]	F, B	b	b		b	e	132
flavescens	F, B	b	b		b	e	132
	F, B	b	b		b	c	195
gonorrhoeae	F, B	b	b	e	b		253
	F, B	b	b		b	e	132
	F	b	b		b		341
	P	b	b				413
	P	b	b				341
lactamicus[f]	F, B	b	b		b	b	195
meningitidis	F, B	b	b	e	b		253
	F, B	b	b		b	b	132

Table 2 (continued)
FATTY ACIDS OF GRAM-NEGATIVE BACTERIA[a]

Organism	Types of lipids analyzed	S	U	Br	C	OH	UK	Reference
mucosa	F, B	b	b			b		195
ovis[f]	F, B	b	b			b	b	195
	F, B	b	b			b	c	132
sicca	F, B	b	b					101
	F,B	b	b			b		195
subflava	F, B	b	b			b	b	195
(flava)	F, B	b	b			b	c	195
(perflava)	F, B	b	b			b		195
TAXA OF UNKNOWN AFFILIATION								
Achromobacter								
xylosoxidans	F	b	b		b			415
	B					b		415
spp.	F	b	b	b				121
Agrobacterium								
stellatum[f]	F	b	b	e				271
Methylosinus								
trichosporium	P	e	b					398
Paracoccus								
denitrificans	F	b	b					101, 407
halodenitificans	F	b	b					101

Note: B = bound; Br = branched; C = cyclopropane; F = free; G = glycolipids; N = neutral; OH = hydroxy; OL = ornithine lipid; P = phospholipids; S = saturated; Sp = sphingolipids; U = unsaturated; UK = unknown fatty acid(s); v = variable. Names in parentheses are those under which the strain was reported; they are given where this lends clarity or important information.

[a] Classification according to *Bergey's Manual, 8th ed.*[476]
[b] *>5% of total fatty acids.*
[c] *1—5% of total fatty acids.*
[d] *Taxonomic position unknown.*
[e] *<1% of total fatty acids.*
[f] *Species not recognized.*
[g] *Major amounts of unusual methoxy fatty acids are present.*

Table 3
FATTY ACIDS OF GRAM-POSITIVE BACTERIA[a]

Organism	Types of lipids analyzed	S	U	Br	C	OH	UK	Reference
COCCI								
Micrococcaceae								
Micrococcus								
caseolyticus[b]	F	c	c	c			d	405
conglomeratus[b]	F	d		c				405
cryophilus[b]	F	d	c					318
	B	e	c			c		317
diversus[b]	F	c	c		c		c	101
luteus	F, B	c	d	c			c	131
	F	c		c				250
(lysodeikticus)	F	c	d	c				101
	F	d		c				405
(Sarcina flava)	F	d		c				405
	P	c		c				125
(S. lutea)	F			c				405
mucilaginosus[b]	F, B	c	d	c				131
radiodurans[b]	F	c	c					101
roseus	F	c		c				250
	F, B	c	d	c			c	131
	F	c	c	c			c	101
(agilis)	F	c	e	c				101
	F	c		c				405
(tetragenus)	F			c				405
varians	F, B	c	d	c			c	131
	F	d	c	c				405
spp.	F	c		c				250
sp.	F	c		c				405

Table 3 (continued)
FATTY ACIDS OF GRAM-POSITIVE BACTERIA[a]

Organism	Types of lipids analyzed	S	U	Br	C	OH	UK	Reference
Staphylococcus								
aureus	F	c		c				35
	F	c	c	c				310
	F, B	c		c				77
	F, B	c	d	c				131
	P	c	d	c				310, 311, 316
epidermidis	F, B	c	d	c				131
(halotolerant)	P	c		c				176
saprophyticus	F, B	c	d	c				131
	F, B	c		c			c	74
Aerococcus								
viridans	F, B	c	c				d	379
Pediococcus								
acidilactici	F, B	c	c		e		e	379
cerevisiae	F, B	c	c					408
	F, B	c	c		c		d	379
halophilus	F, B	c	c		c		d	379
homari	F, B	c	c				c	379
							f	378
pentosaceus	F, B	c	c		e		e	379
urinae-equi	F, B	c	c				d	379
Streptococcus								
faecalis	F	c	c		c			102
	B	c	c		c			102
	F, B	c	c		c		e	379
	F, B	c	c	e			c	74
	F, B	c	c		c		c	10
faecalis var. *faecalis*	P	c	c		c			89, 90
faecalis var. *zymogenes*	P	c	c		c			89, 90

faecium	F, B	c	c		c		c	10
faecium var. *casseliflavus*	F, B	c	c		c		c	10
faecium var. *durans*	F, B	c	c		c		c	10
hemolyticus	P	c	c					89
lactis	P	c	c		c			90
	F	c	c		c		e	379
mitis	F, B	c	c					194
mutans	F, B	c	c	d				75
	F, B	c	c		g			194
salivarius	F, B	c	c					194
sanguis	F, B	c	c				c	194
uberis	F	c	c					35
	F, B	c	c					194
spp.	F, B	c	c					194
sp.	F, B	c	c	d				75
	F, B	c	c	c	d			76
sp. (motile) *Planococcus*?	F, B	c	c					102
Peptococcaceae								
Gemella								
haemolysans	F, B	c	c			c	c	195
Sarcina								
maxima	F	c		c		c		405
ventriculi	F	c		c		c		405
ENDOSPORE-FORMING RODS AND COCCI								
Bacillaceae								
Bacillus								
acidocaldarius	F, B	c		c			h	67
alvei	F	c	d	c				47c
caldolyticus[b]	F, B	c		c				400
caldotenax[b]	F	c		c				400
cereus	P	c	c	c				147
	P	c	c	c				29

Table 3 (continued)
FATTY ACIDS OF GRAM-POSITIVE BACTERIA[a]

Organism	Types of lipids analyzed	S	U	Br	C	OH	UK	Reference
coagulans	F, B	c		c				380
globiporus	F, B	c	c	c				144
insolitus	F, B	c	c	c				144
laevolacticus	F, B	d		c				380
larvae	F	c	e	c				47c
	F, B	c	e	c				143
lentimorbus	F	c	d	c				47c
	F, B	c	e	c				143
licheniformis	F	c		c				329
megaterium	F, B	c	d	c				74
mixolactis[b]	F, B	d		c				380
popilliae	F	c	e	c				47c
	F, B	c	e	c				143
psychrophilus	F, B	c	c	c				144
pumilus	F	c	c	c				329
racemilacticus	F, B	d		c				380
stearothermophilus	F	c	c	c				329
	F, B	c	c	c				425
	F	c		c				280
subtilis	F	c		c				35, 405
	F, B	d		c				380
subtilis var. *niger*	F, B, Spo	c	c	c				47e
	P	c		c				146
thuringiensis	F	c	d	c				47c
Clostridium								
butyricum	P	c	c		c		d, i	105

pasteurianum	F	c	c	d	c		j	53
tartarivorum[b]	F	c	d	c	c		j	53
thermosaccharolyticum	F	c	e	c	c		j	53
	Spo	c	c			c		288
sp. (psychrophile)	F	c	c		c		j	53
Sporolactobacillus								
inulinus	F, B	c		c				380
laevus[b]	F, B	d		c				380
racemicus[b]	F, B	c		c				380
Sporosarcina								
ureae	F	c		c				405
NONSPORING ROD-SHAPED BACTERIA								
Lactobacillaceae								
Lactobacillus								
acidophilus	F	c	c					35
	F	c	c	e	c			388
	F, B	c	c		c			381
bulgaricus	F	c	c	e	c			388
casei	F	c	c	e	c			388
	F, B	c	c		c		d	379—381
delbrueckii	F	c	c	e	c			388
	F, B	c	c		c		e	379
fermentum	F	c	c	e	c			388
	F, B	c	c		c			381
helveticus	F	c	c	e	c			388
heterohiochii	F	c	c		c		d	381
	F, B	c	c		c			377
	P	c	c		c			376
homohiochii	F, B	c	c		c		c	381
lactis	F	c	c	d	c			388
plantarum (arabinosus)	F	c	c	e	c			388
	F, B	c	c		c		e	379—381
sake[b]	F, B	c	c		c		e	379
yamanashiensis[b]	F, B	c	c		c			380

Table 3 (continued)
FATTY ACIDS OF GRAM-POSITIVE BACTERIA[a]

Organism	Types of lipids analyzed	S	U	Br	C	OH	UK	Reference
TAXA OF UNKNOWN AFFILIATION								
Listeria								
monocytogenes	F	c	c	c				139
(4 serotypes)	F	d	d	c				348
	F, B	c	c	c				301
Erysipelothrix								
rhusiopathiae	F	c	c	c				348
	F	c	d	c				348

Note: Spo = spores; see Table 2 for explanation of other letter symbols.

[a] Classification according to *Bergey's Manual*, 8th ed.[47b]
[b] Species not recognized.
[c] >5% of total fatty acids.
[d] 1—5% of total fatty acids.
[e] <1% of total fatty acids.
[f] Major amounts of conjugated dienoic acids.
[g] 2 of 18 strains contained major amounts of Δ-$C_{19:0}$ and Δ-$C_{21:0}$.
[h] Cyclohexyl fatty acids are present in major amounts.
[i] Major amounts of aldehydes are also present (derived from plasmalogens).
[j] Unsaturated cyclopropane fatty acids are also present.

Table 4
FATTY ACIDS OF IRREGULARLY SHAPED GRAM-POSITIVE BACTERIA OF UNKNOWN RELATION TO ACTINOMYCETES

Organism	Types of lipids analyzed	S	U	Br	C	OH	UK	Reference
Arthrobacter								
atrocyaneus[a]	F, B	b	b	c				41
aurescens[a]	F, B	b	b	c				41
citreus	F, B	b	b	c				41
crystallopoietes[a]	F	c		c				182
	P	c		c				182
	P	c		c				327
globiformis	F, B	c	b	c				41
marinus[d]	F	c	c	e	c			271
nicotianae[a]	F, B	b	b	c				41
oxydans[a]	F, B	c	c	c				41
pascens[a]	F, B	c	b	c				41
	F	c		c				327
ramosus[a]	F, B	c	b	c				41
simplex	F	c		c			c	418
	F, B	c	c	c				41
	P	c	c	b			f	422
tumescens	F, B	c	c	b				41
ureafaciens[a]	F, B	b	e	c				41
sp.	F, B	c	b	c				41
Bifidobacterium								
bifidum var. *pennsylvanicus*	F	c	c	c	c			388
spp.	F	c	c	b	b			388
Brevibacterium								
ammoniagenes	F, B	c	c	e				41
linens	F, B	c	b	c				41
sulphureum	F, B	c	c	c				41
Corynebacterium[g]								
alkanolyticum	P	c	c					169
								168

Table 4 (continued)
FATTY ACIDS OF IRREGULARLY SHAPED GRAM-POSITIVE BACTERIA OF UNKNOWN RELATION TO ACTINOMYCETES

Organism	Types of lipids analyzed	S	U	Br	C	OH	UK	Reference
insidiosum	F, B	c	b	c				41
michiganense	F, B	c	c	c				41
poinsettiae	F	b	b	c				348
pyogenes	F, B	c	b	c				252
sepedonicum	F, B	c	b	c				41
Propionibacterium								
acidi-propionici (arabinosum)	F, B	c	b	c				252
acidi-propionici (pentosaceum)	F, B	c	b	c				252
acnes	F, B	c	b	c				252
acnes (anaerobium)	F, B	c	b	c				252
acnes (diphtheroides)	F, B	c	b	c				252
acnes (liquifaciens)	F, B	c	b	c				252
freudenreichii	F, B	c	b	c				252
freudenreichii (shermanii)	F, B	c	b	c				252
granulosum	F, B	c	b	c				252
jensenii (zeae)	F, B	c	b	c				252
jensenii	F, B	c	b	c				252
thoenii	F, B	c	b	c				252

Note: See Table 2 for explanation of upper case letter symbols.

[a] Species may be synonymous with *A. globiformis*.
[b] 1—5% of total fatty acids.
[c] >5% of total fatty acids.
[d] Species of uncertain taxonomic position.
[e] <1% of total fatty acids.
[f] Hydroxy fatty acids present.
[g] Corynebacteria not known to have a cell wall of type IV.[29a,204]

Table 5
FATTY ACIDS OF ACTINOMYCETES[a]

Organism	Types of lipids analyzed	S	U	Br/TB[b]	C	OH[c]	UK	Ref.
Actinomycetaceae								
Actinomyces								
viscosus	G -	d	d					427
Oerskovia								
turbata	F	d	d	d/0				294
	P	d		d/0				294
Rothia								
dentocariosa	G	d		d/0				283
Actinoplanaceae								
Actinoplanes								
philippinensis	F, B	d	e	d/0				23
spp.	F, B	d	e	d/0				23
Streptosporangium								
album	F, B	d	e	d/f				23
amethystogenes	F, B	d	e	d/d				23
roseum	F, B	d	e	d/f				23
	P					c		420
virialbum	F, B	d	e	d/d				23
vulgare	F, B	d	e	d/f				23
Corynebacteriaceae[g]								
Corynebacterium[g]								
diphtheriae	F	d	d	0/0				35
	P	d	d	d/0				44
equi	F	d	d	d/0			d	348
	P	d		d/0				44
fascians	F, B	d	d	f/0				41
ovis	P	d		d/0				44
pseudodiphtheriticum	F	d	d	d/0				348
xerosis	F, B	d	d	h/0			d	74
Micromonosporaceae								
Micromonospora								
chalcea	F, B	d	e	d/0				23
fusca	F, B	d	e	d/h				23
globosa	F, B	d	e	d/f				23
spp.	F, B	d	e	d/d				23
sp.	P	d		d/d				347
Mycobacteriaceae								
Mycobacterium								
album	?	d	d	0/0				33
avium	F, B	d	d	0/d?				358
	F	d	d	0/0			d	133
	F	d	d	h/d				54
intracellulare	F	d	d	0/0			d	133
	F, B	d	d	0/d?				358
kansasii	F, B	d	d	f/d				357, 359
	F	d	d	0/0			h	133
marinum	F, B	d	d	h/d				359
paraffinicum	?	d		0/0				185
rubrum[i]	F	d	d	0/0				33
scrofulaceum	F	d	d	0/0			f	133
smegmatis	F	d	d	d/d				54
tuberculosis	F	d	d	0/d				54
vaccae	F, B	d	d	0/0				391

Table 5 (continued)
FATTY ACIDS OF ACTINOMYCETES[a]

Organism	Types of lipids analyzed	S	U	Br/TB[b]	C	OH[c]	UK	Ref.
Nocardiaceae								
Micropolyspora								
polyspora	F, B	d	e	d/0				23
rectivirgula[i]	F?	d	f	d/0				116
virida[i]	F?	d	d	d/0				116
viridinigra[i]	F?	d	d	d/f?				116
spp.	F?	d	d	d/f?				116
Nocardia								
actinoides[i]	F, B	d	d	d/h				372
asteroides	F, B	d	d	d/d				372
	F, B	d	d	f/d				372
	F	d	d	h/d				294
(leishmanii)	P					d		420
brasiliensis	F, B	d	e	f/d				23
	F, B	d	e	f/f				23
	F, B	d	d	h/d				372
	F, B	d	d	h/h				372
caviae	F	d	d	h/d				294
	P	d	d	f/d				294
convoluta	F, B	d	d	h/h				372
corallina[j]	F, B	d	d	f/h				372
erythropolis	F	d	d	h/d				294
farcinica	F	d	d	h/d				294
	F, B	d	d	f/0				41
globerulus	F, B	d	d	d/d				372
kirovani	CW	d	d			d,k		383
minima	F, B	d	d	d/d				372
opaca	F, B	d	d	0/0				41
pellegrino[j]	F, B	d	e	d/d				23
rhodochrous	F	d	d	d/d				405
ruber[j]	F, B	d	d	f/d				372
rubra[j]	F	d	d	0/0				35
rubropertincta[j]	F, B	d	e	0/d				23
sp.	F, B	d	d	h/O				41
	F, B	d	e	h/d				23
Streptomycetaceae								
Chainia								
olivacea	Sc			d/0		f		207
Streptomyces								
aureofaciens	F	d	d	d?/0			d	191
	F, B	d	e	d/f				23
coelicolor	F, B	d	e	d/f				23
flavovirens	F, B	d	e	d/h				23
gardneri	F, B	d	e	d/h				23
gelaticus	F, B	d	e	d/0				23
griseus	F, B	d	e	d/h				23
lavendulae	F	d		d/0				382
mediterranei	F	d		d/0				294
	P	d		d/0		d		294
olivaceus	P	d	d	d/0				25
toyocaensis	F	d	d	d/0		d		159
	P	d	d	d/0		d		159
venezuelae	F, B	d	e	d/0				23
viridochromogenes	F, B	d	e	d/0				23
spp.	F, B	d	e	d/0				23
	F, B	d	e	d/f				23

Table 5 (continued)
FATTY ACIDS OF ACTINOMYCETES[a]

Organism	Types of lipids analyzed	S	U	Br/TB[b]	C	OH[c]	UK	Ref.
Thermomonosporaceae								
Actinomadura								
dassonvillei	F?	d	d	d/h				4
helvata[i]	F?	d	d	d/h				4
madurae	F?	d	d	f/d				4
	F?	d	d	f/h				4
	F, B	d	d	f/h				372
pelletieri	F, B	d	d	h/f				372
	F, B	d	e	d/f				23
	F?	d	d	d/h				4
	F?	d	d	f/f				4
Actinomadura								
pusilla[i]	F?	d	d	d/h				4
roseoviolacea[i]	F?	d	d	d/h				4
spp.	F?	d	d	d/h				4
Microbispora								
amethystogenes	F, B	d	e	d/d				23
chromogenes	F, B	d	e	d/d				23
	F	d		d/d				282
diastatica	F, B	d	e	d/d				23
parva	F, B	d	e	d/d				23
rosea	F, B	d	e	d/d				23
	P					d		420
Thermomonospora								
curvata	F, B	d	e	d/0			d	23
Saccharomonospora								
viridis								
(Thermomonospora viridis)	F, B	d	e	d/0				23
(Thermoactinomyces glaucus)	F, B	d	e	d/h			h	23
(Thermopolyspora glauca)	F, B	d	e	d/h			h	23

Note: CW = cell wall; G = glycolipid; Sc = sclerotia; for other letter symbols see Table 2.

[a] Classification modified from Reference 60.
[b] Tuberculostearic acid alone or with homologs.
[c] See Table 6 for mycolic acids.
[d] >5% of total fatty acids.
[e] Lipid sample hydrogenated; unsaturated fatty acids calculated with saturated.
[f] 1—5% of total fatty acids.
[g] Corynebacteria having a cell wall composition of type IV.[204]
[h] <1% of total fatty acids.
[i] Species of uncertain taxonomic position.
[j] Member of the *N. rhodochrous* group.
[k] Nocardomycolic acid.

Tuberculostearic Acid (10-Methyloctadecanoic Acid)

Originally isolated from *Mycobacterium tuberculosis* in 1929 by Anderson and Chargaff,[11b] tuberculostearic acid (see Table 1) or its lower homologs have been reported widely among actinomycetes and their relatives, including — besides *Mycobacterium* — members of the genera *Arthrobacter*,[417] *Microbispora*,[23,282] *Micromonos-*

pora,[347] *Nocardia*,[16,294,421] certain *Streptomyces*,[124a,151] and *Streptosporangium*.[23,282] It was reported to be absent from strains of *Actinoplanes, Micropolyspora (Thermopolyspora), Oerskovia, Saccharomonospora (Thermomonospora)*, certain *Streptomyces*, and *Thermomonospora*.[23,282,294] The phosphatidyl inositol mannosides of *Athrobacter simplex* were reported to be acylated with tuberculostearic acid to a much greater extent than the other phospholipids (40% vs. 1 to 5%).[417]

Hydroxy Fatty Acids

Hydroxylated fatty acids (see Table 1) are most frequently reported as occurring in Gram-negative bacteria, where they are found largely as components of lipopolysaccharides, ornithine lipids, or the polymer poly-β-hydroxybutyrate.[325] Members of the genera from which such compounds have been isolated include the pseudomonads, where they are found uniquely in the "bound" lipids, the neisseriae, the moraxellae, and certain species of *Alcaligenes*, where they may be found both in "free" and "bound" form.

Hydroxy fatty acids also occur in certain actinomycete genera, both free and as components of complex lipids, including phospho-, peptido-, glyco-, and ornithine lipids, and even attached to the peptidoglycan. α-Hydroxy fatty acids were reported from *Microbispora* and from *Streptosporangium*, as well as from *Arthrobacter*.[420,422] In *Streptomyces*, both α-hydroxy[200,420] and β-hydroxy forms[294] are produced. None were reported from *Oerskovia*.[294] Unusual α-hydroxy *anteiso* acids were found to be acylated to the phosphatidyl ethanolamine of *Streptomyces sioyaensis*[160] and among the fatty acids of *Nocardia (leishmanii) asteroides* and *Arthrobacter simplex*. None were found in *Microbispora (Waksmania) rosea* or *Streptosporangium roseum*.[420] Of particular interest to actinomycetologists are the mycolic acids (see Table 1) — α-branched, β-hydroxy fatty acids that are produced by members of the *Mycobacterium-Nocardia-Corynebacterium* group.[16,111,211] The differences in the mycolates produced by members of these genera permit one to classify a given strain unambiguously.[206,209] In general, the mycolates ("eumycolates") from mycobacteria have chain lengths centering around 80 carbons (C_{60} to C_{90}), and α-branches having 20 to 24 carbons. Nocardiae produce mycolates (nocardic or nocardomycolic acids) having about 50 carbons (C_{38} to C_{60}), with α-branches ranging from 10 to 16 carbons. The mycolates (corynomycolates) of corynebacteria of the animal types are about 30 carbons in length (C_{28} to C_{36}) and have α-branches like those of nocardomycolates. The main chain of the eumycolates may contain additional carbonyl groups, methyl branches, or cyclopropane rings, and the α-branches are saturated. In general, mycolates from nocardiae and corynebacteria are simpler than those from mycobacteria, and although unsaturations may occur in both the main chain and side branches,[16,128,208] no other functional groups have been reported. For methods of purification and analysis of these compounds, consult References 206 and 209. Other methods that aim at shortening this analysis have been developed by Kanetsuna and Bartoli[149] and by Minnikin et al.[243] "Lipids-characteristic-of-*Nocardia* A" (LCNA), originally reported to be characteristic of members of the genus *Nocardia*,[246] have recently been shown to be nocardomycolates.[108] Analyses of the types of mycolates produced by various mycobacteria, nocardiae, and corynebacteria[206,209] showed that, with certain exceptions, their previous generic assignments were unchanged (Table 6). For a more detailed discussion of these results see Lipids Useful in Taxonomy.

PHOSPHOLIPIDS

Common Bacterial Phospholipids

As previously noted, our largest reservoir of data on lipids of bacteria concerns their fatty acid composition. However, chemotaxonomists have become increasingly aware

Table 6
THE OCCURRENCE OF MYCOLIC ACIDS IN *CORYNEBACTERIUM, MYCOBACTERIUM,* AND *NOCARDIA*

Species analyzed	Mycolate found	Generic assignment	Reference
Corynebacterium			
diphtheriae	Corynomycolate	*Corynebacterium*	209
pseudodiphtheriticum (hofmannii)	Corynomycolate	*Corynebacterium*	401
pseudotuberculosis	Corynomycolate	*Corynebacterium*	401
pyogenes	None	?	401
ulcerans	Corynomycolate	*Corynebacterium*	419
Mycobacterium			
acapulcensis[a]	Eumycolate	*Mycobacterium*	21, 209
avium	Eumycolate	*Mycobacterium*	209, 247
aurum[a]	Eumycolate	*Mycobacterium*	206
bovis	Eumycolate	*Mycobacterium*	206
brevicale[a]	Nocardomycolate	*Nocardia*	206
chitae[a]	Eumycolate	*Mycobacterium*	206, 209
diernhoferi	Eumycolate	*Mycobacterium*	206
flavescens	Eumycolate	*Mycobacterium*	21, 247
fortuitum	Eumycolate	*Mycobacterium*	209, 247
gallinarum[a]	Eumycolate	*Mycobacterium*	206
gastri	Eumycolate	*Mycobacterium*	206
intracellulare	Eumycolate	*Mycobacterium*	206
kansasii	Eumycolate	*Mycobacterium*	*206*
lacticolum var. aliphaticum	Corynomycolate	*Corynebacterium*	186
marinum	Eumycolate	*Mycobacterium*	206
paraffinicum	Eumycolate	*Mycobacterium*	206
	Nocardomycolate[b]	*Nocardia*	184, 185
parafortuitum[a]	Eumycolate	*Mycobacterium*	206
phlei	Eumycolate	*Mycobacterium*	206, 322a
rhodochrous[a]	Nocardomycolate	*Nocardia*	206, 209
scrofulaceum	Eumycoate	*Mycobacterium*	209
sp. (scotochromogen)	Eumycolate	*Mycobacterium*	322a
smegmatis	Eumycolate	*Mycobacterium*	206
terrae	Eumycolate	*Mycobacterium*	206
terrae (novum)	Eumycolate	*Mycobacterium*	206
thamnopheos	Nocardomycolate	*Nocardia*	206
thermoresistibile[a]	Eumycolate	*Mycobacterium*	206
tuberculosis	Eumycolate	*Mycobacterium*	206
vaccae	Eumycolate	*Mycobacterium*	21, 206, 247
Nocardia			
amarae	Nocardomycolate[c]	*Nocardia*	208
asteroides	Nocardomycolate	*Nocardia*	206, 209, 294
autotrophica (spp.)	No mycolates	?	206
brasiliensis	Nocardomycolate	*Nocardia*	199, 206
calcarea	Nocardomycolate[b]	*Nocardia*	244
carnea	Nocardomycolate[c]	*Nocardia*	208
caviae	Nocardomycolate	*Nocardia*	206, 232, 244, 294
corallina	Nocardomycolate[b]	*Nocardia*	21, 27, 247
erythropolis	Nocardomycolate[b]	*Nocardia*	232, 294, 424
farcinica (African)	Eumycolate	*Mycobacterium*	17, 206
farcinica	Nocardomycolate	*Nocardia*	232, 294
kirovani	Nocardomycolate	*Nocardia*	232, 384
pellegrino	Nocardomycolate	*Nocardia*	209
rhodochrous	Nocardomycolate	*Nocardia*	21, 208, 247, 405

Table 6 (continued)
THE OCCURRENCE OF MYCOLIC ACIDS IN *CORYNEBACTERIUM, MYCOBACTERIUM,* AND *NOCARDIA*

Species analyzed	Mycolate found	Generic assignment	Reference
rubra	Nocardomycolate	*Nocardia*	209
sumatra	Nocardomycolate	*Nocardia*	232
vaccinii	Nocardomycolate[c]	*Nocardia*	208

[a] Species *incertae sedis.*
[b] Low range of molecular weight similar to *rhodochrous* type.
[c] Monounsaturated α-branch.

that, although fatty acids are useful, patterns of other lipids — particularly phospholipids — can give a welcome additional perspective on problem groups.

As far as is known, phospholipids occur primarily in association with membranes, playing a vital role in both their structure and function. They are often referred to as "polar" lipids, because, although they are soluble in organic solvents, they have hydrophilic groups that serve to make one end of each molecule "water-loving" or polar. The opposite end is substituted with fatty compounds, such as fatty acids (diacyl phospholipids), fatty aldehydes (plasmalogens), or alcohols (diether phospholipids), and thus "prefers" organic solvents (Table 7).

The most common bacterial phospholipids are illustrated in Table 8. All of these are also found in eukaryotes. Diphosphatidyl glycerol (DPG) and phosphatidyl glycerol (PG) are the most commonly occurring phospholipids in all types of bacteria. Phosphatidic acid (PA), an intermediate in the biosynthesis of phospholipids, occurs widely, but in small amounts. Phosphatidyl ethanolamine (PE) is found in high concentrations in many bacteria, but it is particularly common among Gram-negative bacteria and actinomycetes. Furthermore, it is probably the metabolically most stable phospholipid, usually found in cells at any point in the growth cycle. The N-methylated derivatives of PE — phosphatidyl methylethanolamine (PME) and phosphatidyl dimethylethanolamine (PDME) — are not common compounds, and when they occur, they seem to be limited to Gram-negative bacteria or actinomycetes. Phosphatidyl choline (PC), like PME and PDME (to which it is related), may be taxonomically significant in certain Gram-negative forms; it has been reported only very rarely in Gram-positive bacteria and actinomycetes. Phosphatidyl serine (PS) is restricted in occurrence to Gram-negative bacteria. In many bacteria the aminoacyl phosphatidyl glycerols (AAPG) — also known as lipoamino acids — occur, or their production is appreciably enhanced when growth takes place at low pH. All those reported to date are esters with L-amino acids; D-amino acid esters have been found only in *Mycoplasma.*[9]

Also included in Table 8 is a representative of the phosphatidyl inositol mannosides (PIM). Among prokaryotes, these phospholipids appear to be confined in their occurrence to the actinomycetes and their close allies. The report that PIM were found in *Propionibacterium (shermanii) freudenreichii*[43] has not been borne out by later work.[44] PIM have also been found in yeasts.[349] They may contain from one to six mannose residues; one residue is always at position 2 of the inositol, and the rest are attached as a chain at position 6. The subject has been well reviewed by Brennan and Lehane.[44]

An estimate of the relative frequency of occurrence of all of these common phospholipids in bacteria and actinomycetes is presented in Table 8; recent data on individual taxa are found in Tables 9 to 12. Quantitative data on phospholipids are not given, since the quantity of a phospholipid is a function of the strain examined, its age, growth conditions, etc. In general, qualitative differences are only occasionally ob-

Table 7
GENERAL STRUCTURES OF PHOSPHATIDES

$H_2C{-}O{-}C({=}O){-}R$ $HC{-}O{-}C({=}O){-}R'$ $H_2C{-}O{-}P({=}O)(OH){-}O{-}R''$	Diacyl ester type (most common form)	VII-1
$H_2C{-}OH$ $HC{-}O{-}C({=}O){-}R'$ $H_2C{-}O{-}P({=}O)(OH){-}O{-}R''$	Monacyl ester type (lyso form)	VII-2
$H_2C{-}O{-}CH{=}CH{-}R$ $HC{-}O{-}C({=}O){-}R'$ $H_2C{-}O{-}P({=}O)(OH){-}O{-}R''$	Monoacyl monoalk-1-enyl ether (plasmalogen)	VII-3
$H_2C{-}O{-}CH_2{-}R$ $HC{-}O{-}CH_2{-}R'$ $H_2C{-}O{-}P({=}O)(OH){-}R''$	Diether form (extreme halophiles)	VII-4
$H_2C{-}O{-}C({=}O){-}R$ $HC{-}O{-}C({=}O){-}R'$ $H_2C{-}O{-}P({=}O)(OH){-}R''$	Phosphono form	VII-5
See Table 13 [XIII-9]	Sphingolipid	

Note: R = R′, = saturated, unsaturated, or branched aliphatic chain; R″ — see Table 8.

Table 8
COMMON BACTERIAL PHOSPHOLIPIDS

R″ (Table 4, [IV-1]) =	Name	Abbreviation	Occurrence in bacteria (%)		
			G−	G+	Actinomycetes[a]
$-H$	Phosphatidic acid	PA	0.1—10	0.1—10	0.1—10
$-CH_2CH(OH)CH_2OH$	Phosphatidyl glycerol	PG	>50	>50	>50
$-CH_2CH(OH)(CH_2)$ O $R-CH-C=O$ NH_2	Phosphatidyl-3′-O-aminoacyl glycerol (lipoamino acid)	AAPG	0.1—10	0.1—10	0.1—10
$-CH_2CH(OH)CH_2$ O $HO-P=O$ O $CH_2-CH-CH_2$ O O $O=C$ $C=O$ R R	Diphosphatidyl glycerol (cardiolipin)	DPG	>50	>50	>50
$-CH_2CH_2NH_3^+$	Phosphatidyl ethanolamine	PE	>50	0.1—10	>50
$-CH_2CH_2NH-CH_3$	Phosphatidyl-N-methylethanolamine	PME	0.1—10	0	0.1—10

$-CH_2CH_2N(CH_3)_2$	Phosphatidyl-N-dimethylethanolamine	PDME	0.1—10	0	0.1—10
$-CH_2CH_2N(CH_3)_3^+$	Phosphatidyl choline	PC	10—50	0.1—10	0.1—10
$-CH_2CH(NH_2)COOH$	Phosphatidyl serine	PS	0.1—10	0	0
–O-inositol (ring with OH, OH, OH, HO, OH)	Phosphatidyl inositol	PI	0.1—10	10—50	>50
–O-inositol (ring with OR, OR, OH, HO, OH)	Phosphatidyl inositol mannosides (mono-, di-, tri-, tetra-, and penta-); R = mannose residues (1 to 5)	PIM	0	0	>50

[a] Includes the bacteria of unknown relation to actinomycetes.

Table 9
PHOSPHOLIPIDS OF GRAM-NEGATIVE BACTERIA

Organism	DPG[a]	PG	PE	PME or PDME	PS	PC	PI	PA	Other	Reference
BUDDING OR APPENDAGED[b] BACTERIA										
Hyphomicrobium sp.						c				18
PHOTOTROPHIC BACTERIA										
Chlorobiaceae										
Chlorobium										
limicola	c	c				c?	c?		d	162
limicola var. *thiosulfatophilum*	c	c				c?	c?		d	162
phaeobacteroides	c	c				c?	c?		d	162
phaeovibroides	c	c				c?	c?		d	162
vibrioforme	c	c				c?	c?		d	162
vibrioforme var. *thiosulfatophilum*	c	c				c?	c?		d	162
Pelodictyon										
luteolum	c	c				c?	c?		d	162
Chloroflexus										
aurantiacus[e]		c					c		d	162
Rhodospirillaceae										
Rhodomicrobium										
vanneilii		c	c			c		c	f	262
Rhodopseudomonas										
capsulata		c	c			c				262
gelatinosa		c	c						f	262
palustris	c	c	c			c			f	262
spheroides	c	c	c			c		c	f	262
viridis	c	c	c			c			f	300
Rhodospirillum										
rubrum	c	c	c							46b
	c	c	c			c				262
GLIDING BACTERIA										
Cytophagaceae										
Sporocytophaga sp.		c	c						d	272
CHEMOLITHOTROPHIC BACTERIA										
Nitrobacteriaceae										
Nitrobacter										
agilis						c				18
agilis (winogradsky)						c				18
SULFUR-OXIDIZING BACTERIA										
Thiobacillus										
neopolitanus	c	c	c	c						2
thiooxidans	c	c	c	c					g	31
Sulfolobus										
acidocaldarius							c		h	203
SPIROCHETES										
Spirochaetaceae										
Spirochaeta										
aurantia	c	c								216a
litoralis	c	c						c		216a

Table 9 (continued)
PHOSPHOLIPIDS OF GRAM-NEGATIVE BACTERIA

Organism	DPG[a]	PG	PE	PME or PDME	PS	PC	PI	PA	Other	Reference
stenostrepta	c	c			c			c		216a
zuelzerae	c	c								233
Treponema	c	c								
denticola	c	c	c			c				216a
pallidum	c	c				c			i	233
phagedenis	c	c	c			c				216a
refrigens	c	c			c	c				216a
scoliodontum	c	c	c			c				216a
vincentii	c	c				c				216a
SPIRAL AND CURVED BACTERIA										
Spirillaceae										
Campylobacter										
fetus		c	c		c		c			368
Spirillum										
linum		c	c						d	272
AEROBIC RODS AND COCCI										
Azotobacteriaceae										
Azotobacter										
vinelandii	c	c	c		c					226
Halobacteriaceae										
Halobacterium										
cutirubrum		c, j							j, k, l	192
Halococcus										
morrhuae									j, k	414
Pseudomonadaceae										
Pseudomonas										
aeruginosa	c		c							285
	c	c	c						d	272
	c		c			c			m	39
diminuta		c							n	411
		c								412
fluorescens	c	c	c							100
	c	c	c		tr					61
	c	c	c							237
pavonacea		c	c						d	412
perfectomarinus	c	c	c							272
rubescens[o]		c	c						d	409, 412
stutzeri	c	c	c							409, 412
syncyanea	c	c	c							409, 412
Rhizobiaceae										
Agrobacterium										
stellulatum[p]	c	c	c			c				272
Rhizobium										
japonicum	c	tr?	c		c	c				47g
leguminosarum	c	c	c			c				86
sp. (*Lotus*)	c		c			c				98
FACULATIVELY ANAEROBIC RODS										
Enterobacteriaceae										
Escherichia										
coli	c	c	c					c		310
	c	c	c		tr					114
	c	c	c		tr			tr		59
	c	c	c							403

Table 9 (continued)
PHOSPHOLIPIDS OF GRAM-NEGATIVE BACTERIA

Organism	DPG[a]	PG	PE	PME or PDME	PS	PC	PI	PA	Other	Reference
Salmonella										
typhimurium	c	c	c						q	273
	c	c	c							362
	c	c	c							279
	c	c	c							173
Yersinia										
pestis	c	c	c	c	c?			c		363
pseudotuberculosis	c	c	c	c	c		c	c		363
Vibrionaceae										
Photobacterium										
phosphoreum	c	c	c							78
Vibrio										
alginolyticus	c	c	c							272
algosus[c]	c	c	c					c?		272
cholerae	tr	c	c							272
fischeri (marinus)	c	c	c							272
fischeri	c	c	c							78
	c	c	c							272
marinofulvus[c]	c	c	c							272
parahaemolyticus	c	c	c							272
	c	c	c					c		69
sp.	c	c	c							272
	c	c	c						d	272
	tr	tr	c							272
ANAEROBES										
Bacteroidaceae										
Bacteroides										
fragilis										
var. *fragilis*									s	92
var. *thetaiotamikron*									s	92
	c	c	c		c				s	340
var. *vulgatus*									s	92
putredinis									s	92
ruminicola									s	190
Fusobacterium										
mortiferum	c	c	c						i	119
necrophorum									s	190
COCCI AND COCCOBACILLI										
Neisseriaceae										
Acinetobacter sp.	c	c	c							361
Neisseria										
gonorrhoeae	c	c	c							413
	c	c	c			c				341
Veillonellaceae										
Megasphaera										
elsdenii			c		c				i	298
			c		c				d, i	386
TAXA OF UNCERTAIN POSITION										
Achromobacter										
aquamarinus	c	c	c						d	272
xylosoxidans	c	c	c							415
spp.	c	c	c							121

Table 9 (continued)
PHOSPHOLIPIDS OF GRAM-NEGATIVE BACTERIA

Organism	DPG[a]	PG	PE	PME or PDME	PS	PC	PI	PA	Other	Reference
Alcaligenes										
eutrophus	c	c	c							356
faecalis			c							99
sp.			c							352
Alteromonas										
haloplanktes	c	c	c					c	t	69
MARINE BACTERIA		c	c					tr	t	69
Bordetella										
pertussis	c	tr	c		c				d	353
Brucella										
abortus	c	c	c	c	c	c			d	353
	c		c			c			d	39
melitensis	c	c	c	c	tr	c				353
Desulfovibrio										
desulfuricans	c	c	c		c					223
gigas		c	c							223
vulgaris	c	c	c							223
Haemophilus										
parainfluenzae	c	c	c							373
Methylosinus										
trichosporium	c	c	c		c	c				398
Moderate halophile (rod)	c	c	c						u	286
Paracoccus										
denitrificans	c	c	c	c	tr	c		c	d	407
Thermus										
aquaticus	c	c	c				c	c	v	307

Note: tr = traces.

[a] See Table 8 for abbreviations.
[b] Classification according to *Bergey's Manual*, 8th ed.[476]
[c] Phospholipid reported present.
[d] Unknown phospholipids present.
[e] Taxonomic position uncertain.
[f] Ornithyl PG.
[g] Ornithine-containing phospholipid.
[h] Diether phospholipids only.
[i] Major amounts of plasmalogens (Table 7 [VII-3]).
[j] Diphytanyl ether analogs.
[k] Phosphatidyl glycerol phosphate.
[l] Phosphatidyl glycerol sulfate.
[m] Cell wall preparation.
[n] 3-O-[6′-(*sn*-glycero-3″-phosphoryl)-α-D-glucopyranosyl]-*sn*-glycerol (see Table 13 [XIII-2], *P. diminuta* cell walls).
[o] Not *Pseudomonas;* placement unknown.
[p] Not *Agrobacterium;* placement unknown.
[q] 3-O-Acylphosphatidyl glycerol.
[r] Generic assignment uncertain.
[s] Major amounts of sphingolipids (e.g., Table 13 [XIII-9]).
[t] Bisphosphatidic acid.
[u] Monoglucosyl PG.
[v] Major amounts of a phospholipid containing an unsaturated hydroxyundecylamine.

Table 10
PHOSPHOLIPIDS OF GRAM-POSITIVE BACTERIA

Organism	DPG[a]	PG	PE	AAPG	PI	PA	Other	Reference
COCCI[b]								
Micrococcaceae								
Micrococcus								
caseolyticus[c]	d	d		d			e	405
conglomeratus[c]	d	d		d			e	405
cryophilus[c]	d	d	d				e	318
luteus group								
M. luteus	d	d					e	177
M. lysodeikticus	d	d			d		e	405
	d	d			d			70, 360
Sarcina flava	d	d			d			405
S. lutea	tr	d			d			405
	d	d					e	177
roseus group								
M. roseus	d	d						405
M. tetragenus	d	d			d		e	405
varians	d	d					e	405
Staphylococcus								
aureus	d	d	tr	d			g	331
	d	d					e	177
	d	d		d		d		310, 311
	d	d		d				148
	d	d		d				316
	d	d		d				292
	d	d					g	32
epidermidis	d	d					e	177
epidermidis (halotolerant)	d	d				d	e, f	176
Planococcus								
citreus	d	d	d				e	177
sp.	d	d	d				e	177
Peptostreptococcaceae								
Sarcina								
maxima	d	d					e	405
ventriculi	d	d					e	405
Streptococcaceae								
Streptococcus								
faecalis	d	d					f	90
faecalis var. *faecalis*							f	8, 9, 88, 94a
faecalis var. *zymogenes*							f	90
hemolyticus							f	89
lactis							f	90
Pediococcus								
cerevisiae	d	d				+	e, g?	408
ENDOSPORE-FORMING RODS								
Bacillaceae								
Bacillus								
alvei	d	d	d					47c
cereus		d	d					147
	d	d	d					238
	d	d	d					29
		d	d	d			e	229
larvae	d	d	d					47c
lentimorbus	d	d	d					47c

Table 10 (continued)
PHOSPHOLIPIDS OF GRAM-POSITIVE BACTERIA

Organism	DPG[a]	PG	PE	AAPG	PI	PA	Other	Reference
ENDOSPORE-FORMING RODS (continued)								
Bacillus (continued)								
megaterium		d	d	d			h	229
							h	290
popilliae	d	d	d					47f
stearothermophilus	d	d	d				e	49
	d	d	d				e	242, 276
	d	d						242
subtilis		d	d	d			e	229
		d	d					146
	d	d	d, i	d			e	405
	d	d	d	d				30
subtilis var. *niger*	d	d	d	d				47e
thuringiensis	d	d	d					47f
Clostridum								
butyricum							j	231
Sporosarcina								
urae	d	d	d				e	177
	d	d	d					405
NONSPORING ROD-SHAPED BACTERIA								
Lactobacillaceae								
Lactobacillus								
acidophilus	d	d		d				85
arabinosus	d	d		d				85
bulgaricus	d	d		d				85
casei	d	d		d				85
delbruckii	d	d					e	85
fermentum	d	d		d				85
helveticus	d	d						85
heterohiochii	d	d					e	376
lactis	d	d		d				85
plantarum	d	d		d				85
GENUS OF UNCERTAIN POSITION								
Listeria								
monocytogenes	d	d					e, k	181

[a] See Tables 8 and 9 for abbreviations.
[b] Classification according to *Bergey's Manual,* 8th ed.[47b]
[c] Species of uncertain taxonomic position.
[d] Phospholipid present.
[e] Unknown phospholipids present.
[f] Phosphatidyl diglucosyl diglycerides (see Table 13 [XIII-5]).
[g] Phosphatidyl glucose (see Table 13 [XIII-1]).
[h] Glucosaminyl phosphatidyl glycerol (see Table 13 [XIII-3]).
[i] PE and PG not separated in this system.
[j] Aldehydogenic derivative of phosphatidyl ethanolamine (see Table 13 [XIII-10]).
[k] Bisphosphatidyl glycerol phosphate (see Table 13 [XIII-7]).

Table 11
PHOSPHOLIPIDS OF IRREGULARLY SHAPED GRAM-POSITIVE BACTERIA OF UNKNOWN RELATION TO ACTINOMYCETES

Organism	DPG[a]	PG	PE	PME or PDME	PC	PS	PI	PIM_x	Other	Ref.
Arthrobacter										
citreus	b	b					b	b		178
crystallopoietes	b	b					b	b		182
	b	b					b			327
globiformis	b	b						b		178
	b	b					b			327
marinus	tr	b	b							272
pascens	b	b					b			327
simplex	b	b								422
	b	b	b					b		417
	b	b						b		178
tumescens	b	b						b		178
ureafaciens	b	b						b		178
Bifidobacterium										
asteroides	b	b							d	85
bifidum		b							d, e	85
bifidum var. *pennsylvanicus*	b	b							d, e	85
breve	b	b							d, e	85
indicum	b	b							d	85
coryneforme	b	b							d	85
globosum	b	b							d, e	85
liberorum	b	b							d, e	85
ruminale	b	b							d	85
Brevibacterium										
albidum	b	b					tr	tr		178
ammoniagenes	b	b	tr				b	b		178
citreum	b	b					b	b		178
herolum	b	b					tr	tr		178
lipolyticum	b	b					tr	tr		178
testaceum	b	b					tr	tr		178
Cellulomonas										
biazotea	b	b						b		178
fimi	b	b						b		178
Corynebacterium										
alkanolyticum	b	b	b				b			168, 169
aquaticum	b	b	b				b	b		117,165
	b	b								178
coelicolor	b	b					b			405
flaccumfaciens	b	b					tr	tr		178
lilium	b	b	b				b	b		178
michiganense	b	b								178
poinsettiae	b	b					tr	tr		178
Microbacterium										
flavum	b	b	tr				b	b		178

[a] See Table 8 for abbreviations of phospholipids, and see Tables 9 and 12 for other abbreviations.
[b] Phospholipid present.
[c] Unknown phospholipids present.
[d] Galactosyl phosphatidyl glycerol diglyceride (Compound #15; Table 13 [XIII-4]).
[e] Alanyl phosphatidyl glycerol.

Table 12
PHOSPHOLIPIDS OF ACTINOMYCETES

Organism	DPG	PG	PE	PME or PDME	PC	PS	PI	PIM_x	Other	Ref.
Actinomycetaceae[a]										
Actinomyces										
viscosus	b		b				b			427
Oerskovia										
turbata	b	b					b	b		178
Corynebacteriaceae[c]										
Corynebacterium										
diphtheriae	b	b	tr				b	b		178
	b	b?					b	b		44
equi		b?						b		44
	b	b	b				b	b		178
fascians	b	b	b				b	b		178
ovis	b	b	b					b		44
xerosis	b							b		44
Micromonosporaceae										
Micromonospora sp.	b		b				b	b		347
Mycobacteriaceae										
Mycobacterium										
avium	b		b					b		55
phlei	b		b					b		72
	b	b	b				b	b		178
smegmatis	b		b					b		72
	b		b					b		55
tuberculosis	b	b	b				b	b		178
	b		b					b	d	55
vaccae	b		b			b				391
Nocardiaceae										
asteroides	b		b				b	b		423
	b	b	b				b	b		178
	b		b				b	b		294
brasiliensis	b	b	b				b	b		178
calcarea	b	b	b				b	b		178
caviae	b	b	b				b	b		178
	b		b				b	b		294
coeliaca	b		b		b		b	tr	e	164
corallina	b		tr				b	b		423
	b	b	b				b	b		178
eppingerii	b		tr				b	b		423
Nocardia										
erythropolis	b		tr				b	b		423
	b		b				b	b		294
	b	b	b				b	b		178
farcinica	b	b	b				b	b		178
	b		b				b	b		294
opaca	b	b	b				b	b		178
polychromogenes	b		b				b	b	e	164
	b		tr				b	b		423
rhodochrous	tr								d	405
rubra	b	b	tr				b	b		423
	b	b	b				b	b		178
rugosa	b	b	b				b	b		178

Table 12 (continued)
PHOSPHOLIPIDS OF ACTINOMYCETES

Organism	DPG	PG	PE	PME or PDME	PC	PS	PI	PIM_x	Other	Ref.
Streptomycetaceae										
Streptomyces										
canosus	b	b	b	b	b		b		d	183
mediterranei	b		b					b		294
olivaceus	b		b					b	f	180
toyocaenisis	b		b					tr		159
Thermomonosporaceae										
Actinomadura										
madurae	b	b								178

[a] Classification modified from the system of Cross and Goodfellow.[60]
[b] Phospholipid present.
[c] Restricted to corynebacteria having a cell wall of Type IV.[29a,204]
[d] Unknown phospholipids present.
[e] Phosphatidyl monoglucosyl diglyceride.
[f] Phosphatidyl-2,3-butanediol.

served under growth conditions where essential nutrients are lacking (see Effect of Medium Constituents). Lyso forms (see Table 7 [VII-2]) of the phospholipids frequently have been reported to occur in bacteria. It is likely that they really exist in the cell, although they were formerly thought to be degradation products resulting from isolation procedures. They do not appear to have any taxonomic significance and have not been included in the tables. Phospholipid fatty acids generally reflect the make-up of the cellular "pool" of fatty acids. There are, however, frequent pronounced differences in the fatty acids of individual phospholipids and between the two acylated positions of the glycerol.

Unusual Bacterial Phospholipids

The structures of some bacterial phospholipids of apparently rather limited occurrence are illustrated in Table 13. These have been included in this review on the chance that further studies may show some of them to be of taxonomic significance. It is, in fact, more likely that such unusual compounds will be useful in taxonomy than those of more widespread occurrence. In many cases the original reports were limited to the examination of one strain (e.g., Table 13 [XIII-1, -2, -6, -7, -10]). Glucosaminyl phosphatidyl glycerol (Table 13 [XIII-3]) has been found in both the Gram-positive *Bacillus megaterium* and the Gram-negative *Pseudomonas putida (ovalis)*. Other compounds are produced by more than one species of a given genus. A case in point is the phosphogalactolipid known as "Compound 15" (Table 13 [XIII-4]), which was isolated from six different species of the genus *Bifidobacterium*.[85] Its structure, which has only recently been determined,[390] contains an unusual galactofuran moiety. It was not found in the nine species of *Lactobacillus* examined.

Thus far reported solely from *Streptococcus* are the acylated derivatives of glycerophosphoryl diglucosyl diglycerides (GDD), two of which are shown in Table 13 [XIII-5].[8,9,88-90] It has recently been found that the membrane lipoteichoic acid of *S. faecalis* is linked to this type of phospholipid via a phosphodiester bond within the membrane; thus, GDD may play a significant structural role in the streptococcal membrane, and possibly in the taxonomy of the group.

Diether phospholipids (Table 7 [VII-4]), such as the diphytanyl glycerol ether analog of phosphatidyl glycerophosphate shown in Table 13 [XIII-8], have been shown to be present in large quantities in extreme halophiles of the genera *Halobacterium* and *Hal-*

Table 13
UNCOMMON BACTERIAL PHOSPHOLIPIDS

Shorthand formula	Example (producer)	References
FA FA Glycerol–Phosphate Glucose	Structure XIII-1 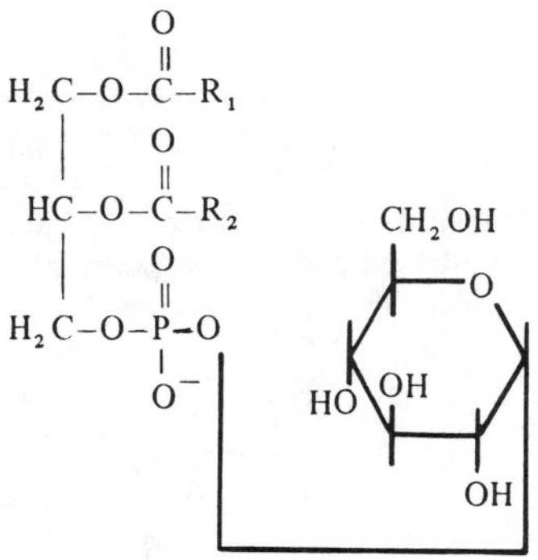 $H_2C-O-C(=O)-R_1$ $HC-O-C(=O)-R_2$ $H_2C-O-P(=O)(O^-)-O$ — glucose (CH_2OH, HO, OH, OH) Phosphatidyl glucose (3-*sn*-phosphatidyl-1′-glucose) *Staphylococcus aureus*	330
Glycerol–Phosphate Glucose–Glycerol	Structure XIII-2 H_2C-OH $HC-OH$ $H_2C-O-P(=O)(O^-)-O-CH_2$ — glucose (HO, OH, OH) — $O-CH_2$, $HO-CH$, $HO-CH_2$ 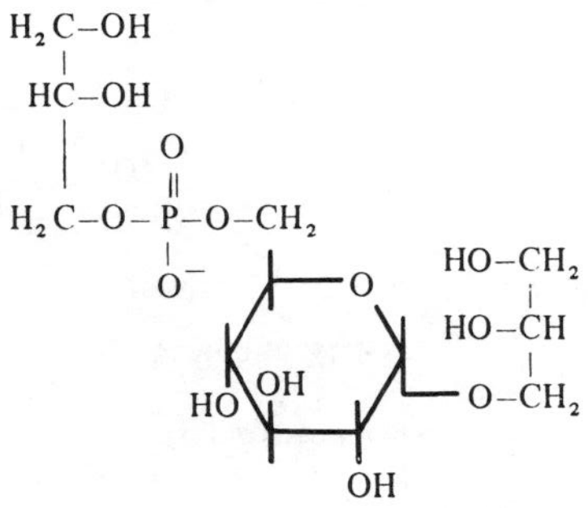 3-*O*-(6′-(*sn*-glycero-3″-phosphoryl)-α-D-glucopyranosyl)-*sn*-glycerol (Tetraacylation not shown; exact positions unknown)	411

Table 13 (continued)
UNCOMMON BACTERIAL PHOSPHOLIPIDS

Shorthand formula	Example (producer)	References
FA FA / Glycerol–Phosphate / Glycerol / Glucosamine	Structure XIII-3	277, 289

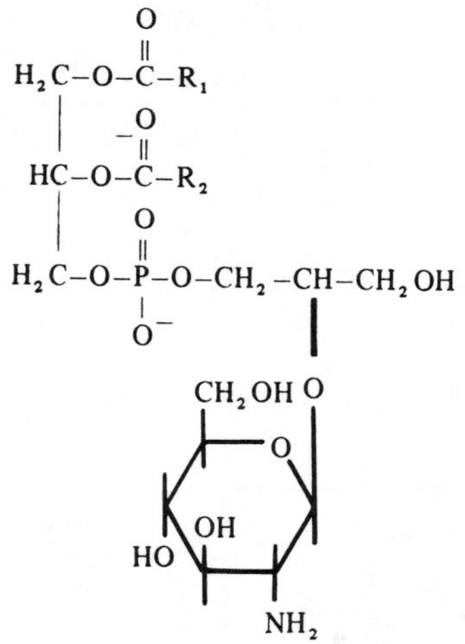

```
FA FA
|  |
Glycerol–Phosphate
          |
          Glycerol
          |
          Glucosamine
```

```
           O
           ||
H2C–O–C–R1
 |         O
 |         ||
HC–O–C–R2
 |         O
 |         ||
H2C–O–P–O–CH2–CH–CH2OH
           |           |
           O⁻          O
     CH2OH             |
   (glucopyranose ring: O, OH, HO, NH2)
```

Glucosaminyl phosphatidyl glycerol
1′-(1,2-diacyl-*sn*-glycero-3-phosphoryl)-2′*-*O*-β-(2-amino-2-deoxy-D-glucopyranosyl)-*sn*-glycerol (* = or 3′)
Bacillus megaterium
Pseudomonas putida (*ovalis*)

Structure XIII-4

390

Glycerol–Phosphate
|
Galactose (Furan)
|
Glycerol
| |
FA FA

$H_2C{-}OH$
|
$HC{-}OH$
|
$H_2C{-}O{-}P(=O)(O^-){-}O{-}CH_2{-}C(OH)(H){-}$ (furanose ring: O; OH; OH) ${-}O{-}CH_2$
|
$HC{-}O{-}C(=O){-}R_1$
|
$H_2C{-}O{-}C(=O){-}R_2$

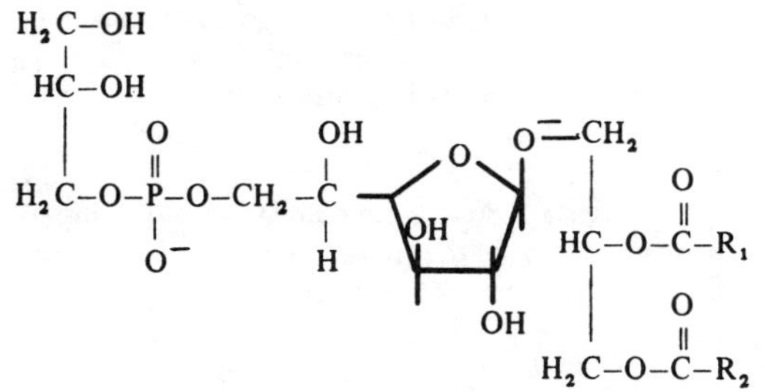

"Compound 15"

3-*O*-(6′-(*sn*-glycero-*1*-phosphoryl)-β-D-galactofuranosyl)-*sn*-1,2-diglyceride
Bifidobacterium

Table 13 (continued)
UNCOMMON BACTERIAL PHOSPHOLIPIDS

Shorthand formula	Example (producer)	References
X_1 – Glucose – Glucose(X_2) – Glycerol(FA)(FA)	Structure XIII-5 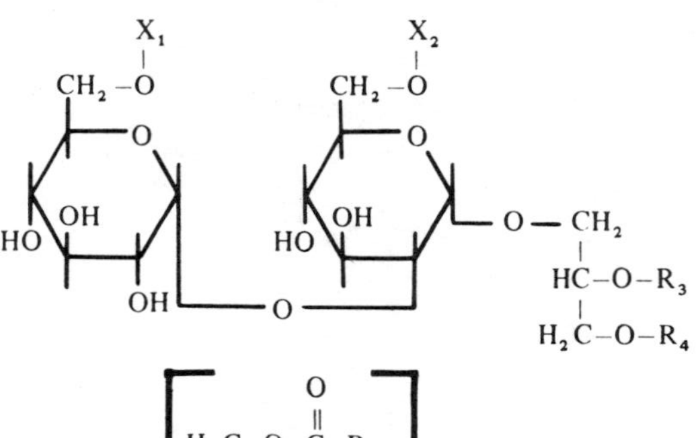 Phosphatidyl kojibiosyl diglyceride (Acylated derivatives of glycerophosphoryl diglucosyl diglycerides) Type I: $X_2 = R_1 = R_2 = H$ $X_1 = Y$ R_3, R_4 = Various fatty acids in ester linkage 1(3)-*O*-(6″-(*sn*-glycero-1-phosphoryl)-2′-*O*-(α-D-glucopyranosyl)-α-D-glucopyranosyl)-diglyceride Type II: $X_1 = H$ $X_2 = Y$ R_3, R_4 = Various fatty acids in ester linkage. 3(1)-*O*-(6′-(1,2-diacyl-*sn*-glycero-3-phosphoryl)-2′-*O*-(α-D-glucopyranosyl)-α-D-glucopyranosyl)-1(3),2-diacylglycerol *Streptococcus faecalis* var. *faecalis* (I and II) *S. faecalis* var. *zymogenes* (I and II) *S. hemolyticus* (Serogroup A) (I) *S. lactis* (II)	8, 9, 89, 90

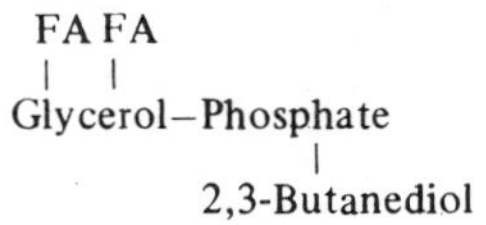

```
FA FA
|  |
Glycerol–Phosphate
               |
           2,3-Butanediol
```

Structure XIII-6 — 26, 180

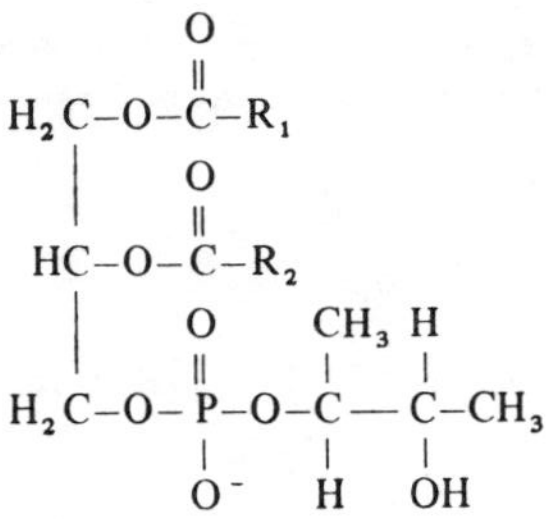

```
           O
           ||
H2C–O–C–R1
  |        O
  |        ||
 HC–O–C–R2
  |        O      CH3  H
  |        ||      |    |
H2C–O–P–O–C — C–CH3
           |       |    |
           O-      H    OH
```

1,2-di-*O*-acyl-*sn*-glycerophosphoryl butane-2,3-diol
Streptomyces (Actinomyces) olivaceus

```
FA  FA
|   |
Glycerol–Phosphate–Glycerol
                      |
                  Phosphate
                      |
Glycerol–Phosphate–Glycerol
|   |
FA  FA
```

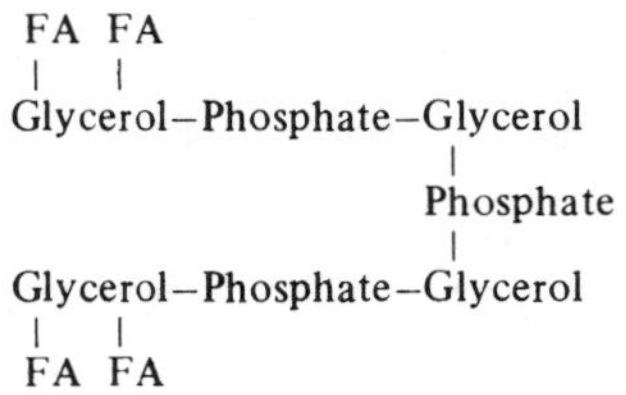

Structure XIII-7 — 181

```
           O                          O
           ||                         ||
H2C–O–C–R1                 R4—C–O–  CH2
  |        O                  O        |
  |        ||                 ||       |
 HC–O–C–R2                 R3C–O–CH
  |        O                  O        |
  |        ||                 ||       |
H2C–O–P–O–CH2        H2C–O–P–O–CH2
           |     |                |
           O-  HC–OH  HO–CH       O-
                 |      O
                 |      ||
               H2C–O–P–O–CH2
                        |
                        O-
```

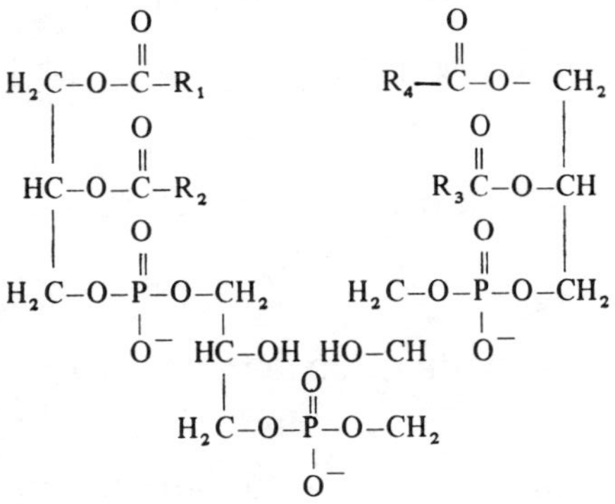

Bisphosphatidyl glyceryl phosphate
Listeria monocytogenes

Table 13 (continued)
UNCOMMON BACTERIAL PHOSPHOLIPIDS

<table>
<tr><th>Shorthand formula</th><th>Example (producer)</th><th>References</th></tr>
<tr><td><pre>
DA   DA
| |
Glycerol–Phosphate–Glycerol
                   |
                   Phosphate
</pre></td><td>Structure XIII-8<pre>
H2C–O–(CH2–CH2–CH–CH2)3CH2CH2CHCH3
 | | |
 | CH3   CH3
HC–O–(CH2–CH2–CH–CH2)3CH2CH2CHCH3
 | | |
 | CH3   CH3
 | O   O
 | ‖   ‖
H2C–O–P–O–CH2CHCH2–O–P–O⁻
        | | |
        O⁻   OH   O⁻
</pre>Diphytanyl glycerol ether analog of phosphatidylglycerophosphate
(2,3-di-*O*-(3′,7′,11′15′-tetramethyl hexadecyl)-*sn*-glycero-1-phosphoryl-3″-*sn*-glycerol-*1*″-phosphate
Halobacterium (*cutirubrum*) *salinarum*
Halococcus (*Sarcina*) *morrhuae*</td><td>136, 158, 414</td></tr>
<tr><td><pre>
Branched sphingosine–Phosphate
         | |
         FA   Ethanolamine
</pre></td><td>Structure XIII-9<pre>
                    H H   O⁻
                    | | |
CH3CH (CH2)11 –C–C–CH2O– P –OCH2CH2NH3⁺
  | | | ‖
 CH3   HO NH   O
                      /
                   C–O–R
</pre>Ceramide phosphinicoethanolamine (phosphorylethanolamine ester of 15-methylhexadecasphinganine)
Bacterioides</td><td>190, 340, 404</td></tr>
</table>

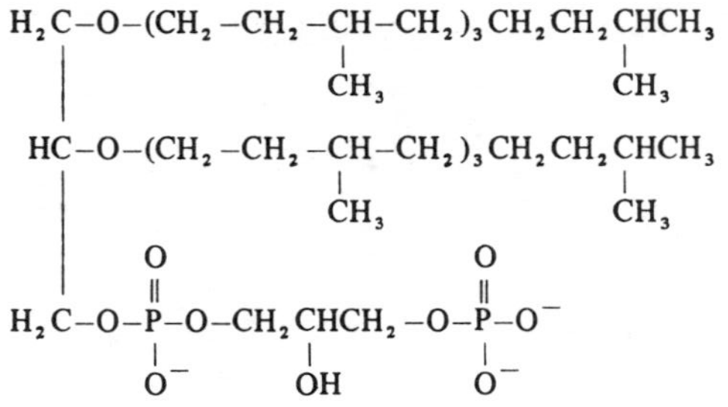

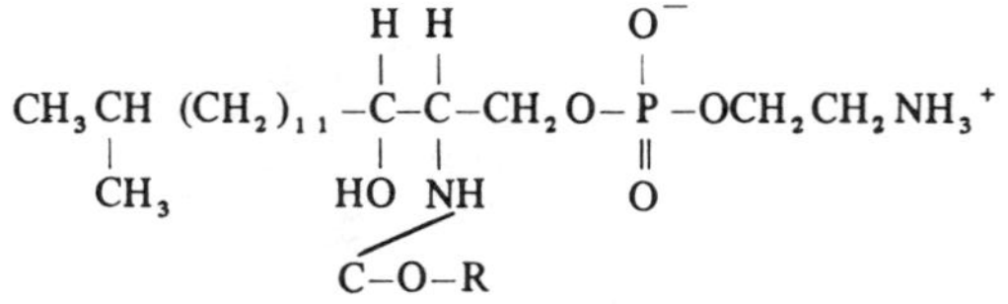

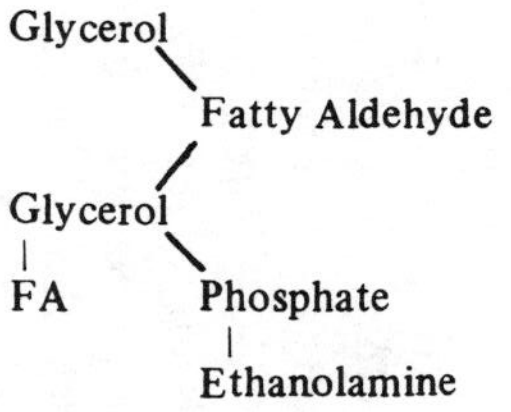

Structure XIII-10 231

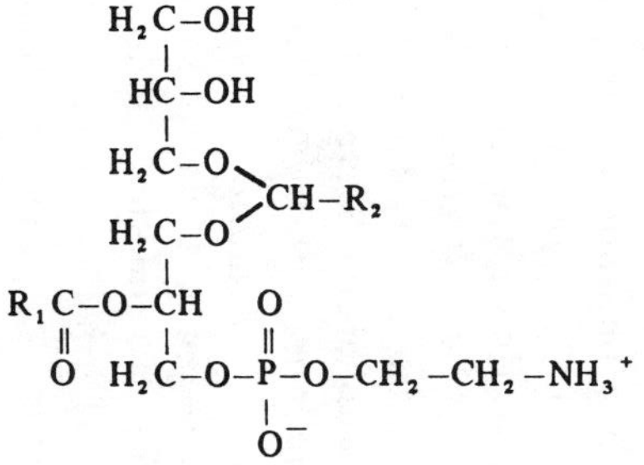

```
Glycerol
        \
         Fatty Aldehyde
        /
Glycerol
 |      \
FA       Phosphate
          |
         Ethanolamine
```

```
          H2C-OH
            |
           HC-OH
            |
          H2C-O
                 \
                  CH-R2
                 /
          H2C-O
            |
R1C-O-CH         O
  ||      |           ||
  O   H2C-O-P-O-CH2-CH2-NH3+
              |
              O-
```

Aldehydogenic analog of phosphatidylethanolamine
Clostridium butyricum

ococcus (Sareina).[136,156,158,414] For a review of these lipids in halophiles, consult Reference 154. Moderate halophiles of the genera *Micrococcus* and *Vibrio* have normal fatty acid ester-containing phospholipids.[156,263] A large number of marine isolates[272] had phospholipids similar to those of nonmarine isolates, and no diether lipids. On the other hand, the nonhalophile *Sulfolobus acidocaldarius* produced glyco- and phospholipids that were almost exclusively of the diether type. Instead of dihydrophytanyl alcohol groups, these lipids contain $C_{40}H_{80}$ isopranol chains, which were thought to contain a ring and to be branches.[66,68,203] Since *Sulfolobus* is an acidophilic extreme thermophile, it would seem that bacteria coming from extreme environments appear to be equipped with lipids of a highly unusual nature.

Phosphonolipids (see Table 7 [VII-5]) are phospholipids in which the phosphorus is directly linked to the first carbon of a nitrogenous base, such as ethanolamine or choline, without the usual intermediate oxygen. These compounds have been reported in minor amounts (0.8 to 3.1%) in *Mycobacterium smegmatis* and *M. phlei*, where they were tentatively identified as the phosphonic acid analogs of phosphatidyl ethanolamine.[320]

Other rare phospholipids, known as sphingolipids (see Table 13 [XIII-9]), appear to be produced only by certain anaerobes. They are also referred to as ceramide phospholipids (ceramide = N-acyl sphingosine) — e.g., ceramide phosphorylethanolamine, ceramide phosphorylglycerol, or ceramide phosphinicoethanolamine.[340] They differ chemically from the common glycerophospholipids in that they are phosphoryl esters of branched or straight-chain N-acyl dihydrosphingosine; the acyl group is a normal or hydroxy fatty acid. White and his colleagues[404] have reported the isolation and characterization of such compounds from the Gram-negative anaerobe *Bacteroides melaninogenes*. More recently, *B. ruminicola* and *Fusobacterium necrophorum*,[190] as well as *B. fragilis* subsp. *fragilis*, *B. putredinis*, *B. fragilis* subsp. *thetaiotomicron*, and *B. fragilis* subsp. *vulgatus*,[92] were also found to contain these compounds. The predominant ceramide bases were 15-methylhexadecasphinganine or 17-methyloctadecasphinganine. These bases have been reported to occur in free form in the cells.[340] Neither *B. amylophilus* (4 strains), *B. succinogenes* (1 strain), nor *B.* sp. (1 strain) had any sphingolipids whatsoever, nor did other anaerobes examined. These include *Butyrivibrio fibrisolvens*, *Selenomonas ruminantium*, *Megasphaera (Peptostreptococcus) elsdenii*, *Ruminococcus flavefaciens*, and *R. albus*.[190] Strains of *Fusobacterium mortiferum (Sphaerophorus freundii)*, *F. necrophorum (S. necrophorus)*, and *F. varium (S. varius)* were also found to lack sphingolipids.[92] Thus, the data on *F. necrophorum* seem to conflict.

It may be that most fusibacteria contain only plasmalogens, another type of phospholipid widely reported in anaerobic bacteria.[104] These alk-1-enyl ethers (Table 7 [VII-3]) contain an α-β unsaturated (alk-1-enyl) ether on the first carbon (C-1) of the glycerol, a fatty acid esterfied to C-2, and the usual phosphate ester on C-3. Genera whose members have been reported to contain plasmalogens include *Bacteroides*, *Clostridium*, *Desulfovibrio*, *Megasphaera (Peptostreptococcus)*, *Propionibacterium*, *Ruminococcus*, *Selenomonas*, *Treponema*, and *Veillonella*.[103,309] Bacteria that were found to contain no plasmalogens include representatives of the genera *Bacillus*, *Corynebacterium*, *Escherichia*, *Lactobacillus*, *Leuconostoc*, *Pseudomonas*, *Rhodospirillum*, *Streptococcus*, and *Streptomyces*.[103] A report that *Treponema pallidum* has plasmalogens, whereas *T.* (now *Spirochaeta*) *zuelzerae* does not,[233] would seem to confirm their recent placement in two different genera.[47b] However, a word of caution is in order. In the above studies the presence of plasmalogens was actively sought, but it should not be assumed that plasmalogens are not present merely because a report on the phospholipids of a given organism does not mention them. They usually exist in mixtures with the normal diacyl phospholipids and may be overlooked if the organic solvent-soluble residues are not examined for phosphorus-containing compounds following strong alkaline deacylation. For further information, consult Reference 63.

OTHER COMPLEX LIPIDS

Glycolipids

Glycolipids are compounds that contain "carbohydrates in combination with long-chain aliphatic acids or alcohols and which are extracted . . . into organic solvents without prior use of hydrolytic procedures".[324] This definition excludes (1) lipopolysaccharides that are also composed of sugars and lipids, but are water-soluble, (2) degradation products of larger molecules arising through hydrolysis, and (3) the sugar-containing phospholipids. Representative types of bacterial glycolipids are given in Table 14.

Shaw's review[324] on glycolipids from bacteria does not cover those from mycobacteria and their close relatives; however, the literature through 1971 on the glycolipids from this group has been well covered by Asselineau's comprehensive text[16] and by Lederer's[210] and Goren's[111] reviews.

The most widespread glycolipids among bacteria are the glycosyl diglycerides. These are *sn*-1,2-diglycerides glycosidically linked with various carbohydrates at the 3-position (see Table 14 [XIV-1 and XIV-2]). Shaw[324] states that this class of glycolipid is most widespread among Gram-positive bacteria; however, a perusal of Sastry's tables[321] and of Table 15 shows that this is no longer true. Thus, we can now list the following Gram-negative genera as having members reported to contain glycosyl diglycerides: *Bacteroides, Butyrivibrio, Chlorobium, Chloroflexus, Chloropseudomonas, Chromatium, Flavobacterium, Halobacterium, Pelodictyon, Pseudomonas, Thermus, Treponema,* and *Yersinia.* Among Gram-positive bacteria, we may cite *Actinomyces, Arthrobacter, Bacillus, Bifidobacterium, Corynebacterium, Lactobacillus, Listeria, Microbacterium, Micrococcus, Nocardia, Rothia, Staphylococcus, Streptococcus,* and *Streptomyces.*

From a taxonomic viewpoint, there appears to be some value in the observation that members of some genera seem to contain identical glycolipids. Thus, based on the available (and often limited) data to date, arthrobacters contain mono- and digalactosyl and dimannosyl diglycerides; treponemas, except *T. zuelzerae,* contain monogalactosyl diglycerides; pseudomonads contain glucosyl and glucuronosyl diglycerides; lactobacilli contain galactosylglucosyl diglycerides or diglucosyl galactosyl diglycerides; staphylococci contain diglucosyl diglycerides; and streptococci contain glucosyl and diglucosyl diglycerides. Tornabene[363] found some mono- and diglucosyl diglycerides in *Yersinia pseudotuberculosis,* but not in *Y. pestis.* Although yersiniae are now classed as Enterobacteriaceae,[47b] no members of this family have been reported to produce glycosyl diglycerides; thus, *Y. pseudotuberculosis* would seem — at least from the point of view of lipid composition — to be something of a maverick. Also, one strain of *Bacillus stearothermophilus* contained mono- and diglucosyl diglycerides, but another (grown under the same conditions) did not.[236] No glycosyl diglycerides have been found in *Rhodopseudomonas* or in *Rhodospirillum,*[321] although many other phototrophs synthesize them. The related phospholipids, glycerophosphoryl diglycosyl diglycerides, have been reported in various bacteria that produce the analogous diglycosyl diglycerides[9] (see Table 13 [XIII-5]).

Acylated sugars, another type of glycolipid, have been widely reported from both Gram-positive and Gram-negative bacteria. These are simple compounds consisting of sugars linked to fatty acids by means of a glycosidic bond (see Table 14 [XIV-4 and XIV-5]). They have been reported to be produced by members of the following genera: *Arthrobacter, Brevibacterium, Corynebacterium, Enterobacter, Escherichia, Micro-*

Table 14
BACTERIAL GLYCOLIPIDS AND SULFOGLYCOLIPIDS

Type	Abbreviated structure	Example (producer)	References
Monoglycosyldiglyceride	Sugar–Glycerol (FA, FA)	Structure XIV-1 Monogalactosyl diglyceride *Arthrobacter globiformis*	396
Diglycosyldiglyceride	FA FA	Structure XIV-2 Galactosylglucosyl diglyceride *Streptococcus pneumoniae*	47h

Structure XIV-1

Monogalactosyl diglyceride
Arthrobacter globiformis

Structure XIV-2

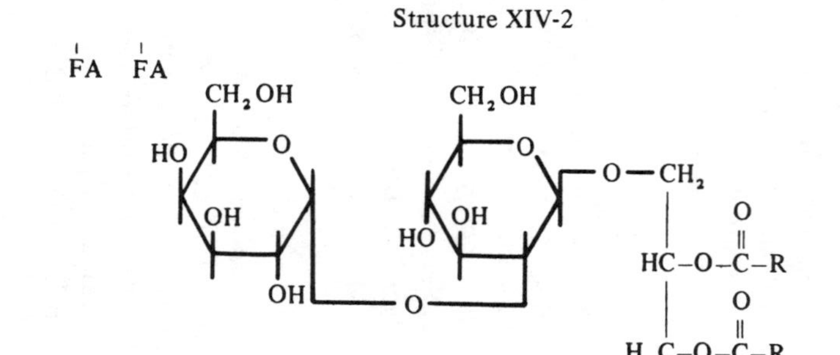

Galactosylglucosyl diglyceride
Streptococcus pneumoniae

Note: FA = fatty acids; MA = mycolic acids; Alc = alcohol; PA = phthioceranic acids.

Lipoglycoside		Structure XIV-3 Rhamnolipid *Pseudomonas aeruginosa*	79
Acyl sugar	$(FA)_x$	Structure XIV-4 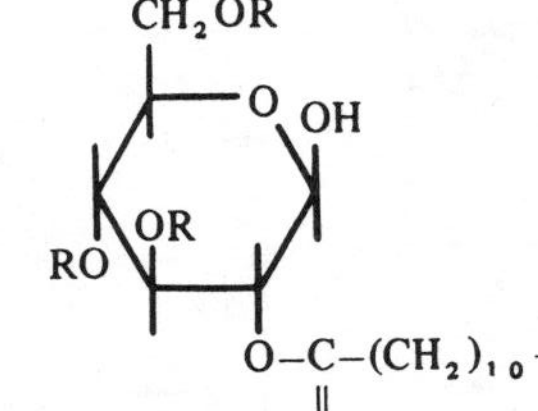 $R = CH_3CO$ Tetraacyl glucose *Escherichia coli* *Pseudomonas fluorescens* *Streptococcus faecalis*	324, 402

Table 14 (continued)
BACTERIAL GLYCOLIPIDS AND SULFOGLYCOLIPIDS

Type	Abbreviated structure	Example (producer)	References
Acyl sugar		Structure XIV-5 $R = C_{87}H_{175}O_2$ 6,6′-Trehalose dimycolate (cord factor) *Mycobacterium tuberculosis*	1
Glycolipid sulfate	Sulfate | Sugar–Sugar–Sugar | Glycerol | | Alc Alc	Structure XIV-6 $R = C_{20}H_{41}$ (phytanyl) Sulfolipid from extreme halophiles	154, 155, 157

Glycolipid Sulfate — Structure XIV-7 — 110, 111

Trehalose
|
Sulfate

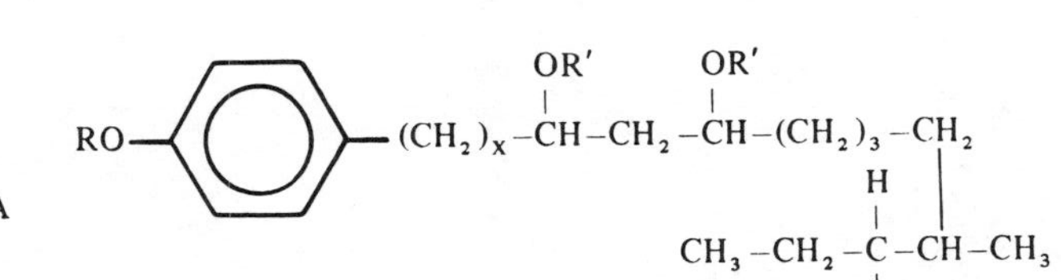

$$R_1 = C_{15}H_{31}-\underset{}{\overset{OH}{CH}}-(\overset{CH_3}{CH}-CH_2)_7-\overset{CH_3}{CH}-CO-$$

$$R_2 = C_{15}H_{31}-CH_2(\overset{CH_3}{CH}-CH_2)_6-\overset{CH_3}{CH}-CO-{}_{\delta}$$

$$R_3 = C_{15}H_{31}CO-$$

Sulfolipid I (provisional)
Mycobacterium tuberculosis

Mycoside — Structure XIV-8 — 96

Mycoside

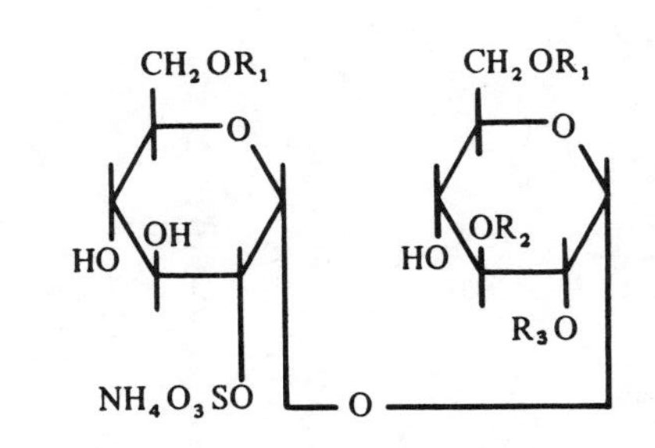

$$RO-C_6H_4-(CH_2)_x-\overset{OR'}{CH}-CH_2-\overset{OR'}{CH}-(CH_2)_3-CH_2-\overset{}{CH}(CH_3)-\underset{OCH_3}{\overset{H}{C}}-CH_2-CH_3$$

R = 2-*O*-Methyl Rhamnose
x = 14-18
R_1 = n $C_{16:0}$ or
Mycocerosic acids
Mycoside B
Mycobacterium bovis

Table 15
BACTERIAL GLYCOLIPIDS

Microorganism	Type of glycolipid	Formula, if known	Ref.
GRAM-POSITIVE BACTERIA			
Bacillus			
cereus	None	—	238
megaterium	Monoglycosyl diglyceride	1-(O-β-Glucosaminyl)-2,3-Di	290
stearothermophilus A	Mono- and diglucosyl diglycerides	—	242
stearothermophilus B	None	—	242
Clostridium			
butyricum	Monoglycosyl diglyceride	1-(O-β-Glu*p*)-2,3-Di	231
Lactobacillus			
heterohiochii	Diglycosyl diglyceride	Footnote a	376
Micrococcus			
luteus (lysodeikticus)	Glycolipid	—	70
Staphylococcus			
aureus	Mono- and diglucosyl diglycerides	—	32
epidermidis	Mono- and diglucosyl diglycerides	3(1)-O-β-D-Glu*p*-1(3),2-Di	176
		3(1)-O-β-D-Glu*p*-(1 → 6)-O-β-D-Glu*p*)-1(3),2-Di	176
Streptococcus			
faecalis	Diglycosyl diglyceride	Contains 3-O-α-kojibiosylglycerol	90
ACTINOMYCETES AND OTHERS			
Arthrobacter		3-O-β-D-Gal*p*-1,2-Di	
crystallopoietes	Mono- and diglycosyl diglycerides	3-(O-β-D-Gal*p*-(1 → 6)-O-β-D-Gal*p*)-1,2-Di	327
		3-(O-α-D-Mann*p*-(1 → 3))-O-α-D-Mann*p*)-1,2-Di	182
		3-O-β-D-Gal*p*-1,2-Di	
globiformis	Mono- and diglycosyl diglycerides	3-(O-β-D-Gal*p*-(1 → 6)-O-β-D-Gal*p*)-1,2-Di	182
		3-(O-α-D-Mann*p*-(1 → 3)-O-α-D-Mann*p*)-1,2-Di	
pascens	Mono- and diglycosyl diglycerides	3-O-β-D-Gal*p*-1,2-Di	182
		3-(O-β-D-Gal*p*-(1 → 6)-O-β-D-Gal*p*)-1,2-Di	
		3-(O-α-D-Mann*p*-(1 → 3)-O-α-D-Mann*p*)-1,2-Di	
Actinomyces			
viscosus	Monoglycosyl diglyceride	1-Monogalactosyl-Di	427
Bifidobacterium			
bifidum var. *pennsylvanicus*	Mono-, di-, and tri-, glycosyl diglycerides	2,3-O- diacyl-β-D-Gal*f*-1,2-Di	389
		3′-O-acyl-β-D-Gal*f*-1,2-Di	
		Gal*p*(1 → 2)Gal*p*(1 → 2)Gal*p*-1,2-Di	
		O-β-D-Gal*p*-1,2-Di	
		3′-O-acyl-β-D-Gal*f*(1 → 2)-β-D-Gal*f*-1,2-Di	
		Gal*p*(1 → 2)Gal*p*-1,2-Di	
Cornyebacterium			
aquaticum[b]	Diglycosyl diglyceride	Dimannosyl diglyceride	117, 165
diphtheriae	Acyl sugar	Acyl glucose	44
	Cord factor	Diacyl trehalose	
ovis	Cord factor	Dicorynomycolyl trehalose	44
Micromonospora sp.	Cord factor	Diacyl trehalose	347

Table 15 (continued)
BACTERIAL GLYCOLIPIDS

Microorganism	Type of glycolipid	Formula, if known	Ref.
Mycobacterium			
avium	Mycoside C type	Mycoside C_2	394
farcinogenes var. *senegalense*	Mycoside C type	Mycoside C′	198
sp. (nonphotochromogen)	Mycoside C type	Mycoside C	198
marinum	Mycoside B type	Mycoside G	95
Nocardia			
asteroides	Cord factor	Dimycolyl trehalose	423
caviae	Acyl sugar-polyol	Acid polyol glucose acylated with fatty acids	293
coeliaca	Acyl sugar	Acyl glucose	164
polychromogenes	Acyl sugar	Triacyl glucose	164
	Diglycosyl diglyceride	Diglucosyl diglyceride	
Rothia			
dentocariosa	Diglycosyl diglyceride	Dimannosyl diglyceride	283
GRAM-NEGATIVE BACTERIA			
Chlorobium sp.	Mono- and diglycosyl diglycerides	Galactosyl diglyceride Galactosyl-rhamnosyl diglyceride	162
Chloroflexus[c] sp.	Mono- and diglycosyl diglycerides	Galactosyl diglyceride Diaglactosyl diglyceride	162
Flavobacterium			
thermophilum	Tetraglycosyl diglyceride	Gal*f*-(1 → 2)-Gal*p*(1 → 6)GluN-(acyl)-(1 → 2)-Glu*p*-Di	281
Halobacterium			
salinarium (cutirubrum)	Triglycosyl diether	—	192
	Sulfated glycolipid	—	
Moderate halophile	Monoglycosyl diglyceride	Glucuronosyl diglyceride	286
Pelodictyon sp.	Mono- and diglycosyl diglycerides	Galactosyl diglyceride Galactosyl-rhamnosyl diglyceride	162
Pseudomonas			
rubescens[c]	Monoglycosyl diglycerides	Glucosyl diglyceride	409
Selenomonas			
ruminantium	Acylated amino sugars	O-and N-Acyl-β-glucosaminyl-1,6-glucosamine	141
Spirochaeta			
zuelzerae	Monoglycosyl digycerides	Glucosyl diglyceride	233
Sulfolobus			
acidocaldarius	Monoglycosyl polyol diglyceride Diglycosyl diglyceride	Glucosyl polyol diglyceride Glucosyl galactosyl diglyceride	203
Treponema			
pallidum	Monoglycosyl diglycerides	Galactosyl diglyceride	233
Yersinia			
pestis	None	—	363
pseudotuberculosis	Mono- and diglycosyl diglycerides	Glucosyl diglyceride Diglucosy diglyceride	363

Note: Di = diglyceride; *f* = furanosyl; Gal = galactose; Gal*f* = Galactofuran; Glu = glucose; GluN = glucosamine; *p* = pyranosyl.

[a] Contained long-chain fatty acids not found in other cellular lipids.
[b] Lacks acyl glucose characteristic of other corynebacteria.
[c] Taxonomic position unknown.

monospora, Mycobacterium, Nocardia, Pseudomonas, and *Streptococcus.*[324] An unusual triglycosyl diether was found in the lipids of *Halobacterium cutirubrum,*[192] adding still another ether lipid to the list produced by members of this taxon.

A glycolipid from *Nocardia caviae,*[293] which was never fully characterized, was said to contain an acidic polyol in glycosidic linkage, with glucose acylated by normal fatty acids. A compound characterized as 6-glucose corynomycolate was isolated from the lipids of glucose-grown *Corynebacterium diptheriae* and *Mycobacterium smegmatis.*[45] A similar substance, 6-glucose β-hydroxy-α-oleyl oleate, was reported from an oleic acid-requiring strain of *Brevibacterium thiogenitalis.*[264] According to Hackett and Brennan,[117] absence of acyl glucose in *Corynebacterium aquaticum* sets this species apart from the other corynebacteria.

Among the most well-studied types of acylated sugars are the "cord factors" and their various analogs. These are acylated trehaloses in which the acyl groups may be fatty acids, mycolic acids (q.v.), or very-long-chain polyunsaturated acids (phleic acids). They occur exclusively in actinomycetes and related forms. Cord factor, or 6, 6′- trehalose dimycolate[261] (see Table 14 [XIV-5]), was originally so named because it was believed to be the substance responsible for the cord formation in virulent strains of *Mycobacterium tuberculosis.*[37] As very similar substances are isolated from other non-cord-forming mycobacteria, this is no longer considered to be true. A glycolipid thought to be trehalose monomycolate was also isolated from *M. tuberculosis* H37Rv.[221] Other cord factors have been reported from *M. bovis* and *M. butyricum* (both acylated at the 6,6′-positions, with eumycolates having 77 to 89 carbons),[1] from *M. fortuitum* (acylated at positions 4 and 6 with tuberculostearic, palmitic, or stearic acids),[392,393] and from *M. phlei* and *M. smegmatis* (acylated at eight positions with phleic acids).[14] Some were also isolated from *Corynebacterium diphtheriae* (corynomycolic and corynomycolenic acids at the 6,6′-positions),[323] from *Nocardia asteroides* (trehalose corynomycolate, positions of acyls unknown),[128,423] and from *N. rhodochrous* (acylated at the 6,6′-positions with nocardomycolates having 38 to 46 carbons).[128] *Arthrobacter paraffineus* was said to produce a trehalose dicorynomycolate not substituted at the usual 6,6′-positions when it was grown on an alkane-containing medium. *Corynebacterium alkanum, C. hydrocarboclastus, C. pseudodiphtheriae, Brevibacterium butanicum, Nocardia butanica,* and *N. convoluta* also produced trehalose lipids on such a medium.[129,344,345] No trehalose lipids were found in alkane-utilizing strains of *Micrococcus, Microbacterium,* or *Pseudomonas.*

Lipomannans (LM) have recently been isolated from membranes of *Micrococcus luteus* (strains of *M. "lysodeikticus", "M. flavus",* and *"M. sodonensis"*). The LM from the *"lysodeikticus"* strain has been partially characterized as having a long chain of (about 60) hexose residues with two branch points, linked at one end to the C-1 of a glycerol moiety acylated with two fatty acids.[295]

Among the complex lipids of mycobacteria that must be mentioned are the so-called "mycosides." Their history and chemistry have been covered by a recent review by Goren.[111] These glycolipids were originally thought to be species-specific; unfortunately, this no longer seems to be true. Mycosides A and B (see Table 14 [XIV-8]) are glycolipids containing a phenolic moiety; the mycosides of type C (see Table 16 [XVI-3]) are really peptidoglycolipids.

A second group of peptidoglycolipids, known collectively as "waxes D", are found in certain actinomycetes with a cell wall of Type IV.[16,204] Wax D was formerly a term employed by Anderson (see Reference 16) to refer to an inhomogeneus mycobacterial lipid fraction extracted by chloroform and insoluble in boiling acetone. However, it has come to refer to a part of this fraction — specifically, certain complex glycolipids containing mycolic acids, arabinose, galactose, and, in many cases, amino acids and amino sugars, which are produced by certain mycobacteria, corynebacteria, and nocardiae (see Table 16 [XVI-4]). Waxes D are usually found in "bound" form, but sometimes they are also "free". For excellent reviews of their chemistry, biological properties, and biosynthesis, consult References 111 and 210. The occurrence of waxes D in the *Mycobacterium-Corynebacterium-Nocardia* complex (see References 16, 20, 150, 199, 211, 228, 234, 383, 416, and 423) underlines the close taxonomic relationship of these genera that has been proposed on the basis of cell wall composition.[206]

Amino-Acid-Containing Lipids

Over the years, a growing number of reports on bacterial lipids have been concerned aith amino-acid-containing lipids (nonphospholipids), especially those with an ornithine moiety. A review by Thiele and Schwinn[354] has recently appeared on this subject. Found in Gram-negative bacteria and in actinomycetes, ornithine lipids may be divided into two principal types (see Table 16 [XVI-1 and XVI-2]), depending on the identity and placement of the fatty substituents and on the presence or absence of a polyol. Ornithine-containing lipids whose structures have not been elucidated have been reported from *Rhodospirillum rubrum*[46b,314] and from *Mycobacterium* sp.[202]

Peptidolipids

A small number of peptidolipids (peptide-containing lipids) have been isolated from mycobacteria and from nocardiae.[24,115,196] Considering how few have been isolated and the immense amount of work that has been carried out on lipids of this group of organisms, it seems that this type of lipid is too rare to have any taxonomic importance.

Sulfolipids

To date, all sulfolipids reported from bacteria are glycolipid sulfates. These compounds are relatively unstable, undergoing spontaneous desulfation under mild conditions.[111] *Mycobacterium tuberculosis* produces several 2,3,6,6′-tetraacyl-α,α'-trehalose-2′-sulfates,[110-112] the acyl groups being principally fatty acids of the phthioceranic acid type (see Table 14 [XIV-7]). Minor amounts of other sulfoglycolipids were also found. *Halobacterium cutirubrum* produces a sulfated diether triglycosyl diglyceride. As in other types of lipids from this species, the ether-linked moieties were phytanyl alcohol residues (q.v.).[155,157,192] A diether glucosyl polyol diglyceride monosulfate is produced by *Sulfolobus acidocaldarius,*[203] and an unidentified sulfolipid was reported from *Rhodopseudomonas sphaeroides.*[262] Their instability and the difficulties attendant to their isolation and characterization make it unlikely that sulfolipids will play much of a role in bacterial taxonomy.

Waxes

Waxes are compounds composed of fatty acids linked to long-chain fatty alcohols. Their apparent widespread occurrence in *Acinetobacter* spp. may be taxonomically

Table 16
ORNITHINE-CONTAINING LIPIDS AND PEPTIDOGLYCOLIPIDS

Abbreviated structure	Structure	Producers	References
Ornithine – OH Fatty Acid (FA) – FA	$NH_2-CH_2-(CH_2)_2-CH(COOH)-NH-C(=O)-CH_2-CH(R)-O-C(=O)-R$ (bracket labels R_1, R_2) Type I [XVI-1] R_1 = 3-OH-FA R_2 = FA's	*Pseudomonas rubescens*	409
		Thiobacillus thiooxidans	174
		Rhodospirillum rubrum	46b
		Rhodopseudomonas spheroides	109
		Desulfovibrio gigas	224
Ornithine–Polyol–FA – FA	$NH_2-CH_2-(CH_2)_2-CH(C(=O)-O-(CH_2)_x-O-C(=O)-R)-NH-C(=O)-R$ (bracket labels R_1, R_2) Type II [XVI-2] R_1 = OH, cyclopropane, or branched-chain FA's R_2 = β-OH or n-FA's x = 2 or 3	*Mycobacterium bovis*	299
		Streptomyces sp.	26
		Stm. toyocaensis	159
		Stm. sioyaensis	159
		Brucella melitensis	353, 354
		B. abortus	353, 354
		Bordetella pertussis	353, 354

Note: Ala = alanine; Arab = arabinose; Dap = diaminopimelic acid; Glu = glutamic acid; GluNAc = N-acetyl glucosamine; MurN-Gly = N-glycoyl muramic acid.

<table>
<tr><td>FA–Oligopeptide
|
Sugar
Nitrogenous Compound
|
Sugar</td><td>R–Phe–allo Thr–Ala–Alaninol
| |
S_1 S_2
[X-3]
R = -OH FA
S_1 = O-diacetyl-6-deoxytalose
S_2 = 2,3,4-tri-O-methyl rhamnose
Mycoside C_{1217}</td><td>Mycobacterium sp.</td><td>197</td></tr>
<tr><td></td><td>Mycolic Acid $\overset{5}{-}$ Arab(1 → 5)Arab
| 1 or 3
Arab(1 → 5)Arab(1 → 5)Arab(1 → 5)Gal(1 → 4)Gal
|
O
|
HO–P=O
|
O
|
GluNAc–MurN–Gly
|
Tetrapeptide
(Ala-Glu-Dap-Ala)
Wax D
[XVI-4]</td><td>Various actinomycetes</td><td>211, 228</td></tr>
</table>

significant.[94,225,361] Waxes have also been found in *Micrococcus cryophilus,*[318] in *Clostridium* spp.,[123] and possibly in *Brucella melitensis.*[353]

Hydrocarbons

Except for the fact that they lack a carboxyl group, hydrocarbons have strong structural analogies with fatty acids, occurring in normal (straight-chain), monounsaturated, or branched forms. They may be important in the taxonomy of Gram-positive cocci. Morrison et al.[250] reported hydrocarbons in all micrococci examined, except in staphylococci. *Sarcina lutea* and *S. flava* (now synonomous with *Micrococcus luteus*) were also reported to produce these compounds.[367] Other bacteria from which hydrocarbons have been isolated include members of the genera *Clostridium,*[123] *Corynebacterium,*[193] *Enterobacter,*[370] *Escherichia,*[260] *Rhizobium* (infective on *Lotus*),[98] and *Nocardia.*[64] A review on this subject appeared in 1970.[6]

Lipid-Containing Iron-Binding Compounds

The "bactins" mycobactin and nocobactin are organic solvent-soluble, iron-chelating agents produced by mycobacteria and nocardiae. These relatively large molecules contain hydroxamic groups similar to those found in water-soluble, iron-binding agents, known as siderochromes, which are produced by ordinary bacteria; however, in other aspects they are chemically different from them. Bactins are produced maximally only in iron-deficient media. For reviews of their history, chemistry, and biosynthesis, consult References 305 and 335. Of taxonomic interest is the finding that single strains of mycobacteria representing nine different "clusters", obtained by phenetic methods, each produced different mycobactins.[335] It would be of value to know whether all strains within each cluster would produce the same mycobactin type. Recently[304] an examination of nocobactins from various nocardiae showed that those of *Nocardia asteroides, N. brasiliensis,* and *N. caviae* represented the three principal chemical types. No nocobactins were found in any of the strains of the *"rhodochrous"* group examined.

Bound Lipids

The genera in which the "bound" lipids (those that require acid or alklaline hydrolysis for their release) appear to differ in a marked manner from those that are free include *Pseudomonas,*[255] *Alcaligenes,*[65,99] and possibly also *Moraxella*[131] and *Neisseria.*[101,131,195,253] There are probably many others, especially among Gram-negative bacteria, which will come to light as more data are accumulated. A large proportion of the lipids of *Leptospira* is found in bound form.[179]

Use of Patterns of Unknown Lipids in Taxonomy

Jenkins et al.[134] found that crude extracts from whole cells could be separated by two-dimensional thin-layer chromatography to yield reproducible patterns characteristic of certain groups of mycobacteria. The lipids and lipid-like substances were not identified. Others have employed this system with success.[113] It has also been used in the numerical taxonomy of this group.[187] A second, similar technique has recently been described by Tsukamura and Mizuno.[371]

STABILITY OF LIPID COMPOSITION

No serious consideration can be given to the utility of lipid composition in taxonomy without a hard look at the stability of this character. It is, of course, true that bacterial lipid profiles may change through variation within the natural population or as a consequence of induced mutation.[332] Thus, in order to assay the limits of variability and mutability, it is not only desirable, but necessary, to accumulate as large a reservoir of lipid data on as many different strains as possible. This is true for all taxonomic characters.

Genetic and environmental factors must be weighed. At present there is a substantial amount of data on the influence of medium components, pH, age, growth temperature, and aeration on lipid composition. Previous reviews of this subject include those of Kates,[152] O'Leary,[267] Veerkamp,[388] and Ambron and Pieringer.[9]

Effect of Growth Temperature

Observations going back more than 50 years[350] show that the growth temperature may have a profound effect on both qualitative and quantitative lipid composition.

FATTY ACIDS

The most commonly reported effect of growth temperature change is an increase in unsaturated fatty acids with a corresponding decrease in normal and, sometimes, cyclopropane fatty acids in the total lipids of some mesophilic strains grown at temperatures below their optima. To this we may now add a shortening of fatty acid chain length with decreasing temperature. Both types of change are presumed to contribute to the maintenance of the "fluidity" of the lipids by lowering their melting points.

To *Escherichia coli* and *Serratia marcescens*, which are already known to undergo the first type of response,[152] we may now add *Bacillus licheniformis*,[56] *Listeria monocytogenes*,[348] *Neisseria* sp.,[131] an unidentified marine bacterium,[271] and *Pseudomonas fluorescens*.[100] In *P. fragi*, dropping the temperature from 25° to 7° increased the total unsaturated fatty acids by about 10% in the "free" lipids, but only about 2.6% in the "bound" ones.[235] In contrast, lowering the temperature of growth had little or no effect on the fatty acids of *Lactobacillus lactis*,[388] 35 strains of *Mycobacterium kansasii*,[357] *Staphylococcus aureus*,[138] or *Thiobacillus novellus*.[215] Likewise, the polyunsaturated phleic acids of *M. phlei* were unchanged by a switch from 37° to 30°; however, the $C_{18:1}$ content of this strain increased.[14]

The chain-shortening response was evident in the free lipids of *Bifidobacterium bifidum* var. *pennsylvanicus*[388] and of the psychrophile *Micrococcus cryophilus*.[317] In the latter, cells grown at both 20° and 0° contain 95% $C_{16:1}$ and $C_{18:1}$; however, there is a fourfold increase in the $C_{16:1}$:$C_{18:1}$ ratio when the growth temperature is dropped to 0°. The same response was observed to a lesser extent in the bound fatty acids of this organism. Chain length shortening plus a transitory shift in one medium to increased unsaturation were found in the fatty acids of *Listeria monocytogenes* when the growth temperature was dropped from 37° to 4°.[348] In the extreme thermophile *Flavobacterium thermophilum*, a growth temperature decrease from 82° to 49° was reflected in increased percentages of fatty acids of shorter chain lengths: *iso*-$C_{17:0}$ decreased from 64% to 31% of the total, *anteiso*-$C_{15:0}$ increased from 2% to 13%, and *iso*-$C_{15:0}$ increased from 12% to 31%.[280] A similar response was also observed in the thermophiles *Bacillus caldolyticus* and *B. caldotenax* grown at lower than optimal temperatures.[400] In *Thermus aquatius*, total lipids increased 70% as the temperature was increased from 50° to 75°. The only marked qualitative change was the loss of *anteiso*-$C_{17:0}$ at temperatures greater than 55°, along with a modest trend toward higher melting fatty acids.[306,307]

The fatty acids of the phospholipids generally reflect the overall profile of the cellular fatty acids; however, there are some exceptions. When the fatty acids of the phospholipids of *Escherichia coli* were analyzed, Aibara et al.[5] and Kito et al.[172] found that following a shift in growth temperature from ca. 40° to ca. 20° there was an overall increase in unsaturation in the fatty acid moieties. The most rapid changes occurred in PG; only minor changes were found in the fatty acids of DPG. In *Clostridium butyricum,*[166] a change of the growth temperature from 37° to 25° was reflected in an increase of unsaturated as well as cyclopropane fatty acids of all the phospholipids, along with a decrease in the saturated fatty acids. The alk-1-enyl chains of the PE and PME plasmalogens of this organism were affected in exactly the reverse way, becoming more saturated when the temperature was shifted from 37° to 30°. At 25° the trend was inexplicably reversed toward the higher temperature profile. When they examined the effect of dropping the temperature to 25° on the type of acylations found at the two positions of the glycerol moieties of PE and PME of the diacyl type, these workers found a general trend toward shorter chain lengths, fewer cyclopropane fatty acids, and greater unsaturation on C-1, whereas on C-2 these changes were less marked.

In *Pseudomonas fluorescens,*[61] a drop in the growth temperature from 33° to 5° was accompanied by an increase of only certain unsaturated and cyclopropane phospholipid fatty acids, namely, $C_{16:1}$ and $\Delta C_{17:0}$, but not $C_{18:1}$ or $\Delta C_{19:0}$.

The effect of growing an organism at higher than its optimal temperature is not clear-cut. In some cases an increase in the saturated fatty acids occurs, whereas in others different or no effects are noted. For example, the expected increase of saturated fatty acids occurs in *Escherichia coli* and in *Bacillus subtilis,* but not in *Staphylococcus aureus.*[152] In *Lactobacillus heterohiochii,* the ratio of unsaturated/saturated fatty acids decreases with increasing temperature (25° to 39°), and the fatty acids having a chain length greater than C_{20}, normally occurring in amounts up to 36.5% in this organism, also decline. In contrast, in *Lactobacillus lactis* a change in growth temperature from 37° to 45° had little effect on the fatty acids in the log phase, but a marked increase in $C_{18:1}$ at the expense of $C_{16:0}$ in the stationary phase. However, little change in the average chain lengths was found.[377,388]

OTHER LIPIDS

At present, there is no general rule concerning the effect of growth temperature on complex lipids. Because so little data have been accumulated on this subject, the results seem to be conflicting, as well as confusing. However, it may be true that certain types of changes — for example, in major types of phospholipids — in response to temperature may be specific for a given taxon, and therefore may become a useful taxonomic character. A strain of *Bacillus stearothermophilus* grown at its optimal temperature of 55° showed a 15% increase of phospholipids and a substantial decrease of other lipids, compared to cells grown at 37°.[275] The major phospholipid at 55° was DPG, and at 37° it was PG; PE was not affected.[212] In *Salmonella typhimurium,* a decrease of growth temperature from 37° to 32° was accompanied by a loss of PG and DPG and an increase of PE.[273] In *Vibrio marinus,* a decrease in growth temperature resulted in a decrease in the proportions of the major phospholipids, PG and PE, accompanied by an increase in their lyso forms.[272] The extreme thermophile *Thermus aquaticus,* grown at 50° instead of its optimum of 75°, showed an overall loss of total lipids (26% to 10%), phospholipids (37 moles to 19 moles of lipid P), and glycolipids (16 moles to 5 moles of lipid carbohydrate). The proportion of each lipid within each class remained the same.[307] In *Clostridium butyricum* a decrease in the growth temperature from 37° to 25° resulted in a 12.3% increase in the plasmalogen glycerol phosphoglycerides and a decrease of 8% in the plasmalogen ethanolamine and N-methylethanolamine phospholipids. However, the glycerol hemiacetal form of ethanolamine and N-methylethanolamine plasmalogens increased. The diacyl phospholipids did not change.[166]

In contrast, no change was observed in the proportions of neutral to polar lipids or in the types and quantities of phospholipids in the psychrophile *Micrococcus cryophilus* when grown at 5° vs. 20°.[318] Likewise, the types and quantities of phospholipids of yersiniae were unchanged when the growth temperature was dropped from 37° to 27°.[363] This was also true for *Pseudomonas fluorescens,* whether the cells were grown at 5°, 22°, or 30°.[61]

In view of the obvious influence growth temperature has on the composition of most bacterial lipids, one may well ask whether all strains to be compared should be grown at their optimal temperature or at a single common temperature, even if the latter may be far from the optimum for many of them. The former alternative seems preferable, since it will yield the "natural" lipid profile for a given strain, and thereby, as the sum of many strains, a true lipid profile for the entire genus. It is also desirable to grow bacteria at their optimal temperature from the point of view of reproducibility, since bacteria grown at other temperatures may be expected to show abnormalities in their metabolism under what for many of them should be considered conditions of stress. Despite their finding of reduced amounts of saturated and increased unsaturated fatty acids in certain low-temperature-requiring marine bacteria, as compared to the fatty acid patterns of similar nonmarine bacteria growing at higher temperatures, Oliver and Colwell[271] concluded that all strains examined could be considered the typical *"Vibrio"* type. They also compared the fatty acid composition of two *V. marinus* strains, one a mesophile and the other a psychrophile. Their profiles were more similar when both were grown at their optimal temperatures (25° and 15°) than when both were grown at 15°. Thus, these authors recommend growing bacteria at their optimal temperatures.

A comparison of the fatty acid profiles of members of the genus *Bacillus* with different temperature optima shows that in almost every case the major fatty acids are branched-chain with major or minor amounts of saturated fatty acids. Other components are present in some species. The psychrophiles *Bacillus psychrophilus, B. insolitus,* and *B. globisporus* have, in addition to the typical profile of blanched- and straight-chain fatty acids, large amounts of unusual Δ^5 unsaturated fatty acids.[144] The thermophiles *B. caldolyticus* and *B. caldotenax* have major amounts of branched-chain fatty acids, although those of cells grown at higher temperatures tend toward longer chain lengths.[400] In *B. stearothermophilus,* Shen et al.[329] found branched fatty acids with longer chain lengths at higher temperatures, whereas Yao et al.[425] reported that growth at higher temperatures caused a decrease in the total amounts of branched-chain fatty acids (although they remained the major components) and an increase in n-$C_{16:0}$ and $C_{16:1}$, a change that also raised the lipid melting points. Only when one examines the fatty acids from *B. acidocaldarius,* an acidotolerant thermophile, does the profile change. Here the branched-chain fatty acids are minor components, replaced by major quantities of unusual cyclohexyl fatty acids.[67] Thus, again we have an example of unusual lipids produced by cultures from extreme environments.

Chan et al.[53] reported that members of the genus *Clostridium* grown at their different temperature optima showed qualitative and quantitative differences in their fatty acid content. The thermophiles *Clostridium thermosaccharolyticum* and *C. tartarivorum,* grown at 55°, had more branched-chain fatty acids than the mesophile *C. pasteurianum;* the psychrophile had none at all. Both the thermophiles and the mesophile had more saturated cyclopropane fatty acids than the psychrophile. Both strains with (or having) lower temperature optima had more unsaturated fatty acids than the thermophiles and, in particular, contained certain unique unsaturated cyclopropane fatty acids absent in the latter. Previously published data[152,268,325] reveal that the fatty acids of all of these strains, except for the branched-chain types, are within the range of variation expected for the genus. The data of Chan and colleagues[53] simply broadened

Table 17
EFFECT OF CARBON SOURCE ON MAJOR FATTY ACIDS OF *MYCOBACTERIUM CONVOLUTUM*

		Fatty acids		
Substrate	Major (>15%)	Odd carbon, %	Even carbon, %	None detected
Acetate	$C_{16:0}$, $C_{16:1}$, $C_{18:1}$, $BrC_{19:0}$	14	86	$C_{17:0}$, $C_{18:0}$
Propionate	$C_{17:1}$, $C_{15:0}$, $BrC_{18:0}$	96	4	$BrC_{19:0}$, $C_{16:1}$, $C_{18:1}$, $C_{18:0}$
Isopropanol.	$C_{16:0}$, $BrC_{19:0}$, $C_{16:1}$	23	77	$C_{17:0}$, $C_{18:0}$
Isopropylamine	$C_{16:0}$, $C_{16:1}$	28	72	$C_{17:0}$, $C_{18:0}$
n-Propylamine	$C_{15:0}$, $C_{17:0}$, $BrC_{18:0}$	95	5	$BrC_{19:0}$, $C_{16:1}$, $C_{18:1}$
1,3-Propanediamine	$C_{16:0}$, $BrC_{19:0}$, $C_{18:1}$	9	90	

Adapted from Cerniglia, C. E. and Perry, J. J., *J. Bacteriol.*, 124, 285, 1975. With permission.

the data base on the genus; in so doing, they have modified its definition in terms of lipids. Thus, genera that harbor strains with diverse temperature requirements will, in many cases, vary widely in types and quantities of fatty acids.

Effect of Medium Composition

CARBON SOURCE

In an extensive study of the effect of different carbon sources on the fatty acids of *Mycobacterium convolutum*, Cerniglia and Perry[50] found that by switching substrates they could obtain major quantitative and qualitative changes. The cells were preadapted to each substrate in a basal salts medium, then grown in the same medium and analyzed for fatty acids. Their results are summarized in Table 17. A study using *Mycobacterium vaccae*[391] also showed that with isopropanol or acetate as a carbon source the major fatty acids were those with an even number of carbons, whereas uneven carbon fatty acids predominated with propionate. Changing the carbon source from acetate to propionate, however, has no effect on the phospholipids. In *Listeria monocytogenes*, replacement of glucose by glycerol in a tryptose-NaCl medium resulted in an increase in *iso*-fatty acids and a reduction of the *anteiso*-types.[139] A change from 1% glucose to 3% sucrose in the growth medium of *Streptococcus mutans* caused a marked increase of branched-chain fatty acids in the strains examined.[76] In contrast, either carbon dioxide or glycerol as a carbon source yielded the same fatty acid profile in *Mycobacterium kansasii;* addition of citrate, glucose, or carbonate to the other medium constituents had no effect.[357] Glucose and glycerol both yielded the same phospholipid pattern in *Nocardia asteroides*.[423] Increasing the concentration of glucose or substituting glycerol in yeast extract medium had no effect on the fatty acids of *Lactobacillus heterohiochii;* however, increasing the yeast extract concentration gave rise to an increase of the fatty acids with chain lengths greater than 20 carbons. The ethanol concentration in the medium of the alcoholophilic organism had a direct relationship to the unsaturated fatty acid content.[377] A comparison of acetate with pyruvate or glucose as carbon sources showed little effect on the fatty acid profiles of the thermophilic bacilli *Bacillus caldolyticus* and *B. caldotenax*. However, nitrogren sources were extremely critical (see below).[399] Addition of small amounts of glucose or acetate to the growth medium had no effect on the fatty acids of *Thiobacillus thiooxidans*.[214]

The presence of acylated sugars in the cell may depend to a singular degree on the growth conditions used. Thus, in many cases their production may be totally suppressed, or the sugar moiety may be changed by the addition or withdrawal of a single constituent in a medium.[129,324,344,423]

NITROGEN SOURCE

Depletion of the utilizable nitrogen in both batch and chemostat culture has been reported to cause accumulation of total lipids in Gram-negative bacteria.[82] Ammonium sulfate substituted for asparagine in the growth medium of *Mycobacterium kansasii* produced no change in the fatty acid composition of the cells.[357] Changes following the omission of brain-heart infusion from the medium pointed up the fact that certain amino acids (including leucine, isoleucine, and valine) were the critical components in obtaining the "typical" fatty acid profile of *Bacillus caldotenax* and *B. caldolyticus*. Their omission caused an increase of branched-*anteiso* and straight-chain, even-carbon fatty acids and a decrease in branched-chain odd-carbon fatty acids.[399] In *B. cereus*, the addition of isoleucine and other precursors of branched-chain fatty acids (such as L-α-aminobutyrate, L-homoserine, and D,L-α-methylbutyrate) increased the *anteiso* types of fatty acids at the expense of other branched types. In contrast, addition of L-leucine or isovalerate to the medium strongly enhanced *iso*-odd fatty acids, and L-valine and isobutyrate increased the *iso*-even types.[145] The presence or absence of branched-chain precursors had no significant effect on the fatty acids of *Thiobacillus novellus*.[215]

Azotobacter vinelandii cells grown with nitrogen gas or ammonium ions as nitrogen sources had the same types of phospholipids and phospholipid fatty acids; however, in the N_2-grown cells, phospholipids accounted for 90 to 94% of the total lipids, whereas in the NH_4^+-grown cells they accounted for only 67%. It was surmised that the large amounts of specialized membranes elaborated during nitrogenase induction were responsible for this difference.[226]

In continuous culture under nitrogen-limiting conditions and with decreasing temperature, *Pseudomonas fluorescens* showed a loss of both PG and PE, but a marked increase of DPG. In contrast, when carbon was limiting, PG and DPG decreased, whereas PE increased.[100]

COMPLEX MEDIUM CONSTITUENTS

The withdrawal of an essential constituent (milk) from the growth medium of *Bifidobacterium bifidum* var. *pennsylvanicus* caused an increase in total lipids, with neutral lipids unchanged and total phospholipids increased by 10% (DPG and PG), an increase of about 10% in monogalactosyl diglyceride and a decline of 44% in phosphogalactolipid, 13% in digalactosyl diglyceride, and a drop in other glycolipids.[84,387]

A change from one complex medium (GC Medium Base®) to another (Trypticase®-soy) did not change the overall fatty acid profile of most neisseriae tested, except that one component, $C_{17:1}$, was completely absent from cells grown in the latter medium.[195]

Analyses of cells grown on an agar medium vs. its broth counterpart show that substantially different fatty acid patterns may be obtained in certain organisms. *Neisseria lactamicus* grown in Trypticase®-soy broth had $C_{16:0}$, $C_{14:0}$, and $C_{18:1}$ as major fatty acids; in cells grown on the same medium solidified with agar, the major fatty acids were hydroxy $C_{12:0}$, $C_{16:0}$, and certain longer-chain unknowns.[195] In *Pseudomonas rubescens,* glucosyl diglyceride, glucuronosyl diglyceride, and the ornithine lipid were isolated only from cells grown on nutrient agar; in those grown in nutrient broth, these components were missing. In *P. fluorescens,* however, the ornithine lipid was produced under both conditions of growth.[409]

The fatty acid profile of *Bacillus subtilis* var. *niger* grown in a nutrient broth of French origin was compared to that of cells grown in an equivalent medium of American manufacture. In the latter, branched-chain fatty acids accounted for 53% of the total, whereas straight-chain ones accounted for 47%; in the former (French manufacture), the corresponding figures were 35% and 65%.[47e]

On the other hand, Eberhard and Rouser[78] found that major changes in medium constituents did not change the gross phospholipid composition of *Vibrio (Photobacterium) fischeri* or of *P. phosphoreum;* Beebe[30] found very little difference in the phospholipids of *Bacillus subtilis,* whether grown on Trypticase®-soy broth or on a semisynthetic medium. Likewise, little change was found in the fatty acids or phospholipids of yersiniae or of *Escherichia coli,* whether the cells were grown on Trypticase®-soy broth or on a defined medium.[260,363] No change in the fatty acid profile of a strain of *Neisseria* was brought about by changing the blood source or by omitting the blood from the agar used to grow the cells.[131]

AUTOTROPHIC VS. HETEROTROPHIC GROWTH

Autotrophically grown *Thiobacillus novellus* had more straight-chain fatty acids than heterotrophically-grown cells,[215] whereas *"T." trautweinii* had fatty acids of longer chain lengths.[3] The ornithine lipid of *Rhodospirillum rubrum* was produced whether cells were grown auto- or heterotrophically.[46b]

BIOTIN

Addition of biotin to the medium in which *Mycobacterium kansasii* was cultivated had no effect on the fatty acid patterns,[357] but in *Thiobacillus novellus* $\Delta C_{19:0}$ was increased at the expense of $C_{18:1}$.[215]

PHOSPHATE, SULFATE, AND MAGNESIUM

Using chemostat-grown cultures, Minnikin and colleagues[238] tested the effects of limiting both phosphate and magnesium (Mg^{2+}) on the phospho- and glycolipids of a number of Gram-negative and Gram-positive bacteria. They found that in one strain of *Bacillus subtilis* batch-grown cells had the following percentages of lipids in the early stationary phase: 5% DPG, 12% PE, 25% PG, trace lysyl PG, 38% diglucosyl diglyceride (DGD), and 20% unknown glycolipid (UG). Corresponding figures in the early log phase were 28% DPG, 17% PE, 16% PG, 0% lysyl PG, 22% DGD, and 8% UG. In Mg^{2+}-limited chemostat-grown cultures, the lipid pattern was simplified to 54% PG, 39% DGD, and 7% UG. When phosphate was limiting (chemostat), the cells contained 19% DPG, 10% PG, 56% DGD, 15% UG, and no PE. Minnikin et al.[240] concluded that in this organism PE and DGD were interchangeable. In other studies, however, these same workers, using a different strain of *B. subtilis* as well as strains of *B. subtilis* var. *niger* and *B. cereus,* found that phosphate- and magnesium-limited cells showed only quantitative, rather than qualitative, changes in their lipid compositions.[236,239]

Bacillus subtilis grown under sulfate- or magnesium-limiting conditions has a very high lipid content, as compared to controls.[82] In Mg^{2+}-limited cells of *Pseudomonas diminuta,* PG and phosphatidyl glucosyl diglyceride were unchanged from the control; however, glucosylglucuronosyl diglyceride (GGD) increased, whereas glucuronosyl diglyceride (GD I) and glucosyl diglyceride (GD II) both decreased. In cultures limited in phosphate, the phospholipids decreased markedly, whereas GGD increased, and GD I and GD II did not change.[241] In chemostat-grown *P. fluorescens,* limitation of magnesium resulted in little change in the phospholipid pattern, but complete disappearance of the ornithine lipid was observed. In contrast, the authors reported that in phosphate-limited cultures the cells were completely lacking in phospholipids, the ornithine lipid being the sole component. This remarkable finding has not been substantiated by others, but an examination of the morphology and physiology of the cells formed under these conditions would be of great interest. Data were not given on the relative growth rate of cells under these conditions, and there was no indication as to whether the cell morphology was grossly normal. However, the organism was isolated from a low-phosphorus marine environment and may very well be capable of normal metabolic activity without phospholipids in its membranes.[237]

Guenther and colleagues[114] found that lack of magnesium caused an increase of the total phospholipid content of *Escherichia coli,* largely due to an increase in PE and DPG; PG and PS were unaffected. Depletion of phosphate in batch or chemostat culture caused an accumulation of lipids in Gram-negative cells,[82] but a change of 0.05 to 1.25% in phosphate concentration had no effect on the fatty acid content of *Thiobacillus novellus.*[215]

SALTS AND METALS

The production of the iron-complexing mycobactins by *Mycobacterium smegmatis* is dependent on the level of iron in the medium: high levels of Fe^{2+} repress production. Zn^{2+} is also toxic. Maximum production required 1.8 μM Fe^{2+}, ca. 0.5 μM Mn^{2+}, ca. 0.5 μM Zn^{2+}, and at least 0.17 mM Mg^{2+}.[303]

Addition of copper as $CuCl_2 \cdot 2H_2O$ to the medium at 1.2 to 6.3 μg/ml has been reported to increase the production of the trehalose lipid of *Arthrobacter paraffineus* by about 40%.[346] Peterson et al.[287] demonstrated the toxicity of lead to *Micrococcus luteus* as measured by a decrease in the total cellular lipids. At 0.6 mg of lead per ml of Trypticase®-soy broth, diglycerides and ketones increased in proportion to other lipids, whereas aliphatic hydrocarbons and DPG decreased. After more than six transfers of the culture in lead-containing medium, however, the lipids were restored to their normal levels.

High salt content is frequently the basis for selective isolation media for cocci (e.g., mannitol-salt). A study of the influence of 10% NaCl in the growth medium on the lipids of *Staphylococcus aureus* showed that PG and lyso-PG decreased and that DPG increased with increasing NaCl content. The fatty acids of DPG showed an increase of branched-chain types with a drop in straight-chain fatty acids; PG and lyso-PG showed exactly the reverse trend.[148] In contrast, a comparison of complete, half-, and quarter-strength sea salts in the growth medium of marine bacteria (unidentified) showed only very slight decreases of unsaturated fatty acids and increases of saturated fatty acids with decreasing salinity.[271]

FATTY ACIDS AND ALKANES

Addition of serum to media in which fastidious (and not so fastidious) organisms are grown is common practice. In some cases members of certain genera, such as *Leptospira,* are dependent on the preformed fatty acids that the serum supplies for their normal growth. In all cases the lipids of cells grown in serum-containing media may be expected to reflect the fatty acids provided. Growth of *Staphylococcus aureus*[310] in serum-containing media increased the actual cellular content of phospholipids, free fatty acids, and glycerides, but the relative proportion of the phospholipids decreased in relation to the other two types of lipids. The quantity of monounsaturated fatty acids increased, diunsaturated fatty acids were directly incorporated into the cells, and branched- and long-chain ($>C_{20}$) fatty acids declined. Phosphatidic acid and an unknown phospholipid appeared, and the other phospholipids (PG, lysyl-PG, and DPG) decreased proportionally. In *Escherichia coli* cells grown on serum-containing medium, Redai et al.[310] found that extractable lipids increased twofold, with neutral lipids representing one half of this amount. As in *S. aureus*, new phospholipids (including lyso-PE and four unknowns) appeared.

Yin et al.[426] also demonstrated that *S. aureus* would take up whatever fatty acids were added to the medium. In this case the added $C_{17:0}$, $C_{18:1}$, and $C_{18:2}$ could be recovered from the cells, although none of these fatty acids was found in cells grown in conventional low-lipid media. In some cases other types of ellular fatty acids were influenced by such a supplement. The addition of fatty acids in the form of Tweens® to the medium in which *Lactobacillus lactis* was grown changed the fatty acid profiles markedly from a quantitative viewpoint.[388]

Many mycobacteria, nocardiae, and arthrobacters are known to metabolize hydrocarbons. Their resulting fatty acid composition has been shown to be influenced by the chain length of the hydrocarbon supplied.[33,171,185,222,391,417,422] Other bacteria may be influenced as well.[56,225,284,333,334]

Other Factors

pH

In many bacteria aminoacyl phosphatidyl glycerols (AAPG) occur only at, or are enhanced by, low pH of the growth medium. Other phospholipids may also be affected by low pH.

For example, *Staphylococcus aureus* cells harvested at pH 4.7 at mid-log phase contained 79% lysyl-PG (LPG) and 14% PG. Following incubation for 2 hr at pH 7.0, LPG declined to 50%, PG increased to 42%, and DPG was unaffected. In contrast, cells harvested at pH 6.5 and incubated at pH 5.0 for 2 hr showed an increase from 38% to 83% in LPG with a corresponding decrease from 57% to 12% in PG. The fatty acid content of the phospholipids was not affected.[118]

The phospholipids of *Streptococcus faecalis* cells grown in a tryptone-yeast extract-K_2HPO_4 medium containing no glucose were 42% DPG, 0% PG, and 58% unknown. When 1% glucose was added to the medium and the pH was permitted to drop normally to ca. 4.2, AAPG appeared, and the phospholipid pattern became quite complex: 12% DPG, 44% PG plus alanyl-PG, 4% unknown, 33% LPG plus arginyl-PG (APG), and 7% 2′,3′-dilysyl-PG plus diglucosylphosphatidyl glycerol. If, on the other hand, 1% glucose was added to the medium, but the pH was maintained at 7.2, the phospholipids isolated were 12% DPG, 58% PG plus unknown, 20% LPG plus APG, and 10% 2′,3′-dilysylphosphatidyl glycerol plus diglucosylphosphatidyl glycerol.[257]

Phosphatidyl ethanolamine metabolism is also influenced by pH. In *Bacillus subtilis* var. *niger*, raising the pH from 5.1 to 8.1 eliminated PE in magnesium-limited cells, but not in phosphate-limited cells.[236] When the pH of the medium was low, the PE content of *B. stearothermophilus* was increased over that of cells harvested at neutral pH.[212]

In *Bifidobacterium bifidum* var. *pennsylvanicus*, if the pH of the milk-containing medium was permitted to drop naturally to 5.2 instead of being maintained artificially at 6.8, DPG was replaced as the major phospholipid by the phosphogalactolipids glycerophosphoryl galactosyl monoglyceride and glycerophosphoryl galactosyl diglyceride. No other qualitative changes were reported. When milk was absent from the medium, pH changes had little effect.[387]

Siolipin (Table 16 [XVI-2]) production was also affected by pH. *Streptomyces sioyaensis* produced more the lysine homolog of siolipin at pH 3.0 than at pH 8.0, where the ornithine-type predominated.[170]

Little has been reported on the effect of pH on fatty acids. Drucker et al.,[75] using chemostat-grown *Streptococcus mutans,* found an increase of shorter straight-chain and unsaturated fatty acids when the pH of the medium rose from 5.8 to 8.4.

OSMOLARITY

An increase of 1% in the concentration of sucrose in the growth medium of *Streptococcus mutans* gave rise to a major increase of unsaturated fatty acids.[75] *Escherichia coli* responded to media of higher osmolarity by increased PG production. PE was unaffected.[258]

AERATION

Aerated cultures of strains representing seven genera showed substantial qualitative and quantitative differences in their fatty acid content when compared to cultures grown under nonaerated conditions.[74] Under nonaerated conditions (unshaken), these bacteria had remarkably similar fatty acid profiles, dominated by saturated and unsaturated fatty acids, including n-$C_{16:0}$, $C_{18:1}$, and $C_{16:1}$. When grown under aerated (shaken) conditions, highly individual patterns emerged in six or seven strains with the appearance of major new fatty acid types: *Bacillus megaterium* iso-$C_{15:0}$; *Corynbacterium xerosis,* $C_{17:0}$, $C_{17:1}$, $C_{18:0}$; *Klebsiella pneumoniae (aerogenes),* unknown fatty acid; *Branhamella (Neisseria) catarrhalis,* $\Delta C_{19:0}$ plus a marked shift to major amounts of $C_{18:1}$; *Serratia marcescens,* unknown fatty acid, anteiso-$C_{19:0}$; *Staphylococcus saprophyticus,* anteiso-$C_{15:0}$; *Streptococcus faecalis,* no change. Thus, proper aeration appears to be very critical to the expression of individual bacterial profiles.

Escherichia coli K-12 cells grown under anaerobic conditions showed a 3.5-fold increase in free fatty acids (particularly unsaturated types) over those grown aerobically in the same medium. There were no qualitative changes in the bound fatty acids or hydrocarbons.[260] In contrast, insaturated fatty acids (especially $C_{18:1}$) increased in *Streptococcus mutans* with increased rates of oxygenation in chemostat culture.[75] In *Thiobacillus denitrificans,* growth under anaerobic conditions altered the fatty acid profile profoundly; there were fewer and shorter acids, as compared to aerobic controls.[3]

Increased aeration by shaking decreased the amount of the polyunsaturated phleic acids of *Mycobacterium phlei.*[14] In *M. kansasii, Thiobacillus novellus,* and *Yersinia,* agitation had no effect on the fatty acid profile.[215,357,363] If aeration was stopped in the log phase of growth of *Bacillus stearothermophilus,* total phospholipid synthesis was reduced, DPG increased, and PG decreased. Reaeration reversed these effects.[49]

GROWTH RATE

A slowing of the growth rate of *Bacillus subtilis* var. *niger* (through nutrient exhaustion, lowered temperature, addition of antibiotics, or slowly metabolizable amino acids) caused an accumulation of branched-chain fatty acids at the expense of straight-chain types.[47d]

AGE

The age of the cells analyzed can be an important factor in the lipid profile of certain bacteria. The greatest changes, when they occur, are usually during the exponential phase of growth; however, there are also important shifts during the stationary phase. One of the most common observations concerning fatty acids is the conversion of $C_{18:1}$ to $\Delta C_{19:0}$ with age. This has been observed in the Gram-positive genus *Lactobacillus,*[388] but appears to be largely a characteristic of Gram-negative bacteria, such as members of the genera *Escherichia, Serratia, Agrobacterium, Pseudomonas,* and *Yersinia.*[152,363]

A variation of this theme was found in *Bifidobacterium bifidum* var. *pennsylvanicus,* where there was a decrease of saturated and an increase of unsaturated and cyclopropane fatty acids in the stationary phase. On the average, the chain lengths of the fatty acids increased.[388]

In a study on the effect of growth phase on lipids of *E. coli,* it was found that the above transition from unsaturated to cyclopropane fatty acids took place in the free lipids in the early stationary phase; on longer incubation, however, an increase of branched-chain fatty acids occurred at the expense of the others. In contrast, the bound fatty acids changed little in the early stationary phase, except for slight chain-lengthening; on further aging, the dominant branched-chain fatty acids diminished, whereas the cyclopropane and especially the unsaturated fatty acids became dominant.[7]

In three Gram-negative marine isolates, little change was observed in the fatty acids of stationary-phase cells, compared to those harvested in the log phase.[271]

In Gram-positive bacteria — especially in actinomycetes — the effect of aging on fatty acids is usually more difficult to characterize. In *Bacillus cereus,* iso-odd, branched-chain fatty acids rose somewhat in proportion to the other fatty acids, whereas anteiso types declined.[145] They increased in *B. subtilis* var. *niger,* with anteiso-$C_{15:0}$ and iso-$C_{17:0}$ appearing only after the early log phase.[47e] Normal fatty acids were unaffected.

In *Lactobacillus, Sporolactobacillus,* and *Bacillus,* the fatty acid profile became completely stable after 48 hr;[380,381] in *Pediococcus,* the pattern stabilized 48 hr after the stationary phase was reached. Reproducibility of lipid patterns under these conditions was reported to be excellent.[379] In *Micrococcus cryophilus,* no effect of aging was found on the fatty acids at either 0° or 20°.[317]

Early reports that the lipid content of mycobacteria and corynebacteria continues to increase long after the stationary phase is reached[16] have been qualified by the finding that *Mycobacterium tuberculosis* continues to accumulate lipid in media containing glycerol (such as those used in earlier studies, namely Sauton's[322b] and Long's[216b]), whereas lipid content is stable throughout the stationary phase in the same medium made with glucose.[13] Bordet and co-workers[40] found that even in a glycerol medium the total lipid content of *Nocardia asteroides* does not increase in the stationary phase; however, the concentration of the wall lipids in the form of mycolic acids continues to rise, apparently at the expense of other fatty acids.

Mycobacteria are often difficult to grow, and long periods of static incubation (up to 1 month or more) are frequently needed to obtain sufficient biomasses for analysis. Nevertheless, it was found that the fatty acids of *M. kansasii* did not change after 14, 28, and 60 days of incubation.[357] Likewise, the fatty acids, including tuberculostearic acid (TBA), of *M. tuberculosis* var. *hominis* H37Rv and *M. avium* changed little from the early log into the late stationary phase. In the nonpathogenic *M. smegmatis,* however, almost all strains showed a decrease in $C_{18:1}$ and a coressponding increase in TBA.[54] In *Nocardia asteroides,* aging gave rise to a similar increase in TBA (along with certain fatty acids of chain length greater than 20 carbons) and to a decline of $C_{16:0}$, $C_{18:0}$, and $C_{18:1}$.[28]

When the membranes of *Micrococcus luteus (Sarcina flava)* were analyzed, it was found that in the free lipids $BrC_{15:0}$ dropped from 46.3% in the log phase to 20.6% in stationary phase, whereas the principal fatty acids that increased were n-$C_{17:0}$, n-$C_{18:0}$, Br$C_{23:0}$, and an unknown. The bound lipids released by acid or alkaline hydrolysis showed a completely different picture: the major fatty acid released in both cases,n-$C_{17:0}$, remained unchanged from the log into the stationary phase, whereas the concentration of the branched-chain fatty acids declined in the lipids hydrolyzed by acid and rose in those hydrolyzed by alkali.[125]

The phospholipid fatty acids of *Clostridium butyricum* showed an increase in saturation and a concomitant decline in unsaturation when the cells from the early stationary phase were compared with those from mid-log phase.[105] As far as the phospholipids themselves are concerned, no general rule can be made; however, the qualitative aspect of the phospholipid profile is rarely modified with aging. In *Alcaligenes (Hydrogenomonas) eutrophus* grown chemolithotrophically, PE increased from 45% to 63% from the late log to the stationary phase, whereas PG and DPG declined (38% to 27% and 16% to 10%, respectively).[356]

Although in most bacteria PE was said to be constant with age, Oliver and Colwell[272] found that in certain marine *Vibrio* strains it increased in the stationary phase, whereas PG and DPG declined. In *Pseudomonas diminuta,* PG also declined with age; however, phosphatidyl glucosyl diglyceride increased, as did the glycolipid glucosyl glucuronosyl diglyceride. It is interesting that the glucuronosyl diglyceride and glucosyl diglycerides of this strain dropped initially, but recovered their original levels upon aging.[241] In *Neisseria gonorrhoeae,*[341] PC increased two- to fivefold, whereas PG decreased in the stationary phase; in *Thiobacillus neopolitanus,* PME and DPG increased at the expense of PE and PG.[2]

Micrococcus luteus (lysodeikticus) growing in a peptone-yeast extract medium at 30° was reported to show the greatest changes in lipid composition in the mid- to late log phase. Total lipids decreased from 10% to 2%, but the proportion of the phospholipids rose. DPG increased to a maximum in the log phase, then declined in the stationary phase, whereas PG did exactly the reverse. PI continued to rise throughout, but PA was unchanged.[71] Essentially the same findings were reported by Thomas and Ellar.[360]

Two different groups — Short and White[331] and Polonovski et al.[292] — working with *Staphylococcus aureus* found that DPG accumulated during the stationary phase, whereas PG declined; phosphatidyl glucose and lysyl PG remained essentially constant. Komura et al.[177] reported that the phospholipid composition of members of the genera *Micrococcus* (also as *Sarcina*), *Staphylococcus, Sporosarcina,* and *Planococcus* was stabilized after 16 hr of incubation in nutrient broth.

In *Bacillus stearothermophilus,* the phospholipid concentration was stable throughout the log phase, but increased in both the log-to-stationary transition (largely PE) and throughout the stationary phase (mostly DPG).[49] Pitel and Gilvarg[291] reported that phospholipid synthesis increased during the initiation of sporulation of an amino acid auxotroph of *B. megaterium,* but they did not specify the proportions of component phospholipids. *B. thuringiensis* and *B. popilliae* both showed an increase of PG and a decline of DPG from the log to the stationary phase. PE declined in *B. popilliae,* but was stable in *B. thuringiensis.* By the time of sporulation, DPG had disappeared completely in the latter.[47f]

In contrast, phospholipids increased in the log phase of *Bifidobacterium bifidum* var. *pennsylvanicus,* but declined in the stationary phase, largely through losses of PG. DPG and the phosphogalactolipids increased proportionally.[387]

Phosphatidyl inositol mannosides have been found to vary with the age of the culture. In *Corynebacterium aquaticum,* PIM_2 and PIM_3 were the dominant forms in the log phase. During the stationary phase, these declined, whereas PIM, PIM_4, and PI increased.[117] In *Mycobacterium bovis* and in *M. smegmatis,* PIM was the major form in log phase, whereas PIM_2 and PIM_x increased proportionally in the stationary phase. The other two major phospholipids, PE and DPG, declined.[167]

Chandramouli and Venkitasubramanian[55] found that the PIMs increase with age in the saprophytic species (*Mycobacterium smegmatis*), but decrease slightly in the virulent species (*M. tuberculosis* and *M. avium*). As in the previous study, PIM was most common in young cells of all strains, with PIM_2 and PIM_3 increasing in the older saprophyte cells. PIM_2 was equal to PIM in the older virulent strains; no PIM_3 was found. DPG and PE decreased with age in the saprophytes; in the virulent strains they rose in concentration and then dropped. The same authors reported that glycerides were found to decline upon aging in *M. avium* and in *M. smegmatis,* whereas there was a twofold increase of these compounds in *M. tuberculosis.*

MORPHOLOGICAL STAGE

Variation in bacterial morphology is usually a function of age. In *Campylobacter (Vibrio) fetus,* the young vibrioform cells contain PE, PG, PI and PS in that order of concentration. The coccoform cells that appeared upon aging showed a marked decline of PE, disappearance of PS, and appearance of PA.[368] The coccoform cells also contained 60% neutral lipids, as compared to 32% in the vibrioform cells. This increase was largely due to increased free fatty acids; glycerides, glycolipids, and phospholipids showed a drop in concentration.

Filamentous *Rothia dentocariosa* cells had twice as much glycolipid and phospholipid as the older coccal forms.[283] In contrast, Chow and Schmidt[58] found that the stalked cells and swarmers of *Caulobacter crescentus* had the same overall composition. Various species of *Arthrobacter* showed no differences when the rod and coccal stages were compared.[41,327] The spores of *Clostridium thermosaccharolyticum* were found to contain hydroxy-$C_{18:0}$, a fatty acid that had not been reported from vegatative cells.[288]

ORGANELLE ANALYSIS

Analysis of the lipids of a specific organelle of a microorganism is a challenging process; the principal difficulty is the isolation of enough pure material for analysis. Nevertheless, the potential rewards for the taxonomist might seem worth the trouble, since the isolation and analysis of inner or outer membranes, cell walls, spores, and the like might reasonably yield data of much greater taxonomic value than the "garbage can" approach, where all the cellular lipids are studied. Interestingly, this does not seem to be borne out by recently published results.

For example, a comparison of the simple and complex lipids of both whole cells and cell envelopes of *Enterobacter (Aerobacter) aerogenes*[370] showed no important differences in the major or minor fatty acids associated with them. Only the proportions of the types of lipids differed: in the whole cells, hydrocarbons represented 58% of the total lipids, aromatic compounds 12%, simple lipids 11%, and complex lipids 19%; in contrast, in the cell walls the percentages were 19, 1, 4, and 76, respectively. Unfortunately, the types of specific lipids within each broad group were not studied. However, in *Escherichia coli* an examination of the phospholipids of the inner and outer membranes showed them to be qualitatively and quantitatively similar.[59] This was also true for a *Pseudomonas* sp.[73] and for *Proteus mirabilis.*[315] Osborn et al.[279] isolated the inner and outer membranes of *Salmonella typhimurium,* but here again the gross composition of the phospholipids was similar in that only the proportions of each component varied. Likewise, the plasma and mesosomal membranes of *Staphylococcus aureus* contained the same phospholipids[32] and fatty acids[351] as did those of *Micrococcus luteus (lysodeikticus).*[360]

Goldman[106] reported finding PIM_2 as the major phosphoinositol mannoside in whole cells, cytoplasm, cytoplasmic membrane, and cell walls of *Mycobacterium tuberculosis.* The cell wall had the highest PIM_2/PIM_5 ratio, whereas PIM was found in greater amounts in the cytoplasm than in the membrane or wall.

Analysis of the cell wall fatty acids of *Lactobacillus plantarum*[374] showed them to be qualitatively the same as those from whole cells, with one exception — a substantial quantity (12.5%) of $C_{14:1}$ was found in the walls, whereas none was detectable in the whole-cell extracts. Qualitatively, the cell walls and membranes of *Escherichia coli* had the same phospholipid content; quantitatively, lyso PE was found in much greater quantities in the cell wall; some quantitative differences in fatty acids were also detected.[403]

It was found that dimannosyl diglyceride was located primarily in the cell walls of *Rothia dentocariosa,* and mannolipids were found almost exclusively in the walls of *Actinomyces viscosus* and *A. bovis;*[283] however, both compounds can be visualized in whole-cell extracts.

MISCELLANEOUS FACTORS

Thoen et al.[358] found that the lipids of freshly isolated strains of mycobacteria did not differ from similar strains that had been maintained on laboratory media for 2 years. This was also the case for a strain of *Mycobacterum bovis* that had been maintained by transferring for several years; however, another similar strain showed a decrease in total lipid production following lyophilization.[342]

In general, except at lysis, little change occurred in the lipid composition of cells of *Escherichia coli* or *Salmonella typhimurium* following infection with phage.[57,173]

Antibiotic resistance, earlier thought to be a function of a changed balance of cellular lipids,[408] does not appear to change the overall lipid pattern from a qualitative point of view. However, quantitative differences in the content of various lipids between parent strains and antibiotic-resistant mutants have frequently been reported.[34,77,285,316,413]

PHOSPHOLIPID FATTY ACIDS VS. POSITION ON GLYCEROL

Differentiation of the fatty acids that occupy position 2 (C-2) of the glycerol of phospholipids from those on the next carbon (usually C-1, sometimes C-3) may be carried out by the use of phospholipases — i.e., enzymes that specifically attack the ester bond at one or the other position. The taxonomic significance of such data must wait until more strains are studied.

At the moment there appears to be a reasonable correlation between the length of the fatty acid chains at a given position and the Gram reaction of the strain. Most Gram-negative bacteria for which there are suitable data (*Agrobacterium tumefaciens, Escherichia coli,* and *Serratia marcescens*) have the shorter-chain fatty acids on C-1 and the longer-chain fatty acids on C-2.[122] The reverse is true for Gram-positive strains (*Bacillus cereus, B. subtilis, Clostridium butyricum, Mycobacterium tuberculosis, M. phlei, M. butyricum,* and *Streptococcus faecalis* — both var. *faecalis* and *zymogenes*), which have the longer fatty acids on C-1 and the shorter ones on C-2. In *Azotobacter agilis* and *S. hemolyticus* there is no substantial difference between the two positions.

There are also positional preferences, which also follow "Gram boundaries" for the various types of fatty acids among bacteria. Gram-negative bacteria having predominantly saturated acids at C-1 and cyclopropane at C-2 include *A. tumefaciens, E. coli,* and *S. marcescens.*[122] According to this report, *A. agilis* did not contain cyclopropane fatty acids; thus, its pattern is C-1 saturated and C-2 unsaturated. *Brucella abortus* and *B. melitensis* had predominantly hydroxy fatty acids at C-1 and cyclopropane at

C-2.[355] Among the following Gram-positive bacteria, we find largely saturated fatty acids at C-2 (the dominant type at C-1, if discernible, is given in parentheses): *B. cereus, B. subtilis, C. butyricum* (cyclopropane), *M. tuberculosis* (unsaturated, TBA), *M. phlei* (TBA), *M. butyricum* (TBA), and *S. faecalis* (var. *faecalis* and var. *zymogenes*).[89,122,146,147,265]

ISOLATION AND CHARACTERIZATION OF LIPIDS

Production of Bacterial Cells for Lipid Analysis

As we have seen, lipids found in a given bacterial strain depend, among other factors, largely on the medium in which the organism is grown, the temperature of incubation, the aeration, the pH, the age of the cells at harvest, and similar considerations. It is of the utmost importance that anyone wanting to utilize lipid spectra to aid in the classification of unknown strains should analyze in parallel (and under the same conditions) authentic representatives of the taxon to which he feels the unknowns belong. These results will be more valuable in resolving a taxonomic problem than any attempt to duplicate "literature results" by trying to imitate the growth and analytical conditions of other investigators.

In general, the growth medium to be used should support the growth of all strains to be compared; it should contain few to no lipid constituents, unless the organism is dependent on these for growth. Temperature of growth should be the optimum for each strain to be studied, and growth should take place under well-aerated (agitated, sparged, or on agar) conditions when the organisms are aerobic. The cells may be harvested at any phase of growth, but there is some evidence that more reproducible results may be obtained from older cultures that are well into the stationary phase. At harvest the cells should exhibit normal morphology; abnormally shaped cells reflect a distorted metabolism.

Extraction, Purification, and Analysis of Bacterial Lipids

Microbial lipids may be associated with other cellular constituents in three basic ways, with the type of the association dictating the extraction method to be used:

1. Binding by hydrophobic or Vander Waals' forces (neutral lipids, such as glycerides, waxes, and hydrocarbons).
2. By hydrogen bonding (polar lipids, such as phospholipids and glycolipids).
3. Covalent bonding (bound lipids, such as fatty acids linked to peptides, polysaccharides, and the like by ester, ether, amide, or glycosidic bonds).

For safety, prior to extraction all microorganisms should be autoclaved in the liquid media in which they were grown. It has been our experience that autoclaving of the cells under these conditions does not appreciably alter the lipid constitution; in fact, the inactivation of the cellular lipases may be beneficial.

In extraction, a dilemma is presented for the biochemist. If the extraction technique is gentle enough, the lipids obtained will probably have undergone only relatively minor degradation; however, the biochemist will probably have missed a certain proportion of the total lipids. On the other hand, if the extraction technique is rigorous enough to obtain all or most of the total cellular lipids, he will undoubtedly have caused the degradation of many of them. The taxonomist is free of such limitations. He may autoclave, extract, and hydrolyze to his heart's content. Provided that he keeps all parameters constant, he will probably arrive at a constant pattern of products, albeit many of them degraded, at least in part. The drawback to the latter method is that, if an unknown product is found, it will be more difficult to guess which of the known lipids it is, since most — if not all — of these will have been isolated under milder conditions.

Neutral lipids may be extracted with nonpolar solvents, such as chloroform, ether, petroleum ether, methylene chloride, or benzene; polar lipids may be extracted with polar solvents, such as methanol, usually admixed with chloroform and/or water; and bound lipids may be obtained by acid or alkaline hydrolysis followed by extraction with appropriate solvents.

A technique widely used for quantitative extraction of free lipids is the technique of Bligh and Dyer.[36] In this procedure, which appears to enjoy universal approbation among lipidologists, the cells are extracted at room temperature with chloroform:methanol:water (1:2:0.8 v/v). Chloroform and water are added to this uniphase extract until two phases form. The lipids are then recovered from the chloroform layer, leaving most of the nonlipid contaminants behind in the aqueous layer. Another widely used method is that of Folch et al.,[91] which employs chloroform:methanol (2:1) as extractant. In general, all solvents should be freshly distilled; all procedures, especially those involving heat, should be carried out *in vacuo* or under nitrogen.

Moss and colleagues[254] have developed a procedure for the isolation of both free and bound fatty acids. Sterilized, fresh, or lyophilized cells are heated with 5% sodium hydroxide in 50% (aqueous) methanol for 15 min at 100° in tubes with Teflon®-lined caps. The extract is cooled and acidified to pH 2.0; boron trichloride-methanol is added and heated at 100° for 5 min. The entire reaction mixture is added to a saturated sodium chloride solution, and the whole is extracted twice with an equal volume of chloroform:hexane (1:4). The extracts are combined, reduced in volume, and gas-chromatographed. Attempts at transesterification using boron trihalides caused extensive degradation of the cyclopropane fatty acids and were less effective than the above technique in releasing the hydroxy fatty acids.

Another technique uses whole-cell methanolysis to obtain the free and bound fatty acids. In this procedure dry cells are heated with methanolic hydrochloric acid, and the resulting methyl esters of the fatty acids are gas-chromatographed.[93] Unfortunately, it has been reported that this technique tends to degrade 3-hydroxy and cyclopropane fatty acids and gives rise to artifacts.[254,395,410]

Crude lipids may be fractionated in various ways. For example, one may take advantage of differences in their solubilities in various solvents. A case in point is the classic separation by the use of solubility in excess cold acetone: neutral lipids remain in solution, whereas the phospho- and glycolipids are precipitated out. Lipids may also be fractionated by column chromatography on silica gel or other adsorbants and eluted with increasingly polar solvents to fractionate the sample into lipid classes (neutral lipids, glycolipids, and phospholipids). This method may also be modified to separate each lipid class into its constituent parts. Other absorbants include alumina, magnesium oxide, magnesium silicate (Florisil®), DEAE or TEAE cellulose, or ion-exchange resins.

Thin-layer chromatography (TLC) is one of the most useful techniques for the analysis of lipid mixtures. Thus, mono-, di, and triglycerides may be separated from free fatty acids, methyl esters of fatty acids, waxes, and hydrocarbons; hydroxy fatty acids may be separated from nonhydroxylated ones; saturated fatty acids may be separated from unsaturated fatty acids; phospholipids may be separated from glycolipids, and so forth. All compounds may be visualized by spraying with specific reagents for preliminary characterization, and their relative mobilities may be compared with known standards. Lipids may be prepared in sufficient quantities for further analysis by preparative TLC by using thick layers of adsorbant, with the various fractions being scraped off and eluted by appropriate solvents. Fatty acids in ester linkage may be released from their parent compounds by saponification, and the fatty acids thus freed may be methylated and then analyzed by gas—liquid chromatography (GLC). Based on their retention times, compared with authentic compounds, on various columns with different liquid phases, and using appropriate methods of derivatization or degradation, the individual fatty acids may be identified with reasonable certainty.

A rapid method for the isolation and tentative identification of bacterial phospholipids has recently been described; this procedure uses the direct ion-exchange paper chromatography of trichloroacetic acid-precipitated solids.[177] Unfortunately, the identification of phospholipids, even if pure, solely on the basis of R_f on paper or thin-layer chromatography and of reactions with various reagents is hazardous. Only the isolation and characterization of the purified phospholipids, followed by degradative analysis, can assure proper identification of such compounds.

Kates' excellent handbook,[153] covering all of these techniques as they apply to lipids, is especially recommended. Useful specialized guides include the volume edited by Marinetti[227] (phospholipids and glycolipids) and the text on phospholipids edited by Ansell, Hawthorne, and Dawson.[12] Other detailed descriptions of various aspects of these techniques include those of Randerath,[302] Stahl,[337] Renkonen,[312] Snyder,[336] Johnson and Davenport,[135] Hanahan,[120] and Mitruka.[245]

LIPIDS USEFUL IN TAXONOMY

The following sections, arranged alphabetically, summarize the areas in which recent data on lipid composition may be valuable in taxonomy, either in defining a given genus or in differentiating amoung certain species. Any genus name not given here has been omitted because of insufficient useful data. Specific references not given may be found by referring to the appropriate tables (Tables 2 to 6, 9 to 13, and 15).

Gram-negative Bacteria

Alcaligenes

The fatty acid data show that members of this genus contain hydroxy fatty acids in their bound lipids. It is not known whether they exist in the free lipids also. Cyclopropane fatty acids may also be a useful taxonomic marker. The aglycolytic pseudomonads (*P. alcaligenes, P. pseudoalcaligenes,* and *P. putrefaciens*), not easily distinguishable from *Alcaligenes* by other means, were readily differentiated from them on the basis of fatty acid profile. 3-OH-$C_{10:0}$ and 3-OH-$C_{12:0}$ were found in *P. alcaligenes* and in *P. pseudoalcaligenes,* whereas *Alcaligenes* contained 2-OH-$C_{12:0}$ and 3-OH-$C_{14:0}$. The last compound is found in certain members of the Enterobacteriaceae, but has only been found in two other pseudomonads, *P. cepacia* and *P. pseudomallei,* both of which may be distinguished from *Alcaligenes* species on other grounds. *Pseudomonas putrefaciens* lacked hydroxy acids altogether; it also contained major amounts of branched-chain fatty acids and a unique $C_{17:1}$ not found in other pseudomonads or *Alcaligenes* species.[65]

AGROBACTERIUM

See Achromobacter.

ACHROMOBACTER

Achromobacter xylosoxidans appears to have a fatty acid profile similar to that of pseudomonads, namely, hydroxy fatty acids are found only in the bound lipids. Achromobacters are now distinguished from rhizobia and from agrobacteria, which they otherwise resemble, on the basis of the relationships that members of the latter two have with plants. Phospholipid composition also appears to be useful in their differentiation: rhizobia and agrobacteria contain PC; the achromobacters tested to date do not. Hendrie et al. (see Reference 47b) proposed the placement of achromobacters in *Alcaligenes,* but this is not supported by phospholipid data; in fact, on this basis, *Alcaligenes eutrophus* might more properly be placed in *Achromobacter.*

ARIZONA

See Enterobacteriaceae.

AZOTOBACTER

Although none of the lipids reported from *Azotobacter* seem to be genus-specific, certain phospholipids may serve as markers for some species: PME for *A. agilis,* and PS for *A. vinelandii.* A survey of additional strains seems warranted.

BACTEROIDES (see also *Fusobacterium*)

Most *Bacteroides* species contain only normal and unsaturated fatty acids in their free lipids. Many contain large amounts of unknown fatty acids, possibly β-hydroxy types, which are absent in fusobacteria.[92] *Bacteroides melanogenicus, B. succinogenes,* and *B. ruminicola* have been found to contain branched-chain and normal fatty acids, but not unsaturated ones; thus, they form a distinct group.

BORDETELLA

See *Brucella-Bordetella.*

BRUCELLA

All recent information on the phospholipid composition of this genus has been due to the work of Thiele and his colleagues.[355] Their findings show that the genus is relatively homogeneous, as judged by this criterion.

BRUCELLA/BORDETELLA

According to Thiele and Schwinn,[353] fatty acid and phospholipid patterns show that these two genera should not be in the same family. Nevertheless, they have very similar ornithine-containing lipids (q.v.).

CAMPYLOBACTER (see also *Vibrio*)

The report of PI in *C. fetus* shows this taxon to be unusual among vibroid bacteria and among Gram-negative bacteria in general.

CHLOROBIACEAE (*Chlorobium, Pelodictyon,* and *Chloroflexus*)

The fatty acids of the Chlorobiaceae are very similar, except for *Chloroflexus,* which not only has no cyclopropane fatty acids, but contains unusual unsaturated odd-chain length fatty acids, unsaturated largely at Δ^9 and Δ^{11}.[162] The phospholipids of Chlorobiaceae are unlike those of other photosynthetic bacteria in that they do not contain PE. Kenyon and Gray[162] concluded that, if one considers all the lipids they contain (including the monogalatosyl diglycerides), these bacteria are closer to blue-green algae than to other photosynthetc bacteria. This relationship is underlined by the fact that ring structures of the chlorophylls are on the same oxidation level.

CITROBACTER

See Enterobacteriaceae.

DESULFOVIBRIO

According to Makula and Finnerty,[223] *D. gigas* is set apart from the two other species they examined (*D. desulfuricans* and *D. vulgaris*) by its low PE content. It was not grown in the same medium as the other strains, and no details of the pH reached during growth were given. As previously noted (see Stability of Lipid Composition), pH in the production of PE can be very important; thus, the validity of this difference must be clarified.

ENTEROBACTERIACEAE

All members of the family Enterobacteriaceae, with the exception of *Proteus vulgaris,* have been reported to contain cyclopropane fatty acids in addition to unsaturated and saturated fatty acids (see Tables 2 to 5).[268,325] Many early reports do not mention cyclopropanic acids, probably because these compounds were destroyed or went unrecognized. As a general rule, the concentration of cyclopropane fatty acids exceeds that of unsaturated fatty acids in the stationary phase in this group.

ESCHERICHIA

See Enterobacteriaceae.

FUSOBACTERIUM (see also *Bacteroides*)

Except among the related *Bacteroides,* the presence of branched-chain fatty acids in *Fusobacterium* is essentially unique among Gram-negative bacteria. *Fusobacterium mortiferum (Sphaerophorus ridiculosus)* has been reported to contain plasmalogens, whereas *F. necrophorum,* like the other members of the Bacteroidaceae, has been reported to contain sphingolipids. As this very basic difference does not appear to be echoed in other taxonomic differences, further work on the lipids is obviously needed.

GEMELLA

See *Neisseria.*

HALOCOCCUS and *HALOBACTERIUM*

The phospholipids of members of both of these genera are of the dialkyl ether type. These unusual lipids appear to be absent from moderate halophiles and halotolerant bacteria.

HYPHOMICROBIUM

Like its photosynthetic counterpart *Rhodomicrobium,* from which it is now separated,[47b] *H. vulgare* contains a very high level (greater than 90%) of Δ^{11}-$C_{18:1}$.[18] Their phospholipids differ, however.[325] The unusual *Hyphomicrobium* sp. reported in Table 9 differs profoundly in its phospholipid composition (PC only) from *H. vulgare,* which contains DPG, PG, PE, and PDME.[325] However, there appears to be substantial heterogeneity by other criteria among other species that have been assigned to this genus in the past.[47b]

KLEBSIELLA

See Enterobacteriaceae.

LEPTOSPIRA (see also *Treponema*)

Like treponemas, leptospirae require preformed fatty acids for growth; thus, their fatty acid patterns reflect those supplied in the medium. Reports of polyunsaturated fatty acids in these organisms undoubtedly stem from this fact.

MEGASPHAERA

Megasphaera elsdenii has a very unusual phospholipid pattern; it has no DPG or PG, and only PE, PS, and plasmalogens are found.

MORAXELLA

The presence of hydroxy fatty acids in the lipids of moraxellae may be another case, like *Pseudomonas*, of characteristic differences of possible taxonomic importance between free and bound lipids. It has been reported that species of moraxellae may be reliably recognized on the basis of GLC analysis of whole-cell methanolysates. The following groups were distinguished: (1) *M. nonliquifaciens, M. bovis, M. lacunata;* (2) *M. osloensis;* (3) *M. phenylpyrouvica;* (4) *M. kingii;* and (5) *M. urethralis.*[132]

MORAXELLA/NEISSERIA

Members of these genera were reported to be separable on the basis of their fatty acid profiles only with computer assistance.[132]

NEISSERIA

Several studies have shown that the compound 3-OH-$C_{12:0}$ may be a taxonomic marker for *Neisseria*.[195,253,341] Thus, Lambert et al.[195] found this compound in all but *"N." denitrificans* and *"N." haemolysans*. Based on this and other evidence, they proposed that these two species might be placed elsewhere, pending the evaluation of more strains. The current placement of *N. haemolysans* is within the Streptococcaceae (as *Gemella*), since sections of its cell wall show it to be of the Gram-positive type. *Neisseria denitrificans* is currently relegated to *species incertae sedis.*[47b] Other groups among the neisseriae that could be distinguished on the basis of fatty acid composition included the following:

1. *"N." catarrhalis, N. caviae, N. ovis,* and *N. cinerea,* because they contained unique n-$C_{10:0}$ and $C_{17:1}$ fatty acids.
2. *N. lactamicus,* because, like *Gemella haemolysans,* it contained substantial amounts of unknown fatty acids having more than 20 carbons.
3. *N. flavescens, N. canis, N. subflava (N. flava, N. perflava), N. sicca, N. cuniculi, N. mucosa,* and one *N. cinerea,* which could not be distinguished from each other.

These groupings are interesting in view of the fact that *"N." catarrhalis* has been shifted to the genus *Branhamella* on the basis of its pigments, physiology, and GC%, and *N. caviae* and *N. ovis* were considered to be closely allied to it.[47b] Based on lipid composition, *N. lactamicus* might also fit in *Branhamella.* Since the third group contains most of the "recognized" species of *Neisseria,* the several *"species incertae sedis"* (*N. canis* and *N. cuniculi*) also in that group might be considered true neisseriae as far as fatty acid composition is concerned.

NITROBACTERIACEAE

The genera belonging to this family, except for one report of cyclopropane fatty acids in *Nitrobacter winogradskyi,* seem to have the same simple fatty acid profile: saturated and unsaturated fatty acids usually confined to two or four principal compounds.[38] *Nitrobacter winogradskyi* is also set apart from other Nitrobacteriaceae that have been analyzed because of its PC content; the others contan only DPG, PG, and PE.

PARACOCCUS

The statement that *Paracoccus* is closely related to *Pseudomonas, Alcaligenes,* and certain thiobacilli[47b] does not seem to be borne out by the phospholipid composition of *P. halodenitrificans,* which is considerably more complex than these other genera, as it contains DPG, PG, PE, PME, PC, and PA.

PROTEUS

See Enterobacteriaceae.

PROVIDENCIA

Providencia stuarti, now a member of the Enterobacteriaceae as *Proteus inconstans,*[47b] showed the typical increase of cyclopropane over unsaturated fatty acids in a stationary-phase characteristic of members of this family. In contrast, *P. alcalifaciens,* also now reduced to synonomy with *P. inconstans,* did not show this characteristic shift.[220]

PSEUDOMONAS (see also *Alcaligenes*)

It is perhaps not an unexpected finding in this extraordinary genus of metabolic gymnasts that a remarkable diversity of fatty acids is produced by practically all species. These include normal, unsaturated, branched (some strains), cyclopropane (some strains), and hydroxy fatty acids. The last type of fatty acid appears to be characteristic of the "bound" lipids of most of the recognized species examined to date. Hydroxy fatty acids are also widely distributed among members of the genera *Alcaligenes, Moraxella,* and *Neisseria.*

According to Moss and colleagues,[255,256] these hydroxy fatty acids may also be useful in distinguishing among *Pseudomonas* species (see Table 18). *Pseudomonas maltophila* also contains some very unusual branched hydroxy fatty acids, which were considered to be a taxonomic character of primary importance.[255] Fatty acid data supported proposals based on other criteria that *P. cepacia* is identical to *P. multivorans* and *P. kingii,*[319] and that *P. rubescens* be excluded from the genus.[412]

Pseudomonas diminuta, which has already been set apart from other members of the genus on the basis of DNA homology, also differs from the others in its lack of PE and DPG and in its production of glycosyl diglycerides. Other "pseudomonads" that show this unusual lack of DPG include *P. rubescens,* which also contains glycosyl diglycerides (and has been discussed above), and *P. pavonacea,* which differs from other members in its DNA composition.[412]

RHIZOBIUM

The difficulties inherent in the present system of classifying rhizobia on the basis of their association with plants may be the cause of the differences in fatty acid profiles reported for *R. japonicum* in Table 2 and in a previous analysis.[268] A certain *Rhizobium* sp. was found to have some very unusual fatty acids, including 11-methyloctadec-11-enoic, 12-methoxy-11-methyl- and 11-methoxy-12-methyloctadecanoic, and 11-methoxy- and 13-methoxynonadecanoic acids, amounting to 20 to 30% of the total fatty acids.[97]

RHODOSPIRILLACEAE (*RHODOMICROBIUM, RHODOPSEUDOMONAS,* AND *RHODOSPIRILLUM*)

Like the other phototrophs, members of the Rhodospirillaceae contain major amounts (up to 90%) of unsaturated fatty acids.[162] PC and ornithyl-PG have been reported for certain species in the family; these and DPG may be species-specific. The unidentified non-phosphorus-containing ornithine lipids may be potentially useful in classification.[46b,262]

Table 18
FATTY ACIDS USEFUL IN DISTINGUISHING AMONG *PSEUDOMONAS* SPECIES

Taxon	3-OH $C_{10:0}$	2-OH $C_{12:0}$	3-OH $C_{12:0}$	3-OH $C_{14:0}$	2-OH $C_{16:0}$
P. acidovorans	+	−	−	−	−
P. testosteroni	+	−	−	−	−
P. alcaligenes	+	−	−	+	+
P. stutzeri	+	−	+	−	−
P. aeruginosa	+	+	+	−	−
P. putida	+	+	+	−	−
P. cepacia	−	−	−	+	+
(multivorans)					
(kingii)					
P. maltophila[a]	−	−	+	−	−

[a] Contains also *i*-3-OH$C_{11:0}$ (2%) and *i*-3-OH $C_{13:0}$ (3%).

Modified from Moss, C. W., Samuels, S. B., and Weaver, R. E., *Appl. Microbiol.*, 24, 596, 1972, and from Moss, C. W., Samuels, S. B., Liddle, J., and McKinney, R. M., *J. Bacteriol.*, 14, 1018, 1973. With permission.

SALMONELLA

See Enterobacteriaceae.

SERRATIA

See Enterobacteriaceae.

SPIROCHAETA (see also *Treponema*)

Spirochaetae contain saturated branched-chain fatty acids, whereas treponemas and leptospirae contain only saturated and unsaturated fatty acids. *Spirochaeta stenostrepta* was reported to contain more bound than free fatty acids.[137] Both members of the Spirochaetaceae examined (*S. zuelzerae* and *Treponema pallidum*) lack PE and share the unusual character of producing large amounts of free aldehydes.[233]

THERMUS

The fatty acids of *Thermus aquaticus* contain large amounts of branched-chain fatty acids, an uncommon characteristic among Gram-negative bacteria.

THIOBACILLUS

Thiobacillus is one of the few genera in the 8th edition of *Bergey's Manual*[47b] to have been characterized by the fatty acid profiles of the various species. Agate and Vishniac,[3] who grouped various species together on this basis, found a correlation of their groupings with GC%. Group I was comprised of *T. neopolitanus* and *T. ferrooxidans* (56 to 57 GC%), with fatty acids equally distributed between those larger than C_{14} and those smaller; Group II contained *T. thioparus, T. thiocyanooxidans, T. denitrificans, T. novellus,* and *T. intermedius* (62 to 68 GC%), with fatty acids mostly longer than C_{14}; Group III consisted of *T. thiooxidans* and *T. concretivorus* (50 to 52 GC%), with fatty acids mostly shorter than C_{14}. Among the phospholipids, it appears that a content of PME may be a good marker at the generic level.

TREPONEMA (see also Spirochaeta)

The requirement of these organisms for preformed fatty acids for growth makes their characterization by lipid analysis more difficult. Thus, such parasitic treponemes as *T. minutum, T. refringens, T. reiteri,* and *T. kazan* have fatty acid profiles that resemble those of *T. pallidum.*[233] When *T. refringens* was grown in a medium without serum as the source of long-chain fatty acids, only n-$C_{16:0}$ and $C_{18:1}$ were found in the cellular fatty acids. When grown in a medium with serum, other fatty acids, including $C_{18:2}$, were found, having been taken up directly or transformed from the serum fatty acids.[216] Despite these difficulties, it was possible to differentiate as follows among the genera *Spirochaeta, Treponema,* and *Leptospira.*[216a]

	Spirochaeta	*Treponema*	*Leptospira*
Branched-chain fatty acids	+	−	−
Unsaturated fatty acids	−	+	+
Monoglycosyl digylceride	−	+	−
Phosphatidyl choline	−	+	−
De novo fatty acid synthesis	+	−	±

VIBRIO

Although otherwise unremarkable, the fatty acid profile of *Vibrio* spp. shows a minor but almost constant complement of branched-chain fatty acids. *Spirillum linum* and *Vibrio (Photobacterium) fischeri* also had "vibrioid" fatty acid profiles, underlying their close relationship with *Vibrio,* a similarity that has been recognized in the latest edition of *Bergey's Manual* (see Reference 271).

Removal of *V. fetus* to *Campylobacter* seems to be confirmed by its lack of DPG and the presence of PS and PI. A previous report of PI in *V. (comma) cholerae*[126] requires reexamination. Oliver and Colwell[272] considered *V. parahaemolyticus, V. cholerae, V. alginolyticus,* and *V. marinofulvus* to have typical "*Vibrio*" phospholipid patterns. *Vibrio marinus* was unlike the rest, with low PE and high PG. Remarkable among the marine vibrio strains studied was the detection of lyso-phosphatides in large amounts.

YERSINIA

Yersinae differ from other members of the Enterobacteriaceae in several ways:

1. They contain some free hydroxy fatty acids that are rare in the others.
2. Glycosyl glycerides, not reported from any other members of the Enterobacteriaceae, were found in *Y. pseudotuberculosis.*
3. They contain PME and PS, also not reported in major amounts from other Enterobacteriaceae.

The phospholipids may serve as a good generic marker. More data are needed to confirm the present status of the genus.

Gram-positive Bacteria

AEROCOCCUS

See *Pediococcus.*

BACILLUS

All *Bacillus* species contain major amounts of branched-chain plus small amounts of saturated fatty acids. Certain mesophilic species (*B. anthracis. B. alvei, B. cereus, B. megaterium, B. pumilus, B. stearothermophilus,* and *B. thuringiensis*) also contain substantial amounts of unsaturated fatty acids. In most cases the position of the unsaturation has not been determined, but in certain species (*B. cereus, B. anthracis,* and *B. thuringiensis*) it is of the *cis*-Δ^{10} type. In three psychrophiles (*B. insolitus, B. globisporus,* and *B. psychrophilus*), substantial amounts of unusual Δ^5 isomers were found.[144] Not surprisingly, three thermophiles (*B. acidocalderius, B. caldotenax,* and

B. caldolyticus) contained no unsaturated fatty acids, although all had major amounts of the characteristic *Bacillus* branched-chain types.[67,399,400] *Bacillus acidocaldarius* had two cyclohexyl fatty acids (11-cyclohexylundecanoic and 13-cyclohexyltridecanoic), which accounted for as much as 65% of the total fatty acids under certain conditions.[67]

Bacilli are unusual among Gram-positive bacteria in that they contain PE (the others being *Planococcus* and the actinomycetes). In fact, the published data to date show that all but two strains have been reported to contain this phospholipid; the two exceptions are a strain of *B. stearothermophilus,* presumably a wild type,[242] and a *B. subtilis* mutant.[30]

According to Minnikin et al.,[242] most strains of *Bacillus* contain glycosyl diglycerides, although some exceptions have been noted — i.e., certain *B. stearothermophilus* strains[242,276] and a strain of *B. cereus.*[238]

CLOSTRIDIUM

Chan et al.[53] have recently reported the occurrence of unsaturated cyclopropane fatty acids in representative clostridia, including meso-, thermo-, and psychrophiles. All strains examined contained 12,13-methylene-9-tetradecenoic acid. This is the first report of such unsaturated cyclopropane fatty acids from bacteria. It would be interesting to know whether some of the unknown fatty acids reported in the fatty acid patterns of other clostridia would be of this type.[268,325] Previous work by Moss and Lewis[251] showed that *C. perfringens, C. sporogenes, C. bifermentans,* and *C. butyricum* could be differentiated both from other clostridia and from each other. Ten other species had less individual profiles.

ERYSIPELOTHRIX

See *Listeria.*

GEMELLA

Like *Sarcina* (the other genus of the Peptococcaceae about which we have information), *Gemella* contains hydroxy fatty acids, a type that is unusual in Gram-positive bacteria.

LACTOBACILLUS

Among aerobic Gram-positive rods that have been studied, lactobacilli are unique in their fatty acid profile, which is characterized by negligible amounts of branched-chain and by major amounts of cyclopropane fatty acids. Within the genus itself, however, no subgroups could be distinguished on this basis. *Lactobacillus heterohiochii* strains, alone of eight species of alcoholophilic or alcoholotolerant lactobacilli, had large amounts of fatty acids of chain lengths greater than 20.[381] All strains except *L. delbruckii* had major amounts of lactobacillic acid.[388] Two species of lactobacilli that had diaminopimelic acid in their cell walls (none of the others do) had fatty acids similar to other lactobacilli.[380]

AAPG has been reported from *L. acidophilus, L. arabinosus, L. bulgaricus, L. casei, L. fermentum, L. lactis,* and *L. plantarum,* but not from *L. delbruckii, L. helveticus,* or *L. heterohiochii;* thus, it may have some utility in speciation. All lactobacilli reported to date have contained galactosylglucosyl or diglucosyl galactosyl diglycerides.

LACTOBACILLUS/BIFIDOBACTERIUM

In general, bifidobacteria have much lower amounts of lactobacillic acid than lactobacilli, although their fatty acid profiles are otherwise quite similar.[388] Phospholipid composition gives added weight to the separatin of these two genera. All bifidobacteria contain a unique phospholipid, galactosyl phosphatidyl glycerol diglyceride, also known as "Compound 15" (see Table 13 [XIII-4]), that was absent in lactobacilli. Other differences include GC%, cell wall composition, and metabolic pathways.[85]

LACTOBACILLUS/SPOROLACTOBACILLUS

On the basis of their fatty acid profiles, Uchida and Mogi[380] were able to place the catalase-negative, spore-forming homolactic fermenters of the genus *Sporolactobacillus* within the Bacillaceae rather than within the Lactobacillaceae. Sporolactobacilli had branched-chain fatty acids similar to those of *Bacillus* species and no cyclopropane fatty acids.

LISTERIA

Strains representing serotypes I, II, III, and IVB of *Listeria monocytogenes* have essentially the same fatty acid profiles, except that IVB types lack fatty acids having 18 carbons. The others have minor amounts of these (2.3 to 6.8%).[348] Raines et al.[301] and Julak and Mara[139] found no relationship of serotype to fatty acid patterns.

LISTERIA/ERYSIPELOTHRIX

These two genera are often placed side by side in taxonomic textbooks. Their fatty acid types are alike, but quantitatively *Listeria* has predominantly branched-chain types, and *Erysipelothrix* has predominantly unsaturated and saturated fatty acids.[348]

MICROCOCCUS

In recognized species,[47b] branched-chain and saturated fatty acids are characteristic. Unsaturated fatty acids are minor components, and cyclopropane types are absent. Whiteside et al.[405] distinguished three broad groups:

1. The "*M. lysodeikticus*" type, which included *M. luteus (Sarcina flava)* and *M. tetragenus.*
2. The *"M. roseus"* type, which also included *M. conglomeratus.*
3. The *"M. varians"* type.

These three groups undoubtedly correspond to the three recognized species of the genus.[47b]

In contrast, compared to the above, the species of uncertain affiliation show many anomalies:

1. *M. caseolyticus* contains major amounts of $C_{18:1}$ and does not appear to belong among the micrococcaceae.
2. *M. cryophilus,* although reported to be a Gram-positive organism, has phospholipid and fatty acid patterns that are more like those of Gram-negative bacteria. Sections of the cell wall show this species to be quite unlike micrococci and staphylococci.
3. *M. diversus,* a Gram-negative coccus, has a fatty acid pattern more typical of the rods of the Enterobacteriaceae than of the Gram-negative cocci of *Neisseria* or *Paracoccus.*
4. *M. radiodurans,* which resembles *M. roseus* physiologically and biochemically, differs in its unusual cell wall ultrastructure. It is possible that its resistance to radiation may be related to its unusually high content of unsaturated fatty acids.[101]

MICROCOCCUS/STAPHYLOCOCCUS

Jantzen et al.[131] considered the key components in the differentiation of these two genera to be certain fatty acids present in one or the other genus. Micrococci contained n-$C_{15:0}$. Staphylococci contained *iso*- and *anteiso*-$C_{19:0}$ and $C_{20:0}$. In certain cases, however, these compounds were produced in only minor amounts. Hydrocarbons were found in all micrococci examined, whereas none were found in staphylococci.[250]

PEDIOCOCCUS

Fatty acid patterns of pediococci are like those of streptococci, including the fact that some species contain cyclopropane fatty acids, principally lactobacillic acid (LA), and some do not. *Pediococcus cerevisiae* and *P. halophilus* both contain large amounts of LA, whereas *P. pentosaceus* and *P. acidolactici* contain only small amounts of other cyclopropane fatty acids. *Pediococcus urinae-equi, P. homari,* and *Aerococcus viridans* contain no cyclopropane fatty acids, but they do contain several unknown unsaturated types.[379] These authors suggested that the lactobacillic-acid-containing species may be related to the lactic streptococci.

PLANOCOCCUS

The phospholipid pattern of planococci is unusual for Gram-positive bacteria and differs from other Micrococcaceae in its content of PE.

SARCINA

This genus is currently reserved for anaerobic packet-forming cocci and thus has been set apart from morphologically similar aerobic forms. Although the phospholipid content of sarcinae is similar to that of their aerobic counterparts, their fatty acids show them to be quite unlike the Micrococcaceae.

SPOROLACTOBACILLUS

See *Lactobacillus.*

SPOROSARCINA

The inclusion of this genus in the Bacillaceae rather than in the Micrococcaceae is supported by the presence of PE in *S. urea.* PE is absent from the aerobic sarcinae, now included in *Micrococcus.*

STAPHYLOCOCCUS (see also *Micrococcus*)

The occurrence of AAPG in this genus may be species-specific: *S. aureus* contains lysyl-PG, whereas *S. epidermidis* does not. All strains thus far reported contain diglucosyl diglycerides.[321]

STREPTOCOCCUS

A comparison of 31 strains of human, animal, and plant origin in the *S. faecalis* and *S. faecium* groups showed no qualitative differences in fatty acid content. However, these two species can be grouped with *S. lactis* and *S. cremoris* on the basis of the presence of cyclopropane fatty acids (largely lactobacillic acid) in their fatty acid profiles. Species in which none of these compounds occurred include *S. bovis, S. equinus, S. hemolyticus, S. mutans, S. pyogenes,* and *S. uberis.*[10,87,268,325]

Lambert and Moss[194] found that the fatty acids of both *S. mutans* and *S. salivarius* were characterized by large amounts (13 to 26%) of eicosenoic acid ($C_{20:1}$). This compound was absent or present only in traces in *S. sanguis, S. mitis, S. uberis,* and *S.* spp. The fact that Drucker et al.[75,76] had not reported this unusual fatty acid in *S.*

mutans could be explained by the Drucker group's use of a boron-trihalide reagent in the direct esterification of the cells, a technique known to destroy unsaturated and cyclopropane fatty acids.[254] The differences in the branched-chain fatty acid content of *S. mutans* reported by the two groups could also be accounted for on this basis.

All streptococci thus far reported contain glucosyl and diglucosyl diglycerides. They also contain major amounts of unusual phosphatidyl diglucosyl diglycerides (Table 13 [XIII-5]).

Actinomycetes and Irregularly Shaped Bacteria of Uncertain Affinity (ISBUA)

Most actinomycetes that have been analyzed show a typical Gram-positive lipid profile, consisting of branched-chain fatty acids of the *iso/antieiso* type and of minor amounts of monounsaturated fatty acids. Mycobacteria, nocardiae, and actinomadurae of the *madurae* type stand apart from this main group in that they contain only minor amounts of branched-chain fatty acids, but major amounts of unsaturated types. Both groups contain saturated fatty acids, but not cyclopropane fatty acids. Mycobacteria often contain long-chain fatty acids (greater than 20 carbons) not found in other actinomycetes.[133]

It is unfortunate that one of the major reports on the fatty acids of various genera of actinomycetes[23] examined lipid samples only after hydrogenation; thus, all unsaturated fatty acids were measured as saturated fatty acids. Also, the data of Efimova and Tsyganov[80] and many of those of Guzeva et al.[116] have not been included, since no effort was made to distinguish branched-chain from normal fatty acids.

Among the fatty acids that are unique to actinomycetes are tuberculostearic acid (TBA) and its homologs and the mycolic acids (see previous sections for discussion of these compounds). Mycolic acids are probably found only in strains of the *Nocardia-Mycobacterium-Corynebacterium* complex. As previously discussed, the data on the mycolic acid type, for the most part, support the generic assignments made by other criteria. Among the taxa that showed anomalous results were *Mycobacterium brevicale, M. rhodochrous* and *M. thamnopheos.* Since these contained nocardomycolates,[201,206] they should be assigned to the genus *Nocardia. "Nocardia" farcinica* strains recently isolated from African cases of bovine farcy turned out to contain eumycolates similar to those of mycobacteria.[52,206] As *Nocardia farcinica* is the type species of the genus, it was proposed[206] that this epithet be considered a *nomen dubium* and that *N. asteroides* be considered the type species. No mycolates of any kind were found in four strains of *N. autotrophica* (published as *N.* spp.) or in *Corynebacterium pyogenes,* suggesting the need for taxonomic reassessments.[206] Mycolates are also absent in *Actinomadura madurae, A. pelletieri, A. dassonvillei, Micropolyspora faeni, N. aerocolonigenes, Oerskovia turbata, Pseudonocardia thermophila, Saccharomonospora (Thermomonospora) viridis,* and *Streptomyces* spp.;[205b,294] however, nocardomycolates were found in *Micropolyspora brevicatena.*[124b] *Nocardomycolates* (as LCNA, q.v.), were not found in *Actinobifida chromogena, Thermoactinomyces (Actinobifida) dichotomica, M. mesophilica, M. rectiirgula, M. virida,* or *Streptomyces griseus.*[249] Twenty-three *N. pellegrino* strains contained LCNA having the same mobilities as LCNA of *N. rhodochrous* strains on thin-layer chromatography; thus, the authors concluded that the organisms were properly placed in the genus *Nocardia* rather than in *Mycobacterium.*[248]

For the first time that the reviewer is aware of, lipids have been included as a character in a numerical taxonomic study.[107] The authors used not only morphological and physiological characters, but also chemotaxonomic and serological characters. The mobility of nocardomycolic acids on TLC plates, using the LCNA technique,[247] was found to be one of 14 characters with the "greatest resolving power" in distinguishing clusters of the *N. rhodochrous* group.

At first glance, the array of complex mycobacterial lipids seems bewildering in its variety. In particular, O'Leary[267,269] has expressed grave reservations about the reality of many of the lipids isolated from mycobacteria, as he says, based on "no experimental evidence, only intellectual unease."[269] Although it is true that mycobacterial lipids appear to represent an extraordinary summit of complexity among microbial lipids, indirect evidence would seem to favor their genuineness. Taken as a whole, the broad group of the actinomycetes, to which the mycobacteria belong, is an extraordinarily productive group biosynthetically, producing complex antibiotics, complex pigments, volatile substances, and many other compounds.[205a] Furthermore, the structures of cord factors, mycolic acids, mycosides, waxes D, and similar compounds have been confirmed by lipid chemists in many different laboratories in different countries. In addition, the capacity of actinomycetes to transform and break down many molecules that resist attack by other microorganisms gives evidence of a remarkably diverse battery of enzymes. Mycobacteria, in particular, possess a unique (for bacteria) fatty acid synthetase, which appears to be of the multienzyme complex type.[111,127] Why, then, should they not produce complex lipids?

ACTINOMADURA

The unusual phospholipid pattern of *A. madurae* (lack of PE) supports the previous proposal, made on the basis of cell wall composition, to remove members of this group from the genus *Nocardia*.[204]

Agre et al.[4] found that the genus contains two groups in terms of fatty acid composition, confirming the original observation of Lechevalier and Lechevalier[204] that it was heterogeneous in terms of morphology and whole-cell sugar type. Major amounts of unsaturated fatty acids were found in *A. madurae* and in *A. pelletieri*, whereas in *A. dassonvillei*, *A. helvata*, *A. pusilla*, *A. roseoviolacea*, and *Actinomadura* spp., branched-chain fatty acids were dominant.

ACTINOPLANES

The members of this genus examined to date lack tuberculostearic acid.

ARTHROBACTER (see also *Corynebacterium*)

Most arthrobacters contain major amounts of branched-chain and normal fatty acids and minor amounts of unsaturated fatty acids. Bowie et al.[41] proposed removal of *A. simplex* and *A. tumescens* from the genus on the basis of their high content of unsaturated fatty acids, as well as on the basis of their different cell composition and the results of a numerical taxonomic study.

Most of the species analyzed to date show an affinity for the broad group of the actinomycetes in their content of PI and/or PIM. Only *A. marinus* seems to have an appreciably different pattern, and thus it should probably be placed elsewhere. The three strains that have been examined to date all contain mono- and digalactosyl and dimannosyl diglycerides.

BIFIDOBACTERIUM

Despite their seemingly strong morphological affinity for the actinomycetes, bifidobacteria do not resemble them in regard to fatty acid or phospholipid profiles. Not only do bifidobacteria lack PI and PIM, but they all contain galactosyl phosphatidyl glycerol diglyceride (Compound 15 — see Table 13 [XIII-4]), and many contain alanyl-PG; neither compound has been reported from the actinomycetes. No important differences were found in the phospholipid composition of bifidobacteria from various sources, including ten strains of *B. bifidum* var. *pennsylvanicus* and one strain each of *B. asteroides*, *B. coryneforme*, *B. indicum*, *B. ruminale*, and *B. globosum*.[85]

BREVIBACTERIUM

Patterns of phospholipids in the brevibacteria that have been analyzed to date confirm the placement of these species close to the actinomycetes. On the basis of both fatty acid and cell wall composition, as well as on the basis of GC%, it has been proposed that *B. ammoniagenes* be placed among the corynebacteria of the *diptheriae* or animal-associated type.[41]

CELLULOMONAS (see also *Oerskovia*)

The new location of this genus among actinomycete-like organisms[47b] is supported by its phospholipid patterns, in particular the content of PIM.

CORYNEBACTERIUM

Although this genus, as presently constituted, is heterogeneous from many points of view, the phospholipid patterns show that many species are closely related to the actinomycetes. For purposes of this review, those species that are known to have a cell wall composition of Type IV[204] have been placed with the actinomycetes; the others have been classed among the ISBUA until further studies may be carried out. Komura et al.[178] divided the genus into two groups: members of the first group were characterized by a low GC% (52 to 60) and contained little or no trace of PE; the second group had a high GC% (62 to 71) and contained PE.

Most corynebacteria having a cell wall of Type IV appear to differ from the mycobacteria and nocardiae (which they resemble in this and many other ways, including mycolic acid production) by the fact that they contain substantial amounts of branched-chain fatty acids. Bowie et al.[41] found that three species of plant-pathogenic corynebacteria (*C. insidiosum, C. michiganese,* and *C. sepedonicum*) were similar to *Arthrobacter* in fatty acid and DNA composition; the fourth species (*C. fascians*) was more like the animal pathogens represented by *C. diphtheriae.* Previous work had also shown that the first three species were similar to *Arthrobacter* on the basis of numerical taxonomy and enzymatic and serological characterstics, although their cell walls differed from arthrobacters in composition. Similar data supported the conclusion that *C. fascians* belonged with the *diptheriae* types. The authors concluded that, if one accepts the current definition of *Arthrobacter* as having diaminopimelic acid in the cell wall, plant corynebacteria will have to be placed in another generic slot. On the basis of their results they proposed new definitions for *Arthrobacter, Corynebacterium,* and *Mycobacterium,* which would differentiate among these genera on the basis of GC%, cell wall composition, lipid composition, and morphology. Thus, *Arthrobacter* would show a morphological cycle of rods to cocci on aging when grown in a rich medium, have a GC% of 60 to 80, and contain *anteiso*-$C_{15:0}$ as the principal fatty acid component; *Corynebacterium* spp. would show a tendency to pleomorphism, no branching or mycelium, a GC% of 50 to 60, a cell wall composition of Type IV, and n-$C_{16:0}$ as the major fatty acid; and *Mycobacterium* spp. may show branching, would have a GC% of 62 to 70%, a cell wall of Type IV, and n-$C_{16:0}$ as the principal fatty acid.

MICROBACTERIUM

Microbacterium flavum, the only member of this heterogeneous genus that has been analyzed for lipids, shows an affinity for corynebacteria of the animal type in its cell wall composition and GC%,[47b] but also in its phospholipid composition.

MICROBISPORA

TBA and its homologs have been found in all examined species of this genus.

MICROMONOSPORA

Tuberculostearic acid has been reported in some species.

MICROPOLYSPORA

The genus is heterogeneous in terms of mycolic acid production (see general comments at the beginning of this section).

MYCOBACTERIUM (see also *Corynebacterium*)

Mycobacteria may be distinguished from other closely related actinomycetes on the basis of their mycolic acid type (q.v.). Besides unsaturated and normal fatty acids, many mycobacteria also contain tuberculostearic acid.

Mycobacterium kansaii strains can be differentiated from strains of *M. avium, M. intracellulare* (Battey), *M. marinum, M. tuberculosis,* and various scotochromogens by their content of a saturated, branched-chain fatty acid with the methyl group on position 2, eluting between n-$C_{14:0}$ and n-$C_{15:0}$ on nonpolar columns, and between $C_{12:0}$ and $C_{14:0}$ on polar columns.[357]

NOCARDIA

Nocardiae may be distinguished from other closely related actinomycete genera on the basis of the type of mycolic acids they produce. Except for *Nocardia coeliaca,* which contains PC, all members of this genus have a typical actinomycete pattern of phospholipids (DPG, PE, PI, and PIM). It is interesting that *N. coeliaca* is one of the species of *Nocardia* that lacks nocardomycolic acids. The African isolates of *"N." farcinica,* now classed in *Mycobacterium,*[51] contain mycosides — glycolipids found only in mycobacteria.[198]

OERSKOVIA

Like *Actinomadura,* this genus differs in its phospholipid composition from most of the other genera of actinomycetes by its lack of PE. Oerskoviae have been thought to be similar to cellulomonads,[296] but this relationship is not borne out by either fatty acid or phospholipid composition. From only limited data, it appears that members of the Actinomycetaceae, at least *Oerskovia* and *Rothia,* may lack unsaturated fatty acids altogether.

PROPIONIBACTERIUM

As the genus is presently constituted, the fatty acid patterns appear remarkably consistent from species to species and very similar to members of the other genera that have been placed in the ISBUA, including those that were formerly classed as anaerobic corynebacteria. The only differences found were in the isomer of the branched-chain $C_{15:0}$ fatty acids produced. In the first group (*Propionibacterium freudenreichii,* and also *P. shermanii,* now a synonym), mostly *anteiso*-$C_{15:0}$ and little *iso*-$C_{15:0}$ were found; in the second group (*P. acidi-propionici* [(*arabinosum, pentosaceum*], *P. jensenii* [also as *P. zeae*], and *P. thoenii*), the *iso*-$C_{15:0}$ isomer predominated. The *Corynebacterium acnes* strains analyzed were all similar to the secon group.[252] Thus, the transfer of the latter to *Propionibacterium,* as well as some of the other species transfers that have been made in the current *Bergey's Manual,*[47b] do not appear to conflict with the fatty acid profiles reported here.

ROTHIA

See *Oerskovia.*

SACCHAROMONOSPORA

Members of this genus have a typical actinomycete fatty acid profile; they may contain TBA and its homologs.

STREPTOMYCES

In most cases, streptomycetes have a typical actinomycete lipid spectrum. The exception is the Russian *Streptomyces canosus,* which is reported to contain PC and PDME. As there is no information on the cell wall composition of this strain, its placement is uncertain. In the second case, an interesting phospholipid produced by *S. olivaceus* is phosphatidyl-2,3-butanediol (see Table 13 [XIII-6]). These results indicate that a wider look at the many members of this enormous genus might bring to light some important taxonomic distinctions based on phospholipid composition.

STREPTOSPORANGIUM

All strains examined to date contain TBA.

THERMOMONOSPORA

Some strains contain TBA.

Blue-green Algae (Cyanobacteria)

Stanier et al.[338] have proposed that blue-green algae be included within the bacterial domain on the basis of their prokaryotic nature; this has been endorsed by the editors of the 8th edition of *Bergey's Manual.*[47b] As Stanier et al.[338] originally pointed out, although they have cell walls similar to those of Gram-negative bacteria, there are fundamental differences between the pigments and overall photosynthetic processes of blue-greens and those of the phototrophic bacteria. In many cases there are also differences in their lipid compositions. Although some of the simpler unicellular blue-greens have bacteria-like fatty acid profiles, other unicellular forms and the "higher" filamentous forms contain polyunsaturated fatty acids, including tri- and tetra-unsaturated types.[161,163,274] Also, the mono- and diglycosyl diglycerides isolated from *Agmenellum quadruplicatum* were bacteria-like, whereas the sulfolipid, sulfoquinovosyl diglyceride, was like those of plants. The phospholipids were limited to PG, which is not a common bacterial profile.[274] Thus, it appears that the "cyanobacteria" are truly an intermediate group between the higher algae and plants and the bacteria. Since classifications in taxonomy should be as rational and useful as possible, it would seem reasonable to place these organisms in a kingdom of their own, distinct from either the bacteria or the higher algae.

CONCLUSIONS

At present, the usefulness of lipid composition in the taxonomy of bacteria appears to be limited to phospholipids and fatty acids, including mycolic acids, and possibly hydrocarbons and waxes. There is not sufficient data presently available from which to draw conclusions concerning the utility of glycolipid, sulfolipid, and peptidoglycolipid composition as taxonomic criteria. Furthermore, it may be that the production of these substances is, in many cases, too dependent on growth conditions to be fully reliable. All of these lipids may be extracted from whole cells; the study of the lipid composition of specific organelles (such as cell walls or mesosomal, cytoplasmic, or outer membranes) does not now seem to hold promise of yielding data more useful that those obtainable from whole-cell extracts.

The identification of the simpler lipids (such as fatty acids, mycolic acids, phospholipids, and glycolipids), for which standards are available, is well within the competence of the taxonomist. However, one might ask what value such data really have. Is it worth the trouble? First, it is evident that only in rare cases (such as the resolution of the problem of defining the generic boundaries of the *Nocardia-Mycobacterium-Corynebacterium* group) can one look to lipid composition as a primary taxonomic character; it is, however, an important auxiliary one. The data show that, for the most part, where enough information has been compiled under standard conditions, qualitative bacterial fatty acid and phospholipid compositions support many of the familial and generic groupings presently recognized in the 8th edition of *Bergey's Manual*.[47b] This is also true for actinomycetes, using the classification system of Cross and Goodfellow[60] as modified in this review. In other cases, lipid composition data may point up heterogeneity among the species of a given genus. Such evidence, taken together with other taxonomic characters, may help delineate these taxa, as has been done with *Thiobacillus*,[47b] or it may help to push out the misfits, such as *Pseudomonas rubescens*.

In situations where morphological, physiological, or immunological data fail to lead to an unequivocal decision, information on lipid composition may tend to tip the taxonomic balance in one direction or another. Such is the case with *Sporolactobacillus*, where fatty acid composition showed a clear alliance with *Bacillus* rather than with *Lactobacillus*.[380]

Lipid composition was also able to cast light on the problem of whether lepromatous leprosy is caused by a mycobacterium or by a mixture of corynebacteria, mycobacteria, and other bacteria, as has recently been proposed. Etemadi and Convit,[83] using gas chromatography and high-resolution mass spectroscopy, were able to show that the lipids from tissues of leprous patients contained eumycolates of the types that have been isolated from *Mycobacterium tuberculosis* and *M. kansasii*. Thus, they were able to conclude that the disease is probably caused by a mycobacterium.

Finally, it is likely that lipids may serve as a useful means of characterizing some of the enormous numbers of faceless saprophytes that lie haplessly within genera such as *Brevibacterium*, *Acinetobacter*, or *Achromobacter*. Most bacteria that are beyond the clinicomedical pale (including those from soil, sewage, and water) are in great need of definitive taxonomic study and could undoubtedly be broken down by such means into broad groups. Those strains that are lumped together because they produce the same type of reaction (''corroding bacilli''), are isolated from the same environment (''farcy bacillus''), or have seemingly similar morphological characters (''coryneform bacteria'') require critical taxonomic evaluation. Such groupings are often shown to stand up when other characters of the strains involved are evaluated. More often, closer examination brings to light basic differences, as in the cases of the groups cited above, where lipid, cell wall, or DNA composition show them to be quite heterogeneous.

Are there pitfalls in the analysis of bacterial lipids? As in any scientific technique, there are a few. A perusal of the section on the effect of medium constituents, age of cells, growth temperature and the like on lipid composition should make it clear that whatever conditions are chosen for growth and analysis should be standardized for all strains to be compared. Authentic reference strains and chemical standards are indispensable. Given adequate and appropriate nutrients, preferably those free of fatty acids, bacteria that have been grown at their optimal temperature and harvested for analysis at a given growth phase will give reproducible lipid profiles, both qualitatively and, usually (within limits), quantitatively.

Quantitative discrepancies between analyses of representatives of a given species are easily explained by differences in the strains employed, growth conditions, and methods of extraction. Less easily explained are the occasional profound qualitative differences that appear in the literature. These are more likely to lie in (1) incorrect identification or impurity of strains, or (2) the sole use of retention times in gas chromatography for the identification of fatty acids, or of relative mobility of a phospholipid on thin-layer chromatography, without suitable complementary degradative or derivative analytical procedures.

Other problems occur when a chemist analyzes a biomass that has been grown for him (antifoam oils, Tweens® and the like, or unsuitable growth conditions causing abnormal lipid metabolism may have been used, or mixed or contaminated cultures may have been extracted). Thus, the chemist whose only interest is in the novel chemistry of the lipids he finds can leave the taxonomists in a quandary when they examine his results. It is far better for the taxonomist to take the plunge into the great chemical sea, preferably with a chemist nearby to fish him out if he goes over his head. The combination is much more likely to yield accurate and taxonomically useful results.

REFERENCES

1. **Adam, A., Senn, M., Vilkas, E., and Lederer, E.,** Spectrométrie de masse de glycolipides. 2. Diesters de trehalose naturels et synthetiques, *Eur. J. Biochem.*, 2, 460, 1967.
2. **Agate, A. D. and Vishniac, W.,** Changes in phospholipid composition of *Thiobacillus neapolitanus* during growth, *Arch. Mikrobiol.*, 89, 247, 1973.
3. **Agate, A. D. and Vishniac, W.,** Characterization of *Thiobacillus* species by gas—liquid chromatography of cellular fatty acids, *Arch. Mikrobiol.*, 89, 257, 1973.
4. **Agre, N. S., Efimova, T. P., and Guzeva, L. N.,** Heterogeneity of the genus *Actinomadura,* Lechevalier et Lechevalier, *Microbiology USSR*, 44, 220, 1975.
5. **Aibara, S., Kato, M., Ishinaga, M., and Kito, M.,** Changes in positional distribution of fatty acids in the phospholipids of *Escherichia coli* after shift-down in temperature, *Biochim. Biophys. Acta,* 270, 301, 1972.
6. **Albro, P. W. and Dittmer, J. C.,** Bacterial hydrocarbons: Occurrence, structure and metabolism, *Lipids*, 5, 320, 1970.
7. **Alimova, E. K. and Gurskii, E. V.,** Changes in fatty acid composition of free and bound lipids in growth phases of *Escherichia coli, Microbiology USSR*, 41, 575, 1972.
8. **Ambron, R. T. and Pieringer, R. A.,** The metabolism of glyceride glycolipids. V. The identification of the membrane lipid formed from diglucosyl diglyceride in *Streptococcus faecalis* ATCC 9790 as an acylated derivative of glyceryl phosphoryl diglucosyl glycerol, *J. Biol. Chem.*, 246, 4216, 1971.
9. **Ambron, R. T. and Pieringer, R. A.,** Phospholipids in microorganisms, in *Form and Function of Phospholipids,* Ansell, G. B., Hawthorne, J. N., and Dawson, R. M. C., Eds., Elsevier, New York, 1973, p. 289.
10. **Amstein, C. F. and Hartman, P. A.,** Differentiation of some enterococci by gas chromatography, *J. Bacteriol.*, 113, 38, 1973.

11a. **Anderes, E. A., Sandine, W. E., and Elliker, P. R.,** Lipid in antibiotic-sensitive and -resistant strains of *Pseudomonas aeruginosa, Can. J. Microbiol.*, 17, 1357, 1971.

11b. **Anderson, R. J. and Chargaff, E.,** The chemistry of the lipoids of tubercle bacilli. VI. Tuberculostearic acid and phthioic acid from the acetone-soluble fat, *J. Biol. Chem.*, 85, 77, 1929.

12. **Ansell, G. B., Hawthorne, J. N., and Dawson, R. M. C. (Eds.),** *Form and Function of Phospholipids,* Elsevier, New York, 1973.
13. **Antoine, A. D. and Tepper, B. S.,** Environmental control of glycogen and lipid content of *Mycobacterium tuberculosis, J. Bacteriol.*, 100, 538, 1969.
14. **Asselineau, C. P., Montrozier, H. L., Promé, J.-C., Savagnac, A. M., and Welby, M.,** Etude d'un glycolipide polysinaturé synthétisé par *Mycobacterium phlei, Eur. J. Biochem.*, 28, 102, 1972.
15. **Asselineau, J.,** *Les Lipides Bacteriéns,* Hermann, Paris, 1962.
16. **Asselineau, J.,** *The Bacterial Lipids,* Hermann, Paris, 1966.
17. **Asselineau, J., Lanéelle, M.-A., and Chamoiseau, G.,** De L'étiologie du farcin de zébus tchadiens: Nocardiose ou mycobactériose? II. Composition lipidique, *Rev. Elev. Méd. Vét. Pays Trop.*, 22, 205, 1969.
18. **Auran, T. B. and Schmidt, E. L.,** Similarities between *Hyphomicrobium* and *Nitrobacter* with respect to fatty acids, *J. Bacteriol.*, 109, 450, 1972.

19. **Azuma, I.**, Glycolipids of mycobacterial cell envelopes, *Tampakushitsu Kakusan Koso*, 16, 680, 1971.
20. **Azuma, I., Kanetsuna, F., Tanaka, Y., Mera, M., Yanagihara, Y., Mifuchi, I., and Yamamura, Y.**, Partial chemical characterization of cell wall of *Nocardia asteroides* strain 131, *Jpn. J. Microbiol.*, 17, 154, 1973.
21. **Azuma, I., Ohuchida, A., Taniyama, T., Yamamura, Y., Shoji, K., Hori, M., Tanaka, Y., and Ribi, E.**, The mycolic acids of *Mycobacterium rhodochrous* and *Nocardia corallina, Biken J.*, 17, 1, 1974.
22. **Azuma, I., Yamamura, Y., Tanaka, Y., Kohsaka, K., Mori, T., and Itoh, T.**, Cell wall of *Mycobacterium lepraemurium* strain Hawaii, *J. Bacteriol.*, 113, 515, 1973.
23. **Ballio, A. and Barcellona, S.**, Relations chimiques et immunologiques chez les actinomycetales. I. Les acides gras de 43 souches d'actinomycètes aérobies, *Ann. Inst. Pasteur Paris*, 114, 121, 1968.
24. **Barber, M., Jolles, P., Vilkas, E., and Lederer, E.**, Determination of amino acid sequences in oligopeptides by mass spectrometry. I. The structure of fortuitine, an acyl-nonapeptide methyl ester, *Biochem. Biophys. Res. Commun.*, 18, 469, 1965.
25. **Batrakov, S. G., Panosyan, A. G., Konova, I. V., and Bergelson, L. D.**, Diol lipids. XXVI. Identification of a *threo*-butane-2,3-diol phospholipid from *Actinomyces olivaceus, Biochim. Biophys. Acta*, 337, 29, 1974.
26. **Batrakov, S. G., Pilipenko, T. V., and Begel'son, L. D.**, New ornithine-containing lipid from actinomycetes, *Dokl. Akad. Nauk SSSR*, 200, 226, 1971.
27. **Batt, R. D., Hodges, R., and Robertson, J. G.**, Gas chromatography and mass spectrometry of the trimethylsilyl ether methyl ester derivatives of long-chain hydroxy acids from *Nocardia corallina, Biochim. Biophys. Acta*, 239, 368, 1971.
28. **Beaman, B. L., Burnside, J., and O'Donnell, B.**, Ultrastructural and biochemical analysis of *Nocardia asteroides* 14759 cell walls during its growth cycle, *1st Int. Conf. on the Biology of the Nocardiae*, Merida, Venezuela, 1974, p. 54.
29. **Beaman, T. C., Pankratz, H. S., and Gerhardt, P.**, Chemical composition and ultrastructure of native and reaggregated membranes from protoplasts of *Bacillus cereus, J. Bacteriol.*, 117, 1335, 1974.

29a. **Becker, B., Lechevalier, M. P., and Lechevalier, H. A.**, Chemical composition of cell-wall preparations from strains of various form-genera of aerobic actinomycetes, *Appl. Microbiol.*, 13, 236, 1965.

30. **Beebe, J. L.**, Isolation and characterization of a phosphatidyl-ethanolamine-deficient mutant of *Bacillus subtilis, J. Bacteriol.*, 107, 704, 1971.
31. **Beebe, J. L. and Umbriet, W. W.**, Extracellular lipid of *Thiobacillus thiooxidans, J. Bacteriol.*, 108, 612, 1971.
32. **Beining, P. R., Huff, E., Prescott, B., and Theodore, T. S.**, Characterization of the lipids of mesosomal vesicles and plasma membranes from *Staphylococcus aureus, J. Bacteriol.*, 121, 137, 1975.
33. **Bekhtereva, M. N., Gerasimova, N. M., and Donets, A. T.**, Lipids of saprophytic mycobacteria grown on media with fatty acids, *Microbiology USSR*, 40, 712, 1971.
34. **Bishop, E. A. and Bermingham, M. A. C.**, Lipid composition of Gram-negative bacteria, sensitive and resistant to streptomycin, *Antimicrob. Agents Chemother.*, 4, 378, 1973.
35. **Blaschy, H. and Zimmermann, W.**, Gaschromatographische Untersuchungen von Fettsäuren verschiedener Bakterienarten, *Zentralbl. Bakteriol. Parasitenkd. Infektionskr. Hyg. Abt. I. Orig. Reihe A*, 218, 468, 1971.
36. **Bligh, E. G. and Dyer, W. J.**, A rapid method of total lipid extraction and purification, *Can. J. Biochem. Physiol.*, 37, 911, 1959.
37. **Bloch, H.**, Studies on the virulence of tubercle bacilli. Isolation and biological properties of a constituent of virulent organisms, *J. Exp. Med.*, 91, 197, 1950.
38. **Blumer, M., Chase, T., and Watson, S. W.**, Fatty acids in the lipids of marine and terrestrial nitrifying bacteria, *J. Bacteriol.*, 99, 366, 1969.
39. **Bobo, R. A. and Eagon, R. G.**, Lipids of cell walls of *Pseudomonas aeruginosa* and *Brucella abortus, Can. J. Microbiol.*, 14, 503, 1968.
40. **Bordet, C., Vacheron, M. J., and Michel, G.**, Evolution de la composition lipidique de *Nocardia asteroides* au cours de la croissance, *FEBS Lett.*, 5, 253, 1969.
41. **Bowie, I. S., Grigor, M. R., Dunckley, G. G., Loutit, M. W., and Loutit, J. S.**, The DNA base composition and fatty acid constitution of some Gram-negative pleomorphic soil bacteria, *Soil Biol. Biochem.*, 4, 397, 1972.
42. **Bradley, T. J. and Khan, N. H.**, Extracellular lipid of *Pseudomonas aeruginosa* NCTC 2000 obtained by *n*-butanol extraction of stationary liquid cultures, *Microbios*, 9, 15, 1974.
43. **Brennan, P. and Ballou, C. E.**, Phosphatidylmyoinositol monomannoside in *Propionibacterium shermanii, Biochem. Biophys. Res. Commun.*, 30, 69, 1968.
44. **Brennan, P. J. and Lehane, D. P.**, The phospholipids of corynebacteria, *Lipids*, 6, 401, 1971.

45. **Brennan, P. J., Lehane, D. P., and Thomas, D. W.**, Acylglucoses of the corynebacteria and mycobacteria, *Eur. J. Biochem.*, 13, 117, 1970.
46a. **Brian, B. L. and Gardner, E. W.**, A simple procedure for detecting the presence of cyclopropane fatty acids in bacterial lipids, *Appl. Microbiol.*, 16, 549, 1968.
46b. **Brooks, J. L. and Benson, A. A.**, Studies on the structure of an ornithine-containing lipid from *Rhodospirillum rubrum, Arch. Biochem. Biophys.*, 152, 347, 1972.
47a. **Brooks, J. B., Kellogg, D. S., Thacker, L., and Turner, E. M.**, Analysis by gas chromatography of hydroxy acids produced by several species of *Neisseria, Can. J. Microbiol.*, 18, 157, 1972.
47b. **Buchanan, R. E. and Gibbons, N. E. (Eds.)**, Bergey's Manual of Determinative Bacteriology, 8th ed., Williams & Wilkins, Baltimore, 1974.
47c. **Bulla, L. A., Jr., Bennett, G. A., and Shotwell, O. L.**, Physiology of spore-forming bacteria associated with insects, II, *J. Bacteriol.*, 104, 1246, 1970.
47d. **Bureau, G.**, Effet du ralentissement de la croissance sur le métabolisme des acides gras de *Bacillus subtilis* var. *niger, C. R. Acad. Sci.*, 274, 468, 1972.
47e. **Bureau, G. and Mazliak, P.**, Métabolisme des acides gras au cours de la sporulation de *Bacillus subtilis* var. *niger, Ann. Inst. Pasteur Paris*, 120, 144, 1971.
47f. **Bulla, L. A. and St. Julian, G.**, Lipid metabolism during bacterial growth and sporulation: Phospholipid pattern in *Bacillus thuringiensis* and *Bacillus popilliae*, in *Spores V*, Halvorson, H. O., Hanson, R., and Campbell, L. L., Eds., American Society for Microbiology, Washington, D.C., 1972, p. 191.
47g. **Bunn, C. R. and Elkan, G. H.**, The phospholipid composition of *Rhizobium japonicum, Can. J. Microbiol.*, 17, 291, 1971.
47h. **Brundish, D. E., Shaw, N., and Baddiley, J.**, The glycolipids from the non-capsulated strain of *Pneumococcus* I-192R, A.T.C.C. 12213, *Biochem. J.*, 97, 158, 1965.
48. **Button, G. L., Miller, M. A., and Tsang, J. C.**, Antibiogram and lipid analysis of a pigmented strain of *Serratia marcescens* and its non-pigmented variants, *Antimicrob. Agents Chemother.*, 7, 219, 1975.
49. **Card, G. L.**, Metabolisms of phosphatidyl glycerol, phosphatidylethanolamine and cardiolipin of *Bacillus stearothermophilus, J. Bacteriol.*, 114, 1125, 1973.
50. **Cerniglia, C. E. and Perry, J. J.**, Metabolism of *n*-propylamine, isopropylamine, and 1,3-propanediamine by *Mycobacterium convolutum, J. Bacteriol.*, 124, 285, 1975.
51. **Chamoiseau, G.**, *Mycobacterium farcinogenes* agent causal du farcin du boeuf en Afrique, *Ann. Microbiol. Paris*, 124A, 215, 1973.
52. **Chamoiseau, G. and Asselineau, J.**, Examen des lipides d'une souche de *Nocardia farcinica* : Présence d'acides mycoliques, *C. R. Acad. Sci., Ser. D*, 270, 2603, 1970.
53. **Chan, M., Himes, R. H., and Akagi, J. M.**, Fatty acid composition of thermophilic, mesophilic, and psychrophilic clostridia, *J. Bacteriol.*, 106, 876, 1971.
54. **Chandramouli, V. and Venkitasubramanian, T. A.**, Effect of age on fatty acids C_{14} to C_{19} of mycobacteria, *Am. Rev. Respir. Dis.*, 108, 387, 1973.
55. **Chandramouli, V. and Venkitasubramanian, T. A.**, Effect of age on the lipids of mycobacteria, *Indian J. Chest Dis.*, Suppl. 16, 199, 1974.
56. **Chang, N. C. and Fulco, A. J.**, The effects of temperature and fatty acid structure on lipid metabolism in *Bacillus licheniformis* 9259, *Biochim. Biophys. Acta*, 296, 287, 1973.
57. **Chattopadhyay, P. K. and Dutta, J.**, Effects of bacteriop hage M13 infection upon phospholipid and fatty acid compositions of *Escherich ia coli, Lipids*, 10, 497, 1975.
58. **Chow, T. C. and Schmidt, J. M.**, Fatty acid composition of *Caulobacter crescentus, J. Gen. Microbiol.*, 83, 369, 1974.
59. **Cronan, J. E., Jr. and Vagelos, P. R.**, Metabolism and function of the membrane phospholipid of *Escherichia coli, Biochim. Biophys. Acta*, 265, 25, 1972.
60. **Cross, T. and Goodfellow, M.**, Taxonomy and classification of the actinomycetes, in *Actinomycetales: Characteristics and Practical Importance*, Sykes, G. and Skinner, F. A., Eds., Academic Press, New York, 1973, p. 11.
61. **Cullen, J., Phillips, M. C., and Shipley, G. G.**, The effects of temperature on the composition and physical properties of the lipids of *Pseudomonas fluorescens, Biochem. J.*, 125, 733, 1971.
62. **Cummins, C. S.**, Bacterial cell wall structure, in *Handbook of Microbiology*, Vol. 2, Laskin, A. I. and Lechevalier, H. A., Eds., CRC Press, Cleveland, 1973, p. 167.
63. **Dawson, R. M. C.**, Analysis of phosphatides and glycolipids by chromatography of their partial hydrolysis or alcoholysis products, in *Lipid Chromatographic Analysis*, Marinetti, G. V., Ed., Marcel Dekker, New York, 1967, p. 163.
64. **Dedyukhina, E. G., Andreev, L. V., Popkov, G. P., and Eroshin, V. K.**, Biosynthesis of hydrocarbons by alkane-oxidizing microorganisms, *Microbiology USSR*, 41, 584, 1972.

65. **Dees, S. B. and Moss, C. W.,** Cellular fatty acids of *Alcaligenes* and *Pseudomonas* sp. isolated from clinical specimens, *J. Clin. Microbiol.,* 1, 414, 1975.
66. **deRosa, M., Gambacorta, A., and Bu'lock, J. D.,** Extremely thermophilic acidophilic bacteria convergent with *Sulfolobus acidocaldarius, J. Gen. Microbiol.,* 86, 156, 1975.
67. **deRosa, M., Gambacorta, A., and Minale, L.,** Cyclohexane fatty acids from a thermophilic bacterium, *J. Chem. Soc. D,* p. 1334, 1971.
68. **deRosa, M., Gambacorta, A., and Minale, L.,** Cyclic diether lipids from very thermophilic acidophilic bacteria, *Chem. Commun.,* p. 543, 1974.
69. **DeSiervo, A. J. and Reynolds, J. W.,** Phospholipid composition and cardiolipin synthesis in fermentative and non-fermentative marine bacteria, *J. Bacteriol.,* 123, 294, 1975.
70. **DeSiervo, A. J. and Salton, M. R. J.,** Biosynthesis of cardiolipin in the membranes of *Micrococcus lysodeikticus, Biochim. Biophys. Acta,* 239, 280, 1971.
71. **DeSiervo, A. J. and Salton, M. R. J.,** Changes in phospholipid composition of *Micrococcus lysodeikticus* during growth, *Microbios,* 8, 73, 1973.
72. **Dhariwal, K. R., Chander, A., and Venkitasubramanian, T. A.,** Phospholipids of *Mycobacterium phlei, Experientia,* 31, 776, 1975.
73. **Diedrich, D. L. and Cota-Robles, E. H.,** Heterogeneity in lipid composition of the outer membrane and cytoplasmic membrane of *Pseudomonas* BAL-31, *J. Bacteriol.,* 119, 1006, 1974.
74. **Drucker, D. B. and Owen, J.,** Chemotaxonomic fatty acid fingerprints of bacteria grown with and without aeration, *Can. J. Microbiol.,* 19, 247, 1973.
75. **Drucker, D. B., Griffith, C. J., and Melville, T. H.,** Fatty acid fingerprints of *Streptococcus mutans* grown in a chemostat, *Microbios,* 7, 17, 1973.
76. **Drucker, D. B., Griffith, C. J., and Melville, T. H.,** Fatty acid fingerprints of streptococci: Variability due to carbohydrate source, *Microbios,* 9, 187, 1974.
77. **Dunnick, J. K. and O'Leary, W. M.,** Correlation of bacterial lipid composition with antibiotic resistance, *J. Bacteriol.,* 101, 892, 1970.
78. **Eberhard, A. and Rouser, G.,** Quantitative analysis of the phospholipids of some marine bioluminescent bacteria, *Lipids,* 6, 410, 1971.
79. **Edwards, J. R. and Hayashi, J. A.,** Structure of a rhamnol ipid from *Pseudomonas aeruginosa, Arch. Biochem. Biophys.,* 111, 415, 1965.
80. **Efimova, T. P. and Tsyganov, V. A.,** Fatty acids contained in the lipopolysaccharides of actinomycetes, *Microbiology USSR,* 38, 571, 1969.
81. **Elin, R. J. and Wolff, S. M.,** Bacterial endotoxins, in *Handbook of Microbiology,* Vol. 2, Laskin, A. I. and Lechevalier, H. A., Eds ., CRC Press, Cleveland, 1973, p. 215.
82. **Ellwood, D. C. and Tempest, D. W.,** Effects of environment on bacterial wall content and composition, in *Advances in Microbial Physiology,* Vol. 7, Rose, A. H. and Tempest, D. W., Eds., Academic Press, New York, 1972, p. 83.
83. **Etemadi, A. H. and Convit, J.,** Mycolic acids from "noncultivable" mycobacteria, *Infect. Immun.,* 10, 236, 1974.
84. **Exterkate, E. and Veerkamp, J. H.,** Biochemical changes in *Bifidobacterium bifidum* var. *pennsylvanicus* after cell wall inhibition, *Biochim. Biophys. Acta,* 231, 545, 1971.
85. **Exterkate, F. A., Otten, B. J., Wassenberg, H. W., and Veerkamp, J. H.,** Comparison of the phospholipid composition of *Bifidobacterium* and *Lactobacillus* strains, *J. Bacteriol.,* 106, 824, 1971.
86. **Faizova, G. K., Borodulina, Yu. S., and Samsonova, S. P.,** Lipids in nodular bacteria *(Rhizobium leguminosarium), Microbiology USSR,* 40, 411, 1971 (English transl.).
87. **Field, C. R., Keeler, R., and Carter, P. B.,** Cellular fatty acids of group D streptococci, *Abstr. Annu. Meet. Am. Soc. Microbiol.,* p. 51, 1972.
88. **Fischer, W. and Landgraf, H. R.,** Glycerophosphoryl phosphatidyl kojibiosyl diacylglycerol, a novel phosphoglucolipid from *Streptococcus faecalis, Biochim. Biophys. Acta,* 381, 227, 1975.
89. **Fischer, W., Ishizuka, I., Landgraf, H. R., and Herrmann, J.,** Glycerophosphoryl diglucosyl diglyceride, a new phosphoglycolipid from *streptococci, Biochim. Biophys. Acta,* 296, 527, 1973.
90. **Fischer, W., Landgraf, H. R., and Herrmann, J.,** Phosphatidyldiglucosyl diglyceride from *streptococci* and its relationship to other polar lipids, *Biochim. Biophys. Acta,* 306, 353, 1973.
91. **Folch, J., Lees, M., and Sloane-Stanley, G. H.,** A simple method for the isolation and purification of total lipides from animal tissues, *J. Biol. Chem.,* 226, 497, 1957.
92. **Fritsche, D. and Thelen, A.,** Die Abgrenzung der Genera *Bacteroides* und *Sphaerophorus* auf Grund der Struktur ihrer komplexen Lipoide, *Zentralbl. Bakteriol. Parasitend. Infektionskr. Hyg. Abt. I Orig. Reihe A,* 223, 356, 1973.
93. **Frøholm, L. O., Jantzen, E., Hytta, R., and Bøyre, K.,** Gas chromatography of bacterial whole-cell methanolysates, *Acta Pathol. Microbiol. Scand.* 80, 672. 1972.
94. **Gallagher, I. H. C.,** Occurence of waxes in *Acinetobacter, J. Gen. Microbiol.* 68, 245, 1971.

94a. **Ganfield, M.-C. and Pieringer, R.A.**, Phosphatidyl kojibiosyl diglyceride. The covalently linked lipid constituent of the membrane lipoteichoic acid from *Streptococcus faecalis (faecium)* ATCC 9790, *J. Biol. Chem.*, 250, 702, 1975.

95. **Gastambide-Odier, M** Variantes de mycosides caractérisées par des résidus glycosidiques substitués par des chaines acyles, *Eur. J. Biochem.* 33, 81, 1973.

96. **Gastambide-Odier, M. and Sarda, P.**, Contributions à l'étude de la structure et la biosynthèse de glycolipides spécifiques isolés de mycobactéries: Les mycosides A et B, *Pneumologie,* 142, 241, 1970.

97. **Gerson, T., Patel, J. J., and Nixon, L. N.**, Some unusual fatty acids of *Rhizobium, Lipids,* 10, 134, 1975.

98. **Gerson, T. and Patel, J. J.**, Neutral lipids and phospholipids of free-living and bacteroid forms of two strains of *Rhizobium* infective on *Lotus pedunculatus, Appl. Microbiol.*, 30, 193, 1975.

99. **Ghanekar, A. S. and Nair, P. M.**, Lipids of *Alcaligenes faecalis, Indian J. Biochem. Biophys.*, 11, 233, 1974.

100. **Gill, C. O.**, Effect of growth temperature on the lipids of *Pseudomonas fluorescens, J. Gen. Microbiol.*, 89, 293, 1975.

101. **Girard, A. E.**, A comparative study of the fatty acids of some micrococci, *Can. J. Microbiol.*, 17, 1503, 1971.

102. **Girard, A. E. and Cosenza, B. J.**, Fine structure and fatty acid composition of a motile streptococcus, *Antonie van Leeuwenhoek J. Microbiol. Serol.*, 37, 65, 1971.

103. **Goldfine, H.**, Comparative aspects of bacterial lipids, in *Advances in Microbial Physiology,* Rose, A. H. and Tempest, D. W., Eds., Academic Press, New York, 1972, p. 1.

104. **Goldfine, H. and Hagen, P.-O.**, Bacterial plasmalogens, in *Ether Lipids,* Snyder F., Ed., Academic Press, New York, 1972 p. 329.

105. **Goldfine, H. and Panos, C.**, Phospholipids of *Clostridium butyricum* IV, *J. Lipid Res.*, 12, 214, 1971.

106. **Goldman, D. S.**, Subcellular localization of individual mannose-containing phospholipids in *Mycobacterium tuberculosis, Am. Rev. Respir. Dis.*, 102, 543, 1970.

107. **Goodfellow, M., Lind, A., Mordarska, H., Pattyn, S., and Tsukamura, M.**, A cooperative numerical analysis of cultures considered to belong to the "rhodochrous" taxon, *J. Gen. Microbiol.*, 85, 291, 1974.

108. **Goodfellow, M., Minnikin, D. E., Patel, P. V., and Mordarska, H.**, Free nocardomycolic acids in the classification of nocardias and strains of the *"rhodochrous"* complex, *J. Gen. Microbiol.*, 74, 185, 1973.

109. **Gorchein, A.**, Structure of the ornithine-containing lipid from *Rhodopseudomonas spheroides, Biochim. Biophys. Acta,* 306, 137, 1973.

110. **Goren, M. B.**, Sulfolipid I of *Mycobacterium tuberculosis,* Strain H37Rv. II. Structural studies, *Biochim. Biophys. Acta,* 210, 127, 1970.

111. **Goren, M. B.**, Mycobacterial lipids: Selected topics, *Bacteriol. Rev.*, 36, 33, 1972.

112. **Goren, M. B., Brokl, O., Das, B. C., and Lederer, E.**, Sulfolipid I of *Mycobacterium tuberculosis,* Strain H37Rv. Nature of the acyl substituents, *Biochemistry,* 10, 72, 1971.

113. **Grange, J. M. and Stanford, J. L.**, Reevaluation of *Mycobacterium fortuitum* (synonym: *Mycobacterium ranae*), *Int. J. Syst. Bacte riol.*, 24, 320, 1974.

114. **Guenther, T., Richter, L., and Schmalbeck, J.**, Phospholi pids of *Escherichia coli* in magnesium deficiency, *J. Gen. Microbiol.*, 86, 191, 1975.

115. **Guinand, M. and Michel, G.**, Structure d'un peptidolipid e isolé de *Nocardia asteroides,* la peptidolipine NA, *Biochim. Biophys. Acta,* 125, 75, 1966.

116. **Guzeva, L. N., Efimova, T. P., Agre, N. S., and Krasil'niko v, N. A.**, Fatty acids in the mycelia of actinomycetes that form catenate spo res, *Microbiology USSR,* 42, 19, 1973.

117. **Hackett, J. A. and Brennan, P. J.**, The mannophosphoinositides of *Corynebacterium aquaticum, Biochem. J.*, 148, 253, 1975.

118. **Haest, C. W. M., deGier, J., Op den Kamp, J. A. F., Bartels, P., and Van Deenen, L. L. M.**, Changes in permeability of *Staphylococcus aureus* and derived liposomes with varying lipid composition, *Biochim. Biophys. Acta,* 255, 720, 1972.

119. **Hagen, P.-O.**, Lipids of *Sphaerophorus ridiculosis*: Plasmalogen composition, *J. Bacteriol.*, 119, 643, 1974.

120. **Hanahan, D. J.**, Ether-linked lipids: Chemistry and methods of measurement, in *Ether Lipids, Chemistry and Biology,* Snyder, F., Ed., Academic Press, 1972, p. 25.

121. **Hardy, R., Hobbs, G., Mackie, P. R., and Horsley, R. W.**, Characterization of closely related *Achromobacter* spp. by means of chromatographic techniques, *J. Gen. Microbiol.*, 68, iii, 1971.

122. **Hildebrand, J. G. and Law, J. H.**, Fatty acid distribution in bacterial phospholipids. The specificity of the cyclopropane synthetase reaction, *Biochemistry,* 3, 1304, 1964.

123. **Hobbs, G., Hardy, R., and Mackie, P.,** Characterization of *Clostridium* species by means of their lipids, *J. Gen. Microbiol.*, 68, ii, 1971.

124a. **Hofheinz, W. and Grisebach, H.,** The biogenesis of macrolides and fatty acids of *Streptomyces erythraeus* and *Streptomyces halstedii, Z. Naturforsch. Teil B*, 20, 43, 1965.

124b. **Horan, A. C.,** unpublished results.

125. **Hunter, M. I. S. and Thirkell, D.,** Variation in fatty acid composition of *Sarcina flava* membrane lipid with the age of the bacterial culture, *J. Gen. Microbiol.*, 65, 115, 1971.

126. **Ikawa, M.,** Bacterial phosphatides and natural relationships, *Bacteriol. Rev.*, 31, 54, 1967.

127. **Ilton, M., Jevans, A. W., McCarthy, E. D., Vance, D., White, H. B., III, and Bloch, K.,** Fatty acid synthetase activity in *Mycobacterium phlei*: regulation by polysaccharides, *Proc. Natl. Acad. Sci. U.S.A.*, 68, 87, 1971.

128. **Ioneda, T., Lederer, E., and Rozanis, J.,** Sur la structure des diesters de tréhalose ("cord factors") produits par *Nocardia asteroides* et *Nocardia rhodochrous, Chem. Phys. Lipids*, 4, 375, 1970.

129. **Itoh, S. and Suzuki, T.,** Fructose-lipids of *Arthrobacter, Corynebacteria, Nocardia* and *Mycobacteria* grown on fructose, *Agric. Biol. Chem.*, 38, 1443, 1974.

130. **Jackson, T. J., Ramaley, R. F., and Meinschein, W. G.,** Fatty acids of a non-pigmented, thermophilic bacterium similar to *Thermus aquaticus, Arch. Mikrobiol.*, 88, 127, 1973.

131. **Jantzen, E., Bergan, T., and Bovre, K.,** Gas chromatography of bacterial whole cell methanolysates. VI. Fatty acid composition of strains within *Micrococcaceae, Acta Pathol. Microbiol. Scand. Sect. B*, 82, 785, 1974.

132. **Jantzen, E., Bryn, K., Bergan, T., and Bøvre, K.,** Gas chromatography of bacterial whole-cell methanolysates. V., *Acta Pathol. Microbiol. Scand. Sect. B*, 82, 767, 1974.

133. **Javora, J. and Bacílek, J.,** Kvantitativní analýza mykobakteriálních mastných kyselin C_{10}—C_{26} plynovou chromatografií, *Stud. Pneumol. Phtiseol. Cech.*, 32, 241, 1972.

134. **Jenkins, P. A., Marks, J, and Schaefer, W. B.,** Lipid chromatography and seroagglutination in the classification of rapidly growing mycobacteria, *Am. Rev. Respir. Dis.*, 103, 179, 1971.

135. **Johnson, A. R. and Davenport, J. B.,** *Biochemistry and Methodology of Lipids*, Interscience, New York, 1971.

136. **Joo, C. N. and Kates, M.,** Synthesis of the naturally occurring phytanyl diether analogs of phosphatidyl glycerophosphate and phosphatidyl glycerol, *Biochim. Biophys. Acta*, 176, 278, 1969.

137. **Joseph, R.,** Fatty acid composition of *Spirochaeta stenostrepta, J. Bacteriol.*, 112, 629, 1972.

138. **Joyce, G. H., Hammond, R. K., and White, D. C.,** Changes in membrane lipid composition in exponentially growing *Staphylcoccus aureus* during the shift from 37 to 25°C, *J. Bacteriol.*, 104, 323, 1970.

139. **Julák, J. and Mara, M.,** Effect of glucose or glycerin in cultivation media on the fatty acid composition of *Listeria monocytogenes, J. Hyg. Epidemiol. Microbiol. Immunol.*, 17, 329, 1973.

140. **Kaltenbach, C. M., Moss, C. W., and Weaver, R. E.,** Cultural and biochemical characteristics and fatty acid composition of *Pseudomonas diminuta* and *Pseudomonas vesiculare, J. Clin. Microbiol.*, 1, 339, 1975.

141. **Kamio, Y., Kim, K. C., and Takahashi, H.,** Identification of the basic structure of a glycolipid from *Selenomonas ruminantium* as β-glucosaminyl-1,6-glucosamine, *Agric. Biol. Chem.*, 36, 2195, 1972.

142. **Kanai, K. and Kondo, E.,** Chemistry and biology of mycobacteria grown *in vivo, Jpn. J. Med. Sci. Biol.*, 27, 135, 1974.

143. **Kaneda, T.,** Fatty acids of *Bacillus larvae, Bacillus lentimorbus* and *Bacillus popilliae, J. Bacteriol.*, 98, 143, 1969.

144. **Kaneda, T.,** Major occurrence of *cis*-Δ^5 fatty acids in three psychrophilic species of *Bacillus, Biochem. Biophys. Res. Commun.*, 43, 298, 1971.

145. **Kaneda, T.,** Factors affecting the relative ratio of fatty acids in *Bacillus cereus, Can. J. Microbiol.*, 17, 269, 1971.

146. **Kaneda, T.,** Positional distribution of fatty acids in phospholipids from *Bacillus subtilis, Biochim. Biophys. Acta*, 270, 32, 1972.

147. **Kaneda, T.,** Positional preference of fatty acids in phospholipids of *Bacillus cereus* and its relation to growth temperature, *Biochim. Biophys. Acta*, 280, 297, 1972.

148. **Kanemasa, Y., Yoshioka, T., and Hayashi, H.,** Alteration of the phospholipid composition of *Staphylococcus aureus* cultured in medium containing NaCl, *Biochim. Biophys. Acta*, 280, 444, 1972.

149. **Kanetsuna, F. and Bartoli, A.,** A simple chemical method to differentiate *Mycobacterium* from *Nocardia, J. Gen. Microbiol.*, 70, 209, 1972.

150. **Kanetsuna, F. and San Blas, G.** Chemical analysis of a mycolic acid-arabinogalactan-mucopeptide complex of mycobacterial cell wall, *Biochim. Biophys. Acta*, 208, 434, 1970.

151. **Kataoka, T. and Nojima, S.,** The phospholipid composition of some actinomycetes, *Biochim. Biophys. Acta*, 114, 681, 1967.

152. **Kates, M.**, Bacterial lipids, *Adv. Lipid Res.*, 2, 17, 1964.
153. **Kates, M.**, *Techniques of Lipidology. Isolation, Analysis and Identification of Lipids*, Elsevier, New York, 1972.
154. **Kates, M.**, Ether-linked lipids in extremely halophilic bacteria, in *Ether Lipids, Chemistry and Biology*, Snyder, F., Ed., Acdemic Press, New York, 1972, p. 351.
155. **Kates, M. and Deroo, P. W.**, Structure determination of the glycolipid sulfate from the extreme halophile *Halobacterium cutirubrum*, *J. Lipid Res.*, 14, 438, 1973.
156. **Kates, M., Palameta, B., Joo, C. N., Kushner, D. J., and Gibbons, N. E.**, Aliphatic diether analogs of glyceride-derived lipids. IV. The occurrence of di-O-dihydrophytyl glycerol ether-containing lipids in extremely halophilic bacteria, *Biochemistry*, 5, 4092, 1966.
157. **Kates, M., Palameta, B., Perry, M. P., and Adams, G. A.**, A new glycolipid sulfate ester in *Halobacterium cutirubrum*, *Biochim. Biophys. Acta*, 137, 213, 1967.
158. **Kates, M., Yengoyan, L. S., and Sastry, P. S.**, A diether analog of phosphatidyl glycerophosphate in *Halobacterium cutirubrum*, *Biochim. Biophys. Acta*, 98, 252, 1965.
159. **Kawanami, J.**, Lipids of *Streptomyces toyocaensis*. On the structure of siolipin, *Chem. Phys. Lipids*, 7, 159, 1971.
160. **Kawanami, J., Kimura, A., Nakagawa, Y., and Otsuka, H.**, Lipids of *Streptomyces sioyaensis*. V. On the 2-hydroxy-13-methyltetradecanoic acid from phosphatidylethanolamine, *Chem. Phys. Lipids*, 3, 29, 1969.
161. **Kenyon, C. N.**, Fatty acid composition of unicellular strains of blue-green algae, *J. Bacteriol.*, 109, 827, 1972.
162. **Kenyon, C. N. and Gray, A. M.**, Preliminary analysis of lipids and fatty acids of green bacteria and *Chloroflexus aurantiacus*, *J. Bacteriol.*, 120, 131, 1974.
163. **Kenyon, C. N., Rippka, R., and Stanier, R. Y.**, Fatty acid composition of physiological properties of some filamentous blue-green algae, *Arch. Mikrobiol.*, 83, 216, 1972.
164. **Khuller, G. K. and Brennan, P. J.**, The polar lipids of some species of *Nocardia*, *J. Gen. Microbiol.*, 73, 409, 1972.
165. **Khuller, G. K. and Brennan, P. J.**, Further studies on the lipids of corynebacteria. The mannolipids of *Corynebacterium aquaticum*, *Biochem. J.*, 127, 369, 1972.
166. **Khuller, G. K. and Goldfine, H.**, Phospholipids of *Clostridium butyricum*. V. Effects of growth temperature on fatty acid, alk-1-enyl ether group, and phospholipid composition, *J. Lipid Res.*, 15, 500, 1974.
167. **Khuller, G. K., Banerjee, B., Sharma, B. V. S., and Subrahmanyan, D.**, Effect of age on the composition of major phospholipids of mycobacteria, *Indian J. Biochem. Biophys.*, 9, 274, 1972.
168. **Kikuchi, M. and Nakao, Y.**, Relation between cellular phospholipids and the excretion of L -glutamic acid by a glycerol auxotroph of *Corynebacterium alkanolyticum*, *Agric. Biol. Chem.*, 37, 515, 1973.
169. **Kikuchi, M., Karamaru, T., and Nakao, Y.**, Relation between the extracellular accumulation of L - glutamic acid and the excretion of phospholipids by penicillin-treated *Corynebacterium alkanolyticum*, *Agric. Biol. Chem.*, 37, 2405, 1973.
170. **Kimura, A. and Otsuka, H.**, The changes of lysine- and ornithine-lipids in *Streptomyces sioyaensis*, *Agric. Biol. Chem.*, 33, 781, 1969.
171. **King, D. H. and Perry, J. J.**, The origin of fatty acids in the hydrocarbon-utilizing microorganism *Mycobacterium vaccae*, *Can. J. Microbiol.*, 21, 85, 1975.
172. **Kito, M., Aibara, S., Kato, M., Ishinaga, M., and Hata, T.**, Effect of changes in fatty acid composition of phospholipid species on the β-galactoside transport system of *Escherichia coli* K-12, *Biochim. Biophys. Acta*, 298, 69, 1973.
173. **Knipprath, W. G., Cohen, L. W., and Allen, C. F.**, Lipid changes in *Salmonella typhimurium* on infection with bacteriophage P_{22}, *Biochim. Biophys. Acta*, 231, 107, 1971.
174. **Knoche, H. W. and Shively, J. M.**, The structure of an ornithine-containing lipid from *Thiobacillus thiooxidans*, *J. Biol. Chem.*, 247, 170, 1972.
175. **Kojima, R.**, Fatty acid compositions of bacteria and their significance in bacterial studies, *Nippon Saikingaku Zasshi*, 26, 311, 1971.
176. **Komaratat, P. and Kates, M.**, The lipid composition of a halotolerant species of *Staphylococcus epidermidis*, *Biochim. Biophys. Acta*, 398, 464, 1975.
177. **Komura, I., Yamada, K., and Komagata, K.**, Taxonomic significance of phospholipid composition in aerobic gram-positive cocci, *J. Gen. Appl. Microbiol.*, 21, 97, 1975.
178. **Komura, I., Yamada, K., Otsuka, S.-I., and Komagata, K.**, Taxonomic significance of phospholipids in coryneform and nocardioform bacteria, *J. Gen. Appl. Micribiol.*, 21, 251, 1975.
179. **Kondo, E. and Ueta, N.**, Composition of fatty acids and carbohydrates in *Leptospira*, *J. Bacteriol.*, 110, 459, 1972.

180. **Konova, I. V., Batrakov, S. G., Panosyan, A. G., Gerasimova, N. M., and Kislova, L. M.,** Lipid composition of *Actinomyces olivaceus,* a vitamin B_{12} synthesizer, *Prikl. Biokhim. Mikrobiol.,* 11, 306, 1975; *Chem. Abstr.,* 83, 41422w, 1975.
181. **Kosaric, N. and Carroll, K. K.,** Phospholipids of *Listeria monocytogenes, Biochim. Biophys. Acta,* 239, 428, 1971.
182. **Kostiw, L. L., Boylen, C. W., and Tyson, B. J.,** Lipid composition of growing and starving cells of *Arthrobacter crystallopoietes, J. Bacteriol.,* 111, 103, 1972.
183. **Koval'chuk, L. P., Donets, A. T., and Razumovski, P. N.,** Phospholipids of *Actinomyces canosus, Izv. Akad. Nauk Mold. SSR Ser. Biol. Khim Nauk,* 6, 49, 1973; *Biol. Abstr.,* 58, 45335, 1974.
184. **Krasil'nikov, N. A., Koronelli, T. V., and Rozynov, B. V.,** Mycolic acids of *Mycobacterium paraffinicum, Dokl. Akad. Nauk SSSR,* 201, 1248, 1971.
185. **Krasil'nikov, N. A., Koronelli, T. V., and Rozynov, B. V.,** Aliphatic and mycolic acids of *Mycobacterium paraffinicum, Microbiology USSR,* 41, 715, 1972.
186. **Krasil'nikov, N. A., Koronelli, T. V., Rozynov, B. V., and Kalyuzhnaya, T. V.,** Mycolic acids of pigmented paraffin-oxidizing mycobacteria, *Microbiology USSR,* 42, 213, 1973.
187. **Kubica, G. P., Baess, I., Gordon, R. E., Jenkins, P. A., Kwapinski, J. B. G., McDurmont, C., Pattyn, S. R., Saito, H., Silcox, V., Stanford, J. L., Takeya, K., and Tsukamura, M.,** A co-operative numerical analysis of rapidly growing mycobacteria, *J. Gen. Microbiol.,* 73, 55, 1972.
189. **Kunsman, J. E.,** Chracterization of the lipids of six strains of *Bacteroides ruminicola, J. Bacteriol.,* 113, 1121, 1973.
190. **Kunsman, J. E. and Caldwell, D. R.,** Comparison of the sphingolipid content of rumen *Bacteroides* species, *Appl. Microbiol.,* 28, 1088, 1974.
191. **Kurylowicz, W., Malinowski, K., and Kurzatkowski, W.,** Fatty acids of the mycelium of *Streptomyces aureofaciens* during tetracycline biosynthesis, *Acta Microbiol. Pol. Ser. B,* 3, 179, 1971.
192. **Kushwaha, S. C., Kates, M., and Martin, W. G.,** Characterization and composition of the purple and red membrane from *Halobacterium cutirubrum, Can. J. Biochem.,* 53, 284, 1975.
193. **Lacave, C., Asselineau, J., and Toubiana, R.,** Sur quelques constituents lipidiques de *Corynebacterium ovis, Eur. J. Biochem.,* 2, 37, 1967.
194. **Lambert, M. A. and Moss, C. W.,** Cellular fatty acid composition of *Streptococcus mutans* and related streptococci, *J. Dent. Res.,* 55, A96, 1976.
195. **Lambert, M. A., Hollis, D. G., Moss, C. W., Weaver, R. E., and Thomas, M. L.,** Cellular fatty acids of nonpathogenic *Neisseria, Can. J. Microbiol.,* 17, 1491, 1971.
196. **Lanéelle, G., Asselineau, J., Wolstenholme, W. A., and Lederer, E.,** Determination de séquences d'acides aminés dans des oligopeptides par la spectrométrie de masse. III. Structure d'un peptidolipide de *Mycobacterium johnei, Bull. Soc. Chim. Fr.,* p. 2133, 1965.
197. **Lanéelle, G. and Asselineau, J.,** Structure d'un glycoside de peptidolipide isolé d'une mycobactérie, *Eur. J. Biochem.,* 5, 487, 1968.
198. **Lanéelle, G., Asselineau, J., and Chamoiseau, G.,** Presence de mycosides C' (formes simplifiees de mycoside C) dans les bacteries isolees de bovins atteints du farcin, *FEBS Lett.,* 19, 109, 1971.
199. **Lanéelle, M.-A. and Asselineau, J.,** Caracterisation de glycolipides dans une souche de *Nocardia brasiliensis, FEBS Lett.,* 7, 64, 1970.
200. **Lanéelle, M. -A., Asselineau, J., and Castelnuovo, G.,** Relations chimiques et immunologiques chez les actinomycetales. IV. Composition chimique des lipides de quatres souches de *Streptomyces* et d'une souche de *N. (Str.) gardneri, Ann. Inst. Pasteur Paris,* 114, 305, 1968.
201. **Lanéelle, M.-A., Asselineau, J., and Castelnuovo, G.,** Etudes sur les mycobacteries et les nocardiae. IV. Composition des lipides de *Mycobacterium rhodochrous, M. pellegrino* sp. et de quelques souches de nocardiae, *Ann. Inst. Pasteur,* 108, 69, 1969.
202. **Lanéelle, M.-A., Lanéelle, G., Bennet, P., and Asselineau, J.,** Chimie des microorganismes. V. Les lipides d'une mycobactérie non-photochromogène, *Bull. Soc. Chim. Biol.,* 47, 2047, 1965.
203. **Langworthy, T. A., Mayberry, W. R., and Smith, P. F.,** Long-chain glycerol diether and polyol dialkyl glycerol triether lipids of *Sulfolobus acidocaldarius, J. Bacteriol.,* 119, 106, 1974.
204. **Lechevalier, H. A. and Lechevalier, M. P.,** A critical evaluation of the genera of aerobic actinomycetes, in *The Actinomycetales,* Prauser, H., Ed., G. Fischer-Verlag, Jena, 1970, p. 393.

205a. **Lechevalier, H. A., Lechevalier, M. P., and Gerber, N. N.,** Chemical composition as a criterion in the classification of actinomycetes, *Adv. Appl. Microbiol.,* 14, 47, 1971.

205b. **Lechevalier, M. P.,** unpublished data.

206. **Lechevalier, M. P., Horan, A. C., and Lechevalier, H. A.,** Lipid composition in the classification of nocardiae and mycobacteria, *J. Bacteriol.,* 105, 313, 1971.
207. **Lechevalier, M. P., Lechevalier, H. A., and Heintz, C. E.,** Morphological and chemical nature of the sclerotia of *Chainia olivacea* Thirumalachar and Sukapure of the order Actinomycetales, *Int. J. Syst. Bacteriol.,* 23, 157, 1973.

208. **Lechevalier, M. P. and Lechevalier, H. A.**, *Nocardia amarae* sp. nov., an actinomycete common in foaming activated sludge, *Int. J. Syst. Bacteriol.*, 24, 278, 1974.
209. **Lechevalier, M. P., Lechevalier, H. A., and Horan, A. C.**, Chemical characteristics and classification of nocardiae, *Can. J. Microbiol.*, 19, 965, 1973.
210. **Lederer, E.**, Glycolipids of mycobacteria and related microorganisms, *Chem. Phys. Lipids*, 1, 294, 1967.
211. **Lederer, E.**, The mycobacterial cell wall, *Pure Appl. Chem.*, 25, 135, 1971.
212. **Lee, Y. H. and Oo, K. C.**, Metabolism of phospholipids in *Bacillus stearothermophilus*, *J. Biochem.*, 74, 615, 1973.
213. **Lennarz, W. J.**, Bacterial lipids, in *Lipid Metabolism*, Wakil, S. J., Ed., Academic Press, New York, 1970, p. 155.
214. **Levin, R. A.**, Fatty acids of *Thiobacillus thiooxidans*, *J. Bacteriol.*, 108, 992, 1971.
215. **Levin, R. A.**, Effect of cultural conditions on the fatty acid composition of *Thiobacillus novellus*, *J. Bacteriol.*, 112, 903, 1972.
216a. **Livermore, B. P. and Johnson, R. C.**, Lipids of Spirochaetales: Comparison of the lipids of several members of the genera *Spirochaeta*, *Treponema* and *Leptospira*, *J. Bacteriol.*, 120, 1268, 1974.
216b. **Long, E. R.**, The chemical composition of the active principle of tuberculin. I. A non-protein medium suitable for the production of tuberculin in large quantity, *Am. Rev. Tuberc. Pulm. Dis.*, 13, 393, 1926.
217. **Lüderitz, O.**, Recent results on the biochemistry of the cell wall lipopolysaccharides of *Salmonella* bacteria, *Angew. Chem. Int. Ed. Engl.*, 9, 649, 1970.
218. **MacFarlane, M. G.**, Phosphatidylglycerols and lipoamino acids, *Adv. Lipid Res.*, 2, 92, 1964.
219. **MacFarlane, M. G.**, Lipids of bacterial membranes, in *Metabolism and Physiological Significance of Lipids*, Dawson, R. M. C. and Rhodes, D. N., Eds., John Wiley & Sons, New York, 1964, p. 399.
220. **Machtiger, N. A. and O'Leary, W. M.**, Fatty acid compositions of paracolons: *Arizona*, *Citrobacter* and *Providencia*, *J. Bacteriol.*, 114, 80, 1973.
221. **Maeda, J.**, Isolation and characterization of toxic glycolipid from the firmly bound lipids of human tubercle bacilli, *Jpn. J. Bacteriol.*, 27, 469, 1972.
222. **Makula, R. A. and Finnerty, W. R.**, Microbial assimilation of hydrocarbons: cellular distribution of fatty acids, *J. Bacteriol.*, 112, 398, 1972.
223. **Makula, R. A. and Finnerty, W. R.**, Phospholipid composition of *Desulfovibrio* species, *J. Bacteriol.*, 120, 1279, 1974.
224. **Makula, R. A. and Finnerty, W. R.**, Isolation and characterization of an ornithine-containing lipid from *Desulfovibrio gigas*, *J. Bacteriol.*, 123, 523, 1975.
225. **Makula, R. A., Lockwood, P. J., and Finnerty, W. R.**, Comparative analysis of the lipids of *Acinetobacter* species grown on hexane, *J. Bacteriol.*, 121, 250, 1975.
226. **Marcus, L. and Kaneshiro, T.**, Lipid composition of *Azotobacter vinelandii* in which the internal membrane network is induced or repressed, *Biochim. Biophys. Acta*, 288, 296, 1972.
227. **Marinetti, G. V., Ed.**, *Lipid Chromatographic Analysis*, Vol. 1, Marcel Dekker, New York, 1967.
228. **Markovits, J., Vilkas, E., and Lederer, E.**, Sur la structure chimique des cires D, peptidoglycolipides macromoléculaires des souches humanes de *Mycobacterium tuberculosis*, *Eur. J. Biochem.*, 18, 287, 1971.
229. **Mastroeni, P., Nacci, A., Teti, D., and Teti, M.**, Sul significato tassonomico dei fosfolipidi di *Bacillus cereus*, *Bacillus subtilis*, *Bacillus megaterium*, *G. Batteriol. Virol. Immunol Ann. Osp. Maria Vittoria Torino Parte I Sez Microbiol.*, 64, 20, 1971.
230. **Matsumoto, M. and Miwa, M.**, Bacterial lipids, *Yukagaku*, 20. 678, 1971.
231. **Matsumoto, M., Tamiya, K., and Koizumi, K.**, Studies on the neutral lipids and a new type of aldehydrogenic ethanolamine phospholipid in *Clostridium butyricum*, *J. Biochem. Tokyo*, 69, 617, 1971.
232. **Maurice, M. T., Vacheron, M. J., and Michel, G.**, Isolement d'acides nocardiques de plusieures espèces de *Nocardia*, *Chem. Phys. Lipids*, 7, 9, 1971.
233. **Meyer, H. and Meyer, F.**, Lipid metabolism in the parasitic and free-living spirochetes *Treponema pallidum* (Reiter) and *Treponema zuelzerae*, *Biochim. Biophys. Acta*, 231, 93, 1971.
234. **Migliore, D., Augier, J., Boisvert, H., and Jollès, P.**, Wax D from different bovine strains of *Mycobacterium*, *J. Bacteriol.*, 107, 548, 1971.
235. **Miller, A., III, Sandine, W. E., and Elliker, P. R.**, Fatty acid composition of lipid synthesized by *Pseudomonas fragi*, *J. Dairy Sci.*, 54, 919, 1971.
236. **Minnikin, D. E. and Abdolrahimzadeh, H.**, Effect of pH on the proportions of polar lipids in chemostat cultures of *Bacillus subtilis*, *J. Bacteriol.*, 120, 999, 1974.

237. **Minnikin, D. E. and Abdolrahimzadeh, H.,** The replacement of phosphatidylethanolamine and acidic phospholipids by an ornithine-amide lipid and a minor phosphorus-free lipid in *Pseudomonas fluorescens* NCMB 129, *FEBS Lett.*, 43, 257, 1974.
238. **Minnikin, D. E., Abdolrahimzadeh, H., and Baddiley, J.,** The interrelation of phosphatidylethanolamine and glycosyl diglycerides in bacterial membranes, *Biochem. J.*, 124, 447, 1971.
239. **Minnikin, D. E., Abdolrahimzadeh, H., and Baddiley, J.,** The interrelation of polar lipids in bacterial membranes, *Biochim. Biophys. Acta*, 249, 651, 1971.
240. **Minnikin, D. E., Abdolrahimzadeh, H., and Baddiley, J.,** Variation of polar composition of *Bacillus subtilis* (Marburg) with different growth conditions, *FEBS Lett.*, 27, 16, 1972.
241. **Minnikin, D. E., Abdolrahimzadeh, H., and Baddiley, J.,** Glycolipids and biological membranes: the replacement of phospholipids by glycolipids in *Pseudomonas diminuta, Biochem. Soc. Trans.*, 1, 431, 1973.
242. **Minnikin, D. E., Abdolrahimzadeh, H., and Baddiley, J.,** The occurrance of phosphatidylethanolamine and glycosyl diglycerides in thermophilic bacilli, *J. Gen. Microbiol.*, 83, 415, 1974.
243. **Minnikin, D. E., Alshamaony, L., and Goodfellow, M.,** Differentiation between mycobacteria, nocardiae and related taxa by analysis of whole-cell methanolysates, *1st Int. Conf. on the Biology Nocardiae*, Merida, Venezuela, 1974, p. 36.
244. **Minnikin, D. E., Patel, P. V., and Goodfellow, M.,** Mycolic acids of representative strains of *Nocardia* and the "rhodochrous" complex, *FEBS Lett.*, 39, 322, 1974.
245. **Mitruka, B. M.,** *Gas Chromatographic Applications in Microbiology and Medicine*, John Wiley & sons, New York, 1975.
246. **Mordarska, H. and Rethy, A.,** Preliminary studies on the chemical character of the lipid fraction in *Nocardia, Arch. Immunol. Ther. Exp.*, 18, 455, 1970.
247. **Mordarska, H., Mordarski, M., and Goodfellow, M.,** Chemotaxonomic character and classification of some nocardioform bacteria, *J. Gen. Microbiol.*, 71, 77, 1972.
248. **Mordarska, H., Mordarski, M., and Pietkiewicz, D.,** Chemical analysis of hydrolysates and cell extracts of *Nocardia pellegrino, Int. J. Syst. Bacteriol.*, 23, 274, 1973.
249. **Mordarskaya, G., Guzeva, L. N., and Agre, N. S.,** Lipids from the mycelia of thermophilic actinomycetes, *Microbiology USSR*, 42, 142, 1973, (English transl).
250. **Morrison, S. J., Tornabene, T. G., and Kloos, W. E.,** Neutral lipids in the study of relationships of members of the family Micrococcaceae, *J. Bacteriol.*, 108, 353, 1971.
251. **Moss, C. W. and Lewis, V. J.,** Characterization of clostridia by gas chromatography. I. Differentiation of species by cellular fatty acids, *Appl. Microbiol.*, 15, 390, 1967.
252. **Moss, C. W., Dowell, V. R., Jr., Farshtchi, D., Raines, L. J., and Cherry, W. B.,** Cultural characteristcs and fatty acid composition of propionibacteria, *J. Bacteriol.*, 97, 561, 1969.
253. **Moss, C. W., Kellogg, D. S., Farshy, D. C., Lambert, M. A., and Thayer, J. D.,** Cellular fatty acids of pathogenic *Neisseria, J. Bacteriol.*, 104, 63, 1970.
254. **Moss, C. W., Lambert, M. A., and Merwin, W. H.,** Comparison of rapid methods for analysis of bacterial fatty acids, *Appl. Microbiol.*, 28, 80, 1974.
255. **Moss, C. W., Samuels, S. B., Liddle, J., and McKinney, R. M.,** Occurrence of branched-chain hydroxy fatty acids in *Pseudomonas maltophila, J. Bacteriol.*, 114, 1018, 1973.
256. **Moss, C. W., Samuels, S. B., and Weaver, R. E.,** Cellular fatty acid composition of selected *Pseudomonas* species, *Appl. Microbiol.*, 24, 596, 1972.
257. **Mota, J. S., Silva, M. T., and Guevra, F. C.,** Variations in the membranes of *Streptococcus faecalis* related to different cultural conditions, *Arch. Mikrobiol.*, 83, 293, 1972.
258. **Munro, G. F. and Bell, C A.,** Effects of external osmolarity on phospholipid metabolism in *Escherichia coli* B, *J. Bacteriol.*, 116, 257, 1973.
259. **Muramatsu, T.,** Plasmalogens, *Yukagaku*, 22, 646, 1973.
260. **Naccarato, W. F., Gilbertson, J. R., and Gelman, R. A.,** Effects of different culture media and oxygen upon lipids of *Escherichia coli* K-12, *Lipids*, 9, 322, 1974.
261. **Noll, H., Bloch, H., Asselineau, J., and Lederer, E.,** The chemical structure of the cord factor of *Mycobacterium tuberculosis, Biochim. Biophys. Acta*, 20, 299, 1956.
262. **Oelze, J. and Drews, G.,** Membranes of photosynthetic bacteria, *Biochim. Biophys. Acta*, 265, 209, 1972.
263. **Ohno, Y., Yano, I., and Masui, M.,** Separation and identification of molecular species of phospholipids from a Gram-negative moderately halophilic bacterium, *FEBS Lett.*, 50, 50 1975.
264. **Okazaki, H., Sugino, H., Kanzaki, T., and Fukuda, H.,** L-Glutamic acid fermentation. VI. Structure of a sugar lipid produced by *Brevibacterium thiogenitalis, Agric. Biol. Chem.*, 33, 764, 1969.
265. **Okuyama, H., Kankura, T., and Nojima, S.,** Positional distribution of fatty acids of phospholipids from *Mycobacteria, J. Biochem. Tokyo*, 61, 732, 1967.
266. **O'Leary, W. M.,** The fatty acids of bacteria, *Bacteriol. Rev.*, 26, 421, 1962.

267. **O'Leary, W. M.,** *The Chemistry and Metabolism of Microbial Lipids,* World Publ., Cleveland, 1967.
268. **O'Leary, W. M.,** Lipoidal contents of specific microorganisms, in *Handbook of Microbiology,* Vol. 2, Laskin, A. I. and Lechevalier, H. A., Eds., CRC Press, Cleveland, 1973, p. 275.
269. **O'Leary, W. M.,** Microbial lipids, in *Molecular Microbiology,* John Wiley & Sons, New York, 1974, p. 185.
270. **O'Leary, W. M.,** The chemistry of microbial lipids, *CRC Crit. Rev. Microbiol.,* 4, 41, 1975.
271. **Oliver, J. D. and Colwell, R. R.,** Extractable lipids of Gram-negative marine bacteria: fatty acid composition, *Int. J. Syst. Bacteriol.,* 23, 442, 1973.
272. **Oliver, J. D. and Colwell, R. R.,** Extractable lipids of Gram-negative marine bacteria: phospholipid composition, *J. Bacteriol.,* 114, 897, 1973.
273. **Olsen, R. W. and Ballou, C. E.,** Acylphosphatidyl glycerol, a new phospholipid from *Salmonella typhimurium, J. Biol. Chem.,* 246, 3305, 1971.
274. **Olson, G. J. and Ingram, L. O.,** Effects of temperature and nutritional changes on the fatty acids of *Agmenellum quadruplicatum, J. Bacteriol.,* 124, 373, 1975.
275. **Oo, K. C. and Lee, K. L.,** The lipid content of *Bacillus stearothermophilus* at 37° and 55°, *J. Gen. Microbiol.,* 69, 287, 1971.
276. **Oo, K. C. and Lee, Y. H.,** The phospholipids of a facultatively thermophilic strain of *Bacillus stearothermophilus, J. Biochem.,* 71, 1081, 1972.
277. **Op den Kamp, J. A. F., Bonsen, P. P. M., and Van Deenen, L. L. M.,** Structural investigations of glucosaminylphosphatidyl glycerol from *Bacillus megaterium, Biochim. Biophys. Acta,* 176, 298, 1969.
278. **Op den Kamp, J. A. F. and Van Deenen, L. L. M.,** Bacterial phospholipids and membranes, in *Structural and Functional Aspects of Lipoproteins in Living Systems,* Tria, E. and Scanu, A. M., Eds., Academic Press, New York, 1969, p. 227.
279. **Osborn, M. J., Gander, J. E., Parisi, E., and Carson, J.,** Mechanism of assembly of the outer membrane of *Salmonella typhimurium, J. Biol. Chem.,* 247, 3962, 1972.
280. **Oshima, M. and Miyagawa, A.,** Comparative studies on the fatty acid composition of moderately and extremely thermophilic bacteria, *Lipids,* 9, 476, 1974.
281. **Oshima, M. and Yamakawa, T.,** Chemical structure of a novel glycolipid from an extreme thermophile, *Flavobacterium thermophilum, Biochemistry,* 13, 1140, 1974.
282. **Oshima, M., Ueta, N., and Yamakawa, T.,** Gas-chromatographic study of microbial component. IV. Branched-chain fatty acids in *Microbispora chromogenes* M-22, *Jpn. J. Exp. Med.,* 39, 77, 1969.
283. **Pandhi, P. N. and Hammond, B. F.,** A glycolipid from *Rothia dentocariosa, Arch. Oral Biol.,* 20, 399, 1975.
284. **Patrick, M. A. and Dugan, P. R.,** Influence of hydrocarbons and derivatives of the polar lipid fatty acids of an *Acinetobacter* isolate, *J. Bacteriol.,* 119, 76, 1974.
285. **Pechey, D. T., Yau, A. O. P., and James, A. M.,** Total and surface lipid of cells of *Pseudomonas aeruginosa* and their relationship to gentamycin resistance, *Microbios,* 11, 77, 1974.
286. **Peleg, E. and Tietz, A.,** Glycolipids of a halotolerant moderately halophilic bacterium, *FEBS Lett.,* 15, 309, 1971.
287. **Peterson, S. L., Bennett, L. G., and Tornabene, T. G.,** Effects of lead on the lipid composition of *Micrococcus luteus* cells, *Appl. Microbiol.,* 29, 669, 1975.
288. **Pheil, C. G. and Ordal, Z. J.,** Fatty acid composition of spores of "thermophilic anaerobes", *J. Bacteriol.,* 93, 1727, 1967.
289. **Phizackerley, P. J. R., MacDougall, J. C., and Francis, M. J. O.,** Phosphatidylglycerylglucosamine, *Biochem. J.,* 99, 21C, 1966.
290. **Phizackerley, P. J. R., MacDougall, J. C., and Moore, R. A.,** 1-(O-β-Glucosaminyl)-2,3-diglyceride in *Bacillus megaterium, Biochem. J.,* 126, 499, 1972.
291. **Pitel, D. W. and Gilvarg, C.,** Timing of mucopeptide and phospholipid synthesis in sporulating *Bacillus megaterium, J. Biol. Chem.,* 246, 3720, 1971.
292. **Polonovski, J., Wald, R., Paysant, M., Rampini, C., and Barbu, E.,** Métabolisme du phosphatidylglycérol et du cardiolipide chez *Staphylococcus aureus, Ann. Inst. Pasteur Paris,* 120, 589, 1971.
293. **Pommier, M.-T. and Michel, G.,** Isolement et caractéristiques d'un nouveau glycolipide de *Nocardia caviae, C. R. Acad. Sci. Ser. C,* 275, 1323, 1972.
294. **Pommier, M.-T. and Michel, G.,** Phospholipid and acid composition of *Nocardia* and nocardoid bacteria as criteria of classification, *Biochem. Syst.,* 1, 3, 1973.
295. **Powell, D. A., Duckworth, M., and Baddiley, J.,** A membrane-associated lipomannan in micrococci, *Biochem. J.,* 151, 387, 1975.
296. **Prauser, H.,** DAP-freie, gelbe Actinomyceten mit Tendenz zur Beweglichkeit, *Z. Allg. Mikrobiol.,* 7, 81, 1967.
297. **Prefontaine, G. and Jackson, F. L.,** Cellular fatty acid profiles as an aid to the classification of "corroding bacilli" and certain other bacteria, *Int. J. Syst. Bacteriol.,* 22, 210, 1972.

298. **Prins, R. A., Akkermans-Kruyswijk, J., Franklin-Klein, W., Lankhorst, A., and Van Golde, L. M. G.,** Metabolism of serine and ethanolamine plasmalogens in *Megasphaera elsdenii, Biochim. Biophys. Acta,* 348, 361, 1974.
299. **Promé, J.-C., Lacave, C., and Lanéelle, M.-A.,** Sur les structures de lipides à ornithine de *Brucella melitensis* et de *Mycobacterium bovis* (B.C.G.), *C. R. Acad. Sci.,* 269, 1664, 1969.
300. **Pucheu, N. L., Kerber, N. L., and Garcia, A. F.,** Comparative studies on membranes isolated from *Rhodopseudomonas viridis* grown in presence and absence of yeast extract, *Arch. Mikrobiol.,* 101, 259, 1974.
301. **Raines, L. J., Moss, C. W., Farshtchi, D., and Pittman, B.,** Fatty acids of *Listeria monocytogenes, J. Bacteriol.,* 96, 2175, 1968.
302. **Randerath, K.,** *Thin-Layer Chromatography,* Academic Press, New York, 1968.
303. **Ratledge, C. and Hall, M. J.,** Influence of metal ions on the formation of myobactin and salicylic acid in *Mycobacterium smegmatis* grown in static culture, *J. Bacteriol.,* 108, 314, 1971.
304. **Ratledge, C. and Patel, P. V.,** Lipid-soluble, iron-binding compounds in *Nocardia, "M." rhodochrous* and *Mycobacterium, 1st Int. Conf. on the Biology of the Nocardiae,* Merida, Venezuela, 1974, p. 58.
305. **Ratledge, C. and Snow, G. A.,** Isolation and structure of nocobactin NA, a lipid-soluble iron-binding compound from *Nocardia asteroides, Biochem. J.,* 139, 407, 1974.
306. **Ray, P. H., White, D. C., and Brock, T. D.,** Effect of temperature on the fatty acid composition of *Thermus aquaticus, J. Bacteriol.,* 106, 25, 1971.
307. **Ray, P. H., White, D. C., and Brock, T. D.,** Effect of growth temperature on the lipid composition of *Thermus aquaticus, J. Bacteriol.,* 108, 227, 1971.
308. **Reaveley, D. A. and Burge, R. E.,** Walls and membranes in bacteria, in *Advances in Microbial Physiology,* Vol. 7, Rose, A. H. and Tempest, D. W., Eds., Academic Press, New York, 1972, p. 1.
309. **Rebel, G.,** Alkenyl and alkyl phosphatides in *Proteus* and its L forms, *Arch. Mikrobiol.,* 81, 333, 1972.
310. **Rédai, I., Réthy, A., Sebessi-Gönczy, P., and Váczi, L.,** Lipids in *Staphylococcus aureus* and *Escherichia coli* cultured in the presence of human serum, *Acta Microbiol. Acad. Sci. Hung.,* 18, 297, 1971.
311. **Rédai, I., Sebessy-Gönczy, P., and Váczi, L.,** Effect of chloramphenicol on phospholipid synthesis in sensitive *Staphylococcus aureus* strains, *Acta Microbiol. Acad. Sci. Hung.,* 19, 187, 1972.
312. **Renkonen, O.,** Thin-layer chromatographic analysis of subclasses and molecular species of polar lipids, *Prog. Thin-Layer Chromatogr. Relat. Methods,* 2, 143, 1971.
313. **Rietschel, E. T.,** Chemical structure and biological activity of endotoxins (lipopolysaccharides) and lipid A, *Naunyn Schmiedeberg's Arch. Pharmacol.,* 287, 73, 1975.
314. **Rivas, E., Kerber, N. L., Viale, A. A., and García, A. F.,** Isolation of a "basic membrane" fraction enriched in an ornithine-containing lipid from a blue-green mutant of *Rhodospirillum rubrum, FEBS Lett.,* 11, 37, 1970.
315. **Rottem, S., Hasin, M., and Razin, S.,** The outer membrane of *Proteus mirabilis* II, *Biochim. Biophys. Acta,* 375, 395, 1975.
316. **Rozgonyi, F., Váczi, L., Sebessy-Gönczy, P., and Rédai, I.,** Phospholipid content of staphylococci sensitive and resistant to methicillin, *Contrib. Microbiol. Immunol.,* 1, 172, 1973.
317. **Russell, N. J.,** Alteration in fatty acid chain length in *Micrococcus cryophilus* grown at different temperatures, *Biochim. Biophys. Acta,* 231, 254, 1971.
318. **Russell, N. J.,** The lipid composition of the psychrophilic bacterium *Micrococcus cryophilus, J. Gen. Microbiol.,* 80, 217, 1974.
319. **Samuels, S. B., Moss, C. W., and Weaver, R. E.,** The fatty acids of *Pseudomonas multivorans (Pseudomonas cepacia)* and *Pseudomonas kingii, J. Gen. Microbiol.,* 74, 275, 1973.
320. **Sarma, G. R., Chandramouli, V., and Venkitasubramanian, T. A.,** Occurrence of phosphonolipids in mycobacteria, *Biochim. Biophys. Acta,* 218, 561, 1970.
321. **Sastry, P. S.,** Glycosyl diglycerides, *Adv. Lipid Res.,* 12, 251, 1974.

322a. **Sato, H., Arai, H., Yokozawa, A., Motomiya, M., Konno, J., and Yano, I.,** Mycolic acid of scotochromogenic mycobacteria, P6, *Kekkaku,* 49, 53, 1974.

322b. **Sauton, B.,** Sur la nutrition mineralé du bacille tuberculeux, *C. R. Acad. Sci.,* 155, 860, 1912.

323. **Senn, M., Ioneda, T., Pudles, J., and Lederer, E.,** Spectrométrie de masse de glycolipides. I. Structure du "cord factor" de *Corynebacterium diphtheriae, Eur. J. Biochem.,* 1, 353, 1967.
324. **Shaw, N.,** Bacterial glycolipids, *Bacteriol. Rev.,* 34, 365, 1970.
325. **Shaw, N.,** Lipid composition as a guide to the classification of bacteria, *Adv. Appl. Microbiol.,* 17, 63, 1974.
326. **Shaw, N.,** Bacterial glycolipids and glycophospholipids, *Adv. Microb. Physiol.,* 12, 141, 1975.
327. **Shaw, N. and Stead, D.,** Lipid composition of some species of *Arthrobacter, J. Bacteriol.,* 107, 130, 1971.

328. **Shaw, N. and Stead, A.,** Bacterial glycophospholipids, *FEBS Lett.*, 21, 249, 1972.
329. **Shen, P. Y., Coles, E., Foote, J. L., and Stenesh, J.,** Fatty acid distribution in mesophilic and thermophilic strains of the genus *Bacillus, J. Bacteriol.*, 103, 479, 1970.
330. **Short, S. A. and White, D. C.,** Metabolism of glucosyl diglycerides and phosphatidyl glucose of *Staphylococcus aureus, J. Bacteriol.*, 104, 126, 1970.
331. **Short, S. A. and White, D. C.,** Metabolism of phosphatidylglycerol, lysylphosphatidylglycerol and cardiolipin of *Staphyloccus aureus, J. Bacteriol.*, 108, 219, 1971.
332. **Silbert, D. F.,** Genetic modification of membrane lipid, *Annu. Rev. Biochem.*, 44, 315, 1975.
333. **Silbert, D. R., Ulbright, T. M., and Honegger, J. L.,** Utilization of exogenous fatty acids for complex lipid biosynthesis and its effect on *de novo* fatty acid formation in *Escherichia coli K-12, Biochemistry*, 12, 164, 1973.
334. **Sinensky, M.,** Temperature control of phospholipid biosynthesis in *Escherichia coli, J. Bacteriol.*, 106, 449, 1971.
335. **Snow, G. A.,** Mycobactins: iron-chelating growth factors from mycobacteria, *Bacteriol. Rev.*, 34, 99, 1970.
336. **Snyder, F.,** The chemistry, physical properties and chromatography of lipids containing ether bonds, *Prog. Thin-Layer Chromatogr. Relat. Methods*, 2, 105, 1971.
337. **Stahl, E.,** *Thin-Layer Chroatography, A Laboratory Handbook*, Springer-Verlag, New York, 1969.
338. **Stanier, R. Y., Kunisawa, R., Mandel, M., and Cohen-Bazire, G.,** Purification and properties of unicellular blue-green algae (ordor Chroococcales), *Bacteriol. Rev.*, 35, 171, 1971.
339. **Steinhauer, J. E., Flentge, R. L., and Lechowich, R. V.,** Lipid patterns of selected microorganisms as determined by gas—liquid chromatography, *Appl. Microbiol.*, 15, 826, 1967.
340. **Stoffel, W., Dittmar, U. K., and Wilmes, R.,** Sphingolipid metabolism in Bacteroidaceae, *Hoppe-Seyler's Z. Physiol. Chem.*, 356, 715, 1975.
341. **Sud, I. J. and Feingold, D. S.,** Phospholipids and fatty acids of *Neisseria gonorrhoeae, J. Bacteriol.*, 124, 713, 1975.
342. **Šula, L., Galliová, J., and Mára, M.,** A sudden decrease in the lipid content of *Mycobacterium bovis* BCG and *Mycobacterium microti* MP. A preliminary communication, *Zentralbl. Bakteriol. Parasitenkd. Infektionskr. Hyg. Abt. I. Orig. Reihe A*, 219, 264, 1972.
343. **Suto, T. and Ryozo, A.,** Chemotaxonomy of microorganisms, *Nippon Saikingaku Zasshi*, 29, 811, 1974.
344. **Suzuki, T., Tanaka, H., and Itoh, S.,** Sucrose lipids of *arthrobacteria, corynebacteria* and *nocardiae* grown on sucrose, *Agric. Biol. Chem.*, 38, 557, 1974.
345. **Suzuki, T., Tanaka, K., Matsubara, I., and Kinoshita, S.,** Trehalose lipid and α-branched-β-hydroxy fatty acid formed by bacteria grown on n-alkanes, *Agric. Biol. Chem.*, 33, 1619, 1969.
346. **Suzuki, T., Yamaguchi, K., and Tanaka, K.,** Effects of cupric ion on the production of glutamic acid and trehalose by an n-paraffin-grown bacterium, *Agric. Biol. Chem.*, 35, 2135, 1971.
347. **Tabaud, H., Tisnovska, H., and Vilkas, E.,** Phospholipides et glycolipides d'une souche de *Micromonospora, Biochimie*, 53, 55, 1971.
348. **Tadayon, R. A. and Carroll, K. K.,** Effect of growth conditions on the fatty acid composition of *Listeria monocytogenes* and comparison with the fatty acids of *Erysipelothrix* and *Corynebacterium, Lipids*, 6, 820, 1971.
349. **Tanner, W.,** The function of *myo*-inositol glycosides in yeasts and higher plants, *Ann. N. Y. Acad. Sci.*, 165, 726, 1969.
350. **Terroine, E.-F., Bonnet, R., Kopp, G., and Véchot, J.,** Sur la signification physiologique des liasons éthyléniques des acides gras, *Bull. Soc. Chim. Biol.*, 9, 605, 1927.
351. **Theodore, T. S. and Panos, C.,** Protein and fatty acid composition of mesosomal vesicles and plasma membranes of *Staphylococcus aureus, J. Bacteriol.*, 116, 571, 1973.
352. **Thiele, O. W.,** The lipids of hydrogen-oxidizing bacteria: occurrence of *cis*-9,10-methylene hexadecanoic acid in *Hydrogenomonas* H16, *Experientia*, 27, 1268, 1971.
353. **Thiele, O. W. and Schwinn, G.,** The free lipids of *Brucella melitensis* and *Bordetella pertussis, Eur. J. Biochem.*, 34, 333, 1973.
354. **Thiele, O. W. and Schwinn, G.,** Bakterielle Ornithinlipide, *Z. Allg. Mikrobiol.*, 14, 435, 1974.
355. **Thiele, O. W., Busse, D., and Schwinn, G.,** Phosphatide der Brucellen, *Z. Allg. Mikrobiol.*, 11, 249, 1971.
356. **Thiele, O. W., Dreysel, J., and Hermann, D.,** The "free" lipids of two different strains of hydrogen-oxidizing bacteria in relation to their growth phases, *Eur. J. Biochem.*, 29, 224, 1972.
357. **Thoen, C. O., Karlson, A. G., and Ellefson, R. D.,** Fatty acids of *Mycobacterium kansasii, Appl. Microbiol.*, 21, 628, 1971.
358. **Thoen, C. O., Karlson, A. G., and Ellefson, R. D.,** Comparison by gas—liquid chromatography of the fatty acids of *Mycobacterium avium* and some other nonphotochromogenic mycobacteria, *Appl. Microbiol.*, 22, 560, 1971.

359. **Thoen, C. O., Karlson, A. G., and Ellefson, R. D.,** Differentiation between *Mycobacterium kansasii* and *Mycobacterium marinum* by gas—liquid chromatographic analysis of cellular fatty acids, *Appl. Microbiol.*, 24, 1009, 1972.
360. **Thomas, T. D. and Ellar, D. J.,** Properties of plasma and mesosomal membranes isolated from *Micrococcus lysodeikticus*: rates of synthesis and characterization of lipids, *Biochim. Biophys. Acta*, 316, 180, 1973.
361. **Thorne, K. J. I., Thornley, M. J., and Glauert, A. M.,** Chemical analysis of the membrane and other layers of the cell envelope of *Acinetobacter* sp., *J. Bacteriol.*, 116, 410, 1973.
362. **Tomlins, R. I., Vaaler, G. L., and Ordal, Z. J.,** Lipid biosynthesis during the recovery of *Salmonella typhimurium* from thermal injury, *Can. J. Microbiol.*, 18, 1015, 1972.
363. **Tornabene, T. G.,** Lipid composition of selected strains of *Yersinia pestis* and *Yersinia pseudotuberculosis, Biochim. Biophys. Acta*, 306, 173, 1973.
367. **Tornabene, T. G. and Markey, S. P.,** Characterization of branched monounsaturated hydrocarbons of Sarcina lutea *and Sarcina flava, Lipids*, 6, 190, 1971.
368. **Tornabene, T. G. and Ogg, J. E.,** Chromatographic studies of the lipid components of *Vibrio fetus, Biochim. Biophys. Acta*, 239, 133, 1971.
369. **Tria, E. and Scanu, A. M. (Eds.),** *Structural and Functional Aspects of Lipoproteins in Living Systems*, Academic Press, New York, 1969.
370. **Truby, C. P. and Bennett, E. O.,** Fatty acid composition of *Aerobacter aerogenes, Antonie van Leeuwenhoek J. Microbiol. Serol.*, 37, 101, 1971.
371. **Tsukamura, M. and Mizuno, S.,** Differentiation among mycobacterial species by thin-layer chromatography, *Int. J. Syst. Bacteriol.*, 25, 271, 1975.
372. **Tsyganov, V. A., Efimova, T. P., and Soboleva, L. V.,** Fatty acids produced by proactinomycetes, *Microbiology USSR*, 42, 705, 1973.
373. **Tucker, A. N. and White, D. C.,** Detection of a rapidly metabolizing portion of the membrane cardiolipin in *Haemophilus influenzae, J. Bacteriol.*, 108, 1058, 1971.
374. **Turujman, N., Jabr, I., and Durr, I. F.,** Amino acid and fatty acid composition of membranes of *Lactobacillus plantarum, Int. J. Biochem.*, 5, 791, 1974.
375. **Uchida, K.,** Cellular fatty acids and taxonomy of bacteria, I, *Hakko Kyokaishi*, 32, 471, 1973.
376. **Uchida, K.,** Lipids of alcoholophilic lactobacilli, II, *Biochim. Biophys. Acta*, 369, 146, 1974.
377. **Uchida, K.,** Effects of cultural conditions on the cellular fatty acid composition *of Lactobacillus heterohiochii*, an alcoholophilic bacterium, *Agric. Biol. Chem.*, 39, 837, 1975.
378. **Uchida, K.,** Occurrence of conjugated dienoic fatty acids in the cellular lipids of *Pediococcus homari, Agric. Biol. Chem.*, 39, 561, 1975.
379. **Uchida, K. and Mogi, K.,** Cellular fatty acid spectra of *Pediococcus* species in relation to their taxonomy, *J. Gen. Appl. Microbiol.*, 18, 109, 1972.
380. **Uchida, K. and Mogi, K.,** Cellular fatty acid spectra of *Sporolactobacillus* and some other *Bacillus-Lactobacillus* intermediates as a guide to their taxonomy, *J. Gen. Appl. Microbiol.*, 19, 129, 1973.
381. **Uchida, K. and Mogi, K.,** Cellular fatty acid spectra of hiochibacteria, alcohol-tolerant lactobacilli, and their group separation, *J. Gen. Appl. Microbiol.*, 19, 233, 1973.
382. **Ueta, N. and Yamakawa, T.,** Gas-chromatographic studies of microbial components. III. Research on precursor of branched-chain fatty acids of *Staphylococcus aureus* and *Streptomyces lavendulae, Jpn. J. Exp. Med.*, 38, 347, 1968.
383. **Vacheron, M.-J., Guinand, M., Michel, G., and Ghuysen, J.-M.,** Structural investigation on cell walls of *Nocardia* sp. The wall lipid and peptidoglycan moieties of *Nocardia kirovani, Eur. J. Biochem.*, 29, 156, 1972.
384. **Vacheron, M.-J. and Michel, G.,** Présence d'acides de poids moléculare élevé dans les triglycerides isolés des parois cellulaires de *Nocardia kirovani, C. R. Acad. Sci. Ser. C*, 273, 1778, 1971.
385. **Váczi, L.,** *The Biological Role of Bacterial Lipids*, Akadémiai Kiadó, Budapest, 1973.
386. **Van Golde, L. M. G., Prins, R. A., Franklin-Klein, W., and Akkermans-Kruyswijk, J.,** Phosphatidylserine and its plasmalogen analogue as major lipid constituents in *Megasphaera elsdenii, Biochim. Biophys. Acta*, 326, 314, 1973.
387. **Van Schaik, F. W. and Veerkamp, J. H.,** Biochemical changes in *Bifidobacterium bifidum* var. *pennsylvanicus* after cell wall inhibition. VIII. Composition and metabolism of phospholipids at different stages and conditions of growth, *Biochim. Biophys. Acta*, 388, 213, 1975.
388. **Veerkamp, J. H.,** Fatty acid composition of *Bifidobacterium* and *Lactobacillus* strains, *J. Bacteriol.*, 108, 861, 1971.
389. **Veerkamp, J. H.,** Biochemical changes in *Bifidobacterium bifidum* var. *pennsylvanicus* after cell wall inhibtion. V. Structure of the galactosyldiglycerides, *Biochim. Biophys. Acta*, 273, 359, 1972.
390. **Veerkamp, J. H. and Van Schaik, F. W.,** Biochemical changes in *Bifidobacterium bifidum* var. *pennsylvanicus* after cell wall inhibition. VII. Structure of the phosphogalactolipids, *Biochim. Biophys. Acta*, 348, 370, 1974.

391. **Vestal, J. R. and Perry, J. J.**, Effect of substrate on the lipids of the hydrocarbon-utilizing *Mycobacterium vaccae, Can. J. Microbiol.*, 17, 445, 1971.
392. **Vilkas, E. and Rojas, A.**, Les lipides de *Mycobacterium fortuitum, Bull. Soc. Chim. Biol.*, 46, 689, 1964.
393. **Vilkas, E., Adam, A., and Senn, M.**, Isolation of a new type of trehalose diester from *Mycobacterium fortuitum, Chem. Phys. Lipids*, 2, 11, 1968.
394. **Voiland, A., Bruneteau, M., and Michel, G.**, Etude du mycoside C_2 de *Mycobacterium avium, Eur. J. Biochem.*, 21, 285, 1971.
395. **Vulliet, P., Markey, S. P., and Tornabene, T. G.**, Identification of methoxyester artifacts produced by methanolic HCl solvolysis of the cyclopropane fatty acids of the genus *Yersinia, Biochim. Biophys. Acta*, 348, 299, 1974.
396. **Walker, R. W. and Bastl, C. P.**, The glycolipids of *Arthrobacter globiformis, Carbohydr. Res.*, 4, 49, 1967.
397. **Ware, J. C. and Dworkin, M.**, Fatty acids of *Myxococcus xanthus, J. Bacteriol.*, 115, 253, 1973.
398. **Weaver, T. L., Patrick, M. A., and Dugan, P. R.**, Whole-cell and membrane lipids of the methylotrophic bacterium *Methylosinus trichosporium, J. Bacteriol.*, 124, 602, 1975.
399. **Weerkamp, A. and Heinen, W.**, The effect of nutrients and precursors on the fatty acid composition of two themophilic bacteria, *Arch. Microbiol.*, 81, 350, 1972.
400. **Weerkamp, A. and Heinen, W.**, Effect of temperature on the fatty acid composition of the extreme thermophiles *Bacillus caldolyticus* and *Bacillus caldotenax, J. Bacteriol.*, 109, 443, 1972.
401. Welby-Gieusse, M., Lanéelle, M.-A., and Asselineau, J., Structure des acides corynomycoliques de *Corynebacterium hofmanii* et leur implication biogenétique, *Eur. J. Biochem.*, 13, 164, 1970.
402. **Welsh, K., Shaw, N., and Baddiley, J.**, The occurrence of acylated sugar derivatives in the lipids of bacteria, *Biochem. J.*, 107, 313, 1968.
403. **White, D. A., Lennarz, W. J., and Schnaitman, C. A.**, Distribution of lipids in the wall and cytoplasmic membrane subfractions of the cell envelope of *Escherichia coli, J. Bacteriol.*, 109, 686, 1972.
404. **White, D. C., Tucker, A. N., and Sweeley, C. C.**, Characterization of the *iso*-branched sphinganines from the ceramide phospholipids of *Bacteroides melaninogenes, Biochim. Biophys. Acta*, 187, 527, 1969.
405. **Whiteside, T. L., de Siervo, A. J., and Salton, M. R. J.**, Use of antibody to membrane adenosine triphosphatase in the study of bacterial relationships, *J. Bacteriol.*, 105, 957, 1971.
406. **Wicken, A. J. and Knox, K. W.**, Lipoteichoic acids: a new class of bacterial antigen, *Science*, 187, 1161, 1975.
407. **Wilkinson, B. J., Morman, M. R., and White, D. C.**, Phospholipid composition and metabolism of *Micrococcus denitrificans, J. Bacteriol.*, 112, 1288, 1972.
408. **Wilkinson, B. J., White, D. C., and White, P. J.**, Lipids of *Pediococcus cerevisiae* and some methicillin-resistant substrains, *Antonie van Leeuwenhoek J. Microbiol. Serol.*, 40, 39, 1974.
409. **Wilkinson, S. G.**, Composition and structure of the ornithine-containing lipid from *Pseudomonas rubescens, Biochim. Biophys. Acta*, 270, 1, 1972.
410. **Wilkinson, S. G.**, Artifacts produced by acidic hydrolysis of lipids containing 3-hydroxy-alkanoic acids, *J. Lipid Res.*, 15, 181, 1974.
411. **Wilkinson, S. G. and Bell, M. E.**, The phosphoglucolipid from *Pseudomonas diminuta, Biochim. Biophys. Acta*, 248, 293, 1971.
412. **Wilkinson, S. G., Galbraith, L., and Lightfoot, G. A.**, Cell walls, lipids and lipopolysaccharides of *Pseudomonas* species, *Eur. J. Biochem.*, 33, 158, 1973.
413. **Wolf-Watz, H., Elmros, T., Normark, S., and Bloom, G. D.**, Cell envelope of *Neisseria gonorrhoeae*: outer membranes and peptidoglycan composition of penicillin-sensitive and -resistant strains, *Infect. Immun.*, 11, 1332, 1975.
414. **Woodrow, G. C., Cheung, H. T., and Cho, K. Y.**, Phospholipid of an extremely halophilic bacterium, *Sarcina morrhuae, Aust. J. Biol. Sci.*, 26, 787, 1973.
415. **Yabuuchi, E., Yano, I., Goto, S., Tanimura, E., Tomiyoshi, I., and Ohyama, A.**, Description of *Achromobacter xylosoxidans*, Yabuuchi and Ohyama, 1971, *Int. J. Syst. Bacteriol.*, 24, 470, 1974.
416. **Yamamura, Y., Misaki, A., and Azuma, I.**, Chemical and immunological studies on polysaccharide antigens of mycobacteria, nocardiae and corynebacteria, *Bull. Int. Union Tuberc.*, 47, 181, 1971.
417. **Yanagawa, S., Fujii, K., Tanaka, A., and Fukui, S.**, Lipid composition and localization of 10-methyl branched-chain fatty acids in *Corynebacterium simplex* grown on n-alkanes, *Agric. Biol. Chem.*, 36, 2123, 1972.
418. **Yanagawa, S., Tanaka, A., and Fukui, S.**, Fatty acid compositions of *Corynebacterium simplex* grown on 1-alkenes, *Agric. Biol. Chem.*, 36, 2129, 1972.
419. **Yano, I. and Saito, K.**, Gas-chromatographic and mass-spectrometric analysis of molecular species of corynomycolic acids from *Corynebacterium ulcerans, FEBS Lett.*, 23, 352, 1972.

420. **Yano, I., Furukawa, Y., and Kusunose, M.,** Occurrence of α-hydroxy fatty acids in Actinomycetales, *FEBS Lett.*, 4, 96, 1969.
421. **Yano, I., Furukawa, Y., and Kusunose, M.,** Phospholipids of *Nocardia coeliaca, J. Bacteriol.*, 98, 124, 1969.
422. **Yano, I., Furukawa, Y., and Kusunose, M.,** Fatty acid composition of *Arthrobacter simplex* grown on hydrocarbons, *Eur. J. Biochem.*, 23, 220, 1971.
423. **Yano, I., Furukawa, Y., and Kusunose, M.,** Occurrence of acylated trehalose in *Nocardia, J. Gen. Appl. Microbiol.*, 17, 329, 1971.
424. **Yano, I., Saito, K., Furukawa, Y., and Kusunose, M.,** Structural analysis of molecular species of nocardomycolic acids from *Nocardia erythropolis* by the combined system of gas chromatography and mass spectrometry, *FEBS Lett.*, 21, 215, 1972.
425. **Yao, M., Walker, H. W., and Lillard, D. A.,** Fatty acids from vegetative cells and spores of *Bacillus stearothermophilus, J. Bacteriol.*, 102, 877, 1970.
426. **Yin, E. J., Simon, I., and Simon, H. J.,** Study of *Staphylococcus aureus* and its stable L-form variant: Changes in cellular composition as a function of long-chain fatty acid supplementation of culture medium, *J. Gen. Appl. Microbiol.*, 21, 87, 1975.
427. **Yribarren, M., Vilkas, E., and Rozanis, J.,** Galactosyl-diglyceride from *Actinomyces viscosus, Chem. Phys. Lipids*, 12, 172, 1974.

BACTERIAL CELL BREAKAGE OR LYSIS

E. H. Cota-Robles and S. M. Stein

The primary goal of breaking microbial cells is the isolation of cellular components, so that more detailed analysis can be made of the properties of these components. The ultimate goal of such investigations is to develop an understanding of how these components function in an integrated state in the intact cell. Thus, the investigator who wishes to study in detail the permeability properties of a bacterial membrane finds it necessary to isolate the membrane or portions of the membrane in order to make the experimental inquiries desired. Similarly, the investigator wishing to study the organization of ribosomal subunits finds it necessary to disrupt cells.

Occasionally, however, investigators utilize cell breakage to bring a halt to certain biochemical reactions. In particular, bacterial virologists have used cell breakage to isolate viral components, so that a temporal as well as biochemical analysis of viral replication can be undertaken.

In 1960, Marr* published a review on the localization of enzymes in bacteria, which succinctly described the principal methods available at that time for effective breakage of bacterial cells. Marr notes that the most frequently used methods were the following: (1) grinding with an abrasive, (2) sonication, (3) ballistic disintegration, and (4) osmotic shock. Amazingly, in this era of technological development few significant innovations have been introduced for disruption of bacterial cells. It is true that the lysozyme-EDTA [(ethylenedinitrilo)tetraacetic acid, edetic acid] method of Repaske with its infinite variations is the most frequently used method currently applied. In particular, EDTA treatment has been found to effect release of periplasmic substances as well as of cell wall components. Repaske's system, supplemented with both ionic and nonionic detergents, has found extensive utility in molecular biology.

The methods cited in Table 1 include physical, chemical, and biological systems that have been used with some success to disrupt bacterial cells. These methods range from application of explosive decompression to the utilization of direct microbial activity, such as the action of *Bdellovibrio bacteriovirus* to lyse cells. The method of choice is controlled by the experimental goals of the investigator. Generally the investigator must select a method that permits his particular experimental system to be preserved. Thus, for the isolation of a fairly stable cytoplasmic enzyme, such as hexokinase, one need have little worry about inactivation of the enzyme by the use of the methods described. However, when an experimentor desires to investigate a multienzyme system, such as the electron transport system, he must use gentle means to disrupt the cell. Similarly, if an investigator wishes to isolate native and undegraded nucleic acids, he must select a method that involves fairly gentle means of disruption.

* *The Bacteria*, Vol. 1, Gunsalus, I. C. and Stanier, R. Y., Eds., Academic Press, New York, 1960, pp. 443—468.

Table 1
METHODS FOR BACTERIAL CELL BREAKAGE OR LYSIS

Method of lysis or breakage	Activity factor(s)	Sensitive organism(s)	Reference
	Physical Methods		
Ballistic disintegration			1—13
Explosive decompression			14, 15
Grinding			16, 17
Osmotic shock	Sudden dilution with water	Halophiles and halotolerant organisms	18
	Low salt concentration	*Halobacterium cutirubrum*	19
	Decreased salt concentration	*Halobacterium salinarum*	20
	Brief exposure to 1 *M* glycerol	*Azotobacter agilia; Rhodospirillum rubrum; Serratia plymuthica*	21
	Low salt concentration	*Micrococcus halodenitrificans*	22
Pressure			23—29
Sonication			30—33
		Actinomycetes	34—36
Temperature extremes		Red halophiles	37
	25°C	*Escherichia coli; Aerobacter aerogenes*	38
	Heat treatment	*Escherichia coli; Salmonella pullorum; Pseudomonas fluorescens; P. pyocyanea*	39
	25°C	*Staphylococcus aureus*	40
	Temperatures above optimum	Psychrophiles	41
	Chemical Methods		
Acids	Dichlorophenoxybutyric acid	*Bacillus subtilis*	42
	Short-chain fatty acids	*Escherichia coli*	43
	Sodium salts of acetic, butyric, formic and propionic acid	*Escherichia coli*	43
	p-Chloromercurobenzoic acid	*Escherichia coli*	44
Buffers	Tris® buffer	*Bacillus cereus*	45
	Sodium acetate buffer	*Escherichia coli*	45
	Tris® + EDTA	*Vibrio cholerae*	46
Cations	Addition of monovalent cations; also heat, urea, and ionic detergents	Red halophiles	47
	Depletion of divalent cations	*Pseudomonas aeruginosa*	48
		Marine pseudomonads	49
Cystamine	Cystamine and closely related derivatives	*Bacillus subtilis* and other *Bacillus* strains	50, 51
Detergents	Anionic and nonionic detergents; also succinylation, phospholipase A, alkaline phosphatase, trypsin, and chymotrypsin	*Mycoplasma laidlawii*, strain B	52
EDTA		*Pseudomonas aeruginosa*	53—55
Glycine		*Aerobacter aerogenes; Bacillus mesentericus; Escherichia coli*	56
		Escherichia coli	57

Table 1 (Continued)
METHODS FOR BACTERIAL CELL BREAKAGE OR LYSIS

Method of lysis or breakage	Activity factor(s)	Sensitive organism(s)	Reference
	Chemical Methods (continued)		
		Mycobacterium smegmatis	58
		Salmonella typhi	59
		Vibrio cholerae; V. El Tor; *Salmonella typhi; S. typhimurium; S. paratyphi* B	60
Physiological solutions	Ringer's solution	*Bacillus anthracis*	61, 62
Zeolite, synthetic		Most bacteria	63
	Biological Methods		
	Metabolic disturbance		
Addition of amino acids	D-Methionine or D-alanine	*Alcaligenes faecalis*	64
	D-Galactose	*Salmonella enteritidis*	65, 66
	Glycine, aspartate, or arginine	*Salmonella typhosa*	67
Antibiotics	Amphomycin	Gr^+ bacteria	68
	Aspartocin	Gr^+ bacteria	69, 70
	Bacitracin	Gr^+ bacteria; meningococcus; gonococcus	71—73
	Bacitracin; also novobiocin, ristocetin, and Vancomycin®	Staphylococci	74—77
	Cephalosporin C	Wide range	78, 79
	Circulin®	Gr^+ and Gr^- bacteria; Gr^- bacteria are more sensitive than Gr^+ bacteria	80
	D-Cycloserine	*Escherichia coli*	81
		Streptococcus faecalis	82
	Megacin	*Bacillus megaterium*	83
	Penicillin	*Bacillus megaterium*	84
		Escherichia coli	85—88
		Staphylococcus; *Escherichia coli*	89—96
	Polymyxin	Gr^- bacteria; some Gr^+ bacteria	97, 98
	Pyocin	*Pseudomonas aeruginosa*	99
	Ristocetin	Gr^+ bacteria; mycobacteria	100, 101
	Subtilin	Gr^+ bacteria	102, 103
	Vancomycin®	Gr^+ bacteria, especially *Staphylococcus aureus*	104—107
Nutritional deficiency	D-Glutamic acid limitation	*Bacillus subtilis*	108
	Glucose depletion	*Bacillus subtilis*	109
	Diaminopimelic acid deprivation	Mutant of *Escherichia coli* requiring diaminopimelic acid	110—113
	N-Acetyl-D-glucosaminidase deficiency	*Lactobacillus bifidus* var. *pennsylvanicus*	114, 115
	Medium depletion	*Streptococcus faecalis*	116—119
	L-Lysine deprivation	*Streptococcus faecalis;* vitamin B-deficient mutant of *S. faecalis*	120, 121
	D-Alanine deprivation	*Streptococcus faecalis;* vitamin B-deficient mutant of *S. faecalis*	120, 121
Oxygen	Aeration during active growth	Anaerobes	122

Table 1 (continued)
METHODS FOR BACTERIAL CELL BREAKAGE OR LYSIS

Method of lysis or breakage	Activity factor(s)	Sensitive organism(s)	Reference
	Biological Methods (continued)		
	Metabolic disturbance (continued)		
	Halt of aeration and establishment of semiaerobic conditions	*Bacillus megaterium*	123
	Sudden removal of oxygen from aerated cultures in the log phase	*Bacillus subtilis*	124, 125
	Enzymes, microbial		
Aeromonas	Enzyme from *Aeromonas hydrophila*	*Staphylococcus aureus*	126
Bacillus	Lytic factor induced in *Bacillus cereus* 569	*Bacillus cereus* 130	127
	Filtrate of *Bacillus cereus*	Aerobic spore formers	128
	Enzyme from spores of *Bacillus cereus*	*Bacillus cereus*	129
	Enzyme from spores of *Bacillus megaterium*	*Bacillus megaterium*	130
	Enzyme from sporulating cells of *Bacillus megaterium*	*Bacillus megaterium*	131
	Autolysin from *Bacillus stearothermophilus*	*Bacillus stearothermophilus*	132
	Autolysin from *Bacillus subtilis*	*Bacillus subtilis*	133, 134
	Filtrate of *Bacillus cereus*	*Micrococcus lysodeikticus*	135
Chalaropsis	Enzyme from the fungus *Chalaropsis*	*Staphylococcus aureus*	136—138
Escherichia coli	Enzyme from a defective lysogenic mutant of *Escherichia coli* K_{12} (λ)	Gr^- bacteria; some Gr^+ bacteria, i.e., *Bacillus megaterium* and *Staphylococcus aureus*	139
	Autolysin from *Escherichia coli*	*Escherichia coli*	140—142
Flavobacterium	L_{11} enzyme from *Flavobacterium*	*Staphylococcus aureus; Micrococcus lysodeikticus*	143
Myxobacterium	Enzyme from *Myxobacterium*	*Arthrobacter crystallopietes; Staphylococcus aureus; Micrococcus lysodeikticus; Sarcina lutea; Rhodospirillum rubrum*	144
		Aerobacter aerogenes; Escherichia coli; Pseudomonas fluorescens; Serratia marcescens; Sarcina lutea; Bacillus subtilis; B. megaterium	145—150
		Staphylococcus aureus, strain Copenhagen	151
Pseudomonas	Enzyme from *Pseudomonas*	*Staphylococcus aureus; S. roseus; Gaffkya tetragena; Sarcina lutea*	152
Sorangium	Enzyme from *Sorangium*	*Arthrobacter globiformis; Micrococcus lysodeikticus*	153, 154
Staphylococcus	Enzyme from *Staphylococcus aureus*	*Micrococcus lysodeikticus*	155, 156

Table 1 (Continued)
METHODS FOR BACTERIAL CELL BREAKAGE OR LYSIS

Method of lysis or breakage	Activity factor(s)	Sensitive organism(s)	Reference
	Biological Methods (continued)		
	Enzymes, microbial (continued)		
	Autolysin from *Staphylococcus aureus*	*Staphylococcus aureus*	157, 158
	Lysostaphin from *Staphylococcus staphylolyticus*	Staphylococci	159
	Enzyme from *Staphylococcus epidermidis*	Staphylococci	160
Streptococcus	Autolysin from *Streptococcus faecalis*	*Streptococcus faecalis*	161—164
	Muralytic enzyme from Group C *Streptococcus*	Group A *Streptococcus*	165
Streptomyces	Enzyme from *Streptomyces albus* G	*Staphylococcus aureus; Micrococcus lysodeikticus; M. roseus; Corynebacterium diphtheriae*	166
		Heat-killed Gr⁻ species; many living Gr⁺ species	167, 168
	R1 enzyme fraction from *Streptomyces albus* G	*Micrococcus radiodurans*	169
	L_3 enzyme from *Streptomyces*	*Corynebacterium diphtheriae*	170
	Enzyme from *Streptomyces* F_1	Gr⁺ organisms	171
	Enzymes, bacteriophage-induced		
Bacillus			172, 173
Escherichia coli			174—178
Klebsiella pneumoniae			179
Staphylococcus			180—182
Streptococcus			183—190
Lysozyme		Bacteria from stationary-phase cells are less sensitive to lysozyme than actively dividing cells	191
	Lysozyme	*Bacillus* sp.	192
		Bacillus megaterium	193, 194
		Bacillus subtilis	195, 196
		Escherichia coli	197
		Micrococcus lysodeikticus	198—205
		Sarcina lutea	206—208
		Streptococcus faecalis F24; *S. faecalis* var. *liquefaciens* 31; S. *faecalis* E1	209
		Sarcina flava	210
	Lysozome + EDTA	*Azotobacter vinelandii*	211
		Salmonella typhi	212
		Escherichia coli	213—217
		Samonella paratyphi B	218, 219
	Lysozyme + anions	Enterococci	220
	Lysozyme + Brij®-58	*Escherichia coli*	221, 222
	Lysozyme + phagocytin	*Escherichia coli*	223
	Lysozyme + quaternary ammonia compounds	Gr⁻ bacteria	224
	Lysozyme + serum or heat treatment or thioglycolic acid	Gr⁻ bacteria	225, 226

Table 1 (continued)
METHODS FOR BACTERIAL CELL BREAKAGE OR LYSIS

Method of lysis or breakage	Activity factor(s)	Sensitive organism(s)	Reference
	Biological Methods (continued)		
	Lysozyme + EDTA or lipase or freezing and thawing	Gr^- bacteria	227
	Lysozyme + trypsin or trypsin and butanol	Gr^+ bacteria	228
	Lysozyme + EDTA + trypsin	*Lactobacillus casei*	229
	Polymyxin B sulfate followed by lysozyme	*Neisseria catarrhalis*	230
	Lysozyme + lipase (steapsin)	*Listeria monocytogenes*	231
	n-Butanol-saturated buffer + lysozyme	*Micrococcus radiodurans*	232
	Lysozyme + sodium dodecylsulfate	*Micrococcus radiodurans*	233
	Lysozyme or glycine	*Mycobacterium smegmatis*	234
	Lysozyme + EDTA + Tris®	*Pseudomonas aeruginosa*	235 236
	Lysozyme + EDTA + lack of Ca^{++}	*Spirillum serpens*	
	Nafcillin (a semisynthetic penicillin) followed by lysozyme or trypsin	*Staphylococcus aureus*	237, 238
	Nafcillin, ampicillin, or cloxacillin and cephalothin followed by lysozyme or trypsin	*Staphylococcus aureus*	239
	Tris® + EDTA + lysozyme or sodium lauryl sulfate	*Vibrio* El Tor	240, 241
Other enzymes	Leucozyme C	*Escherichia coli*	242
	Phagocytin	*Escherichia coli*	243
	Anionic and nonionic detergents; also succinylation, phospholipase A, alkaline phosphatase, trypsin, and chymotrypsin	*Mycoplasma laidlawii,* strain B	244
Immune substances	Complement system + lysozyme	Gr^- bacteria that are susceptible to immune bactericidal reaction	245—249
Direct microbial activity	*Bdellovibrio bacteriovorus*	Gr^- bacteria	250—254

REFERENCES

1. **Cummins, C. S. and Harris, H.,** *J. Gen. Microbiol.,* 14, 583, 1956.
2. **Huff, E. Oxley, H., and Silverman, C. S.,** *J. Bacteriol.,* 88, 1155, 1964.
3. **King, H. K. and Alexander, H.,** *J. Gen. Microbiol.,* 2, 315, 1948.
4. **Mandelstam, J.,** *Biochem. J.,* 84, 294, 1962.
5. **Merkenschlager, M., Schlossmann, K., and Kurz, W.,** *Biochem. Z.,* 329, 332, 1957.

6. Mickle, H., *J. R. Microsc. Soc.*, 68, 10, 1948.
7. Nossal, P. M., *Aust. J. Exp. Biol. Med., Sci.*, 31, 583, 1953.
8. Pickering, B. T., *Biochem. J.*, 100, 430, 1966.
9. Sagniez, G., LeCam, M., Madec, Y., and Bernard, S., *Ann. Inst. Pasteur Paris*, 177, 663, 1969.
10. Salton, M. R. J. and Horne, R. W., *Biochim. Biophys. Acta*, 7, 177, 1951.
11. Sharon, N. and Jeanloz, R. W., *Experientia*, 20, 253, 1962.
12. Shockman, G. D., *Biochim. Biophys. Acta*, 59, 234, 1962.
13. Shockman, G. D., Kalb, J. J., and Toennies, G., *Biochim. Biophys. Acta*, 24, 203, 1957.
14. Foster, J. A., Cowan, R. M., and Maag, T. A., *J. Bacteriol.*, 83, 330, 1962.
15. Van Eseltine, W. P., Jones, R. W., and Gilliard, F. E., *Proc. Soc. Exp. Biol. Med.*, 131, 1446, 1969.
16. McIlwain, H., *J. Gen. Microbiol.*, 2, 288, 1948.
17. Wiggert, W. P., Silverman, M., Utler, M. F., and Werkman, C. H.,, *Iowa State J. Sci.*, 14, 179, 1939.
18. Ingram, M., in *Microbial Ecology*, Spicer, R. E. O. and Spicer, C. C., Eds., Cambridge University Press, New York and London, 1957, p. 90.
19. Kushner, D. J., *J. Bacteriol.*, 87, 1147, 1964.
20. Mohr, V., and Larsen, H., *J. Gen. Microbiol.*, 31, 267, 1963.
21. Robrish, S. A., and Marr, A. G., *Bacteriol. Proc.*, p. 130, 1957.
22. Takahashi, L. and Gibbons, N. E., *Can. J. Microbiol.*, 5, 25, 1959.
23. Edebo, L., *J. Biochem. Microbiol. Technol. Eng.*, 2, 453, 1960.
24. French, C. S. and Milner, H. W., in *Methods in Enzymology*, Vol. I, Colowick, S. P. and Kaplan, N. O., Eds., Academic Press, New York, 1955, p. 64.
25. Hughes, D. E., *Br. J. Exp. Pathol.*, 32, 97, 1950.
26. Milner, H. W., Lawrence, N. S., and French, C. S., *Science*, 111, 633, 1950.
27. Perrine, T. D., Ribi, E., Maki, W., Miller, B., and Oertli, E., *Appl. Microbiol.*, 10, 93, 1962.
28. Ribi, E., Perrine, T., List, R., Brown, W., and Goode, G., *Proc. Soc. Exp. Biol. Med.*, 100, 647, 1959.
29. Vanderheiden, G. J., Fairchild, A. C., and Jago, E. G. R., *Appl. Microbiol.*, 19, 875, 1970.
30. Bosco, G., *J. Infect. Dis.*, 99, 270, 1956.
31. Harvey, E. N., *Biol. Bull. Woods Hole, Mass.*, 59, 306, 1930.
32. Ikawa, M., and Snell, E. E., *J. Biol. Chem.*, 235, 1376, 1960.
33. Marr, A. G. and Cota-Robles, E. H., *J. Bacteriol.*, 74, 79, 1957.
34. Becker, B., Lechevalier, M. P., and Lechevalier, H. A., *Appl. Microbiol.*, 13, 236, 1965.
35. Roberson, B. S., and Schwab, J. H., *Biochim. Biophys. Acta*, 44, 436, 1960.
36. Salton, M. R. J., *J. Gen. Microbiol.*, 9, 512, 1953.
37. Abram, D., and Gibbons, N. E., *Can. J. Microbiol.*, 7, 741, 1961.
38. Mitchell, P. and Moyle, J., *Nature*, 178, 993, 1956.
39. Salton, M. R. J. and Horne, R. W., *Biochim. Biophys. Acta*, 7, 19, 1951.
40. Mitchell, P. and Moyle, J., *J. Gen. Microbiol.*, 16, 184, 1957.
41. Hagan, P. O., Kushner, D. J., and Gibbons, N. E., *Bacteriol. Proc.*, p. 42, 1964.
42. Roman, M. and Gonzales, C., *Microbiol. Esp.*, 20, 63, 1967.
43. Mayo, J. A. and Church, B. D., *Bacteriol. Proc.*, p. 84, 1966.
44. Schaechter, M. and Santomassino, K. A., *J. Bacteriol.*, 84, 318, 1962.
45. Mohan, R. R., Kronish, D. P., Pianotti, R. S., Epstein, R. L., and Schwartz, B. S., *J. Bacteriol.*, 90, 1355, 1965.
46. Adhikari, P. C., Raychaudhuri, C., and Chatterjee, S. N., *J. Gen. Microbiol.*, 59, 91, 1969.
47. Abram, D. and Gibbons, N. E., *Can. J. Microbiol.*, 7, 741, 1961.
48. Brown, M. R. and Melling, J., *J. Gen. Microbiol.*, 59, 263, 1969.
49. Buckmire, F. L. A. and MacLeod, R. A., *Can. J. Microbiol.*, 11, 677, 1965.
50. Weinberg, E. D., *Exp. Cell Res.*, 13, 175, 1957.
51. Weinberg, E. D., Saz, A. K., and Pilgren, E. Y., *J. Gen. Microbiol.*, 19, 419, 1958.
52. Smith, P. F., Koostra, W. L., and Mayberry, W. R., *J. Bacteriol.*, 100, 1166, 1969.
53. Carson, K. J. and Eagon, R. G., *Bacteriol. Proc.*, p. 32, 1964.
54. Gray, G. W. and Wilkinson, S. G., *J. Appl. Bacteriol.*, 28, 153, 1965.
55. Eagon, R. G., Simmons, G. P., and Carson, K. J., *Can. J. Microbiol.*, 11, 1041, 1965.
56. Maculla, E. S. and Cowels, P. W., *Science*, 107, 376, 1948.
57. Gordon, J., Hall, R. A., and Strickland, L. H., *J. Pathol. Bacteriol.*, 64, 299, 1962.
58. Adamek, L. P., Misson, H., Mohelska, H., and Trinka, L., *Arch. Mikrobiol.*, 69, 227, 1969.
59. Diena, B. B., Wallace, R., and Greenberg, L., *Can. J. Microbiol.*, 10, 543, 1964.
60. Jeynes, M. H., *Nature*, 180, 867, 1957.
61. Stahelin, H., *Schweiz. Z. Allg. Pathol. Bakteriol.*, 17, 296, 1954.
62. Stahelin, H., *Schweiz. Z. Allg. Pathol. Bakteriol.*, 16, 111, 1953.

63. Wistreich, G., Lechtman, M. D., Bartholomew, J. W., and Bils, R. F., *Appl. Microbiol.*, 16, 1269, 1968.
64. Lark, C. and Schichtel, R., *J. Bacteriol.*, 84, 1241, 1962.
65. Fukasawa, T. and Mikaido, H., *Nature*, 183, 1131, 1959.
66. Fukasawa, T. and Mikaido, H., *Biochim. Biophys. Acta*, 48, 470, 1961.
67. Nasier, M. M. R. and Ghatak, S., *Indian J. Microbiol.*, 7, 91, 1967.
68. Heinemann, B., Kaplan, M. A., Muir, R. D., and Hooper, I. R., *Antibiot. Chemother.*, 3, 1239, 1953.
69. Shay, A. J., Adam. J., Martin, J. H., Hausmann, W. K., Shu, P., and Bohonos, N., in *Antibiotics Annual 1959—60*, Marti-Ibanez, F., Ed., Antibiotica, New York, 1960, p. 194.
70. Kirsch, E. J., Dornbush, A. C., and Backus, E. K., in *Antibiotics Annual 1959—60*, Marti-Ibanez, Ed., Antibiotica, New York, 1960, p. 205.
71. Abraham, E. P., in *CIBA Lectures in Microbiol. Biochemistry*, John Wiley & Sons, New York, 1957, p. 1.
72. Abraham, E. P., and Newton, G. G. F., in *CIBA Foundation Symposium on Amino Acids and Peptides with Antimetabolic Activity*, Wolstenholme, G. E. W. and O'Connor, C. M., Eds., Little, Brown & Co., Boston, 1958, p. 205.
73. Johnson, B. A., Anker, H., and Meleney, F. L., *Science*, 102, 376, 1945.
74. Park, J. T., *J. Biol. Chem.*, 194, 897, 1952.
75. Jordan, D. C., *Biochem. Biophys. Res. Commun.*, 6, 167, 1961.
76. Reynolds, P. E., *Biochim. Biophys. Acta*, 52, 403, 1961.
77. Wallas, C. H. and Strominger, J. L., *J. Biol. Chem.*, 238, 2264, 1963.
78. Newton, G. G. F. and Abraham, E. P., *Biochem. J.*, 62, 651, 1956.
79. Newton, G. G. F. and Abraham, E. P., *Biochem. J.*, 58, 103, 1954.
80. Murray, F. J., Tetrault, P. A., Kaufman, O. W., and Koffler, H., *J. Bacteriol.*, 57, 305, 1949.
81. Ciak, J. and Hahn, F. E., *Antibiot. Chemother.*, 9, 47, 1959.
82. Shockman, G. D., *Proc. Soc. Exp. Biol. Med.*, 101, 693, 1959.
83. Ivanovics, G. L., Alfoldi, L., and Nagy, E., *J. Gen. Microbiol.*, 21, 51, 1959.
84. Fedorova, G. I., *Antibiotiki*, 14, 880, 1969.
85. Chargaff, E., Schidman, H. M., and Shapiro, H. S., *Nature*, 180, 851, 1957.
86. Hahn, F. E., and Ciak, J., *Science*, 125, 119, 1957.
87. Lederberg, J., *Proc. Natl. Acad. Sci. U.S.A.*, 42, 574, 1956.
88. Liebermeister, K. and Kellenberger, E., *Z. Naturforsch. Teil B*, 118, 200, 1956.
89. Mandelstam, J. and Rogers, H. J., *Biochem. J.*, 72, 654, 1959.
90. Nathenson, S. G. and Strominger, J. L., *J. Pharmacol. Exp. Ther.*, 131, 1, 1961.
91. Park, J. T., *Biochem. J.*, 70, 2P, 1958.
92. Park, J. T., and Strominger, J. L., *Science*, 125, 99, 1957.
93. Roberts, J. and Johnson, M. J., *Biochim. Biophys. Acta*, 59, 458, 1962.
94. Rogers, H. J. and Jeljaszewicz, J., *Biochem. J.*, 84, 576, 1962.
95. Rogers, H. J., in *Resistance of Bacteria to the Penicillins*, (CIBA Foundation Group #13), deReuck, A. V. S. and Cameron, M. P., Eds., Little, Brown & Co., Boston, 1962, p. 25.
96. Wylie, E. B. and Johnson, M. J., *Biochim. Biophys. Acta*, 59, 450, 1962.
97. Newton, B. A., *Bacteriol. Rev.*, 20, 14, 1956.
98. Newton, B. A., *J. Gen. Microbiol.*, 9, 54, 1953.
99. Jacob, F., *Ann. Inst. Pasteur Paris*, 86, 149, 1954.
100. Graudy, W. E., Sinclair, A. C., Theriault, R. J., Goldstein, A. W., Rickher, C. J., Warren, H. B., Jr., Oliver, T. J., and Sylvester, J. C., in *Antibiotics Annual 1956—57*, Welch, H. and Marti-Ibanez, F., Eds., Medical Encyclopedia, New York, 1957, p. 680.
101. Philip, J. E., Schenck, J. R., and Hargie, M. P., in *Antibiotics Annual 1956—1957*, Welch, H. and Marti-Ibanez, F., Eds., Medical Encyclopedia, New York, 1957, p. 699.
102. Bricas, E. and Fromageot, C. L., *Adv. Protein Chem.*, 4, 57, 1953.
103. Jansen, E. F. and Hirschmann, D. J., *Arch. Biochem.*, 4, 297, 1944.
104. Jordan, D. C. and Inniss, W. E., *Nature*, 184, 1894, 1961.
105. Jordan, D. C., *Biochem. Biophys. Res. Commun.*, 6, 167, 1961.
106. McCormick, M. H., Stark, W. M., Pittenger, G. E., Pittenger, R. C., and McGuire, J. M., in *Antibiotics Annual 1955—56*, Welch, H. and Marti-Ibanez, F., Eds., Medical Encyclopedia, New York, 1956, p. 606.
107. Reynolds, P. E., *Biochim. Biophys. Acta*, 52, 403, 1961.
108. Momose, H. J., *Gen. Appl. Microbiol.*, 7(Suppl. 1), 359, 1961.
109. Hadjipetrou, L. P. and Stouthamer, A. H., *Antonie van Leeuwenhoeck J. Microbiol. Serol.*, 29, 256, 1963.
110. Davis, B. D. and Bauman, N., *Science*, 126, 170, 1957.

111. **Davis, B. D.,** *Nature,* 169, 534, 1952.
112. **Meadow, P., Hoare, D. S., and Work, E.,** *Biochem. J.,* 66, 270, 1957.
113. **Rhuland, L. E.,** *J. Bacteriol.,* 73, 778, 1957.
114. **O'Brien, P. J., Glick, M. G., and Zilliken, F.,** *Biochim. Biophys. Acta,* 37, 357, 1960.
115. **Glick, M. G., Sall, T., Zilliken, F., and Mudd, S.,** *Biochim. Biophys. Acta,* 37, 361, 1960.
116. **Shockman, G. D., Conover, M. J., Kolb, J. J., Phillips, P. M., Riley, L. S., and Toennies, G.,** *J. Bacteriol.,* 81, 36, 1961.
117. **Shockman, G. D.,** *Bacteriol. Rev.,* 29, 345, 1965.
118. **Shockman, G. D.,** *Trans. N. Y. Acad. Sci. Ser. II,* 26, 182, 1963.
119. **Toennies, G. and Shockman, G. D.,** *Proc. Int. Congr. Biochem.,* 4, 365, 1958.
120. **Shockman, G. D., Kolb, J. J., and Toennies, G.,** *J. Biol. Chem.,* 230, 961, 1958.
121. **Toennies, G., and Gallant, D. L.,** *Growth,* 13, 7, 1949.
122. **Stolp, H.,** *Arch. Mikrobiol.,* 21, 293, 1955.
123. **Kawata, T., Asaki, K., and Takagi, A.,** *J. Bacteriol.,* 81, 160, 1961.
124. **Kaufmann, W., and Bauer, K.,** *J. Gen. Microbiol.,* 18, 11, 1958.
125. **Nomura, M., and Hosoda, J.,** *J. Bacteriol.,* 72, 573, 1956.
126. **Coles, N. W., Gilbo, C. M., and Broad, A. J.,** *Biochem, J.,* 111, 7, 1969.
127. **Csuzi, S. and Kramer, M.,** *Acta Microbiol Acad. Sci. Hung.,* 9, 297, 1962.
128. **Norris, J. R.,** *J. Gen. Microbiol.,* 16, 1, 1957.
129. **Strange, R. E. and Dark, F. E.,** *J. Gen. Microbiol.,* 16, 236, 1957.
130. **Strange, R. E. and Dark, F. E.,** *J. Gen. Microbiol.,* 17, 525, 1957.
131. **Dark, F. E. and Strange, R. E.,** *Nature,* 180, 759, 1957.
132. **Welker, N. E. and Campbell, L. L.,** *Bacteriol. Proc.,* p. 126, 1966.
133. **Young, F. E. and Spizizen, J. J.,** *J. Biol. Chem.,* 238, 3126, 1963.
134. **Young, F. E., Tipper, D. J., and Strominger, J. L.,** *J. Biol. Chem.,* 239, 3600 1964.
135. **Richmond, M. H.,** *Biochim. Biophys. Acta,* 33, 78, 1959.
136. **Hash, J. H., Wishnick, M., and Miller, P. A.,** *J. Bacteriol,* 87, 432, 1964.
137. **Hash, J. H.** *Arch. Biochem. Biophys.,* 102, 379, 1963.
138. **Tipper, D. J., Strominger, J. L., and Ghuysen, J. G.,** *Science,* 146, 781, 1964.
139. **Jacob, F. and Fuerst, C. R.,** *J. Gen. Microbiol.,* 18, 518, 1958.
140. **Pelzer, H.,** *Z. Naturforsch. Teil B,* 18, 950, 1963.
141. **Weidel, W. and Peltzer, H.,** *Adv. Enzymol.,* 26, 193, 1964.
142. **Weidel, W., Frank, H., and Leutgeb, W.,** *J. Gen. Microbiol.,* 30, 127, 1963.
143. **Kato, K., Kotani, S., Matsubara, T., Kogami, J., Hashimoto, S., Chimori, M., and Kazekawa, I.,** *Biken J.,* 5, 155, 1962.
144. **Ensign, J. C. and Wolfe, R. S.,** *Bacteriol. Proc.,* p. 33, 1964.
145. **Kuhliwein, H.,** *Zentralbl. Bakteriol. Parasitenkd. Infektionskr. Hyg. Abt. I Orig.,* 162, 296, 1955.
146. **Noren, B.,** *Svensk. Bot. Tidskr.,* 47, 309, 1953.
147. **Noren, B.,** *Svensk. Bot. Tidskr.,* 49, 282, 1955.
148. **Noren, B.,** *Svensk. Bot. Tidskr.,* 54, 550, 1960.
149. **Noren, B.,** *Bot. Not.,* 113, 320, 1960.
150. **Noren, B. and Raper, K. B.,** *J. Bacteriol.,* 84, 157, 1962.
151. **Tipper, D. J., Strominger, J. L., and Ensign, J. C.,** *Biochemistry,* 6, 906, 1967.
152. **Zyskind, J. W., Pattee, P. A., and Lache, M.,** *Science,* 147, 1458, 1965.
153. **Tsai, C. S., Whitaker, D. R., Jurasek, L., and Gillespie, D. C.,** *Can. J. Biochem.,* 43, 1971, 1965.
154. **Whitaker, D. R.,** *Can. J. Biochem.,* 43, 1935, 1965.
155. **Arvidson, S., Holme, T., and Wadstrom, T.,** *J. Bacteriol.,* 104, 227, 1970.
156. **Richmond, M. H.,** *Biochim. Biophys. Acta,* 31, 564, 1959.
157. **Huff, E., Silverman, C. S., Adams, N. J., and Woodruff, S. A.,** *J. Bacteriol,* 103, 761, 1970.
158. **Welsh, M. and Salmon, J.,** *Ann. Inst. Pasteur Paris,* 79, 802, 1950.
159. **Schindler, C. A. and Schuhardt, V. T.,** *Proc. Natl. Acad. Sci. U.S.A.,* 51, 414, 1964.
160. **Suginaka, H., Kashiba, S., Amano, T., Kotani, S., and Imanishi, T.,** *Biken J.,* 10, 109, 1967.
161. **Bleiweis, A. S., and Krause, R. M,.** *J. Exp. Med.,* 122, 237, 1965.
162. **Conover, M. J., Thompson, J. S., and Shockman, G. D.,** *Biochem. Biophys. Res. Commun.,* 23, 713, 1966.
163. **Shockman, G. D. and Cheney, M. C.,** *J. Bacteriol.,* 98, 1199, 1969.
164. **Montague, M. D.,** *Biochim. Biophys. Acta,* 86, 588, 1964.
165. **Gooder, H. and Maxted, W. R.,** *Nature,* 182, 808, 1958.
166. **Ghuysen, J. M., Petit, J. F., Munoz, E., and Kato, K.,** *Fed. Proc.,* 25, 410, 1966.
167. **McCarty, M.,** *J. Exp. Med.,* 96, 555, 1952.
168. **Welsch, M.,** *Phenomene d'antibiose chez les Actinomycetes,* J. Duculot, Gembloux, France, 1947.
169. **Dean, C. J., Feldschreiber, P., and Lett, J. T.,** *Nature,* 209, 49, 1966.

170. Mori, Y., Kato, K., Matsubara, T., and Kotani, S., *Biken J.*, 3, 139, 1960.
171. Munoz, E., Ghuysen, J. M., Leyh-Bouille, M., Petit, J. F., and Tinelli, R., *Biochemistry*, 5, 3091, 1966.
172. Girard, P. and Sertic, V., *C. R. Seances Soc. Biol.*, 118, 1286, 1935.
173. Murphy, J. S., *Virology*, 4, 563, 1957.
174. Bradley, D. E., *J. Gen. Virol.*, 3, 141, 1968.
175. Katz, W. E. and Weidal, W., *Z. Naturforsch. Teil B*, 16, 363, 1961.
176. Panijel, J. and Happert, J., *C. R. Seances Soc. Biol.*, 245, 240, 1957.
177. Streisinger, G., Mukai, F., Dreyer, W. J., Miller, B., and Horiuchi, S., *Cold Spring Harbor Symp. Quant. Biol.*, 26, 5, 1961.
178. Weidal, W. and Katz, W., *Z. Naturforsch. Teil B*, 16, 156, 1961.
179. Humphries, J. C., *J. Bacteriol.*, 56, 683, 1948.
180. Gratia, A. and Rhodes, B., *C. R. Seances Soc. Biol.*, 89, 171, 1923.
181. Ralston, D. J., Baer, B., Lieberman, M., and Krueger, A. P., *J. Gen. Microbiol.*, 24, 313, 1961.
182. Ralston, D. J., *Bacteriol. Proc.*, p. 69, 1966.
183. Barkulis, S. S., Smith, C., Boltralik, J. J., and Heymann, H., *J. Biol. Chem.*, 239, 4025, 1964.
184. Fox, E. N., *J. Bacteriol.*, 85, 536, 1963.
185. Freimer, E. H., *J. Exp. Med.*, 117, 377, 1963.
186. Freimer, E. H., Krause, R. M., and McCarty, M., *J. Exp. Med.*, 110, 853, 1959.
187. Krause, R. M., *J. Exp. Med.*, 106, 365, 1957.
188. Markowitz, A. and Dorfman, A., *J. Biol. Chem.*, 237, 273, 1962.
189. Smith, D. G. and Shattok, R. G. E., *J. Gen. Microbiol.*, 34, 165, 1964.
190. Zeleznick, L. D., Boltralik, J. J., Barkulis, S. S., Smith, C., and Heymann, H., *Science*, 140, 400, 1963.
191. Chaloupka, J., Kreckova, P., and Rihova, L., *Folia Microbiol.*, 7, 269, 1962.
192. Tomcsik, J. and Guex-Holzer, S., *Schweiz. Z. Allg. Pathol. Bacteriol.*, 15, 517, 1952.
193. Fedorova, G. I., *Antibiotiki*, 15, 880, 1969.
194. Weibull, C., *J. Bacteriol.*, 66, 688, 1953.
195. Mutsaars, W., *Ann. Inst. Pasteur Paris*, 89, 166, 1955.
196. Wiame, J. M., Storch, R., and Vanderwinkel, E., *Biochim. Biophys. Acta*, 18, 353, 1955.
197. Spizizen, J., *Proc. Natl. Acad. Sci. U.S.A.*, 43, 694, 1957.
198. Fedorova, G. I., *Antibiotiki*, 14, 880, 1969.
199. Fleming, A., *Proc. R. Soc. London Ser. B*, 93, 306, 1922.
200. Fleming, A. and Allison, V. D., *Br. J. Exp. Pathol.*, 3, 252, 1922.
201. McQuillen, K., *Biochim. Biophys. Acta*, 17, 382, 1955.
202. Mitchell, P. and Moyle, J., *J. Gen. Microbiol.*, 15, 512, 1956.
203. Salton, M. R. J., *Nature*, 170, 746, 1952.
204. Saint-Blancard, J., Chuzel, P., Mathieu, Y., Perrot, J., and Jolles, P., *Biochim. Biophys. Acta*, 220, 300, 1970.
205. Tomcsik, J. and Guex-Holzer, S., *Schweiz. Z. Allg. Pathol. Bacteriol.*, 15, 517, 1952.
206. McQuillen, K., *Biochim. Biophys. Acta*, 17, 382, 1955.
207. Mitchell, P. and Moyle, J., *J. Gen. Microbiol.*, 15, 512, 1956.
208. Tomcsik, J. and Guex-Holzer, S., *Schweiz. Z. Allg. Pathol. Bacteriol.*, 15, 517, 1952.
209. Gooder, H., in *Microbial Protoplasts, Spheroplasts and L-Forms*, Guze, L. B., Ed., Williams & Wilkins, Baltimore, 1968, p. 40.
210. Colobert, L. and Lenoir, J., *Ann. Inst. Pasteur Paris*, 92, 74, 1957.
211. Repaske, R., *Biochim. Biophys. Acta*, 22, 189, 1956.
212. Colobert, L., *Ann. Inst. Pasteur Paris*, 95, 156, 1958.
213. Fraser, D. and Mohler, H. R., *Arch. Biochem. Biophys.*, 69, 166, 1957.
214. Mahler, H. R. and Fraser, D., *Biochim. Biophys. Acta*, 22, 197, 1956.
215. Miura, T. and Mizushima, S., *Biochim. Biophys. Acta*, 193, 268, 1969.
216. Repaske, R., *Biochim. Biophys. Acta*, 22, 189, 1956.
217. Rickenberg, H. V., *Biochim. Biophys. Acta*, 25, 206, 1957.
218. Colobert, L., *Ann. Inst. Pasteur Paris*, 95, 156, 1958.
219. Colobert, L., *C. R. Seances Soc. Biol.*, 115, 1904, 1957.
220. Metcalf, R. and Deibel, R. H., *J. Bacteriol.*, 99, 674, 1969.
221. Birdsell, D. C. and Cota-Robles, E. H., *Biochem. Biophys. Res. Commun.*, 31, 438, 1968.
222. Godson, G. N., Nigel, G., and Sinsheimer, R. L., *Biochim. Biophys. Acta*, 149, 476, 1967.
223. Zinder, N. and Arndt, W. P., *Proc. Natl. Acad. Sci. U.S.A.*, 42, 586, 1956.
224. Ceglowski, W. S. and Lear, S. A., *Appl. Microbiol.*, 10, 458, 1962.
225. Gould, G. W. and Hitchins, A. D., *Nature*, 197, 622, 1963.
226. Gould, G. W. and Hitchins, A. D., *J. Gen. Microbiol.*, 33, 413, 1963.

227. **Kohn, A.,** *J. Bacteriol.,* 79, 697, 1960.
228. **Noller, E. C. and Hartsell, S. E.,** *J. Bacteriol.,* 81, 482, 492, 1961.
229. **Barker, D. C. and Thorne, K. J. I.,** *J. Cell Sci.,* 7, 755, 1970.
230. **Warren, G. H., Gray, J., and Yurchenko, J. A.,** *J. Bacteriol.,* 74, 788, 1957.
231. **Ghosh, B. K. and Murray, R. G. E.,** *J. Bacteriol.,* 93, 411, 1967.
232. **Driedger, A. A. and Grayston, M. J.,** *Can. J. Microbiol.,* 16, 889, 1970.
233. **Kitayama, S. and Matsuyama, A.,** *Biochem. Biophys. Res. Commun.,* 33, 418, 1968.
234. **Adamek, L. P., Mison, H., Mohelska, H., and Trnka, L.,** *Arch. Mikrobiol.,* 69, 227, 1969.
235. **Repaske, R.,** *Biochim. Biophys. Acta,* 22, 189, 1956.
236. **Murray, R. G. E.,** in *Microbial Protoplasts, Spheroplasts, and L-Forms,* Guze, L. B., Ed., Williams & Wilkins, Baltimore, 1968, p. 1.
237. **Warren, G. H. and Gray, J.,** *Proc. Soc. Exp. Biol. Med.,* 114, 439, 1963.
238. **Warren, G. H. and Gray, J.,** *Proc. Soc. Exp. Biol. Med.,* 128, 776, 1968.
239. **Warren, G. H. and Gray, J.,** *Proc. Soc. Exp. Biol. Med.,* 126, 15, 1967.
240. **Adhikari, P. C., Raychaudhuri, C., and Chatterjee, S. N.,** *J. Gen. Microbiol.,* 59, 91, 1969.
241. **Birdsell, D. C. and Cota-Robles, E. H.,** *Biochem. Biophys. Res. Commun.,* 31, 438, 1968.
242. **Amano, T.,** in *Microbial Protoplasts, Spheroplasts, and L-Forms,* Guze, L. B., Ed., Williams & Wilkins, Baltimore, 1968, p. 30.
243. **Hirsch, J. G.,** *J. Exp. Med.,* 103, 598, 1956.
244. **Smith, P. F., Koostra, W. L., and Mayberry, W. R.,** *J. Bacteriol.,* 100, 1166, 1969.
245. **Crombi, L. B.,** M.S. thesis, University of Minnesota, Minneapolis, 1966.
246. **Inoue, K., Tanigawa, Y., Takubo, M., Satani, M., and Amano, T.,** *Biken J.,* 2, 1, 1959.
247. **Muschel, L. H.,** in *CIBA Foundation Symposium on Complement,* Wolstenholme, W. and Knight, J., Eds., Churchill, London, 1965, p. 153.
248. **Muschel, L. H.,** in *Microbial Protoplasts, Spheroplasts, and L-Forms,* Guze, L. B., Ed., Williams & Wilkins, Baltimore, 1968, p. 19.
249. **Wilson, L. and Spitznagel, K., Jr.,** *J. Bacteriol.,* 96, 1339, 1968.
250. **Stolp, H. and Petzold, H.,** *Phytopathol. Z.,* 45, 364, 1962.
251. **Stolp, H. and Starr, M. P.,** *Bacteriol. Proc.,* p. 47, 1963.
252. **Stolp, H. and Starr, M. P.,** *Antonie Van Leeuwenhoek J. Microbiol. Serol.,* 29, 217, 1963.
253. **Shilo, M. and Bruff, B.,** *J. Gen. Microbiol.,* 40, 317, 1965.
254. **Starr, M. P. and Bargent, N. L.,** *J. Bacteriol.,* 91, 2006, 1966.

ISOLATION OF FUNGAL CELL WALLS

W. J. Nickerson

In general, the preparation of clean cell walls suitable for chemical analysis or electron microscopy entails some form of mechanical agitation of a washed cell mass. Relatively fragile cells may be disrupted in a sonic oscillator,[1] but most yeasts and filamentous fungi require agitation with glass beads. The Mickle disintegrator[2] was one of the first such devices. It operates as an electrically driven "tuning fork", but the cups for cell paste and glass beads are of limited capacity; cooling can be achieved only by operating in a cold room. The Waring blender can accommodate much larger volumes of cell paste and glass beads,[3,4] but again cooling cannot be accomplished efficiently; long periods (more than 1 hour) are required for substantial cell breakage.

More recently, the Braun cell homogenizer[5-7] has come into widespread usage. It also involves shaking a mixture of glass beads and cell paste, but provision has been made for cooling the mixture during operation. A stream of liquid carbon dioxide (obtained by fitting a tank of carbon dioxide with an adductor tube) cools the shaking vessel. Times for essentially complete cell breakage are short (1 to 2 minutes).

A novel type of disintegrator that employs a smooth spinning disk has recently been described for the preparation of yeast cell walls.[8] A combination of shear forces produced by laminar flow, collision, and rolling glass beads causes cell breakage. Little heat is produced in the operation of this device, but breakage times are longer than noted above.

A completely different method for disrupting microorganisms entails the sudden release of high pressures. In the Ribi refrigerated cell fractionator,[7,9,10] a cell suspension is forced through a narrow orifice under a pressure of approximately 35,000 lb/in^2. Cell walls of yeasts and fungi have been prepared with this apparatus and have been found to possess antigenic activity.

After the cells are broken, the walls are separated from cellular debris (and glass beads) by centrifugation. Usually a number of centrifugations — alternating water, buffer, and denser supporting medium (such as 10% sucrose or mannitol) — are required before a preparation is obtained that can be termed an isolated clean cell wall. Light and electron microscopy of stained and unstained preparations are employed to follow the progress of the "cleaning-up" procedure. Infrared spectral analysis[8] has been found to be useful in following the course of purification of yeast cell walls. A decrease in the slope of the lines connecting peaks at 7.1 to 8.1 μm and at 6.05 to 7.1 μm accompanies increasing purification of the wall preparation. The absorption band at 8.1 μm (P = O) is essentially absent in a clean cell wall preparation, indicating the removal of cytoplasmic contamination of nucleic acids and polyphosphates.

REFERENCES

1. **Tokunaga, J. and Bartnicki-Garcia, S.,** *Arch. Mikrobiol.*, 79, 293, 1971.
2. **Mickle, H.,** *J. R. Microsc. Soc.*, 68, 10, 1948.
3. **Lamanna, C. and Mallette, M. F.,** *J. Bacteriol.*, 67, 503, 1954.
4. **Falcone, G. and Nickerson, W. J.,** *Science*, 124, 272, 1956.
5. **Young, F. E., Spizizen, J., and Crawford, I. P.,** *J. Biol. Chem.*, 238, 3119, 1963.
6. **Mill, P. J.,** *J. Gen. Microbiol.*, 44, 329, 1966.
7. **Novaes-Ledieu, M., Jiménez-Martinez, A., and Villanueva, J. R.,** *J. Gen. Microbiol.*, 47, 237, 1967.
8. **Řehaček, J., Beran, K., and Bičik, V.,** *Appl. Microbiol.*, 17, 462, 1969.
9. **Ribi, E., Perrine, T., List, R., Brown, W., and Goode, G.,** *Proc. Soc. Exp. Biol. Med.*, 100, 647, 1959.
10. **Gerhardt, P. and Judge, J. A.,** *J. Bacteriol.*, 87, 945, 1964.

DIAGNOSTIC MEDICAL MICROBIOLOGY

Herbert N. Prince

INTRODUCTION

Diagnostic medical microbiology deals with the isolation and identification of infectious agents from clinical specimens. It involves laboratory technicians and senior microbiologists in hospitals and public health laboratories. The techniques used in this discipline are not restricted to the medical laboratory. It is frequently essential for public health reasons to determine the presence or absence of infectious or potentially infectious microorganisms from nonclinical materials. The diagnostic techniques applicable here (e.g., foods, drugs, water, etc.) are also discussed in certain ''official'' and ''semi-official'' regulatory compendia.[1]

This review will attempt to give a synopsis of an expanding subject. It is hoped that the reader will gain a general appreciation of the techniques involved in the microbiological diagnosis of disease. In addition, this chapter will stress the importance of microbial physiology as it reduced to practice in the identification of microorganisms. It is to be emphasized that direct experience in the clinical or industrial laboratory under the supervision of an experienced diagnostician is the only way to teach the identification of bacteria. A review of this nature is, therefore, an attempt to explain the theoretical basis of the culture media employed, as well as an attempt to describe the microscopic, colonial, and biochemical tests most current in the art. The reader is encouraged to study more advanced texts and articles if he is interested in the more specialized areas of medical mycology, anaerobic bacteriology, antibiotic sensitivity testing, serology, and virological techniques.[2-4]

PATHOGENICITY

Pathogens

Pathogens are defined as disease-producing microorganisms. They are bacterial, mycotic, protozoan, chlamydial, or viral in nature. No one is sure what a pathogen is. Pathogens can lose virulence or become attenuated and become ''nonpathogenic''. ''Nonpathogens'' can gain entry in high numbers during states of reduced resistance and produce disease. The diseases that they produce can be transmitted vertically, horizontally, or via a zig-zag route. Organisms can also be transmitted directly from the environment to an individual (coccidiomycosis, Legionaire's disease, food poisoning). Transmission of infection from mother-to-child (e.g., congenital syphilis, herpes type 2) is an example of ''vertical'' transmission. Contraction of infection by one individual directly from another is an example of ''horizontal'' transmission (e.g., streptococcal sore throat, V.D., influenza, etc.). ''Zig-zag'' transmission is a viral or rickettsial phenomenon associated with insects or ticks which serve as intermediaries between man and other mammals or birds (e.g., an arbovirus disease such as Eastern or Western encephalitis).

There are some organisms which can be regarded as (1) ''obligate pathogens'', organisms that are extremely invasive at a relatively low multiplicity input regardless of age or state of health (e.g., arboviruses, rabies virus, *Treponema pallidum, Neisseria gonorrhoeae, N. meningitis, Coccidioides immitis, Francisella tularensis, Pasteurella pestis* (Yersinia), herpesvirus. Enter their environment and you've got the disease. Organisms regarded as (2) ''pathogens'' can be isolated very frequently from diseased sites (e.g., *Staphylococcus aureus, Streptococcus pyogenes, Escherichia coli, Proteus, Streptococcus pneumoniae, Candida, Salmonella,* and *Nocardia.* Every laboratory technician recognizes these as the organisms that you are likely to find. Organisms regarded as (3) ''opportunistic pathogens'' (e.g., Gram-negative rods such as member of the genus *Serratia, Acinetobacter Pseudomonas, Enterobacter*), are thought to comprise the normal or associated flora or to be part of the inanimate environment. They react to alterations in host resistance and produce disease. In short, the proposition put simply is either:

1. A pathogen is a microorganism that *can* produce disease (parasite-oriented genotypic view).
2. A pathogen is an organism that *has* produced a disease (host-oriented phenotypic view).

Associated Flora

The normal or associated flora has been considered as a latent reservior of potential infection.[5,6] Associated organisms that exist on or in specific body sites are known to possess specific surface organelles (adhesions) which fit as in a lock and key into specific receptor sites on the host cell.[6] Once adhesion occurs, colonization by the associated flora begins. This may even be considered ''infection''; when the equilibrium is destroyed and invasion occurs, the normal ''flora'' has produced a ''disease''. This newer concept of associated flora as opposed to normal flora implies the potential pathogenicity of every organism residing in or on the human body. There is no evidence to disprove this theory except, perhaps, that we know that the oral, enteric, and vaginal lactobacilli do not produce infections in man. The key to pathogenicity is attachment, invasiness, and multiplication. Physiological and structural defenses develop with and are a part of the human animal from birth. This defense against invasion begins to falter throughout life; during injury, during concurrent disease, during drug therapy, during old age, and culminates with death. Health lies somewhere amidst the skirmishes of disease, the retreat of obligate pathogens, and the containment of opportunistic associated flora. It is a constant war.

Infectious diseases, therefore, represent a complicated interaction between host and parasite. Accordingly, the medical microbiologist should have a basic knowledge of human anatomy and physiology, immunology and pathology, as well as a knowledge of the ''normal'' and ''abnormal'' flora. In this interaction the nature of the parasite may be fixed, a function of its intrinsic genotype. The nature of the host is variable, depending upon age, previous illness, presence of concurrent disease, and previous exposure to drugs. There might then exist a mosaic of host phenotypic states upon which and into which a sufficient number of virulent microbial genotypes can propagate and cause disease. Interposed between the phenotypic fluidity of the host and the genotypic rigidity of the microorganisms is the clinical microbiologist. He must help the physician to diagnose disease or he must help the industrial or public health laboratory to prevent disease.

A monograph on diagnostic microbiology can be indexed according to disease, organism, or specimen. Since the disease or organism is frequently in question, the arrangement of procedures and tests on the basis of specimen is most common and most useful.

A disease-oriented index would cover such topics as shown in Tables 1 and 2. The techniques employed in identifying bacteria related to these diseases rely heavily upon cytology, colonial morphology, microbial physiology, nutritional requirements, and selective inhibition by antimicrobial agents. Advances in our knowledge of metabolic pathways and mechanisms of electron transport have allowed the development of tests that were nonexistent a few years ago. New tests are continually being developed.

Useful Cytological Characteristics

In some poorly supervised laboratories microscopic examinations are totally replaced by biochemical tests. To prevent errors, simple and special staining techniques routinely should

Table 1
CATEGORIES OF DISEASE FOR WHICH A MICROBIOLOGICAL DIAGNOSIS IS REQUESTED

Target Organ Index

Disease: body system	Types	Examples of syndrome
Respiratory	Upper respiratory, lower respiratory, bacterial, viral, fungal	"Cold", sore throat, bronchitis, pneumonia, TB, histoplasmosis, candidiasis
Ophthalmic	Acute bacterial, chronic bacterial, viral, or chlamydial	Conjunctivitis, blephlaritis, herpes (type I), keratitis, APC fever, pink eye, trachoma
Otic	Otitis externa, otitis media bacterial fungal	Swimmer's ear, common ear ache
Febrile systemic	Bacteremia, virema	Scarlet fever, septicemias, meningitis, viral fever and rashes
Gastrointestinal	Bacterial, viral, fungal, acute, chronic	Diarrhea, vomiting, typhoid fever, amebic dysentery, general enteritis, viral fevers and rashes
Venereal	Bacterial, viral, chlamydial, protozoan	Syphilis, gonorrhea, chancroid, herpes type II, (genitalis), lymphogranuloma venereum
Genitourinary	Bacterial, yeasts, protozoan	Urethritis, vaginitis, epididymitis, cystitis, cervicitis, prostatitis
Dermatologic	Bacterial, fungal, viral	Acne, boils, warts, wound infection, dermatomycoses, (athletes foot, ring worm, *T. cruris)*, smallpox, herpes

Table 2
CATEGORIES OF ORGANISMS ASSOCIATED WITH COMMON INFECTIOUS DISEASES

Etiological Index

Organism	Disease
Gram-positive cocci	Pyogenic eye-ear-nose infections, skin infections, "strep" throats, meningitis, boils wound infections, subacute bacterial endocartitis, urethritis, nephritis, prostatitis, septicemia
Acid fast bacilli	Tuberculosis, leprosy (Hansen's disease)
Gram-negative rods	Urethritis, vaginitis, cystitis, nephritis, otitis media, septicemia, diarrhea, appendicitis, peritonitis, burn infections, typhoid fever, dysentery
Gram-positive spore-forming bacilli (a) aerobic (b) anaerobic (c) Gram-positive anaerobic diphtheroids	(a) Anthrax, diarrhea (Bacillus cereus), conjunctivitis (Bacillus subtilis) (b) tetanus, gas gangrene, wound infections, botulism (c) acne (*Propionibacterium acnes*)
Yeasts and molds	Meningitis, aspergillosis, histoplasmosis, vaginitis, otitis externa, coccidioidomycosis, cryptococcosis, candidiasis, nail and skin infections (onychomycosis and dermatomycosis)
Viruses	DNA vs RNA; ether sensitive vs. ether resistant; polio, diarrhea, fever and rash, influenza, measles, herpes (eye or genitals), mumps, gastritis, common cold, rabies, meningitis, smallpox, myocarditis, enteritis, Bornholm's disease, APC fever (adenovirus syndrome), infectious mononucleosis (E-B virus).
Gram-negative cocci	Conjunctivitis, gonorrhea, meningitis
Gram-positive nonsporulating rods	Actinomycosis (cutaneous and systemic); nocardiosis (cutaneous and pulmonary)

Note: Table 2 is designed to show, from a different perspective, diseases as they are associated with specific groups of organisms, and is, accordingly, an etiology-oriented index.

be employed. For example, an entire battery of tests for pathogenic enteric Gram-negative rods may be needlessly performed on a harmless spore-forming member of the genus *Bacillus*. To avoid this, Gram stains should be performed on all atypical or sparse growth on EMB agar to rule out the presence of Gram-positive bacilli. It is a common misconception that EMB only supports the growth of Gram-negative bacteria. *Hemophilus hemolyticus* can be misdiagnosed as *Streptococcus pyogenes* (both β-hemolytic) by failure to perform a Gram stain on all small-sized colonies detected on blood agar. Similarly, confusion between a diagnosis of *Enterobacter* or *Klebsiella* can be resolved by a test for motility. The microscope is frequently the most under-used piece of equipment in the clinical laboratory. The most useful tests are (wet mount or stain):

1. Shape (rods, spheres, chains, clusters, filamentous, branching)
2. Gram reaction (proper attention to age and source of culture or specimen)
3. Acid fast reaction
4. Spore formation (endospores for bacteria, conidia and chlamydospores for fungi)
5. Motility
6. Type of flagellation

Morphology is especially important in the identification of fungi and actinomycetes. These should be observed in undisturbed cultures in order to detect the formation and arrangements of conidia and other sporulating structures. True fungi and actinomycetes are identified by the microscopic appearance of spores (conidia) and the spore-bearing structures (conidiophores). Dermatophytic fungi are identified by the size of the spores (microconidia vs. macroconidia) as well as by hyphal structure (racquet hyphae, spiral hyphae, pectinate bodies). It is important to recognize that the carbohydrate or nitrogen content of the medium can affect spore production.

Some Useful Biochemical Reactions

Biochemial reactions elicited by organisms grown on specific substrates can be determined by a variety of conventional test media or by more recently developed multiple-test rapid diagnostic kits. They are especially useful in the identification of Gram-negative rods (fermenters and nonfermentors). Rapid diagnostic kits are also available for yeasts and staphylococci. A partial list of biochemical reactions used as phenotypic traits appears below:

1. Production of acid and gas from glucose and other hexoses*
2. Cleavage of disaccharides and fermentation of corresponding hexoses
3. Oxidation of hexoses
4. Oxidation of disaccharides
5. Amino acid deaminases and decarboxylases
6. Cytochrome oxidase
7. Catalase production
8. Production of H_2S
9. Production of indole from tryptophane
10. Dissimilation of glucose to acetyl methyl carbinol (Voges-Proskauer test)
11. Utilization of citrate as a sole source of carbon
12. Coagulase production

* The acids detected through phenol red or brom thymol blue pH changes are usually lactic, pyruvic, acetic, or formic acid, or combinations of these. If an organism produces stoichiometric amounts of lactic acid from glucose, it is called "homofermentative". If it produces a mixture of acids, is it called "heterofermentative". The gases produced are for the most part CO_2, H_2, or a mixture of these. Oxidative organisms generally produce weaker organic acids such as gluconic or 2-keto-gluconic acid.

Table 3
SELECTIVE INHIBITORS IN CULTURE MEDIA

Inhibitor	Organisms inhibited
KCN	Certain GNR and GP
Boric Acid	Certain GNR, GPC, GPB at 43C
Na lauryl sulfate	GPB, GPC
Na azide	GNR
Crystal violet	Gram positive
Na desoxycholate	Gram positive
B-Phenethylalcohol	GNR
Cetyl trimethyl ammonium bromide (cetrimide)	GNR, except *Pseudomonas* and *Serratia* and some *Enterobacter*, GPC, GBP
Brilliant green	GPC, GPB
NaCl	GNR, Streptococci (concentration-dependant)
Neomycin SO_4	General bacteria as required
Sodium sulfite and basic fuchsin	GP
Eosin Y, methylene blue	GP
Chloramphenicol	General bacteria as required
Malachite green	General bacteria as required (TB culture)
Cyclohexamide	Nonpathogenic fungi
Na selenite	Certain GNR (e.g., coliforms)
Na Heptadecyl SO_4 (Tergitol)	GPC, GPB
Nitrothiazole acetyl piperazine or nitroimidazoles	Anaerobes
Tetracycline	General bacteria as required

Note: GNR, Gram-negative rods;
GP, Gram-positive (C) cocci, (B) or bacilli.

Some Useful Inhibitory Media

Inhibitory media are essential when working with a mixed flora. Such media are described as "selective" or "differential". In most cases, an inhibitory medium is selective and differential with respect to occurrence and appearance of growth. Selective media are designed to encourage the growth of certain organisms and to inhibit the growth of others (e.g., general taxons such as Gram-positive vs. Gram-negative bacteria), or to allow visual discrimination among colony types within on generally selected taxon (e.g., *Salmonella* vs. *Escherichia*). Since most clinical specimens contain a mixed flora, selective and differential media are routinely employed.

The incorporation of antimicrobial agents into culture media is essential for the selective growth of organisms from a mixed population. In the early part of the 20th century microbiologists empirically tested the effect of natural substances and dyes. In more recent times antibiotic and/or synthetic chemicals have been added to broth or agar to shift a mixed flora in the direction of growth of a single taxon or group of bacteria. Table 3 gives examples of some of the more useful agents currently employed in clinical microbiology. As can be noted, basic dyes, desoxycholic acid, antibiotics, and inorganic salts are effective agents. Most of these antimicrobial agents were detected by chance or empirical screening.

Media used in diagnostic bacteriology are designed to support and/or selectively inhibit growth with production of a visual effect. The visual effect is detected by incorporation of pH indicators in the medium or is developed by the addition of colorimetric reagents to the culture tubes. The biochemical reactions found useful are predicated upon the detection of threshold amounts in assays with limited sensitivity. Lysine decarboxylase is significant in the detection of *Salmonella* but one must not assume that lysine decarboxylase is absent in *Escherichia coli*. Cytochrome oxidase is significant for *Pseudomonas* but one must not assume that it is absent in *Acinetobacter*. It is a question of degree and not an absolute presence or absence of any of these enzymes. In clinical medicine, a more quantitative assessment is made within a range of normal values (e.g., alkaline phosphatase, blood glucose, glutamine-oxalacetate transaminase, etc.). The point is that the technician performing blood chemistries on humans and the technician performing biochemical tests on cultures are studying the same thing, outlying values detectable by quantitative tests in humans or quick ''go-no-go'' qualitative tests in bacteria.

As the microbiologist gains experience, prudent use of a few selective and differential media will give rapid presumptive answers as to genus, so as to quickly aid the physician (see Table 4). Extensive biochemical tests are required for definitive speciations. The exigencies of patient care, however, frequently place a greater premium on determinations of genus and antibiotic sensitivity than on complete identification. It is important to identify the source of the organism and/or the suspected disease so that the microbiologist can gain an early diagnostic lead.

Table 4 is a practical guide to the diagnostician insofar as it suggests specific culture media for specific specimens. At the same time it gives a synopsis of the organisms to be expected and the diseases for which a diagnosis is being made. The great majority of these culture media are available commercially and require little time to prepare. Differences exist with respect to stability, storage, and the need to add supplements, such as vitamin mixtures, carbohydrates, serum, blood, etc. The diagnostic microbiologist should be aware of the basic composition of all culture media employed as well as be aware of some of the theoretical considerations which allow the components of the medium to differentiate and select.

Inoculation of Specimens

All specimens received in the laboratory should be properly identified with the patient's name, a number, source of specimen, date of specimen, attending physician, and tests required, and then promptly inoculated. Samples can be refrigerated, but the time should, in general, not exceed 6 hr (24 to 48 hr for urine). Ancillary information on clinical diagnosis or current antibacterial therapy is often helpful. All samples, except stools, should be collected in a sterile container. Environmental samples (floors, water, equipment) should be identified so as to exclude the use of inappropriate diagnostic media.

The choice of media for primary isolation and the decision to perform or not perform a Gram stain on the specimen is important. The chief objective of a clinical microbiology laboratory is to recover organisms and then to accurately identify them. Accordingly, the type of media and the method of inoculation will determine to a large extent the success of the laboratory in making a bacteriological diagnosis (see Table 4). The use of Gram stains, enrichment tests, differential or selective media depends upon a variety of factors, not the least of which is the speed with which a diagnosis is desired. In certain cases, Gram stains and all three types of media are used together. Selective media are the most inhibitory and care should be exercised in attempting primary isolations in the presence of the various antimicrobial agents. Some of the more common specimens will be described along with the primary methods of inoculation and the organisms most likely to be encountered.

Normally Sterile Sites

Specimens from normally sterile sites should be cultured on media capable of supporting the growth of any organism likely to be present, including the most fastidious pathogens. Differential and/or selective media should never be used alone. Any organism isolated from blood, bone marrow, tissue, cerebrospinal, synovial, pleural, or ascitic fluid should be carefully evaluated. An organism normally regarded as ''nonpathogenic'' should be evaluated in the light of the numbers present and the consistency of isolation before being discarded as a contaminant. The final decision is that of the attending physician. *Staphylococcus epidermidis* is of no significance on the skin but is significant in the blood.

Table 4
SPECIMEN GUIDE TO CHOICE OF MEDIA

Specimen	Source	Common organisms suspected	Disease	Some culture media
Throat, nose, ear or eye swab	Pharynx or nasopharynx or draining ear	*S. aureus, S. pyogenes, H. influenza, S. pneumonia,* meningococcus, *C. albicans,* myxoviruses, rhinoviruses	Pharyngitis, sinusitis, thrush, sore throat, otitis media	Blood agar under 10% CO_2 and anaerobic, gram stain, Choc. agar TSI, M-S
Sputum	Lower respiratory tract (not saliva) (1st am specimen)	*S. aureus, Klebsiella, S. pyogenes, S. pneumonia, M. tuberculosis, Histoplasma, C. albicans, Cryptococcus neoformans*	Various pneumonias, tracheitis, tuberculosis, various mycoses, bronchitis, bronchiolitis	As above (special treatment for TB)
Urine	AM, 3-glass specs, catheterization or midstream	*S, aureus, E. coli, Klebsiella Ent./Serratia, S. pyogenes, M. tuberculosis, Ps. sp., Proteus*	Cystitis, nephritis, urethritis, vaginitis, prostatitis	Blood agar, Mac, EMB, M-S quantitative loop onto plates or broth dilution for bacteriuria, TSI
Blood	Venous	A variety of aerobic or anaerobic GP and GN organisms	Febrile disease, subacute bacterial endocarditis, etc.	TSB, BHI (aerobic and anaerobic) FTM
Gastrointestinal	Feces	*S. aureus, E. coli, Salmonella arizona, Shigella, Cholera vibrio, B. cereus, C. perfingers, Vibrio parahemolyticus*	Typhoid fever, dysentery, diarrhea, enteritis, "food poisoning", gastritis	Selenite cysteine, EMB lactose, Mac, XLD, TSI, FTM, Blood agar
Uro-genital swabs	Urethral exudate Vaginal exudate Prostatic secretion Cervical swab	*Staph., Strp., Hemophilus, (corynebacterium) Candida, E. coli, gonococcus, Trichomonas, Proteus, Pseudomonas*	Urethritis, vaginitis prostatitis, vulvitis	Blood agar, Thayer-Martin CO_2, Choc. Mac FTM, cooked meat if anaerobes suspected EMB, FTM
Wounds	Surgical specimens, tissues, skin swabs, etc.	A variety of aerobic and anaerobic GP and GN organisms	Superficial or deep infection	Blood agar, CO_2 anaerobic, MAC, M-S, cooked meat

Note: TSI, triple sugar iron agar; M-S, mannitol salts agar; FTM, fluid thioglycollate medium; GN Gram negative; EMB Eosin methylene blue agar; MAC, MacConkeys agar; XLD, xylose lysine decarboxylase agar; GNR, Gram-negative rod (s); LIA, lysine iron agar; SS, Salmonella Shigella; GP, Gram-positive; BHI, brain-heart infusion; Choc., Chocolate agar.

Heavy Normal Flora Sites

Specimens from the skin and various body orifices are cultured as above but differential and selective media are also employed to prevent overgrowth of organisms normally present (e.g., chloramphenicol to inhibit skin bacteria in isolating dermatophytes). The question of what is a contaminant and what is a potential pathogen is a difficult one when dealing with heavily contaminated sites. The bacteriologist, from his retrospective knowledge of the "normal" or associated "flora" (e.g., throat or skin or vagina), must decide whether a shift to a predominant flora has occurred. This would indicate that the relative balance of the normal flora has been altered in favor of a potential pathogen. Specimens from the throat, lungs, and intestinal tract pose special problems. Under normal conditions certain organisms are not pathogens; under special circumstances they can produce infections (phenotypic vs. genotypic view). Unfortunately, the bacteriologist does not have a "normal flora" reference for a specific patient as a cardiologist has a prior EKG, or a radiologist has a prior lung X-ray. A "heavy normal" flora should be reported, as it may indirectly indicate a viral infection which has compromised the normal host tissue and altered the resident ecological balance. Organisms that might be encountered from sites of heavy resident flora are: staphylococci, fermenting and nonfermenting Gram-negative rods, streptococci, members of the genus *Neisseria,* diphtheroids, lactobacilli, and a variety of Gram-positive and Gram-negative anaerobes.

Delivery of Specimen to the Lab

Specimens are most frequently inoculated to solid or broth medium by the use of swabs. Swabs that must be sent through the mail should be coated with a nonnutritive protective medium such as Stuart's transport medium, where viability at room temperature can be extended. Fluids are best inoculated with a bacteriological loop instead of a swab since small numbers of cells may be absorbed to the fiber and not be recovered. In some cases, a sterile Pasteur pipette or calibrated loop can be used.

Tissue specimens from biopsy or autopsy should be minced in saline in a sterile mortar and pestle to give a 10 to 20% suspension. All combinations of primary, differential, and selective media should be inoculated, since it is seldom possible to obtain a second specimen. The homogenate should be inoculated by loop or pipette as with a fluid.

EENT Specimens — Eye, Ear, Nasal, Throat, and Pharyngeal Swabs and Exudates

The material submitted to the laboratory is streaked onto blood agar and incubated under 10% CO_2 (candle jar) and in a anaerobic jar to aid in the isolation of beta-hemolytic streptococci. In addition, inoculate plates of EMB and MacConkey's agar for the detection of Gram-negative rods and tubes of FTM to detect anaerobes. If *Hemophilus influenza* is suspected either clinically or as a result of Gram stain, rabbit blood agar is preferred since sheep blood may inhibit *Hemophilus* species. The hemolytic reactions of streptococci on blood agar are described later in this chapter. Differentiation of beta-hemolytic streptococci *(Streptococcus pyogenes)* from *Hemophilus hemolyticus* is made by Gram stain and by failure to grow well on sheep blood agar. Differentiation of hemolytic streptococci from beta-hemolytic staphylococci is made on the basis of colonial morphology, pigment, and by the catalase test, one of the simplest biochemical tests performed by the diagnostic microbiologist. Add a drop of 3% H_2O_2 to some growth rubbed onto the surface of a microscope slide or add 0.5 mℓ of 3% H_2O_2 to 0.5 mℓ of a broth culture. Staphylococci but not streptococci cause evolution of oxygen. This test is also useful in differentiating between nonhemolytic streptococci and catalase positive nonhemolytic diphtheroid organisms, which are frequently part of the upper respiratory flora. The principle strains of *Streptococcus* producing communicable disease in man belong to Lancefield's Group A (pharyngitis, rhinitis, tonsillitis, pneumonia, scarlet fever, as well as pyodermas and septicemia). Susceptibility to the antibiotic bacitracin, 0.04 unit (disc), helps differentiate type A from beta-hemolytic B,C,D,F, and G, the latter serological types being insensitive to this antibiotic.

Streptococcus pyogenes colonies at 18 to 24 hr on blood agar (reduced O_2 tension) are about 0.5 mm in diameter, transparent to opaque and domed with a smooth surface and entire edge. The organism is virulent for albino mice, 0.5 mℓ intraperitoneally of a 10^{-3} dilution of an overnight broth culture killing mice in 24 to 48 hr.

Streptococcus pneumoniae (formerly *Diplococcus pneumoniae*) produces alpha-hemolysis of sheep blood agar and is also extremely virulent for white mice by the i.p. route and for rabbits s.c. This organism is found in 30 to 70% of normal humans and from this site can invade the sinuses or middle ear (otitis media), eventually producing mastoiditis, meningitis, purulent conjunctivitis, or pneumonia. Colonies on blood agar under CO_2 are alpha-hemolytic, round with entire edges, domed, mucoid in appearance, and about 1 mm in diameter. Anaerobically the colonies show beta-hemolysis. Differentiation from other alpha- and beta-hemolytic organisms can be made on the basis of the bile solubility test (flood the plate with a solution of 10% Na desoxycholate; colonies of the pneumonococcus will dissolve). Pneumococci are also sensitive to the ethylhrocuprein hydrochloride (EHC) and will show a zone of inhibition around a 5 mcg paper disc containing this diagnostic agent.

5. Genitourinary Tract (GU) Specimens — Exudates and Urine

The detection by pour plates, tube dilution, or surface streaking of a calibarated loop of 100,000/mℓ or more organisms in a clean-catch midstream sample or in a catherized specimen is considered a significant bacteriuria. Direct surface streaking with a 0.001-mℓ loop onto MacConkey's agar allows more rapid enumeration of Gram-negative forms (cystitis, nephritis, pyelitis). Isolated colonies can be picked directly to TSI slants and speciations rendered as per the appropriate biochemical tests. The physician should be kept closely informed of the numbers and types isolated.

Urethreal and vaginal exudates or cervical swabs should be Gram stained and streaked onto a suitable agar as rapidly as possible (sheep blood (CO_2) agar, MacConkey's agar, Chocolate agar (CO_2) and a plate of Thayer-Martin (CO_2) medium if the gonococcus is suspected). Incubate all plates at 35°C and examine at 24 and 48 hr. Examine for staphylococci, streptococci, anaerobes, and Gram-negative rods as described above for EENT and GU cultures.

SPECIAL TESTS FOR GRAM-POSITIVE COCCI

Test	Staph aureus	Staph epidermidis	*Micrococcus*	*Strept. pneumoniae*	Beta-strep. sp.	Other streptococci: Fecal	Other streptococci: Other
Hemolysis (Beta)	+	0	0	0	+	0	0
7.5% salt	+	+	0 to var.	0	0	0	0
Mannitol	+	0 to var.	0	0 to var.	0	0	0 to var.
Coagulase	+	0	0	0	0	0	0
DNAase	+	0	0	0	0	0	0
Anaerobic growth	+	+	0	+	+	+	+
Catalase	+	+	+	0	0	0	0
Sensitivity to bacitracin	NA	NA	NA	NA	+	0	0
Bile solubility	0	0	NA	+	0	0	0
Sensitivity (EHC)	NA	NA	NA	+	0	NA	NA

NA = not applicable.

Blood Cultures

When a physician requests a blood culture the technician must appreciate that a critically ill patient is involved. Extreme care must be exercised at all stages of the analysis. Detection of true bacteremia can only be accomplished under conditions which minimize contamination during collection and by the use of enriched and osmotically stabilized aerobic and anerobic culture media. Osmotic stabilization of organisms having undergone cell wall damage (L-forms) can be achieved by the addition of 10% sucrose to trypticase soy broth, brain-heart infusion broth, or fluid thioglycollate medium. The inclusion of sodium polyanethol sulfonate (Liquoid®) will prevent clotting of the blood and will neutralize the bactericidal effect of human serum and inactivate certain antibiotics. Sodium citrate, 0.5% in any of the above broths is also an effective anticoagulant. Although procedures will vary from hospital to hospital, a generally acceptable technique is to withdraw 10 mℓ of blood and inoculate 5 mℓ into 100 mℓ of supplemented FTM and 5 mℓ into 100 mℓ supplemented TSB. These techniques will allow propagation of aerobic and anaerobic streptococci, pneumonococci, staphylococci, meningococci, coliform bacilli, and members of the genera *Salmonella, Hemophilus, Bacteroides, Clostridium, Acinetobacter, Pseudomonas, Alcaligenes, Candida, Brucella,* and species of *Francisella (Pasteurella tularensis) Yersinia (Pasteurella pestis)* and *Pasteurella multocida.*

Blood cultures are incubated at 35 to 37°C and examined daily for turbidity up to 10 days. When signs of growth appear (turbidity or gas) Gram stains and subcultures should be made on a variety of media (blood agar, chocolate agar, MacConkey's agar, EMB agar) set up in pairs and incubated aerobically and anaerobically. If all cultures, subcultures, and Gram stains are negative, the blood culture may be reported as ''no growth detected''. If the physician suspects viremia, appropriate inoculations into tissue culture, chick embryos, and suckling mice must be considered, as well as appropriate serological studies on paired acute and convalescent sera (see Table 5).

Viral cultures — A complete discussion of virus isolation and identification is beyond the scope of this chapter. A brief synopsis of appropriate tests is shown in Table 5. Tissue cultures must be supplemented with penicillin and streptomycin and in some cases amphotericin to prevent overgrowth by bacteria and fungi. Neomycin, gentimycin, or other combinations can also be employed. Balanced salt solutions that are used to obtain throat washings or fecal suspensions must also contain appropriate antimicrobial agents. All attempts at viral isolation and identification must be performed by a properly trained and apprenticed virologist.

SPECIFIC CULTURE MEDIA

The list presented here is not exhaustive, but attempts to portray typical reductions of theory to practice by describing the chemical composition and the microbial reactions on media frequently used in the diagnostic laboratory. In all cases, the ingredients are listed in grams per liter of reconstituted broth. Most of those media are commercially available (DIFCO, BBL, etc.) and are easily reconstituted with deionized or distilled water. After preparation, all media should be quality controlled by challenge with appropriate sensitive as well as responsive organisms. All of the media listed can be obtained from the commercial sources cited in the references.[7,8]

Blood Agar: pH 7.3 ±

Any all-purpose medium such as nutrient agar, brain heart infusion agar, or soy-bean casein digest agar to which 5% defibrinated blood has been added to is called blood agar. The most commonly employed supplement is whole citrated or defibrinated sheep blood. Goat, horse, or rabbit blood may be employed. Human blood can be employed but care must be taken to test for the presence of streptococcal antistreptolysin antibodies. The secretion of extracellular hemolysins produces complete lysis of the erythrocytes around the colony (beta-hemolysis), or incomplete hemolysis, as shown by a green discoloration of the

Table 5
CLINICAL ISOLATION AND IDENTIFICATION OF VIRUSES

Virus	Clinical material	Susceptible host	Cell culture and response	Other tests	Type blood for HI
Adeno	Eye, throat swab	H. Ep 2, HEK, HeLa	CPE (rounded cells)	SNT-CC, HI, FA, CF	Rat or rhesus
Arbo	Autopsy, serum, CSF	S. mice(death), duck embryo	Plaques-BHK and vero cells	SNT-CC mice, CF, HI	Goose
Corona	Throat and nasal	HETOC, HEK, S. mice (death)	Cilia, inhibition, CPE	HI-OC 43, EM, SNT-CC, or mice	Chick
Coxsackie A	Feces and throat	S. mice (death), PMK	Flaccid paralysis, CPE	SNT-mice,SNT-CC	
Coxsackie B	Feces and throat	PMK, WI-38, S. mice (death)	S. mice (death) (spastic paralysis), CPE (R/S/D)	SNT-CC	
Cytomegalo	Urine and throat	HEL	CPE (giant cells, focal lesions)	IHA, CF, FA	Sheep
ECHO	Feces	PMK, HEL	CPE(R/S/D)	SNT-CC	
Epstein Barr	Throat and blood	Lymphoid cells	Experimental transformation	SNT, IFA	
Herpes simplex	Throat and vesicles, buccal, eye, genital	HEL, H.Ep2, WI-38, MRC-5	CAM (Pocks), CPE, S. mice (death)	IHA, CF, FA, SNT-CC	Sheep
Influenza	Nasal and throat	AMN, all eggs, PMK	Amniotic fluid (HA test), HAD on inoculated cell cultures	HI, CF, DID	Guinea pig, chick, human O
LCM	Blood or CSF	3 week mice	BHK cells	FA, Brain smears, SNT-mice	
Measles	Throat and urine	PMK	CPE	SNT-CC, FA, HA	Vervet
Mumps	Throat	PMK	HAd on inoculated cell cultures, syncytial CPE on some cultures	HI, HAdI	Guinea pig
Parainfluenza	Throat	PMK	HAd on inoculated cultures	HI, HAdI	Guinea pig
Polio	Feces and CSF	PMK, WI-38, H.Ep2	CPE	SNT-CC	
Rabies	Brain	3 week mice	Death (IC inoc.)	FA, SNT-mice	
Reo	Throat and feces	PMK, HEK	CPE	SNT, HI	
Respiratory syncytial	Nose and throat	H.Ep2, PMK	CPE	SNT-CC, CF, FA	
Rhino	Nasal and throat	HEL, WI-38	CPE	SNT-CC	
Rubella	Throat	PAGMK, RK-13	No CPE in PAGMK use interference test for ID	Detect. by INT. (PAGMK) SNT-CPE (RK-13)	1 day old chick
Vaccinia	Vesicles	CAM, PMK, WI-38	Pocks on CAM, CPE (rounding, CB)	EM, AGP, FA	Human O, chick
Varicella-Zoster	Vesicles	HEL	CPE	SNT-FA	
Variola	Vesicles	CAM	CPE, Pocks on CAM	EM, AGP, pocks reduction	

Table 5 (continued)
CLINICAL ISOLATION AND IDENTIFICATION OF VIRUSES

Key:

AGP	= agar gel precipitin	HI	= hemagglutination inhibition
ALL	= allantoic cavity	HEK	= human embryonic kidney
AMN	= amniotic cavity	HEL	= human embryonic lung
BSS	= balanced salt solution	HETOCH	= human embryonic tracheal organ culture
BHK	= baby hamster kidney cells	H.Ep 2	= human epithelioma number 2
CB	= cytoplasmic bridging	IFA	= indirect fluorescent antibody
CC	= cell culture	IHA	= indirect hemagglutination
CAM	= chorioallantoic membrane	IC	= intracerebral
CF	= complement fixation	INT	= interference test
CNS	= central nervous system	LCM	= Lymphocytic choriomeningitis
CPE	= cytopathic effect	MIT	= metabolic inhibition test
CSF	= cerebral spinal fluid	OC-43	= coronavirus strain
D. emb.	= duck embryo	PAGMK	= primary African green monkey
DID	= double immunodiffusion	PMK	= primary Rhesus monkey kidney
EM	= electron microscopy	RFIT	= rapid fluorescent focus inhibition test
FA	= fluorescent antibody	RK-13	= rabbit kidney cells
HA	= hemagglutination test	R/S/D	= rounding, shrinking, degeneration
HAd	= hemadsorption	RSV	= respiratory syncytial virus
HAdI	= hemadsorption inhibition	SNT	= serum neutralization test
		S. mice	= suckling mice

hemoglobin around the colonies (alpha-hemolysis). Failure to affect the cellular integrity of erythrocytes, as seen by no change in the color of the agar surrounding of the colony, is also of diagnostic value (gamma-hemolysis). All blood agar should be freshly prepared, preincubated at 35 to 37°C for sterility, and quality control checked with human strains (*Streptococcus pyogenes* for beta-hemolysis, *Streptococcus viridans*, or *Streptococcus pneumoniae* for alpha-hemolysis).

The presence or absence of beta-hemolysis is useful not only for the streptococci and staphylococci but is a useful diagnostic aid for speciations of Gram-negative rods and spore-forming members of the genus *Bacillus* and other groups. Typical hemolytic reactions can be inhibited by the presence of carbohydrates; accordingly, any blood agar base employed must not contain glucose.

Chocolate Agar

Chocolate agar is essentially a blood agar heated to obtain lysis of the erythrocytes and coagulation of the serum proteins. It was originally described early in this century to detect *Hemophilus pertussis*. Whole defribrinated blood alone, or in combination with dehydrated hemoglobin, may be employed, as well as other supplements, e.g., 1% Iso Vitalex® (BBL mixture of vitamins, amino acids, purines, and pyrimidines). Lysing of the blood cells causes the di- and triphosphopyridine nucleotides and heme to become available to fastidious organisms. Chocolate agar is especially useful for the isolation and maintenance of species of *Hemophilus*, *Neisseria* and *Corynebacterium vaginale* (formerly *Hemophilus vaginitis*) (*Gardnerella*). Species of *Staphylococcus* and *Streptococcus* will also grow luxuriantly but the typical hemolytic reactions seen on blood agar will not be noted.

Eosin-Methylene Blue Agar (EMB) pH 7.1 to 7.2

	grams per liter
Peptone	10
Lactose	10
Sucrose	5 (optional[a])
Dipotassium phosphate	2.0

Agar	15.0
Eosin Y	0.4
Methylene blue	0.065
Distilled water	1000 mℓ

[a] When present, lactose is at 5 g/ℓ.

This medium is primarily used to presumptively identify *Escherichia coli,* which produces a blue-black colony with a greenish-metallic sheen. Although differentiation depends upon pH changes, the medium does not contain a conventional pH indicator. The dark color results from fermentation of lactose or sucrose. Nonlactose fermenters produce colorless or high pink colonies *(Salmonella, Shigella, Pseudomonas).*

The metallic green-black coloration produced by the lactose and/or sucrose fermentations is due to concentration of the insoluble eosin-methylene blue complex precipitated out at low intracolonial pH (similar to the metallic precipitate seen with Wright's stain when staining blood smears). Caution should be exercised in identifying *E. coli* merely on the basis of the metallic sheen, since members of the genera *Enterobacter* and *Citrobacter* can produce the same effect, as can *Proteus* if sucrose is present. Lactose-fermentating members of the *Klebsiella-Enterobacter* group typically produce large mucoid colonies with blue-purple centers. The combination of eosin and methylene blue acts as an inhibitor of Gram-positive organisms. The inhibition is not total, however, and colonies of *Bacillus* and *Staphylococcus* will appear, although small and atypical in size. EMB is also useful in the differentiation of *Candida* species: (1) *C. albicans* = spidery colonies, (2) other *Candida* or yeast = smooth, round colonies.

MacConkey's Agar: pH 7.1 +

	grams per liter
Peptone	17.0
Proteose (polypeptides)	3.0
Lactose	10.0
Bile salts	1.5
NaCl	5.0
Agar	13.5
Neutral red (vital dye)	0.03
Crystal violet	0.001
Distilled water	1000 mℓ

MacConkey's agar shares two things in common with EMB: (1) Gram-negative rods are selected and differentiated with concurrent inhibition of Gram-positive organisms and (2) the diagnostic color changes are related to fermentation of lactose, but are not related to the inclusion of a conventional pH indicator. The medium is more selective than EMB and relies upon a bile salt mixture and crystal violet to inhibit the Gram-positive forms. It is especially useful for the detection of enteric organisms; lactose-fermentating strains produce brick red colonies while lactose-negative forms (*Salmonella, Shigella, Pseudomonas, Acinetobacter, Proteus*) produce pink or colorless colonies. Typical strains of *Escherichia coli* also show a zone of opaque red precipitation around the colony, whereas, other lactose-fermenters will show the red colony with a pale pink periphery and no pericolonial precipitate. As with EMB agar, extremely glycolytic strains may produce intracolonial acidity that concentrates the neutral red dye (red colonies). On this medium, lactose-negative colonies often produce a pale yellow discoloration of the agar. Bacteriuria (10^5 organisms/mℓ in urine) is diagnosed on this medium by direct streak with a calibrated loop.

Endo Agar: pH 7.4 ±

	grams per liter
Dipotassium phosphate	3.5
Peptone	10.0
Agar	15.0
Lactose	10.0
Sodium sulfite	2.5
Basic fuchsin	0.5

Endo agar is a solid medium for the detection of coliform and other enteric forms. It is also used commonly in public health laboratories to detect coliforms in drinking water. Gram-positive bacteria are inhibited by the combination of sodium sulfite and basic fuchsin. Lactose-fermentating bacteria produce pink to red colonies with or without a brick red metallic sheen; as with MacConkey's agar, pericolonial reddening of the agar may occur (but by an entirely different organic reaction). Strains of *Salmonella, Shigella, Pseudomonas* and other lactose-negative species produce colorless to faint pink colonies. The classical Feulgen reaction (Schiff's reagent) is employed here. Strains such as *E. coli* cleave lactose and ferment glucose with production of acetaldehyde. Acetaldehyde fixes the bisulfite from the decolorized bisulfite-fuchsin reagent (bisulfite adds across the double bond of acetaldehyde) and the regenerated red fuchsin dye concentrates inside the colony.

Cetrimide Agar: pH 7.2 ±

	grams per liter	
Peptone	20.0	supplement with 10 mℓ of glycerol per liter
Magnesium chloride	1.4	
Potassium sulfate	10.0	
Agar	13.6	
Cetrimide[a]	0.3	
Distilled water	1000 mℓ	

[a] Cetyl trimethyl ammonium bromide.

This medium is employed for presumptive identification of *Pseudomonas aeruginosa,* which is not inhibited by the quaternary salt (cetyl trimethyl ammonium bromide). The medium promotes the development of the water-soluble pigment pyocyanin as well as the pigment fluorescein, giving the colonies and the surrounding agar a characteristic blue-green color (sometimes reddish or lavender).* Confirmation of *P. aeruginosa* should be made by testing for cytochrome oxidase, reduction of NO_3^- to N_2, growth at 42°C, growth on SS, oxidation but not fermentation of glucose, motility and, ideally, a falgellar stain. Not all strains produce pyocyanin and so this battery of tests should be also run on the nonchromogenic strains (5 to 10%). Other pseudomonads that will grow and produce the typical greenish-yellow water-soluble pigment on cetrimide agar are *P. putida, P. fluorescens* (caution: also *Serratia*). *P. maltophilia,* and *P. stutzeri* do not grow on this medium and furthermore, *P. maltophilia* is cytochrome oxidase negative. *Ps. cepacia* produces variable usually negative growth on this medium but is cytochrome oxidase positive. This organism also oxidizes mannitol and lactose and is esculin positive but does not produce DNAase.

* If the pigment pyorubin is formed.

Mannitol Salt Agar: pH 7.4 ±

Beef extract	1.0
Proteose peptone No. 3	10.0
Sodium chloride	75.0
Mannitol	10.0
Agar	15.0
Phenol red	0.025
Distilled water	1000 mℓ

This medium is recommended for the selective isolation of pathogenic staphylococci, which are tolerant of the high-salt concentration. Strains of *S. aureus* that are putatively pathogenic produce luxuriant butyrous colonies surrounded by a yellow zone signifying fermentation of mannitol. Such colonies must be confirmed as pathogenic strains by the coagulase test (ability of a loopful of the colony to solidify or produce a precipitate in 1 mℓ of rabbit, human, or horse plasma at 35 to 37°C in 4 to 24 hr). Nonpathogenic cocci (e.g., *Staphylococcus epidermidis*) produce small to medium butyrous colonies without a yellow zone. It should be stressed that spore-forming members of the genus *Bacillus,* yeasts, and certain molds also grow on this medium. The bacilli generally produce a more mucilagenous or mucoid type of growth. Gram stains should be performed on all acidogenic colonies. If there is doubt as to whether the strains are fermentating or oxidizing the mannitol, plates can be seeded and then overlayed with 2 to 3 mℓ of plain agar. Media such as Baird-Parker agar or Vogel-Johnson agar are frequently used in place of Mannitol-Salt agar. Details can be obtained from the manufacturer.

Kligler Iron Agar Slant (KIA): pH 7.4 ±

	grams per liter
Peptone	20.0
Lactose	10.0
Glucose	1.0
Sodium chloride	5.0
Ferric ammonium citrate	0.5
Sodium thiosulfate	0.5
Agar	15.0
Phenol Red	0.025

KIA is an extremely useful medium for differentiating among members of the enterobacteriaceae on the basis of their ability to either ferment glucose alone or ferment glucose and lactose. Fermentation of glucose is shown by a yellow butt (with or without a separation of the agar in the butt due to gas formation) and a red or alkaline slant (e.g., *Salmonella, Shigella,* and *Proteus* or any other lactose-negative form). A yellow butt plus a yellow slant indicates fermentation of both glucose and lactose. Production of H_2S is shown by a blackening in any part of the medium resulting from precipitation of iron sulfide in the presence of the ferric salt indicator. It is important that the butt be of the proper depth and the slant of the proper length. Inoculate from the center of a single colony out of EMB, MacConkey's, etc. by streaking the slant and stabbing the butt so as to get aerobic and relatively anaerobic conditions. One notes that the medium contains ten times more lactose than glucose. Stoichiometric fermentation of glucose in the butt does not release sufficient monocarboxylic acids to the upper slant to effect a pH change in the phenol red indicator when all of the glucose is fermented. Lactose-positive organisms continue to ferment the second sugar and then produce an upward migration of acid from the butt to acidify the slant. Thus, coliform organisms (*E. coli, Citrobacter, Enterobacter, Klebsiella,* etc.) produce a yellow butt and yellow slant.

Readings should be made at 24 and 48 hr. Frequently an acid slant will revert to alkaline by oxidation of the acetate and pyruvate via the citric acid cycle. KIA is an exceedingly useful medium in the clinical lab and is an example of biochemical theory reduced to practice.

Triple Sugar Iron Agar Slant (TSI): pH 7.3 ±

TSI is supplemented KIA agar to which sucrose has been added to permit separation of lactose-negative *Proteus* strains from lactose-negative *Salmonella* strains that cannot be accomplished with the double sugar medium of Kliger.

	grams per liter
Peptone	20.0
Lactose	10.0
Sucrose	10.0
NaCl	5.0
Glucose	1.0
Ferrous ammonium sulfate	0.20
Sodium thiosulfate	0.20*
Agar	13.0
Phenol red	0.025

* Present at more than twice this concentration in KIA.

One notes the same ratio of disaccharides to glucose as in KIA to explain acidification of the slant and/or butt region (stoichiometric fermentation of sugars in fixed proportions and upper migration of acid). Since Fe_2S_3 is yellow and FeS is black, the presence of ferrous iron (TSI) instead of ferric (KIA) requires less reducing potential, and thus the $Na_2S_2O_3$ concentration is lower in TSI. Thus, a blackening precipitate (FeS) appears as an indicator of H_2S production (the H_2S is derived from amino acids in the peptone and not from reduction of inorganic $SO_4^=$). A summary of TSI reactions is shown in Table 6.

Lysine Iron Agar LIA Agar: pH 6.7 ±

LIA agar was developed for the detection of Arizona strains, especially those which ferment lactose rapidly. It is especially useful in the diagnosis of typhoid fever and dysentery and other febrile enteric diseases. *Salmonella* and Arizona cultures produce lysine decarboxylase and rapidly form H_2S in large amounts. Decarboxylation of lysine is seen by purple butt and slant (alkaline reactions).

	grams per liter
Peptone	5.0
Yeast extract	3.0
Glucose	1.0
L-Lysine	10.0
Ferric ammonium citrate	0.5
Sodium thiosulfate	0.04
Brom Cresol purple	0.02
Agar	13.50
H_2O	1 ℓ

LIA agar can detect fermentation of glucose, decarboxylation of lysine, and oxidative deamination of lysine, although it is not a substitute for the standard lysine decarboxylase test. One notes the 10:1 ratio of lysine to glucose. Only those organisms that fail to decarboxylate lysine will produce an acid (yellow) butt. Those that ferment the 1% glucose will continue with decarboxylation of lysine converting it from an amphoteric amino acid to a basic substance. The alkalinity of the diamine diffuses throughout to yield a purple color in both butt and slant. A summary of LIA reactions is shown in Table 7.

Table 6
SUMMARY OF TSI REACTIONS OF THE ENTEROBACTERIACEAE

Inoculate slant and stab butt from an isolated colony on EMB or MacConkey's agar or from TSA slant made from such an isolated colony

Organism	Slant	Butt	Gas	H_2S	CO test
	Ewings Section I[a]				
Escherichia	A(K)	A	+ (−)	−	Negative
Shigella	K	A	−	−	Negative
Salmonella typhi	K	A	−	+ (−)	Negative
Other Salmonellae	K	A	+	+ + +(−)	Negative
Arizona	K(A)	A	+	+ + +	Negative
	Ewing's Section II[b]				
Citrobacter	K(A)	A	+	+ + +	Negative
Edwardsiella	K	A	+	+ + +	Negative
Klebsiella	A	A	+ +	−	Negative
Enterobacter	A	A	+ +	−	Negative
E. hafniae	K	A	+	−	Negative
Serratia	K(A)	A	−	−	Negative
Proteus vulgaris	A(K)	A	+	+ + +	Negative
P. mirabilis	K(A)	A	+	+ + +	Negative
P. morganii	K	A	− (+)	−	Negative
P. rettgeri	K	A	−	−	Negative
Providencia	K	A	+(−)	−	Negative

Note: Cultures resembling *Salmonella* can be tested directly with polyvalent O antiserum (one drop of heavy saline suspension from the slant plus one drop of antiserum on a microscopic slide; agglutination can be seen within a few minutes grossly). Further typing can be then rendered with specific group O antiserum. Details on the serological tests performed as slide or tube agglutination with somatic (O) or flagella (H) antisera are provided in the selected references. ( ), occasional reaction; K, alkaline; A, acid.

[a] Produce epidemic and endemic enteric disease.
[b] Occasionally enteric pathogens which also may produce serious infections in extraintestinal sites.

Simmon's Citrate Agar and the IMVIC test

Simmon's Citrate Agar: pH 6.9 ±	grams per liter
Ammonium dihydrogen phosphate	1.0
Dipotassium phosphate	1.0
Sodium chloride	5.0
Sodium citrate	2.0
Magnesium sulfate	0.2
Agar	15.0
Bromthymol blue	0.08

Simmon's citrate agar is the medium of choice for performing the citrate test, one of the four components of the classical IMVIC series of reactions (an acronym derived from Indole, Methyl red, Voges-Proskauer, and Inorganic Citrate).

Table 7
SUMMARY OF LIA REACTIONS OF ENTEROBACTERIACEAE

This medium is to be used in connection with TSI especially for Salmonella detection

LD conventional test[a]	Organism	Slant	Butt[b]	Gas	H_2S
d	*Escherichia*	K (purple)	K, Nor A	− or +	−
−[a]	*Shigella*	K	A (yellow)	−	−
+	*Salmonella*	K	K or N	−	+ (−)
+	*S. typhi*	K	K	−	+ or −
+	*Paratyphi A*	K	A (yellow)	+ or −	− or +
+	*Arizona*	K	K or N	−	+ (−)
−[a]	*Citrobacter*	K	A (yellow)	− or +	+ or −
+	*Edwardsiella*	K	K	− or +	+
+	*Klebsiella*	K or N	K or N	+ or −	−
	Enterobacter cloacae	K or N	A (yellow)	+ or −	−
+	*aerogenes*	K	K or N	+ or −	−
+	*hafniae*	K	K or N	− or +	−
+	*Serratia*	K or N	K or N	−	−
	Proteus				
−[a]	*vulgaris*	R[c]	A (yellow)	−	− (+)
−[a]	*mirabilis*	R[c]	A (yellow)	−	− (+)
−[a]	*morganii*[d]	K or R[c]	A (yellow)	−	−
−[a]	*rettgeri*	R[c]	A (yellow)	−	−
−[a]	*Providencia*	R[c]	A (yellow)	−	−

Note: ( ), occasional reaction; K, alkaline; A, acid; N, no change; d, different biochemical types.

[a] Note the correlation between a failure to decarboxylate lysine (without subsequent production of basic diamine) and acidification of the butt due to glucose fermentation (N slant, A butt).

[b] Theoretically all should be acid by definition of the Enterobacteriaceae, but lysine decarboxylase alkalinity masks the acid produced by glucose.

[c] R, red color due to deamination of lysine.

[d] *Morganella morganii*.

The production of indole from tryptophane is detected by the addition of Kovac's reagent (*p*-dimethyl amino benzaledehyde dissolved in amyl alcohol and concentrated HCl) to peptone broth with the rapid visualization of a cherry red color in the amyl alcohol layer.

The methyl red test is an assay for acid production from glucose in a buffered broth. A few drops of an alcoholic solution of methyl red is added to a 48-hr culture and an immediate red color indicates a positive reaction (extensive glycolysis). Negative tests produce a yellow color. Equivocal red-orange colors should be repeated with tests incubated for 4 to 5 days.

The Voges-Proskauer test is performed on an aliquot of the same tube of buffered peptone glucose broth as was used for the methyl red test (MR-VP broth). To 1 mℓ of MR-VP broth add the VP reagent.* A positive test is indicated by the development of a pink color with the final reading taken at 4 hr. This test is based upon the formation of acetylmethyl-carbinol from glucose, which in turn is oxidized in an alkaline medium to diacetyl. Diacetyl reacts with the creatine present in peptone to form a pink color.

The citrate test refers to the ability of an organism to utilize sodium citrate as a sole source of carbon. Those organisms capable of assimilating Na citrate will grow on the surface of the agar slant or in the broth and produce an alkaline pH (blue color). Most bacteria, of course, can utilize citrate as a sole source of carbon including *E. coli* but the utilization is pH dependent. At low pH tricarboxylic acids such as citrate are not ionized and pass freely

* 15 drops 5% alpha-naphthol in absolute EtOH followed by 10 drops of 40% KOH.

Table 8
TYPICAL IMVIC REACTIONS

	I	M	V	IC
E. coli	d[a]	+	−	−
Enterobacter	−	−	+	+
Klebsiella	−	−	+	+
Salmonella	−	+	−	d
Arizona	−	+	−	+
Citrobacter freundii	−	+	−	+
Serratia marcescens	−	−	+	+
Shigella	−	+	−	−
Proteus vulgaris	+	+	−	d
Proteus mirabilis	−	+	−	+
Providencia	+	+	−	+

[a] d, different biochemical types.

through the cell walls and cytoplasmic membrane to be metabolized via the Krebs cycle. The citrate test involves a special set of conditions buffered at pH 6.9 to 7.0 so as to discriminate between those organisms which are less susceptible to the steric hinderance associated with the transport of anions above the isoelectric point of the cell. A summary of IMVIC reactions is shown in Table 8.

Sabouraud Dextrose Agar: pH 5.6 ±

	grams per liter
Dextrose	40.0
Polypeptone peptone	10.0
Antibiotics	As required
Agar	15.0

Sabouraud dextrose agar may be used for the isolation, identification, and maintenance of pathogenic and nonpathogenic fungi. The low pH tends to partially repress growth of bacteria. Most fungi grow as well at pH 7 as at pH 5 to 6. Sabouraud dextrose agar can be prepared at pH 6.9 to 7.2, but antibiotics must be added to make it selective. Isolation of pathogenic fungi from heavily contaminated clinical material can be obtained by addition of penicillin G (20 units/mℓ) and streptomycin (40 μg/mℓ) immediately before use. For a more stable medium, chloramaphenicol alone is added (0.05 g/ℓ) or tetracycline at 100 μg/mℓ. If one is interested in isolation of pathogenic fungi only, the antibiotics cycloheximide (0.40 g/ℓ) and chloramphenicol (0.05 g/ℓ) are included (Mycosel Agar). Cycloheximide inhibits the common nonpathogenic strains of *Penicillium, Aspergillus, Trichoderma, Alternaria, Neurospora,* and other strains associated with clinical samples, but dermatophytes of the genera *Trichophyton, Microsporum,* and *Epidermophyton* are insensitive to this antibiotic. A quick screen for dermatophytes can be obtained by the use of DTM agar, which selectively inhibits bacteria and nonpathogenic fungi; dermatophytes produce a deep red color in the agar by alkaline reaction.

Casein-Soybean Digest Broth (Trypticase Soy Broth): pH 7.3

	grams per liter
Peptone (tryptic digest of USP casein)	17.0

Papaic digest of soy meal USP (phytone peptone)	3.0
NaCl	5.0
Dipotassium phosphate	2.5
Glucose	2.5
Distilled water	1 ℓ

This is an excellent all-purpose medium for the growth of fastidious organisms. Swabs from material with scanty exudates (e.g., some eye cultures) can be inoculated directly into the broth and then subcultured to selective media for speciation. Quantitative enumeration of bacteria in urine can also be determined by planting decimal portions (1.0 mℓ of undiluted urine to 1.0 mℓ of urine diluted 10^{-5}) for detection of bacteriuria. Casein-soybean digest broth is also useful for sterility tests in a quality control program designed to assay the sterility of parenterals, broth media, sterile reagents (saline, blood, transport media, etc.) autoclaved hospital supplies, and sterile devices. It is one of the media designated for the sterility testing of drugs and devices in the U.S. Pharmacoepia. A laminar flow-hood should be employed for all sterility tests, but is otherwise unnecessary.

Casein-Soybean Digest Agar (Trypticase Soy Agar): pH 7.3

	grams per liter
Peptone (tryptic digest of USP casein)	15.0
Papaic digest of soy meal (phytone peptone)	5.0
NaCl	5.0
Agar	15.0
Distilled water	1 ℓ

Casein-soybean digest agar will support growth of the same range of organisms as the corresponding broth. It can be used as a base for blood agar or chocolate agar. With or without blood, it is a useful all-purpose agar medium to be used wherever total aerobic plate counts are desired (e.g., urine cultures) either by the pour plate method or by surface streaking of calibrated loops to tempered plates. Casein-soybean digest agar plates with or without blood are also useful as settling plates to determine the number of bacteria in the air either while working at a laminar flow hood (e.g., when doing tissue culture work or performing an aseptic fill) or for general monitoring of the microbial environment for contamination control in the lab, ward, or operating room. Addition of glucose must be considered if it is to be considered nutritionally similar to TSB.

Carbohydrate Fermentation Media

Even though rapid diagnostic kits are becoming increasingly popular to determine carbohydrate fermentation, a variety of older basal media are still available to which sterile solutions of filtered carbohydrates are added with or without addition of an inverted test tube (Durham) to detect gas (e.g., phenol red broth which contains peptone, beef extract, NaCl, and the appropriate carbohydrate). Lactose fermentation is frequently used for detection of coliform organisms in water and dairy products as well as for the identification of fecal coliforms and *E. coli,* in which case gas is produced from lactose in the presence of boric acid at 42°C. The above type of medium can be employed without a pH indicator since gas production is the criterion measured. Lactose is a relatively heat stable carbohydrate and can be added directly to the broth before autoclaving. Production of gas automatically indicates acid production since the H_2 and CO_2 produced are derived from dissimilation of the pyruvate and lactate.

One of the most effective means of determining carbohydrate fermentation is to inoculate Cystine Trypticase agar (CTA) stabs. To the CTA semisolid medium weighed out Q.S. for 1 ℓ, add a sterile filtered stock solution to give a 0.5 to 1.0% final concentration of the desired carbohydrate, bring to a boil for 1 min, dispense in screw cap tubes (1/2 full), autoclave and store at room temperature for use (*Note:* without added carbohydrate CTA is an excellent maintenance media and can be used to detect motility). Fermentation can be noted by yellowing of a deep stab if the sugar has been incorporated into the CTA.

Employing CTA base without added sugar, fermentation can be noted by placing a commercially available filter paper carbohydrate disc into the tube and then stabbing the disc at the same time into the medium. Acid and gas can be seen in the area of the disc; the bottom of the tube serves as a carbohydrate-free control. Fermentation can also be detected by employing tubes of phenol red broth base (disc immersed) or phenol red agar pour plates (steak surface with organisms; place the carbohydrate disc on the surface, press firmly, and incubate-yellow zone around the disc indicates fermentation). However, only use of the stab culture or the Durham tube (inverted vial) can reveal gas production

CTA Agar: pH 7.3 ±

	grams per liter
Cystine	0.50
Trypticase peptone	20.0
Agar	2.5
NaCl	5.0
Sodium sulfite	0.5
Phenol red	0.017
Water	1 ℓ

FTM (Fluid Thioglycollate Medium): pH 7.1 ±

About 0.1 g of $CaCO_3$ should be placed into each tube if lactic or propionic acid bacteria or members of the genus *Clostridium* are to be maintained in this medium.

	grams per liter
Pancreatic digest of Casein (USP)	15.0
L-Cystine	0.5
Dextrose	5.0
Yeast extract	5.0
Sodium chloride	2.5
Sodium thioglycollate	0.5
Resazurin (redox indicator)	0.001
Agar	0.750

Note: Store at room temperature in screw cap tubes; boil up fresh before use if more than the upper 1/3 of the tube is oxidized. Do not boil again if broth becomes reoxidized.

FTM will support the growth of fastidious organisms such as streptococci and lactobacilli without addition of ascitic fluids. Swabs can be inoculated directly into tubes of FTM and subcultures made to selective media. The growth of organisms in the lower reduced zone indicates the possible presence of an obligate anaerobe. Facultative organisms or aerobes will grow throughout the medium and in the upper oxidized zone (pink resazurin layer). Direct inoculation of a specimen is recommended whenever the clinical picture or Gram stain suggests an obligate anaerobe. This broth should be used in conjunction with TSB for sterility work on parenterals, broth media, and sterile reagents (see USP 20). The presence of sulfhydryl groups not only lowers the redox potential but also inactivates mercurials. Thus, FTM was originally designed for sterility tests for formulations containing Hg preservatives and not for the propagation of anaerobes. FTM is useful in isolating organisms from blood. The pathogenic protozoan *Trichomonas vaginalis* can be grown in this medium if the pH is adjusted to 6.0 and human, calf, or horse serum is added to a final concentration of 5.0% v/v.

Cooked Meat Medium pH 7.2 ±

	grams per liter
Dextrose	2.0
Polypeptone peptone	20.0
Beef heart	454
Sodium chloride	5.0

Cooked meat medium has been used for over 30 years for the propagation of obligate anaerobes. These organisms are able to sequester themselves in sulfhydryl-rich crevices within the meat surface. The low redox potential is maintained if the meat is cooked up fresh prior to use. In food laboratories cooked meat medium is useful for propagation of *Clostridium botulinum*. In the hospital laboratory it is useful whenever an anaerobe is suspected (wounds, genital tract, blood, etc.). Aerobic organisms grow equally well but frequently do not digest the meat to the same extent as members of the genus *Clostridium*. Any growth must be confirmed as an anaerobe by Gram stain and by demonstration that a colony picked from an anaerobic jar does not grow aerobically. Thus, cooked meat tubes must be subcultured to blood agar aerobically and anaerobically; subculture to FTM and TSB tubes is also useful in deciding if an obligate anaerobe is present, but only if one is dealing with a pure culture.

Hugh and Leifson OF Medium (Oxidation Fermentation Media)

If a Gram-negative rod produces no change on a TSI slant, it is considered to be a nonfermenter. One must next determine if it is capable of oxidizing carbohydrates. OF medium is aseptically dispensed into tubes as 5 mℓ semisolid stabs. It is useful in the study of nonfermenting Gram-negative rods which must be bifurcated into oxidizers or nonoxidizers. The culture is stabbed to the bottom into duplicate tubes for each sugar (add the desired carbohydrate solution to 1.0% final concentration prior to sterilization; autoclave only 10 min). Cover one tube of each pair with 5 mm of sterile mineral oil and incubate for 48 hr or longer. Record presence of acid (yellowing of medium) in both tubes. Fermentative organisms will produce color changes in both the open and closed tubes. Oxidative organisms will produce yellowing of the open tube at the surface but no change in the closed tube (even though growth in the closed tube may occur). The low concentration of agar in this semisolid medium also allows one to note motility of the organisms by migration from the vertical stab line (similar to CTA stab). Hugh and Leifson OF medium takes advantage of our knowledge of glycolysis and the direct oxidation of hexoses for those organisms not possessing an Embden-Meyerhoff system. It is important to include the closed control, lest rapid and extensive oxidation be confused with true fermentation. A more complete appreciation of the value of the OF system can be obtained from the following differential chart, where one notes its value in identifying *Pseudomonas, Acinetobacter, Alcaligenes, Achromobacter*, and *Flavobacterium* (Table 9). Included in Table 9 are data on the cytochrome oxidase test and growth on SS agar so as to give a more complete synopsis of the tests for

Table 9
OF MEDIUM GLUCOSE OXIDATION BY NONFERMENTERS AND OTHER PERTINENT REACTIONS

	CO	Glucose open	Glucose sealed	Motility	SS agar
Pseudomonas	+	A	0	+	+
Alcalignees	+	0	0	+	0
Moraxella	+	0	0	0	0
Acinetobacter	0	A, 0	0	0	0
Achromobacter	+	A	0	+	+
Flavobacterium	+	A	0 (+)	0 (+)	0
Fermenter control	0	A	A		

Note: +, positive; A, acid; 0, no change; (+), delayed or occasional.

nonfermenting bacilli (NFB). Miniaturized rapid diagnostic kits have also been developed for this group of organisms (e.g., API).

Candle Jar

Carbon dioxide behaves as a growth supplement for many bacteria. The simplest way of producing an atmosphere of carbon dioxide is to light a candle inside a large wide-mouthed jar. The petri dishes are placed within the jar, the candle is lit, and the lid screwed in place. Within about 1 min the candle will extinguish due to the blanketing effect of CO_2, which is heavier than air. Accordingly, one has produced an atmosphere of approximately 10% CO_2; the microaerophilic environment encourages growth for some species by the combined removal of oxygen and the addition of CO_2. The biochemical basis for CO_2 enrichment is not clearly understood, but it is known that CO_2 fixation occurs regularly in certain metabolic pathways. The candle jar enhances the growth of Gram-positive and Gram-negative cocci as well as strains of *Hemophilus*.

ANAEROBIC ORGANISMS

Anaerobes in pure or mixed cultures are present in over 80% of the clinical abscesses associated with human or veterinary tissues. Successful isolations can be accomplished in most (but not all) cases with the use of commercially available anaerobe jars, such as the BBL Gas-Pak® system, which is available from most bacteriological supply houses. In this system, water is added to an envelope which then generates a mixture of H_2 and CO_2 into a sealed chamber. The O_2 passes through a dry catalyst and "cold combustion" occurs. A strip of filter paper, impregnated with the methylene blue, is included in the kit as an anaerobic indicator. It will slowly change from blue (oxidized methylene blue) to colorless (reduced methylene blue). For best results, pour plates of blood agar from tubes of prereduced agar are used, or a medium that has been freshly rehydrated and autoclaved the day of use. Some obligate anaerobes will not grow in the Gas-Pak® if prereduced media is not employed.

Prereduced media generally contain L-cysteine (0.05%) and are made under a stream of CO_2 in the absence of oxygen. When made in this manner and tightly sealed, an anaerobic environment is maintained. Every time the culture is opened for examination or transfer a CO_2 gas "flush" is required to wash out oxygen. Details on the preparation of a variety of prereduced media have been published.[9,10]

Specimens to be examined for anaerobes include (1) abscessess or drainage from any part of the body, (2) postsurgical or other infections, (3) wounds, (4) burn infections, and (5) blood. Little is to be gained in taking routine anaerobic cultures of throat swabs, urine, or fecal samples because obligate anaerobes are normal to these sites. Oral, bronchial, and brain abscesses should also be checked for *Actinomyces* (candle jar). Genital, oral, or bronchial infections should be observed with phase or dark filed microscopy for spirochetes. About 80% of all specimens from abscesses, wounds, or infected sites will contain obligate anaerobes if properly cultured in a prereduced system. About 65% of these specimens will also contain facultative bacteria which can be distinguished by (1) careful attention to colonial morphology, (2) Gram stain, and (3) regrowth after aerobic subculture. Anaerobes may not be isolated if the specimen is cultured in broth medium before direct streaking (broth enrichment); the more fastidious obligate anaerobes will be overgrown by less fastidious facultative bacteria. Abscesses and wounds can yield up to seven different kinds of anaerobes and for plural and unogenital infections up to four may be isolated.

The genera of obligately anaerobic bacteria are as follows:

Rods

Form spores	*Clostridium* Gram positive)	
Do not form spores	*Proprionibacterium*	motile and Gram positive
	Arachnia	
	Bifidobacterium	
	Lactobacillus	
	Actinomyces	
	Eubacterium	
	Fusobacterium	motile and Gram negative
	Leptotrichia	
	Bacteroides	
	Butyrivibro	
	Succinivibrio	
	Succinimonas	
	Vibrio	
	Selenomonas	
	Treponema	

Cocci

Gram positive	*Peptostreptococcus*
	Peptococcus
Gram negative	*Veillonella*
	Acidaminococcus

The following is a list of pathogenic obligate anaerobes isolated from clinical specimens:

* *P. acnes* (skin, intestinal tract)
* *P. avidum*
* *Lactobacillus catanaforme* (pleural)
* *Bifidobacterium eriksonii* (intestinal tract)
*** *Eubacterium lentum* (frequently in abscesses)
*** *E. limosum* (frequently in infections)
** *E. alactolyticum* (pleural)
*** *Bacteroides fragilis* (blood, wounds, abscesses but rare in pleural specimens)
*** *B. vulgatus*
** *B. oralis* (infections of mouth and URI)
** *B. corrodens* (mouth)
** *Fusobacterium fusiforme* (mouth or UR abscesses)
** *F. mortiferm* (abscesses)
** *F. naviforme* (abscesses)

** *F. nucleatuim* (mouth, URI)
** *F. necrophorum*
** *F. russii* (abscesses)
** *F. varuim* (abscesses)

Note: * infrequent, ** occasionally, *** frequent, URI, upper respiratory infection.

Treponema

The incidence of many types of treponemes in clinical material is poorly established (except for positive dark-filed examination of a syphilitic lesion). *T. vincentii* is isolated from the oral cavity and *T. refringens* (formerly called *Borrelia refringens* or *T. genitalis*) is isolated from oral or genital material.

Clostridium

Clostridium perfringens usually makes up 50% of the clostridia isolated from clinical material. This organism sporulates poorly. Other clostridia can be shown to be sporogenous by Gram stain after 3 to 7 days on a slant of cooked media agar at 30°C. Clinical specimens will frequently harbor a group of nonpathogenic clostridia as follows: *C. inocuum, C. sordelli, C. bifermentans, C. sporogenes, C. septicuum, C. ramosum, C. sphenoides. C. sporogenes* is regarded as a nontoxigenic variant of *C. botulinum.*

The pathogenic clostridia, as judged by toxic production in culture (mouse i.p. assay) or pathogenicity in guinea pigs (i.m.) are as follows (t = produces toxin, n = necrosis): *C. histolyticum* (n), *C. limosum* (n), *C. perfringens* (t), *C. septicum* (t), *C. chauvoei* (t), *C. botulinum* (t), *C. carnis* (n), *C. novyi* (t), *C. haemolyticum* (t), *C. sordelli* (t), *C. difficule* (n).

The isolation and identification of obligate anaerobes requires special skill and patience. Unlike aerobic Gram-negative rods, a simple battery of biochemical tests is inadequate. When working with these organisms the bacteriologist should pay great attention to the source of the organisms, the Gram stain, and colonial morphology. As with all clinical isolates, the performance of appropriate antibiotic sensitivity tests is invaluable.

ACKNOWLEDGMENTS

The author wishes to thank Richard N. Prince, Rutgers University, for assistance. Many thanks to Joseph Rubino, Assistant Scientific Director, Gibraltar Biological Laboratories, and to Mary Jones, St. Michael's Medical Center, Newark, N.J., for discussions and review of data and concepts and for their invaluable assistance.

REFERENCES

1. *United States Pharmacoepia,* 20th revision, Mack Publishing, Easton, Pa., 1980.
2. *Manual of Clinical Microbiology,* 3rd ed., American Society of Microbiology, Washington, D.C., 1980.
3. **Lorian, V.,** *Antibiotics in Laboratory Medicine,* Williams & Wilkins, Baltimore, 1980.
4. **Waterson, A. P.,** *Recent Advances in Clinical Virology,* Churchill Livingstone, London, 1980.
5. **Rosebury, T.,** *Microorganisms Indigenous to Man,* Maple Press, York, Pa., 1962.
6. **Mackowiak, P. A.,** The normal microbial flora, *New Engl. J. Med.,* 307 (2), 83, 1982.
7. DIFCO Manual, 9th ed., Difco Laboratories, Detroit, Mich., 1969.
8. **BBL,** *Manual of Laboratory Procedures,* 5th ed., Becton-Dickinson Co., Cockeysville, Md., 1973.
9. Manual of Methods for General Bacteriology, American Society of Microbiology, Washington, D.C., 1981.
10. *Outline for Clinical Methods in Anaerobic Bacteriology,* Virginia Polytechnic Institute, Blacksburgh, Va., 1970.

MECHANISMS OF ACTION OF ANTIMICROBIAL AGENTS

Steven M. Pogwizd and Stephen A. Lerner

INTRODUCTION

This table attempts to summarize current information about, and working models for, the mechanisms of action of a variety of antimicrobial agents and to provide references to more extensive information. We have limited the list to agents with antibacterial or antifungal properties. We have also included some antitumor agents which are active against prokaryotic as well as eukaryotic cells. The drugs listed below are classified according to their primary site of action in the cell.

Antimicrobial agent	Mechanism of action	Ref.
	Inhibition of Cell Wall Synthesis	
Phosphonomycin (fosfomycin)	Phosphonomycin acts as an analog of phosphoenolpyruvate and binds covalently to a cysteinyl residue of the phosphoenolpyruvate: UDP-GlcNAc-3-enolypyruvyl transferase; this complex interferes with the formation of UDP-*N*-acetyl muramic acid (UDP-MurNAc), early in the synthesis of bacterial cell wall	1, 2
Cycloserine	D-Cycloserine, a structural analog of D-alanine, competitively inhibits alanine racemase and D-analyl-D-alanine synthetase; this blocks the formation of D-alanine from L-alanine and the synthesis of the D-alanyl-D-alanine dipeptide prior to its addition to the UDP-MurNAc tripeptide, the rigidly planar D-cycloserine has greater affinity for alanine racemase than does either D- or L-alanine	3—5
O-Carbamyl-D-Serine	*O*-Carbamyl-D-serine competitively inhibits alanine racemase and leads to accumulation of the UDP-MurNAc tripeptide; unlike D-cycloserine, *O*-carbamyl-D-serine does not inhibit D-alanyl-D-alanine synthetase	5—7
Vancomycin Ristocetin	Vancomycin and ristocetin are glycopeptide antibiotics that interfere with bacterial cell wall synthesis; although the structure and action of vancomycin have been studied in greater detail, the mechanism of action of ristocetin is believed to be similar to that of vancomycin; these drugs are bactericidal, and may even kill cell wall-deficient bacteria (L forms), suggesting secondary effects not limited to cell wall synthesis; the structure of vancomycin has recently been elucidated: a 1450-dalton tricyclic glycopeptide, in which are linked two chlorinated tyrosines, three substituted phenylglycines, glucose, vancosamine (a unique amino sugar), *N*-methyl leucine, and aspartic acid amide; this structure can undergo hydrogen bonding to the acyl-D-alanyl-D-alanine terminus of various peptidoglycan precursors (UDP-MurNAc-peptapeptide, MurNAc-pentapeptide pyrophospholipid, GlcNAc-MurNAc-pentapeptide pyrophospholipid, and the growing point of the peptidoglycan); the amino acid in the first position of the pentapeptide chain may also affect this interaction; the affinity of vancomycin binding to murein precursors and their accessibility to vancomycin, in vivo may determine the major site of inhibition of cell wall synthesis, which has not yet been ascertained; it has been	8—10, 10a, 10b

MECHANISMS OF ACTION OF ANTIMICROBIAL AGENTS (continued)

Antimicrobial agent	Mechanism of action	Ref.
	Inhibition of Cell Wall Synthesis	
	shown that neither the formation of a vancomycin-cell wall acceptor complex nor inhibition of peptidoglycan synthetase contribute to the antibiotic effect of vancomycin	
Tunicamycins	The tunicamycins are a family of glucosamine-containing nucleoside antibiotics which interfere with the assembly of peptidoglycan, teichoic acid, and teichuronic acid in Gram-positive bacteria and with the glycosylation of proteins in eukaryotic cells; these antibiotics act upon the synthesis of "lipid intermediates" which participate in the formation of the macromolecules by specifically inhibiting reactions of the type: polyprenyl phosphate + UDP-*N*-acetylhexosamine → polyprenyl *N*-acetylhexosaminyl pyrophosphate + UMP; the *N*-acetylhexosamine that is involved in the case of teichoic and teichuronic acids and the eukaryotic glycoproteins is *N*-acetylglucosamine; in the case of the peptidoglycan, it is *N*-acetylmuramyl pentapeptide; an analogous reaction, i.e., the transfer of galactosyl phosphate from UDP-galactose to polyprenyl-phosphate to form polyprenyl galactosyl pyrophosphate, the first intermediate in O-antigen synthesis, is not sensitive; mycospocidin, streptovirudins, and antibiotic 24010 have chemical properties similar to those of the tunicaymycins	11—15
Diumycin Moenomycin Macarbomycin Prasinomycin 11,837RP Janiemycin Enduracidin	Both of these groups of lipid-containing antibiotics inhibit bacterial cell wall synthesis and lead to accumulation of "lipid intermediate" (GlcNAc-MurNAc-pentapeptide-P-P-lipid); transfer of the disaccharide-pentapeptide unit from this lipid intermediate to an acceptor molecule is inhibited by these antibiotics, and may be due to their structural similarity to the C_{55}-isoprenylmonophosphate	16—18
Gardimycin	Gardimycin, a new sulfur-containing peptide antibiotic, inhibits the transfer of disaccharide-pentapeptide units from the C_{55}-isoprenyl phosphate carrier to the cell wall acceptor; its mechanism of action is thus similar to diumycin, janiemycin, and related antibiotics	19
Bacitracin	In the presence of divalent cations, bacitracin complexes with the membrane-bound pyrophosphate form of the undecaprenyl (C_{55}-isoprenyl) carrier molecule that remains after the disaccharide-peptide unit is transferred to the nascent peptidoglycan chain; this binding inhibits the enzymatic dephosphorylation of this carrier to its monophosphate form, which is required for another round of synthesis and transfer of the disaccharide-peptide unit; similar complex formation with intermediates of sterol biosynthesis in animal tissues (e.g., farnesyl pyrophosphate and mevalonic acid) may account in part for the toxicity of bacitracin in animal cells	20—22
Penicillins Cephalosporins	Penicillins and cephalosporins interfere with the terminal step in the cross-linking of peptidoglycan chains (i.e., transpeptidation of the N-terminus of the amino acid or peptide bridge onto the penultimate D-alanine of the pentapeptide chain with release of the terminal D-alanine); it has been suggested that penicillin is a structural analog of a transition state of the D-alanyl-D-alanine terminus of the peptidoglycan strand which is involved in transpeptidation; whereas a transpeptidase might normally form	23—27

MECHANISMS OF ACTION OF ANTIMICROBIAL AGENTS (continued)

Antimicrobial agent	Mechanism of action	Ref.
	Inhibition of Cell Wall Synthesis	
	an acyl-enzyme intermediate with the penultimate D-alanine, in the presence of a penicillin the highly reactive CO–N bond of its β-lactam ring might acylate the transpeptidase irreversibly, thereby inactivating it; this view is complicated by the presence in bacteria of a multiplicity of penicillin-binding proteins (PBPs), which may include not only penicillin-sensitive enzymes such as transpeptidases and D-alanine carboxypeptidases, but also other enzymatic activities such as β-lactamases and endopeptidases; the possibility of multiple target sites for the actions of penicillins and cephalosporins is supported by the observations that various β-lactam antibiotics produce differential effects on cell division, elongation, and lysis; furthermore, some PBPs bind specific β-lactam antibiotics preferentially; the lytic and killing action of penicillin may not result simply from the inhibition of peptidoglycan synthesis, but may also involve the action of bacterial murein hydrolases; it has been observed that lipoteichoic acids, which inhibit these autolytic enzymes, are released from bacteria upon treatment with penicillin in vitro; it has thus been postulated that the release of these endogenous inhibitors upsets the regulation of murein hydrolase activity in the cell and may result in killing and osmotic lysis	
Isoniazid (INH)	Isoniazid is an antituberculous agent which blocks synthesis of mycolic acids in the cell walls of mycobacteria; the resulting defective cell envelope, which is low in mycolic acid, permits leakage of essential cell components; since the biosynthetic pathway of mycolic acids and the enzymes involved have not been elucidated, the exact nature of the biochemical effect of isoniazid remains to be clarified	28—30
Polyoxins	The polyoxins are a family of antifungal peptide nucleosides; they inhibit cell wall chitin synthesis in some filamentous fungi by competitive inhibition of chitin synthetase, the enzyme which catalyzes the transfer of *N*-acetylglucosamine (GlcNAc) from UDP-GlcNAc to an acceptor molecule; kinetic studies have indicated that the action of the polyoxins results from the competition between nucleoside moieties of polyoxin and of UDP-GlcNAc for the binding site of chitin synthetase	31—33
	Interference with Membrane Integrity	
Tyrocidins Gramicidin S	The tyrocidins and the related gramicidin S are cyclic peptides which promote leakage of cytoplasmic solutes and may cause uncoupling of oxidative phosphorylation as a secondary effect; the interaction between gramicidin S and phospholipids appears to be electrostatic, so it is unlikely that the peptide ring is incorporated into the lipid region of the membrane	34, 35
Polymyxins	Polymyxins initiate disorganization of the cytoplasmic membrane, causing leakage of intracellular components; as a result of studies with liposome systems, it has been postulated that the mechanism of action of polymyxin involves an ionic interaction between the phospholipid phosphate and the ammonium group of polymyxin; simultaneous proton transfers between these two groups and also between an ammonium group on the	36—38

MECHANISMS OF ACTION OF ANTIMICROBIAL AGENTS (continued)

Antimicrobial agent	Mechanism of action	Ref.
	Interference with Membrane Integrity	
	phospholipid and a γ-amino group on polymyxin result in neutralization of charge on the polar head of the phospholipid which can alter the electrostatic and hydrophobic stabilizing forces of the membrane; liposomes derived from methylated phospholipids are not sensitive to polymyxin; methyl groups may prevent proper alignment of phospholipid and antibiotic or they may increase the basicity of the phospholipid amino group which would make proton transfer less favored	
Polyenes Amphotericin B Nystatin	The integrity of sterol-containing membranes of eukaryotes is impaired by the binding of polyene antibiotics; studies with liposome systems have shown that interaction between these antibiotics and sterols causes formation of pores with radii of approximately 5—7 Å; the antibiotic is anchored at the membrane-water interface by its polar end and extends its full length halfway through the lipid bilayer; the antibiotic molecules may orient themselves into a cylindrical half-pore with their polyhydroxylic surfaces interacting with each other and with water, and their lipophilic surfaces facing outward; molecular model building suggests that sterol molecules fit snugly betwen each pair of polyene molecules and stabilize these half-pores; pore formation results from hydrogen bonding between the rings of hydroxyl groups at the ends of two opposing half-pores	39, 40
Valinomycin Enniatin	Valinomycin, a cyclic depsipeptide, promotes leakage of K^+ from prokaryotic and eukaryotic cells by formation of a lipid-soluble complex with these ions; the free antibiotic molecule has both a hydrophilic and a hydrophobic side, so that it becomes localized at membrane surfaces; after the K^+ ion enters the molecule, shedding its normal hydration shell, valinomycin's carbonyl groups move toward the center of the molecule and stabilize the complex by a system of induced dipoles between K^+ and the carbonyl oxygen atoms; the resulting increase in lipid solubility facilitates passage of the complex through the lipid membrane; although valinomycin can also form a complex with sodium, the smaller sodium ions fit less tightly in the valinomycin structure than do potassium ions, resulting in decreased stability of the complex; the enniatins, cyclic hexadepsipeptides related to valinomycin, appear to have similar mechanisms of action	41—43
Nonactin	The cyclic macrotetralide, nonactin, complexes with a potassium ion within its ring of eight oxygen atoms (four carbonyl and four tetrahydrofuran); complex formation is accompanied by folding into a conformation resembling the seam of a tennis ball, with nonpolar functional groups projecting outward to form a lipophilic shell; this leads to leakage of potassium ions across membranes in a manner similar to that with valinomycin	43—45
Nigericin Monensin	Although the antibiotics nigericin and monensin are not cyclic molecules, they can form complexes with K^+ and Na^+ which are similar to those of valinomycin; in these complexes, the antibiotic molecule assumes a cyclic conformation as a result of hydrogen bonding between a carboxyl group at one end of the molecule and alcohol groups at the other; the complexes	43, 45, 46

MECHANISMS OF ACTION OF ANTIMICROBIAL AGENTS (continued)

Antimicrobial agent	Mechanism of action	Ref.
	Interference with Membrane Integrity	
	contain an ion surrounded by ether and hydroxyl oxygen atoms of the drug; their lipophilic exterior facilitates passage of the complexes through membranes; unlike valinomycin, these antibiotics promote a 1:1 exchange of potassium and hydrogen atoms, since the carboxyl groups of the molecule must be deprotonated before complex formation can take place	
Gramicidin A	The mechanism of action of the linear peptide gramicidin A is different from that of other ionophores; the cation permability of membranes is increased by the formation of transient transmembrane channels of high conductivity; it has been suggested (1) that these channels consist of two antibiotic molecules in a head-to-head dimer oriented in a left-handed helix, and (2) that the diameter of the channel is about 4 Å and its length is about 25—30 Å (just less than the lower limits of hydrophobic thickness of the lipid bilayer)	47—49
	Inhibition of Nucleic Acid Synthesis	
Azaserine DON (6-diazo-5-oxo-L-norleucine) Azotomycin	Azaserine and DON are analogs of glutamine which inhibit purine biosynthesis; they compete with glutamine and bind irreversibly to a cysteinyl residue of the biosynthetic enzyme which converts formylglycinamide ribonucleotide to its corresponding amidine derivative; these drugs inhibit less effectively a variety of other enzymes which utilize glutamine as a substrate (e.g., XMP aminase and 5-phosphoribosyl-1-pyrophosphate amidotransferase); the hydrolysis of azotomycin to DON in vivo may account for its similar biochemical effects	50—52
Hadacidin	Hadacidin, an analog of L-aspartic acid, inhibits purine biosynthesis by competing with aspartate in the conversion of inosinic acid to adenylosuccinic acid; although adenylosuccinate synthetase is readily inhibited by hadacidin, other enzymes which utilize aspartate as substrate are not significantly inhibited	53
Psicofuranine Decoynine	Psicofuranine and decoynine inhibit bacterial nucleic acid synthesis by the noncompetitive inhibition of amination of xanthylic acid (XMP) to guanylic acid (GMP)	54, 55
Bromodeoxyuridine Iododeoxyuridine	Bromodeoxyuridine (5-Bromo-2′-deoxyuridine) and iododeoxyuridine (5-iodo-2′-deoxyuridine) are structural analogs of deoxythymidine. These compounds may be incorporated into DNA and lead to errors in replication and transcription due to faulty base pairing.	56
5-Fluorocytosine	5-Fluorocytosine is taken up and deaminated by yeast cells but not by mammalian cells; the resulting 5-fluorouracil may block thymidylate synthetase, which normally forms thymidylate from uridylate, and other steps in pyramidine biosynthesis; the 5-fluorouracil may also be incorporated into mRNA, causing errors in translation	57, 58
Hydroxyurea	Hydroxyurea inhibits DNA synthesis in prokaryotic and eukaryotic cells; it appears that hydroxyurea inhibits ribonucleoside diphosphate reductase, the enzyme which converts ribonucleotides to deoxyribonucleotides; the drug may require conversion to a reactive derivative, *N*-carbamyloxyurea, before it can exert its inhibitory effect	59—61

MECHANISMS OF ACTION OF ANTIMICROBIAL AGENTS (continued)

Antimicrobial agent	Mechanism of action	Ref.
	Inhibition of Nucleic Acid Synthesis	
Intercalating agents Acridines Quinacrine Proflavine Chloroquine Phenanthridines Ethidium Propidium Anthracyclines Daunomycin Nogalomycin Thiaxanthenones Hycanthone Miracil D	Intercalating agents interfere with DNA and RNA synthesis to various extents in numerous microbial systems, as well as inhibit DNA and RNA polymerases in vitro; these planar inhibitors are intercalated (inserted) between adjacent stacked base pairs in the DNA, which remain planar and perpendicular to the helical axis but move apart to accommodate the inhibitor molecule; the configuration of the intercalated molecule is stabilized by electrostatic interaction between the rings and the DNA bases above and below it; intercalation reduces strand separation of the DNA, as shown by its increased thermal stability; since strand separation is essential for DNA replication, increased stabilization may account for inhibition of DNA synthesis; the intercalated molecule may hinder the attachment of RNA polymerase to the DNA template, and may inhibit both initiation and elongation phases of RNA synthesis; the intercalating agents exhibit a weaker secondary binding to the phosphate groups of the double helix; however, it appears that the primary binding (i.e., intercalation) is mainly responsible for the ability of these drugs to interfere with nucleic acid synthesis; although the action of these various agents is essentially the same, they may differ in the specificity of their binding to particular bases in DNA and in the fine details of DNA complex formation	62—66
Actinomycin D	Actinomycin D binds to double helical DNA and specifically inhibits DNA-dependent RNA polymerase at concentrations much lower than those which inhibit DNA synthesis; complex formation between actinomycin D and DNA apparently depends on intercalation of the phenoxazone ring of the drug between two adjacent stacked guanine-cytosine base pairs which are in opposite orientation, while the peptide subunit of the drug lies in the minor groove of the DNA helix; the complex is stabilized by a hydrogen bond between the 2-amino group of guanine and the carbonyl oxygen of the L-threonine residue of actinomycin D; the inhibition of RNA chain elongation, which is more sensitive to actinomycin D than is initiation, may be due to prevention of RNA polymerase movement along the template by the bulkiness of the drug's cyclic pentapeptide rings	52, 67
Chromomycin Mithramycin Olivomycin	These agents are potent inhibitors of RNA synthesis (and to a much lesser extent, DNA synthesis) both in vitro and in vivo; like actinomycin D, they form stable complexes which require the presence of helical DNA and the free amino group of guanine; however, complex formation does not cause unwinding of the DNA helix, which suggests these agents do not intercalate into DNA; the ease of dissociation of DNA-drug complexes by organic solvents suggests that the interaction is not covalent in nature; the binding of these drugs to DNA prevents the movement of the RNA polymerase along the DNA template	52, 68, 69
Quinoxaline Echinomycin Triostin A	Quinoxaline antibiotics inhibit RNA synthesis (and to a lesser extent, DNA synthesis) by binding to the DNA template; this binding appears to be specific for helical DNA, but lacks base-pair specificity	70

MECHANISMS OF ACTION OF ANTIMICROBIAL AGENTS (continued)

Antimicrobial agent	Mechanism of action	Ref.
	Inhibition of Nucleic Acid Synthesis	
Hedamycin Rubiflavin Pluramycin	Hedamycin, rubiflavin, and pluramycin preferentially inhibit DNA synthesis in vivo, with less effect on RNA synthesis; complex formation with DNA is essentially irreversible, but does not involve covalent binding; from the available information, it is not possible to decide whether these antibiotics act by intercalating between the DNA base pairs or whether they interact with functional groups in the grooves of the DNA template; it has been proposed that the antibiotics inhibit DNA replication in vivo by preventing separation of the DNA strands	45, 68
Distamycin Netropsin	The structurally related antibiotics, distamycin and netropsin, inhibit DNA and RNA synthesis; they act as DNA template poisons and inhibit the initiation of RNA synthesis by RNA polymerase; the antibiotics also inhibit the DNA polymerase I reaction, thought to catalyze repair synthesis of DNA but not DNA replication in vivo; the antibiotics bind to both native and denatured DNA, and show a preference for AT-rich segments of DNA; one fundamental difference between the two drugs exists: distamycin produces a decrease in DNA viscosity while netropsin increases the viscosity of DNA; this difference presents problems in the designing of models of the DNA-antibiotic complexes	52, 71, 72
Kanchanomycin	Kanchanomycin inhibits RNA synthesis in bacterial and mammalian cells; it requires divalent cations for formation of two types of complexes with DNA (an initial complex is formed and converted over time to a second complex); the inhibition of RNA synthesis does not appear to be due solely to binding of the antibiotic to DNA template, but may involve inactivation of the RNA polymerase; it has been suggested that a ternary complex between polymerase, DNA, and bound antibiotic may be formed, from which the antibiotic can be released slowly or not at all; binding of kanchanomycin to DNA template may also impair DNA synthesis	73—75
Mitomycin Porfiromycin	Mitomycin and porfiromycin inhibit DNA synthesis and lead to massive degradation of preexisting DNA in microbial and mammalian cells; mitomycin in its reduced form is capable of forming covalent cross-links between complementary strands of DNA; it appears likely that its aziridine ring may take part in the formation of one covalent link with DNA; it is possible that the second link may involve the methylene group of the methylurethane side-chain; the target site in the DNA appears to be the 0(6) atom of guanine, although no direct evidence has been obtained; as a result of the interaction between mitomycin and DNA: (1) DNA synthesis stops when the replicating fork reaches a mitomycin cross-link, and (2) degradation of DNA results from the excision of DNA in the region of the mitomycin cross-linking and the appearance of nucleases which are associated with lysogenic phages induced by mitomycin	45, 52
Carzinophillin	Like mitomycin, carzinophillin introduces covalent cross-linking between DNA strands; however, its action does not lead to single-stranded breaks in DNA	76, 77
Anthramycin	Anthramycin inhibits DNA and RNA synthesis in bacteria and	68, 78

MECHANISMS OF ACTION OF ANTIMICROBIAL AGENTS (continued)

Antimicrobial agent	Mechanism of action	Ref.
	Inhibition of Nucleic Acid Synthesis	
	mammalian cells; RNA synthesis appears to be inhibited to a slightly greater degree than is DNA synthesis; anthramycin reacts with DNA (specifically with DNA containing guanine) to form a nearly irreversible covalent complex. The result is the inactivation of a DNA template for DNA and RNA polymerization reactions	
Sibiromycin	Sibiromycin binds covalently to template DNA of both prokaryotic and eukaryotic cells, inhibiting DNA synthesis and DNA-dependent elongation of the RNA; the complexing of the antibiotic with DNA increases the melting temperature of the DNA but does not affect its viscosity in solution; it appears that DNAs with a high GC content bind more antibiotic than DNAs with a low GC content	52, 79
Streptonigrin	Streptonigrin leads to the production of single-stranded breaks in DNA at concentrations below those which inhibit DNA synthesis; this degradation, which is the main lethal action of streptonigrin, requires the interaction of oxygen with a reduced form of the drug; despite its structural resemblance to mitomycin, streptonigrin does not cause formation of covalent cross-links in DNA	45, 52
Bleomycin Phleomycin	Bleomycin and phleomycin inhibit DNA synthesis in intact bacterial and mammalian cells; in its copper-free form, bleomycin: (1) causes single-stranded scissions of DNA, (2) releases thymine (as well as other bases), (3) inhibits DNA ligase, and (4) releases DNA from the membrane-DNA complex; as yet, it is not known whether one (or all) of these effects is responsible for inhibition of nucleic acid synthesis in vivo	80—82
Neocarzinostatin	Neocarzinostatin specifically inhibits DNA synthesis in bacterial and mammalian cells by introducing breaks into single-stranded and double-stranded DNA; these breaks cannot be sealed by polynucleotide ligase, nor can the degraded DNA act as substrate for DNA polymerase I; the fact that little adenine or guanine is released suggests that there is some base specificity in the reaction	83
Ansamycins Rifamycins Streptovaricins	The rifamycins specifically inhibit bacterial DNA-dependent RNA polymerases by blocking the RNA chain initiation step; these antibiotics bind very tightly to the β subunit of the RNA polymerase in a molar ratio of 1:1; they do not bind well to β subunits of RNA polymerase isolated from rifamycin-resistant mutants; the mechanism of action of streptovaricins is probably identical to that of the rifamycins; however, they inhibit bacterial RNA polymerases to a lesser extent than do the rifamycins, because they form less stable complexes with these enzymes	52, 84, 85
Streptolydigin Tirandamycin	Streptolydigin, like the rifamycins and streptovaricins, inhibits bacterial RNA polymerases, by binding to the β subunit of these enzymes; however, there are important differences: (1) streptolydigin inhibits the RNA chain elongation step and (2) the binding of streptolydigin is much weaker than that of the rifamycins; the mode of action of tirandamycin is identical to that of streptolydigin, but it is much less potent	86—88
Lomofungin	Lomofungin inhibits RNA synthesis by a direct effect on bacterial and yeast DNA-dependent RNA polymerases without inter-	89, 90

MECHANISMS OF ACTION OF ANTIMICROBIAL AGENTS (continued)

Antimicrobial agent	Mechanism of action	Ref.
	Inhibition of Nucleic Acid Synthesis	
	acting with template DNA; in addition, lomofungin has been shown to interfere with DNA synthesis in yeast	
Novobiocin Coumermycin	Novobiocin and coumermycin inhibit replicative DNA synthesis in bacteria by their interference with the ATP-dependent DNA-melting activity of DNA gyrase, the enzyme which catalyzes the introduction of negative superhelical turns into covalently closed circular DNA; mutations conferring resistance to either of these antibiotics in *E. coli* lie in the *cou* locus and result in formation of a drug-resistant DNA gyrase (see Nalidixic Acid and Oxolinic Acid)	91—93
Nalidixic acid Oxolinic acid	Nalidixic acid and oxolinic acid cause a reversible inhibition of bacterial DNA synthesis by their effect on a nicking-closing activity that can relax positively supercoiled DNA; mutations conferring resistance to either of these antibiotics in *E. coli* lie in the *nal A* locus and result in resistance of this nicking-closing activity as well as of DNA gyrase, the enzyme that catalyzes the introduction of negative superhelical turns into covalently closed circular DNA; it has been postulated that the activity of DNA gyrase, which is critical for DNA replication, has two components: (1) a coumermycin-sensitive enzyme that utilizes ATP hydrolysis to produce a twisting stress on the DNA (DNA-melting), and (2) a nalidixic acid-sensitive protein (Pnal) that catalyzes nicking and closing; the degree of independence of these components has not yet been clarified (see Novobiocin and Coumermycin)	92, 93
Edeines	Edeines A and B are linear oligopeptides which inhibit bacterial DNA synthesis and, at a tenfold higher concentration, protein synthesis; although DNA polymerase I is insensitive, a crude mixture of DNA polymerases II and III was reported to be inhibited by edeines; initial studies have found no binding of edeines to DNA; inhibition of protein synthesis is due to interference with messenger-directed binding of aminoacyl-tRNA to both the A and P sites of the ribosome	83, 94
Hydroxyphenylazo-pyrimidines	Arylhydrazinopyrimidines, the active forms of the hydroxyphenylazopyrimidines, reversibly inhibit bacterial DNA synthesis by interacting with DNA polymerase III; it has been proposed that inhibition results from reversible formation of a ternary complex involving hydroxyphenylhydrazinouracil (or hydroxyphenylhydrazinoisocytosine), DNA template-primer, and DNA polymerase III with 1:1:1 stoichiometry	83, 95
Arabinosyl cytosine Arabinosyl adenine	Arabinosyl cytosine and arabinosyl adenine are analogs of their corresponding ribonucleosides and deoxyribonucleosides; these drugs inhibit prokaryotic and eukaryotic cells; the primary target of inhibition for both these drugs is believed to be replicative DNA polymerase, since resistant bacterial mutants appear to have altered DNA polymerase III; arabinosyl nucleosides must be converted into the nucleoside 5′-triphosphate form in order to be active; two models of antibiotic action have been proposed: (1) the drugs competitively inhibit the incorporation of deoxynucleoside triphosphates into the polymerase reaction, and (2) the analogs are incorporated into DNA, thereby impeding extension from the 3′-terminus of the primer	55, 83

MECHANISMS OF ACTION OF ANTIMICROBIAL AGENTS (continued)

Antimicrobial agent	Mechanism of action	Ref.
	Inhibition of Protein Synthesis	
Althiomycin	Althiomycin inhibits protein synthesis in prokaryotes but not in eukaryotic cells; the antibiotic blocks peptidyl transferase activity of the 50S ribosomal subunit	96, 97
Aminoglycosides	The prototype aminoglycoside, streptomycin, binds irreversibly to a single site on the 30S subunit of streptomycin-sensitive bacterial ribosomes; as a result, the initiation complexes which form with these ribosomes are nonproductive; presumably, aminoacyl-tRNA cannot bind to the distorted acceptor site, and fMet-tRNA is released; elongation is also retarded, but less drastically; streptomycin causes misreading of mRNA in vitro and in vivo, but these effects do not have a major influence on the bactericidal action of this drug; studies with ribosomes from streptomycin-resistant mutants have shown that a particular protein, S12, of the 30S ribosomal subunit is important for the binding and effects of streptomycin; the other aminoglycosides have similar mechanisms of action, but they appear to bind to multiple sites on the 30S ribosome	98, 99
Aurintricarboxylic acid	Aurintricarboxylic acid (ATA), a triphenylmethane dye, inhibits initiation of polypeptide synthesis by preventing the attachment of mRNA to prokaryotic and eukaryotic ribosomal subunits; at higher concentrations of ATA, other steps in protein synthesis are affected	100—102
Borrelidin	Borrelidin inhibits prokaryotic and eukaryotic protein synthesis by specific inhibition of threonyl-tRNA synthetase; the inhibition is noncompetitive and may be due to the structural similarity of borrelidin to the 3′-*O*-threonyl-adenosine-5′-phosphoryl moiety of threonyl-tRNA	103, 104
Chloramphenicol	Chloramphenicol inhibits peptide bond formation on bacterial ribosomes by binding to the 50S ribosomal subunit at low and high affinity sites; this inhibition seems to depend on interference with the binding of the aminoacyl end of the aminoacyl-tRNA with the ribosomes	105, 106
Erythromycin	Erythromycin, a macrolide antibiotic, inhibits prokaryotic, but not eukaryotic, protein synthesis as a result of its binding to the 50S ribosomal subunit; ribosomal protein L4 may be involved in this interaction, since some erythromycin-resistant mutants have exhibited an altered L4 protein; it has been postulated that erythromycin inhibits translocation by preventing the positioning of peptidyl-tRNA in the donor site; other studies have indicated that the peptidyl transferase reaction may also be involved	107—109
Fusidic acid	Fusidic acid is a steroidal antibiotic which inhibits protein synthesis in prokaryotic and eukaryotic subcellular systems; the antibiotic binds to the ternary complex of prokaryotic EF–G (or EF–2 in eukaryotes) and GDP with the ribosome, preventing release of the elongation factor for further rounds of translocation; it has been proposed that this sequestration of EF–G may not explain the inhibitory action of fusidic acid in vivo, since growing cells appear to contain an excess of elongation factors; alternative explanations include (1) interference with subsequent complex formation between the ribosome and aminoacyl-tRNA—EF–Tu—GTP, and (2) possible effects of fusidic acid on both ribosomal A and P sites	110—112

MECHANISMS OF ACTION OF ANTIMICROBIAL AGENTS (continued)

Antimicrobial agent	Mechanism of action	Ref.
	Inhibition of Protein Synthesis	
Kirromycin	Kirromycin interacts with bacterial elongation factor Tu (EF–Tu), preventing the release of EF–Tu–GDP from the ribosome after GTP hydrolysis and thereby blocking subsequent peptidyl transfer, it has been postulated that kirromycin affects the conformation of the EF–Tu molecule, reducing allosteric control between its sites which interact with GTP, aminoacyl-tRNA, and the ribosome	113
Lincomycin Clindamycin	Lincomycin and its chlorinated derivative, clindamycin, inhibit the peptidyl transferase function of the 50S ribosomal subunit; however, a recent report suggests a secondary interaction of these antibiotics at lower concentrations with the process of peptide chain initiation in cell-free systems; interaction with the 50S ribosomal subunit results in alteration of the P site to produce an enhanced binding capacity for the fMet–tRNA	100, 114
Pactamycin	Pactamycin binds to smaller ribosomal subunits from prokaryotes and eukaryotes, but only when they are free from mRNA; as a result of this binding, the altered initiation complexes which are formed are less stable; it is possible that the interaction of this antibiotic with the initiator region also interferes with the binding of aminoacyl-tRNA to the acceptor site; this would account for the inhibition by pactamycin of chain elongation	115—117
Puromycin	Puromycin acts as an analog of the aminoacyl-adenosine end of aminoacyl-tRNA and competes with aminoacyl-tRNA for an acceptor site in the peptidyl transferase reaction of both prokaryotic and eukaryotic ribosomes; puromycin can then serve as an acceptor of the nascent peptide chain of ribosome-bound peptidyl-tRNA, causing premature release of the incomplete polypeptide chain; as a secondary effect of the release of nascent polypeptides from ribosomes, polysomes break down	45, 55, 100, 118
Sparsomycin	Sparsomycin inhibits the peptidyl transferase reaction of prokaryotic and eukaryotic ribosomes; it interferes with the acceptor recognition site, thereby blocking its interaction with the incoming aminoacyl-tRNA and fixing peptidyl-tRNA in the donor site	116, 119
Spectinomycin Kasugamycin	Unlike aminoglycosidic aminocyclitols, those which lack an aminosugar, such as spectinomycin and kasugamycin, (1) are bacteriostatic rather than bactericidal, (2) interact reversibly with the 30S ribosomal subunit, and (3) do not cause miscoding; spectinomycin appears to inhibit the translocation reaction and require cytidylic or guanylic acid residues in messenger RNA; mutational alteration of protein S5 of the 30S ribosomal subunit leads to spectinomycin resistance; kasugamycin inhibits the binding of fMet-tRNA to 30S ribosomes in the presence of initiation factors and the GTP analog, GMPPCP; the analysis of kasugamycin-resistant mutants suggests that the action of kasugamycin requires methylated 16S RNA and an intact ribosomal protein S2	120—122
Streptogramin group Virginiamycin Ostreogrycin Synergistin Mikamycin	These antibiotics are a mixture of two or more compounds which can be classified into two major groups, A and B; antibiotics of group A bind to the peptidyl transferase center of the 50S ribosomal subunit, preventing the binding of the 3′-terminal ends of the peptidyl-tRNA and aminoacyl-tRNA substrates,	123, 124

MECHANISMS OF ACTION OF ANTIMICROBIAL AGENTS (continued)

Antimicrobial agent	Mechanism of action	Ref.
	Inhibition of Protein Synthesis	
Pristinamycin Vernamycin	and thereby block peptide bond formation as well as initiation and elongation; it has been postulated that antibiotics in group B block translocation of the growing polypeptide chain from the A-site to the P-site (although EF–G-dependent GTPase activity is unaffected by these antibiotics); group A antibiotics exhibit a marked synergism with antibiotics of group B against Gram-positive bacteria due to an increased affinity of group A antibiotics for the ribosome when in the presence of group B antibiotics	
Tetracycline	Tetracycline inhibits protein synthesis on both 70S and 80S ribosomes, although 70S ribosomes are more sensitive; thus, tetracycline is more effective against protein synthesis in intact prokaryotic cells than against eukaryotic cells; tetracycline inhibits the binding of aminoacyl-tRNA to the ribosomal acceptor site but has little effect in binding to the donor site except at high drug concentrations; although the drug binds to most nucleoproteins, it binds preferentially to the 30S and presumably, 40S subunits; in intact cells, the drug at low concentrations causes breakdown of polyribosomes	45, 100
Thiostrepton Group Thiostrepton Siomycin Thiopeptin Micrococcin Sporangiomycin	The thiostrepton group of sulfur-containing peptide antibiotics inhibit protein synthesis in Gram-positive bacteria; they bind to the 50S ribosomal subunit and inhibit in vitro: (1) EF–G-dependent GTP hydrolysis, (2) the formation of ribosome-EF–G–GTP complexes, and (3) EF–Tu-dependent binding of aminoacyl-tRNA to the ribosomal A site; however, in vivo these drugs specifically inhibit aminoacyl-tRNA binding but do not affect translocation	125—127
	Inhibition of Synthesis of Essential Small Molecules Miscellaneous Effects	
Sulfonamides	The sulfonamides are structural analogs of paraaminobenzoic acid (PABA) which compete with this natural substrate for dihydropteroate synthetase in the bacterial biosynthesis of dihydrofolate; in bacterial and mammalian cells, dihydrofolate is reduced by a reductase to tetrahydrofolate, which serves as a cofactor for one-carbon transfer in the biosynthesis of essential metabolites such as purines, thymidylate, glycine, and methionine; the susceptibility of bacterial dihydrofolate reductases to inhibition by trimethroprim (see Trimethoprim) thus permits sequential blockade of the biosynthesis of tetrahydrofolate in bacteria by the combination of a sulfonamide plus trimethoprim; mammalian cells, which cannot synthesize dihydrofolate from PABA, derive preformed folates (folate, dihydrofolate, and also tetrahydrofolate) from the medium (or diet), while bacteria are unable to take up these compounds and are thus dependent on their biosynthesis of tetrahydrofolate	128, 129
Trimethoprim	Trimethoprim is a potent inhibitor of bacterial dihydrofolate reductases, which catalyze the reduction of dihydrofolate to the active tetrahydrofolate; cofactors of tetrahydrofolate function as donors of one-carbon fragments in the synthesis of thymidylate (TMP) and other essential compounds such as purines, methionine, pantothenate, and *N*-formyl methionyl-tRNA$_f$; the synthe-	128, 130, 131

MECHANISMS OF ACTION OF ANTIMICROBIAL AGENTS (continued)

Antimicrobial agent	Mechanism of action	Ref.
	Inhibition of Synthesis of Essential Small Molecules Miscellaneous Effects	
	sis of TMP from uridylate involves not only C_1 transfer, but also the oxidation of tetrahydrofolate to dihydrofolate, which must be reactivated by dihydrofolate reductase; the toxic effect of the limitation of tetrahydrofolate is primarily due to the deprivation of TMP which results in thymineless death; mammalian cells are protected by the extraordinarily low affinity of their dihydrofolate reductases for trimethoprim	
Pyrimethamine	The antimalarial agent pyrimethamine is an inhibitor of protozoal dihydrofolate reductases; as in the use of trimethoprim, the selective action of pyrimethamine against plasmodial parasites is due to a greater affinity for their reductases than for mammalian reductases	131, 132
Melinacidin	Melinacidin inhibits the growth of Gram-positive bacteria by blocking their biosynthesis of nicotinic acid, nicotinamide, and their derivative, the essential cofactor, nicotinamide adenine dinucleotide (NAD)	133
Cerulenin	Cerulenin inhibits the growth of a variety of fungi and bacteria by inhibiting their biosynthesis of fatty acids and sterols; the two catalytic steps affected by cerulenin are: (1) the conversion of acetyl-CoA to acetoacetyl-CoA by acetoacetyl-CoA thiolase, and (2) the conversion of acetoacetyl-CoA to a 3-hydroxy-3-methyl-glutaryl-CoA (HMG–CoA) by HMG-CoA synthetase	134
Griseofulvin	The antifungal agent, griseofulvin, inhibits only those fungi which possess chitin in their cell wall; however, long before the hyphae become noticeably distorted; demonstrable changes occur in the nuclei; griseofulvin causes breakdown of spindles and development of irregular masses of chromatin; although the exact mechanism of action of griseofulvin is unknown, it appears likely that its primary effect on fungi is due to its effect on the cell division process	135—138

REFERENCES

1. **Kahan, F. M., Kahan, J. S., Cassidy, P. J., and Kropp, H.,** The mechanism of action of fosfomycin (phosphonomycin), *Ann. N.Y. Acad. Sci.*, 235, 364, 1974.
2. **Wu, H. C. and Venkateswaran, P. S.,** Fosfomycin-resistant mutant of *Escherichia coli, Ann. N.Y. Acad. Sci.*, 235: 587, 1974.
3. **Strominger, J. L., Ito, E., and Threnn, R. H.,** Competitive inhibition of enzymatic reactions by oxamycin, *J. Am. Chem. Soc.*, 82, 998, 1960.
4. **Roze, U. and Strominger, J. L.,** Alanine racemase from *Staphylococcus aureus:* conformation of its substrates and its inhibitor, D-cycloserine, *Mol. Pharmacol.*, 2, 92, 1966.
5. **Neuhaus, F. C.,** D-Cycloserine and *O*-carbamyl-D-serine, in *Antibiotics*, Vol. 1, Gottlieb, D. and Shaw, P. D., Eds., Springer-Verlag, New York, 1967, 40.
6. **Lynch, J. L. and Neuhaus, F. C.,** On the mechanism of action of the antibiotic *O*-carbamyl–D-serine in *Streptococcus faecalis, J. Bacteriol.*, 91, 449, 1966.
7. **Neuhaus, F. C.** Selective inhibition of enzymes utilizing alanine in the biosynthesis of peptidoglycan, *Antimicrob. Agents Chemother.*, 1967, 304, 1968.
8. **Hammes, W. P. and Neuhaus, F. C.,** On the mechanism of action of vancomycin: inhibition of peptidoglycan synthesis in *Gaffkya homari, Antimicrob. Agents Chemother.*, 6, 722, 1974.

9. **Perkins, H. R. and Nieto, M.,** The chemical basis for the action of the vancomycin group of antibiotics, *Ann. N.Y. Acad. Sci.*, 235, 348, 1974.
10. **Jordan, D. C. and Reynolds, P. E.,** Vancomycin, in *Antibiotics*, Vol. 3, Corcoran, J. W. and Hahn, F. E., Eds., Springer-Verlag, New York, 1975, 704.
10a. **Williams, D. H. and Kalman, J. R.,** Structural and mode of action studies on the antibiotic vancomycin. Evidence from 270-MH proton magnetic resonance, *J. Am. Chem. Soc.*, 99, 2768—2774, 1977.
10b. **Sheldrick, G. M., Jones, P. G., Kennard, O., Williams, D. H., and Smith, G. A.,** Structure of Vancomycin and Its Complex with Acetyl-D-alanyl-D-alanine, *Nature*, 271, 223—225, 1978.
11. **Takatsuki, A., Kawamura, K., Okina, M., Kodama, Y., Ito, T., and Tamura, G.,** The structure of tunicamycin, *Agric. Biol. Chem.*, 41, 2307, 1977.
12. **Ward, J. B.,** Tunicamycin inhibition of bacterial wall polymer synthesis, *FEBS Lett.*, 78, 151, 1977.
13. **McArthur, H., Roberts, F., Hancock, I., and Baddiley, J.,** Lipid intermediates in the biosynthesis of the linkage unit between teichoic acids and peptidoglycan, *FEBS Lett.*, 86, 193, 1978.
14. **Tkacz, J. and Lampen, J. O.,** Tunicamycin inhibition of polyisoprenyl N-acetyl-glycosaminyl pyrophosphate formation in calf-liver microsomes, *Biochem. Biophys. Res. Commun.*, 65, 248, 1975.
15. **Eckardt, K., Thrum, H., Bradler, G., Tonew, E., and Tonew, M.,** Streptovirudins, new antibiotics with antibacterial and antiviral activity, *J. Antibiot.*, 28, 274, 1975.
16. **Lugtenberg, E. J. J., van Schijndel-Van Dam, A., and van Bellegam, T. H. M.,** *In vivo* and *in vitro* action of new antibiotics interfering with the utilization of *N*-acetyl-glucosamine-*N*-acetyl-muramyl-pentapeptide, *J. Bacteriol,*. 108, 20, 1971.
17. **Linnett, P. E. and Strominger, J. L.,** Additional antibiotic inhibitors of peptidoglycan synthesis, *Antimicrob. Agents Chemother.*, 4, 231, 1973.
18. **Brown, W. E., Seinerova, V., Chan, W. M., Laskin, A. I., Linnett, P., and Strominger, J. L.,** Inhibition of cell wall synthesis by the antibiotics diumycin and janiemycin, *Ann. N.Y. Acad. Sci.*, 235, 399, 1974.
19. **Somma, S., Merati, W., and Parenti, F.,** Gardimycin, a new antibiotic inhibiting peptidoglycan synthesis, *Antimicrob. Agents Chemother.*, 11, 396, 1977.
20. **Siewert, G. and Strominger, J. L.,** Bacitracin, an inhibitor of the dephosphorylation of lipid pyrophosphate, an intermediate in biosynthesis of the peptidoglycan of bacterial cell walls, *Proc. Natl. Acad. Sci. U.S.A.*, 57, 767, 1967.
21. **Stone, K. J. and Strominger, J. L.,** Mechanism of action of bacitracin: complexation with metal ion and C_{55}-isoprenyl pyrophosphate, *Proc. Natl. Acad. Sci. U.S.A.*, 68, 3223, 1971.
22. **Stone, K. J. and Strominger, J. L.,** Inhibition of sterol biosynthesis by bacitracin, *Proc. Natl. Acad. Sci. U.S.A.*, 69, 1287, 1972.
23. **Blumberg, P. M. and Strominger, J. L.,** Interaction of penicillin with the bacterial cell: penicillin-binding proteins and penicillin-sensitive enzymes, *Bacteriol. Rev.*, 38, 291, 1974.
24. **Strominger, J. L., Willoughby, E., Kamiryo, T., Blumberg, P. M., and Yocum, R. R.,** Penicillin-sensitive enzymes and penicillin-binding components in bacterial cells, *Ann. N.Y. Acad. Sci.*, 235, 210, 1974.
25. **Tomasz, A.,** Novel aspects of penicillin action. Introduction, in *Microbiology—1977*, Schlesinger, D., Ed., American Society for Microbiology, Washington, D.C., 1977, 175.
26. **Strominger, J. L.,** How penicillin kills bacteria: a short history, in *Microbiology—1977*, Schlesinger, D., Ed., American Society for Microbiology, Washington, D.C., 1977, 177.
27. **Tomasz, A. and Höltje, J. V.,** Murein hydrolases and the lytic and killing action of penicillin, in *Microbiology—1977*, Schlesinger, D., Ed., American Society for Microbiology, Washington, D.C., 1977, 209.
28. **Takayama, K., Wang, L., and David, H. L.,** Effect of isoniazid on the *in vivo* mycolic acid synthesis, cell growth, and viability of *Mycobacterium tuberculosis*, *Antimicrob. Agents Chemother.*, 2, 29, 1972.
29. **Takayama, K.,** Selective action of isoniazid on the synthesis of cell wall mycolates in mycobacteria, *Ann. N.Y. Acad. Sci.*, 235, 426, 1974.
30. **Krishna Murti, C. R.,** Isonicotinic acid hydrazide, in *Antibiotics*, Vol. 3, Corcoran, J. W. and Hahn, F. E., Eds., Springer-Verlag, New York, 1975, 623.
31. **Endo, A., Kakiki, K., and Misato, T.,** Mechanism of action of the antifungal agent polyoxin D, *J. Bacteriol.*, 104, 189, 1970.
32. **Hori, M., Kakiki, K., Suzuki, S., and Misato, T.,** Studies on the mode of action of polyoxins. III. Relation of polyoxin structure to chitin synthetase inhibition, *Agric. Biol. Chem.*, 35, 1280, 1971.
33. **Hori, M., Kakiki, K., and Misato, T.,** Further study on the relation of polyoxin structure to chitin synthetase inhibition, *Agric. Biol. Chem.*, 38, 691, 1974.
34. **Hunter, F. E., Jr., and Schwartz, L. S.,** Tyrocidines and gramicidin S, in *Antibiotics*, Vol. 1, Gottlieb, D. and Shaw, P. D., Eds., Springer-Verlag, New York, 1967, 636.
35. **Pache, W., Chapman, D., and Hillaby, R.,** Interaction of antibiotics with membranes: polymyxin B and gramicidin S, *Biochim. Biophys. Acta*, 255, 358, 1972.

36. **HsuChen, C. C. and Feingold, D. S.,** The mechanism of polymyxin B action and selectivity toward biological membranes, *Biochemistry,* 12, 2105, 1973.
37. **Feingold, D. S., HsuChen, C. C., and Sud, I. J.,** Basis for the selectivity of action of the polymyxin antibiotics on cell membranes, *Ann. N.Y. Acad. Sci.,* 235, 480, 1974.
38. **Lounatmaa, K., Makela, P. H., and Sarvas, M.,** Effect of polymyxin on the ultrastructure of the outer membrane of wild-type and polymyxin-resistant strains of *Salmonella, J. Bacteriol.,* 127, 1400, 1976.
39. **Andreoli, T. E.,** The structure and function of amphotericin B — cholesterol pores in lipid bilayer membranes, *Ann. N.Y. Acad. Sci.,* 235, 448, 1974.
40. **Holz, R. W.,** The effects of the polyene antibiotics nystatin and amphotericin B on thin lipid membranes, *Ann. N.Y. Acad. Sci.* 235, 469, 1974.
41. **Shemyakin, M. M., Ovchinnikov, Y. A., Ivanov, U. T., Antonov, V. K., Vinogradova, E. I., Shkrob, A. M., Malenkov, G. G., Evstratov, A. V., Laine, I. A., Melnik, E. I., and Rabova, I. D.,** Cyclodepsipeptides as chemical tools for studying ionic transport through membranes, *J. Membrane Biol.,* 1, 402, 1969.
42. **Pinkerton, M., Steinrauf, L. K., and Dawkins, P.,** The molecular structure and some transport properties of valinomycin, *Biochem. Biophys. Res. Commun.,* 35, 512, 1969.
43. **Grell, E., Funck, T., and Eggers, F.,** Structure and dynamic properties of ion-specific antibiotics, *Membranes,* 3, 1, 1975.
44. **Kilbourn, B. T., Dunitz, J. D., Pioda, L. A. R., and Simon, W.,** Structure of the K^+ complex with nonactin, a macrotetralide antibiotic possessing highly specific transport properties, *J. Mol. Biol.,* 30, 559, 1967.
45. **Gale, E. G., Cundliffe, E., Reynolds, P. E., Richmond, M. H., and Waring, M. J.,** *The Molecular Basis of Antibiotic Action,* John Wiley & Sons, New York, 1972.
46. **Pinkerton, M. and Steinrauf, L. K.,** Molecular structure of monovalent metal cation complexes of monensin, *J. Mol. Biol.,* 49, 533, 1970.
47. **Urry, D. W.,** The gramicidin A transmembrane channel: a proposed $\pi_{(L,D)}$Helix, *Proc. Natl. Acad. Sci. U.S.A.,* 68, 672, 1971.
48. **Hladsky, S. B. and Haydon, D. A.,** Ion transfer across lipid membranes in the presence of gramicidin A. I. Studies of the unit conductance channel, *Biochim. Biophys. Acta,* 274, 294, 1972.
49. **Urry, D. W.** A molecular theory of ion-conducting channels: a field-dependent transition between conducting and nonconducting conformations, *Proc. Natl. Acad. Sci. U.S.A.,* 69, 1610, 1972.
50. **Pittillo, R. F. and Hunt, D. E.,** Azaserine and 6-diazo-5-oxo-L-norleucine (DON), in *Antibiotics,* Vol. 1, Gottlieb, D. and Shaw, P. D., Eds., Springer-Verlag, New York, 1967, 481.
51. **Brockman, R. W., Pittillo, R. F., Shaddix, S., and Hill, D. L.,** Mode of action of azotomycin, *Antimicrob. Agents Chemother.,* 1969, 56, 1970.
52. **Kursten, H. and Kursten, W.,** *Inhibitors of Nucleic Acid Synthesis,* Springer-Verlag, New York, 1974.
53. **Shigeura, H. T.,** Hadacidin, in *Antibiotics,* Vol. 1, Gottlieb, D. and Shaw, P. D., Eds., Springer-Verlag, New York, 1967, 451.
54. **Hanka, L. J.,** Psicofuranine, in *Antibiotics,* Vol. 1, Gottlieb, D. and Shaw, P. D., Eds., Springer-Verlag, New York, 1967, 457.
55. **Suhadolnik, R. J.,** *Nucleoside Antibiotics,* John Wiley & Sons, New York, 1970, 96.
56. **Roy-Burman, P.** *Analogues of Nucleic Acid Components,* (Recent Results in Cancer Research, 25), Springer-Verlag, New York, 1970.
57. **Heidelberger, C.,** Fluorinated pyrimidines, *Prog. Nucl. Acid. Res. Mol. Biol.,* 4, 1, 1965.
58. **Polak, A. and Scholer, H. J.,** Mode of action of 5-fluorocytosine and mechanisms of resistance, *Chemotherapy,* 21, 113, 1975.
59. **Elford, H. L.,** Effect of hydroxyurea on ribonucleotide reductase, *Biochem. Biophys. Res. Commun.,* 33, 129, 1968.
60. **Sinha, N. K. and Snustad, D. P.,** "Mechanism of inhibition of deoxyribonucleic acid synthesis in *Escherichia coli* by hydroxyurea, *J. Bacteriol.,* 112, 1321, 1972.
61. **Timson, J.,** Hydroxyurea, *Mutat. Res.,* 32, 115, 1975.
62. **Hahn, F. E.,** Chloroquine (resochin), in *Antibiotics,* Vol. 3, Corcoran, J. W. and Hahn, F. E., Eds., Springer-Verlag, New York, 1975, 58.
63. **DiMarco, A., Arcamone, F., and Zunino, F.,** Daunomycin (daunorubicin) and adriamycin and structural analogues: biological activity and mechanism of action, in *Antibiotics,* Vol. 3, Corcoran, J. W. and Hahn, F. E., Eds., Springer-Verlag, New York, 1975, 101.
64. **Waring, M.,** Ethidium and propidium, in *Antibiotics,* Vol. 3, Corcoran, J. W. and Hahn, F. E., Eds., Springer-Verlag, New York, 1975, 141.
65. **Wolfe, A. D.,** Quinacrine and other acridines, in *Antibiotics,* Vol. 3, Corcoran, J. W. and Hahn, F. E., Eds., Springer-Verlag, New York, 1975, 203.

66. **Hirschberg, E.,** Thiaxanthenones: miracil D and hycanthone, in *Antibiotics,* Vol. 3, Corcoran, J. W. and Hahn, F. E., Eds., Springer-Verlag, New York, 1975, 274.
67. **Sobell, H. M.,** The stereo chemistry of actinomycin binding to DNA and its implication in molecular biology, *Prog. Nucl. Acid. Res. Mol. Biol.,* 13, 153, 1973.
68. **Goldberg, I. H. and Friedman, P. A.,** Antibiotics and nucleic acids, *Ann. Rev. Biochem.,* 40, 775, 1971.
69. **Gause, G. F.,** Olivomycin, chromomycin and mithramycin, in *Antibiotics,* Vol. 3, Corcoran, J. W. and Hahn, F. E., Eds., Springer-Verlag, New York, 1975, 197.
70. **Katigiri, K., Yoshida, T., and Sato, K.,** Quinoxaline antibiotics, in *Antibiotics,* Vol. 3, Corcoran, J. W. and Hahn, F. E., Eds., Springer-Verlag, New York, 1975, 234.
71. **Hahn, F. E.,** Distamycin A and netropsin, in *Antibiotics,* Vol. 3, Corcoran, J. W. and Hahn, F. E., Eds., Springer-Verlag, New York, 1975, 79.
72. **Zimmer, C.,** Effects of the antibiotics netropsin and distamycin A on the structure and function of nucleic acids, *Prog. Nucl. Acid Res. Mol. Biol.,* 15, 285, 1975.
73. **Friedman, P. A., Joel, P. B., and Goldberg, I. H.,** Interaction of kanchanomycin with nucleic acids. I. Physical properties of the complex, *Biochemistry,* 8, 1535, 1969.
74. **Friedman, P. A., Li, T., and Goldberg, I. H.,** Interaction of kanchanomycin with nucleic acids. II. Optical rotatory dispersion and circular dichroism, *Biochemistry,* 8, 1545, 1969.
75. **Goldberg, I. H.,** Kanchanomycin, in *Antibiotics,* Vol. 3, Corcoran, J. W. and Hahn, F. E., Eds., Springer-Verlag, New York, 1975, 166.
76. **Terawaki, A. and Greenberg, J.,** Inactivation of transforming deoxyribonucleic acid by carzinophillin and mitomycin C, *Biochim. Biophys. Acta,* 119, 59, 1966.
77. **Terawaki, A. and Greenberg, J.,** Effect of carzinophillin on bacterial deoxyribonucleic acid: formation of inter-strand cross-links in deoxyribonucleic acid and their disappearance during post-treatment incubation, *Nature (London),* 209, 481, 1966.
78. **Kohn, K. W.,** Anthramycin, in *Antibiotics,* Vol. 3, Corcoran, J. W. and Hahn, F. E., Eds., Springer-Verlag, New York, 1975, 3.
79. **Gause, G. F.,** Sibiromycin in *Antibiotics,* Vol. 3, Corcoran, J. W. and Hahn, F. E., Eds., Springer-Verlag, New York, 1975, 269.
80. **Umezawa, H.,** Chemistry and mechanism of action of bleomycin, *Fed. Proc.,* 33, 2296, 1974.
81. **Umezawa, H.,** Bleomycin, in *Antibiotics,* Vol. 3, Corcoran, J. W. and Hahn, F. E., Eds., Springer-Verlag, New York, 1975, 21.
82. **Sleigh, M. and Grigg, G. W.,** The mechanism of sensitivity to phloemycin in growing *Escherichia coli* cells, *Biochem. J.,* 155, 87, 1976.
83. **Cozzarelli, N. R.,** The mechanism of action of inhibitors of DNA synthesis, *Ann. Rev. Biochem.,* 46, 641, 1977.
84. **Wehrli, W. and Staehelin, M.,** Actions of the rifamycins, *Bacteriol. Rev.,* 35, 290, 1971.
85. **Wehrli, W. and Staehelin, M.,** Rifamycins and other ansamycins, in *Antibiotics,* Vol. 3, Corcoran, J. W. and Hahn, F. E., Eds., Springer-Verlag, New York, 1975, 252.
86. **Siddhikol, C., Erbstoeszer, J. W., and Weisblum, B.,** Mode of action of streptolydigin, *J. Bacteriol.,* 99, 151, 1969.
87. **Cassani, C., Burgess, R. R., Goodman, H. M., and Gold, L.,** Inhibition of RNA polymerase by streptolydigin, *Nature New Biol.,* 230, 197, 1971.
88. **Reusser, F.,** Tirandamycin, an inhibitor of bacterial ribonucleic polymerase, *Antimicrob. Agents Chemother.,* 10, 618, 1976.
89. **Cano, F. R., Kuo, S.-C., and Lampen, J. O.,** Lomofungin, an inhibitor of deoxyribonucleic acid-dependent ribonucleic acid polymerases, *Antimicrob. Agents Chemother.,* 3, 723, 1973.
90. **Cannon, M. and Jiminez, A.,** Lomofungin as an inhibitor of nucleic acid synthesis in *Saccharomyces cerevisiae, Biochem. J.,* 142, 457, 1974.
91. **Gellert, M., O'Dea, M. H., Itoh, T., and Tomizawa, J.,** Novobiocin and coumermycin inhibit DNA supercoiling catalyzed by DNA gyrase, *Proc. Natl. Acad. Sci. U.S.A.,* 73, 4474, 1976.
92. **Sugino, A., Peebles, C. L., Kreuzer, K. N., and Cozzarelli, N. R.,** Mechanism of action of nalidixic acid: purification of *Escherichia coli nal A* gene product and its relationship to DNA gyrase and a novel nicking-closing enzyme, *Proc. Natl. Acad. Sci. U.S.A.,* 74, 4767, 1977.
93. **Gellert, M., Mizuuchi, K., O'Dea, M. H., Itoh, T., and Tomizawa, J.,** Nalidixic acid resistance: a second genetic character involved in DNA gyrase activity, *Proc. Natl. Acad. Sci. U.S.A.,* 74, 4772, 1977.
94. **Kurylo-Borowska, Z.,** Edeines, in *Antibiotics,* Vol. 3, Corcoran, J. W. and Hahn, F. E., Eds., Springer-Verlag, New York, 1975, 129.
95. **Mackenzie, J. M., Neville, M. M., Wright, G. E., and Brown, N. C.,** Hydroxyphenylazopyrimidines: characterization of the active forms and their inhibitory action on a DNA polymerase from *Bacillus subtilis, Proc. Natl. Acad. Sci. U.S.A.,* 70, 512, 1973.
96. **Pestka, S.,** Studies on transfer ribonucleic acid — ribosome complexes. XIX. Effect of antibiotics on peptidyl puromycin synthesis on polyribosomes from *Escherichia coli, J. Biol. Chem.,* 247, 4669, 1972.

97. **Pestka, S.,** Althiomycin, in *Antibiotics,* Vol. 3, Corcoran, J. W. and Hahn, F. E., Eds., Springer-Verlag, New York, 1975, 323.
98. **Lewin, G.,** *Gene Expression,* Vol. 1, John Wiley & Sons, London, 1974.
99. **Schlessinger, D. and Medoff, G.,** Streptomycin, dihydrostreptomycin, and the gentamicins, in *Antibiotics,* Vol. 3, Corcoran, J. W. and Hahn, F. E., Eds., Springer-Verlag, New York, 1975, 535.
100. **Pestka, S.,** Inhibitors of ribosome function, *Ann. Rev. Microbiol.,* 25, 487, 1971.
101. **Tai, P., Wallace, B. J., and Davis, B. D.,** Action of aurintricarboxylate, kasugamycin, and pactamycin on *Escherichia coli* polysomes, *Biochemistry,* 12, 616, 1973.
102. **Apirion, D. and Dohner, D.,** Aurintricarboxylic acid. A non-antibiotic organic molecule that inhibits protein synthesis, in *Antibiotics,* Vol. 3, Corcoran, J. W. and Hahn, F. E., Eds., Springer-Verlag, New York, 1975, 327.
103. **Nass, G., Poralla, K., and Zahner, H.,** Effect of the antibiotic borrelidin on the regulation of threonine biosynthetic enzymes in *E. coli, Biochem. Biophys. Res. Commun.,* 34, 84, 1969.
104. **Poralla, K.,** Borrelidin, in *Antibiotics,* Vol. 3, Corcoran, J. W. and Hahn, F. E., Eds., Springer-Verlag, New York, 1975, 365.
105. **Lessard, J. L. and Pestka, S.,** Studies on the formation of transfer ribonucleic acid — ribosome complexes. XXIII. Chloramphenicol, aminoacyl-oligonucleotides, and *Escherichia coli* ribosomes, *J. Biol. Chem.,* 247, 6909, 1972.
106. **Pestka, S.,** Chloramphenicol, in *Antibiotics,* Vol. 3, Corcoran, J. W. and Hahn, F. E., Eds., Springer-Verlag, New York, 1975, 370.
107. **Cannon, M. and Burns, K.,** Mode of action of erythromycin and thiostrepton as inhibitors of protein synthesis, *FEBS Lett.,* 18, 1, 1971.
108. **Tanaka, S., Otaka, T., and Kaji, A.,** Further studies on the mechanism of erythromycin action, *Biochim. Biophys. Acta,* 331, 128, 1973.
109. **Oleinick, N. L.,** The erythromycins, in *Antibiotics,* Vol. 3, Corcoran, J. W. and Hahn, F. E., Eds., Springer-Verlag, New York, 1975, 396.
110. **Cundliffe, E. and Burns, D.,** Long term effects of fusidic acid on bacterial protein synthesis *in vivo, Biochem. Biophys. Res. Commun.,* 49, 766, 1972.
111. **Burns, K., Cannon, M., and Cundliffe, E.,** A resolution of conflicting reports concerning the mode of action of fusidic acid, *FEBS Lett.,* 40, 219, 1974.
112. **Tanaka, N.,** Fusidic acid, in *Antibiotics,* Vol. 3, Corcoran, J. W. and Hahn, F. E., Eds., Springer-Verlag, New York, 1975, 436.
113. **Wolf, H., Chinali, G., and Parmeggiani, A.,** Kirromycin, an inhibitor of protein biosynthesis that acts on elongation factor Tu, *Proc. Natl. Acad. Sci. U.S.A.,* 71, 4910, 1974.
114. **Reusser, F.,** Effect of lincomycin and clindamycin on peptide chain initiation, *Antimicrob. Agents Chemother.,* 7, 32, 1975.
115. **Cohen, L. B., Herner, A. E., and Goldberg, I. H.,** Inhibition by pactomycin of the initiation of protein synthesis. Binding to N-acetylphenylalanyl transfer ribonucleic acid and polyuridylic acid to ribosomes, *Biochemistry,* 8, 1312, 1969.
116. **Goldberg, I. H., Stewart, M. L., Ayuso, M., and Kappen, L. S.,** On the mechanism of inhibition of polypeptide synthesis by the antibiotics sparsomycin and pactamycin, *Fed. Proc.,* 32, 1688, 1973.
117. **Goldberg, I. H.,** Pactamycin, in *Antibiotics,* Vol. 3, Corcoran, J. W. and Hahn, F. E., Eds., Springer-Verlag, New York, 1975, 498.
118. **Tanaka, S., Igarashi, K., and Kaji, A.,** Studies on the action of tetracycline and puromycin, *J. Biol. Chem.,* 247, 45, 1972.
119. **Herner, A. E., Goldberg, I. H., and Cohen, L. B.,** Stabilization of N-acetylphenylalanyl transfer ribonucleic acid binding to ribosomes by sparsomycin, *Biochemistry,* 8, 1335, 1969.
120. **Okuyama, A., Machiyama, N., Kinoshita, T., and Tanaka, N.,** Inhibition by kasugamycin of initiation complex formation on 30S ribosomes, *Biochem. Biophys. Res. Commun.,* 43, 196, 1971.
121. **Tanaka, N.,** Aminoglycoside antibiotics, in *Antibiotics,* Vol. 3, Corcoran, J. W. and Hahn, F. E., Eds., Springer-Verlag, New York, 1975, 340.
122. **Yoshikawa, M., Okuyama, A., and Tanaka, N.,** A third kasugamycin resistance locus, ksgC, affecting ribosomal protein S2 in *Escherichia coli* K-12, *J. Bacteriol.,* 122, 796, 1975.
123. **Vazquez, D.,** The streptogramin family of antibiotics, in *Antibiotics,* Vol. 3, Corcoran, J. W. and Hahn, F. E., Eds., Springer-Verlag, New York, 1975, 521.
124. **Barbacid, M., Contreras, A., and Vazquez, D.,** The mode of action of griseoviridin at the ribosomal level, *Biochim. Biophys. Acta,* 395, 347, 1975.
125. **Cundliffe, E.,** The mode of action of thiostrepton *in vivo, Biochem. Biophys. Res. Commun.,* 44, 912, 1971.
126. **Pestka, S. and Bodley, J. W.,** The thiostrepton group of antibiotics, in *Antibiotics,* Vol. 3, Corcoran, J. W. and Hahn, F. E., Eds., Springer-Verlag, New York, 1975, 551.

127. **Cundliffe, E. and Dixon, P. D.,** Inhibition of ribosomal A site functions by sporangiomycin and micrococcin, *Antimicrob. Agents Chemother.*, 8, 1, 1975.
128. **Hitchings, G. H.,** Mechanism of action of trimethoprim-sulfamethoxazole — I, *J. Infect. Dis.*, 128, S433, 1973.
129. **Anand, N.,** Sulfonamides and Sulfones, in *Antibiotics,* Vol. 3, Corcoran, J. W. and Hahn, F. E., Eds., Springer-Verlag, New York, 1975, 668.
130. **Burchall, J. J.,** Mechanism of action of trimethoprim-sulfamethoxazole — II, *J. Infect. Dis.*, 128, S437, 1973.
131. **Burchall, J. J.,** Trimethoprim and pyrimethamine, in *Antibiotics,* Vol. 3, Corcoran, J. W. and Hahn, F. E., Eds., Springer-Verlag, New York, 1975, 304.
132. **Ferone, R., Burchall, J. J., and Hitchings, G. H.,** *Plasmodium berghei* dihyrodofolate reductase: isolation, properties, and inhibition of antifolates, *Mol. Pharmacol.*, 5, 49, 1969.
133. **Reusser, F.,** Mode of action of melinacidin, an inhibitor of nicotinic acid biosynthesis, *J. Bacteriol.*, 96, 1285, 1968.
134. **Omura, S.,** The antibiotic cerulenin, a novel tool for biochemistry as an inhibitor of fatty acid synthesis, *Bacteriol. Rev.*, 40, 681, 1976.
135. **Crackower, S. H. B.,** The effects of griseofulvin on mitosis in *Aspergillus nidulans, Can. J. Microbiol.*, 18, 683, 1972.
136. **Gull, K. and Trinci, A. P. J.,** Griseofulvin inhibits fungal mitosis, *Nature (London),* 244, 292, 1973.
137. **Kappas, A. and Georgopoulds, S. G.,** Interference of griseofulvin with the segregation of chromosomes at mitosis in diploid *Aspergillus nidulans, J. Bacteriol.*, 119, 334, 1974.
138. **Huber, F. M.,** Griseofulvin, in *Antibiotics,* Vol. 3, Corcoran, J. W. and Hahn, F. E., Eds., Springer-Verlag, New York, 1975, 606.

METHODS FOR BIOASSAY OF ANTIBIOTICS

David M. Isaacson and T. B. Platt

Table 1
TEST ORGANISMS USED FOR THE BIOASSAY OF SELECTED ANTIBIOTICS[a]

Test organism number	Test organism
1	*Bacillus cereus* var. *mycoides* (ATCC 11778)
2	*Bacillus subtilis* (ATCC 6633)
3	*Bordetella bronchiseptica* (ATCC 4617)
4	*Candida tropicalis* (ATCC 13803)
5	*Corynebacterium xerosis* (NTCC 9755)
6	*Escherichia coli* (ATCC 10536)
7	*Klebsiella pneumoniae* (ATCC 10031)
8	*Micrococcus flavus* (ATCC 10240) (*Micrococcus luteus*)
9	*Micrococcus flavus* resistant to dihydrostreptomycin (ATCC 10240A) (*Micrococcus luteus*)
10	*Micrococcus flavus* resistant to neomycin (ATCC 14452) (*Micrococcus luteus*)
11	*Microsporum gypseum* (ATCC 14683)
12	*Mycobacterium smegmatis* (ATCC 607)
13	*Paecilomyces varioti* (ATCC 22319)
14	*Proteus vulgaris* (ATCC 21100)
15	*Pseudomonas aeruginosa* (ATCC 25619)
16	*Pseudomonas aeruginosa* (ATCC 29336)
17	*Saccharomyces cerevisiae* (ATCC 2601)
18	*Saccharomyces cerevisiae* (ATCC 9763)
19	*Sarcina lutea* (ATCC 9341) (*Micrococcus luteus*)
20	*Sarcina subflava* (ATCC 7468) (*Micrococcus luteus*)
21	*Sarcina subflava* resistant to dihydrostreptomycin (ATCC 7468/d) (*Micrococcus luteus*)
22	*Staphylococcus aureus* (ATCC 6538P)
23	*Staphylococcus aureus* (ATCC 9144)
24	*Staphylococcus epidermidis* (ATCC 12228)
25	*Streptococcus faecalis* (ATCC 10541) (*Streptococcus faecium*)

[a] See Table 5 for the specific microbiological assay methods that employ these test organisms.

Table 2
METHODS FOR PREPARATION OF SUSPENSIONS OF THE TEST ORGANISMS[a]

Method A

Maintain the test culture by weekly transfers on slants of medium 1. Suspend a 24-hr slant culture (grown at 35°C) with 3 mℓ of sterile solution K. Transfer the suspension onto the surface of 250 mℓ of medium 1 in a Roux bottle. Incubate for 24 hr at 32 to 37°C. Wash the resulting cell growth from the agar surface with 50 mℓ of solution K. Store the cell suspension at 5°C.

Method B

Maintain the test culture by weekly transfers on slants of medium 1. Suspend a 24-hr slant culture (grown at 35°C) with 3 mℓ of sterile solution K. Transfer the suspension onto the surface of 250 mℓ of medium 32 in a Roux bottle. Incubate for 24 hr at 32 to 37° C. Wash the resulting cell growth from the agar surface with 50 mℓ of sterile solution K. Transfer the suspension to a sterile centrifuge bottle. Centrifuge at 2000 × *g* for 10 min. Decant the supernatant. Resuspend the cells in 60 mℓ of sterile solution K. Heat the suspension for 30 min at 70°C. Store the spore suspension at 5°C.

Method C

Maintain the test culture by weekly transfers on slants of medium 1. Suspend a 24-hr slant culture (grown at 35°C) with 3 mℓ of sterile solution K. Transfer suspension onto the surface of 250 mℓ of medium 1 in a Roux bottle. Incubate for 24 hr at 32 to 37°C. Wash the resulting cell growth from the agar surface with 50 mℓ of sterile medium K. Transfer the suspension to a sterile centrifuge bottle. Heat for 30 min at 70°C, then centrifuge at 2000 × *g* for 10 min. Decant the supernatant. Resuspend the cells with 30 mℓ of sterile distilled water. Repeat centrifugation and wash two additional times. Resuspend the cells in 60 mℓ of sterile distilled water. Store the spore suspension at 5°C.

Method D

Same as Method A, except incubate the slants for 24 hr at 30°C and incubate the Roux bottles for 48 hr at 30°C.

Method E

Maintain the test organisms in 100-mℓ portions of medium 3. To prepare the test broth culture, transfer a loopful of the stock culture to 100 mℓ of medium 3. Incubate for 18 hr at 37°C. Store the test broth culture at 5°C.

Method F

Same as Method C, except incubate the Roux bottle for 7 days at 37°C. Suspend the spores in 100 mℓ of sterile distilled water.

Method G

Maintain the test culture by weekly transfers on slants of medium 6. Incubate for 24 hr at 37°C. Suspend a 24-hr slant culture with 10 mℓ of medium 39. Adjust the cell density to 80% light transmission at 660 nm.

Method H

Maintain the test culture by biweekly transfers on slants of medium 30. Incubate at 25°C. Suspend growth with 10 mℓ of sterile solution K. Transfer the slant wash to 3-ℓ conical flasks containing 200 mℓ of medium 22. Incubate for 6 to 8 weeks at 25°C. If sporulation is 80%, harvest the spores on the mycelial layer with a sterile spatula. Place the harvested spores in 50 mℓ of sterile solution K. Store up to 2 months at 5°C.

Method I

Maintain the test culture by biweekly transfers on slants containing 20 mℓ of medium 12. Incubate at 25°C. Wash the spores from the slants with 5 mℓ of solution K per slant. Store the spore suspension at 5°C.

Table 2 (continued)
METHODS FOR PREPARATION OF SUSPENSIONS OF THE TEST ORGANISMS[a]

Method J

Same as Method A, except use medium 28.

Method K

Maintain the test culture by weekly transfers on slants of medium 1. Incubate for 20 hr at 37°C. To prepare test inoculum, transfer a loopful of the stock culture to 200 mℓ medium 3. Incubate 18 hr with shaking at 30°C.

Method L

Maintain the test culture on slants of medium 8. Incubate for 24 hr at 37°C. To prepare the test inoculum, transfer a loop of the slant culture to medium 3. Incubate for 20 hr at 37°C without shaking. Store the test broth culture at 5°C.

Method M

Maintain the test culture on slants of medium 6. Incubate for 24 hr at 30°C. To prepare the test inoculum, transfer a loop of the slant culture to 100 mℓ of medium 3. Incubate for 18 hr at 30°C.

Method N

Maintain the test culture on slants of medium 31. Incubate for 7 days at 35°C. To prepare the test inoculum, wash a slant culture with 5 mℓ of medium 17 and transfer to 1000 mℓ of medium 14. Incubate, with shaking, for 16 hr at 35°C. Store the test broth culture for one week at 5°C.

Method O

Maintain the test culture on slants of medium 36. Transfer weekly. To prepare the test inoculum, incubate slant for 2 days at 37°C, wash the slant culture with 3 mℓ of sterile solution K and transfer to 100 mℓ of medium 34. Add 50 g glass beads. Incubate on rotary shaker, 130 cycles/min and radius of 3.5 cm, for 5 days at 27°C. Store the test broth culture for 2 weeks at 5°C.

Method P

Maintain the test culture by weekly transfer on slants of medium 36. Suspend a 24-hr slant culture (grown at 37°C) with 3 mℓ of sterile solution K. Transfer to 250 mℓ of medium 36 in a Roux bottle. Incubate 24 hr at 37°C. Wash the resulting cell growth from the agar surface with 50 mℓ of medium 37. Store the cell suspension at 5°C for 1 week.

Method Q

Maintain the test culture by weekly transfer on slants of medium 16. Incubate for 18 hr at 37°C. To prepare the test inoculum, transfer a loopful of the stock culture to 100 mℓ of medium 39. Incubate for 18 hr at 37°C. Store the test broth culture at 5°C.

Method R

Same as Method A except use medium 27 and incubate for 48 hr at 37°C.

[a] Refer to Table 3 for descriptions of the media cited above, and to Table 4 for descriptions of the solutions cited above.

Table 3
COMPOSITION OF CULTURE MEDIA USED FOR MICROBIOLOGICAL ASSAYS OF ANTIBIOTICS[a]

Ingredient / Number	1	2	3	4	5	6	7	8	9	10	11	12	13	14	15	16	17	18	19	20	21
(FDA Number)	(1)	(2)	(3)	(4)	(5)	—	—	(8)	(9)	(10)	(11)	—	(13)	—	—	—	—	(18)	(19)	(20)	(21)
Beef extract	1.5	1.5	1.5	1.5	1.5	—	—	1.5	—	—	1.5	—	—	1.5	3.0	3.0	1.5	1.5	2.4	—	—
Yeast extract	3.0	3.0	1.5	3.0	3.0	—	—	3.0	—	—	3.0	—	—	1.5	—	2.0	6.5	3.0	4.7	—	—
Peptic digest of animal tissue	6.0	6.0	5.0	6.0	6.0	—	10.0	6.0	—	—	6.0	—	10.0	5.0	5.0	5.0	5.0	6.0	9.4	10.0	10.0
Pancreatic digest of casein	4.0	—	—	—	—	15.0	—	—	17.0	17.0	4.0	—	—	—	—	—	10.0	4.0	—	—	—
Papaic digest of soybean meal	—	—	—	—	—	5.0	—	—	3.0	3.0	—	—	—	—	—	—	—	—	—	—	—
Infusion from horse meat	—	—	—	—	—	—	454	—	—	—	—	—	—	—	—	—	—	—	—	—	—
Glucose	1.0	—	1.0	1.0	—	—	—	—	2.5	2.5	1.0	20.0	20.0	11.0	—	—	11.0	1.0	10.0	40.0	40.0
Glycerol	—	—	—	—	—	—	—	—	—	—	—	—	—	—	—	—	—	—	—	—	—
L-Cystine	—	—	—	—	—	—	—	—	—	—	—	—	—	—	—	—	—	—	—	—	—
Potassium dihydrogen phosphate	—	—	1.32	—	—	—	—	—	—	—	—	—	—	1.32	—	—	—	—	—	—	—
Dipotassium phosphate	—	—	3.68	—	—	—	—	—	2.5	2.5	—	—	—	3.68	—	—	—	—	—	—	—
Sodium chloride	—	—	3.5	—	—	5.0	5.0	—	5.0	5.0	—	—	—	3.5	—	—	—	—	10.0	—	—
Sodium sulfite	—	—	—	—	—	—	—	—	—	—	—	—	—	—	—	—	—	—	—	—	—
Chloramphenicol[b]	—	—	—	—	—	—	—	—	—	—	—	—	—	—	—	—	—	—	—	0.05	0.05
Cycloheximide[b]	—	—	—	—	—	—	—	—	—	—	—	—	—	—	—	—	—	—	—	—	0.2
Dihydrostreptomycin[b]	—	—	—	—	—	—	—	—	—	—	—	—	—	—	—	—	—	—	—	—	—
Erythromycin[b]	—	—	—	—	—	—	—	—	—	—	—	—	—	—	—	—	—	—	—	—	—
Neomycin[b]	—	—	—	—	—	—	—	—	—	—	—	—	—	—	—	—	—	—	—	—	—
Novobiocin[b]	—	—	—	—	—	—	—	—	—	—	—	—	—	—	—	—	—	—	—	—	—
Tetracycline[b]	—	—	—	—	—	—	—	—	—	—	—	—	—	—	—	—	—	—	—	—	—
Polysorbate 80 (mℓ)	—	—	—	—	—	—	—	—	—	10.0	—	—	—	—	—	—	—	20.0	—	—	—
Agar	15.0	15.0	—	15.0	15.0	15.0	20.0	15.0	20.0	12.0	15.5	17.0	—	—	15.0	15.0	—	15.0	23.5	15.0	15.0
pH ± 0.05 (after sterilization)	6.55	6.55	7.0	6.55	7.9	7.3	7.1	5.9	7.25	7.25	7.9	6.5	5.65	6.0	7.9	6.8	6.8	7.9	6.1	5.65	5.65

Ingredient / Number	22	23	24	25	26	27	28	29	30	31	32	33	34	35	36	37	38	39	40	41
(FDA Number)	(22)	(23)	(24)	(25)	(26)	(27)	(28)	(29)	—	—	(32)	(33)	(34)	(35)	(36)	(37)	(38)	—	—	—
Beef extract	—	1.5	1.5	1.5	1.5	1.5	1.5	6.0	—	3.0	1.5	1.5	10.0	10.0	—	—	—	3.0	—	1.5
Yeast extract	—	3.0	3.0	3.0	3.0	3.0	3.0	—	—	—	3.0	3.0	—	—	—	—	—	—	5.0	3.0
Peptic digest of animal tissue	10.0	6.0	6.0	6.0	6.0	6.0	6.0	10.0	10.0	5.0	6.0	6.0	5.0	5.0	—	—	15.0	5.0	—	6.0
Pancreatic digest of casein	—	4.0	4.0	4.0	4.0	4.0	4.0	—	—	—	4.0	4.0	—	—	15.0	17.0	—	—	15.0	—
Papaic digest of soybean meal	—	—	—	—	—	—	—	—	—	—	—	—	—	—	5.0	3.0	5.0	—	—	—
Infusion from horse meat	—	—	—	—	—	—	—	—	—	—	—	—	—	—	—	—	—	—	—	—
Glucose	40.0	1.0	1.0	1.0	1.0	1.0	1.0	—	40.0	—	1.0	1.0	—	—	—	2.5	5.5	—	10.0	—
Glycerol	—	—	—	—	—	—	—	—	—	—	—	—	10.0	10.0	—	—	—	—	—	—
L-Cystine	—	—	—	—	—	—	—	—	—	—	—	—	—	—	—	—	0.7	—	—	—
Potassium dihydrogen phosphate	—	—	—	—	—	—	—	—	—	—	—	—	—	—	—	—	—	—	—	—
Dipotassium phosphate	—	—	—	—	—	—	—	—	—	—	—	—	—	—	—	2.5	—	—	—	—
Sodium chloride	—	—	—	—	—	—	—	—	—	—	—	—	3.0	3.0	5.0	5.0	4.0	—	2.5	10.0
Sodium sulfite	—	—	—	—	—	—	—	—	—	—	—	—	—	—	—	—	0.2	—	—	—
Chloramphenicol[b]	—	—	—	—	—	—	—	—	—	—	—	—	—	—	—	—	—	—	—	—
Cycloheximide[b]	—	—	—	—	—	—	—	—	—	—	—	—	—	—	—	—	—	—	—	—
Dihydrostreptomycin[b]	—	—	—	—	—	—	0.5	—	—	—	—	—	—	—	—	—	—	—	—	—
Erythromycin[b]	—	0.6	—	—	—	—	—	—	—	—	—	—	—	—	—	—	—	—	—	—
Neomycin[b]	—	—	—	—	—	0.1	—	—	—	—	—	—	—	—	—	—	—	—	—	—
Novobiocin[b]	—	—	—	—	—	—	—	—	—	—	—	0.01	—	—	—	—	—	—	—	—
Tetracycline[b]	—	—	—	—	0.1	—	—	—	—	—	—	—	—	—	—	—	—	—	—	—
Polysorbate 80 (mℓ)	—	—	—	—	—	—	—	—	—	—	—	—	—	—	—	—	—	—	—	—
Agar	—	15.0	30.0	15.0	15.0	15.0	15.0	15.0	15.0	15.0	15.0	15.0	—	17.0	15.0	—	15.0	—	24.0	15.0
pH ± 0.05 (after sterilization)	5.65	6.55	6.55	6.55	6.55	6.55	6.55	7.9	5.65	6.8	6.55	6.55	7.0	7.0	7.3	7.3	7.0	6.8	6.0	9.0

[a] Refer to Tables 2 and 5 for specific applications of the media listed in this table. Each value in the table equals the concentration in grams per liter of the ingredient in the numbered culture medium.

[b] Activity.

Table 4
COMPOSITION OF SOLVENTS, BUFFERS, AND DILUENTS USED FOR MICROBIOLOGICAL ASSAYS OF ANTIBIOTICS

Solution	Name	Ingredients[a]	Amount
A	1% Phosphate buffer pH 6.0 ± 0.05	Dipotassium phosphate	2.0 g
		Potassium dihydrogen phosphate	8.0 g
		Distilled water to make	1000 mℓ
B	0.1 *M* Phosphate buffer pH 7.9 ± 0.1	Dipotassium phosphate	16.73 g
		Potassium dihydrogen phosphate	0.523 g
		Distilled water to make	1000 mℓ
C	0.1 *M* Phosphate buffer pH 4.5 ± 0.05	Potassium dihydrogen phosphate	13.6 g
		Distilled water to make	1000 mℓ
D	10% Phosphate buffer pH6.0 ± 0.05	Dipotassium phosphate	20.0 g
		Potassium dihydrogen phosphate	80.0 g
		Distilled water to make	1000 mℓ
E	0.2 *M* Phosphate buffer pH 10.5 ± 0.1	Dipotassium phosphate	35.0
		10 *N* Potassium hydroxide	2.0 mℓ
		Distilled water to make	1000 mℓ
F	0.1 *M* Phosphate buffer pH 7.0 ± 0.1	Dipotassium phosphate	13.6 g
		Potassium dihydrogen phosphate	4.0 g
		Distilled water to make	1000 mℓ
G	1% Phosphate buffer with 0.1% sodium nitrate pH 6.0 ± 0.05	Dipotassium phosphate	2.0 g
		Potassium dihydrogen phosphate	8.0 g
		Sodium nitrate	1.0 g
		Distilled water to make	1000 mℓ
H	0.01 *N* Methanolic hydrochloric acid	1.0 *N* Hydrochloric acid	10.0 mℓ
		Methyl alcohol to make	1000 mℓ
I	80% Isopropyl alcohol solution	Isopropyl alcohol	800 mℓ
		Distilled water to make	1000 mℓ
J	1% Sodium carbonate	Sodium carbonate	10.0 g
		Distilled water to make	1000 mℓ
K	0.9% Saline	Sodium chloride	9.0 g
		Distilled water to make	1000 mℓ
M	Dimethylformamide solution pH 6.7 ± 0.3	Dimethylformamide	1000 mℓ
		Adjust pH with 10 *N* potassium hydroxide and 10 *N* hydrochloric acid immediately before use	
N	80% Dimethylsulfoxide	Dimethylsulfoxide	800 mℓ
		Distilled water to make	1000 mℓ
O	25% Dimethylsulfoxide	Dimethylsulfoxide	250 mℓ
		Distilled water to make	1000 mℓ
P	0.05 *M* Tris buffer	Tris(hydroxymethyl) nitromethane	7.56 g
		Distilled water to make	1000 mℓ

[a] Adjust phosphate buffers to correct pH with 18 *N* phosphoric acid or 10*N* potassium hydroxide

Table 5
SELECTED METHODS FOR THE MICROBIOLOGICAL ASSAY OF ANTIBIOTICS

		Test organism			Assay media										
					Agar-diffusion assays				Turbidimetric assays		Preparation of standard solutions				
					Base layer		Seed Layer								
Antibiotic	Test method	Test organism number (refer to Table 1)	Method of preparation of test inoculum (refer to Table 2)	Milliliters of test inoculum per liter of assay medium	Medium number (refer to Table 3)	Volume per plate (mℓ)[c]	Medium Number (refer to Table 3)	Volume per plate (mℓ)[d]	Medium number (refer to Table 3)	Volume per tube (mℓ)	Initial concentration and initial solvent (refer to Table 4)	Diluent employed for subsequent dilutions (refer to Table 4)	Assay range, (units or µg/mℓ)[e]	Incubation temperature (°C)[f]	Ref.
Amoxicillin	AD[a]	19	A	5.0	11	21	11	4	—	—	1000 µg/mℓ in distilled water	B	0.064—0.156(0.100)	32—35	1
Amphomycin	AD	10	R	5.0	2	21	1	4	—	—	100 µg/mℓ in B	B	6.4—15.6(10.0)	37	1
Amikacin Sulfate	TU[b]	22	A	1.0	—	—	—	—	3	9	1000 µg/mℓ in distilled water	Distilled water	8.0—12.5(10.0)	37	1
Amphotericin B	AD	18	D	2.0	—	—	19	8	—	—	1000 µg/mℓ in dimethyl sulfoxide	E	0.64—1.56(1.0)	30	2, 29
Ampicillin	AD	19	A	10.0	11	21	11	4	—	—	100 µg/mℓ in distilled water	B	0.64—1.56(1.0)	32—35	1, 2, 29
Bacitracin	AD	20 or 21[g]	A[j]	2.0	2	21	1	4	—	—	100 units/mℓ in A	A	0.64—1.56(1.0)	32—35	1, 2, 29
	AD	8 or 9[h]	A[j]	2.0	2	21	1	4	—	—	100 units/mℓ in A	A	64—156(100)	32—35	1, 2, 29
Bleomycin	AD	12	O	10.0	35	10	35	6	—	—	2.0 units/mℓ in F	F	0.01—0.16(0.04)	32—35	1
Candicidin	TU	18	A	2.0	—	—	—	—	13	9	1000 µg/mℓ in dimethyl sulfoxide	Distilled water	0.03—0.12(0.06)	25	1, 2
Capreomycin	TU	7	A	1.0	—	—	—	—	3	9	1000 µg/mℓ in distilled water	Distilled water	64—156(100)	37	1—4, 29
Carbenicillin	AD	15	A	0.5	9	21	10	4	—	—	1000 µg/mℓ in A	A	64—156(100)	37	1, 2
Carbomycin	AD	19	A	4.0	1	21	1	4	—	—	1000 µg/mℓ in methanol	B	0.6—1.5(1.0)	30—32	5, 6
Cefaclor	AD	19	A	0.2	—	—	1	10	—	—	1000 µg/mℓ in C	C	0.025—0.20(0.1)	30	34, 35
	TU	23	K	30.0	—	—	—	—	3	10	1000 µg/mℓ in C	C	0.05—0.25(1.5)	37	34
Cefadroxil	AD	22	A	0.5	2	21	1	4	—	—	1000 µg/mℓ in A	A	12.8—31.2(20.0)	37	26
Cefazaflur	AD	2	B	2.0	—	—	1	7	—	—	1000 µg/mℓ in A	A	0.5—4.0(1.5)	30	15
Cefazolin	AD	22	A	0.5	2	21	1	4	—	—	10,000 µg/mℓ in D	A	0.64—1.56(1.0)	32—35	1
Cephacetrile	AD	22	A	5.0	2	21	1	4	—	—	1000 µg/mℓ in A	A	6.4—15.6(10.0)	32—35	1
Cephalexin	AD	22	A	0.5	2	21	1	4	—	—	1000 µg/mℓ in A	A	12.8—31.2(20.0)	32—35	1, 2
Cephaloglycin	AD	22	A	2.0	2	21	1	4	—	—	100 µg/mℓ in distilled water	C	6.4—15.6(10.0)	32—35	1, 2
Cephaloridine	AD	22	A	3.0	2	21	1	4	—	—	1000 µg/mℓ in A	A	0.64—1.56(1.0)	32—35	1, 2
Cephalothin	AD	22	A	1.0	2	21	1	4	—	—	1000 µg/mℓ in A	A	0.64—1.56(1.0)	32—35	1, 2, 29
Cephanone	AD	2	B	2.0	—	—	8	10	—	—	1000 µg/mℓ in K	K	1.2—20.0(5.0)	32—35	36
Cephapirin	AD	22	A	0.8	2	21	1	4	—	—	1000 µg/mℓ in A	A	0.64—1.56(1.0)	32—35	1
Cephradine	AD	19	A	0.5	—	—	1	10	—	—	1000 µg/mℓ in A	A	0.05—0.50(0.1)	30—32	32[n]
Chloramphenicol	AD	19	A	20.0	1	21	1	4	—	—	10,000 µg/mℓ in ethanol	A	32.0—78.1 (50.0)	32—35	2
	TU	6	A	1.0	—	—	—	—	3	9	10,000 µg/mℓ in ethanol	A	2.0—3.12(2.5)	37	1, 2, 29
Chlortetracycline	AD	1	C	4.0	4	21	4	4	—	—	1000 µg/mℓ in 0.01 N HCl	C	0.064—0.156(0.100)	30	2, 30
	TU	22	A	1.0	—	—	—	—	3	9	1000 µg/mℓ in 0.01 N HCl	C	0.038—0.094(0.060)	37	1, 2
Clindamycin	AD	19	A	15.0	11	21	11	4	—	—	1000 µg/mℓ in distilled water	B	0.64—1.56(1.0)	37	1, 2
Cloxacillin	AD	22	A	2.0	2	21	1	4	—	—	1000 µg/mℓ in A	A	3.2—7.81(5.0)	32—35	1, 2, 29
Colistimethate, sodium	AD	3	A	0.4	9	21	10	4	—	—	10,000 µg/mℓ in distilled water	D	0.64—1.56(1.0)	37	1, 2, 29
Colistin	AD	3	A	0.4	9	21	10	4	—	—	10,000 µg/mℓ in distilled water	D	0.64—1.56(1.0)	37	1, 2, 30
Cycloserine	AD	22	A	0.4	2	10	1	4	—	—	1000 µg/mℓ in distilled water	A	32.0—78.1(50.0)	30	1, 2, 29
	TU	22	A	1.0	—	—	—	—	3	9	1000 µg/mℓ in distilled water	Distilled water	32.0—78.1(50.0)	37	1, 2
Dactinomycin	AD	2	B	0.2	5	10	5	4	—	—	10,000 µg/mℓ in methanol	B	0.5—2.0(1.0)	37	1, 2, 29
Demeclocycline	AD	1	C	4.0	8	21	5	4	—	—	1000 µg/mℓ in 0.1 N HCl	C	0.064—0.156(0.10)	37	2
	TU	22	A	1.0	—	—	—	—	3	9	1000 µg/mℓ in 0.1 N HCl	C	0.064—0.156(0.10)	37	1, 2

Dicloxacillin	AD	22	A	2.0	2	21	1	4	—	—	1000 μg/mℓ in A	A	3.2—7.81(5.0)	32—35	1, 2
Dihydrostreptomycin	AD	2	B	0.5	5	21	5	4	—	—	1000 μg/mℓ in B	B	0.64—1.56(1.0)	37	1, 2, 30
	TU	7	A	1.0	—	—	—	—	3	9	1000 μg/mℓ in distilled water	Distilled water	24.0—37.5(30.0)	37	1, 2
Doxycycline	AD	1	C	4.0	8	21	8	4	—	—	1000 μg/mℓ in 0.1 N HCl	B	0.064—0.156(0.10)	30	2, 31
	TU	22	A	1.0	—	—	—	—	3	9	1000 μg/mℓ in 0.1 N HCl	C	0.064—0.156(0.10)	37	1, 2
Erythromycin	AD	19	A	10.0	11	21	11	4	—	—	10,000 μg/mℓ in methanol	B	0.64—1.56(1.0)	37	1, 2, 29, 30
Fusidate, sodium	AD	5	M	10.0	40	10	40	5	—	—	1000 μg/mℓ in A	A	0.10—3.0(1.8)	32—35	8—10
	TU	22	A	2.5	—	—	—	—	3	9	1000 μg/mℓ in J	Distilled water	80—125(100)	37	2
Gentamicin	AD	24	A	0.3	11	21	11	4	—	—	1000 μg/mℓ in B	B	0.064—0.156(0.10)	37	1, 2, 29
Gramicidin	TU	25	E	10.0	—	—	—	—	3	9	1000 μg/mℓ in 95% ethanol	95% ethanol[j]	0.028—0.057(0.10)	37	1, 2
Griseofulvin	AD	11	H	10.0	20	6	21	4	—	—	1000 μg/mℓ in dimethylformamide[i]	B	3.2—7.81(5.0)	30[k]	1, 2, 11, 29
Hamycin	AD	13	I	60.0	—	—	7	20	—	—	1000 μg/mℓ in dimethylsulfoxide	K[l]	0.05—0.10(0.075)	30	12
Hetacillin	AD	19	A	5.0	11	21	11	4	—	—	1000 μg/mℓ in B	B	0.064—0.156(0.10)	32	2, 13
Kanamycin	AD	22	A	0.1	11	21	11	4	—	—	1000 μg/mℓ in B	B	3.2—7.81(5.0)	32—35	2, 29
Kanamycin B	AD	2	B	0.5	5	21	5	4	—	—	1000 μg/mℓ in B	B	0.64—1.56(1.0)	32—35	1, 2
Lincomycin	AD	19	A	10.0	11	21	11	4	—	—	1000 μg/mℓ in distilled water	B	1.28—3.12(2.0)	37	1, 2, 29
Methacycline	AD	1	C	4.0	8	21	8	4	—	—	1000 μg/mℓ in H	C	0.064—0.156(0.10)	30	2, 17
	TU	22	A	1.0	—	—	—	—	3	9	1000 μg/mℓ in H	C	0.038—0.094(0.066)	37	1, 2
Methicillin	AD	22	A	2.0	2	21	1	4	—	—	1000 μg/mℓ in A	A	6.4—15.6(10.0)	32—35	1, 2, 29
Minocycline	TU	22	A	2.0	—	—	—	—	3	9	1000 μg/mℓ in 0.1 N HCl	C	0.070—0.143(0.100)	37	1, 2, 18
Mithramycin	AD	22	A	1.0	8	10	8	4	—	—	1000 μg/mℓ in distilled water	A	0.5—2.0(1.0)	32	1, 2
Mitomycin	AD	2	B	5.0	2	21	15	4	—	—	200 μg/mℓ in distilled water[l]	F	0.5—4.0(1.5)	32—35	19
Nafcillin	AD	22	A	2.0	2	21	1	4	—	—	1000 μg/mℓ in A	A	1.28—3.12(2.0)	32—35	1, 2, 20, 29
Neomycin	AD	22	A	0.4	11	21	11	4	—	—	1000 μg/mℓ in B	B	6.4—15.6(10.0)	32—35	1, 2, 30
	AD	24	A	2.0	11	21	11	4	—	—	1000 μg/mℓ in B	B	0.64—1.56(1.0)	37	2, 29
Netilmicin	AD	24	A	0.3	11	21	11	4	—	—	1000 μg/mℓ in B	B	0.064—0.156(0.10)	37	14
Novobiocin	AD	24	A	20.0	2	21	1	4	—	—	10,000 μg/mℓ in ethanol	D	0.32—0.781(0.5)	35	1, 2, 30
Nystatin	AD	17	D	10.0	—	—	19	8	—	—	1000 μg/mℓ in dimethylformamide[i]	D[j]	12.8—31.2(20.0)	30	2, 29
	TU	4	N	30.0	—	—	—	—	35	5	360 μg/mℓ in dimethylsulfoxide	O[j, o]	0.05—0.12(0.09)	30	33[n]
Oleandomycin	AD	24	A	2.0	11	21	11	4	—	—	10,000 μg/mℓ in ethanol	B	3.2—7.81(5.0)	37	1, 2, 30
	TU	22	A	1.0	—	—	—	—	3	9	1000 μg/mℓ in B	B	2.0—3.12(2.5)	37	2
Oxacillin	AD	22	A	1.0	2	21	1	4	—	—	1000 μg/mℓ in A	A	3.2—7.81(5.0)	32—35	1, 2, 29
Oxytetracycline	AD	1	C	4.0	8	21	8	4	—	—	1000 μg/mℓ in 0.1 N HCl	C	0.64—1.56(1.0)	30	1, 2
	TU	22	A	1.0	—	—	—	—	3	9	1000 μg/mℓ in 0.1 N HCl	C	0.16—0.39(0.25)	37	2, 29
Paromomycin	AD	24	A	20.0	11	21	11	4	—	—	1000 μg/mℓ in B	B	0.64—1.56(1.0)	37	1, 2
Penicillin G	AD	22	A	2.0	2	21	1	4	—	—	1000 units/mℓ in A	A	0.64—1.56(1.0)	37	1, 2, 29
Phenethicillin	AD	19	A	4.0	11	21	11	4	—	—	1000 units/mℓ in distilled water	A	0.064—0.156(0.1)	32—35	1, 2
Phenoxymethyl-penicillin	AD	22	A	2.0	2	21	1	4	—	—	1000 units/mℓ in methanol	A	0.64—1.56(1.0)	37	1, 2, 21, 29
Phosphonomycin	AD	14	Q	30.0	—	—	16	5	—	—	1000 μg/mℓ in P	P	0.22—7.0(1.75)	37	22
Polymyxin B	AD	3	A	0.4	9	21	10	4	—	—	20,000 units/mℓ in distilled water	D	6.4—15.6(10.0)	37	1, 2, 29, 30
Rifampin	AD	2	B	1.0	2	21	2	4	—	—	1000 μg/mℓ in methanol	A	3.2—7.81(5.0)	30	1, 2
Rolitetracycline	AD	1	C	4.0	8	21	8	4	—	—	1000 μg/mℓ in methanol	C	0.64—1.56(1.0)	30	2
	TU	22	A	1.0	—	—	—	—	3	9	1000 μg/mℓ in methanol	A	0.16—0.39(0.25)	37	1, 2
Sisomicin	AD	22	A	0.3	11	21	11	4	—	—	1000 μg/mℓ in B	B	0.064—0.156(0.10)	37	24
Spectinomycin	AD	7	A	2.0	—	—	16	8	—	—	1000 μg/mℓ in B	B	31.25—500(150)	37[m]	25
	TU	6	A	0.2	—	—	—	—	3	9	1000 μg/mℓ in B	Distilled water	24.0—37.5(30.0)	37	1
Streptomycin	AD	2	B	0.5	5	21	5	4	—	—	1000 μg/mℓ in B	B	0.64—1.56(1.0)	30	1, 2, 29
	TU	7	A	1.0	—	—	—	—	3	9	1000 μg/mℓ in distilled water	Distilled water	24.0—37.5(30.0)	37	1, 2, 29

Table 5 (continued)
SELECTED METHODS FOR THE MICROBIOLOGICAL ASSAY OF ANTIBIOTICS

Antibiotic	Test method	Test organism: Test organism number (refer to Table 1)	Test organism: Method of preparation of test inoculum (refer to Table 2)	Test organism: Milliliters of test inoculum per liter of assay medium	Agar-diffusion assays, Base layer: Medium number (refer to Table 3)	Agar-diffusion assays, Base layer: Volume per plate (mℓ)[c]	Agar-diffusion assays, Seed Layer: Medium Number (refer to Table 3)	Agar-diffusion assays, Seed Layer: Volume per plate (mℓ)[d]	Turbidimetric assays: Medium number (refer to Table 3)	Turbidimetric assays: Volume per tube (mℓ)	Preparation of standard solutions: Initial concentration and initial solvent (refer to Table 4)	Preparation of standard solutions: Diluent employed for subsequent dilutions (refer to Table 4)	Assay range, (units or μg/mℓ)[e]	Incubation temperature (°C)[f]	Ref.
Tetracycline	AD	1	C	4.0	8	21	8	4	—	—	1000 μg/mℓ in 0.1 *N* HCl	C	0.64—1.56(1.0)	32—35	1, 2
	TU	22	A	1.0	—	—	—	—	3	9	1000 μg/mℓ in 0.1 *N* HCl	C	0.160—0.390(0.250)	37	1, 2, 29, 30
Thiostrepton	AD	22	A	1.0	—	—	19	8	—	—	1000 μg/mℓ in M	B	0.64—1.56(1.0)	32—37	2
	AD	22	L	1.0	41	10	41	5	—	—	1000 μg/mℓ in dimethylsulfoxide	N	1.25—5.0(2.5)	37	27
Ticarcillin	AD	16	P	15.0	38	21	38	4	—	—	1000 μg/mℓ in A	A	3.20—7.81(5.00)	37	1
Tobramycin	AD	2	B	5.0	5	10	5	5	—	—	1000 μg/mℓ in B	B	0.025—0.50(0.20)	30—37	23
	TU	22	A	1.0	—	—	—	—	3	9	1000 μg/mℓ in distilled water	Distilled water	2.00—3.125(2.50)	37	1
Troleandomycin	AD	24	A	2.0	18	21	18	4	—	—	1000 μg/mℓ in I	E	9.6—23.4(15.0)	37	2
	TU	8	A	0.5	—	—	—	—	3	9	1000 μg/mℓ in I	A	16.0—39.0(25.0)	37	1, 2
Tylosin	AD	19	A	5.0	11	21	11	4	—	—	10,000 μg/mℓ in methanol	B	3.2—7.81(5.0)	32[n]	2
Tyrothricin	TU	25	E	10.0	—	—	—	—	3	9	1000 μg/mℓ in ethanol	Ethanol[i]	0.14—0.285(0.20)	37	1, 2
Vancomycin	AD	2	B	2.0	8	10	8	4	—	—	400 μg/mℓ in distilled water	C	6.4—15.6(10.0)	30	2, 29
Viomycin	AD	2	B	0.5	5	10	5	4	—	—	1000 μg/mℓ in distilled water	B	32.0—78.1(50.0)	37	2
	TU	7	A	1.0	—	—	—	—	3	9	1000 μg/mℓ in distilled water	Distilled water	64.0—156(100)	37	1, 2, 29

[a] AD, agar-diffusion assay.
[b] TU, turbidimetric assay.
[c] Volume of uninoculated agar per 90-mm petri plate.
[d] Volume of inoculated agar per 90-mm petri plate, layered onto the solidified base layer.
[e] The value in parentheses is the reference concentration; samples are diluted to this concentration. In the turbidimetric assays, 1 mℓ of each concentration is added to each tube unless otherwise noted.
[f] Incubation time: 18 to 24 hr for agar-diffusion assays; 3 to 6 hr for turbidimetric assays.
[g] Test organism 21 is used for tests of solutions containing bacitracin and dihydrostreptomycin.
[h] Test organism 9 is used for tests of solutions containing bacitracin and dihydrostreptomycin.
[i] The solution is further diluted in solvent to 20 times the final concentration of each standard curve point, so that each final dilution is 1 + 19 with the indicated buffer.
[j] Add only 0.1 mℓ of solution to each tube.
[k] Incubation time: 48 hr.
[l] Solutions are prepared and stored in low-actinic glassware.
[m] Prediffuse for 5 hr at 5°C before incubation.
[n] Modified.
[o] Dilute to 30 μg/mℓ in diluent O, then dilute to the assay range with medium 14.

REFERENCES

1. **Anon.,** Code of Federal Regulations, Title 21, Food and Drugs, Part 436, Subpart D-Microbiological Assay Methods, 1982, 240.
2. **Arret, B., Johnson, D. P., and Kirshbaum, A.,** Outline of details for microbiological assays of antibiotics: second revision, *J. Pharm. Sci.*, 60, 1689, 1971.
3. **Stark, W. M., Higgens, C. E., Wolfe, R. N., Hoehn, M. M., and McGuire, J. M.,** Capreomycin, a new antimycobacterial agent produced by *Streptomyces capreolus*, sp. n., in *Antimicrobial Agents and Chemotherapy 1962*, Sylvester, J. C., Ed., American Society for Microbiology, Ann Arbor, Mich., 1963, 596.
4. **Black, H. R., Griffith, R. S., and Peabody, A. M.,** Absorption, excretion and metabolism of capreomycin in normal and diseased states, *Ann. N.Y. Acad. Sci.*, 135, 974, 1966.
5. **Grove, D. C. and Randall, W. A.,** *Assay Methods of Antibiotics, A Laboratory Manual*, Medical Encyclopedia, New York, 1955, 104.
6. **Dony, J.,** Technique of determination of magnamycin, *Ann. Pharm. Fr.*, 12, 307, 1954.
7. **Bran, J. L., Levison, M. E., and Kaye, D.,** Clinical and *in-vitro* evaluation of cephapirin, a new cephalosporin antibiotic, *Antimicrob. Agents. Chemother.*, 1, 35, 1972.
8. **Hilson, G. R. F.,** *In-vitro* studies of a new antibiotic (fucidin), *Lancet*, 1, 932, 1962.
9. **Godtfresden, W., Roholt, K., and Tybring, L.,** Fucidin, a new orally active antibiotic, *Lancet*, 1, 928, 1962.
10. **Saggers, B. A. and Lawson, D.,** *In-vivo* penetration of antibiotics into sputum in cystic fibrosis, *Arch. Dis. Child.*, 43, 404, 1968.
11. **Knoll, E. W., Bowman, F. W., and Kirshbaum, A.,** Plate assays for griseofulvin in pharmaceutical preparations and body fluids, *J. Pharm. Sci.*, 52, 586, 1968.
12. **Piggott, W. R., Williams, T. W., Jr., Witorsch, P., and Emmons, C. W.,** Bioassay for hamycin, in *Antimicrobial Agents and Chemotherapy — 1965*, Hobby, G. L., Ed., American Society for Microbiology, Washington, D.C., 1966, 353.
13. **Sutherland, R. and Robinson, O. P. W.,** Laboratory and pharmacological studies in man with hetacillin and ampicillin, *Br. Med. J.*, 2, 804, 1967.
14. **Oden, E. M.,** personal communication, Schering Corp., 1978.
15. **Actor, P., Uri, J. V., Guarini, J. R., Zajac, I., Phillips, L., Sachs, C. S., De Marinis, R. M., Hoover, J. R. E., and Weisbach, J. A.,** A new parenteral cephalosporin, SK&F 59962: *in-vitro* and *in-vivo* antibacterial activity and serum levels in experimental animals, *J. Antibiot. (Tokyo)*, 28, 471, 1975.
16. Minimum Requirements of Antibiotic Products, Ministry of Health and Welfare, Tokyo, Japan, 1961, 400.
17. **English, A. R., McBride, T. J., and Riggio, R.,** Biological studies of 6-methylene oxytetracycline, a new tetracycline, in *Antimicrobial Agents and Chemotherapy — 1961*, Sylvester, J. C., Ed., American Society for Microbiology, Ann Arbor, Mich., 1962, 462.
18. **Redin, G. S.,** Antibacterial activity in mice of minocycline, a new tetracycline, in *Antimicrobial Agents and Chemotherapy — 1961*, Sylvester, J. C., Ed., American Society for Microbiology, Ann Arbor, Mich., 1962, 371.
19. Minimum Requirements of Antibiotic Products, Ministry of Health and Welfare, Tokyo Japan, 1961, 481.
20. **Anon.,** *Fed. Reg.*, 37 (Part 149d), 4907, 1972.
21. Biological assays, in *Pharmacopoeia of India*, Government of India Press, Nasik, 1966, 1012.
22. **Stapley, E. O., Hendlin, D., Mata, J. M., Jackson, M., Wallick, H., Hernandez, S., Mochales, S., Currie, S. A., and Miller, R. M.,** Phosphonomycin. I. Discovery and *in-vitro* biological characterization, in *Antimicrobial Agents and Chemotherapy — 1969*, Hobby, G. L., Ed., American Society for Microbiology, Washington, D.C., 1970, 284.
23. **Lamb, J. W., Mann, J. M., and Simmons, R. J.,** Factors influencing the microbiological assay of tobramycin, *Antimicrob. Agents Chemother.*, 1, 323, 1972.
24. **Weinstein, M. J., Marquez, J. A., Testa, R. T., Wagman, G. H., Oden, E. M., and Waitz, J. A.,** Antibiotic 6640 a new *Micromonospora*-produced aminoglycoside antibiotic, *J. Antibiot. (Tokyo)*, 23, 551, 1970.
25. **Hanka, L. J., Mason, D. J., and Sokolski, W. T.,** Actinospectacin, a new antibiotic, II. Microbiological assay, *Antibiot. Chemother.*, 11, 123, 1961.
26. **Anon.,** *Fed. Reg.*, 43, 20976, 1978.
27. **Levin, J. D. and Pagano, J. F.,** Thiostrepton, in *Analytical Microbiology*, Kavanagh, F., Ed., Academic Press, New York, 1963, 365.
28. **Horwitz, W., Ed.,** in *Official Methods of Analysis of the Association of Official Analytical Chemists*, 11th ed., Association of Official Analytical Chemists, Washington, D.C., 1970, 261 and 752.
29. *The Pharmacopeia of the United States of America*, 19th revision, U.S. Pharmacopeial Convention, Inc., Bethesda, Md., 1975, 595.

30. **Gavin, J. J.,** Biological methods, in *Handbook of Analytical Chemistry,* Sect. 9, Meites, L., Ed., McGraw-Hill, New York, 1963.
31. **von Wittenau, M. S. and Twomey, T. M.,** The disposition of doxycycline by man and dog, *Chemotherapy,* 16, 217, 1971.
32. **Grove, D. C. and Randall, W. A.,** in *Assay Methods of Antibiotics, A Laboratory Manual,* Medical Encyclopedia, Inc., New York, 1961, 14.
33. **Gerke, J. R. and Madigan, M. E.,** Amphotericin B and other polyene antifungal antibiotics: photometric assay and factors influencing activity, *Antibiot. Chemother.,* 11, 225, 1961.
34. **Lamb, J. W.,** personal communication, Eli Lilly & Co., 1978.
35. **Foglesong, M. A., Lamb, J. W., and Dietz, J. V.,** Stability and blood level determinations of cefaclor, a new oral cephalosporin antibiotic, *Antimicrob. Agents Chemother.,* 13, 49, 1978.
36. **Wick, W. E. and Preston, D. A.,** Biological properties of three 3-heterocyclic thiomethyl cephalosporin antibiotics, *Antimicrob. Agents Chemother.,* 1, 221, 1972.

METHODS FOR BIOASSAY OF ANTIFUNGAL AGENTS IN BIOLOGIC FLUIDS

Smith Shadomy and Ana Espinel-Ingroff

INTRODUCTION

Using graded concentrations of a standard preparation of the agent being assayed and an agar layer seeded with a suitable indicator organism, it is possible to perform agar radial diffusion bioassays of antifungal agents in serum or spinal fluid in a manner identical to that used for bioassay of antibacterial agents. Four antifungal agents used in the treatment of serious or systemic mycotic disease will be discussed in this chapter. These include 5-fluorocytosine (Ancobon®, flucytosine, 5-FC), amphotericin B (Fungizone®), miconazole (Monistat®, i.v.) and ketoconazole (Nizoral®).

Nystatin, an antifungal agent used topically and orally in treatment of *Candida* infections, is rarely employed in situations where knowledge of serum concentrations is required. However, if such information is required, the bioassay to be described for amphotericin B is fully applicable using a nystatin standard in lieu of the amphotericin B standard. Knowledge of serum levels is not required for topical agents such as haloprogin and tolnaftate used in the treatment of dermatophytic infections or candicidin used in treatment of vaginal candidosis. Serum levels for potassium iodide, used orally in treatment of lymphatic sporotrichosis, and griseofulvin, used orally in treatment of dermatophytic infections, cannot be determined by bioassay.

Based upon procedures developed in our laboratory and by others five bioassays using essentially the same basic method will be described. These include individual bioassays for amphotericin B and 5-FC alone,[7,10,11] a bioassay for amphotericin B in the presence of 5-FC,[7,13] a bioassay for miconazole, a drug now used both topically in treatment of dermatophytic infections and systemically in treatment of life threatening fungal infections[5] and a bioassay for ketoconazole, an orally active imidazole now used in treatment of a number of mycotic infections.[2] The presence of amphotericin B in clinical specimens does not interfere in bioassays for 5-FC because of the much larger zones of inhibition produced by the latter compound and the ease with which the former compound is inactivated in such clinical specimens by moderate heating.[7,13]

INDICATOR ORGANISMS

1. *Paecilomyces variotii* ATCC 22319 (MSCC 5605, NIAID) is used for assay of amphotericin B in the absence of 5-FC.
2. *Chrysosporium pruinosum* ATCC 36374 (MCVC 69.2; MSCC 2452, NIAID) being resistant to 5-FC is used for assay of amphotericin B in the presence of 5-FC.
3. *Saccharomyces cerevisiae* ATCC 36375 (ATCC 9763) is used when assaying 5-FC.
4. *Candida stellatoidea* ATCC 36232 (MCVC 52.10; MO-010, NCDC Mycology Proficiency Testing Program) is used in bioassays for miconazole.
5. *Candida pseudotropicalis* ATCC 28838 (Carshalton strain) is used in the bioassay for ketoconazole.

MEDIA

Five media are used in bioassays of these four drugs: (1) antibiotic medium M-12 (Difco, BBL) for assay of amphotericin B in specimens free of other antifungal agents; (2) antibiotic medium M-12 supplemented with 10 μg/mℓ of cytosine for assay of amphotericin B in specimens containing 5-FC; (3) yeast morphology agar (YMA, Difco) adjusted to a pH of 6.5 for assay of 5-FC; (4) Emmon's modification of Sabouraud's dextrose agar (Difco, BBL) for assay of miconazole; (5) Kimmig's agar (EM Reagents) for assay of ketoconazole.

ASSAY PLATES

Large, 23 × 23 × 1.8 cm, rigid transparent polystyrene assay plates (A/S Nunc Bioassay Plate, Vangard International Inc.) are used. Because of their tendency to warp, regular disposable plastic petri dishes are unsuited for bioassay purposes.

SUSPENSIONS OF INDICATOR ORGANISMS

Fresh, pure cultures of the indicator organisms should be maintained and transferred weekly on slants of Sabouraud's agar free of antibiotics. *P. variotii,* which must be incubated under completely aerobic conditions, can be used only when the culture is mature enough for the spores to show a tan color when examined microscopically; this normally is after 3 to 4 days incubation. Cultures of *S. cerevisiae, C. stellatoidea* and *C. pseudotropicalis* are ready for harvest after 48 to 96 hr incubation; *C. pruinosum* requires 4 to 6 days growth before harvesting. Mature cultures of the yeasts can be harvested directly from agar slants using sterile physiological saline. Cultures of the filamentous organisms are harvested by washing mycelial mats to remove mature spores. The resulting suspensions then are allowed to stand to permit mycelial fragments to settle to the bottom of the tubes. Homogenous suspensions with transmissions of 70 to 80% as measured at 530 μm then are prepared from the harvested cultures in sterile physiological saline; 1 mℓ of such a suspension is required for each assay plate. Suspension of all five organisms will remain viable at 4°C for up to 1 week.

STOCK DRUG SOLUTIONS

Amphotericin B — Prepare a 1000 μg/mℓ (actual activity) stock solution of drug standard (secondary Amphotericin B standard, E. R. Squibb & Sons, Inc. or U.S.P. Reference Standard) in either dimethylsulfoxide (DMSO) or 50% alkaline isopropyl alcohol (prepared by adding 6 drops of 10% NaOH to 100 mℓ of alcohol). With either solution, wait 30 min for self-sterilization before using. This stock solution can be used for 1 week if kept at 4°C.

5-Fluorocytosine — A stock solution containing 10,000 μg/mℓ of 5-FC (5-fluorocytosine powder, Hoffman-La Roche, Inc.) is prepared in distilled water and sterilized by filtration. It can be kept indefinitely at −20°C.

Miconazole — Stock solutions of standard material are difficult to prepare because of solubility problems; further, the drug will not remain dissolved in aqueous solutions when added to culture media. A solution containing 1000 μg/mℓ (miconazole base, Janssen Pharmaceutica) is prepared in alkalinized isopropyl alcohol using the I.V. preparation of the drug (Janssen Pharmaceutica). This may be stored at 4°C.

Ketoconazole — A stock solution containing 1000 μg/mℓ (ketoconazole base, Janssen Pharmaceutica) is prepared in 0.2 *N* HCl. This is stored at −20°C.

CONCENTRATIONS OF DRUGS USED IN PREPARATION OF STANDARD CURVES

While suitable standard curves can be prepared using only three or four concentrations of drug, one of the methods to be presented for processing bioassay data requires a minimum of five separate concentrations prepared in a doubling, geometric progression. Thus, in all the following methods, a minimum of six concentrations of drug are used. Further, all of the bioassays to be described require pooled, normal human sera as final diluents. Individual lots of such sera must be pretested against the various indicator organisms to determine that they are free of any preexisting inhibitory factors. Finally, it is important to note that all test concentrations of drug should produce zones of inhibition which are in the linear regions of the dose response curves.

Amphotericin B — Drug concentrations: 1.0, 0.5, 0.25, 0.125, 0.063, and 0.031 μg/mℓ. Pipette 0.25 mℓ of the 1000 μg/mℓ stock solution into a 5-mℓ, sterile, graduated volumetric flask and bring to volume with alkalinized isopropyl alcohol. Mix thoroughly. This will give a concentration of 50 μg/mℓ. Dilute this solution 1:10 in pooled, normal human serum to obtain a 5 μg/mℓ concentration. Either serum or cerebrospinal fluid may be used in preparation of the standard curve in bioassays of spinal fluid specimens. Sterile filtered urine should be used in assays of urine specimens. For the final concentration of 1 μg/mℓ, dilute 1.0 mℓ of the latter concentration in 4 mℓ of serum. Then serially dilute 1:2 through 5 more dilutions using 1.0 mℓ volumes of pooled serum as diluent. Use a new pipette for each dilution.

5-Fluorocytosine — Drug concentrations: 80, 40, 20, 10, 5, and 2.5 μg/mℓ. These are prepared by first diluting the 10,000 μg/mℓ stock solution 1:10 in sterile saline to give a 1000 μg/mℓ solution which is then diluted 1:10 in pooled normal human serum to give a solution containing 100 μg/mℓ. The latter is then added to volumes of pooled, normal human serum as shown below to give the six final concentrations.

Pooled serum (mℓ)	**Drug (100 μg/mℓ)**	**Final concn. (μg/mℓ)**
0.2	0.8	80
0.6	0.4	40
0.8	0.2	20
0.9	0.1	10
0.95	0.05	5
0.975	0.025	2.5

The same twofold dilution procedure described for amphotericin B also may be used with 5-FC after obtaining the initial 80 μg/mℓ solution. This is easily achieved by making 2 mℓ of this latter solution (0.4 mℓ of serum + 1.6 mℓ of 100 μg/mℓ) and then diluting down to the sixth tube in 1.0 mℓ volumes.

Miconazole — Drug concentrations: 8, 4, 2, 1, 0.5, 0.25 μg/mℓ. Prepare a 100 μg/mℓ solution by diluting the 1000 μg/mℓ stock solution 1:10 in pooled, normal human serum. Add 0.4 mℓ of this solution to 4.6 mℓ of pooled, normal human serum. This results in a 16 μg/mℓ solution which is then serially diluted in 1.0 mℓ volumes, through 5 dilutions.

Ketoconazole — Drug concentrations: 32, 16, 8, 4, 2, 1, 0.5, 0.25 μg/mℓ. Prepare a 100 μg/mℓ solution by diluting the 1000 μg/mℓ stock solution 1:10 in pooled, normal human serum. Add 0.64 mℓ of this solution to 1.36 mℓ of pooled, normal human serum. This results in a 32 μg/mℓ solution which then is serially diluted, in 1.0 mℓ volumes, through seven dilutions.

PREPARATION OF THE PLATES

A 100 mℓ volume of sterile, molten medium at about 45 to 50°C is seeded with 1 mℓ of an adjusted suspension of the appropriate indicator organism. The pH of YMA is adjusted from the normal value of 4.5 to 6.5 by adding 0.3 mℓ of 1 *N* NaOH to each 100-mℓ flask of medium; this prevents formation of serum precipitates which interfere with measurements of the resulting zones of inhibition. When assaying for amphotericin B in the presence of 5-FC, the medium (M-12) should be supplemented with 10 μg/mℓ of cytosine. This is obtained by adding 0.1 mℓ of a 10,000 μg/mℓ sterile (by filtration) stock suspension of cytosine. Stock solutions of cytosine can be kept indefinitely in small aliquots at −20°C. The flask of seeded medium should be immediately and thoroughly mixed and then poured into the plate which is placed on a level surface to obtain a uniform layer of agar (about 4 mm). Allow the agar to harden.

When the hardened plates, standard drug dilutions, and specimens to be tested are ready, gently place sterile, stainless steel cylinders (Fisher, ''Pen assay'') on the hardened agar. Arrange them to provide 12 or 16 cylinders per plate. If cylinders are not available, alternate procedures include either use of sterile paper discs or cutting of holes in the agar layer. The small zones of inhibition obtained in the bioassay for amphotericin B will reduce the sensitivity of the assay if paper discs are used.

Each specimen and standard curve dilution are tested in duplicate. When assaying for 5-FC no more than two to four samples should be placed on each plate as the diameters of the resulting zones of inhibition can be 50 to 60 mm or more. If 5-FC concentrations in excess of 100 μg/mℓ or ketoconazole concentrations in excess of 10 μg/mℓ are anticipated, such specimens *must be* diluted in pooled serum prior to assay. Otherwise, the nonlinearity of the dose response curves invalidates the results for such specimens. If amphotericin B is present in specimens being assayed for 5-FC, the specimens should be heated at 56°C for 30 min prior to assay.

Introduce 0.1 mℓ of the material being tested into each cylinder. Be sure not to spill the sample or tip the cylinder. Samples of the pooled, human serum used in preparation of the standard curve as well as pretreatment specimens from the patients, if available, should be tested for possible nonspecific inhibitory activity against the indicator organisms. The presence of any zone of inhibition around the pooled, human serum or pretreatment controls indicates that the assay may not be valid.

In bioassays for amphotericin B, 5-FC and ketoconazole incubate the plates at 30°C for 24 to 48 hr or until clear zones of inhibition can be seen. The bioassay for miconazole requires a 24-hr period for prediffusion of the drug at 4°C before overnight incubation at 30°C.

Reading of the Plates

After the incubation period, remove the cylinders and measure the diameters of the zones of inhibition to the nearest 1.0 mm using a metric ruler or to the nearest 0.1 mm using a metric caliper-micrometer. Record three separate measurements from three different angles. Thus, each patient's sample or each drug dilution should have six readings made from two zones of inhibition. Determine the average of these measurements.

Plotting the standard curve and determination of drug levels in specimens

Three different methods can be used for plotting the standard curve and determining serum drug levels. These include visual inspection, use of a simplified mathematical formula, and regression analysis. Visual inspection is the simplest but is most subject to human error and bias. The mathematical formula eliminates bias from the standard curve but still requires visual inspection for determination of drug levels in specimens. Regression analysis depends upon the availability of a calculator or microcomputer; however, this is compensated for by elimination of bias and possible error in both evaluation of data and interpretation of results. It also provides for quality control checks as well as means of evaluating the statistical reliability of the data for the standard curve.

1. Visual Inspection

Using the final averaged zone size values for the six standard drug concentrations, plot the curve on 2 or 3 cycle semilogarithmic paper as required by the concentration range. Drug concentrations (in μg/mℓ) are plotted, as log values, from the ordinate and the calculated average zone size values (in mm) are plotted from the abscissa. The "best line of fit" is then selected by visual inspection. Levels of drug in the specimen samples are determined by plotting the specimen zone size values in mm from the abscissa to the dose response curve. Results in μg/mℓ are then read directly on the ordinate.

2. Mathematical Determination of the Standard Curve

As noted above, this method requires a minimum of five equally spaced concentrations of drug in a doubling or geometric sequence. All four standard curves are prepared using a minimum of six drug concentrations. Thus, if either the highest or lowest concentrations of drug gave questionable or uncertain results, data for five doubling concentrations still are available.

In order to draw the calculated dose response line, instead of plotting by inspection, zone size values for "Low" and "High" plotting points on the curve are calculated as follows:[3]

$$L = \frac{3e + 2d + c - a}{5} \qquad H = \frac{3a + 2b + c - e}{5}$$

where L = calculated zone of inhibition, lowest concentration plot point; H = calculated zone of inhibition, highest concentration plot point; a = diameter of zone of inhibition, dilution I (highest concentration); b = dilution II zone size diameter; c = dilution III zone size diameter; d = dilution IV zone size diameter; and e = dilution V (lowest concentration) zone size diameter. Plot the high and low points on 2 or 3 cycle semilogarithmic paper as before and draw a straight line between these two points. The H value usually corresponds with the 1 μg/mℓ concentration (dilution I) and the L value with the 0.063 concentration (V) in the amphotericin B standard curve; corresponding values for the 5-FC standard curve are 80 and 5 μg/mℓ. These calculated high and low values should be close to the actual measured zone sizes for dilutions I and V, respectively, and can be used as quality control checks of the accuracy of the dose response curve; measured zone size values for the remaining four concentrations of drug should plot closely to this straight line. As before, drug levels in specimens are determined by plotting the specimen zone size values in mm from the abscissa to the dose response line and reading corresponding drug concentrations in μg/mℓ from the ordinate.

The following is an example of the "High" and "Low" calculations obtained in a bioassay for 5-FC:

Value	5-Fluorocytosine concn. (μg/mℓ)	Avg. zone size diam. (mm)
a	80	73.2
b	40	64.9
c	20	56.8
d	10	50.2
e	5	41.8

$$L = \frac{3(41.8) + 2(50.2) + 56.8 - 73.2}{5} = 41.88$$

$$H = \frac{3(73.2) + 2(64.9) + 56.8 - 41.8}{5} = 72.88$$

The above values show excellent agreement with the actual measured zone size values for 5 and 80 μg/mℓ. As already noted, this provides for a quality control check of the linearity of the response curve.

3. Regression Analysis

A third method for processing of bioassay data involves regression analysis and use of the estimating equation. Statistically, regression analysis is a measurement of both the relationship between two variables, the coefficient of regression, and of the changes caused in one variable by changes in the second variable. Applied to bioassay data, regression analysis provides for both a measurement of the relationship between paired drug concentrations and corresponding sizes of zones of inhibition as well as a means for calculation of expected values for one variable at any given value of the second variable.

Three series of calculations are required in application of regression analysis to bioassay data. These include (1) data summations and calculation of the coefficient of correlation, r; (2) calculation of statistics describing the "Least Squares" regression line; and (3) use of the estimating equation for calculation of serum drug levels in specimens based upon their measured zones of inhibition.

Statistically, the coefficient of correlation, r, is a measurement of the degree of reliability or significance of the relationship between two series of related or unrelated variables. Linear regression measures the degree of change in one variable, the dependent variable, brought about by any given change in the second or independent variable. In processing of bioassay data, drug concentrations are treated as the independent variable, X, while the diameters of zones of inhibition are the dependent variable, Y. The limiting value of r is $\pm$ 1.0; the closer to 1.0 the value of r, the more reliable or statistically significant is the relationship between the paired data for the standard curve. The coefficient of correlation is calculated using the formulae:

$$r = \frac{\Sigma xy}{\sqrt{\Sigma x^2 \times \Sigma y^2}} \tag{1}$$

$$\Sigma xy = \Sigma XY - \frac{(\Sigma X) \times (\Sigma Y)}{n} \tag{2}$$

$$\Sigma x^2 = \Sigma X^2 - \frac{(\Sigma X)^2}{n} \tag{3}$$

$$\Sigma y^2 = \Sigma Y^2 - \frac{(\Sigma X)^2}{n} \tag{4}$$

In the above, ΣX is the sum of all values of X, or of all drug concentrations employed in the standard curve, ΣX^2 is the sum of all values of X^2, ΣY is the sum of all values of Y, or of zones diameters from the standard curve, ΣY^2 is the sum of all values Y^2, ΣXY is the sum of the cross-products of each pair of corresponding values for X and Y and n is the number of data pairs of X and Y. Results from Equations 2 and 3 also will be required for calculation of statistics describing the "Least Squares" or standard curve regression line. This represents the mathematical relationship between the two variables. This relationship can be used to predict values of Y for any corresponding value of X or calculate X from any measured value of Y. Values for X are expressed as logarithms in order to provide a linear rather than a curvilinear dose response curve or line. In processing data from standard curves employing twofold increasing increments of drug in relatively high concentrations X should be expressed in terms of $\log_2$ values:

$$\log_2 X = \log X \times 3.32193 \tag{5}$$

where log X is the common log of X.

The equation for statistics describing the regression line is derived as follows:

$$Y_c = \overline{Y} = \beta(X - \overline{X}) \qquad (6)$$

where Y_c is the calculated value of Y for any corresponding value for X; $\overline{X}$ and $\overline{Y}$ are the means of the measured values of X and Y and β is the regression coefficient describing the slope of the line. β is calculated using the results of equations 2 and 3 above as follows:

$$\beta = \frac{\Sigma xy}{\Sigma x^2} \qquad (7)$$

One more statistic is required to describe the regression line. This is α or intercept, the value for Y when X equals zero. In all calculations using log values of X, α will be the value of Y for a drug concentration of 1.0 μg/mℓ. α is calculated from the equation

$$\alpha = \overline{Y} - \beta \cdot \overline{X} \qquad (8)$$

Once α and β have been calculated, the estimating equation is applied for calculation of serum drug concentrations from corresponding measured diameters of zones of inhibition. It also can be used to calculate zone size values for corresponding drug concentrations if a graphic display of the results for the standard curve is desired. The estimating equation is expressed as

$$Y_c = \alpha + \beta \cdot X \qquad (9)$$

where Y_c is the calculated value for any corresponding value of X, and

$$X_c = \frac{Y - \alpha}{\beta} \qquad (10)$$

where X_c is the calculated value for any corresponding value of Y.

A sample application of the above calculations used in processing of bioassay data follows. In this sample, representative data from the bioassay of 5-FC concentrations in several clinical specimens will be used. Data for the standard curve are the same as in the previous example; $\log_2$ values of X will be used in the statistical analyses.

1. The Standard Curve

5-FC concn. μg/mℓ (X)		**Average diameter of resulting zones of inhibition, mm (Y)**
X	**$\log_2$X**	
—	—	
80	6.322	73.2
40	5.322	64.9
20	4.322	56.8
10	3.322	50.2
5	2.322	41.8

2. Correlation Analysis and Calculation of r

X	X^2	Y	Y^2	XY
6.322	39.9677	73.2	5358.24	462.7704
5.322	28.3237	64.9	4212.01	345.3978
4.322	18.6797	56.8	3226.24	245.4896
3.322	11.0357	50.2	2520.04	166.7644
2.322	5.3917	41.8	1747.24	97.0596
Σ = 21.610	103.3985	286.9	17,063.77	1,317.4818

$$n = 5$$

$$\Sigma xy = \Sigma XY - \frac{(\Sigma X) \times (\Sigma Y)}{n}$$

$$= 1{,}317.4818 - \frac{21.61 \times 286.90}{5}$$

$$\Sigma xy = 77.500$$

$$(3) \qquad \Sigma x^2 = \Sigma X^2 - \frac{(\Sigma X)^2}{5}$$

$$= 103.3985 - \frac{(21.61)^2}{5}$$

$$\Sigma x^2 = 10.0001$$

$$(4) \qquad \Sigma y^2 = \Sigma Y^2 - \frac{(\Sigma Y)^2}{n}$$

$$= 17{,}063.77 - \frac{(286.9)^2}{5}$$

$$\Sigma y^2 = 601.448$$

$$(1) \qquad r = \frac{\Sigma xy}{\sqrt{\Sigma x^2 \times \Sigma y^2}}$$

$$= \frac{77.500}{\sqrt{10.0001 \times 601.448}}$$

$$= \frac{77.500}{77.5535}$$

$$r_{[3]} = 0.9993$$

3. Calculation of β and α for the "Least Squares" Regression Line

$$(7) \qquad \beta = \frac{\Sigma xy}{\Sigma x^2}$$

$$= \frac{77.50}{10.0}$$

$$\beta = 7.75$$

$$(8) \quad \alpha = \overline{Y} - \beta \cdot \overline{X}$$

$$\alpha = \frac{286.9}{5} - \beta \left[\frac{21.61}{5}\right]$$

$$= 57.38 - 33.4955$$

$$\alpha = 23.8845$$

4. Application of the Estimating Equation

Calculation of the zones of inhibition for given upper and lower concentrations of drug provide both for plotting points for graphic display of the standard dose response curve and for quality control checks of the actual measured zone size values. Thus, calculation of two values, the statistically predicted zone sizes for the 80 and 5 μg/mℓ concentrations are performed using the estimating equation (Equation 9):

$$(9) \quad Y_c = \alpha + \beta \cdot X$$

$$Y_{c(80)} = 23.8845 + 7.75 \times 6.322$$

$$= 72.88$$

$$Y_{c(5)} = 23.8845 + 7.75 \times 2.322$$

$$= 41.88$$

The above calculated zone size values agree very well with the actual measured zone sizes of 73.2 and 41.8 mm. These quality control checks together with the highly significant value of r, 0.9993, prove the reliability of the standard curve. This can be confirmed by reference to statistical tables which would show that the significance of the above value of r with 3 degrees of freedom is <0.001 (1). If desired, the t statistic can be calculated:

$$(11) \quad t = r\frac{\sqrt{N - 2}}{1 - r^2} \qquad df = N - 2$$

$$t_{[3]} = 46.267, P < 0.001$$

Serum drug concentrations in clinical specimens being assayed are now calculated, first as $\log_2$ values and then converted to arithmetic values.

Three specimens were assayed; the results were as follows:

Spec. no.	Measured zone of inhibition (mm)	5-FC concn. (μg/mℓ)
2015	52.6	—
2019	73.7	—
2021	79.5	—

The estimating equation is used as follows (Equation 10):

$$(10) \quad X_c = \frac{Y - \alpha}{\beta}, \quad \text{thus}$$

$$\log_2 X_{2015} = \frac{52.6 - 23.88}{7.75}$$

$$= 3.7058$$

$$\log_2 X_{2019} = \frac{73.7 - 23.88}{7.75}$$

$$= 6.4284$$

$$\log_2 X_{2021} = \frac{79.5 - 23.88}{7.75}$$

$$= 7.1768$$

Conversion of the above $\log_2$ values to arithmetic values is accomplished by dividing the $\log_2$ of X by 3.32193 and then taking the antilogarithm of the resulting value. Thus, the serum drug concentrations were as follows: 2015 = 13.05 μg/mℓ; 2019 = 86.13 μg/mℓ; 2021 = 144.69 μg/mℓ.*

The above calculations were performed using a hand calculator. Today, there are available a variety of modestly priced programmable desk top calculators or minicomputers which can be used for processing bioassay data. Most models feature built-in programs which will perform most if not all of the above calculations.

ANTICIPATED RESULTS

Peak amphotericin B serum concentrations usually will be in the range of 1.0 to 1.5 μg/mℓ; 0.3 μg/mℓ is considered to be an effective concentration in treatment of most fungal infections.[12] Concentrations in cerebrospinal fluids will be much less, 0.02 to 0.08 μg/mℓ; however, concentrations in the range of 1.5 to 2.0 μg/mℓ can be obtained with intrathecal therapy. Therapeutic serum concentrations of 5-FC generally are in the range of 60 to 80 μg/mℓ;[9] paired cerebrospinal fluid concentrations will be about 40 to 60 μg/mℓ. Serum concentrations of 5-FC in excess of 125 μg/mℓ should be regarded as excessive and potentially toxic.[8] Serum concentrations of miconazole will rarely exceed 2 to 4 μg/mℓ.[5] Ketoconazole serum levels are both dose dependent and subject to wide patient-to-patient variability. On an average, however, sera obtained 2 to 4 hr after ingestation of a 400-mg dose will contain approximately 4 to 6 μg/mℓ of active drug.

* As already noted, a value of this magnitude may be in error because of nonlinearity of the dose response curve in the higher concentration range. A specimen giving such a value should be reassayed after dilution 1:2 or 1:5 in normal serum or cerebrospinal fluid.

REFERENCES

1. **Batson, H. C.,** *An Introduction to Statistics in the Medical Sciences,* Burgess Publishing, Minneapolis, Minn., 1958, 80.
2. **Borelli, D., Bran, J. L., Fuentes, J., Legendre, R., Leiderman, E., Levine, H. B., Restrepo, M. A., and Stevens, D. A.** Ketoconazole, an oral antifungal: laboratory and clinical assessment of imidazole drugs, *Postgrad. Med. J.,* 55, 657, 1979.
3. **Deutschberger, J. and Kirshbaum, A.,** Simplified equations for fitting least squares lines to data, *Antibiot. Chemother.,* 9, 752, 1959.
4. **Duncan, R. D., Knapp, R. G., and Miller, M. C., III,** *Introductory Biostatistics for the Health Sciences,* John Wiley & Sons, New York, 1977, 97.
5. **Espinel-Ingroff, A., Shadomy, S. and Fisher, J. F.,** Bioassay for miconazole, *Antimicrob. Agents Chemother.,* 11, 365, 1977.
6. **Finney, D. J., Hazlewood, T., and Smith, M. H.,** Logarithms to Base 2, *J. Gen. Microbiol.,* 12, 222, 1955.
7. **Kasper, R. L. and Drutz, D. J.,** Rapid simple bioassay for 5-fluorocytosine in the presence of amphotericin B, *Antimicrob. Agents Chemother.,* 7, 462, 1975.
8. **Kaufman, C. A. and Frame, P. T.,** Bone marrow toxicity associated with 5-fluorocytosine therapy, *Antimicrob. Agents Chemother.,* 11, 244, 1977.
9. **Schönebeck, J., Polak, A., Fernex, M., and Scholer, H. J.,** Pharmacokinetic studies on the oral antimycotic agent 5-fluorocytosine in individuals with normal and impaired kidney function, *Chemotherapy,* 18, 321, 1973.
10. **Shadomy, S.,** *In vitro* studies with 5-fluorocytosine, *Appl. Microbiol.,* 17, 871, 1969.
11. **Shadomy, S., McCay, J. A., and Schwartz, S. I.,** Bioassay for hamycin and amphotericin B in serum and other biological fluids, *Appl. Microbiol.,* 17, 497, 1969.
12. **Utz, J. P. and Buechner, H. A.,** Mucormycosis (phycomycosis), in *Management of Fungus Diseases of the Lungs,* Buechner, H. A., Ed., Charles C Thomas, Springfield, Ill., 1971, 175.
13. **Utz, J. P., Garriques, I. L., Sande, M. A., Warner, J. F., Mandell, G. L., McGehee, R. F., Duma, R. J., and Shadomy, S.,** Therapy of cryptococcosis with a combination of flucytosine and amphotericin B, *J. Inf. Dis.,* 132, 368, 1975.

APPENDIX: TABLES OF CONTENTS OF THE MULTI-VOLUME CRC HANDBOOK OF MICROBIOLOGY*

Volume I
BACTERIA

* Editors: A. I. Laskin and H. A. Lechevalier

Volume II
FUNGI, ALGAE, PROTOZOA, AND VIRUSES

Volume III
MICROBIAL COMPOSITION:
Amino Acids, Proteins, and Nucleic Acids

Volume IV
MICROBIAL COMPOSITION: Carbohydrates, Lipids, and Minerals

Volume V
MICROBIAL PRODUCTS

SUBSTANCES RELATED TO CARBOHYDRATES

ALIPHATIC AND RELATED COMPOUNDS

ALICYCLIC COMPOUNDS

AROMATIC COMPOUNDS

NITROGEN-CONTAINING COMPOUNDS

HETEROCYCLIC COMPOUNDS

MISCELLANEOUS COMPOUNDS

INDEXES

Volume VI
GROWTH AND METABOLISM

Volume VII
MICROBIAL TRANSFORMATION

Volume VIII
TOXINS AND ENZYMES

TOXINS

ENZYMES

INDEX

Volume IX: Part A
ANTIBIOTICS

Volume IX: Part B
ANTIMICROBIAL INHIBITORS

INDEX

A

B

D

F

G

H

I

J

K

L

M

N

O

P

Q

R

T

U

V